The Rhizosphere

Biochemistry and Organic Substances at the Soil-Plant Interface

Second Edition

Soil Biochemistry, Volume 1, edited by A. D. McLaren and G. H. Peterson

Soil Biochemistry, Volume 2, edited by A. D. McLaren and J. Skujins

Soil Biochemistry, Volume 3, edited by E. A. Paul and A. D. McLaren

Soil Biochemistry, Volume 4, edited by E. A. Paul and A. D. McLaren

Soil Biochemistry, Volume 5, edited by E. A. Paul and J. N. Ladd

Soil Biochemistry, Volume 6, edited by Jean-Marc Bollag and G. Stotzky

Soil Biochemistry, Volume 7, edited by G. Stotzky and Jean-Marc Bollag

Soil Biochemistry, Volume 8, edited by Jean-Marc Bollag and G. Stotzky

B2311595

The Rhizosphere

Biochemistry and Organic Substances at the Soil-Plant Interface

Second Edition

edited by

Roberto Pinton
Zeno Varanini
Paolo Nannipieri

CRC Press
Taylor & Francis Group
Boca Raton   London   New York

CRC Press is an imprint of the
Taylor & Francis Group, an **informa** business

CRC Press
Taylor & Francis Group
6000 Broken Sound Parkway NW, Suite 300
Boca Raton, FL 33487-2742

© 2007 by Taylor & Francis Group, LLC
CRC Press is an imprint of Taylor & Francis Group, an Informa business

No claim to original U.S. Government works
Printed in the United States of America on acid-free paper
10 9 8 7 6 5 4 3 2 1

International Standard Book Number-10: 0-8493-3855-7 (Hardcover)
International Standard Book Number-13: 978-0-8493-3855-7 (Hardcover)

Visit the Taylor & Francis Web site at
http://www.taylorandfrancis.com

and the CRC Press Web site at
http://www.crcpress.com

Preface to the First Edition

The research on plant – soil interaction is focused on the processes that take place in the rhizosphere, the soil environment surrounding the root. Many of these processes can control plant growth, microbial infections, and nutrient uptake.

The rhizosphere is dominated by organic compounds released by plant roots and microorganisms. Furthermore, stable components of soil organic matter, namely, humic and fulvic substances, can influence both plant and microorganism. A variety of compounds are present in the rhizosphere, and they range from low-molecular-weight root exudates to high-molecular-weight humic substances. The chemistry and biochemistry of these substances is becoming increasingly clear, and their study promises to shed light on the complex interactions between plant and soil microflora.

The aim of the book is to provide a comprehensive and updated overview of the most recent advances in this field and suggest further lines of investigation. As an interdisciplinary approach is necessary to study such a complex subject, the book provides a good opportunity to summarize information concerning agronomy, soil science, plant nutrition, plant physiology, microbiology, and biochemistry. The book is therefore intended for advanced students, and researchers in agricultural, biological, and environmental sciences interested in deepening their knowledge of the subject and developing new experimental approaches in their specific field of interest.

The first chapter defines the spatial and functional features of the rhizosphere that make this environment the primary site of interaction between soil, plant, and microorganisms. Among the multitude of organic compounds present in the rhizosphere, those released by plant roots are the most important from a qualitative and quantitative point of view; furthermore, the relationships with soil components of any released compound need to be considered (Chapter 2). The release of these compounds strongly depends on the physiological status of the plants and is related to the ability of plant roots to modify the rhizosphere in order to cope with unfavorable stress-inducing conditions. These aspects are discussed in Chapter 3, with particular emphasis on water and physical, and nutritional stresses. A thorough analysis of how root exudates may influence the dynamics of microbial populations at the rhizosphere will be provided in Chapter 4. However, the importance of the role played by biologically active substances produced by microbial populations cannot be underrated, and the organic compounds acting as signals between plants and microorganisms must be identified and characterized (Chapter 7). In this context the biochemistry of the associations between mycorrhizae and plants (Chapter 9) and the interaction between rhizobia and the host plant (Chapter 10) is also considered.

It has been long recognized that both roots and microorganisms compete for iron at the rhizosphere; a wealth of literature is already available on this subject, and many studies on the production of siderophores by microbes are being carried out. Their potential use by plants and the relationship with other plant-borne iron chelating substances is still a matter of debate (Chapter 8). The fulfillment of the nutritional requirement of plants and microorganisms also depends on the processes leading to mineralization and humification of organic residues (Chapter 6). The presence of humic and fulvic substances can have a considerable effect on root habitability, plant growth, and mineral nutrition (Chapter 5). Knowledge of these aspects needs to be reconsidered at the rhizosphere (Chapter 5 and Chapter 6). The development of specific models can shed light on the events taking place at the rhizosphere (Chapter 11). Validation of the models and a better understanding of these phenomena may come from the correct use and development of new experimental approaches (Chapter 12).

We realize that the information in the book is still largely descriptive and that the interdisciplinary view of the causal relationships in the rhizosphere is still in its infancy. Nevertheless, we do hope that our efforts and these high-quality scientific contributions will stimulate further interest in and work on this fascinating topic.

Roberto Pinton
Zeno Varanini
Paolo Nannipieri
Udine, Florence

Preface to the Second Edition

After the first edition of this book published in 2001, interest in rhizosphere research very much increased, becoming a major area of scientific research, and some important steps ahead have been taken in the area covered by the book. This is testimonied by the vast array of publications and the number of meetings focusing on this subject. On this basis we developed the idea to edit a second edition of the book. The book structure remained essentially the same, reflecting the multidisciplinary approach of the first edition. Most of the original chapters were maintained, expanded, and updated. The second edition contains new information obtained since the first edition was published, which integrates material coming from the first edition that still remains valid.

Furthermore, some new chapters have been included that deal with areas gaining increasing importance for understanding the complex biochemistry of soil–microbe–plant interactions. Subjects have been added that describe the role of nutrient availability in regulating root morphology and architecture (Chapter 5), and the involvement of root membrane activities, besides the well-known release of exudates, in determining (and responding to) the nutritional conditions at the rhizosphere (Chapter 6). Molecular signals between root–root (including allelopathy) and root–microbe, excluding those involving rhizobia and mychorriza, have been discussed (Chapter 10). Manipulation of microbial population for biocontrol and rhizosphere management has been also considered (Chapter 11). Gene flow in the rhizosphere has been discussed for its important role in the evolution of rhizosphere microorganisms and their coevolution with plants (Chapter 14), and in relation to the fate of genetically modified organisms added to the soil–plant system.

All the chapters contain new information deriving from a molecular approach, which contribute to a better understanding of the biochemical processes occurring in the rhizosphere.

We do hope that the efforts made by the editors and by the different contributors can help to make more vigorous the interconnection among scientists interested in rhizosphere biochemistry and molecular biology.

Roberto Pinton
Zeno Varanini
Paolo Nannipieri
Udine, Florence

The Editors

Roberto Pinton is full professor of plant nutrition and head of the Department of Agricultural and Environmental Sciences at the University of Udine, Italy. The author or coauthor of over 160 articles, abstracts, proceedings, and book chapters, he is a research fellow of the Alexander von Humboldt Foundation and a member of the steering committee for the International Symposium on Iron Nutrition and Interactions in Plants. He is topic editor for the section on root and rhizosphere and a member of the editorial advisory board to the *Encyclopedia of Plant and Crop Science* (Dekker).

Zeno Varanini is full professor of agricultural biochemistry at the University of Verona, Italy. The author or coauthor of over 180 articles, abstracts, proceedings, and book chapters, he was a member of the Advisory Committee for Agricultural Sciences of the National Research Council of Italy. He is president of the Italian Society of Agricultural Chemistry and of the Italian Association of Scientific Societies in Agriculture.

Paolo Nannipieri is full professor of agricultural biochemistry and head of the Department of Soil Science and Plant Nutrition at the University of Florence, Italy. The author or coauthor of over 200 articles, abstracts, proceedings, and book chapters, he is the editor of four international books, editor-in-chief of *Biology and Fertility of Soils*, and a member of the editorial boards of *Archives of Agronomy and Soil Science* and *Arid Soil Research and Rehabilitation*. He has been president of the Italian Society of Agricultural Chemistry, and he is chairman of soil biology division in the International Union of Soil Sciences.

Contributors

Élan R. Alford
Graduate Degree Program in Ecology
Colorado State University
Fort Collins, CO

Luigi Badalucco
Dipartimento di Ingegneria e Tecnologie
  Agro-Forestali
Università di Palermo
Palermo, Italy

Paola Bonfante
Dipartimento di Biologia Vegetale
Università di Torino
Torino, Italy

Melissa J. Brimecombe
Trek Diagnostic Systems
West Sussex, England

David E. Crowley
Department of Environmental Sciences
Riverside, California

Peter R. Darrah
Department of Plant Science
University of Oxford
Oxford, England

Frans A.A.M. De Leij
School of Biomedical and Molecular Sciences
University of Surrey, Guildford
Surrey, England

Junichiro Horiuchi
Department of Horticulture and Landscape
  Architecture
Colorado State University
Fort Collins, CO

Elisabeth Kay
Claude Bernard University
Center of Microbial Ecology
Lyon, France

Stephan M. Kraemer
Institute of Terrestrial Ecology
ETH Zurich
Schlieren, Switzerland

Johan Leveau
Netherlands Institute of Ecology
  (NIOO-KNAW)
Heteren, The Netherlands

Bettina Linke
Humboldt University
Institute of Biology—Applied Botany
Berlin, Germany

Ben Lugtenberg
Leiden University
Clusius Lab
Leiden, The Netherlands

James M. Lynch
Alice Holt Lodge—Forest Research
University of Surrey, Farnham
Surrey, England

Francis M. Martin
INRA
UMR INRA-UHP
Nancy, France

Anne Mercier
Claude Bernard University
Center of Microbial Ecology
Lyon, France

J. Alun W. Morgan
Horticulture Research International
Warwick University, Wellesbourne
Warwick, England

Paolo Nannipieri
Dipartimento di Scienza del Suolo e Nutrizione
  della Pianta
University of Florence
Firenze, Italy

Günter Neumann
Institute of Plant Nutrition
Hohenheim University
Stuttgart, Germany

Mark W. Paschke
Department of Forest, Rangeland,
 and Watershed Stewardship
Colorado State University
Fort Collins, CO

Silvia Perotto
Dipartimento di Biologia Vegetale
Università di Torino
Torino, Italy

Laura G. Perry
Department of Forest, Rangeland, and
 Watershed Stewardship and Department of
 Horticulture and Landscape Architecture
Colorado State University
Fort Collins, CO

Roberto Pinton
Dipartimento di Scienze Agrarie e Ambientali
Università di Udine
Udine, Italy

Volker Römheld
Institute of Plant Nutrition
Hohenheim University
Stuttgart, Germany

Tiina Roose
Mathematical Institute
University of Oxford
Oxford, England

Wolfgang Schmidt
Institute of Plant and Microbial Biology, Stress
 and Development Lab
Academia Sinica
Nankang
Taipei, Taiwan

Pascal Simonet
Claude Bernard University
Center of Microbial Ecology
Lyon, France

Nicholas C. Uren
Honorary Senior Research Fellow
Department of Agricultural Sciences
La Trobe University
Victoria, Australia

Zeno Varanini
Dipartimento Scientifico e Tecnologico
Università di Udine,
Verona, Italy

Johannes A. van Veen
Netherlands Institute of Ecology
 (NIOO-KNAW)
Heteren, The Netherlands

Jorge M. Vivanco
Department of Horticulture and Landscape
 Architecture
Colorado State University
Fort Collins, CO

Timothy M. Vogel
Claude Bernard University
Center of Microbial Ecology
Lyon, France

Dietrich Werner
Department of Biology
Philipps-Universität Marburg
Marburg, Germany

John M. Whipps
Horticulture Research International
Warwick University
Wellesbourne
Warwick, England

Table of Contents

1 Types, Amounts, and Possible Functions of Compounds Released into the Rhizosphere by Soil-Grown Plants

Nicholas C. Uren

CONTENTS

I. INTRODUCTION

The rhizosphere is defined here as that volume of soil affected by the presence of the roots of growing plants. The overall change may be deemed biological, but chemical, biological, and physical properties of the soil in turn are affected to varying degrees. A multitude of compounds are released into the rhizosphere of soil-grown plants, most of which are organic compounds and are normal plant constituents derived from photosynthesis and other plant processes (Table 1.1). The relative and absolute amounts of these compounds produced by plant roots vary with the plant species, cultivars, plant's age, and environmental conditions including soil properties, particularly, the level of physical, chemical, and biological stress and so on [1,2,15–18].

An impression one gets from the voluminous literature on the rhizosphere and related topics is that the rhizosphere is deemed by many as a feature of not only soil-grown plants but also of those plants grown *in vitro* in any sort of medium. If that is the case, then a new name other than *rhizosphere* is required for those media other than soil. Similarly, one gets the impression that each and every compound released has a specific role or function, but the reality is that very few proposed effects are established; some are feasible, and some, probably the majority, must remain speculative and unproven. From the time this chapter was first completed in 1999 [19], nothing seems to have changed in the sense that there is no shortage of enthusiastic speculative reviews or of fanciful titles (for example, see Reference 20 and Reference 21).

It is salutary to read the review of Rovira [22], a renowned rhizosphere researcher, where he records significant progress in the understanding of the rhizosphere and how his optimism, born in the 1960s, has transformed into frustration in the 1990s. Admittedly, the complexity of the system

TABLE 1.1
Organic Compounds Released by Plant Roots

Sugars and Polysaccharides

Arabinose, desoxyribose, fructose, galactose, glucose, maltose, mannose, mucilages of various compositions, oligosaccharides, raffinose, rhamnose, ribose, sucrose, xylose

Amino Acids[a]

α-alanine, β-alanine, α–amino adipic, γ-amino butyric, arginine, asparagine, aspartic, citrulline, cystathionine, cysteine, cystine, deoxymugineic, 3-epihydroxymugineic, glutamine, glutamic, glycine, homoserine, histidine, isoleucine, leucine, lysine, methionine, mugineic, ornithine, phenylalanine, proline, serine, threonine, tryptophan, tyrosine, valine

Organic Acids[a]

Acetic, aconitic, aldonic, ascorbic, benzoic, butyric, caffeic, citric, p-coumaric, erythonic, ferulic, formic, fumaric, glutaric, glycolic, glyoxilic, lactic, malic, malonic, oxalacetic, oxalic, p-hydroxy benzoic, piscidic, propionic, pyruvic, succinic, syringic, tartaric, tetronic, valeric, vanillic

Fatty Acids[a]

Linoleic, linolenic, oleic, palmitic, stearic

Sterols

Campesterol, cholesterol, sitosterol, stigmasterol

Growth Factors

p-amino benzoic acid, biotin, choline, n-methyl nicotinic acid, niacin, pantothenic, vitamins B_1 (thiamine), B_2 (riboflavin), and B_6 (pyridoxine)

Enzymes

Amylase, invertase, peroxidase, phenolase, phosphatases, polygalacturonase, protease

Flavonones and Nucleotides

Adenine, flavonone, guanine, uridine or cytidine

Miscellaneous

Auxins, p-benzoquinone, scopoletin, hydrocyanic acid, 8-hydroxyquinoline, glucosides, hydroxamic acids, luteolin, unidentified ninhydrin-positive compounds, unidentified soluble proteins, reducing compounds, ethanol, glycinebetaine, inositol and myo-inositol-like compounds, Al-induced polypeptides, dihydroquinone, quercetin, quercitrin, sorgoleone

[a] The pKa's of these acids and the pH of the solution will determine the form that these acids adopt.

Source: From Rovira, A.D., *Bot. Rev.*, 35, 35, 1969; Hale, M.G. et al., *Interactions between Non-Pathogenic Soil Micro-Organisms and Plants,* Dommergues, V.R. and Krupa, S.V., Eds., Elsevier, Amsterdam, 1978, p. 163; Kraffczyk, I. et al., *Soil Biol. Biochem.*, 16, 315, 1984; Curl, E.A. and Truelove, B., *The Rhizosphere*, Springer-Verlag, New York, 1986; Schönwitz, R. and Ziegler, H., *Z. Pflanzenernähr. Bodenk.*, 152, 217, 1989; Einhellig, F.A. and Souza, I.F., *J. Chem. Ecol.*, 18, 1, 1992; Basu, U. et al., *Plant Physiol.,* 106, 151, 1994; Fan, T.W.M. et al., *Anal. Biochem.*, 251, 57, 1997; Shinmachi, F., *Plant Nutrition for Sustainable Food Production and Environment,* Ando, T., Fujita, K., Mae, T., Matsumoto, H., Mori, S., and Sekiya, J., Eds., Kluwer Academic Publishers, Dordrecht, 1997, p. 277; Bais, H.P. et al., *Plant Physiol.*, 128, 1173, 2002; Dakora, F.D. and Phillips, D.A., *Plant Soil*, 245, 35, 2002; Inderjit, S. and Weston, L.A., in *Root Ecology*, de Kroon, H. and Visser, E.J.W., Eds., Springer-Verlag, Berlin, 2003, chap. 10; Rumberger, A. and Marshner, P., *Soil Biol. Biochem.*, 35, 445, 2003; Vivanco, J.M. et al., *Ecol. Lett.*, 7, 285, 2004. With permission.

is great, daunting at times, and a source of some frustration. But another source of frustration has been born out of unfulfilled expectations — expectations that were, as it turned out, unrealistically high. A period of rationalization may be arising as, for example, Vaughan and Ord [23] prefer to refer to phenolic acids as *phytotoxins* rather than allelochemicals because the proof of allelopathy is difficult to establish; other authors may not be quite so circumspect. Similarly, Jones et al. [24] state: "Root exudates released into the soil surrounding the root have been implicated in many mechanisms for altering the level of soluble ions and molecules within the rhizosphere, however, very few have been critically evaluated." Further similar cautionary statements have been made [25–29] and none more apt than the following by Farrar and Jones [30]:

> Root exudation cannot be simply explained by a single mechanism but is moreover a combination of complex multidirectional fluxes operating simultaneously. While we currently possess a basic understanding of root exudation, its overall importance in plant nutrition, pathogen responses, etc., still remains largely unknown. Future research should therefore be directed at quantifying the significance of root exudates in realistic plant–soil systems.

Many of the problems arise out of the extrapolation of what happens in solution cultures to soils. Although solution cultures have served, and continue to serve, very useful functions in basic research of plant science, they differ from soils in several important ways: (1) the surface area available in soils for processes such as sorption is much greater than in solution cultures, (2) solution cultures are mixed continuously, (3) the microbial ecology differs greatly between the two media, and (4) the status of water and O_2 in the two systems is usually quite different. It is not difficult to find quotes from eminent plant scientists that indicate their support for this general view. For example, "Laboratory studies blind us to the complexity found by careful study of roots in soil" [31] and "The idea, however, that the laboratory control is the norm is false and can lead to misunderstanding and poor predictions of behavior" [32].

In this chapter I have continued to take a skeptical view, as I did previously [19], of the rosy pictures so commonly painted in the current literature. The role of the devil's advocate has been adopted to raise issues that are too readily disregarded or assumed glibly to be true. Although some attention will be given to the quantitative aspects of the release of root products, the main purpose is to classify types of root products on the basis of their known properties and perceived roles in the rhizosphere.

Because most phytoactive compounds released by plants do not persist in soil in a free and active form for very long, it appears implausible that they should be implicated in a process of infection or nutrient acquisition, for example. However, circumstances in the rhizosphere or at the root–soil interface may arise, through normal root growth through soil that preserve the compound and its activity, and thus facilitate the process. Because the theory of Mn uptake in soils of neutral and alkaline pH failed to adequately explain the observed facts, the "right set of circumstances" was invoked by Uren and Reisenauer [25] to explain how labile reducing agents secreted by roots may be protected physically from O_2 and microbial degradation, and thus be preserved to react with insoluble oxides of Mn at the root–soil interface. Provided that the right set of circumstances occurs sufficiently and frequently, then the Mn needs of the plant will be met, but if not, then deficiency prevails. The right set of circumstances may have relevance in other situations where other types of root exudates remain phytoactive, in what appears to be a hostile environment. Such possibilities are discussed later in this chapter.

This chapter considers the various types of root products with a potential functional role in the usually tough environment of soil. Only direct effects of immediate benefit to plant growth, e.g., an increase in nutrient solubility, will be considered here. Although root products of a plant species may have a direct effect on important groups of soil organisms such as rhizobia and mycorrhizae, their effect on the plant is not immediate; these and other aspects related to microbial activity in the rhizosphere will not be considered here (see Chapter 3, Chapter 9, and Chapter 11). For some reviews of the microorganisms in the rhizosphere, the reader is referred to Bowen and Rovira [33] and to more recent works [34–38].

II. ROOT GROWTH, THE RHIZOSPHERE, AND ROOT PRODUCTS

Roots are linear underground organs of plants that grow through soil with a complex architecture, a three-dimensional configuration, which in turn secures the plant and facilitates the exploitation of soil for nutrients and water [39,40]. The complexity of root growth and the architecture of root systems are illustrated in several reviews [41–43] (see also Chapter 5), whereas more recent emphasis has been on the influence of nutrients on root morphology and architecture [44–46]. The pattern of root growth, given adequate supplies of nutrients, is determined largely by the type of plant, the soil water potential, and its interaction with soil structure. The growth of roots is such that in soil under ideal conditions, favoring full exploitation of nutrients and water, roots avoid one another and rarely do neighboring roots interfere with one another; the heterogeneity of soil structure tends to separate roots spatially [47]. Under suboptimal soil conditions where yield is decreased, any configuration of the root system imposed by the conditions cannot make good the deficit.

The successful models of nutrient uptake have included the parameter for root surface area, or proxies for surface area, such as root length and radius or rate of root growth and radius (see Chapter 12). The sensitivity analysis carried out by Silberbush and Barber [48] reinforced what the modelers perceived to be important. Those soil conditions that inhibit root elongation, such as Al toxicity, high bicarbonate activities, and heavy metal toxicity, restrict the uptake of the least mobile nutrients and so, for example, iron chlorosis is a common symptom of heavy metal toxicity.

The rhizosphere forms around each root as it grows because each root changes the chemical, physical, and biological properties of the soil in its immediate vicinity. The rhizosphere along the axis of each root can be described in terms of the longitudinal and radial gradients that develop as a result of root growth, nutrient and water uptake, rhizodeposition, and subsequent microbial growth. In solution cultures most of these gradients tend to be obliterated by the active mixing, such that a phytoactive compound released at one point on the root axis may have an impact on or at another more distal (nearer the root apex) point, a situation that may have little or no relevance for soil-grown plants. Further, although reabsorption of sugars and amino acids occurs in solution-based systems [49], it seems unrealistic that root exudates can play a major role in plant growth, because they are highly biodegradable and assimilated by microbes in the rhizosphere.

The evaluation of rhizodeposition in soil in terms of not only the quantities of compounds released by roots and their identification but also the sites of release, their fate, and their impact remains a major difficulty faced by researchers in this field [50–56], see Chapter 13. When faced with such an apparently intractable situation, one might ask, "Does it matter?" Maybe for most root products it does not matter as it is really quite remarkable that, given adequate light and support, healthy vigorous plants can be grown in sterile nutrient solutions. The point here is that it is possible, that in most respects plants can manage quite well without the majority of effects that have been proposed to occur in the rhizosphere.

Rhizodeposits stimulate microbial growth because of their high energy and C content, and thus they will tend to decrease the availability of nutrients in the region of the rhizosphere where they are released from the root. Because the apical regions of the root (i.e., from the root hair zone to the root apex) have extracted most nutrients that are available for uptake before there is extensive colonization of the rhizosphere by saprophytic microorganisms, there is no impact on plant growth but microbial growth is likely to be limited. Merckx et al. [57] found that microbial growth in the rhizosphere of maize was limited by the depletion of mineral nutrients. Similarly, it was found that microbial respiration was not limited in the rhizosphere of winter wheat by available C [58], which in turn indicates, as one would expect, that perhaps some other nutrient element is limiting. The high C-to-N ratio of root products leads to the suggestion that N may be significant [59], and so it is not a surprise that rhizodeposition by maize plants enhanced microbial denitrification and immobilization of N in the rhizosphere [60]. Immobilization of other nutrients may occur also, but, as suggested earlier, it may not be a problem for the plant because root extension and absorption of nutrients occurs at a more rapid rate than the colonization of the rhizosphere with competing microorganisms. Ultimately, nutrients immobilized by microbial assimilation are likely to become available (recycled) following the death and degradation of microbes and microfauna, but the timing and location of such events in relation to the nutrient-absorbing regions of the root has not been pursued, probably because of the difficulties involved. For example, Mary et al. [61] measured the recycling of C and N during the decomposition of root mucilage, glucose, and roots by simply mixing the substrates with soil at low rates of addition. Although mixing by soil fauna and cultivation may occur in some circumstances such as annual plants, is it realistic where, say, perennial plants are growing?

High root densities may lead to the overlapping of rhizospheres but not in a consistent way, and extensive overlapping is more likely to be the exception rather than the rule. Some poor soil structures, e.g., cloddiness, may cause clumping of roots, but such an arrangement would not appear to be the preferred way, and it is unlikely to confer any benefits upon the plant's growth. Even if there are any benefits, they are not sufficient to make good the decrease in plant yield brought about by poor structure. Although some of the contrived systems of studying plant root exudates utilize clumping of roots in an attempt to mimic the rhizosphere (see Reference 62), the effects measured in those systems may reflect what happens when roots are clumped together, nothing more, and not what happens in more normal and desirable circumstances in soil-grown plants in the field. In some pot and field experiments, the Fe status of peanuts was increased by close intercropping with maize [63], and although the cause is no doubt due to an increase in the availability of Fe in the intermingling rhizospheres, there is no direct evidence to support the contention that phytosiderophores were responsible [64].

A common situation that requires some consideration is the emergence of lateral roots. The apices of these roots must grow through the rhizosphere of the superior axis from which they originate, and thus may experience specific effects related to the type of exudates produced by the main axis and the microbial population. If the situation in nonsterile solution cultures has relevance in soil, then the high level of bacterial colonization of the rhizoplane observed near emerging laterals of maize in solution cultures [65] may be significant. However, any effects on the new lateral root apex, which may be significant and quite specific, have not been investigated.

Root products are all the substances produced by roots and released into the rhizosphere (Table 1.2) [25]. Although most root products are C compounds, the term includes ions, sometimes O_2, and even water. Root products may also be classified on the basis of whether they have either a perceived functional role (excretions and secretions) or a nonfunctional role (diffusates and root debris). Excretions are deemed to facilitate internal metabolism such as respiration, whereas secretions are deemed to facilitate external processes such as nutrient acquisition [25]. Both excretion and secretion require energy, and some exudates may act as either. For example, protons derived from CO_2 production during respiration are deemed *excretions*, whereas those derived from an organic acid involved in nutrient acquisition are deemed *secretions*.

TABLE 1.2
Root Products: A Classification

Product	Compound
Root exudates	
Diffusates	Sugars, organic acids/anions, amino acids, water, inorganic ions, oxygen, riboflavin etc.
Excretions	Carbon dioxide, bicarbonate ions, protons, electrons, ethylene, etc.
Secretions	Mucilage, protons, electrons, enzymes, siderophores, allelochemicals, etc.
Border cells	Root cap cells separated from the root apex
Root debris	Cell contents, lysates, etc.

Source: From Uren, N.C. and Reisenauer, H.M., *Adv. Plant Nutr.*, 3, 79, 1988. With permission.

The name *root border cells* has been adopted for those root cap cells that separate during growth from the root apex [66]. Most plant species, but not all, appear to exhibit this release of border cells *in vitro*, particularly in free water [67]. In soil, maize border cells have been observed to remain intact and alive among root hairs in the rhizosphere [68–70] and to continue secretion of mucilage for up to 3 weeks after separation [71]. In 1988, Uren and Reisenauer [25] suggested that "if these cells had fulfilled their functions of protection, secretion of mucilage, and geotropic response near the root cap, then it appears wasteful of photosynthate that they should not serve another purpose in association with more proximal regions of the root." Although many other roles have been proposed for border cells [67,69,72–74], it would seem, thus far, "to date, research on root cap biology and its relationship with the rhizosphere has raised more questions than it has answered" [67].

Most root debris comes from the senescence of cortical cells, an event which may be the trigger for infection with mycorrhizae, but it is probably of little, real, direct consequence for plant growth in fertile soil. The possibility that phytohormones such as indole acetic acid, cytokinin, and abscisic acid produced by rhizosphere bacteria of field-grown maize [75] do have a consistent effect on plant growth has yet to be established. The production of plant growth-regulating substances in the rhizosphere has been reviewed extensively by Arshad and Frankenberger [76], and they conclude that there are many unresolved aspects when it comes to the consideration of *in situ* events and circumstances.

Root products as defined by Uren and Reisenauer [25] represent a wide range of compounds. Only secretions are deemed to have a direct and immediate functional role in the rhizosphere. CO_2, although labeled an excretion, may play a role in rhizosphere processes such as hyphal elongation of vesicular-arbuscular mycorrhiza [77]. Also, root-derived CO_2 may have an effect on nonphotosynthetic fixation of CO_2 by roots subject to P deficiency and thus contribute to exudation of large amounts of citrate and malate as observed in white lupins [78]. The amounts utilized are very small and, in any case, are extremely difficult to distinguish from endogenous CO_2 derived from soil and rhizosphere respiration.

III. AMOUNTS RELEASED

The bulk of root products are C compounds derived from products of photosynthesis. The root products that are not C compounds are few (H^+, inorganic ions, water, electrons, etc.), but nevertheless they are deemed to be highly significant. Both H^+ and electrons may be secreted as C compounds in the form of undissociated acids and reducing agents, respectively, but plasma membrane-bound entities are believed to be the main sites of H^+ and electron transport [79–81]. The origins of root-mediated pH changes in the rhizosphere have been discussed by Hinsinger et al. [82], whereas Ryan et al. [83] do not exclude the possibility of exudation of undissociated organic acids. The reducing capacity of roots has been known ever since it has been discovered that they require oxygen. Schreiner et al. [84] found that selenite was reduced to metallic Se by wheat roots in solution culture, and Lund and Kenyon [85] showed that onion roots reduced methylene blue, but nonsterile conditions prevailed in these and other similar experiments. Uren [86] found that sterile sunflower roots reduced an insoluble higher oxide of Mn impregnated into filter paper and showed unequivocally that roots in their own right secreted reducing agents.

Estimates of the amounts and proportions of photosynthate committed to roots and to root products vary considerably, and the shortcomings associated with measurements have been critically and realistically reviewed [27,87,88] (see also Chapter 13). The units used, or those which might be used, need closer attention. In relation to the cycling of C rhizodeposition, rates of kg C ha^{-1} y^{-1} are appropriate, whereas micromoles per unit root length per hour might be more appropriate in relation to the secretion of phytosiderophores. Because uptake, for example, may be restricted to a specific region of each root, then the units of micromoles per unit root length per hour might be even more relevant still. Darrah [27] concludes that the major challenge to quantify the individual flux components of rhizodeposition remains. Obviously, further huge challenges exist as so little is known of the timing of the release (in relation to the stage of growth and development), the sites of the release, and other aspects of individual secretion.

TABLE 1.3
Rough Estimates of the Fates of Carbon Fixed by Soil-Grown Plants

Photosynthesis = 100%

           Shoots = 50%

                 Shoots = 45%

                 Respiration = 5%

           Roots = 50%

                 Root biomass = 25%

                 Root products = 25%

                               Respiration = 15%

                               Root debris including border cells = 10%

                               Diffusates <1% (guess)

                               Secretions <1% (guess) includes mucilage — may be more

Note: Amounts and relative proportions depend on species, cultivars, environmental conditions, health, age, level of chemical, physical and biological stress, and so on.

Source: From Darrah, P.R., *Plant Soil,* 187, 265, 1996; Whipps, J.M., *The Rhizosphere*, Lynch, J.M., Ed., John Wiley and Sons, Chichester, U.K., 1990, p. 59; Lynch, J.M. and Whipps, J.M., *The Rhizosphere and Plant Growth,* Keister, D.L. and Creagan, P.B., Eds., Kluwer Academic Publishers, Dordrecht, 1991, p. 15. With permission.

All the variables aside, approximately 50% of fixed C is committed to roots (Table 1.3). Fifty percent of this C is retained as root tissue and the other 50% is root products. Three fifths (15% of the net fixed C) is used in root respiration and two fifths (10%) make up border cells, root debris, diffusates, and secretions. Of the latter, border cells and root debris predominate ahead of secretions (largely mucilage), with diffusates making up the difference [27,87,88]. The contribution of root border cells is difficult to estimate, but as Griffin et al. [89] estimated that for sterile peanuts grown in solution culture "95 to 98% of the sloughed organic matter plus total sugars lost by roots is sloughed organic matter," the contribution might be significant. By comparison, 5 to 10% of C deposited by maize roots in sand culture was attributed to sloughed cap cells [90]. More recent estimates (for example, see Reference 91) are not greatly different, and so at present we have to accept these rough estimates in much the same way that we acknowledge that many factors affect the proportion released by roots and that sick or stressed plants make a larger commitment than healthy plants [25]. It is often construed without much evidence for soil-grown plants that such an extra commitment is a controlled response to stress, which in turn enables the plant to overcome the stress, the so-called *stress response*. Ryan et al. [83] emphasize the need for caution when stressed plants are involved: "It is important to remember that P-deficient plants are stressed plants, and every metabolic perturbation that they display will not necessarily be directed toward increasing P availability in the rhizosphere." In field soil, the environment is not nearly as friendly as in solution cultures, and incontinent roots are likely to encourage infection by pathogens as much as by beneficial or saprophytic microorganisms. There must be a limit to what quantities and proportions of the fixed C can be lost in stress responses but, as with quantitative estimates of exudation, they must remain as uncertain at best.

In an investigation of the phytotoxicity and antimicrobial activity of (±)-catechin exuded by knapweed roots, extraction of the soil with absolute methanol for 24 h gav concentrations ranging from 292 to 390 µg/g [10]. Assuming a bulk density of 1 g cm^{-3} and a volumetric water content of 0.25, the concentration of catechin in the soil solution would be of the order of 1200 µg/ml (approximately, 4 mM), which is to be compared with the 50 to 60 µg/ml of (–)-catechin required to give an allelopathic response in *Arabidopsis* seedlings. The extracted concentrations in the soil would amount to about 1% of the total soil organic matter and as such appear abnormally high. However, as there have been so few studies of the chemical forms and activities of potential allelochemicals in soils, we have little idea of what is normal or abnormal, and thus caution is required in evaluating new data.

For wheat plants grown to maturity under irrigation [92] in a soil of neutral pH [93], one can calculate that a mature wheat plant (yield 13.2 g dm [dry matter]) with a Mn concentration of 42 mg kg^{-1} took up 556 µg of Mn (i.e., $556/55 \times 2$ equivalents of Mn). If ascorbic acid ($M_{AA} = 176$), or another reducing agent of similar equivalent weight and C content, is assumed to reduce insoluble Mn oxides at the root–soil interface, then it can be calculated that an amount of ascorbic acid equivalent to about 0.01% of the total C in the mature plant needs to be secreted to give the concentration of 42 mg Mn/kg in the mature plant. In this calculation it is assumed that all the Mn in the plant comes from the reduction of insoluble Mn oxides and that every molecule of reductant hits its target with concomitant uptake of the Mn^{2+} formed. As the root system is made up only 1.7% of the total dry matter in these mature wheat plants, it is difficult to give a more precise estimate of the proportion of the total C attributed to Mn mobilization. Nevertheless, 0.01% is near the maximum, and values less than 0.01% are realistic.

Similar calculations might be performed for other secretions, which are neither changed nor consumed in their interaction with soil entities. For example, complexation of Fe^{3+} by a ligand secreted by the root involves first the diffusion of the ligand away from the root to the insoluble oxides of Fe, a complex forms between the ligand and Fe^{3+}, and then, if the appropriate activity gradient exists, the complex diffuses to the root. At the plasmalemma, if the Fe^{3+} is complexed by a phytosiderophore [59] it is absorbed by the root, but if the Fe^{3+} is separated from the ligand, then the ligand is free to diffuse back into the soil, and so the process may continue. The quantities of root secretion required in such a "search and fetch" role are likely to be much less than the earlier case for Mn, where the secretion is destroyed, or in cases such as Al toxicity, where it is important that the secretion stays associated with the metal (if not permanently, then at least until the root tip has progressed beyond the point of interaction).

IV. TYPES OF ROOT PRODUCTS: SECRETIONS AND THEIR ROLES

Root products contain probably every type of compound that exists in plants, except for chlorophyll and other specific compounds associated with photosynthesis (Table 1.1). The range of compounds is increasing with the increasing sensitivity and analytical capabilities of modern equipment. For example, Fan et al. [8] analyzed comprehensively the root exudates of iron-stressed barley plants with multinuclear or multidimensional nuclear magnetic resource (NMR) and silylation gas chromatography/mars spectrometry (GC-MS), not only bypassing tedious traditional methods but also detecting unknown and unexpected ligands.

Most root products are by-products, which represent some of the costs of growth and development, and, except in the development of symbiotic relationships, they simply become the substrate for attendant microorganisms. With time the organic C compounds are converted progressively to either CO_2 or into recalcitrant forms of organic matter (e.g., humins). There may be indirect effects associated with heterotrophic activity, which may be either harmful or beneficial, but these will not be discussed here (see Chapter 3). Of the other root products, secretions that facilitate external processes are our primary interest here, and they are discussed in the following text.

Root products may be classified on the basis of their (1) chemical properties such as composition, solubility, stability (e.g., hydrolysis and oxidation), volatility, molecular weight, etc., (2) site of origin, and (3) established, not just perceived, functions. The chemical properties in turn determine their biological activity and how the compounds will behave in soils; their persistence in soil is very much an outcome of their chemical behavior, particularly sorption and biodegradability. Root products as chemical signals, and issues relating to their persistence, etc., are discussed in Chapter 11.

The persistence of a secretion and the likelihood that it will reach an appropriate nutrient source and be effective in its role is a prime concern. A secretion must be free to diffuse through a portion of the rhizosphere, but a sort of tyranny of distance exists. The longer it takes, or the further it must travel, then the greater is the chance that it will be rendered ineffective by either microbial degradation or assimilation, or chemical degradation or reaction, or by sorption, or by a combination of these processes. Low-molecular-weight exudates (sugars, amino acids, and other organic acids), the so-called diffusates, may be more mobile, but usually they are more readily assimilated by a wider range of microorganisms than are high-molecular-weight compounds such as mucilage, although Mary et al. [61] found the mineralization rate of C from mucilage of maize roots to be comparable with that of glucose.

Secretions may be classified also on the basis of their biological activity. Some of the classes are phytohormones, ectoenzymes, phytoalexins, allelochemicals, and phytotoxins, or referred to as "chemical signals" (see Chapter 11). However, whether or not they can exert their potential activity, which so often has been illustrated in solution cultures or under axenic conditions, depends on their survival in the soil, being at the right place, and for long enough at appropriate concentrations. All these preceding classes of compounds, except for phytoalexins and protectors against toxic Al, are usually secreted during normal plant growth in the absence of stress. Phytoalexins are secreted in response to an external stimulus (infecting organisms), which presumably is a chemical compound, whereas in the case of Al toxicity the plant response in soil occurs if the Al is present in a mobile and active form [83]. The possible roles of some different types of root secretions are given in Table 1.4.

The growth of roots through soil is perceived often as improving soil structure for plant growth. In the context of this review, the question is whether or not a plant's root products directly improve the soil structure for the growth of that plant. However, it is a difficult question to answer, because in addition to the release of root products there may be shearing and compression that together, in turn, may tend to destabilize aggregates [94], perhaps with some benefits (e.g., mineralization of physically protected organic N). Accompanying root elongation and radial expansion, there is the

TABLE 1.4
Possible Roles of Some Different Types of Root Secretions

Role	Action
Acquisition of nutrients	
Fetchers	Seek and fetch, e.g., phytosiderophores
Modifiers	Modification of the rhizosphere soil with, e.g., protons, reductants
Ectoenzymes	Convert unusable organic forms to usable ones, e.g., phosphatase
Acquisition of water	Modification of the rhizosphere soil with mucilage
Protection against physical stress	Response to high soil strength through modification of interface through lubrication and amelioration of rhizosphere soil
Protection against pathogens	Defensive response to invasion, e.g., phytoalexins
Protection against toxic elements	Response to toxic entity, e.g., complexation of Al^{3+}
Protection against competition	Modification of rhizosphere soil with phytoactive compounds, e.g., allelochemicals
Establishment of symbiotic relationships	Chemotactic response
Rhizobia	
Endomycorrhizae	
Ectomycorrhizae	

permeation of soil with mucilage that has been shown *in vitro*, and when accompanied by wetting and drying, to confer stability on soil aggregates [95]. Other than permeation of soil with mucilage, most effects on structure would appear to be indirect effects, and thereby of little benefit to the plant whose roots brought about the change in structure.

Whiteley [96] found that remolded soil near (300 to 600 μm from the root surface) the roots of peas showed evidence of orientation of the clay fraction. He argues that the change could only have come about if the soil water potential had increased because of mucilage secretion by the root tip; the lower soil strength then allowed deformation by the growing root. Also, he believes that the lower soil strength may then predispose the remolded soil in the rhizosphere to penetration by root hairs and lateral roots. The secretion of water by roots as observed by McCully [97] and Young [98] adds weight to his argument.

It is impossible to discuss the possible effects of mucilage associated with the root apex without including border cells, although there are some species of plants which do not produce border cells [67]. Bengough and Mckenzie [99] suggested that the mucilage assists root cap cells, or acts in concert with them, to decrease the friction between the growing root tip and soil or, conversely, the mucilage acts as a lubricant. Iijima et al. [100] showed that the border cells of maize roots tips were more effective than mucilage at decreasing soil resistance to root elongation. Presumably, then, those plant species that do not produce either much mucilage or border cells are not at a great disadvantage when it comes to root growth in soil.

The permeation of soil at the root–soil interface by mucilage from the root cap may affect structure, and it may oppose the damaging effects of compression and shearing. Read et al. [101] found that the mucilages of maize, lupin, and wheat contained phospholipids that indirectly were shown to decrease the surface tension of water and thus facilitate possibly the permeation of soil with mucilage. And, attempts to measure the development of water repellency in the rhizosphere of barley, oil-seed rape, potato, and Italian ryegrass were not convincing [102]. Whalley et al. [103] attributed a decrease in the infiltration of water into rhizosphere soil to deformation of the soil, rather than to any other changes due to mucilage or the development of water repellency. Also, they found that neither natural mucilage from maize nor polygalacturonic acid affected the soil water characteristic between 0 and -15 kPa.

It is sometimes claimed that mucilage and similar gels may help to maintain hydraulic conductivity between root and soil [104]. However, the hydraulic conductivity of soils is often substantially decreased when soils are irrigated with wastewater, which is largely due to the production of microbial biomass, particularly extracellular polysaccharides. For example, Wu et al. [105] found that the hydraulic conductivity of sand was decreased by one to one-half orders of magnitude 3 weeks after treatment with a mixture of dextrose and nutrient solution. These extracellular polysaccharides form gels which may store large quantities of water and allow water and ions to diffuse through them at rates not much less than free water, but they could be expected to restrict mass flow of water, and thus some nutrients, to roots [106].

Another apparent paradox exists as mucilage is most easily seen on roots when they are immersed in water, and yet the evidence of its secretion in soil, apart from electron micrographs [71], is the development of rhizosheaths in soil-grown plants, particularly at relatively low water potentials [107]. The explanation may be simple in that at high water potentials the cohesive and adhesive properties of mucilage are low and bonds are easily broken, whereas in drier conditions the bonds are much stronger and so soil sticks more readily to the root. An alternative suggestion is that mucilage and water is secreted as a gel when conditions favor guttation, that is, at night and at water potentials from -120 to -500 kPa [97].

The release of mucilage from the periplasmic space can be triggered by contact with water at high potential [108]. In soil, growing roots are only exposed to high soil–water potential when the soil is saturated after rain or irrigation, but they make most of their growth at water potentials between -10 and -1000 kPa when most intra-aggregate pores are full of water, and aggregate surfaces are covered by a thick water film [106]. When a root tip makes contact with the aggregates, the mucilage will tend to be secreted and to form a gel on the surfaces of aggregates [109] and in those pores that favor accommodation of the macromolecules of mucilage, that is, pores that are full of water, big enough, and do not repel the molecules. If the diameter of the mucilage molecule is taken as 68 nm [110], and diffusion of molecules is severely restricted in pores up to one order of magnitude larger than the molecule [111], then mucilage will move into pores whose diameters are greater than about 680 nm, which is equivalent to a soil water potential of about −500 kPa. Pores of such size are large enough to accommodate bacteria, and so the mucilage molecules may not be safe from microbial degradation. We obviously need to know more about the secretion of mucilage from the periplasmic space, its physical and chemical properties, its interaction with soil, and the consequences.

In soil, the chances that any enzyme retains its activity are very slim, indeed, because inactivation can occur by denaturation, microbial degradation, and sorption [112,113], although it is possible that sorption may protect an enzyme from microbial degradation or chemical hydrolysis and retain its activity. The nature of most enzymes, particularly size and charge characteristics, is such that they would have very low mobility in soils, so that if a secreted enzyme is to have any effect, then it must operate close to the point of secretion, and its substrate must be able to diffuse to the enzyme. Secretory acid phosphatase was found to be produced in response to P-deficiency stress by epidermal cells of the main tap roots of white lupin and in the cell walls and intercellular spaces of lateral roots [114]. Such apoplastic phosphatase is safe from soil but can only be effective when presented with soluble organophosphates, which are often present in the soil solution [115]. However, because the phosphatase activity in the rhizosphere originates from a number of sources [116], mostly microbial, and is much higher in the rhizosphere than in bulk soil [117], it seems curious that plants would have a need to secrete phosphatase at all. The role of phosphatases in the rhizosphere remains uncertain [118].

Of enzymes that appear to acquire or retain their activity in soils, urease is an example. It is a microbial product, bound to soil, and causes very rapid hydrolysis of urea upon its addition to soil.

By contrast when phytase from seedlings of transgenic *Arabidopsis thaliana* was added to soil, it was rapidly immobilized and deactivated by adsorption, which was favored by low soil pH [119]. The phytase could be desorbed by increasing the pH but gradually the overall activity decreased, possibly due to proteinase activity.

Many phytoalexins and allelochemicals have much in common, in that they are designed to protect plants from either injury by other organisms or competition from other plants, and they are usually aromatic compounds. Chemical signals are discussed in Chapter 11. The phytoalexins are usually larger and more complex molecules [120] than those deemed to be allelochemicals [23]. Compounds of this type that are secreted by roots have to run the usual gauntlet in the rhizosphere, and if any of the secretions are likely to survive the trip, then the phenolics implicated in allelopathy have some chance. Not only do the phenolics have low molecular weights and usually have some negative charge, but they also have some antimicrobial activity. In spite of these attributes, benzoic and cinnamic acid derivatives are easily degraded microbially in soils [121], and they are chemically oxidized by Mn oxides [122,123] and adsorbed by soils as well [123,124]. Further, when nine different phenolics (caffeic acid, chlorogenic acid, *p*-coumaric acid, ellagic acid, ferulic acid, gallic acid, *p*-hydroxybenzoic acid, syringic acid, and vanillic acid) were added to three different soils, they had no effect on seed germination, on seedling growth, or on early plant growth of several species, even when added to soils at rates well above the concentrations detected in soils [125]. The case for these types of compounds as having an important role in root secretions is not strong, in spite of the case argued earlier, except perhaps in the right set of circumstances, discussed later. The situation presented here is supported by the statement that "it is important to demonstrate that phenolics, released by the plant, should have enough bioactive concentration and persistence in the rhizosphere in order to argue their probable involvement in allelopathy" [126]. Other potential allochemical compounds, such as momilactone B found in rice exudates [127] and sorgoleone in sorghum spp. [128], appear to be better candidates but very much depends on their fate and behavior in soil.

Inderjit and Duke [129] indicate that there is a lack of consensus in identifying any compound as an allelochemical, and they discuss the requirements for a chemical compound to be deemed an allelochemical. The difficulty associated with the isolation, identity, activity, etc., of suspected allelochemicals is illustrated by the work of Wu et al. [130], who isolated significant bioactive phenolic and hydroxamic acids exuded into agar growth medium by wheat seedlings, and they found that the so-called allelochemicals inhibited the growth of annual ryegrass (*Lolium rigidum*). Annual ryegrass is one of the worst weeds in cereal growing in Australia, and so one can only wonder about the relevance of the finding, particularly in the context of the difficulties faced in the search for allelochemicals.

The roots of canola plants and decaying residues of canola crops release 2-phenylethylisothiocyanate, which is believed to inhibit soilborne pathogens [13], but its microbial degradation is rapid. In a Luvisol (20% clay) the concentration decreased 50% from the initial concentration of 3382 pmol g^{-1} in close to 2 h. Such rapid degradation suggests that if the compound is to have an effect it will be close to the root, and it will need to be secreted in a specific zone and not be released along with cortical degeneration.

Nonaromatic organic acids such as citric have been implicated in nutrient acquisition since the 19th century [131] and, in spite of the certainty with which some authors assert that these acids play an unquestionable role, there is still uncertainty [132,133]. The process involves secretion of the acid in either an undissociated form (H_3X) or a dissociated form (H_2X^-, HX^{2-}, or X^{3-}). Although it would seem that, at the pHs that prevail in the cytoplasm and in the soil solution, the form secreted is anionic and not as the acid [83], it is possible that the undissociated acids are secreted or that there is a concomitant efflux of H^+ [133]. The form of the acid in the rhizosphere depends on the pH and on the availability of metals and their tendency to form complexes of varying stability. The acid *per se* may mobilize metals by dissolution (e.g., $Fe(OH)_3$) or cation exchange, whereas the anion through its tendency to form stable soluble complexes may protect metals from precipitation, or it may cause insoluble sources to dissolve, or it even may precipitate a cation such as Ca^{2+} [134]. These respective reactions for Fe can be represented as follows, where X^{3-} represents an anion such as citrate:

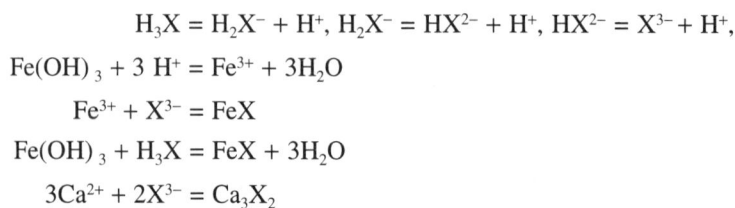

$$H_3X = H_2X^- + H^+, \ H_2X^- = HX^{2-} + H^+, \ HX^{2-} = X^{3-} + H^+,$$
$$Fe(OH)_3 + 3\,H^+ = Fe^{3+} + 3H_2O$$
$$Fe^{3+} + X^{3-} = FeX$$
$$Fe(OH)_3 + H_3X = FeX + 3H_2O$$
$$3Ca^{2+} + 2X^{3-} = Ca_3X_2$$

It is conceived that the soluble complexes once formed diffuse back to the root where the complex is absorbed or the metal is separated from the ligand and then absorbed [29]. The ligand is then released back into the rhizosphere and, as the organic acids are not prone to reabsorption [135], they are fully available to run the gauntlet once more. Because citrate is adsorbed by soil surfaces and rapidly degraded microbially [136], there are serious doubts about the role of citric acid and similar organic acids in the acquisition of nutrients except, perhaps, in the right set of circumstances, where some form of protection is proffered by the spatial arrangement of root and soil surfaces. Further, the soil diffusion rates of soluble exudates such as amino acids are several orders of magnitude less than, say, nitrate [137,138], and so the chances that they are assimilated by microbes in the rhizosphere rather than absorbed by roots are high.

It is curious that white lupins, which have the capacity to produce relatively large quantities of citric acid [134], do not grow as well as other species on calcareous soils [139,140] that cannot produce such large quantities of citric acid [141]. Hinsinger [142] highlights in a table some data from Dinkelaker et al. [134], in which the concentration of citrate in the rhizosphere accompanies an eightfold increase in diethylenetriamine pentaacetic acid (DTPA)-extractable concentration of Fe, and yet white lupins suffer from Fe chlorosis on calcareous soils. Dinkelaker et al. [134] estimated that the quantity of citric acid produced was 23% of the total plant dry weight at harvest. Such a huge release should probably be regarded as abnormal rather than as a result of a stress response mechanism.

The role of the secretion from the root apex of organic acids such as citric and malic in the resistance of maize and wheat respective to Al toxicity [143,144] has emerged recently as one with plausibility [83,145]. These studies have been carried out in solution cultures, but how does the suggestion hold up in soil? The first and probably greatest difficulty is that the toxic species of Al, probably hydrated Al^{3+}, must diffuse to some site in the root apex and stimulate the production and subsequent release of the organic acid. The site may be extracellular or intracellular, but whichever it is, the production of the organic acid must be intracellular, and after its release it must inactivate the toxic Al species in the apoplasm as beyond the apoplasm rapid microbial degradation is likely [146]. The relative freedom of the root apex from microbial colonization and the production of mucilage both help to create the right set of circumstances that allows the detoxification of the Al to take place in the protection provided by the apoplasm. The observation that phosphate was released by the root apex of Al-tolerant cultivars of maize [144] and wheat [147] is of interest in this context, but it may be an unusual situation restricted to solution cultures, because those soils in which the activity of Al in the soil solution is high are usually P deficient as well. As plausible as the role of organic acids may be, there is evidence to suggest that organic acids, and polypeptides, arise out of Al-induced failure of membranes rather than *de novo* synthesis and secretion [7].

Similarly, the pivotal role of the so-called stress response in the acquisition of Fe by plants grown in calcareous soils [59] may have been overrated [118]. The key role of the inhibitory effect of bicarbonate on root growth of calcifuge plant species [148] has been overlooked in most considerations of the acquisition of Fe, the least soluble of all nutrients. Also, it is likely that Fe acquisition in normal healthy plants is a constitutive process, such as Mn uptake in barley [149], so that normal healthy roots acquire their Fe as a matter of course. By the time the Fe stress response is triggered, it is possible that cell membranes are losing their integrity and that compounds normally involved in metabolism requiring or involving Fe are released. It is likely, though, that some of the compounds released as a result of Fe stress are involved in the constitutive process, and their production is related to a species prowess in acquiring Fe from calcareous soils. The inability of the Fe response in so-called iron-efficient sunflowers to overcome Fe stress in a calcareous soil [150] suggests that the Fe stress response in this case may be restricted to culture solutions and of little relevance in calcareous soils. Another reason to question the theory of Fe stress response is that in soil the Fe-active compounds are rapidly decomposed by microorganisms [151–153]. The early reviews on root exudates by Rovira [1] and others [2,15] all drew attention to the numerous factors that affect membrane permeability and cause roots to leak; they are just as relevant today as they were 20 or more years ago [83]. Once again, the right set of circumstances may overcome the problem of microbial decomposition, but it cannot overcome membrane failure.

V. THE RIGHT SET OF CIRCUMSTANCES

Contact reduction was proposed to explain how plants obtain Mn from soils of neutral and alkaline pH [154]. The evidence that sterile roots of sunflower roots could directly reduce insoluble reactive oxides of Mn strengthened the theory [86]. Nevertheless, the idea of the "right set of circumstances" was developed to explain how labile reductants produced by roots may be protected physically from microorganisms and O_2 and be directed toward insoluble oxides of Mn instead [25]. The right set of circumstances are thought to arise where roots contact soil and an interface is created that is saturated with water, and that, with physical blockage, creates a zone with low activity of O_2 [155,156] and, presumably, of microorganisms. If the right set of circumstances arise sufficiently and frequently enough, then the mechanism remains a plausible one [25,157].

The absence of suitable reducing agents among root exudates has been taken as a flaw in this theory, but their absence is explained readily because either they are not looked for nor are adequate precautions taken to exclude O_2; also, the oxidized products are not recognized as derivatives of reducing agents. Evidence that this sort of thing could happen easily is provided by Einhellig and Souza [6], who analyzed for sorgoleone rather than its precursor dihydroquinone, a major root exudate of sorghum seedlings, as dihydroquinone was rapidly oxidized by ambient O_2 to sorgoleone. Also, the right of set circumstances that exist in soil would never occur in well-aerated solution cultures, although ascorbic acid, a suitable but labile reducing agent, has been found in solution cultures of healthy cucumber and tomato [9]. Many phenolic compounds discussed earlier in reference to allelopathy have reducing activity toward Mn oxides. For example, Park et al. [158] found in the root exudates of sunflowers hydroquinone, β-resorcyclic acid, vanillic acid, caffeic acid, salicylic acid, quercetin, gentisic acid, and ferulic acid, some of which readily reduce reactive Mn oxides, e.g., hydroquinone.

The complexity of plant root interactions with soil are such that even the right set of circumstances as described earlier is a simplification. For example, in the case of Mn acquisition, the right set of circumstances must arise at the right time and frequently enough for the plant to acquire sufficient quantities of Mn. It depends on (1) the constitutive properties of the roots to produce reducing compounds, (2) the growth of roots through soil so that the parts of the root producing the reductant and those involved in Mn absorption come into contact with soil, (3) the location of Mn oxides on the soil surfaces contacted by the root, (4) the reactivity of the Mn oxides (their ability to accept electrons) and (5) soil properties such as pH and structure. Although the number of variables involved is high, the probability of the right set of circumstances occurring frequently is also high in most soils: it becomes less so as the opportunities of roots making contact with active oxides becomes less.

VI. CONCLUSIONS

It is likely that of the vast array of compounds released by plant roots very few have a direct effect on the growth of soil-grown plants. One must ask whether or not they serve any useful purpose at all. The fears and uncertainty about what is physiologically normal were expressed by Ayers and Thornton in 1968 [159]; their concerns are as valid today as they were then. The absolute and relative quantities released are at present no more than approximations based largely on a very narrow selection of short-lived agricultural plants.

In attempting to classify types of secretions on the basis of what might happen in the rhizosphere, the foregoing discussion has taken a fairly distrusting view of data derived from *in vitro* experiments such as solution cultures, those using sterilized soil, and those using highly contrived situations where the reality of normal soil-grown plants is disregarded. The discussion has also highlighted the difficulties faced by some secretions and how these difficulties will decrease their likelihood of bringing about the process that they are purported to bring about. However, arguments in cases such as tolerance of Al toxicity and the acquisition of Fe and Mn can be strengthened by invoking the right set of circumstances.

Root products represent a vast array of predominantly organic compounds. Of these, secretions represent a small proportion, but they are deemed the most likely of all root products to have a direct effect on the growth of the plant that produced them. When a secretion is released by a root, all the following are likely to affect its behavior:

1. Site of secretion appropriate or inappropriate
2. Microbial assimilation or degradation
3. Chemical alteration or degradation, for example, oxidation
4. Sorption and persistence with or without activity, for example, ectoenzymes
5. Diffusion and reaction with the target, for example, complex with Al^{3+} or Fe^{3+}
6. Mechanism of uptake, for example, reabsorption by the root

Very few root secretions can be expected to be effective unless the right set of circumstances arise sufficiently often. Further, research involving soil-grown plants is required to establish whether or not the right set of circumstances as discussed earlier do make a real contribution to the well-being of field-grown plants. Similarly, close scrutiny of all research must be made, particularly when results obtained in solution cultures and other contrived situations are believed to be relevant for plants growing in soil.

Finally, if the understanding of the rhizosphere and the functional roles of compounds therein is to increase, then it is important that the enthusiasm shown in recent times continues, but that enthusiasm must be curbed at times by reality checks and frequent reference to what might actually happen in normal soil. One should also heed the words of Tinker and Nye [118]:

It is not surprising that the large literature on the rhizosphere in the past has produced rather little in the way of firm generalizations and mechanisms on the nutritional effects, bearing in mind its complexity, constant variation, and difficulty of access. It is interesting to note the list of processes in the rhizosphere that have been proposed in the past, but that have only slowly and often partially been proven to occur. The critical question is whether they are important for plant growth, and it is only recently that we have gained some insight into this.

ACKNOWLEDGMENT

I thank Dr. C. Tang for his comments and the opportunity to discuss aspects of this chapter.

REFERENCES

1. Rovira, A.D., Plant root exudates, *Bot. Rev.*, 35, 35, 1969.
2. Hale, M.G., Moore, L.D., and Griffin, G.J., Root exudates and exudation, in *Interactions between Non-Pathogenic Soil Micro-Organisms and Plants*, Dommergues, V.R. and Krupa, S.V., Eds., Elsevier, Amsterdam, 1978, p. 163.
3. Kraffczyk, I., Trolldenier, G., and Beringer, H., Soluble root exudates of maize: influence of potassium supply and rhizosphere organisms, *Soil Biol. Biochem.*, 16, 315, 1984.
4. Curl, E.A. and Truelove, B., *The Rhizosphere*, Springer-Verlag, New York, 1986.
5. Schönwitz, R. and Ziegler, H., Interaction of maize roots and rhizosphere microorganisms, *Z. Pflanzenernähr. Bodenk.*, 152, 217, 1989.
6. Einhellig, F.A. and Souza, I.F., Phytotoxicity of sorgoleone found in sorghum root exudates, *J. Chem. Ecol.*, 18, 1, 1992.
7. Basu, U., Basu, A., and Taylor, G.J., Differential exudation of polypeptides by roots of aluminum-resistant and aluminum-sensitive cultivars of *Triticum aestivum* L. in response to aluminum stress, *Plant Physiol.*, 106, 151, 1994.
8. Fan, T.W.M. et al., Comprehensive analysis of organic ligands in whole root exudates using nuclear magnetic resonance and gas chromatography-mass spectroscopy, *Anal. Biochem.*, 251, 57, 1997.
9. Shinmachi, F., Characterization of iron deficiency response system with riboflavin secretion in some dicotyledonous plants, in *Plant Nutrition for Sustainable Food Production and Environment*, Ando, T., Fujita, K., Mae, T., Matsumoto, H., Mori, S., and Sekiya, J., Eds., Kluwer Academic Publishers, Dordrecht, 1997, p. 277.
10. Bais, H.P. et al., Enantriomeric-dependent phytotoxic and antimicrobial activity of (±)-catechin. A rhizosecreted racemic mixture from spotted knapweed, *Plant Physiol.*, 128, 1173, 2002.
11. Dakora, F.D. and Phillips, D.A., Root exudates as mediators of mineral acquisition in low-nutrient environments, *Plant Soil*, 245, 35, 2002.
12. Inderjit, S. and Weston, L.A., Root exudates: an overview, in *Root Ecology*, de Kroon, H. and Visser, E.J.W., Eds., Springer-Verlag, Berlin, 2003, chap. 10.
13. Rumberger, A. and Marshner, P., 2-phenylthiocyanate and microbial community composition in the rhizosphere of canola, *Soil Biol. Biochem.*, 35, 445, 2003.

14. Vivanco, J.M. et al., Biogeographical variation in community response to root allelochemistry: novel weapons and exotic invasion, *Ecol. Lett.,* 7, 285, 2004.
15. Hale, M.G., Foy, C.L., and Shay, F.J., Factors affecting root exudation, *Adv. Agron.,* 23, 89, 1971.
16. Rovira, A.D. and Davey, C.B., Biology of the rhizosphere, in *The Plant Root Environment,* Carson, E.W., Ed., University of Virginia, Charlotteville, VA, 1974, chap. 7.
17. Martin, J.K., Factors influencing the loss of organic carbon from wheat roots, *Soil Biol. Biochem.,* 9, 1, 1977.
18. Hale, M.G. and Moore, L.D., Factors affecting root exudation: 1970–1978, *Adv. Agron.,* 31, 93, 1979.
19. Uren, N.C., Types, amounts, and possible functions of compounds released into the rhizosphere by soil-grown plants, in *The Rhizosphere: Biochemistry and Organic Substances of the Soil-Plant Interface,* Pinton, R., Varanini, Z., and Nannipieri, P., Eds., Marcel Dekker, New York, 2001, chap. 2.
20. Persello-Crtieaux, F., Nussaume, L., and Robaglia, C., Tales from the underground: molecular plant-rhizobacteria interactions, *Plant Cell Environ.,* 26, 189, 2003.
21. Somers, E., Vanderleyden, J., and Srinivasan, M., Rhizosphere bacterial signalling: a love parade beneath our feet, *Crit. Rev. Microbiol.,* 30, 205, 2004.
22. Rovira, A.D., Rhizosphere research — 85 years and frustration, in *The Rhizosphere and Plant Growth,* Keister, D.L. and Creagan, P.B., Eds., Kluwer Academic Publishers, Dordrecht, 1991, p. 3.
23. Vaughan, D. and Ord, B.G., Extraction of potential allelochemicals and their effects on root morphology and nutrient contents, in *Plant Root Growth: an Ecological Perspective,* Atkinson, D., Ed., Blackwell, Oxford, 1991, p. 399.
24. Jones, D.L., Darrah, P.R., and Kochian, L.V., Critical evaluation of organic acid mediated iron dissolution in the rhizosphere and its potential role in root iron uptake, *Plant Soil,* 180, 57, 1996.
25. Uren, N.C. and Reisenauer, H.M., The role of root exudates in nutrient acquisition, *Adv. Plant Nutr.,* 3, 79, 1988.
26. Bowen, G.D., Microbial dynamics in the rhizosphere: possible strategies in managing rhizosphere populations, in *The Rhizosphere and Plant Growth,* Keister, D.L. and Creagan, P.B., Eds., Kluwer Academic Publishers, Dordrecht, 1991, p. 25.
27. Darrah, P.R., Rhizodeposition under ambient and elevated CO_2 levels, *Plant Soil,* 187, 265, 1996.
28. Inderjit, S. and del Moral, R., Plant phenolics in allelopathy, *Bot. Rev.,* 62, 186, 1996.
29. Jones, D.L., Organic acids in the rhizosphere — a critical review, *Plant Soil,* 205, 25, 1998.
30. Farrar, J.F. and Jones, D.L., The control of carbon acquisition by and the growth of roots, in *Root Ecology,* de Kroon, H. and Visser, E.J.W., Eds., Springer-Verlag, Berlin, 2003, chap. 4.
31. McCully, M.E., Roots in soil: unearthing the complexities of roots and their rhizospheres, *Ann. Rev. Plant Physiol. Plant Mol. Biol.,* 50, 695, 1999.
32. Clarkson, D.T., Foreword, in *Plant Physiological Ecology,* Lambers, H., Chapin, F.S., and Pons, T.L. [Authors], Springer-Verlag, New York, 1998.
33. Bowen, G.D. and Rovira, A.D., The rhizosphere and its management to improve plant growth, *Adv. Agron.,* 11, 1, 1999.
34. Kapulnik, Y. and Okon, Y., Plant growth promotion by rhizosphere bacteria, in *Plant Roots: The Hidden Half,* 3rd ed., Waisel, Y., Eshel, E., and Kafkafi, U., Eds., Marcel Dekker, New York, 2002, chap. 48.
35. Dobbelaere, S., Vanderleyden, J., and Okon, Y., Plant growth-promoting effects of diazotrophs in the rhizosphere, *Crit. Rev. Plant Sci.,* 22, 107, 2003.
36. Patterson, E., Importance of rhizodeposition in the coupling of plant and microbial activity, *Eur. J. Soil Sci.,* 54, 741, 2003.
37. Walker, T. et al., Root exudation and rhizosphere biology, *Plant Physiol.,* 132, 44, 2003.
38. Singh, B. et al., Unravelling rhizosphere-microbial intersections: opportunities and limitations, *Trends Microbiol.,* 12, 387, 2004.
39. Barley, K.P., The configuration of the root system in relation to nutrient uptake, *Adv. Agron.,* 22, 159, 1970.
40. Lynch, J., Root architecture and plant productivity, *Plant Physiol.,* 109, 7, 1995.
41. Bernston, G.M., Modelling root architecture: are there tradeoffs between efficiency and potential of resource acquisition, *New Phytologist,* 127, 483, 1994.
42. Spek, L.Y., Generation and visualization of root-like structures in a three-dimensional space, *Plant Soil,* 197, 9, 1997.

43. Thaler, P. and Pagès, L., Modelling the influence of assimilate availability on root growth and architecture, *Plant Soil,* 201, 307, 1998.
44. López-Bucio, J. et al., The role of nutrient availability in regulating root architecture, *Curr. Opin. Plant Biol.,* 6, 280, 2003.
45. Malamy, J.E., Intrinsic and environmental response pathways that regulate root system architecture, *Plant Cell Environ.,* 28, 67, 2005.
46. Ueda, M., Koshino-Kimura, Y., and Okada, K., Stepwise understanding of root development, *Curr. Opin. Plant Biol.,* 8, 71, 2005.
47. Young, I.M., Biophysical interactions at the root-soil interface: a review, *J. Agric. Sci.,* 130, 1, 1998.
48. Barber, S.A. and Silberbush, M., Plant root morphology and nutrient uptake, in *Roots, Nutrient and Water Influx, and Plant Growth,* Barber, S.A. and Bouldin, D.R., Eds., Soil Science Society of America, Madison, WI, 1984, chap. 4.
49. Jones, D.L. and Darrah, P.R., Re-sorption of organic components by roots of *Zea mays* L. and its consequences in the rhizosphere, *Plant Soil,* 143, 259, 1992.
50. Meharg, A.A., A critical review of labelling techniques used to quantify rhizosphere carbon-flow, *Plant Soil,* 166, 55, 1994.
51. Toal, M.E. et al., A review of rhizosphere carbon flow modelling, *Plant Soil,* 222, 263, 2000.
52. Hutsch, B., Augustin, J., and Merbach, W., Plant rhizodeposition — an important source for carbon turnover in soils, *J. Plant Nutr. Soil Sci.,* 165, 397, 2002.
53. Kuzyakov, Y. et al., Qualitative assessment of rhizodeposits in non-sterile soil by analytical pyrolysis, *J. Plant Nutr. Soil Sci.,* 166, 719, 2003.
54. Nguyen, C., Rhizodeposition of organic C by plants: mechanisms and controls, *Agronomie,* 23, 375, 2003.
55. Dilkes, N.B., Jones, D.L., and Farrar, J., Temporal dynamics of carbon partitioning and rhizodeposition in wheat, *Plant Physiol.,* 134, 706, 2004.
56. Jones, D.L., Hodge, A., and Kuzyakov, Y., Plant and mycorrhizal regulation of rhizodeposition, *New Phytologist,* 163, 459, 2004.
57. Merckx, R. et al., Production of root-derived material and associated microbial growth in soil at different nutrient levels, *Biol. Fertil. Soils,* 5, 126, 1987.
58. Cheng, W. et al., Is available carbon limiting microbial respiration in the rhizosphere?, *Soil Biol. Biochem.,* 28, 1283, 1996.
59. Marschner, H., *Mineral Nutrition of Higher Plants,* 2nd ed., Academic Press, London, 1995.
60. Qian, J.H., Doran, J.W., and Walters, D.T., Maize plant contributions to root zone available carbon and microbial transformations of nitrogen, *Soil Biol. Biochem.,* 29, 1451, 1997.
61. Mary, B. et al., C and N cycling during decomposition of root mucilage, roots and glucose in soil, *Soil Biol. Biochem.,* 25, 1005, 1993.
62. Kuchenbuch, R. and Jungk, A., A method for determining concentration profiles at the soil-root interface by thin slicing rhizospheric soil, *Plant Soil,* 68, 391, 1982.
63. Zuo, Y. et al., Studies on the improvement of iron nutrition of peanut by intercropping with maize on a calcareous soil, *Plant Soil,* 220, 13, 2000.
64. Zhang, F.S. and Li., L., Using competitive and facilitative interactions in intercropping systems enhances crop productivity and nutrient-use efficiency, *Plant Soil,* 248, 305, 2003.
65. Schönwitz, R. and Ziegler, H., Quantitative and qualitative aspects of a developing rhizosphere microflora of hydroponically grown maize seedlings, *Z. Pflanzenernähr. Bodenk.,* 149, 623, 1986.
66. Hawes, M.C. and Lin, H.J., Correlation of pectolytic enzyme activity with the programmed release of cells from root caps of pea (*Pisum sativum*), *Plant Physiol.,* 94, 1855, 1990.
67. Hawes, M.C. et al., Root caps and rhizosphere, *J. Plant Growth Regul.,* 21, 352, 2003.
68. Vermeer, J. and McCully, M.E., The rhizosphere of *Zea*: new insight into its structure and development, *Planta,* 156, 45, 1982.
69. McCully, M.E., Cell separation: a developmental feature of root caps which may be of fundamental functional significance, in *Cell Separation in Plants,* Osborne, D.J. and Jackson, M.B., Eds., Springer-Verlag, Berlin, 1989, p. 241.
70. McCully, M., How do real roots work? Some new views of root structure, *Plant Physiol.,* 109, 1, 1995.
71. Foster, R.C., Rovira, A.D., and Cook, T.W., *Ultrastructure of the Root-soil Interface,* American Phytopathological Society, St. Paul, MN, 1983.

72. Hawes, M.C., Living plant cells released from the root cap: a regulator of microbial populations in the rhizosphere?, in *The Rhizosphere and Plant Growth*, Keister, D.L. and Creagan, P.B., Eds., Kluwer Academic Publishers, Dordrecht, 1991, p. 51.

73. Hawes, M.C. et al., Function of root border cells in plant health, *Annu. Rev. Phytopathol.*, 36, 311, 1998.

74. Hawes, M.C. et al., The role of border cells in plant defense, *Trends Plant Sci.*, 5, 128, 2000.

75. Müller, M., Deigele, C., and Ziegler, H., Hormonal interactions in the rhizosphere of maize (*Zea mays* L.) and their effects on plant development, *Z. Pflanzenernähr. Bodenk.*, 152, 247, 1989.

76. Arshad, M. and Frankenberger Jr., W.T., Plant growth-regulating substances in the rhizosphere: microbial production and functions. *Adv. Agro.*, 62, 45, 1998.

77. Balaji, B. et al., Responses of an arbuscular mycorrhizal fungus, *Gigaspora margarita*, to exudates and volatiles from the Ri T-DNA-transformed roots of nonmycorrhizal and mycorrhizal mutants of *Pisum sativum* L. Sparkle, *Exp. Mycol.*, 19, 275, 1995.

78. Johnson, J.F. et al., Root carbon dioxide fixation by phosphorus-deficient *Lupinus albus*, *Plant Physiol.*, 112, 19, 1996.

79. Yan, F. et al., Adaptation of H^+-pumping and plasma membrane H^+ ATPase activity in proteoid roots of white lupin under phosphate deficiency, *Plant Physiol.*, 129, 50, 2002.

80. Zhu, Y.Y. et al., A link between citrate and proton release by proteoid roots of white lupin (*Lupinus albus* L.) grown under phosphorus-deficient conditions, *Plant Cell Physiol.* 46, 892, 2005.

81. Berczi, A. and Moller, I.M., Redox enzymes in the plant plasma membrane and their possible roles, *Plant, Cell Environ.*, 23, 1287, 2000.

82. Hinsinger, P. et al., Origins of root-mediated pH changes in the rhizosphere and their responses to environmental constraints: a review, *Plant Soil*, 248, 43, 2003.

83. Ryan, P.R., Delhaize, E., and Jones, D.L., Function and mechanism of organic anion exudation from plant roots, *Annu. Rev. Plant Physiol. Plant Mol., Biol.*, 52, 527, 2001.

84. Schreiner, O., Sullivan, M.X., and Reid, F.R., Studies in Soil Oxidation, USDA Bur. Soils Bull. No. 73, 1910.

85. Lund, E.J. and Kenyon, W.A., Relation between continuous bio-electric currents and cell respiration. I. Electric correlation potentials in growing root tips, *J. Exp. Zool.*, 48, 333, 1927.

86. Uren, N.C., Chemical reduction of an insoluble higher oxide of manganese by plant roots, *J. Plant Nutr.*, 4, 65, 1981.

87. Whipps, J.M., Carbon economy, in *The Rhizosphere*, Lynch, J.M., Ed., John Wiley and Sons, Chichester, U.K., 1990, p. 59.

88. Lynch, J.M. and Whipps, J.M., Substrate flow in the rhizosphere, in *The Rhizosphere and Plant Growth*, Keister, D.L. and Creagan, P.B., Eds., Kluwer Academic Publishers, Dordrecht, 1991, p. 15.

89. Griffin, G.J., Hale, M.G., and Shay, F.J., Nature and quantity of sloughed organic matter produced by roots of axenic peanut plants, *Soil Biol. Biochem.*, 8, 29, 1976.

90. Iijima, M., Griffiths, B., and Bengough, A.G., Sloughing of cap cells and carbon exudation from maize seedling roots in compacted sand, *New Phytologist*, 145, 477, 2000.

91. Farrar, J. et al., How roots control the flux of carbon to the rhizosphere, *Ecology*, 84, 827, 2003.

92. Hocking, P.J., Dry-matter production, mineral nutrient concentrations, and nutrient distribution and redistribution in irrigated spring wheat, *J. Plant Nutr.*, 17, 1289, 1994.

93. Butler, B.E., A soil survey of the horticultural soils in the Murrumbidgee irrigation areas, New South Wales, Bulletin No. 289, CSIRO, 1979.

94. Goss, M.J., Consequences of the activity of roots on soil, in *Plant Root Growth: An Ecological Perspective*, Atkinson, D., Ed., Blackwell, Oxford, 1991, p. 171.

95. Morel, J.L. et al., Influence of maize root mucilage on soil aggregate stability, *Plant Soil*, 136, 111, 1991.

96. Whiteley, G.M., The deformation of soil by penetrometers and root tips of *Pisum sativum*, *Plant Soil*, 117, 210, 1989.

97. McCully, M.E., Water efflux from the surface of field-grown grass roots. Observations by cryo-scanning electron microscopy, *Physiol. Plant.*, 95, 217, 1995.

98. Young, I.M., Variation in moisture contents between bulk soil and the rhizosheath of wheat (*Triticum aestivum* L. cv. Wembley), *New Phytologist*, 130,135, 1995.

99. Bengough, A.G. and McKenzie, B.M., Sloughing root cap cells decreases the frictional resistance to maize (*Zea mays* L.) root growth, *J. Exp. Bot.*, 48, 885, 1997.

100. Iijima, M., Higuchi, T., and Barlow, P.W., Contribution of root cap mucilage and the presence of an intact root cap in maize (*Zea mays*) to the reduction in soil mechanical impedance, *Ann. Bot.* 94, 473, 2004.

101. Read, D.B. et al., Plant roots release phospholipids surfactants that modify the physical and chemical properties of soil, *New Phytologist,* 157, 315, 2003.

102. Hallet, P.D., Gordon, D.C., and Bengough, A.G., Plant influence on rhizosphere hydraulic properties: direct measurements using a miniaturized infiltrometer, *New Phytologist,* 157, 597, 2003.

103. Whalley, W.R. et al., The hydraulic properties of soil at the root-soil interface, *Soil Sci.,* 169, 90, 2004.

104. Drew, M.C., Properties of roots which affect rates of absorption, in *The Soil-Root Interface,* Harley, J.L. and Russell, J.S., Eds., Academic Press, London, 1979, p. 21.

105. Wu, J. et al., Experimental study on the reduction of soil hydraulic conductivity by enhanced biomass growth, *Soil Sci.,* 162, 741, 1997.

106. Greenland, D.J., The physics and chemistry of the soil-root interface: some comments, in *The Soil-Root Interface,* Harley, J.L. and Russell, J.S., Eds., Academic Press, London, 1979, p. 83.

107. Watt, M., McCully, M.E., and Canny, M.J., Formation and stabilization of rhizosheaths of *Zea mays* L., *Plant Physiol.,* 106, 179, 1994.

108. McCully, M.E. and Boyer, J.S., The expansion of maize root-cap mucilage during hydration. 3. Changes in water potential and water content, *Physiol. Plant.,* 99, 169, 1997.

109. Soileau, J.M., Jackson, W.A., and McCracken, R.J., Cutans (clay films) and potassium availability to plants, *J. Soil Sci.,* 15, 117, 1964.

110. Sealey, L.J., McCully, M.E., and Canny, M.J., The expansion of maize root-cap mucilage during hydration. 1. Kinetics, *Physiol. Plant.,* 93, 38, 1995.

111. Nye, P.H. and Tinker, P.B., *Solute Movement in the Soil-Root System,* University of California Press, Berkeley, CA, 1977.

112. Burns, R.G., Enzyme activity in soil: location and a possible role in microbial ecology, *Soil Biol. Biochem.,* 14, 423, 1982.

113. Nannipieri, P., The potential use of enzymes as indicators of productivity, sustainability and pollution, in *Soil Biota — Management in Sustainable Farming Systems,* Pankhurst, C.E., Doube, B.M., Gupta, V.V.S.R., and Grace, P.R., Eds., CSIRO Australia, 1994, p. 238.

114. Wasaki, J. et al., Properties of secretory acid phosphatase from lupin roots under phosphorus-deficient conditions, in *Plant Nutrition for Sustainable Food Production and Environment,* Ando, T., Fujita, K., Mae, T., Matsumoto, H., Mori, S., and Sekiya, J., Eds., Kluwer Academic Publishers, Dordrecht, 1997, p. 295.

115. Seeling, B. and Jungk, A., Utilization of organic phosphorus in calcium chloride extracts of soil by barley plants and hydrolysis and alkaline phosphatases, *Plant Soil,* 178, 179, 1996.

116. Tarafdar, J.C. and Marschner, H., Phosphatase activity in the rhizosphere and hyposphere of VA mycorrhizal wheat supplied with inorganic and organic phosphorus, *Soil Biol. Biochem.,* 26, 387, 1994.

117. Tarafdar, J.C. and Jungk, A., Phosphatase activity in the rhizosphere and its relation to the depletion of soil organic phosphorus, *Biol. Fertil. Soils,* 3, 199, 1987.

118. Tinker, P.B. and Nye, P.H., *Solute Movement in the Rhizosphere,* Oxford University Press, New York, 2000.

119. George, T.S., Richardson, A.E., and Simpson, R.J., Behavior of plant-derived extracellular phytase upon addition to soil, *Soil Biol. Biochem.* 37, 977, 2005.

120. Bailey, J.A. and Mansfield, J.W., Eds., *Phytoalexins,* John Wiley and Sons, New York, 1982.

121. Haider, K. and Martin, J.P., Decomposition of specifically carbon-14 labeled benzoic and cinnamic acid derivatives in soil, *Soil Sci. Soc. Am. Proc.,* 39, 657, 1975.

122. Lehmann, R.G., Cheng, H.H., and Harsh, J.B., Oxidation of phenolics by soil iron and manganese oxides, *Soil Sci. Soc. Am. J.,* 51, 352, 1987.

123. McBride, M.B., Adsorption and oxidation of phenolic compounds by iron and manganese oxides, *Soil Sci. Soc. Am. J.,* 51, 1466, 1987.

124. Inderjit and Bhowmik, P.C., Sorption of benzoic acid onto soil colloids and its implications for allelopathy studies, *Biol. Fertil. Soils,* 40, 345, 2004.

125. Krogmeier, M.J. and Bremner, J.M., Effects of phenolic acids on seed germination and seedling growth in soil, *Biol. Fertil. Soil,* 8, 116, 1989.

126. Inderjit and del Moral, R., Is separating resource competition from allelopathy realistic?, *Bot. Rev.,* 63, 221, 1997.

127. Kato-Noguchi, H., Allelopathic substance in rice root exudates: rediscovery of momilactone B as an allelochemical, *J. Plant Physiol.,* 161, 271, 2004.
128. Czarnota, M.A. et al., Mode of action, localization of production, chemical nature, and activity of sorgoleone: a potent PSII inhibitor in *Sorghum* spp. root exudates, *Weed Technol.,* 15, 813, 2001.
129. Inderjit and Duke, S.O., Ecophysiological aspects of allelopathy, *Planta,* 217, 529, 2003.
130. Wu, H. et al., Distribution and exudation of allelochemicals in wheat *Triticum aestivum, J. Chem. Ecol.,* 26, 2141, 2000.
131. Dyer, B., On the analytical determination of probably available "mineral" plant food in soils, *J. Chem. Soc. Trans.,* 65, 115, 1894.
132. Jones, D.L. et al., Organic acid behavior in soils — misconceptions and knowledge gaps, *Plant Soil,* 248, 31, 2003.
133. Trolove, S.N. et al., Progress in selected areas of rhizosphere research on P acquisition, *Aust. J. Soil Res.,* 41, 471, 2003.
134. Dinkelaker, B., Römheld, V., and Marschner, H., Citric acid excretion and precipitation of calcium citrate in the rhizosphere of white lupin (*Lupinus albus* L.), *Plant Cell Environ.,* 12, 285, 1989.
135. Jones, D.L. and Darrah, P.R., Influx and efflux of organic acids across the soil-root interface of *Zea mays* L. and its implications in rhizosphere C flow, *Plant Soil,* 173, 103, 1995.
136. Jones, D.L. and Darrah, P.R., Role of root derived organic acids in the mobilization of nutrients from the rhizosphere, *Plant Soil,* 166, 247, 1994.
137. Kuzyakov, Y., Raskatov, A., and Kaupenjohann, M., Turnover and distribution of root exudates of *Zea mays, Plant Soil,* 254, 317, 2003.
138. Owen, A.G. and Jones, D.L., Competition for amino acids between wheat roots and rhizosphere microorganisms and the role of amino acids in plant N acquisition, *Soil Biol. Biochem.,* 33, 651, 2001.
139. White, P.F., Soil and plant factors relating to the poor growth of *Lupinus* species on fine-textured alkaline soils, *Aust. J. Agric. Res.,* 41, 871, 1990.
140. Tang, C. et al., The growth of *Lupinus* species on alkaline soils, *Aust. J. Agric. Res.,* 46, 255, 1995.
141. Zhang, F.S., Ma, J., and Cao, Y.P., Phosphorus deficiency enhances root exudation of low-molecular weight organic acids and utilization of sparingly soluble inorganic phosphates by radish (*Raghanus satiuvs* L.) and rape (*Brassica napus* L.) plants, in *Plant Nutrition for Sustainable Food Production and Environment,* Ando, T., Fujita, K., Mae, T., Matsumoto, H., Mori, S., and Sekiya, J., Eds., Kluwer Academic Publishers, Dordrecht, 1997, p. 301.
142. Hinsinger, P., How do plant roots acquire mineral nutrients? Chemical processes in the rhizosphere, *Adv. Agron.,* 64, 225, 1998.
143. Delhaize, E.P., Ryan, P.R., and Randall, P.J., Aluminium tolerance in wheat (*Triticum aestivum* L.) II. Aluminium-stimulated excretion of malic acid from root apices, *Plant Physiol.,* 103, 695, 1993.
144. Pellet, D.M., Grunes, D.L., and Kochian, L.V., Organic acid exudation as an aluminium-tolerance mechanism in maize (*Zea mays* L.), *Planta,* 196, 788, 1995.
145. Horst, W.J., The role of the apoplast in aluminium toxicity and resistance of higher plants: a review, *Z. Pflanzenernähr. Bodenk.,* 158, 419, 1995.
146. Jones, D.L., Prabowo, A.M., and Kochian, L.V., Kinetics of malate transport and decomposition in acid soils and isolated bacterial populations: the effect of microorganisms on root exudation of malate under Al stress, *Plant Soil,* 182, 239, 1996.
147. Pellet, D.M. et al., Involvement of multiple aluminium exclusion mechanisms in aluminium tolerance in wheat, *Plant Soil,* 192, 63, 1997.
148. Lee, J.A. and Woolhouse, H.W., A comparative study of bicarbonate inhibition of root growth in calcicole and calcifuge plants, *New Phytologist,* 68, 1, 1969.
149. Huang, A. and Graham, R.D., Efficient Mn uptake in barley is a constitutive system, in *Plant Nutrition for Sustainable Food Production and Environment,* Ando, T., Fujita, K., Mae, T., Matsumoto, H., Mori, S., and Sekiya, J., Eds., Kluwer Academic Publishers, Dordrecht, 1997, p. 269.
150. Venkatraju, K. and Marschner, H., Inhibition of iron-stress reactions in sunflower by bicarbonate, *Z. Pflanzenernähr. Bodenk.,* 144, 339, 1981.
151. Bar-Ness, E. et al., Short-term effects of rhizosphere microorganisms on Fe uptake from microbial siderophores by maize and oats, *Plant Physiol.,* 100, 451, 1992.
152. von Wiren, N. et al., Iron inefficiency in maize mutant ys1 (*Zea mays* L. cv. Yellow-stripe) is caused by a defect in uptake of iron phytosiderophores, *Plant Physiol.,* 106, 71, 1994.

153. von Wiren, N. et al., Competition between micro-organisms and roots of barley and sorghum for iron accumulated in the root apoplasm, *New Phytologist,* 130, 511 1995.

154. Leeper, G.W., Relationship of soils to manganese deficiency of plants, *Nature,* 134, 972, 1934.

155. de Willigen, P. and van Noordwjik, M., Mathematical models on diffusion of oxygen to and within plant roots, with special emphasis on effects of soil-root contact: I, *Plant Soil,* 77, 215, 1984.

156. van Noordwjik, M. and de Willigen, P., Mathematical models on diffusion of oxygen to and within plant roots, with special emphasis on effects of soil-root contact: II, *Plant Soil,* 77, 233, 1984.

157. Uren, N.C., Mucilage secretion and its interaction with soil, and contact reduction, *Plant Soil,* 155/156, 79, 1993.

158. Park, K.H. et al., Allelopathic activity and determination of allelochemicals from sunflower (*Helianthus annuus* L.) root exudates. II. Elucidation of allelochemicals from sunflower exudates, *Korean J. Weed Sci.,* 12, 173, 1992.

159. Ayers, W.A. and Thornton, R.H., Exudation of amino acids by intact and damaged roots of wheat and peas, *Plant Soil,* 28, 193, 1968.

2 The Release of Root Exudates as Affected by the Plant Physiological Status

Günter Neumann and Volker Römheld

CONTENTS

I. INTRODUCTION

Apart from the function of plant roots as organs for water and nutrient uptake and anchorage in soils, roots are able also to release a wide range of organic and inorganic compounds into the rhizosphere. Soil–chemical changes related to the presence of these compounds and products of their microbial turnover are important factors affecting microbial populations, availability of nutrients, solubility of toxic elements in the rhizosphere, and thereby, the ability of plants to cope with adverse soil–chemical conditions [1]. Organic rhizodeposition includes lysates, liberated by autolysis of sloughed-off cells and tissues, intact root border cells, as well as root exudates, released passively (diffusates) or actively (secretions) from intact root cells (Table 2.1; see also Chapter 1). In annual plant species, 30 to 60% of the photosynthetically fixed carbon is translocated to the roots, and a considerable proportion of this carbon (up to 70%) can be released into the rhizosphere [2,3] as pointed out in Chapter 1, Chapter 3, and Chapter 13 of this book. This rhizodeposition is affected by multiple factors such as light intensity, temperature, nutritional status of the plants, activity of retrieval mechanisms, various stress factors, mechanical impedance and sorption characteristics of the growth medium, and microbial activity in the rhizosphere. This chapter will focus on the release of root exudates, and highlight effects of the physiological status on root exudation and its significance for adaptations to adverse soil conditions and nutrient efficiency. Because the methods employed for collection and analysis of root exudates play an important role for the qualitative and quantitative interpretation of measured exudate data, methodological aspects will also be discussed in the introductory section.

TABLE 2.1
Root Exudates Detected in Higher Plants

Class of Compounds	Single Components
Sugars	Arabinose, glucose, fructose, galactose, maltose, raffinose, rhamnose, ribose, sucrose, xylose
Amino acids and amides	All 20 proteinogenic amino acids, aminobutyric acid, homoserine, cysrathionine, mugineic acid phytosiderophores (mugineic acid, deoxymugineic acid, hydroxymugineic acid, epi-hydroxymugineic acid, avenic acid, distichonic acid A)
Aliphatic acids	Formic, acetic, butyric, popionic, malic, citric, isocitric, oxalic, fumaric, malonic, succinic, maleic, tartaric, oxaloacetic, pyruvic, oxoglutaric, maleic, glycolic, shikimic, cis-aconitic, trans-aconitic, valeric, gluconic
Aromatic acids	p-Hydroxybenzoic, caffeic, p-coumaric, ferulic, gallic, gentisic, protocatechuic, salicylic, sinapic, syringic
Miscellaneous phenolics	Flavonols, flavones, flavanones, anthocyanins, isoflavonoids
Fatty acids	Linoleic, linolenic, oleic, palmitic, stearic
Sterols	Campestrol, cholesterol, sitosterol, stigmasterol
Enzymes	Amylase, invertase, cellobiase, desoxyruibonuclease, ribonuclease, acid phosphatase, phytase, pyrophosphatase apyrase, peroxidase, protease
Miscellaneous	Vitamins, plant growth regulators (auxins, cytokinins, gibberellins), alkyl sulfides, ethanol, H^+,K^+ Nitrate, Phosphate, HCO_3^-

II. COLLECTION OF ROOT EXUDATES: METHODOLOGICAL ASPECTS

A. COLLECTION TECHNIQUES WITH TRAP SOLUTIONS

Water-soluble root exudates are most frequently collected by immersion of root systems into aerated trap solutions for a defined time period (Figure 2.1A). The technique is easy to perform and allows kinetic studies by repeated measurements using the same plants. Although it is possible to get a first impression about qualitative exudation patterns and even quantitative changes in response to different preculture conditions, the technique also includes several drawbacks, which should be taken into account for the interpretation of experimental data. Application should be restricted to plants grown in nutrient solution, because removal of root systems from solid media (soil, sand) is almost certainly associated with mechanical damage of root cells, resulting in overestimation of exudation rates. On the other hand, it has been frequently demonstrated that the mechanical impedance of solid growth media leads to alterations in root morphology and stimulates root exudation [4,5]. In liquid culture media, simulation of the mechanical forces imposed on roots of soil-grown plants may be achieved by addition of small glass beads [5–7]. Alternatively, exudate collection from plants grown in solid media (sand, vermiculite) may be performed by percolating the culture vessels with the trap solution for a defined time period (Figure 2.1B), after removal of

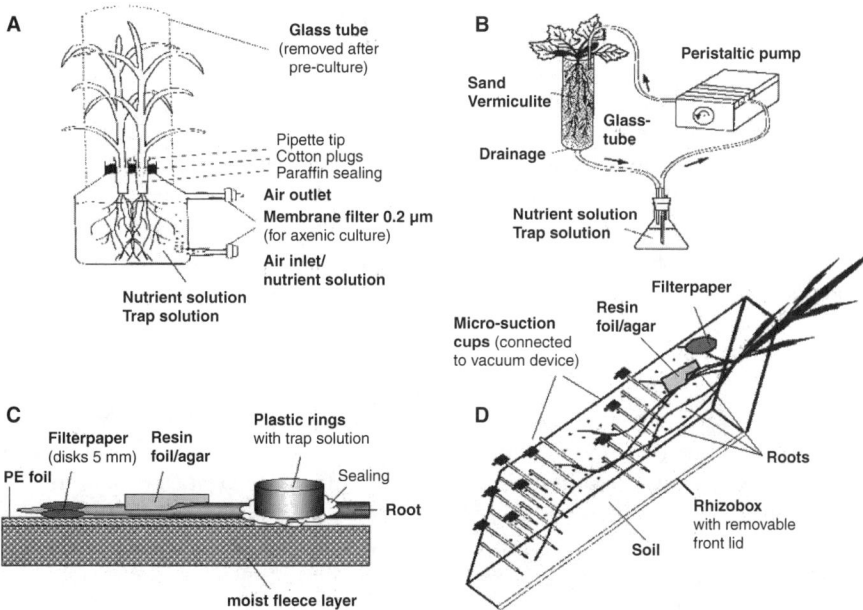

FIGURE 2.1 Techniques for collection of root exudates: (A) Solution culture system [396]; root exudates collected from the whole root system by immersion into aerated trap solutions under sterile conditions (optional). (B) Plant culture in solid media (vermiculite, sand); root exudates collected from the whole root system by percolation of the culture vessels with trap solution [9,11]. (C) Localized root exudate sampling from plants grown in solution culture. Exudates collected into trap solution inside of sealed plastic rings straddling the root [29], or by application of sorption media (filter paper, agar, ion-exchange resins) onto the root surface [18,46]. (D) Localized collection of rhizosphere soil solution from plants grown in soil culture. Rhizoboxes with removable front lids for plant culture (root windows under field conditions). Collection by insertion of microsuction cups (made from HPLC capillaries, 1 mm in diameter) connected to a vacuum collection device [52] or by application of sorption media (filter paper, agar, ion-exchange resins) onto the root surface [28,46].

rhizosphere products accumulated during the preceeding culture period by repeated washing steps [8–11]. For this approach, however, recovery experiments and comparison with results obtained from experiments in liquid culture are essential, because incomplete leaching and sorption of certain exudate compounds to the matrix of solid culture media cannot be excluded [12]. As a modification of the percolation technique, cartridges filled with selective adsorption media (e.g., XAD resin for hydrophobic compounds, anion-exchange resins for carboxylates), which are installed in the draining tube below the plant culture vessel, can be employed for the enrichment of distinct exudate constituents [13,14]. After adsorption to a resin, exudate compounds are also protected to a certain extent against microbial degradation (see Subsection II.C.1).

Trap solutions employed for collection of water-soluble root exudates are nutrient solutions of the same composition as the culture media [8,10,11], solutions of 0.5 to 2.5 mM $CaSO_4$ or $CaCl_2$ to provide Ca^{2+} for membrane stabilization [15] or simply distilled water [9,16–18]. Because the osmotic strength of nutrient solutions is generally low, short-term treatments (1 to 2 h) even with distilled water are not likely to affect membrane permeability by osmotic stress. Accordingly, comparing exudation of amino acids from roots of *Brassica napus* L. into nutrient solution, 20 mM KCl, or distilled water, respectively, revealed no differences during collection periods between 0.5 and 6 h [19]. In contrast, Cakmak and Marschner [20] reported increased exudation of sugars and amino acids from roots of wheat and cotton during a collection period of 6 h when distilled water, instead of 1 mM $CaSO_4$, was applied as trap solution. Thus, for longer collection periods or for repeated measurements, only complete or at least diluted nutrient solutions should be employed as trap solutions to avoid depletion of nutrients and excessive leaching of Ca^{2+}. Long-term exposure of plant roots to external solutions of very low ionic strength is also likely to increase exudation rates due to an increased transmembrane concentration gradient of solutes [21,22]. Prior to further sample preparation, solids, microorganisms, and root border cells in trap solutions should be removed by filtration or centrifugation steps.

Exudate collection in trap solutions usually requires subsequent concentration by vacuum evaporation or lyophilization, due to the low concentration of exudate compounds. Depending on the composition of the trap solution, the reduction of sample volume can lead to high salt concentrations, which may interfere with subsequent analysis, or may even cause irreversible precipitation of certain exudate compounds (e.g., Ca-citrate, Ca-oxalate, proteins). Therefore, if possible, removal of interfering salts by use of ion-exchange resins prior to sample concentration is recommended. Alternatively, solid-phase extraction techniques may be employed for enrichment of exudate compounds from the diluted trap solution [11,23]. High-molecular-weight (HMW) compounds may be concentrated by precipitation with organic solvents (methanol, ethanol, acetone 80% [v/v] for polysaccharides and proteins) or acidification (trichloroacetic acid 10% [w/v], perchloric acid 5% [w/v] for proteins; [24]). Alternatively, ultrafiltration of the trap solutions or even cultivation of plant roots enclosed in dialysis bags is possible [25]. Mucilage polysaccharides adhering to the root surface have been collected by application of vacuum suction [26], by abrasion with a soft brush and subsequent transfer to cellulose acetate filters [27], or simply by collection with forceps [28].

B. LOCALIZED SAMPLING TECHNIQUES IN SOLUTION AND SOIL CULTURE SYSTEMS

In many plants, root exudation is not uniformly distributed over the whole root system. Considerable spatial variation has been reported for the exudation of carboxylates and protons in P-deficient oilseed rape [29,30], the exudation of protons and phenolics in many dicotyledonous plant species in response to Fe deficiency [1,31], or for the release of phytosiderophores in Fe-deficient barley [31]. In all these cases, exudation was mainly confined to apical root zones. Various plant species adapted to low fertile soils, such as members of the Proteaceae and Casuarinaceae, but also white lupin (*Lupinus albus*) are characterized by the formation of cluster roots (proteoid roots; see Figure 2.3A) mainly under conditions of P deficiency or Fe deficiency [32,33]. Exudation of large amounts of carboxylates and protons involved in the mobilization of mineral nutrients such as P and Fe (see

Section IV) is mainly confined to these root clusters [23,34,35], and moreover to distinct stages during cluster root development [18,28,36,37]. Intense root exudation, restricted to distinct root zones (root tips, cluster roots), may enhance the mobilization efficiency due to localized accumulation of exudate compounds in the rhizosphere to a concentration level that is sufficient to mediate desorption of mineral nutrients from the soil matrix [23,38]. This enhanced mobilization effect may be further increased by a low density of microbial colonization [39–41] and a high capacity for nutrient uptake in apical root zones. Longitudinal gradients of exudation along the roots may, however, also reflect gradients in microbial degradation of root exudates, which is frequently more expressed in basal parts of the root sytem than in apical root zones [1,39,42]. This may be attributed to the high intensity of cell division and cell elongation (1 to 2 cm d^{-1}), restricted to the apical parts of the root [43]. This is associated with continuous production of new cell material, which has to be newly colonized by soil microorganisms. However, recent reports suggest, that even in apical root zones, rapid microbial colonization may be possible at least for some species of soil microorganisms [44,45]. Variations in root growth rates and root exudation in different plant species and under different environmental conditions are likely to explain the variability in microbial colonization patterns in different root zones.

Reliable evaluation of root exudation is only possible, considering the spatial variability along the roots. Also the possibility of temporal changes in root exudation, such as transient release of organic acids in cluster rooted plant species [28,46,47], or diurnal variations in exudation of phytosiderophores in Fe-deficient barley [48] (see Subsection IV.C.2), and of citrate in P-deficient white lupin [37] has to be taken into account.

1. Solution Culture Systems

Spatial variation in root exudation has been investigated by separating distinct root zones of plants grown in hydroponic culture with small plastic rings (1.2 cm in diameter), which were sealed with agar [29] or vacuum grease [36,49] and subsequently filled with trap solution for 1 to 2 h (Figure 2.1C). Marschner et al. [50] used agarose sheets that were placed onto the root surface, allowing diffusion of root exudates into the agarose layer. Neumann et al. [18] collected exudates from different root zones with a spatial resolution of 5 mm, by incubating the roots for 3 h between double layers of filter disks (5 mm in diameter) made from moist chromatography paper (preparative quality) with a high soaking capacity (Figure 2.1C). Similarly, Kape et al. [51] used cellulose acetate filters with a high sorption capacity for hydrophobic compounds to investigate spatial variation of flavonoid exudation in soybean seedlings. For the selective adsorption of carboxylate anions released from roots of white lupin, Kamh et al. [46] applied anion-exchange resins enclosed in dialysis bags or agar sheets, which were placed on the surface of different root zones (Figure 2.1C).

2. Soil Culture Systems

Small sheets of chromatography paper [28] as well as resin bags or resin agar sheets [46], applied onto the surface of distinct root zones, were also used for localized collection of compounds released into the rhizosphere soil solution from soil-grown plants, which were cultivated in rhizoboxes (Figure 2.1D). Göttlein et al. [52] reported the construction of microsuction cups made of HPLC capillaries (1 mm outer diameter) and connected to a vacuum collecting device, which were inserted beneath the roots of soil-grown plants for collection of rhizosphere soil solution (Figure 2.1D). This technique was also successfully employed under field conditions by use of root windows, however mainly for analysis of mineral elements. Detection of carboxylates in rhizosphere soil solutions of Norway spruce and silver birch collected with microsuction cups has been recently reported by Sandnes et al. [53].

Extraction of rhizosphere soil [23,35,54,55] is an approach, which can provide information about long-term accumulation of rhizosphere products (root exudates and microbial metabolites) in the soil. Culture systems, which seperate root compartments from adjacent bulk soil compartments by steel or nylon nets [55–57], have been employed to study radial gradients of rhizosphere products in the root environment. The use of different extraction media can account for different adsorption characteristics of rhizosphere products to the soil matrix [22,34]. However, even the extraction with distilled water for extended periods (>10 min) may lead to some contamination from damaged microbial cells and plant residues, which was not observed when centrifugation techniques were used for soil extraction [58].

A new approach to study root exudation of distinct compounds in soil-grown plants uses inoculation of roots with genetically engineered reporter bacteria, which are able to indicate the presence of particular compounds by indicator reactions, such as production of ice-nucleation protein, or activation of bioluminescence genes. This technique has been employed to detect the exudation of amino acids from roots of soil-grown *Avena barbata* [59] or carbon release from roots of *Hordeum vulgare*, depending on the N nutritional status [60].

C. FACTORS AFFECTING THE RECOVERY OF ROOT EXUDATES

1. Microbial Activity

Organic compounds in root exudates are continuously metabolized by root-associated microorganisms at the rhizoplane and in the rhizosphere. Microbial activity results in quantitative and qualitative alterations of the root exudate composition due to degradation of exudate compounds and the release of microbial metabolites. However, ^{14}C-labeling studies with culture systems spatially separating roots and microorganisms, by using Millipore membranes, demonstrated that root exudation can also be stimulated by microbial colonization of plant roots, and by the presence of microbial metabolites [61–63]. Thus, the use of axenic culture systems to avoid microbial degradation of exudate compounds [19,45,64,65] may, on the other hand, underestimate exudation rates compared with nonsterile systems.

Under nonsterile conditions, collection time is an important factor affecting the impact of microbial degradation on recovery of root exudates. Recovery of amino acids in root exudates of *Brassica napus* L. grown in solution culture under nonsterile conditions remained constant during a collection period of 2.5 h but decreased by more than 90% during the next 3.5 h due to microbial metabolization [19]. Investigations on the fate of ^{14}C citrate and ^{14}C malate added to various soils at a realistic concentration level for root exudates (100 to 300 μM) revealed average half-life times of 2 to 3 h, and almost complete mineralization within 48 h [43,66]. Similarly, recovery of various organic acids detected in the rhizosphere soil solution of *Hakea undulata* [28] was little affected by microbial degradation during a 3-h incubation period in soils taken from the culture vessels of the plants, but recovery declined to zero after 20 h (Table 2.2). Rapid microbial degradation in the soil environment

---

TABLE 2.2
Recovery of Water-Extractable Organic Acids Applied to a P-deficient West African Soil Taken from the Culture Vessels of *Hakea undulata* and Incubated under Sterile (chloroform fumigation) and Nonsterile Conditions (organic acid application according to the composition of root exudates)

	Percentage Carboxylate Recovery after Soil Incubation			
Carboxylate	3 h Sterile	3 h Nonsterile	24 h Sterile	24 h Nonsterile
Malic	46.9 ± 4.0	44.5 ± 4.2	34.8 ± 1.7	0
Citric	32.5 ± 3.5	29.5 ± 1.9	12.8 ± 11.1	0
Fumaric	88.1 ± 6.9	64.4 ± 3.1	78.4 ± 1.6	0
trans-Aconitic	71.7 ± 7.4	71.2 ± 7.2	63.1 ± 0.8	0
Lactic	64.6 ± 13.0	69.2 ± 9.5	70.2 ± 5.7	0

has been reported also for other constituents of root exudates, such as amino acids and sugars [30] with similar half-life times in a range between 1 and 5 h [67]. Biodegradation of exudate compounds was found to be most expressed in soils high in organic matter but seems to be inhibited when adsorption to the soil matrix occurs [66,68]. Therefore, sorption materials such as ion-exchange resins [14,46] and reversed-phase materials [13] can be employed to remove certain compounds from the exudate solution to minimize microbial degradation during exudate collection.

The application of various antibiotics such as Rifampicin/Tetracyclin [69], Cefatoxim/Trimethoprim [70], or bacteriostatic compounds such as "Micropur" (Roth, Karlsruhe, Germany) [71] used for root pretreatment or added to collection media is another strategy to prevent biodegradation during root exudate collection. However, depending on dosage and plant species, phytotoxic effects of antibiotics have also been reported (Table 2.3). Antibiotics in the soil environment can rapidly stimulate the release of low-molecular-weight (LMW) organic compounds from dead or damaged microbial cells [72], which may interfere with the determination of root exudates. On the other hand, the function of antibiotics can be affected by adsorption at the soil matrix and by microbial degradation due to the presence of resistant microorganisms [72]. Thus, critical evaluation of root exudate data with application of antibiotics obtained is necessary. Addition of antibiotics to exudate solutions subsequent to the collection period can prevent proliferation of microorganisms and microbial degradation of exudate compounds during further sample processing [73].

TABLE 2.3
Effect of Various Antibiotics on Phytosiderophore (PS) Concentrations in Root Exudates after a 4-h Collection Period and on Fe-PS Uptake in Iron-Deficient Barley and Sorghum

Treatment/Plant Species	Phytosiderophore Concentration in Root Exudates [Relative Values] (%)	Fe-PS Uptake Rate (%)
	Barley	
Control	100	100
(without antibiotics)		
Cefatoxim [30 ppm]	36	74
+Trimethoprim [20 ppm]		
Cefatoxim [60 ppm]	40	57
+Trimethoprim [40 ppm]		
Rifampicin [12.50 ppm]	127	99
+Tetracyclin [6.25 ppm]		
Rifampicin [25.0 ppm]	91	93
+Tetracyclin [12.5 ppm]		
Rifampicin [50 ppm]	n.d.	87
+Tetracyclin [25 ppm]		
Micropur		
1/10 Tablet L^{-1} (1.0 mg L^{-1})	125	100
1/4 Tablet L^{-1} (2.5 mg L^{-1})	112	75
1/1 Tablet L^{-1} (10 mg L^{-1})	75	38
	Sorghum	
Micropur		
Control	100	100
1/10 Tablet L^{-1}	n.d.	n.d.
1/4 Tablet L^{-1}	n.d.	n.d.
1/1 Tablet L^{-1}	<10	15

Note: Fe-PS uptake rates were determined as a parameter for putative phytotoxic effects of the antibiotica treatments; n.d. = not determined.

2. Sorption at the Soil Matrix

Root exudate compounds in soils are differentially affected by adsorption processes, depending on their charge characteristics and on ion-exchange properties of the soil matrix. The lack of charges prevents interactions of sugars with metal ions both in soil solution and at the soil matrix [74]. However, adsorption of more hydrophobic organic exudate compounds such as flavonoids [75] and simple phenolics may be mediated by hydrophobic interactions with humic compounds, and also abiotic oxidation of phenolics and organic acids at Fe and Mn surfaces has been reported [76]. The adsorption of charged compounds such as carboxylic acids, though largely dependent on the soil type and pH, generally tends to increase with the number of negative charges available for anionic interactions with metal surfaces at the soil matrix [77], resulting in rapid removal of certain carboxylate species from the rhizosphere soil solution (Table 2.2). Because metal complexation and ligand-exchange are mechanisms involved in mobilization of mineral nutrients (P, Fe) and exclusion of toxic elements (Al), the most effective organic chelators (e.g., citrate, oxalate, malate) for these elements frequently exhibit the most intense soil adsorption [23,77]. Complex stability of metal carboxylates depends on the number of carboxylic groups and also on their orientation relative to hydroxyl moieties in the molecule [78]. In contrast, sorption of proteinaceous amino acids and the related mobilization of mineral nutrients in soils seem to be comparably low, because of slow reaction kinetics with metal ions [74]. However, the so-called phytosiderophores as non-proteinaceous, tricarboxylic amino acids behave differently and exhibit a fast reaction with amorphous iron (ferrihydrite) in soils [79].

3. Retrieval Mechanisms

Carbon flow in the rhizosphere is not a strictly unidirectional process from root to soil. Active retrieval mechanisms for sugars and amino acids have been identified in plant roots, which were capable of recovering up to 90% of the exudates passively lost into the rhizosphere [22,66,80,81]. Even the preferential uptake of organic nitrogen has been reported for plant species adapted to ecosystems such as arctic tundras, where the rate of nitrogen mineralization is generally low [82]. These findings are in good agreement with reports on the molecular biological characterization of root specific transporters for amino acids [83] and small peptides [84] in higher plants, frequently induced under conditions of N limitation [85,86]. Similarly, induction of a re-uptake system for phytosiderophores as Fe complexes has been reported in graminaceous plant species under Fe deficient conditions [17,40,87]. However, no such retrieval mechanisms could be identified for carboxylic acids [88]. Contrary to sugars and amino acids, strong adsorption of carboxylates in many soils may require an energy investment for remobilization and subsequent uptake, which exceeds the gain of energy by retrieval of carbon originating from carboxylates. The ecological significance of retrieval mechanisms for plants may comprise (1) improved N and Fe acquisition, (2) limitation of passive losses of C and N particularly under stress conditions, and (3) control of microbial colonization at the rhizoplane and in the rhizosphere [89,90]. In this context, the ability of microbial metabolites to stimulate root exudation (e.g., of amino acids) [63] and the expression of retrieval mechanisms for amino acids and peptides by plant roots [83–86] may represent strategies of competition for limiting nutrients between plants and microorganisms in the rhizosphere. However, the contribution of retrieval mechanisms to N and C uptake under real field conditions remains to be established [90].

4. Root Injury

Various techniques for collection of root exudates are associated with the risk of root injury by rupture of root hairs and epidermal cells or rapid changes of the environmental conditions (e.g., temperature, pH, oxygen availability) during transfer of root systems into trap solutions, application of sorption materials onto the root surface, and preparation of root systems for exudate collection.

The possible impact of those stress treatments may be assessed by measuring responses of plant growth in plants either subjected or not subjected to the collection procedure [6], and by comparing exudation patterns after exposure of roots to the handling procedures with different intensities.

III. MECHANISMS OF ROOT EXUDATION

A. DIFFUSION

Release of the major LMW organic constituents of root exudates such as sugars, amino acids, carboxylic acids, and phenolics is a passive process along the steep concentration gradient, which usually exists between the cytoplasm of intact root cells (millimolar range) and the external (soil) solution (micromolar range). Direct diffusion through the lipid bilayer of the plasmalemma (Figure 2.2) is determined by membrane permeability, which depends on the physiological state of the root cell and on the polarity of the exudate compounds, facilitating the permeation of lipophilic exudates [91]. At the cytosolic pH of approximately 7.1 to 7.4 [1], more polar intracellular LMW organic compounds such as amino acids and carboxylic acids usually exist as anions with low plasmalemma permeability. A positive charge gradient, which is directed to the outer cell surface, as a consequence of a large cytosolic K^+ diffusion potential [92] and of plasmalemma ATPase-mediated proton extrusion (Figure 2.3), not only promotes uptake of cations from the external solution, but also the outward diffusion of carboxylate anions. Based on studies on the permeability of lipid bilayers to polar substances by use of synthetic membrane vesicles and the flux density equation of Nobel et al. [93], diffusion-mediated basal exudation of amino acids or malate from plant roots has been calculated at rate of approximately 0.3 nmol h^{-1} cm^{-1} root length or 120 nmol h^{-1} g^{-1} root fresh weight [74,77]. This is in good agreement with experimental data for carboxylate exudation rates of 0.4 to 0.9 nmol h^{-1} cm^{-1} root length (100 to 380 nmol h^{-1} g^{-1} root fresh weight) determined for 5-mm apical root zones of wheat, maize, tomato, potato, and white lupin grown in a hydroponic culture system [94,95]. Jones et al. [74,88] suggested that root exudation of amino acids and sugars generally occurs passively via diffusion and may be enhanced by stress factors affecting membrane integrity such as nutrient deficiency (e.g., K, P, Zn), temperature extremes, or oxidative stress [20,96,97].

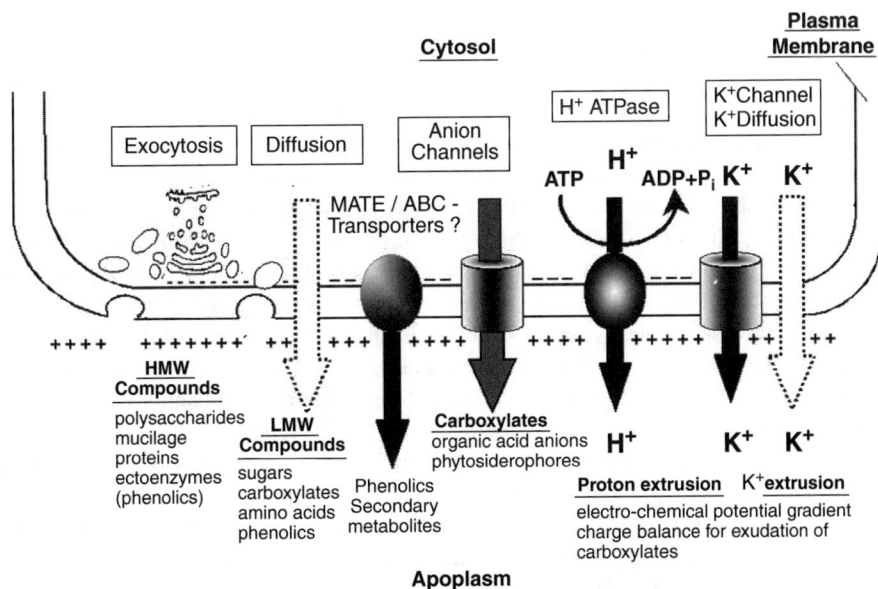

FIGURE 2.2 Model for mechanisms involved in the release of root exudates.

B. Ion-Channels

Root exudation of extraordinary high amounts of specific carboxylates (e.g., citrate, malate, oxalate, phytosiderophores) in response to nutritional deficiency stress or Al toxicity in some plant species cannot simply be attributed to diffusion processes. The controlled release of these compounds, involved in mobilization of mineral nutrients and in detoxification of Al, may be mediated by more specific mechanisms. By use of patch clamp approaches and inhibitor studies, anion channels have been identified in the plasmamembrane of epidermal root cells (Figure 2.2), mediating the release of malate and citrate in wheat and maize under Al stress [98–100] and P-deficiency-induced root exudation of citrate in *Lupinus albus* [18,100]. Anion channels have also been implicated in the release of Fe-, and Zn-mobilizing phytosiderophores in graminaceous plant species [101]. The first identification of a gene encoding for an anion channel responsible for Al-induced root exudation of malate in wheat (*ALMT1*) was provided by Sasaki et al. [102]. The release of carboxylate anions via anion channels seems to be frequently coupled with increased release of protons via plasma-lemma H^+-ATPase or increased K^+ extrusion by K^+ channels, and probably also release of other cations to maintain charge balance (Figure 2.2) [101,103–106].

C. Vesicle Transport

Vesicle transport is involved in root secretion of HMW compounds [107]. The release of mucilage polysaccharides from hypersecretory cells of the root cap is mediated by Golgi vesicles (Figure 2.2). Subsequently, the secretory cells degenerate and are sloughed off. Secretory proteins such as ectoenzymes (e.g., acid phosphatase, phytase, peroxidase, phenoloxidase) are synthesized by membrane-bound polysomes and cotranslationally enter the endomembrane system by vectorial segregation into the ER lumen. While passing through the Golgi apparatus, they are separated from proteins destinated for the vacuolar compartment, and are transported to the plasmalemma by transfer vesicles [108,109]. Processes involved in exocytosis, such as formation of vesicles and their fusion with the plasma membrane strongly depend on extracellular and intracellular calcium levels [1].

Vesicles have also been implicated in storage and release of LMW compounds such as phenolics [110–112] and phytosiderophores [113] in plant roots but the characterization of mechanisms remains to be established.

D. Other Transport Mechanisms

Plant roots are able to secrete a wide range of secondary metabolites, such as phenylpropanoids, quinones, flavonoids, terpenoids, and alkaloids. The large number of different compounds with specific functions in plant–microbial signaling, feeding deterrence, and antimicrobial and allelopathic interactions obviously requires release control in space and time. However, surprisingly little is known concerning the related release mechanisms. Biosynthesis of secondary plant metabolites frequently occurs in close association with the endoplasmic reticulum with subsequent compartmentation in vesicles and vacuoles, to prevent accumulation of toxic levels in the cytosol [112,114,115]. Therefore, vesicle transport has been discussed as a possible release mechanism (Figure 2.2; see Subsection III.C). On the other hand, members of transporter families involved in detoxification and vacuolar compartmentation of pathogen toxins, agrochemicals (e.g., pesticides) and flavonoids, such as multidrug-resistance ATP binding cassette (ABC) transporters [116] and multidrug and toxic compound extrusion (MATE) transporters [117] may also have functions in root exudation of secondary plant metabolites (Figure 2.2; [112]. Accordingly, macroarray analysis for P-deficiency-induced gene expression in *Lupinus albus* revealed an EST with homology to MATE transporters, particularly expressed in cluster roots with high secretory activity for carboxylates and isoflavonoids [118,119].

Apart of release mechanisms responsible for root exudation of LMW and HMW organic compounds, genetically controlled processes are involved even in the liberation of sloughed-off root cells. The number of the so-called root border cells, produced in different plant species is genetically fixed, involves mechanisms of feedback control and can be modified by various environmental factors, such as contact with free water and elevated CO_2 [120]. Root border cells represent the final stage of hypersecretory cells in the root cap, responsible for secretion of mucilage. Embedded into a layer of mucilage polysaccharides, the cells are viable after detachment from the root cap for up to one week, and can be transported during root growth to more basal parts of the root [120,121]. Border cells are able to produce antibiotics and specifically attract root pathogens, such as parasitic nematodes, fungal zoospores, and pathogenic bacteria, thereby counteracting infection of the apical root meristem. In response to infection with pathogenic bacteria and to toxic levels of Al, root border cells exhibit enhanced mucilage excretion, which seems to repel bacteria and alleviates toxic effects of Al on border cell viability [120].

IV. NUTRITIONAL FACTORS

A. PHOSPHORUS (P)

Phosphorus is one of the major limiting factors for plant growth in many soils. Plant availability of inorganic phosphorus (P_i) can be limited by formation of sparingly soluble Ca phosphates, particularly in alkaline and calcareous soils, by adsorption to Fe- and Al-oxide/hydroxide surfaces in acid soils, by formation of Fe/Al-P complexes with humic acids [122,123] and precipitation as Al-, or Fe-phosphates. Phosphorus deficiency can significantly alter the composition of root exudates in a way, which is at least in some plant species related to an increased ability for mobilization of sparingly soluble P sources [30,32,47,77].

1. Role of Carboxylate Exudation for P Mobilization in Soils

Increased root exudation of carboxylates (e.g., citrate, malate, malonate, oxalate) is a P-deficiency response, reported particularly in dicotyledonous plant species [94,95,124]. Mobilization of Pi by exogenous application of carboxylates to various soils with low P availability has been demonstrated in numerous studies [23,122,125–128] and seems to be mediated by mechanisms of ligand exchange, dissolution, and occupation of P sorption sites (e.g., Fe/Al-P and Ca-P) in the soil matrix [23]. Citrate and oxalate were found to be among the most efficient carboxylates with respect to P mobilization in many of these model experiments, according to high stability constants of these carboxylates for complex formation with Fe, Al, and Ca [77]. However, to mediate significant desorption of Pi, carboxylate accumulation of >5 to 10 μmol g^{-1} soil is required for interactions with P sorption sites at the soil matrix [128–131]. A similar carboxylate concentration level in the rhizosphere, has been reported so far only for a very limited number of plant species, such as *Lupinus albus* L. and members of the Proteaceae [23,28,32,46,54] and its still a matter of debate, whether P-deficiency-induced carboxylate exudation observed in many other plants can be really regarded as an adaptive mechanism for chemical P acquisition in soils [6,132].

For 1-cm apical root zones, where the most intense root exudation is frequently located (see Subsection II.B), the amount of rhizosphere soil in a distance of 1 mm from the root surface has been calculated at about 70 mg cm^{-1} root length [133]. Assuming root elongation rates of 1 to 2 cm/d [43], the average residence time of 1-cm apical root zones in a given soil compartment would hardly exceed 12 to 24 h. Thus, accumulation of P-mobilizing carboxylates at rhizosphere concentrations relevant for P desorption (e.g., 10 μmol citrate g^{-1} soil) within 24-h residence time of the root tip in a given soil compartment, would require release rates of at least 30 nmol h^{-1} cm^{-1} root length, and up to 300 nmol h^{-1} cm^{-1} root length, if microbial carboxylate degradation with a half-life time of, for example, 5 h [43] is taken into account. This is far above the reported release rates of carboxylates for many plant species, which in most cases hardly exceed 1 nmol h^{-1} cm^{-1} root length [95]. Stimulation of root exudation by the mechanical impedance of the soil substrate [4,5]

FIGURE 2.3 Clusterlike root structures in different plant species: (A) cluster roots of *Hakea corymbosa* (Proteaceae); (B) clustering of barley roots in response to localized placement of NH_4SO_4; (C) clustering of root hairs (dauciform roots) in apical root zones of sedges (Cyperaceae); (D) clustering of root hairs in the heavy metal hyperaccumulator *Thaspi caerulescens* grown on a Cd-contaminated soil. (Figure 2.3C — courtesy of M. Shane and Hans Lambers, UWA, Perth, Australia.)

or by microbial activity in the rhizosphere [60–62] may contribute to some extent to increased carboxylate exudation [30]. However, a direct involvement in a root-induced strategy for chemical P acquisition seems to be unlikely, because these effects are not specific for P deficiency and strongly depend on environmental conditions. Moreover, the required daily release rates of P-mobilizing carboxylates, such as citrate would easily exceed the total biomass of 1-cm apical root zones (approximately 1 mg dry matter for roots with 1-mm diameter). Therefore, it is not surprising that a general ecological significance of P-deficiency-induced root exudation of carboxylates for P acquisition has been questioned in the recent past [67,77,89].

However, the formation of so-called cluster roots, characteristic for members of the Proteaceae, *Lupinus albus*, and various other plant species [32,47] represents an example for an efficient strategy to increase the concentration of P-mobilizing root exudates in the rhizoshere. The bottlebrush-like clusters of densely-spaced lateral rootlets (50 to 1000 per cm cluster root axis) with limited growth (3 to 5 mm) (Figure 2.3A), densely covered with root hairs are formed mainly in response to low P supply [32,47]. Cluster roots are the sites of intense rhizosphere–chemical changes involved in mobilization of P but also of Ca, Mg, Fe, Mn, Zn, Mo, and Al [23,28,32–35,134] by root-induced pH changes, secretion of carboxylates, phenolics, and acid phosphatase [32,47]. Compared to normal lateral roots, the formation of root clusters increases the radial extension of the rhizosphere by up to 1 cm. Because of the high density of lateral rootlets within this rhizosphere cylinder, the surface area active in root exudation is increased by a factor of 25 to 500 and even higher values may be expected by hyperproliferation of root hairs along the lateral rootlets [32,47,135]. Because lateral rootlets exhibit no more growth activity [32], root exudation into the same soil compartment is possible over extended time periods of 2 to 3 d. The prolonged secretory activity of cluster roots compared with normal lateral roots may further increase the accumulation of root exudates in the rhizosphere by a factor of approximately 3 to 6. Moreover, various modifications in cluster root metabolism are responsible for preferential production and a 5- to 15-fold increase in exudation of root exudates involved in P mobilization (see Subsection IV.A.2.). The combination of these morphological, developmental, and physiological adaptations enables the accumulation of P-mobilizing carboxylates in the rhizosphere of cluster roots at concentration levels up to 45 to 55 μmol g^{-1} soil, reported, for example, for *Lupinus albus*, as a plant species with a proven ability for citrate-mediated P mobilization in the rhizosphere [23,34,130,131]. The modifications in morphology and growth activity of cluster roots mainly increase the quantity of carboxylate accumulation in the rhizosphere, whereas both the composition and the amount of released carboxylates is determined by alterations at the physiological level.

Cluster roots are widely distributed in members of the families of the Fabaceae, Betulaceae, Casuarinaceae, Myricaceae, Cucurbitaceae, Moraceae, Eleagnaceae, Moraceae, and Proteaceae [32,47]. Therefore, Skene [136] postulated, that along with mycorrhizae and N_2-fixing nodules, cluster roots may be regarded as the third major adaptation for nutrient acquisition in terrestrial vascular sporophytes. It remains to be established, whether in other plant species also, "clustering" of root structures by localized proliferation of roots and root hairs or formation of root mats in response to various environmental stimuli (Figure 2.3B to Figure 2.3D) can have a similar impact on accumulation of root exudates and chemical mobilization of nutrients in the rhizosphere. Recently, Lambers et al. [137] demonstrated that so-called dauciform (carrot-shaped) lateral roots in sedges (*Schoenus unispiculatus*, Cyperaceae), formed under P limitation by hyperproliferation of root hairs (Figure 2.3C), release similar amounts of citrate in an exudative burst as observed for cluster roots in *Lupinus albus* or members of the Proteaceae. Intense exudation of carboxylates under P limitation has been reported also in spinach [138], red clover [129], chickpea [94,132], and various *Lupinus* species (*L. luteus, L. angustifolius, L. hispanicus, L. pilosus* [139–141]. Ström et al. [142] suggested that P (and Fe) mobilization, mediated by root exudation of citrate and oxalate, might be related to the ability of various calcicole plant species to grow on calcareous soils.

Although in many soils with low P availability, significant desorption of sparingly soluble Pi forms requires millimolar concentrations of specific carboxylates (e.g., citrate, oxalate) in the extraction solution, much lower application rates were sufficient to reduce soil adsorption of Pi, which was applied simultaneously with carboxylates [133]. Thus, competition of carboxylates with Pi for P-sorption sites in the soil matrix may be a mechanism, which can to some extent prevent soil-fixation of Pi after fertilizer application in agricultural soils even in plant species with moderate exudation rates. This fits well with modeling results of Kirk et al. [143], who identified competition of carboxylates for P-sorption sites as a major mechanism, increasing P availability in the rhizosphere of upland rice, whereas displacement of P from adsorption sites was considered as unimportant.

A possible contribution of rhizosphere-microbial production of carboxylates to chemical P acquisition in plants, driven by root exudation as a carbon source, has been discussed [30]. A wide range of cultivated soil microorganisms are able to mobilise sparingly soluble soil P sources *in vitro* by production of organic chelators and acidification of the growth medium [144]. However, the significance of these processes for P acquisition in plants under the competitive conditions in the rhizosphere remains to be elucidated. Nevertheless, at least for some ectomycorrhizal fungi, significant chemical P mobilization in the myco-rhizosphere by fungal release of carboxylates has been recently demonstrated [145], whereas other studies indicated increased mineralization of carboxylates in presence of ectomycorrhizal associations [146].

A close relationship between P-deficiency-induced exudation of carboxylates in different stages of cluster root development and structural diversity of rhizosphere microbial community structures was reported for *Lupinus albus* [147–149]. Plate counts demonstrated a reduced bacterial abundance associated with a sudden decrease in rhizosphere pH and a pulse of intense citrate exudation in mature root clusters [149,150]. Together with a preceeding release of isoflavonoids [118] and secretion of antifungal cell wall-degrading enzymes, such as chitinase and β-1-3-glucanase [148,150], the transient rhizosphere acidification may represent a plant strategy to counteract microbial degradation of P-mobilizing carboxylates.

In addition to LMW carboxylates, also mucilage can to some extent promote P desorption from clay minerals [151,152], probably mediated by the galacturonate component of the polysaccharide [152,153]. Possible functions of phosphatidylcholine surfactants in root mucilages for P mobilization, inhibition of nitrification and modification soil water retention have been discussed by Read at al. [154].

2. Physiology of Carboxylate Exudation

Apart from root-morphological adaptations (see Subseciton IV.A.1), the potential of certain plant species for mobilization of sparingly soluble soil P sources by root exudation of carboxylates as organic metal chelators is determined also by alterations in root physiology. Major exudate compounds comprise malate, citrate, malonate, and also oxalate, especially in plant species where oxalate replaces malate as the major internal carboxylate anion, for example, members of the Chenopodiaceae [129,138,155].

Preferential root exudation of carboxylates with the highest efficiency in P mobilization under conditions of P limitation seems to be closely related with three groups of physiological alterations: (1) increased biosynthesis of carboxylates in the roots and sometimes also in the shoot tissue, providing exudate compounds or related precursors; (2) reduced metabolic consumption of carboxylates with high potential for P mobilization (e.g., citrate); and (3) expression of specific release mechanisms to mediate controlled root exudation of the respective carboxylates in sufficient amounts. At least the processes involved in biosynthesis and turnover of carboxylates (items 1 and 2 of previous list) can be regarded as more general metabolic responses to P limitation, which have been detected in many plant species, independent of the ability for P-deficiency-induced root exudation of P-mobilizing carboxylates [18,94,124,156–159].

a. Biosynthesis of Carboxylates

Enhanced expression and *in vitro* activities of a specific set of glycolytic enzymes, such as sucrose synthase (SS), phosphoglucomutase (PGM), fructokinase (FK), PPi-dependent phosphofructokinase, and of phosphoenolpyruvate carboxylase (PEPC) represent a widespread metabolic modification detected in P-deficient plant tissues (Figure 2.4). These enzymes may operate as an alternative pathway of carbohydrate catabolism under P-deficient conditions, which facilitates a more economic Pi utilization by Pi recycling, minimization of Pi consumption and utilization of alternative P pools such as PPi [159–163]. Particularly, the enhanced nonphotosynthetic CO_2-fixation via PEPC contributes to carboxylate accumulation in P-deficient plants and can provide a substantial proportion of carbon (>30%) to the biosynthesis of carboxylates (Figure 2.4) [11,94,124,156,157,164]. The cytosolic enzyme catalyzes the carboxylation of phosphoenolpyruvate (PEP) to oxaloacetate, which can be further converted to malate by enhanced expression of cytosolic malate dehydrogenase (MDH) [36,156,157,163,165]. Phosphorus-deficiency-induced upregulation of PEPC is regulated at the transcriptional and also at the posttranslational levels by protein phosphorylation [156,157,165]. PEPC liberates Pi from PEP (Figure 2.4), and may be therefore regarded as an alternative reaction to PEP turnover via pyruvate kinase (PK), which depends on the presence of ADP and Pi as limiting factors under P-deficient conditions [161]. Accordingly, increased activity of PEPC was associated with downregulation of PK in roots of P-deficient *Lupinus albus* (Figure 2.4); [166].

Particularly in dicotyledonous plant species such as tomato, chickpea, and white lupin [94,164] with a high inherent cation/anion uptake ratio, PEPC-mediated biosynthesis of carboxylates may be also linked to excessive net uptake of cations due to inhibition of uptake and assimilation of nitrate under P-deficient conditions (Figure 2.4) [18,164,167]. Excess uptake of cations is balanced by enhanced net release of protons [94,164,168], provided by increased biosynthesis of organic acids via glycolysis and PEPC as a constituent of the intracellular pH-stat mechanism [169,170]. In these plants, P-deficiency-mediated proton extrusion leads to rhizosphere acidification, which can contribute to the solubilization of acid-soluble Ca phosphates in calcareous soils (Figure 2.4) [35,171,172]. In some species (e.g., chickpea, white lupin, oilseed rape, buckwheat), the enhanced

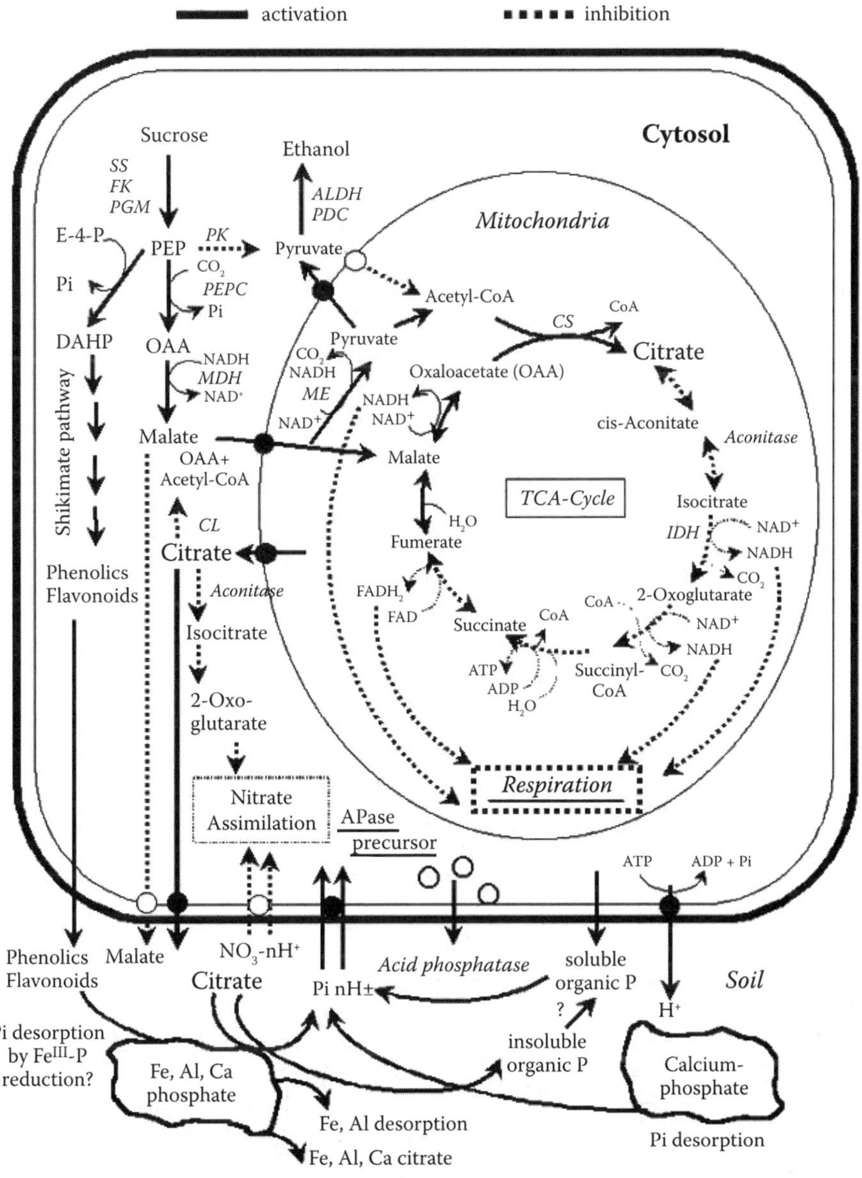

FIGURE 2.4 Model for phosphorus (P)-deficiency-induced physiological changes associated with the release of P-mobilizing root exudates in cluster roots of white lupin. Solid lines indicate stimulation, and dotted lines inhibition of biochemical reaction sequences or metabolic pathways in response to P deficiency. For detailed description, see Section VI.1. Abbreviations: SS = sucrose synthase; FK = fructokinase; PGM = phosphoglu-comutase; PEP = phosphoenolpyruvate; PEPC = PEP-carboxylase; PK = pyruvate kinase; MDH = malate dehydrogenase; ME = malic enzyme; CL = ATP-Citrate lyase, CS = citrate synthase; PDC = pyruvate decarboxylase; ALDH = alcohol dehydrogenase; E-4-P = erythrose-4-phosphate; DAHP = dihydroxyacetone-phosphate; APase = acid phosphatase.

net release of protons is associated with increased exudation of carboxylates, whereas in tomato, carboxylate exudation was negligible despite intense proton extrusion [94,173]. Solution culture experiments revealed that plant species with intense P-deficiency-induced carboxylate exudation, such as oil-seed rape [124], chickpea, and white lupin [94] accumulated organic acids mainly in the root tissue and, moreover, in the root zones where exudation was most intense (e.g., subapical root zones, proteoid roots). In contrast, root exudation of carboxylates even decreased in response to P deficiency in plant species such as Sysimbrium officinale [124], wheat, and tomato [94] and was associated with predominant carboxylate accumulation in the shoots (Figure 2.5). Increased root-to-shoot translocation of carboxylates in P-deficient *Ricinus communis* L. has been reported by Jeschke et al. [125] and may be related to the higher storage capacity for carboxylates in the leaf vacuoles [18].

These findings suggest that increased biosynthesis and intracellular accumulation of carboxylates may be a general metabolic response to P limitation with primary functions in intracellular pH stabilization and more economic P utilization. Depending on plant species, the carboxylates are stored in the vacuoles [94,174], further metabolized [175], or released into the rhizosphere as adaptive response for chemical P acquisition. From an evolutionary point of view, the integration of such already-exisiting structures into novel contexts may be an advantage, offering flexibility for the rapid development of complex adaptations.

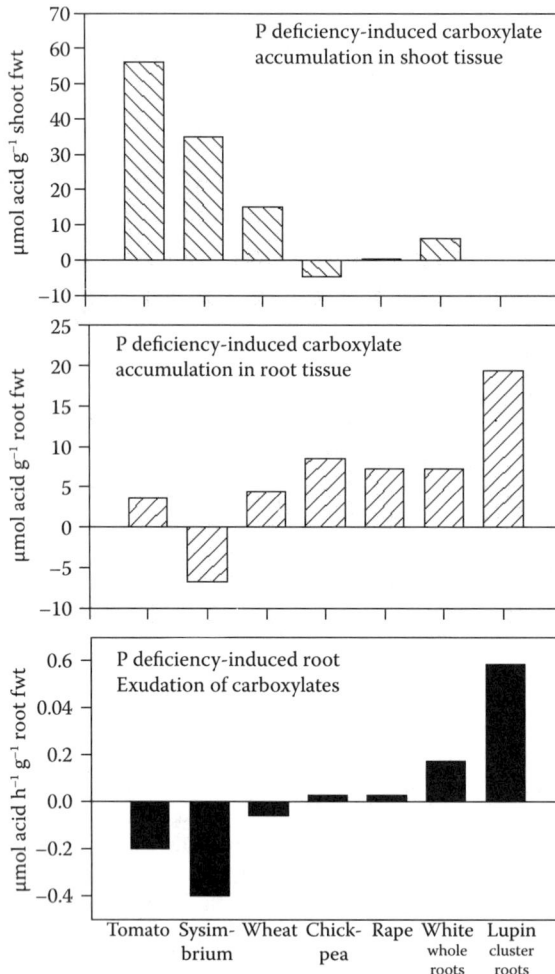

FIGURE 2.5 P-deficiency-induced changes in tissue concentrations and root exudation of carboxylates in different plant species. The zero line represents the P-sufficient control. (Adapted from Neumann, G. and Römheld, V., *Plant Soil*, 211, 121, 1999; Hoffland, E. et al., *New Phytologist*, 122, 675, 1992.)

b.   Reduced Turnover of Carboxylates

The primary products of carboxylate biosynthesis under P limitation are oxaloacetate (PEPC reaction), which is readily converted to malate by the MDH reaction [163]. Accordingly, in many plants, malate is a major carboxylate anion, which accumulates in the root tissue in response to P deprivation but, additionally an overproportional increase in citrate accumulation has been frequently reported [94,124,156,157]. Also cluster roots of *Lupinus albus* with particularly intense expression of P-deficiency-induced citrate exudation exhibit high levels of malate accumulation in growing juvenile root clusters with a high P-nutritional status and high metabolic activity. During ageing of the clusters, declining intracellular P concentrations are associated with a shift to almost exclusive citrate accumulation in the root tissue, which precedes a pulse of intense citrate exudation in mature root clusters [18,37,176]. Redistribution of P from the mature and senescent root clusters to the juvenile, actively growing clusters with a high energy demand may explain declining levels of P_i, ATP, RNA, and respiratory activity during cluster root development in *Lupinus albus* [18,162,176]. Interestingly, similar changes have been observed also in other cluster-rooted plant species such as *Hakea prostrata* [177] and *Hakea undulata* (Neumann, unpublished), not systematically related with *Lupinus albus,* suggesting the influence of a more general metabolic principle, which links citrate accumulation with a low local P status at the tissue level. In face of the high efficiency of citrate for P mobilization in soils (see Subsection IV.A.1), a shift from intracellular accumulation of malate to citrate in the root tissue under P-deficient conditions may be of particular ecological significance.

Oxaloacetate and malate, produced by the PEPC pathway can act as precursors for biosynthesis of citrate via citrate synthase (CS) in the TCA cycle. However, neither measurements of CS activities [94,156,166], nor analysis of differential gene expression under P limitation in various plant species [159,162,163] demonstrated consistent evidence for a distinct upregulation of CS in plant roots under P limitation. Also, attempts to increase production and root exudation of citrate by transgenic overexpression strategies revealed contradictory results [178–181], suggesting that CS may not have a general rate-limiting function for intracellular citrate accumulation. In contrast, there is increasing evidence that reduced activity of various metabolic pathways involved in citrate turnover contributes to intracellular citrate accumulation in plant roots under P limitation and are listed as follows:

1. Downregulation of enzymes involved in citrate degradation such as aconitase, the cytosolic isoform of isocitrate dehydrogenase, and various dehydrogenases of the TCA cycle has been reported for *Lupinus albus* (Figure 2.4), tomato, chickpea, and *Sysimbrium officinale* and was associated with increased citrate accumulation [94,124,166,182]. Accordingly, artificial inhibition of aconitase activity by short-term application of monofluoracetate drastically increased citrate accumulation and root exudation of citrate, even in juvenile root clusters and also in seedling roots of *Lupinus albus* at a rate comparable with mature cluster roots [182]. Intense accumulation and release of citrate was also observed in aconitase mutants of yeast [181]. Declining activities of cytosolic enzymes involved in citrate turnover (aconitase, NADP-isocitrate dehydrogenase) may be related with P-deficiency-induced inhibition of nitrate uptake and assimilation [118,164, 166,167,183], because oxoglutarate as a product of these enzymatic reactions is discussed as important acceptor for amino-N as a product of nitrate reduction [184].
2. Also reduced root respiration, frequently reported under P limitation [18,156,177,183] may contribute to citrate accumulation. It was suggested that limitation of respiration by impairment of the mitochondrial function may induce a feedback inhibition of citrate turnover in the TCA cycle, to prevent excessive production of reducing equivalents (Figure 2.4) [18,183,185].
3. Citrate accumulation in mature root clusters of *Lupinus albus* was associated also with declining expression and activity of ATP-dependent citrate lyase (CL) (Figure 2.4) [186]. The enzyme catalyzes the ATP-dependent cleavage of citrate into oxaloacetate and acetyl-CoA. In juvenile root clusters, high activities of CL may provide acetyl-CoA for the

biosynthesis of lipids and phenolics (Figure 2.4). The production of acetyl-CoA may be limited by downregulation of PK under conditions of P limitation [166], favoring pyruvate conversion to di-, and tricarboxylates via PEPC.

Apart from these P-deficiency-induced metabolic changes determining quantity and the qualitative pattern of carboxylates in root exudates, there is increasing evidence that the composition carboxylates is influenced also by the culture substrate and soil properties [29,187]. These findings are not surprising because plant roots grown in P-deficient soils are frequently exposed to a wide range of additional stress factors with differential impact on organic acid metabolism, such as Al-toxicity, low pH, low levels of nitrate in acid soils, or Fe deficiency, bicarbonate toxicity, high pH and high levels of Ca^{2+} in calcareous soils. Moreover, the plant developmental stage can significantly affect root exudation of carboxylates in an order of magnitude, which, in some cases, even exceeds the responses to P deficiency [132,188].

c. Mechanisms of P-Deficiency-Induced Carboxylate Exudation

Although, metabolic alterations under P limitation are important to provide the most efficient carboxylates for P mobilization in the rhizosphere, the exudation of these compounds in sufficient amounts probably requires the expression of controlled release mechanisms. More detailed investigations on the related transport processes are currently available only for P-deficiency-induced exudation of citrate and malate from cluster roots of *Lupinus albus*: Inhibitor studies [18] and more recently patch clamp approaches [189] revealed that intense pulses of citrate exudation in mature root clusters during 1 to 3 d [18,37] are mediated by anion channels (Figure 2.2). The channels exhibited a highly selective permeability for citrate as well as for malate, and were detected not only in the epidermal plasma membrane of cluster roots but also in lateral roots of P-limited and even P-sufficient lupin plants [189]. This observation may explain the failure of approaches to identify potential target genes by analysis of differential gene expression in cluster roots [119,162,163,190]. The almost exclusive release of citrate during the exudation pulse is probably a consequence of the preferential intracellular accumulation of citrate in mature root clusters [18,189]. Upregulation of the plasma membrane (PM) H^+ ATPase results in a concomittant extrusion of protons (Figure 2.4), which induces intense rhizosphere acidification [35,104,105]. Similar observations have also been recently reported for the cluster-rooted species *Lupinus pilosus* [141]. The release of H^+ and, probably, also of other cations, such as K^+, N^+, and Mg^{2+} is responsible for charge-balance of the secreted carboxylate anions [103,191,192]. Increased H^+ extrusion is also required to balance excess uptake of cations over anions as a consequence of reduced nitrate uptake under P limitation [35,118,164], which complicates stoichiometric calculations. However, a close relationship between citrate exudation and proton extrusion by PM-H^+ATPase is indicated by a stimulation of ATP-dependent transport of H^+ and ^{14}C-labelled citrate into PM vesicles in the inside-out orientation, isolated from cluster roots of P-deficient *Lupinus albus* compared with P-sufficient control plants [105].

The anion channel supposed to be mainly responsible for citrate exudation in cluster roots is activated by hyperpolarization of the plasma membrane [189]. This is in good agreement with the observation that PM hyperpolarization by activation of PM H^+ATPase after fusicoccin treatments stimulates root exudation of citrate also in intact lupin plants [103]. Under natural growing conditions, intracellular accumulation of extraordinary high levels of citrate in mature root clusters (20 to 30 μmol g^{-1} root fwt), probably exceeding the vacuolar uptake capacity [118,176] may contribute to PM hyperpolarization. This effect may be further enhanced by cytosolic acidification as a consequence of enhanced production of organic acids under P limitation, leading to H^+ extrusion by activation of the PM H^+ ATPase. Accordingly, citrate exudation was induced in juvenile root clusters and noncluster roots by increased intracellular citrate accumulation in response to application of aconitase inhibitors [182].

A stimulation of carboxylate exudation (2 to 6 μmol h^{-1} g^{-1} root fresh weight) similar to citrate exudation in *Lupinus albus* (1 to 2 μmol g^{-1} root fresh weight) has been discussed as mechanism for detoxification of lactic acid, which accumulates in root tips of maize under hypoxic conditions [193] and for detoxification of bicarbonate-induced accumulation of citrate and malate in the root tips of rice (see Figure 2.9), with comparable intracellular carboxylate concentrations (20 to 40 μmol g^{-1} root fresh weight).

As a more unspecific effect, enhanced leakiness of membranes in response to P deprivation may also contribute to enhanced release of sugars, amino acids, and organic acids [97] (see also Chapter 1). Schilling et al. [30] reported a shift in the qualitative composition of sugars in root exudates of maize and pea, leading to a higher proportion of pentoses at the expense of glucose and sucrose under P-deficient conditions. Phosphorus-deficiency-induced root exudation of sugars and amino acids has been related to increased mycorrhizal colonization [194,195].

d. Carbohydrate Investments

Root exudation of metal-chelating carboxylates in sufficient amounts to mediate P mobilization in soils can comprise up to 25% of the assimilated carbon [35,196]. This seems to be an extraordinarily high loss of carbon but resembles the carbon investments in various mycotrophic plant species for maintaining mycorrhizal associations [197]. Interestingly, cluster-rooted plant species, characterized by the most intense P-deficiency-induced carboxylate exudation are frequently nonmycorrhizal [47], and cluster roots may be regarded as an alternative strategy for nutrient acquisition [136]. Moreover, nonphotosynthetic, anaplerotic CO_2 fixation via PEPC under P limitation can contribute up to 30% of the exuded carbon [11]. Phosphorus-deficiency-induced downregulation of root respiration by up to 60%, [18,156,177,183] may additionally minimize carbon losses originating from root exudation, because respiratory CO_2 release can comprise 15 to 60% of the net fixed carbon fraction translocated to the roots [198].

3. Exudation of Phenolic Compounds

In many plants, P deficiency also enhances production and root exudation of phenolic compounds (Figure 2.4) [28,32,118,199,200]. Increased biosynthesis of phenolics under P-deficient conditions was suggested as another metabolic bypass reaction involved in liberation and recycling of P_i in P-starved cells [161]. Antibiotic properties of certain phenolic compounds (e.g., isoflavonoids) in root exudates [75] may not only counteract infection by root pathogens, but also prevent the microbial degradation of exudate compounds involved in P mobilization [32]. Certain root flavonoids have been identified as signal molecules for spore germination and hyphal growth of arbuscular mycorrhiza, and flavonoids are likely to be important also as signaling compounds for the establishment of ectomycorrhiza. However, there are also contradictory reports and the nature of root exudates and the related mechanisms are still not clear [89,201,202]. (This subject is reviewed in Chapter 8.) Phenolics may further contribute to P mobilization by reduction of sparingly soluble FeIII phosphates (Figure 2.4) [32]. Extraction experiments with calcareous and acidic soils revealed P extraction efficiencies of various phenolic acids comparable to citrate when supplied in high concentrations up to 100 μmol g^{-1} soil [203]. The specific release of piscidic acid (*p*-hydroxyphenyl tartaric acid) from roots of P-deficient pigeon pea (*Cajanus cajan* L.), which is a strong chelator for FeIII, has been related to enhanced mobilization of Fe-phosphates in Alfisols [204]. However, considering comparatively low exudation rates, phenolic compounds, such as piscidic acid may be more relevant as a signaling compound for the establishment of microbial associations (e.g., AM, rhizobia).

4. Root-Secretory Phosphohydrolases

Enhanced secretion of acid phosphatases [25,55,205] and phytases [206] by plant roots and also by rhizosphere microorganisms [207] under P-deficient conditions may contribute to Pi acquisition by hydrolysis of organic P esters in the rhizosphere (Figure 2.4). Organic P fractions can comprise up to 30 to 80% of the total soil phosphorus, thereby providing a significant proportion of plant available P in natural ecosystems [208].

Phosphorus-deficiency-induced root secretion of acid phosphatases regulated at the transcriptional level [118,209], may involve sensing of external P concentrations in the growth medium [210] and differential induction of isoenzymes [211] with considerable genotypic variation between plant species and even of cultivars [206,212].

In many soils, however, the availability of organic phosphorus seems to be limited mainly by the low solubility of certain P forms such as Ca- and Fe/Al-phytates, which can make up a major proportion of the soil-organic P [213–215]. Moreover, root-secretory acid phosphatases exhibit only limited hydrolytic activity toward phytates, and the release of specific phytases seems to be more abundant in microorganisms than in plants [208]. Another limiting factor for phosphatase-mediated P mobilization is the low mobility of the hydrolytic enzymes (APase, phytase), mainly associated with the root cell wall and with mucilage in apical root zones [216,217]. Phosphatases can be also subjected to adsorption and inactivation on clay minerals and organomineral associations [218,219]. Accordingly, phosphorus acquisition from phytate by *Arabidopsis* and *Trifolium subterraneum* grown on agar medium was significantly increased by transgenic expression of a secretory phytase from *Aspergillus niger*. However, this effect was markedly reduced in soil culture, probably due to strong fixation of phytate and possible inactivation of the enzyme in soils [220,221]. Therefore, limited availability of phytates for enzymatic hydrolysis may explain the accumulation of phytates as dominant organic P fraction in many soils, while sugar-, lipid-, or nucleotide-phosphates exhibit higher solubility and, thus, higher rates of mineralization by enzymes released from plants and microorganisms [208].

Beissner [214] reported that oxalic acid in root exudates can contribute to some extent to phytate mobilization in soils. Similarly, in a P-deficient Arenosol, Pi mobilization by simultaneous application of acid phosphatase and organic acids identified in rhizosphere soil solution of *Hakea undulata* was greater than the additive effect, calculated from separate application of organic acids or acid phosphatase, respectively (Figure 2.6). Hens et al. [222] reported that a substantial proportion of phosphorus, mobilized by soil application of citrate in concentrations characteristic for the rhizosphere of cluster

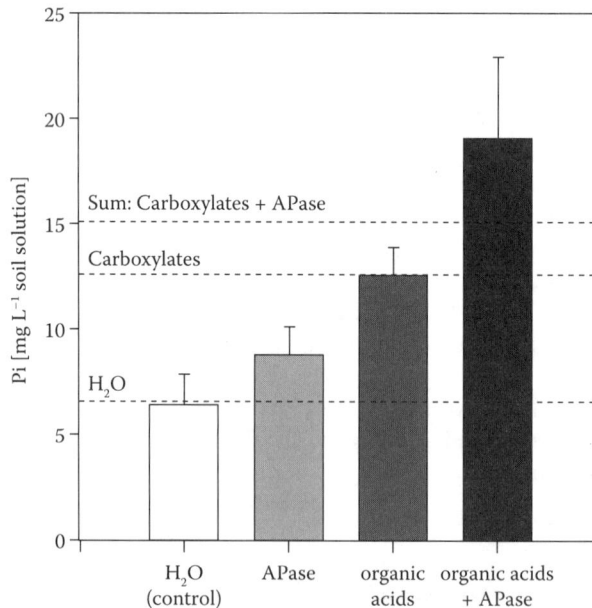

FIGURE 2.6 Water-extractable Pi in a phosphorus-deficient sandy soil from Niger (West Africa) after separate or simultaneous addition of acid phosphatase and of organic acids detected in the proteoid-rhizosphere soil solution of *Hakea undulata*. Organic acids: malic 7.5 mM; citric 2 mM; fumaric 1 mM, t-aconitic 0.6 mM; acid phosphatase: wheat germ APase according to enzyme activity in rhizosphere soil [0.7 U g soil^{-1}].

roots, was hydrolyzable by acid phosphatase. These findings suggest that root exudation of carboxylates, such as citrate and oxalate in sufficient quantities, seems to enhance the solubility not only of inorganic P but also of organic soil P forms, subsequently hydrolyzed by phosphatases in the rhizosphere.

Because there is no evidence for direct uptake of organic P by plant roots in significant amounts, an additional important function of root secretory acid phosphatases may be the rapid retrieval of phosphorus by hydrolysis of organic P, which is permanently lost by diffusion or from sloughed-off and damaged root cells (Figure 2.4) [223]. Accordingly, both high activities of secretory acid phosphatase and a high capacity for Pi uptake were detectable even in senescent cluster roots of *Lupinus albus* in hydroponics and soil culture [18,148].

B. Nitrogen and Potassium

1. Nitrate Assimilation

At least in some plant species, such as maize [224], *Lupinus angustifolius* L. [225] and tomato [226], root exudation of di- and tricarboxylic acids (mainly malate and citrate) seems to be affected by the form of nitrogen supplied as nitrate or ammonium. Generally, exudation of the carboxylates increased with increasing levels of nitrate in the culture medium. This may be related to the function of carboxylates in intracellular pH stabilization. Nitrate reduction in roots and in the shoot is stimulated with increasing nitrate supply, and results in the production of an equivalent amount of OH$^-$, which is neutralized by increased biosynthesis of organic acids or may be released into the rhizosphere when produced in the root tissue [225,226]. The carboxylate anions can be stored in the leaf vacuoles but are also retranslocated to the roots via phloem transport when the leaf storage capacity is limited. In the root tissue, the carboxylate anions are either metabolized by decarboxylation or can be released into the rhizosphere [1,226,227].

2. Ammonium Assimilation

Excess uptake of cations over anions as a consequence of increased ammonium supply is balanced by extrusion of protons. Synthesis of carboxylic acids can contribute to pH stabilization in the root tissue. However, carboxylate anions are required as acceptors for ammonium assimilation in the roots, which is also associated with the production of protons and decarboxylation of organic acids [1]. As a consequence, tissue concentrations and root exudation of carboxylates are declining with increased ammonium supply [226].

High nitrogen concentrations inhibit the production and release of isoflavonoids from lupin roots. Compared with nitrate supply, exudation was strongly enhanced by ammonium application [228]. Similarly, the well-known inhibitory effect of nitrogen on nodulation during establishment of the legume-rhizobium symbiosis is mainly caused by nitrate. [229]. In short-term, intense rhizosphere acidification induced by NH$_4^+$ nutrition or low rates of NO$_3^-$, supply may directly stimulate the release of phenolics and other LMW root exudate compounds as a consequence of an increased electrochemical transmembrane potential gradient and also because of acid-induced impairment of membrane integrity [230] (Table 2.4). Because flavonoids have important functions as chemoattractants, and nod-gene inducers for rhizobia [231], nitrogen effects on nodulation may be explained by differential exudation of these compounds depending on N form supply and the N nutritional status of the plants. Root flavonoids are involved also in pathogen and allelopathic interactions [75,232], and these processes might be similarly affected by nutritional modifications in root exudation. However, the mechanisms involved are largely unknown (see Subsection III.D).

3. Potassium Nutrition

Only limited information is available on effects of potassium (K) supply on root exudation. Increased exudation of sugars, organic acids, and amino acids has been detected in maize as a

TABLE 2.4
Release of Reducing Root Exudates (e.g., phenolics) by Peanut Plants as Affected by Fe Nutritional Status and Short-Term (10-h) Supply of a NH_4^+- N Containing [1 mM] Nutrient Solution

N form in Nutrient Solution	pH of the Nutrient Solution after 10 h	Reducing Substances in the Nutrient Solution [nmol Caffeic Acid Equivalents 10 h^{-1} g^{-1} Root Fresh Weight]	
		+Fe	–Fe
Control (NO$_3^-$-N)	5.5	5	28
NH$_4^+$-N	3.9	9	115

response to K limitation [224]. This may be related to a K-deficiency-induced preferential accumulation of LMW N and C compounds at the expense of macromolecules [1]. In contrast, Gerke [129] reported enhanced extrusion of protons but a reduction in carboxylate exudation in K-deficient wheat, sugar beet, and oil-seed rape. Soil-extraction experiments with carboxylates, amino acids, and sugars revealed that only citrate applied in extraordinary high concentrations [6 mmol g^{-1} soil] was effective in K desorption. Thus, K mobilization by root exudates was suggested to be of minimal importance [129].

C. Iron

Although iron (Fe) is one of the major soil constituents (0.5 to 5%), where it is usually present in the oxidized state (Fe^{III}), plant availability is severely limited by the low solubility of Fe hydroxides at pH levels favorable for plant growth. Therefore, plants need special mechanisms for acquiring Fe from sparingly soluble Fe forms to fit the requirements for growth, especially in neutral and alkaline soils, where the availability of Fe is particularly low [233]. Mechanisms involved in iron acquisition by plants are also discussed in Chapter 1, Chapter 5, and Chapter 7.

1. Strategy I Plants

In dicotyledonous plants and in nongraminaceous monocotyledons (strategy I plants) [234], Fe^{III} solubilization is usually mediated by rhizosphere acidification (Figure 2.7A), by complexation with chelating compounds (Figure 2.7B), and by reduction to Fe^{II} (Figure 2.7C), which is taken up by the roots by a transporter for Fe^{II} (Figure 2.7C) [235]. The respective Fe^{II} transporter genes (IRT1) have been cloned in *Arabidopsis*, tomato, and pea [236] and were identified by functional complementation of a yeast mutant defective in Fe uptake [235].

Root responses to Fe limitation are frequently confined to subapical root zones [31,50], and associated with distinct changes in root morphology, such as proliferation of root hairs, thickening of the root tips, and formation of rhizodermal transfer cells [237,238] to increase the root surface available for root-induced chemical changes and Fe uptake (see also Chapter 5).

Rhizosphere acidification in response to Fe deficiency is mediated by activation of the plasmalemma H$^+$-ATPase (Figure 2.7A) [239,240]. The reductive capacity is increased by enhanced expression of a plasmalemma-bound reductase system with a low pH optimum [241,242], which is further activated by rhizosphere acidification (Figure 2.7C) [230]. A concomitant release of phenolic acids [243,244] and of carboxylates has been discussed as possible mechanism for Fe^{III} complexation [23,245,246], and to some extent also for Fe^{III} reduction in the rhizosphere (Figure 2.7B) [230]. In some plant species, Fe deficiency also stimulates root excretion of flavins [247] with yet unknown functions. Phytohormones such as ethylene and indole acetic acid have been implicated

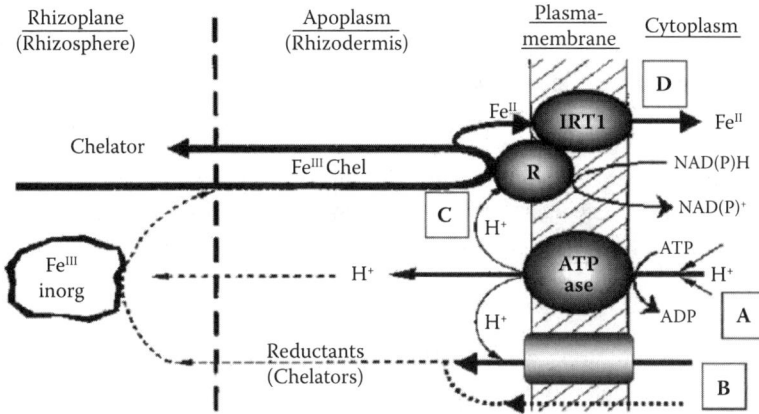

FIGURE 2.7 Model for iron (Fe)-deficiency-induced changes in root physiology and rhizosphere chemistry associated with Fe acquisition in strategy I plants. (Modified from Marschner, H., *Mineral Nutrition of Higher Plants*, 2nd ed., Academic Press, London, 1995.) (A) Stimulation of proton extrusion by enhanced activity of the plasmalemma ATPase→ Fe^{III} solubilization in the rhizosphere. (B) Enhanced exudation of reductants and chelators (carboxylates, phenolics) mediated by diffusion or anion channels → Fe solubilization by Fe^{III} complexation and Fe^{III} reduction. (C) Enhanced activity of plasma-membrane (PM)-bound Fe^{III} reductase further stimulated by rhizosphere acidification (A). Reduction of Fe^{III} chelates, liberation of Fe^{II}. (D) Uptake of Fe^{II} by a plasma-membrane-bound Fe^{II} transporter.

in the signaling of the coordinated strategy I responses to Fe deficiency in dicotyledonous plants [248–250]. However, studies with mutants of *Arabidopsis* and tomato suggest that ethylene-induced modifications of root morphology and the physiological responses to Fe deficiency, such as H^+ extrusion and enhanced reductive capacity at the root surface, may be regulated separately (Chapter 5) [251,252].

Release of carboxylates and phenolics under Fe stress may be stimulated by a steeper electrochemical potential gradient due to enhanced net extrusion of protons and elevated internal carboxylate concentrations. Similar to the P deficiency response, Fe-deficency-induced accumulation of organic acids in the root tissue (citrate, malate) is associated with increased activity of PEPC [253] and probably also with other as yet unknown metabolic modifications involved in enhanced accumulation of carboxylates. However, a reduction in citrate turnover in the TCA cycle due to inhibition of the Fe-dependent aconitase reaction, which has been discussed in earlier studies, seems not to be a limiting factor [254]. The biosynthesis of organic acids in the root tissue in response to Fe limitation does not only provide protons for the H^+-ATPase-mediated rhizosphere acidification but the related metabolism also provides electrons for the Fe deficiency-induced plasma membrane-bound reductase system. Oxidation of citrate via cytosolic aconitase and cytosolic NADP-dependent isocitrate dehydrogenase [255,256] has been reported to be a major direct or indirect electron source for reductase-mediated iron reduction, whereas other studies suggested an important role of NADH [257] and even of ascorbate [258]. Citrate anions are also involved in root-to-shoot translocation of Fe via Fe-citrate in the xylem [253,259,260] and PEPC-mediated increased biosynthesis of carboxylates in the root tissue may contribute to replenishment of carbon losses due to limited photosynthetic CO_2 fixation in leaves affected by Fe-deficiency chlorosis [253].

Increased root exudation of carboxylates in response to Fe deficiency has been reported for chickpea [246], but not for sunflower [261], and seems to be very low compared with the rate of proton extrusion. Based on model calculations, Jones et al. [245] suggested that even the low background exudation of citrate in roots of unstressed oil-seed rape might be sufficient for Fe^{III} solubilization at a rate, which would be high enough to meet the Fe requirements of the plant. However, in face of rapid microbial turnover of carboxylates in the rhizosphere [43], slow formation and low stability of Fe-citrate complexes at soil pH levels above 6.8 [245], the significance of such a mechanism for citrate-mediated Fe mobilization, especially in Fe-deficient calcareous soils with a high buffering capacity, remains questionable. Moreover, humic substances derived from decomposed organic matter, mostly exceeding the concentrations of root-borne chelators in soil solutions, may be involved in contact reduction and transport of Fe to the root surface [262–264].

2. Strategy II Plants

In contrast to strategy I plants, grasses are characterized by a different mechanism for Fe acquisition, with Fe-mobilizing root exudates as main feature. In response to Fe deficiency, graminaceous plants (strategy II plants) [40] are able to release considerable amounts of nonproteinaceous amino acids (Figure 2.8B), so-called phytosiderophores (PS), which are highly effective chelators for Fe^{III} (Figure 2.8) [48,265]. This release takes place predominantly in subapical root zones [31]. Unlike Fe^{III} citrate, Fe^{III}-PS chelates are stable even at high soil-pH levels >7 [245,266,267]. Because of the formation of high-affinity Fe^{III}–PS complexes (Figure 2.8C), there is only minimal competition by chelation with Ca^{2+}, Mg^{2+}, and Al^{3+}, which are usually present in high concentrations in many soils [233]. However, recent studies indicate that sulfate and, particularly, phosphate applied as fertilizers at high rates may inhibit PS-promoted Fe^{III} dissolution, mainly by displacement of PS from the surface of Fe hydroxides [268].

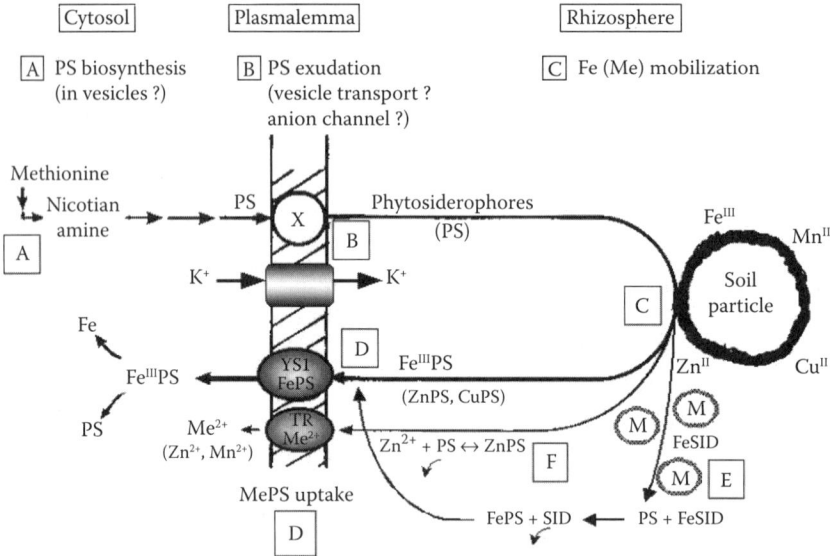

FIGURE 2.8 Model for root-induced mobilization of iron and other micronutrients (Zn, Mn, Cu) in the rhizosphere of graminaceous (strategy II) plants. (Modified from Marschner, H., *Mineral Nutrition of Higher Plants*, 2nd ed., Academic Press, London, 1995.) Enhanced biosynthesis of mugineic acids (phytosiderophores, PS) in the root tissue: (A) biosynthesis of PS; (B) exudation of PS anions by vesicle transport or via anion channels, charge-balanced by concomitant release of K^+; (C) PS-induced mobilization of Fe^{III} (Mn^{II}, Zn^{II}, Cu^{II}) in the rhizosphere by ligand exchange; (D) uptake of metal–PS complexes by specific transporters in the plasma membrane; (E) ligand exchange between microbial (M) siderophores (SID) with PS in the rhizosphere; (F) alternative uptake of microelements mobilized by PS after chelate splitting.

Unlike strategy I plants, where Fe^{III} reduction is a prerequisite for Fe uptake, strategy II involves a specific transport system for Fe^{III}-PS complexes, located at the plasma membrane in roots of graminaceous plants (Figure 2.8D) [269]. The uptake system requires metabolic energy [48], involves H^+ cotransport [270], and is highly specific with respect to Fe^{III} as metal ligand and to PS as organic chelators as well [233]. Although Fe^{III}, chelated by synthetic or microbial siderophores is not recognized by this Fe^{III}-PS transporter [40], Fe^{III} complexes with native or partially degraded microbial siderophores, such as rhizoferrin or dimerum acids [271,272] can improve Fe uptake in graminaceous plant species via exchange chelation with phytosiderophores (Figure 2.8E). However there is also evidence that certain microorganisms (e.g., *Pseudomonas fluorescens*) can profit from PS-mediated Fe mobilization in the rhizosphere [273].

The *YS1* maize mutant, which is defective in Fe^{III}-PS uptake, was recently employed to clone the *ZmYS1* gene by transposon tagging. Heterologous expression of the *YS1* protein restored growth of a yeast mutant defective in Fe uptake after application of Fe^{III}-PS, suggesting that *ys1* codes for the Fe^{III}-PS transporter [87]. *ZmYS1* belongs to a ubiquitous superfamily of oligo-peptide transporters (OPT), mediating the transport of various peptides, including glutathione and enkephalines [274]. *YS1* homologues have also been identified in strategy I plants including gymnosperms and mosses, with possible functions for internal transport of Fe and other metals complexed with nicotianamine (NA) [275], which acts as an ubiquitous metal chelator in higher plants and as precursor for biosynthesis of PS in graminaceous plant species.

After entering the cytosol, the behavior of Fe^{III}-PS complexes is still unknown, but the reduction potential of -102 mV suggests that Fe liberation is possible via reduction by common physiologically available reductants such as NAD(P)H (-320 mV) and glutathione (-230 mV) [233]. Based on speciation analysis, transfer of Fe^{III} and Fe^{II} from PS to nicotinamine has also been discussed with possible functions in intra-, and intercellular Fe trafficking and protection from oxidative damage due to Fe^{II} oxidation at cytosolic pH levels [276,277].

Tolerance to Fe deficiency in different graminaceous plant species is roughly related to the amount of PS exuded under Fe-deficient conditions. High release rates of PS are frequently found in cereals adapted to the calcareous soils of the Fertile Crescent in the Middle East, such as barley, rye, bread wheat, oats, and their wild ancestors. In contrast, rice, maize, sorghum, or millet as plant species originating from the humid tropics with widespread abundance of acid soils not limited in Fe availability, are usually much less efficient in PS release and more susceptible to Fe deficiency chlorosis. However, considerable genotypic variation in Fe tolerance can exist within cultivars of single plant species [232,278,279], suggesting that PS exudation is not the only mechanism determining Fe efficiency. From the ecological point of view, strategy II has advantages over strategy I, especially in well-buffered calcareous soils with high pH, because Fe moblization and Fe uptake by strategy II is less dependent on the external (soil) pH than strategy I [269].

The ability to accumulate high amounts of PS in the rhizosphere (up to 1 mM in Fe-deficient barley) [40] seems to be associated also with a distinct diurnal rhythm of exudation, which is restricted to several hours after onset of the light period [48], and with restriction of PS release to subapical root zones [31]. Temporal and spatial concentration of PS exudation may be a strategy to counteract microbial degradation [40,41,280], dilution by diffusion into the bulk soil [281], and immobilization by sorption of Fe to phospholipids during FePS uptake [282]. In maize and sorghum, with a comparatively high susceptibility to Fe-deficiency chlorosis, PS is continuously released at a relatively low rate [272] in the subapical root zones (Neumann, unpublished).

The molecular mechanism of PS exudation is still not completely clear. Biosynthesis of PS seems to be regulated by the intracellular Fe level [283]. Synthesis in the root tissue increases when Fe supply is limited even before Fe-deficiency chlorosis appears, and rapidly declines after reapplication of Fe [284–288]. Phytosideophores are derived from nicotianamine, with putative functions in regulating the physiological availability of Fe or transport of copper and other micronutrients in the xylem [289, 290]. Nicotianamine is synthesized from L-methionine via trimerization of S-adenosyl-methionine (Figure 2.8A) in a reaction sequence similar to ethylene biosynthesis

(Yang cycle) with continuous recycling of L-methionine [291]. In graminaceous plants, PS formation proceeds by transamination and hydroxylation of nicotianamine to deoxymugineic acid (DMA). DMA is either released as PS into the rhizosphere (e.g., maize) or is converted to higher hydroxylated PS derivatives, such as mugineic acid (MA), hydroxymugineic acid (HMA), epi-hydroxy mugineic acid (epi-HMA), distichonic acid A, and avenic acid A, which were identified as PS in barley, rye, and oat [278,285–288]. The hydroxylation pattern seems to be a factor determining the stability of Fe^{III}-PS complexes [292]. Several genes involved in the biosynthetic pathway have been cloned recently [287,288,293–295] and heterologous expression of the nicotianamine-aminotransferase (NAAT) gene from barley, which represents the key enzyme for PS biosynthesis, stimulated PS exudation, and improved Fe acquisition in rice, with a low inherent capacity for PS release [296].

In roots of Fe-deficient barley, PS biosynthesis proceeds throughout the whole day with a decline of the internal PS levels during the period of release [233,297]. Inhibitory effects of KCN and DCDD and of low root-zone temperatures suggest that PS biosynthesis and the exudation process are highly dependent on metabolic energy [73,267,298]. Low root carbohydrate concentrations under low light conditions, and a stimulation of PS release at high light intensities may indicate the importance of a continuous supply of photosynthates from the shoot to the roots for the biosynthesis of PS in the roots [299]. Accordingly, increased belowground translocation of carbohydrates under elevated atmospheric CO_2 concentrations was also associated with increased root exudation of PS in apical root zones of Fe-deficient barley plants [300].

Ultrastructural investigations in Fe-deficient barley roots revealed, particularly in epidermal cells, the formation of large ER vesicles with attached ribosomes and filled with fibrous materials prior to the release of PS. When PS exudation was terminated, the large vesicles disappeared [113]. Enzymes involved in biosynthesis of PS such as nicotianaminesynthase (NAS) and nicotianamineaminotransferase (NAAT) are associated with these vesicles [301]. Therefore, a function of ER vesicles for storage or transport of PS (Figure 2.8) has been implicated in the mechanism of diurnal PS exudation in barley [113,301,302]. Accordingly, PS release was almost completely inhibited during the period of root exudate collection by short-term (2-h) application of the vesicle transport inhibitor brefeldin-A in barley but not in maize (Table 2.5), where diurnal variation in PS exudation is not detectable [272]. However, in both plant species, PS release was also inhibited by short-term application of various anion-channel antagonists (Table 2.5) [101]. These findings may indicate that after biosynthesis and storage of PS in ER vesicles in barley, the exudation process

TABLE 2.5
Effect of Anion-Channel Antagonists (anthracene-9-carboxylic acid, ethacrynic acid; each 100 μM) and of Brefeldin A (exocytosis inhibitor; 45 μM) on Release of Phytosiderophores from Roots of Fe-Deficient Barley and Maize

	Control (H_2O)	Anthracene-9- Carboxylic Acid	Ethacrynic Acid	Brefeldin-A
Barley				
EpiHMA				
[nmol h^{-1} g^{-1} FW]	87.4 [a]	6.0 [b]	70.9 [a]	8.5 [b]
SE	12.8	4.9	6.8	6.6
Maize				
DMA				
[nmol h^{-1} g^{-1} FW]	30.7 [a]	0.3 [b]	9.7 [c]	52.6 [a]
SE	8.2	0.3	4.4	11.9

Note: Inhibitors were applied during the 2-h period of exudate collection into distilled water, starting at the beginning of the light period.

[a,b,c] Statistically significant differences.

is mediated by anion channels, associated with a concomittant equimolar release of K^+ as counterion (Figure 2.8) [101,267]. Mori [298] suggested that biosynthesis and release of PS in barley might be triggered by diurnal changes in temperature, and not by light signals, but this could not be confirmed by the findings of Kissel [73]. Studies including mutants such as the $ys3$ maize line, which is defective in PS release but not in biosynthesis of PS, could be a powerful tool to elucidate the mechanisms of PS exudation (Basso and Römheld, unpublished).

D. OTHER MICRONUTRIENTS AND HEAVY METALS

Mobilization of micronutrients such as Zn, Mn, Cu, Co, and of heavy metals (Cd, Ni) in soil extraction experiments with root exudates isolated from various axenically grown plants is well documented [66, 303–305] and has been related to the presence of complexing agents.

1. Role of PS

Formation of stable chelates with PS occurs with Fe, but also with Zn, Cu, Co, and Mn (Figure 2.8) [40,306,307] and can mediate the extraction of considerable amounts of Zn, Mn, Cu, and even Cd and Ni in calcareous soils [303,308]. Iron-deficiency-induced PS release in wheat was associated with increased plant uptake of Zn, Ni, Cd, and Cu from contaminated soils [309,310]. However, Cd uptake was rather reduced in wheat and maize grown under Fe-limitation in hydroponics [311,312]. These findings suggest that PS can increase the bioavailability of heavy metals in soils but they are not necessarily involved in heavy metal uptake. Also problems of arsenic (As) contamination in rice have been related with PS-meditated liberation of arsenate from iron plaques at the root surface [313].

There is increasing evidence that PS release in graminaceous plants is also stimulated in response to Zn deficiency [314–316], but possibly also under Mn and Cu deficiency [317]. Similar to Fe deficiency, the tolerance of different graminaceous plant species to Zn deficiency was found to be related to the amount of released PS [315,316], but correlation within cultivars of the same species seems to be low [318]. It is, however, still a matter of debate as to what extent PS release is a specific response to deficiencies of the various micronutrients. Gries et al. [317] reported that exudation of PS in Fe-deficient barley was about 15 to 30 times greater than PS release in response to Zn, Mn, and Cu deficiency. In contrast, PS exudation in Zn-deficient bread wheat was in a similar range as PS release under Fe deficiency in barley [304,315]. Walter et al. [319] demonstrated that Zn-deficiency-induced PS release in bread wheat is probably an indirect response, which is caused by impaired Fe metabolism, supported also by data of Rengel and Graham [320]. Accordingly, upregulation of the metal-PS transporter $ZmYS1$ in maize was induced by Fe deficiency but not by limiting supply of other micronutrients, such as Zn, Mn, or Cu [270,321]. In contrast, Gries et al. [322] suggested that there was a specific response to Cu deficiency in $Hordelymus\ europaeus$ L. Root uptake rates of PS complexes with Cu, Zn, and Co were found to be much lower than uptake of Fe^{III}-PS chelates [233], but may still be sufficient because of a lower demand for micronutrients [322]. Based on studies with the maize $ys1$ mutant, which is defective in Fe-PS uptake, v. Wirén [323] proposed two pathways of Zn uptake in grasses including uptake of the free Zn cation and uptake of the Zn-PS complex via the Fe-PS transport system (Figure 2.8F).

2. Role of Carboxylates, Rhizosphere pH, and Redox State

Mobilization of micronutrients (Mn, Zn, Cu), heavy metals (Cd), and even uranium in the rhizosphere has been also related to rhizosphere acidification and to complexation with organic acids (e.g., citrate) in root exudates [324–329]. This view is further supported by intense mobilization of Mn, Zn, Cu, and Cd observed in soil extraction experiments with leachates from rhizosphere soil or with organic acid mixtures according to the root exudate composition of plant species such as $Lupinus\ albus$, $Hakea\ undulata$, and $Spinacea\ oleracea$ under P-deficient conditions, where exudation of carboxylates and protons is particularly expressed [28,34,138]. However, only limited information exists about the plant availability and uptake of the metal–carboxylate complexes. Solution culture experiments in the presence of complexing agents revealed that plant uptake is correlated mainly with the activity of free uncomplexed metal ions in solution [330–334]. This implies that utilization of chelated metals requires liberation of the metal ligands from the carboxylate complex, which may be mediated by rhizosphere acidification and reduction of metal species such as Mn and Cu by plant or microbial activity [39,129,332].

a.   Manganese, Copper

Similar to Fe acquisition in strategy I plants, Mn mobilization in the rhizosphere of soil-grown plants is a result of the combined effects of rhizosphere acidification, complexation with organic ligands, and reduction of Mn oxides [1] (see also Chapter 1). Phenolics, mucilage, and organic acids in root exudates (especially malate) have been implicated in both complexation and contact reduction of Mn [335–337]. In cluster-rooted plant species such as *Lupinus albus* and members of the Proteaceae, particulary intense exudation of organic acids and phenolics in response to P deficiency is frequently associated also with enhanced Mn mobilization in the rhizosphere, and accumulation of high or even toxic Mn levels in the shoot tissue [28,32,338,339]. Similarly, Mn toxicity was indirectly induced by the Fe deficiency response in flax grown in a calcareous soil high in extractable Mn but low in Fe [340]. However, only a limited number of reports suggests a direct response of root-induced chemical changes to limited Mn supply: Kopittke and Menzies [341] found increased rhizosphere acidification of Rhodes grass (*Chloris gayana*) grown on an alkaline substrate under Mn limitation, whereas Mn-deficiency-induced stimulation of carboxylate exudation was reported for a Mn-efficient genotype of *Medicago sativa* [342]. Besides mobilizing effects of plant root exudates, Mn availability in the rhizosphere is also strongly affected by the activity of microorganisms involved in Mn oxidation and Mn reduction, which, in turn, depend on root exudates as a carbon source [1]. Utilization of Cu complexes with humic acids and citrate has been reported for red clover especially under P-deficient conditions [343]. The authors suggested that this is the result of liberation of complexed Cu in the rhizosphere, mediated by an increased reductive capacity of the roots, which was also identified as an adaptive mechanism to Cu defciency *in Pisum sativum* [332].

b.   Zinc

Despite increased citrate accumulation in roots of Zn-deficient rice plants, root exudation of citrate was not enhanced. However, in distinct adapted rice cultivars, enhanced release of citrate could be observed in the presence of high bicarbonate concentrations in the rooting medium. Bicarbonate toxicity is a stress factor, which is frequently associated with Fe and Zn deficiency in calcareous soils [344,345] and with inhibition of root growth, because of direct rhizotoxic effects [346]. Detrimental effects on root growth have been related to excessive intracellular carboxylate accumulation, particularly in the apical root zones. In contrast, root growth of bicarbonate-tolerant plant species is not affected or even stimulated in response to bicarbonate treatments [344–347]. Bicarbonate-resistant genotypes of rice exhibit stimulation of root growth, lower intracellular concentrations and higher root exudation of carboxylates, associated with higher shoot concentrations of Zn and Fe than bicarbonate-sensitive cultivars [344,345]. These findings suggest that root exudation of carboxylates may act as a mechanism for detoxification of bicarbonate-induced overaccumulation of carboxylates in the root tissue. Moreover, intense carboxylate exudation in response to high bicarbonate concentrations in the soil solution may contribute to mobilization of sparingly soluble nutrients, such as P, Fe, Zn, and Mn in calcareous soils (Figure 2.9).

Increased exudation of sugars, amino acids, and phenolic compounds in response to Zn deficiency has been reported for various dicotyledonous and monocotyledonous plant species, and seems to be related to increased leakiness of membranes [20]. Zinc has essential functions in the stabilization of membranes [348,349] and in preventing oxidative membrane damage as a metal component of superoxide dismutase, which is part of the free-radical scavenging system of higher plants [350]. It is, however, as yet unknown whether this kind of Zn-deficiency-induced root exudation has any impact on mobilization of Zn or other micronutrients in the rhizosphere.

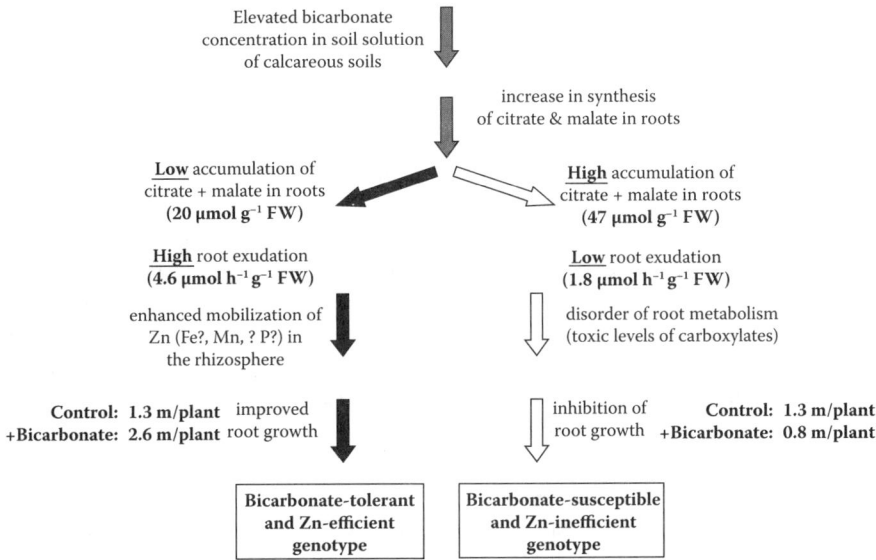

FIGURE 2.9 Proposed role of organic acid metabolism (citrate) in genotypical differences of rice in adaptation to high levels of soil bicarbonate and low Zn availability. (Adapted from Yang, X. et al., *Plant Soil*, 164, 1, 1994.)

c. Heavy Metals

Comparatively high mobility of Cd in soils, associated with high rates of uptake and accumulation in some plant species, is an important aspect from the ecotoxicological point of view. Cadmium mobilization in soils cannot only be mediated by rhizosphere acidification [328] but also to some extent by complexation with carboxylates [129,326] or PS [308]. A comparison of high and low Cd accumulating genotypes of durum wheat, revealed higher levels of carboxylates in the rhizosphere soil of the Cd accumulator [327]. From these findings, it was concluded that plant availability of Cd may be increased by complexation with root-derived carboxylates. Similarly, based on speciation analysis, Wenzel et al. [351] concluded that Ni accumulation in *Thlaspi goesingense* grown on a serpentine soil involved Ni mobilization by complexation with organic ligands in the rhizosphere. In contrast, Gerke [129] suggested that carboxylate complexation of Cd might decrease plant availability, because only free Cd^{2+} seems to be taken up by plant roots [334]. Wallace [330] demonstrated that in soil–plant systems solubility and transport to the root uptake sites are likely to be the limiting steps in uptake of cationic microelements by plants. Intense rhizosphere depletion by expression of specific high-affinity heavy metal uptake systems in roots of accumulator plants, which stimulates the diffusion-mediated delivery of heavy metals from insoluble fractions [352,353] is considered as another important mechanism, increasing the bioavailability of heavy metals in the rhizosphere. Similarly, arsenic hyperaccumulation in ferns *(Pteris vittata)* has been related with multiple mechanisms including mobilization by root-borne carboxylates in the rhizosphere and expression of efficient systems for uptake and translocation to the shoot [354].

V. ALUMINUM TOXICITY

Aluminum toxicity is a major stress factor in many acidic soils. At soil pH levels below 5.0, intense solubilization of mononuclear Al species strongly limits root growth by multiple cytotoxic effects particularly in the zone of transition between cell division and cell elongation in root apices [49,98,355]. There is increasing evidence that Al complexation with carboxylates released in apical root zones in response to elevated external Al concentrations is a widespread mechanism for Al exclusion in many plant species with considerable variation between cultivars (Figure 2.10). Formation of stable Al complexes occurs with citrate, oxalate, tartarate and, to a lesser extent, also with malate [77,78]. The Al–carboxylate complexes are less toxic than free ionic Al species [356] and are not taken up by plant roots [77,78,355]. This explains the well-documented alleviatory effects on root growth in many plant species by carboxylate applications (citric, oxalic, tartaric acids) to the culture media in presence of toxic Al concentrations [8,355,357]. Trapping of Al by carboxylate exudation probably occurs in the apoplastic space of the root apex (Figure 2.10), because Al concentrations and cationic binding sites for carboxylates in the rhizosphere soil would easily exceed the exudation capacity of plant roots. Citrate, malate, and oxalate are the carboxylate anions reported so far to be released from Al-stressed plant roots (Figure 2.10), and Al resistance of species and cultivars seems to be related to the amount of exuded carboxylates [77,78,355, 357,358] but also to the ability to maintain the release of carboxylates over extended periods [359]. In contrast to P-deficency-induced carboxylate exudation, which usually increases after several days or weeks of the stress treatment [18,94], exudation of carboxylates in response to Al toxicity is a fast reaction ocurring within minutes to several hours [106,360] and rapidly drops to base-line levels when Al is removed from treatment solutions [106]. Aluminum-induced carboxylate exudation can reach a level, which is comparable with the extraordinary high exudation of citrate in P-deficent white lupin [74], even in plant species without increased carboxylate exudation under P-deficent conditions such as *Cassia tora* [360], wheat [94], or potato [95]. However, the highly localized pattern of carboxylate release [49,98], restricted to the Al-sensitive zones of the root apex (1 mm in maize) prevents excessive losses of carbon.

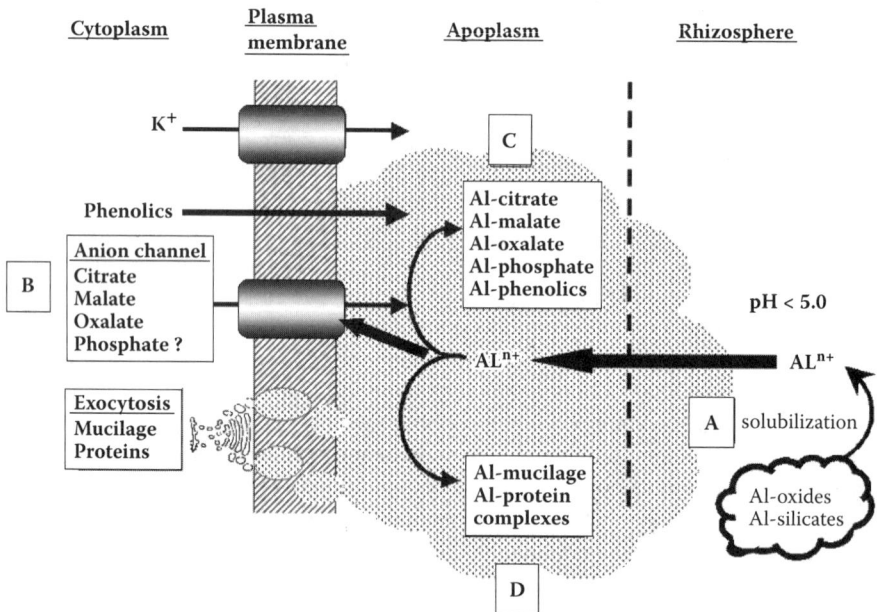

FIGURE 2.10 Model for mechanisms involved in aluminum (Al) exclusion and detoxification at the root apex: (A) Enhanced solubilization of mononuclear Al species from Al oxides and Al silicates in the soil matrix at pH < 5.0. (B) Al-induced stimulation of carboxylate exudation via anion channels, charge-balanced by concomitant release of K^+. (C) Formation of Al-carboxylate complexes in the apoplasm; restricted root uptake and lower toxicity of complexed Al. (D) Al complexation in the mucilage layer (polygalacturonates) and with Al-binding polypeptides. Increased accumulation of Al-chelating carboxylates in the mucilage layer because of limited diffusion.

Patch clamp approaches and inhibitor studies revealed that carboxylate exudation is mediated by Al-induced activation of anion channels in the plasma membrane of epidermal cells in the Al-sensitive zone of the root apex [98–100]. Signaling events leading to channel activation are yet poorly understood. An Al-induced membrane depolarization, which was not caused by malate efflux, was observed in roots of Al-tolerant but not in Al-sensitive genotypes of wheat [361]. The authors suggested that this depolarization together with other yet unknown factors might be involved in gating of a voltage-dependent malate channel in root tips of Al-tolerant wheat lines. Although the release of carboxylates in response to Al treatments occurs almost instantaneously, inhibitory effects of cycloheximide suggest that intact protein synthesis is required for carboxylate exudation [85]. Also involvement of protein phosphorylation, salicylic acid, ABA, and IAA has been discussed [98,354,362,363].

During short-term exposure to Al stress (2 to 3 h) the internal concentrations of carboxylates in root tips of wheat remained at a constant level, and there were no differences in Al-tolerant and Al-sensitive genotypes [77,106]. In root tips of wheat, also the activities of enzymes involved in biosynthesis of malate (PEP carboxylase; NAD malate dehydrogenase) were not changed [106]. It was concluded that the capacity for carboxylate accumulation in the root tissue is not a limiting factor for Al-induced exudation of carboxylates at least in short-term experiments, and genotypic variations cannot be explained on basis of differences in the capacity for biosynthesis of carboxylic acids [106]. However, intense exudation of carboxylates over longer time periods as a prerequisite for an efficient Al detoxification [359], would probably require also increased rates of biosynthesis. In contrast to the instantaneous Al-induced malate release in wheat, release of citrate in response to Al treatments in many other plant species (e.g., soybean, rye, potato) is preceded by a lag phase of several hours associated with modifications of organic acid metabolism [77,364]. Accordingly, Al treatments over 16 h enhanced both the exudation and the internal concentration of citrate in root tips of potato and sunflower (Neumann, unpublished). This is in line with findings of De la Fuente et al. [178] who reported induction of Al tolerance in transgenic tobacco and papaya by constitutive cytosolic expression of a bacterial citrate synthase gene from *Pseudomonas aeruginosa*, which resulted in higher rates of citrate accumulation in the root tissue and higher rates of citrate exudation. Similarly, carboxylate exudation and Al-tolerance were increased by overexpression of citrate synthase in rape [181], of malate dehydrogenase in alfalfa [365], and in aconitase mutants of yeast [181], but there are also contradictory reports [180]. Genetic analysis of the Al-tolerance trait and Al-inducible root exudation of malate in hexaploid wheat, revealed a predominant control by one single gene, which may be involved in the signaling of the exudation process [358]. Using substractive screening for differential gene expression in near-isogenic Al-tolerant wheat genotypes, Sasaki et al. [102] recently identified a candidate gene, encoding the anion channel protein responsible for Al-induced root exudation of malate (*ALMT1*) in wheat.

Aluminum tolerance in different plant species and cultivars is not always correlated with Al-induced carboxylate exudation [366,367], suggesting the presence of additional mechanisms involved in expression of Al resistance. Apart from the well-documented internal detoxification of Al by compartmentation and complexation [355], increased rhizosphere pH and constitutive or Al-induced secretion of other Al-binding agents, such as phosphate, phenolics, polypeptides, hydroxamates, root mucilage, and root border cells may play a role in alleviation of Al stress [27,120,368–370].

There is also increasing evidence that genotypic differences in Al tolerance can be determined by interactions with other stress factors: limitation of Al-induced carboxylate exudation in response to P deficiency has been reported for rape, with differential expression in genotypes of cowpea [371,372]. These findings are particularly important because P limitation frequently occurs simultaneously with Al toxicity on acid mineral soils.

Unlike carboxylic acids, the release and also the production of PS in roots of both Al-sensitive and Al-tolerant wheat cultivars was rapidly inhibited in response to Al treatments and seems to be responsible for Al-induced iron chlorosis in wheat [373].

Increased root exudation of amino acids in response to Cd toxicity has been reported for lettuce and white lupin grown in a hydroponic culture system under axenic conditions [65]. Under similar culture conditions, a transient release of organic acids (citric, maleic) was observed after addition of high Cu concentrations [50 μM] to the culture medium of sunflower [374]. Similarly, Nian et al. [367] reported increased exudation of citrate and malate in response to toxic levels of Al, Cu, and Cd in soybean. These findings suggest that stimulation of root exudation may be triggered by toxic levels of various metal species. It remains to be established whether these responses represent tolerance mechanisms or must be regarded as a consequence of impaired integrity of the plasma membrane.

VI. TEMPERATURE

Diffusion-mediated release of root exudates is likely to be affected by root zone temperature because of temperature dependent changes in the speed of diffusion processes and modifications of membrane permeability [375]. This might explain the stimulation of root exudation in tomato and clover at high temperatures reported by Rovira [376], and also the increase in exudation of sugars and amino acids in maize, cucumber, and strawberry exposed to low temperature treatments (5 to 10°C), which was mainly attributed to a disturbance in membrane permeability [375,376]. A decrease of exudation rates at low temperatures may be predicted for exudation processes that depend on metabolic energy. This assumption is supported by the continuous decrease of phytosiderophore release in Fe-deficient barley by decreasing the temperature from 30 to 5°C [73].

VII. LIGHT INTENSITY

Because a large proportion of the organic carbon released into the rhizosphere is derived from photosynthesis [11], changes in light intensity are likely to modify the intensity of root exudation. Early work of Rovira [376] demonstrated changes in quantity and quality of amino acids in exudates of tomato and clover with decreasing light intensity. High light intensities strongly stimulated the release of PS in roots of Fe- and Zn-deficient barley and wheat cultivars [299]. In P-deficient white lupin, citrate release from proteoid roots followed a diurnal rhythm with exudation peaks during the light period [37]. This behavior might reflect the diurnal variations in carbohydrate (sucrose) supply by the shoot [377] as precursors for citrate biosynthesis.

VIII. WATER SUPPLY

A. DROUGHT

Drought stress increases the soil mechanical impedance on plant roots, which, in turn, can stimulate root exudation [1,4,5]. Increased release of mucilage may contribute to the maintainance of Zn uptake in dry soils by facilitating Zn transport to the root surface in mucilage-embedded soil particles [378]. This effect might be supported by water transfer from the subsoil in the roots, which is subsequently released into the dry top soil layer (hydraulic lift) [379]. Formation of so-called rhizosheaths by inclusion of adhering soil particles into mucilage, which covers the root surface, has been reported particularly for graminaceous plants [121]. Shrinking of mucilage with declining water potentials leads to a more tight association of soil particles within the rhizosheaths and could thereby reduce water losses from the root by approximately 30% [380].

Increased root exudation of carbohydrates under conditions of mild or severe drought stress has been reported for several conifer species [381,382] and was attributed to root damage and increased internal carbohydrate concentrations.

B. EXCESS WATER SUPPLY

High soil moisture levels or flooding are limiting factors for oxygen supply to the roots. Hypoxia causes a shift from aerobic respiration to fermentation of carbohydrates in the root tissue yielding ethanol, lactic acid, and alanine as the main end products, which can accumulate to phytotoxic levels [192,383]. At least in some plant species, the release of substantial amounts of lactic acid [193] and ethanol into the root environment may be involved in detoxification of these metabolites. Also the exudation of sugars and amino acids is frequently enhanced under hypoxic conditions [384,385]. Ethanol is a chemoattractant to plant pathogens, and increased exudation in response to hypoxia may support pathogenic infections especially in presence of other exudate compounds such as amino acids [384].

IX. ELEVATED CO$_2$

The continuous rise of the atmospheric CO_2 concentration during the last decade, mainly as a consequence of anthropogenic CO_2 production, is likely to affect photosynthesis and plant biomass production [386]. Numerous studies demonstrated a stimulation of shoot and root growth in plants exposed to elevated CO_2 concentrations [386,387], but the putative consequences for rhizodeposition and rhizosphere processes are still not clear. Loss of assimilated carbon from maize roots was not affected by increased atmospheric CO_2 levels [388]. Similarly, $^{14}CO_2$ pulse-chase labeling experiments with *Plantago lanceolata* seedlings revealed that carbon loss from the root system was not changed at elevated CO_2 levels, although the belowground assimilate translocation was increased [389]. In P-deficient *Lupinus albus*, elevated CO_2 treatments did not affect the amount of citrate released from individual root clusters, but citrate exudation started in earlier stages of cluster root development, suggesting CO_2-induced modifications of root carbohydrate metabolism, both in hydroponics [37] and in soil culture [148]. This was associated with a higher number of root clusters per root system, which may indicate a CO_2-induced increase in total release of P-mobilizing root exudates per plant, but had no effect on the respective rhizosphere concentrations. Accordingly, the functional and structural diversities of rhizosphere bacteria was closely related with modifications in root exudation during cluster root development but was not affected by CO_2 treatments [148]. A completely different situation was reported for CO_2 effects on Fe-deficiency-induced root exudation of PS in barley, with a stimulation of PS release from indiviual root tips, resulting in increased rhizosphere concentrations [300]. Similarly, phosphatase activity associated with the root surface of *Avena barbata* and *Bromus hordeaceus* was not changed by elevated CO_2 levels [390], but increased on a base of root dry weight and root length in aseptically grown wheat seedlings when the CO_2 treatment was associated with P deficiency [391]. These examples demonstrate that the effects of elevated atmospheric CO_2 concentrations on root exudation are strongly dependent on plant species or even genotypes and also on the culture conditions with consequences for plant-nutrient acquisition, plant microbial interactions and nutrient cycling in the rhizosphere.

X. FUTURE RESEARCH PERSPECTIVES AND PROSPECTS FOR RHIZOSPHERE MANAGEMENT

During the last decade, the functional characterization of plant root exudates involved in rhizosphere processes has attracted increased attention. The role of root exudates in plant–microbial interactions, nutrient acquisition, and plant adaptations to environmental stress or adverse soil–chemical conditions is not only of scientific interest, but also implicates obvious practical aspects associated with the need for production of healthy crop plants and for sustainable agricultural systems.

From the methodological point of view, a critical reevaluation of the techniques employed for collection of root exudates and of the experimental results is indicated for many earlier studies. In the light of more recent findings, spatial variation of exudate release along the roots, effects of mechanical impedance, root injury, microbial turnover in the rhizosphere, and sorption at the soil matrix need to be considered. Further, miniaturization of sampling procedures and analytical techniques (e.g., use of specific microprobes for specific compounds, reporter bacteria, microsuction cups, capillary electrophoresis, image analysis, and videodensitometry) should facilitate nondestructive measurements of rhizosphere processes at a high scale of resolution. Robust routine methods for processing of large sample numbers are necessary to account for heterogenity and variability under field conditions. However, a major critical factor remains the sampling process itself, and it must be kept in mind that even culture techniques with minimum disturbance, such as rhizobox or root window approaches can only approximate the natural growth conditions.

To assess the significance of organic compounds released from plant roots for rhizosphere processes, much more information is necessary about effects of root exudates at realistic rhizosphere concentrations, and on mobility and plant availability of nutrients and toxins in rhizosphere soils under different environmental conditions. This is also a prerequisite for a more precise integration of rhizosphere processes into modeling approaches. It is important to consider the possibility of synergistic effects of simultaneous chemical reactions in the rhizosphere, which have been demonstrated, for example, for the Fe-deficiency response in nongraminaceaous plants including increased redox potential and increased release of reductants and chelating compounds in the apical root zones, stimulated by increased release of protons [1]. Other examples comprise possible effects of transient pH changes in the rhizosphere and release of phenolic compounds and proteins with antimicrobial activity [148–150]; counteracting the microbial degradation of root exudates involved in nutrient mobilization; increased availability of organic P forms for root secretory phosphatases and phytases, mediated by a concomitant release of carboxylates [222]; effects of rhizosphere pH on metal complexation with organic ligands [77,292]; or effects of the P-nutritional status on expression of mechanisms for Al tolerance [371,372].

Flavonoids and other secondary metabolites released by plant roots have important functions in plant–pathogenic interactions, feeding deterrence, nematode resistance, allelopathic interactions, and as signal molecules for the establishment of symbiotic associations [76,231]. However, a detailed analysis of signaling pathways involved in these interactions is currently available only in a limited number of cases (see also Chapter 9). Despite the widespread distribution of mycorrhizal associations in higher plants, the potential impacts of mycorrhizal associations and other plant–microbial interactions on root exudation have not been widely investigated.

Attempts to manipulate root exudation of higher plants by use of genetic engeneering, breeding technologies or modification of culture conditions, to increase the efficiency for nutrient acquisition, the resistance to adverse soil–chemical conditions or designing of plants for phytoremediation and phytomining strategies [392] requires a detailed knowledge of the physiological mechanisms involved in the regulation of root exudation and of the rhizosphere processes as well. During the last decade much progress has been achieved in the characterization of regulatory processes involved in root exudation at the physiological and molecular level. Potential target genes have been identified, responsible for the synthesis and secretion of root exudates involved in alleviation of various environmental stresses, such as P deficiency, Fe deficiency, and Al toxicity. The first transgenic approaches to manipulate root exudation have been tested in the recent past [102,178–181,296,365]. However, there is increasing evidence that simple overexpression strategies are frequently not sufficient to produce consistent effects on root exudation and improvement of stress tolerance: various attempts to increase citrate exudation by overexpression of citrate synthase revealed contradictory results [178–181]. It may be difficult to obtain reproducible effects by interventions into metabolic pathways for central metabolites (e.g., citrate), tightly controlled by multiple internal and external factors. Moreover, an additional requirement for a successful manipulation of root exudation seems to be the tissue-specific expression of suitable and selective export systems [98,189,220].

Promising approaches, now ready for field testing comprise improved Fe acquisition of rice on calcareous soils by expression of the nicotianamine–aminotransferase (NAAT) gene from barley, which represents the key enzyme for biosynthesis of PS [296], and heterologous expression of the Al-inducible malate channel gene (*ALMT1*) from Al-tolerant wheat, which conferred Al tolerance also to Al-sensitive rice, barley, and cell cultures of tobacco [393].

A better understanding of the adaptive functions and the behavior of root exudates in the rhizosphere and the control of release mechanisms may also enable a more directed selection and breeding of plant genotypes adapted to adverse soil conditions [358], as well as improvement of crop rotations and intercropping systems [394,395].

REFERENCES

1. Marschner, H., *Mineral Nutrition of Higher Plants*, 2nd ed., Academic Press, London, 1995.
2. Lynch, J.M. and Whipps, J.M., Substrate flow in the rhizosphere, *Plant Soil*, 129, 1, 1990.
3. Liljeroth, E., Kuikman, P., and Van Veen, J.A., Carbon translocation to the rhizosphere of maize and wheat and influence on the turnover of native soil organic matter at different soil nitrogen levels, *Plant Soil*, 161, 233, 1994.
4. Boeuf-Tremblay, V., Plantureux, S., and Guckert, A., Influence of mechanical impedance on root exudation of maize seedlings at two developmental stages, *Plant Soil*, 172, 279, 1995.
5. Groleau-Renaud, V., Plantureux, S., and Guckert, A., Influence of plant morphology on root exudation of maize subjected to mechanical impedance in hydroponic conditions, *Plant Soil*, 201, 231, 1998.
6. Barber, D.A. and Gunn, K.B., The effect of mechanical forces on the exudation of organic substances by the roots of cereal plants grown under sterile conditions, *New Phytologist*, 73, 39, 1974.
7. Schönwitz, R. and Ziegler, H., Exudation of water soluble vitamins and some carbohydrates by intact roots of maize seedlings (*Zea mays* L.) into a mineral nutrient solution, *Z. Planzenphysiol.*, 107, 7, 1982.
8. Miyasaka, S.C. et al., Mechanism of aluminium tolerance in snapbeans. Root exudation of citric acid, *Plant Physiol.*, 96, 737, 1991.
9. Hülster, A. and Marschner, H., PCDD/PCDF-Transfer in Zuchini und Tomaten, *Veröff. PAÖ*, 8, 579, 1994.
10. Shepherd, T. and Davies, H.V., Effect of exogenous amino acids, glucose and citric acid on the patterns of short-term amino acid accumulation and loss of amino acids in the root-zone of sand-cultured forage rape (*Brassica napus* L.), *Plant Soil*, 158, 111, 1994.
11. Johnson, J.F. et al., Root carbon dioxide fixation by phosphorus deficient *Lupinus albus*. Contribution to organic acid exudation by proteoid roots, *Plant Physiol.*, 112, 19, 1996.
12. Gransee, A. and Wittenmayer, L., Qualitative and quantitative analysis of water-soluble root exudates in relation to plant species and development, *J. Plant Nutr. Soil Sci.*, 163, 381, 2000.
13. Tang, C.S. and Young, C.C., Collection and identification of allelopathic compounds from the undisturbed root system of Bigalte Lompograss (*Hemarthia altissima*), *Plant Physiol.*, 69, 155, 1982.
14. Petersen, W. and Böttger, M., Contribution of organic acids to the acidification in the rhizosphere of maize seedlings, *Plant Soil*, 132, 159, 1992.
15. Ohwaki, Y.H. and Hirata, H., Differences in carboxylic acid exudation among P-starved leguminous crops in relation to carboxylic acid contents in plant tissues and phospholipid levels in roots, *Soil Sci. Plant Nutr.*, 38, 235, 1992.
16. Lipton, D.S., Blanchar, R.W., and Blevins, D.G., Citrate, malate, and succinate concentration in root exudates from P-sufficient and P-stresssed *Medicago sativa* L. seedlings, *Plant Physiol.*, 85, 315 1987.
17. v. Wirén, N. et al., Iron ineffeciency in maize mutant *Ys1* (*Zea mays* L. cv. Yellow-Stripe) is caused by a defect in uptake of iron phytosiderophores, *Plant Physiol.*, 106, 71, 1994.
18. Neumann, G. et al., Physiological adaptations to phosphorus deficiency during proteoid root development in white lupin, *Planta*, 208, 373, 1999.
19. Shepherd, T. and Davies, H.V., Patterns of short-term amino acid accumulation and loss in the root-zone of liquid culture forage rape (*Brassica napus* L.), *Plant Soil*, 158, 99, 1994.
20. Cakmak, I. and Marschner, H., Increase in membrane permeability and exudation in roots of zinc deficient plants, *J. Plant Physiol.*, 132, 356 1988.
21. Prikryl, Z. and Vancura, V., Root exudates in plants. VI. Wheat exudation as dependent on growth, concentration gradients of exudates and the presence of bacteria, *Plant Soil*, 57, 69, 1980.

22. Jones, D.L. and Darrah, P.R., Re-sorption of organic compounds by roots of *Zea mays* L. and its consequences in the rhizosphere II, *Plant Soil*, 153, 47, 1993.
23. Gerke, J., Römer, W., and Jungk, A., The excretion of citric and malic acid by proteoid roots of *Lupinus albus* L.: effects on soil solution concentrations of phosphate, iron, and aluminium in the proteoid rhizosphere samples of an oxisol and a luvisol, *Z. Pflanzenernaehr. Bodenk.*, 157, 289, 1994.
24. Neumann, G., Hülster, A., and Römheld, V., PCDD/PCDF-mobilizing compounds in root exudates of zucchini, *Organohalogen Compd.*, 41, 331, 1999.
25. Sakai, H. and Tadano, T., Characteristics of response of acid phosphatase secreted by the roots of several crops to various conditions in the growth media, *Soil Sci. Plant Nutr.*, 39, 437, 1993.
26. Morel, J.L., Mench, M., and Guckert, A., Measurement of Pb^+, Cu^{2+} and Cd^{2+} binding with mucilage exudates from maize (*Zea mays* L.) roots, *Biol. Fertil. Soils*, 2, 29, 1986.
27. Horst, W.J., Wagner, A., and Marschner, H., Mucilage protects root meristems from aluminium injury, *Z. Pflanzenphys.*, 105, 435, 1982.
28. Dinkelaker, B. et al., Root exudates and mobilization of nutrients, in *Trees — Contributions to modern tree physiology*, Rennenberg, H., Eschrich, W., and Ziegler, H., Eds., Backhuys Publishers, Leiden, The Netherlands, 1997, p. 441.
29. Hoffland, E., Findenegg, G.R., and Nelemans, J.A., Solubilization of rockphosphate by rape. II. Local exudation of organic acids as a response to P-starvation, *Plant Soil*, 113, 161, 1989.
30. Schilling, G. et al., Phosphorus availability, root exudates, and microbial activity in the rhizosphere, *Z. Pflanzenernähr. Bodenk.*, 161, 465, 1998.
31. Marschner, H., Römheld, V., and Kissel, M., Different strategies in higher plants in mobilization and uptake of iron, *J. Plant Nutr.*, 9, 695, 1986.
32. Dinkelaker, B., Hengeler, C., and Marschner, H., Distribution and function of proteoid roots and other root clusters, *Bot. Acta*, 108, 183, 1995.
33. Arahou, M. and Diem, H.G., Iron deficiency induces cluster (proteoid) root formation in *Casuarina glauca*, *Plant Soil*, 196, 71, 1997.
34. Gardner, W.K., Barber, D.A., and Parbery, D.G., The acquisition of phosphorus by *Lupinus albus* L. III. The probable mechanism, by which phosphorus movement in the soil-root interface is enhanced, *Plant Soil*, 70, 107, 1983.
35. Dinkelaker, B., Römheld, V., and Marschner, H., Citric acid excretion and precipitation of calcium citrate in the rhizosphere of white lupin (*Lupinus albus* L.), *Plant Cell Environ.*, 12, 285, 1989.
36. Keerthisinghe, G. et al., Effect of phosphorus supply on the formation and function of proteoid roots of white lupin (*Lupinus albus* L.), *Plant Cell Environ.*, 21, 467, 1998.
37. Watt, M. and Evans, J., Linking development and determinacy with organic acid efflux from proteoid roots of white lupin grown with low phosphorus and ambient or elevated atmospheric CO_2 concentration, *Plant Physiol.*, 120, 705, 1999.
38. Darrah, P.R., Models of the rhizosphere. I. Microbial population dynamics around the root releasing soluble and insoluble carbon, *Plant Soil*, 133, 187, 1991.
39. Schönwitz, R. and Ziegler, H., Quantitative and qualitative aspects of a developing rhizosphere microflora in hydroponically grown maize seedlings, *Z. Pflanzenernaehr. Bodenk.*, 149, 623, 1986.
40. Römheld, V., The role of phytosiderophores in acquisition of iron and other micronutrients in graminaceous species: an ecological approach, *Plant Soil*, 130, 127, 1991.
41. v. Wirén, N. et al., Influence of soil microorganisms on iron acquisition in maize, *Soil Biol. Biochem*, 25, 371, 1993.
42. Foster, R.C., The ultrastructure of the rhizoplane and the rhizosphere, *Annu. Rev. Phytopathol.*, 24, 211, 1986.
43. Jones, D.L., Prabowo, A.M., and Kochian, L., Kinetics of malate transport and decomposition in acid soils and isolated bacterial populations: the effect of microorganisms on root exudation of malate under Al stress, *Plant Soil*, 182, 239, 1996.
44. Bowers, J.H. et al., Infection and colonization of potato roots by *Verticillium dahliae* as affected by *Patylenchus penetrans* and *P. crenatus*, *Phytopathology*, 86, 614, 1996.
45. Wiehe, W., Hecht-Buchholz, C., and Höflich, G., Electronmicroscopic investigations on root colonization of *Lupinus albus* and *Pisum sativum* with two associative plant-growth promoting rhizobacteria, *Pseudomonas fluorescens* and *Rhizobium leguminosarum* bv *trifolii*, *Symbiosis*, 17, 15, 1994.
46. Kamh, M. et al., Mobilization of soil and fertilizer phosphate by cover crops, *Plant Soil*, 211, 19, 1999.

47. Neumann, G. and Martinoia, E., Cluster roots — an underground adaptation for survival in extreme environments, *Trends Plant Sci.*, 7(4), 162, 2002.
48. Tagaki, S., Nimito, K., and Takemoto, T., Physiological aspect of mugineic acid, a possible phytosiderophore of graminaceous plants, *J. Plant Nutr.*, 7, 469, 1984.
49. Ryan, P.R., DiThomaso, J.M., and Kochian, L.V., Aluminium toxicity in roots: an investigation of spatial sensitivity and the role of the root cap, *J. Exp. Bot.*, 44, 437, 1993.
50. Marschner, H., Römheld, V., and Kissel, M., Localization of phytosiderophore release and iron uptake along intact barley roots, *Physiol. Plant.*, 71, 157, 1987.
51. Kape, R. et al., Legume root metabolites and VA-mycorrhiza development, *J. Plant Physiol.*, 141, 54, 1992.
52. Göttlein, A., Hell, U., and Blasek, R., A system for microscale tensiometry and lysimetry, *Geoderma*, 69, 147, 1996.
53. Sandnes, A., Eldhuset, T.D., and Wollebæk, G., Organic acids in root exudates and soil solution of Norway spruce and silver birch, *Soil Biol. Biochem.*, 37, 259, 2005.
54. Grierson, P.F., Organic acids in the rhizosphere of *Banksia integrifolia* L., *Plant Soil*, 144, 259, 1992.
55. Li, M., Shinano, T., and Tadano, T., Distribution of exudates of lupin roots in the rhizosphere under phosphorus deficient conditions, *Soil Sci. Plant Nutr.*, 43, 237, 1997.
56. Helal, H.M. and Sauerbeck, D., A method to study turnover processes in soil layers of different proximity to roots, *Soil Biol. Biochem.*, 15, 223, 1983.
57. Kuchenbuch, R. and Jungk, A., A method for determining concentration profiles at the soil root interface by thin slicing rhizosphere soil, *Plant Soil*, 68, 391, 1982.
58. Zabowski, D., Limited release of soluble organics from roots during the centrifugal extraction of soil solutions, *Soil Sci. Soc. Am. J.*, 53, 977, 1989.
59. Jaeger, C.H., III et al., Interaction of roots and soil microorganisms in rhizosphere N cycling, *Bull. Ecol. Soc. Am.*, 77, (Suppl.), 215, 1996.
60. Darwent, M.J. et al., A Biosensor reporting of root exudation from *Hordeum vulgare* in relation to shoot nitrate concentration, *J. Exp. Bot.*, 54, 325, 2003.
61. Meharg, A.A. and Kilham, K., A novel method of quantifying root exudation in the presence of soil microflora, *Plant Soil*, 133, 111, 1991.
62. Meharg, A.A. and Kilham, K., Loss of exudates from the roots of perennial ryegrass inoculated with a range of microorganisms, *Plant Soil*, 170, 345, 1995.
63. Phillips, D.A. et al., Microbial products trigger amino acid exudation from plant roots, *Plant Physiol.*, 136, 2887, 2004.
64. Sulochana, C.B., Amino acids in root exudates of cotton, *Plant Soil*, 16, 312, 1962.
65. Costa, G., Michaut, J.C., and Guckert, A., Amino acids exuded from axenic roots of lettuce and white lupin seedlings exposed to different cadmium concentrations, *J. Plant Nutr.*, 20, 883, 1997.
66. Jones, D.L. and Darrah, P., Role of root derived organic acids in the mobilization of nutrients from the rhizosphere, *Plant Soil*, 166, 247, 1994.
67. Jones, D.L. et al., Organic acid behaviour in soils: misconceptions and knowledge gaps, *Plant Soil*, 248, 31, 2003.
68. Boudot, J.P., Relative efficiency of complexed aluminium, non-crystalline Al hydroxyde, allophane and imogolite in retarding the biodegradation of citric acid, *Geoderma*, 52, 29, 1992.
69. Schwab, S.M., Menge, J.A., and Leonard, R.T., Quantitative and qualitative effects of phosphorus on extracts and exudates of sudangrass roots in relation to vesicular-arbuscular mycorrhiza formation, *Plant Physiol.*, 73, 761, 1983.
70. Azaizeh, H.A. et al., Effects of vesicular-arbuscular fungus and other soil microorganisms on growth, mineral nutrient acquisition and root exudation of soil-grown maize plants, *Mycorrhiza*, 5, 321, 1995.
71. Amann, C. and Amberger, A., Phosphorus efficiency of buckwheat (*Fagopyrum esculentum*), *Z. Pflanzenernähr. Bodenk.*, 152, 181, 1989.
72. Landi, L. et al., Effectiveness of antibiotics to distinguish the contributions of fungi and bacteria to net nitrogen mineralization, nitrification and respiration, *Soil Biol. Biochem.*, 25, 1771, 1993.
73. Kissel, M., Eisenmangel-induzierte Abgabe von Phytosiderophoren aus Gerstenwurzeln als effizienter Mechanismus zur Eisenmobilisierung, Ph. D. thesis, Hohenheim University, Stuttgart, Germany, 1987.
74. Jones, D.L. et al., Role of proteinaceous amino acids released in root exudates in nutrient acquisition from the rhizosphere, *Plant Soil*, 158, 183, 1994.

75. Rao, A.S., Root flavonoids, *Bot. Rev.*, 56, 1, 1990.
76. Makino, T. et al., Influence of soil chemical properties on adsorption and oxidation of phenolic acids in soil suspension, *Soil Sci. Plant Nutr.*, 42, 867, 1996.
77. Jones, D.L., Organic acids in the rhizosphere — a critical review, *Plant Soil*, 205, 25, 1998.
78. Ryan, P.R., Delhaize, E., and Jones D.L., Function and mechanism of organic anion exudation from plant roots, *Ann. Rev. Plant Physiol. Plant Mol. Biol.*, 52, 527, 2003.
79. Awad, F., Römheld, V., and Marschner, H., Mobilization of ferric iron from a calcareous soil by plant-borne chelators (phytosiderophores), *J. Plant Nutr.*, 11, 701, 1988.
80. Xia, J. and Saglio, P.H., Characterization of the hexose transport system in maize root tips, *Plant Physiol.*, 106, 71, 1988.
81. Darrah, P.R., Rhizodeposition under ambient and elevated CO_2 levels, *Plant Soil*, 187, 265, 1996.
82. Chapin, F.S., III, Moilanen, L., and Kielland, K., Preferential use of organic nitrogen for growth by a non-mycorrhizal arctic sedge, *Nature*, 361, 150, 1993.
83. Fischer, W.N. et al., Amino acid transport in plants, *Trends Plant Sci.*, 3, 188, 1998.
84. Steiner, H.Y. et al., An *Arabidopsis* peptide transporter is a member of a novel family of membrane transport proteins, *Plant Cell*, 6, 189, 1994.
85. Nazoa, P. et al., Regulation of the nitrate transporter gene AtNRT2.1 in *Arabidopsis thaliana*: responses to nitrate, amino acids and developmental stage, *Plant Mol. Biol.*, 52, 689, 2003.
86. Persson, J. and Nashölm, T., Regulation of amino acid uptake in conifers by exogenous and endogenous nitrogen, *Planta*, 215, 639, 2003.
87. Curie, C. et al., Z. Maize yellow stripe 1 encodes a membrane protein directly involved in Fe(III) uptake, *Nature*, 18, 409(6818), 346, 2001.
88. Jones, D.L. and Darrah, P.R., Influx and efflux of organic acids across the soil-root interface of *Zea mays* L. and its implications in rhizosphere C flow, *Plant Soil*, 173, 103, 1995.
89. Jones, D.L., Hodge, A., and Kuzyakov, Y., Plant and mycorrhizal regulation of rhizodeposition, *New Phytologist*, 163, 459, 2004.
90. Jones, D.L. et al., Dissolved organic nitrogen uptake by plants — an important N uptake pathway?, *Soil Biol. Biochem.*, 37, 413, 2005.
91. Guern, J., Renaudin, J.P., and Brown, S.C., The compartmentation of secondary metabolites in plant cell cultures, in *Cell Culture and Somatic Cell Genetics of Plants*, Vol. 4, Constabel, F. and Vasil, I.K., Eds., Academic Press, San Diego, CA, 1987, p. 43.
92. Samuels, A.L., Fernando, M., and Glass, A.D.M., Immunofluorescent localization of plasma membrane H^+-ATPase in barley roots and effects of K nutrition, *Plant Physiol.*, 99, 1509, 1992.
93. Nobel, P.S., *Physiochemical and Environmental Plant Physiology*, Academic Press, London, 1991.
94. Neumann, G. and Römheld, V., Root excretion of carboxylic acids and protons in phosphorus-deficient plants, *Plant Soil*, 211, 121, 1999.
95. Neumann, G. and Römheld, V., Root-induced changes in the availability of nutrients in the rhizosphere, in *Plant Roots The Hidden Half*, 3rd ed., Waisel, Y., Eshel, A., and Kafkafi, U., Eds., Marcel Dekker, New York, 2002, p. 617.
96. Rovira, A.D., Plant root exudates, *Bot. Rev.*, 35, 35, 1969.
97. Ratnayake, M., Leonard, R.T., and Menge, A., Root exudation in relation to supply of phosphorus and its possible relevance to mycorrhizal infection, *New Phytologist*, 81, 543, 1978.
98. Kollmeier, M. et al., Aluminum activates a citrate-permeable anion channel in the aluminum-sensitive zone of the maize root apex. A comparison between aluminium-sensitive and an aluminium-resistant cultivar, *Plant Physiol.*, 126, 397, 2001.
99. Pineros, M.A. and Kochian, L.V., A patch clamp study on the physiology of aluminium toxicity and aluminium tolerance in Zea mays: identification and characterization of Al^{3+}-induced anion channels, *Plant Physiol.*, 125, 292, 2001.
100. Zhang, W.H., Ryan, P.R., and Tyerman, S.D., Malate-permeable channels and cation channels activated by aluminum in the apical cells of wheat roots, *Plant Physiol.*, 125, 1459, 2001.
101. Sakaguchi, T. et al., The role of potassium in the secretion of mugineic acids family phytosiderophores from iron-deficient barley roots, *Plant Soil*, 215, 221, 1999.
102. Sasaki, T. et al., A wheat gene encoding an aluminium-activated malate transporter, *Plant J.*, 37, 645, 2004.
103. Zhu, Y. et al., A link between citrate and proton release by proteoid roots of white lupin (*Lupinus albus* L.) grown under phosphorus-deficient conditions?, *Plant Cell Physiol.*, 46, 892, 2005.

104. Yan, F. et al., Adaptation of H$^+$-pumping and plasma membrane H$^+$ ATPase activity in proteoid roots of white lupin under phosphate deficiency, *Plant Physiol.*, 129, 50, 2002.
105. Kania, A. et al., Use of plasma membrane vesicles for examination of phosphorus deficiency-induced root excretion of citrate in cluster roots of white lupin (*Lupinus albus* L.), in *Plant Nutrition — Food Security and Sustainability of Agro-Ecosystems*, Horst, W.J. et al., Eds., Kluwer Academics, Dortrecht, Boston, London, 2001, p. 546.
106. Ryan, P.R., Delhaize, E., and Randall, P.J., Characterization of Al-stimulated efflux of malate from the apices of Al-tolerant wheat roots, *Planta*, 196, 103, 1995.
107. Battey, N.H. and Blackbourn, H.D., The control of exocytosis in plant cells, *New Phytologist*, 125, 307, 1993.
108. Chrispeels, M., Sorting of proteins in the secretory system, *Ann. Rev. Plant Physiol.*, 42, 21, 1991.
109. Chrispeels, M.J. and Raikhel, N.V., Short peptide domains target proteins to plant vacuoles, *Cell*, 68, 613, 1992.
110. Rougier, M., Secretory activity of the root cap, in *Plant Carbohydrates II, Extracellular Carbohydrates: Encyclopedia of Plant Physiology*, Tanner, W. and Leowus, F.A., Eds., Springer, Berlin, Germany, 1981.
111. Gagnon, H. et al., Biosynthesis of white lupin isoflavonoids from [U-^{14}C]L-phenylalanine and their release into the culture medium, *Plant Physiol.*, 100,76, 1992.
112. Walker, T.S., Root exudation and rhizosphere biology, *Plant Physiol.*, 132, 44, 2003.
113. Nishizawa, N. and Mori, S., The particular vesicles appearing in barley root cells and its relation to mugineic acid secretion, *J. Plant Nutr.*, 10, 1013, 1987.
114. Facchini, P.J., Alkaloid biosynthesis in plants: biochemistry, cell biology, molecular regulation and metabolic engineering applications, *Annu. Rev. Plant Physiol. Plant Mol. Biol.*, 52, 29, 2001.
115. Winkel-Shirley, B., Flavonoid biosynthesis: a colorful model for genetics, biochemistry, cell biology and biotechnology, *Plant Physiol.*, 126, 485, 2001.
116. Martinoia, E., Multifunctionality of plant ABC transporters: more than just detoxifiers, *Planta*, 214, 345, 2002.
117. Brown, M.H., Paulsen, I.T., and Skurray, R.A., The multidrug efflux protein NorM is a prototype of a new family of transporters, *Mol. Microbiol.*, 31, 393 1999.
118. Neumann, G. et al., Physiological aspects of cluster root function and development in phosphorus-deficient white lupin (*Lupinus albus* L.), *Ann. Bot.*, 85, 909, 2000.
119. Uhde-Stone, C. et al., Nylon filter arrays reveal differential gene expression in proteoid roots of white lupin in response to phosphorus deficiency, *Plant Physiol*, 131, 1064, 2003.
120. Hawes, M.C. et al., The role of root border cells in plant defense, *Trends Plant Sci.*, 5, 128, 2000.
121. McCully, M.E., Roots in soil: unearthing the complexities of roots and their rhizospheres, *Annu. Rev. Plant Physiol. Plant Mol. Biol.*, 50, 695, 1999.
122. Gerke, J., Phosphate, aluminium, and iron in the soil solution of three different soils in relation to varying concentrations of citric acid, *Z. Pflanzenernähr. Bodenk.*, 155, 339,1992.
123. Hinsinger, P., Bioavailability of soil inorganic P as affected by root-induced chemical changes — a review, *Plant Soil*, 237, 173, 2001.
124. Hoffland, E. et al., Biosynthesis and root exudation of citric and malic acids in phosphate-starved rape plants, *New Phytologist*, 122, 675, 1992.
125. Fox, T.R., Comerford, N.B., and McFee, W.W., Phosphorus and aluminium release from a spodic horizon mediated by organic acids, *Soil Sci. Soc. Am. J.*, 54, 1763, 1990.
126. Xing-Guo, H. and Jordan, C.F., Mobilization of phosphorus by naturally occuring organic acids in oxisols and Ultisols, *Pedosphere*, 5, 289, 1995.
127. Staunton, S. and Leprince, F., Effect of pH and some organic anions on the solubility of soil phosphate: implications for P bioavailability, *Eur. J. Soil Sci.*, 47, 231, 1996.
128. Ammann, C. and Amberger, A., Verringerung der Phosohatsorption durch Zusatz organischer Verbindungen zu Böden in Abhängigkeit vom pH-Wert, *Z. Pflanzenernähr. Bodenk.*, 151, 41, 1988.
129. Gerke, J., Chemische Prozesse der Nährstoffmobilisierung in der Rhizosphäre und ihre Bedeutung für den Übergang vom Boden in die Pflanze, Cuvillier Verlag, Göttingen, Germany, 1995.
130. Gerke, J., Beißner, L., and Römer, W., The quantitative effect of chemical phosphate mobilization by carboxylate anions on P uptake by a single root. I. The basic concept and determination of soil parameters, *J. Plant Nutr. Soil Sci.*, 163, 207, 2000.

131. Gerke, J., Römer, W., and Beißner, L., The quantitative effect of chemical phosphate mobilization by carboxylate anions on P uptake by a single root. II. The importance of soil and plant parameters for uptake of mobilized P, *J. Plant Nutr. Soil Sci.*, 163, 213, 2000.

132. Wouterlood, M. et al., Carboxylate concentrations in the rhizosphere of lateral roots of chickpea (*Cicer arietinum*) increase during plant development but are not correlated with phosphorus status of soil or plants, *New Phytologist*, 162, 745 2004.

133. Bar-Yosef, B., Root excretions and their environmental effects. Influence on availability of phosphorus, in *Plant Roots: The Hidden Half*, Waisel, Y., Eshel, A., and Kakafi, U., Eds., Marcel Dekker, New York, 1991, p. 503.

134. Tesfamariam, T., Römheld, V., and Neumann, G., Phosphorus-deficiency induced root exudation of carboxylates contributes to molybdenum acquisition in the rhizosphere of leguminous plants, in *Rhizosphere 2004 — A tribute to Lorenz Hiltner*, Hartmann, A., Schmid, M., Wenzel, W., and Hinsinger, P., Eds., GSF-Report, Munich, Neuherberg, Germany, 2005, p. 190.

135. Vance, C.P., Uhde-Stone, C., and Allan, D.L., Phosphorus acquisition and use: critical adaptations by plants for securing a non-renewable resource, *New Phytologist*, 157, 423, 2003.

136. Skene, K., Pattern formation in cluster roots: some developmental and evolutionary considerations, *Ann. Bot.*, 85, 901, 2000.

137. Lambers, H. et al., Acquisition and utilization of phosphorus: lessons from the Australian flora, *Proc. 15th Int. Plant Nutr. Colloq.*, Beijing, 2005.

138. Keller, H. and Römer, W., Ausscheidung organischer Säuren bei Spinat in Abhängigkeit von der P-Ernährung und deren Einfluß auf die Löslichkeit von Cu, Zn und Cd im Boden, in *Pflanzenernährung Wurzelleistung und Exsudation. 8. Borkheider Seminar zur Ökophysiologie des Wurzelraumes*, Merbach, W., Ed., B.G. Teubner Verlagsgesellschaft, Stuttgart, Leipzig, 1998, p. 187.

139. Egle, K., Römer, W., and Keller, H., Exudation of low-molecular weight organic acids by *Lupinus albus* L., *Lupinus angustifolius* L. and *Lupinus luteus* L. as affected by phosphorus supply, *Agronomie*, 23, 511, 2003.

140. Hocking, P.J. and Jeffery, S., Cluster-root production and organic anion exudation in a group of old-world lupins and a new-world lupin, *Plant Soil*, 258, 135, 2004.

141. Ligaba, A. et al., Phosphorus deficiency enhances H+-ATPase activity and citrate exudation in greater purple lupin (*Lupinus pilosus*), *Funct. Plant Biol.*, 31, 1075, 2004.

142. Ström, L., Olsson, T., and Tyler, G., Differences between calcifuge and acidifuge plants in root exudation of low molecular weight organic acids, *Plant Soil*, 167, 239, 1994.

143. Kirk, G.J.D., Santos, E.E., and Santos, M.B., Phosphate solubilization by organic anion excretion from rice growing in aerobic soil: rates of excretion and decomposition, effects on rhizosphere pH and effects on phosphate solubility and uptake, *New Phytologist*, 142, 185, 1999.

144. Banic, S. and Dey, B.K., Phosphate-solubilizing microorganisms of a Lateritic soil, *Zentralbl. Bakteriol. Abt.*, 136, 478, 1981.

145. Casarin, V. et al., Quantification of oxalate ions and protons released by ectomycorrhizal fungi in rhizosphere soil, *Agronomie*, 23, 461, 2003.

146. van Hees, P.A.W. et al., Impact of ectomycorrhizas on the concentration and biodegradation of simple organic acids in a forest soil, *Eur. J. Soil Sci.*, 54, 697, 2003.

147. Marschner, P. et al., Spatial and temporal dynamics of the microbial community structure in the rhizosphere of cluster roots of white lupin (*Lupinus albus* L.), *Plant Soil*, 246, 167, 2002.

148. Wasaki, J. et al., Root exudation, P acquisition and microbial diversity in the rhizosphere of *Lupinus albus* as affected by P supply and atmospheric CO_2 concentration, *J. Environ. Qual.*, 34, 215, 2005.

149. Weisskopf, L. et al., Secretion activity of white lupin's cluster roots influences bacterial abundance, function and community structure, *Plant Soil*, 268, 181, 2005.

150. Weisskopf, L. et al., White Lupin has developed a complex strategy to limit microbial degradation of the excreted citrate required for phosphate nutrition, *Plant Cell Environ.*, 29, 919, 2006.

151. Matar, A.E., Paul, J.L., and Jenny, H., Two-phase experiments with plants growing in phosphate-treated soil, *Soil Sci. Soc. Am. Proc.*, 31, 235, 1967.

152. Grimal, J.Y., Influence de léxsudation racinaire de *Zea mays* L. sur la mobilisation de formes de phosphore difficilement biodisponibles, Ph.D. thesis, Institut National Polytechnique de Lorraine. Ecole Nationale Supérieure dÁgronomie et des Industries Alimentaires. Laboratoire INRA. Agronomie et Environnement, 1994.

153. Nagarajah, S., Posner, A.M., and Quirk, J.P., Competitive absorptions of phosphate with polygalacturonate and other organic anions on kaolinite and oxide surfaces, *Nature,* 228, 83, 1970.

154. Read, D.B. et al., Plant roots release phospholipid surfactants that modify the physical and chemical properties of soil, *New Phytologist,* 157, 315, 2003.

155. Beissner, L., Mobilisierung von Phosphor aus organischen und anorganischen P-Verbindungen durch Zuckerrübenwurzeln, Ph. D. thesis, Georg-August Universität Göttingen, Germany, 1997.

156. Johnson, J.F., Allan, D.L., and Vance, C.P., Phosphorus stress-induced proteoid roots show altered metabolism in *Lupinus albus* L., *Plant Physiol.,* 104, 657, 1994.

157. Johnson, J.F., Vance, C.P., and Allan, D.L., Phosphorus deficiency in *Lupinus albus*: altered lateral root development and enhanced expression of phosphoenolpyruvate carboxylase, *Plant Physiol.,* 112, 31, 1996.

158. Gaume, A. et al., Low-P tolerance by maize (*Zea mays* L.) genotypes: significance of root growth, and organic acids and acid phosphatase root exudation, *Plant Soil,* 228, 253, 2001.

159. Wasaki, J. et al., Transcriptomic analysis of metabolic changes by phosphorus stress in rice plant roots, *Plant Cell Environ.,* 26, 1515, 2003.

160. Theodoru, M.E. and Plaxton, W.C., Metabolic adaptations of plant respiration to nutritional phosphate deprivation, *Plant Physiol.,* 101, 339, 1993.

161. Plaxton, W.C., Metabolic aspects of phosphate starvation in plants, in *Phosphorus in Plant Biology: Regulatory Roles in Molecular, Cellular, Organismic, and Ecosystem Processes,* Lynch, J.P. and Deikman, L., Eds., American Society of Plant Physiologists, 1998, p. 229.

162. Massonneau, A. et al., Metabolic changes associated with proteoid root development in white lupin (*Lupinus albus* L.): relationship between organic acid excretion, sucrose metabolism, and fermentation, *Planta,* 213, 534, 2001.

163. Uhde-Stone, C., Acclimation of white lupin to phosphous deficiency involves enhanced expression of genes related to organic acid metabolism, *Plant Soil,* 248, 99, 2003.

164. Pilbeam, D.J. et al., Effect of withdrawal of phosphorus on nitrate assimilation and PEP carboxylase activity in tomato, *Plant Soil,* 154, 111, 1993.

165. Gilbert, G.A., Vance, C.P., and Allan, D.L., Regulation of white lupin root metabolism by phosphorus availability, in *Phosphorus in Plant Biology: Regulatory Roles in Molecular, Cellular, Organismic, and Ecosystem Processes,* Lynch, J.P. and Deikman, L., Eds., American Society of Plant Physiologists, 1998, p. 157.

166. Kihara, T. et al., Alteration of citrate metabolism in cluster roots of white lupin, *Plant Cell Physiol.,* 44(9), 901, 2003.

167. Rufty, T.W., Jr., MacKnown, C.T., and Israel, D.W., Phosphorus stress effects on assimilation of nitrate, *Plant Physiol.,* 94, 328, 1990.

168. Heuwinkel, H. et al., Phosphorus deficiency enhances molybdenum uptake by tomato plants, *J. Plant Nutr.,* 15, 549, 1992.

169. Davies, D.D., The fine control of cytosolic pH, *Physiol Plant.,* 67, 702, 1986.

170. Sakano, K., Revision of biochemical pH-Stat: involvement of alternative pathway metabolisms, *Plant Cell Physiol.,* 39(5), 467, 1998.

171. Dinkelaker, B., Genotypische Unterschiede in der Phosphateffizienz von Kichererbse (*Cicer arietinum* L.), Ph.D. thesis, Universität Hohenheim, Germany, 1990.

172. Hinsinger, P. and Gilkes, R.J., Dissolution of phosphate rock in the rhizosphere of five plant species grown in an acid, P-fixing mineral substrate, *Geoderma,* 75, 231, 1997.

173. Imas, P. et al., Phosphate induced carboxylate and proton release by tomato roots, *Plant Soil,* 191, 35, 1997.

174. Jeschke, W.D. et al., Effects of P deficiency on assimilation and transport of nitrate and phosphate in intact plants of castor bean (*Ricinus communis* L.), *J. Exp. Bot.,* 48, 75, 1997.

175. Rychter, A.M. and Mikulska, M., The relationship between phosphate status and cyanide-resistant respiration in bean roots, *Physiol. Plant.,* 79, 663, 1990.

176. Peñaloza, E. et al., Spatial and temporal variation in citrate and malate exudation and tissue concentration as affected by P stress in roots of white lupin, *Plant Soil,* 241, 221, 2002.

177. Shane, M.W. et al., Developmental physiology of cluster-root carboxylate synthesis and exudation in Harsh *Hakea*. Expression of phosphoenolpyruvate carboxylase and the alternative oxidase, *Plant Physiol.,* 135, 549, 2004.

178. de la Fuente, J.M. et al., Aluminum tolerance in transgenic plants by alteration of citrate synthesis, *Science*, 276, 1566, 1997.

179. Lopez-Bucio, J. et al., Enhanced phosphorus uptake in transgenic tobacco plants that overproduce citrate, *Nat. Biotechnol.*, 18, 450, 2000.

180. Delhaize, E., Hebb, D.M., and Ryan, P.R., Expression of a *Pseudomonas aeruginosa* citrate synthase gene in tobacco is not associated with either enhanced citrate accumulation or efflux, *Plant Physiol.*, 125, 2059, 2001.

181. Anoop, V.M. et al., Modulation of citrate metabolism alters aluminum tolerance in yeast and transgenic canola overexpressing a mitochondrial citrate synthase, *Plant Physiol.*, 132, 2205, 2003.

182. Kania, A., Römheld, V., and Neumann, G., Role of reactive oxygen species in metabolic changes related with citrate exudation in cluster roots of phosphorus-deficient *Lupinus albus*, in, *Rhizosphere 2004 — A Tribute to Lorenz Hiltner*, Hartmann, A., Schmid, M., Wenzel, W., and Hinsinger, P., Eds., GSF-Report, Munich, Neuherberg, Germany, 2005, p. 190.

183. Kania, A. et al., Phosphorus deficiency-induced modifications in citrate catabolism and in cytosolic pH as related to citrate exudation in cluster roots of white lupin, *Plant Soil*, 248, 117, 2003.

184. Lancien, M. et al., Simultaneous expression of NAD-dependent isocitrate dehydrogenase and other Krebs cycle genes after nitrate resupply to short-term nitrogen starved tobacco, *Plant Physiol.*, 120, 717, 1999.

185. Juszczuk, I.M. and Rychter, A.M., Changes in pyridine nucleotide levels in leaves and roots of bean plants (*Phaseolus vulgaris* L.) during phosphate deficiency, *J. Plant Physiol.*, 151, 399, 1997.

186. Langlade, N.B. et al., ATP citrate lyase: cloning, heterologous expression and possible implication in root organic acid metabolism and excretion, *Plant Cell Environ.*, 25, 1561, 2002.

187. Veneklaas, E.J. et al., Chickpea and white lupin rhizosphere carboxylates vary with soil properties and enhance phosphorus uptake, *Plant Soil*, 248, 187, 2003.

188. Otani, T., and Ae, N., Interspecific differences in the role of root exudates in phosphorus acquisition in *Plant Nutrient Acquisition: New Perspectives*, Ae, N., Arihara, J., Okada, K., and Srinivasan, A., Eds., Springer, Berlin, Heidelberg, New York, 2001, p. 101.

189. Zhang, W.H., Ryan, P.R., and Tyerman, S.D., Citrate-permeable anion channels in the plasma membrane of cluster roots from white lupin, *Plant Physiol.*, 136, 3771, 2004.

190. Penaloza, E. et al., Differential gene expression in proteoid root clusters of white lupin (*Lupinus albus*), *Physiol Plant.*, 116, 28, 2002.

191. Sas, L., Rengel, Z., and Tang, C., Excess cation uptake, and extrusion of protons and organic acid anions by *Lupinus albus* under phosphorus deficiency, *Plant Sci.*, 160(6), 1191, 2001.

192. Roelofs, R., Exudation of carboxylates in Australian Proteaceae: chemical composition, *Plant Cell Environ.*, 24, 891, 2001.

193. Xia, J.H. and Roberts, J.K.M., Improved cytoplasmic pH regulation, increased lactate efflux, and reduced cytoplasmic lactate levels are biochemical traits expressed in root tips of whole maize seedlings acclimated to a low-oxygen environment, *Plant Physiol.*, 105, 651 1994.

194. Graham, J.H., Leonard, R.T., and Menge, J.A., Membrane mediated decrease in root exudation responsible for phosphorus inhibition of vesicular arbuscular mycorrhizae formation, *Plant Physiol.*, 68, 548, 1981.

195. Azcon, R. and Ocampo, J.A., Effect of exudation on VA mycorrhizal infection at early stages of plant growth, *Plant Soil*, 82, 133, 1984.

196. Pate, J.S., Verboom, W.H., and Galloway P.D., Co-occurrence of Proteaceae, laterite and related oligotrophic soils: coincidental associations or causative inter-relationships? *Aust. J. Bot.*, 49, 529, 2001.

197. Lambers, H., Chapin, F.S., III, and Pons, T.L., *Plant Physiological Ecology*, Springer, New York, 1998.

198. Lambers, H., Atkin, O.K., and Millenaar, F.F., Respiratory patterns in roots in relation to their functioning, in *Plant Roots The Hidden Half*, 3rd ed., Waisel, Y., Eshel, A., and Kafkafi, U., Eds., Marcel Dekker, New York, 2002, p. 521.

199. Nair, M.G., Safir, G.R., and Siqueira, J.O., Isolation and identification of vascular-arbuscular mycorrhiza-stimulatory compounds from clover (*Trifolium repens*) roots, *Appl. Environ. Microbiol.*, 57, 434, 1991.

200. Chishaki, N. and Horiguchi, T., Responses of secondary metabolism in plants to nutrient deficiency, *Soil Sci. Plant Nutr.*, 43, 987, 1997.

201. Harrison, M.J., The arbuscular mycorrhizal symbiosis: an underground association, *Trends Plant Sci.*, 2, 54, 1997.

202. Barker, S.J., Tagu, D., and Delp, G., Regulation of root and fungal morphogenesis in mycorrhizal symbioses, *Plant Physiol.*, 116, 120, 1998.
203. Hu, H., Tang, C., and Rengel, Z., Influence of phenolic acids on phosphorus mobilisation in acidic and calcareous soils, *Plant Soil*, 268, 173, 2005.
204. Ae, N. et al., The role of piscidic acid secreted by pigeon pea roots grown on an Alfisol with low soil fertility, in *Genetic Aspects of Plant Mineral Nutrition*, Kluwer Academic, Dordrecht, The Netherlands, 1993, p. 164.
205. Tarafdar, J.C. and Jungk, A., Phosphatase activity in the rhizosphere and its relation to the depletion of organic phosphorus, *Biol. Fertil. Soils*, 3, 199, 1987.
206. Li, M. et al., Secretion of phytase from the roots of several plant species under phosphorus-deficient conditions, *Plant Soil*, 195, 161, 1997.
207. Tarafdar, J.C. and Marschner, H., Phosphatase activity in the rhizosphere and hyphosphere of VA mycorrhizal wheat supplied with inorganic and organic phosphorus, *Soil Biol. Biochem.*, 26, 387, 1994.
208. Richardson, A.E. et al., Utilization of soil organic phosphorus by higher plants, in *Organic Phosphorus in the Environment*, Turner, B.L., Frossard, E., and Baldwin, D.S., Eds., CABI Publishing, Cambridge, MA, 2005, p. 165.
209. Wasaki, J. et al., Properties of secretory acid phosphatase from lupin roots under phosphorus-deficient conditions, *Soil Sci. Plant. Nutr.*, 43, 981, 1997.
210. Wasaki, J., Secreting portion of acid phosphatase in roots of lupin (*Lupinus albus* L.) and a key signal for the secretion from the roots, *Soil Sci. Plant Nutr.*, 45, 937, 1999.
211. Gilbert, G.A., Acid phosphatase in phosphorus-deficient white lupin roots, *Plant Cell Environ.*, 21, 801, 1999.
212. Römer, W., Einfluß von Sorte und Phosphordüngung auf den Phosphorgehalt und die Aktivität der sauren Phosphatasen von Weizen und Gerste — Ein Beitrag zur Diagnose der P-Versorgung von Pflanzen, *Z. Pflanzenernähr. Bodenk.*, 158, 3, 1995.
213. Adams, M.A. and Pate, J.S., Availability of organic and inorganic forms of phosphorus to lupins (*Lupinus* spp.), *Plant Soil*, 145, 107, 1992.
214. Beißner, L. and Römer, W., Improving the availability of phytate-phosphorus to sugarbeet (*Beta vulgaris* L.) by phytase application to soil, in *9th International Kolloquium for the Optimization of Plant Nutrition*, Prague, Czech Republic, 1996, p. 327.
215. Colpaert, J.V., The use of inositol hexaphosphate as a phosphorus source by mycorrhizal and non mycorrhizal scots pine (*Pinus sylvestris*), *Funct. Ecol.*, 11, 407, 1997.
216. Dracup, M.N.H. et al., Effect of phosphorus deficiency on phosphatase activity of cell walls from roots of subterranean clover, *J. Exp. Bot.*, 35, 466, 1984.
217. Eltrop, L., Role of ectomycorrhiza in the mineral nutrition of Norway spruce (*Picea abies* L.), Ph. D. thesis, University of Hohenheim, Stuttgart, Germany, 1993.
218. Sarkar, J., Leonowicz, A. and Bollag, J.M., Immobilization of enzymes on clays and soils, *Soil. Biol. Biochem.*, 21, 223, 1989.
219. Rao, A.M. et al., Interactions of acid phosphatase with clays, organic molecules and organo-mineral complexes, *Soil Sci.*, 161, 751, 1996.
220. Richardson, A.E., Hadobas, P.A., and Hayes, J.E., Extracellular secretion of *Aspergillus* phytase from *Arabidopsis* roots enables plants to obtain phosphorus from phytate, *Plant J.*, 25(6), 641, 2001.
221. George, T.S. et al., Characterization of transgenic *Trifolium subterraneum* L. which expresses *phyA* and releases extracellular phytase: growth and P nutrition in laboratory media and soil, *Plant Cell Environ.*, 27, 1351, 2004.
222. Hens, M., Turner, B.L., and Hocking, P.J., Chemical nature of soil organic phosphorus mobilized by organic anions, in *Proc. 2nd Int. Symp. Phosphorus Dyn. Soil-Plant continuum*, Perth, Western Australia. Uniprint, University of Western Australia, 2003, p. 16.
223. Lefebvre, D.D. et al., Response to phosphate deprivation in *Brassica nigra* suspension cells, *Plant Physiol.*, 93, 504, 1990.
224. Kraffczyk, I., Trolldenier, G., and Beringer, H., Soluble root exudates of maize: influence of potassium supply and rhizosphere microorganisms, *Soil Biol. Biochem.*, 16, 315, 1984.
225. Loss, S.P., Robson, A.D., and Ritchie, G.S.P., Nutrient uptake and organic acid anion metabolism in lupins and peas supplied with nitrate, *Ann. Bot.*, 74, 69, 1994.

226. Imas, P. et al., Carboxylic anions and proton secretion by tomato roots in response to ammonium/nitrate ratio and pH in nutrient solution, *Plant Soil*, 191, 1997.

227. Martinoia, E. and Rentsch, D., Malate compartmentation — responses to a complex metabolism, *Annu. Rev. Plant Physiol. Plant Mol. Biol.*, 45, 447, 1994.

228. Wojtaszek, P., Stobiecki, M., and Gulewicz, K., Role of nitrogen and plant growth regulators in the exudation and accumulation of isoflavonoids by roots of intact white lupin (*Lupinus albus* L.) plants, *J. Plant Physiol.*, 142, 689, 1993.

229. Waterer, J.G., Vessey, J.K., and Raper, C.D., Jr., Stimulation of nodulation in field peas (*Pisum sativum*) by low concentrations of ammonium in hydroponic culture, *Physiol. Plant.*, 86, 215, 1992.

230. Römheld, V. and Marschner, H., Mechanisms of iron uptake by peanut plants: 1. reduction, chelate splitting, and release of phenolics, *Plant Physiol.*, 71, 949, 1983.

231. Phillips, D.A. and Tsai, S.M., Flavonoids as plant signals to rhizosphere microbes, *Mycorrhiza*, 1, 55, 1992.

232. Marschner, H., Soil-root interface: biological and biochemical processes, in *Soil Chemistry and Ecosystem Health*, Huang, P.M., Ed., Soil Science Society of America, Madison, WI, 1998, p. 191.

233. Ma, J.F. and Nomoto, K., Effective regulation of iron acquisition in graminaceous plants. The role of mugineic acids as phytosiderophores, *Physiol. Plant.*, 97, 609, 1996.

234. Römheld, V., Different strategies for iron acquisition in higher plants, *Physiol. Plant.*, 70, 231, 1987.

235. Eide, D. et al., A novel iron-regulated metal transporter from plants identified by functional expression in yeast, *Proc. Natl. Acad. Sci. USA*, 93, 5624, 1996.

236. Schmid, W., Iron solutions: acquisition strategies and signaling pathways in plants, *Trends Plant Sci.*, 8, 188, 2003.

237. Landsberg, E.C., Transfer cell formation in the root epidermis: a prerequisite for Fe-efficiency?, *J. Plant Nutr.*, 5, 415, 1982.

238. Römheld, V. and Kramer, D., Relationship between proton efflux and rhizodermal transfer cells induced by iron deficiency, *Z. Pflanzenphysiol.*, 113, 73, 1983.

239. Bienfait, H.F. et al., Rhizosphere acidification by iron deficient bean plants: the role of trace amounts of divalent metal ions, *Plant Physiol.* 90, 359, 1989.

240. Alcantara, E., de la Guardia, M.D., and Romera, F.J., Plasmalemma redox activity and H^+ extrusion in roots of Fe-deficient cucumber plants, *Plant Physiol.*, 96, 1034, 1991.

241. Brüggemann, W. et al., Plasma membrane-bound NADH-Fe^{3+}-EDTA reductase and iron deficiency in tomato (*Lycopersicon esculentum*). Is there a turbo reductase?, *Physiol. Plant*, 79, 339, 1991.

242. Holden, M.J. et al., Fe^{3+}-chelate reductase activity of plasma membranes isolated from tomato (*Lycopersicon esculentum* Mill.) roots. Comparison of enzymes from Fe-deficient and Fe-sufficient roots, *Plant Physiol.*, 97, 537, 1991.

243. Olsen, R.A. et al., Chemical aspects of the Fe stress response mechanisms in tomatoes, *J. Plant Nutr.*, 3, 905, 1981.

244. Marschner, H. et al., Root-induced changes in the rhizosphere: importance for the mineral nutrition of plants, *Z. Pflanzenernähr. Bodenk.*, 149, 441, 1986.

245. Jones, D.L., Darrah, P., and Kochian, L.V., Critical evaluation of organic acid mediated iron dissolution in the rhizosphere and its potential role in root iron uptake, *Plant Soil*, 180, 57, 1996.

246. Ohwaki, Y. and Sugahara, K., Active extrusion of protons and exudation of carboxylic acids in response to iron deficiency by roots of chickpea (*Cicer arietinum* L.), *Plant Soil*, 189, 49, 1997.

247. Römheld, V. and Marschner, H., Mobilization of iron in the rhizosphere of different plant species, *Adv. Plant Nutr.*, 2, 155, 1986.

248. Landsberg, E.C., Hormonal regulation of iron-stress response in sunflower roots: a morphological and cytological investigation, *Protoplasma*, 194, 69, 1996.

249. Romera, F.J. and Alcantara, E., Iron-deficiency stress responses in cucumber (*Cucumis sativus* L.) roots, *Plant Physiol.*, 105, 1133, 1994.

250. Romera, F.J., Alcantara, E., and de la Guardia, M.D., Ethylene production by Fe-deficient roots and its involvement in the regulation of Fe-deficiency stress responses by strategy I plants, *Ann. Bot.*, 83, 5, 1999.

251. Schmidt, W., Tittel, J., and Schikora, A., Role of hormones in the induction of iron deficiency responses in *Arabidopsis* roots, *Plant Physiol.*, 121, 1109, 2000.

252. Schmidt, W. et al., Hormones induce Fe-deficiency-like root epidermal cell pattern in the Fe-inefficient tomato mutant *fer*, *Protoplasma*, 213, 67, 2000.

253. Abadía, J. et al., Organic acids and Fe deficiency: a review, *Plant Soil*, 241, 75, 2002.
254. De Vos, C.R., Lubberding, H.J., and Bienfait, H.F., Rhizosphere acidification as a response to iron deficiency in bean plants, *Plant Physiol.*, 81, 842, 1986.
255. Bienfait, H.F., Mechanisms in Fe-deficient reactions of higher plants, *J. Plant Nutr.*, 11, 605, 1988.
256. Bienfait, H.F., Is there a metabolic link between H⁺ excretion and ferric reduction by roots of Fe-deficient plants? — A viewpoint, *J. Plant Nutr.*, 19, 1211, 1996.
257. Moog, P.R. and Brüggemann, W., Iron reductase systems on the plant plasma membrane — a review, *Plant Soil*, 165, 241, 1994.
258. Askerlund, P. and Larsson, C., Transmembrane electron transport in plasma membrane vesicles loaded with an NADH-generating system or ascorbate, *Plant Physiol.*, 96, 1178, 1991.
259. Landsberg, E.C., Function of rhizodermal transfer cells in the Fe stress response of *Capsicum annuum* L., *Plant Physiol.*, 82, 511, 1986.
260. Elliott, G.C. and Läuchli, A., Phosphorus efficiency and phosphate-iron interaction in maize, *Agron. J.*, 77, 399, 1985.
261. Venkat Raju, K., Marschner, H., and Römheld, V., Effect of iron nutritional status on iron uptake, substrate pH and production and release of organic acids and riboflavin by sunflower plants, *Z. Pflanzenerenähr. Bodenk.*, 132, 177, 1972.
262. Pinton, R. et al., Water and pyrophosphate-extractable humic substances fractions as a source of iron for Fe-deficient cucumber plants, *Biol. Fert. Soil*, 26, 23, 1998.
263. Pandeya, S.B, Singh, A.C., and Dhar, P., Influence of fulvic acid on transport of iron in soils and uptake by paddy seedlings, *Plant Soil*, 198, 117, 1998.
264. Shen, J. et al., Nutrient uptake, cluster root formation, and exudation of protons and citrate as affected by localized supply of phosphorus in a split-root system, *Plant Sci.*, 168, 837, 2005.
265. Murkami, T. et al., Stabilities of metal complexes of mugineic acids and their specific affinities for iron (III), *Chem. Lett.*, 2137, 1989.
266. Treeby, M., Marschner, H., and Römheld, V., Mobilization of iron and other micronutrient cations from a calcareous soil by plant-borne, microbial, and synthetic metal chelators, *Plant Soil*, 114, 217, 1989.
267. Tagaki, S., The iron acquisition system in graminaceous plants and mugineic acids, in *Nutriophysiology of Metal Related Compounds*, Japanese Society of Soil Science and Plant Nutrition, 1990, p. 6.
268. Hiradate, S. and Inoue, K., Interaction of mugineic acid with iron (hydr) oxides: sulfate and phosphate influences, *Soil Sci. Soc. Am. J.*, 62, 159, 1998.
269. Römheld, V. and Marschner, H., Evidence for a specific uptake system for iron phytosiderophores in roots of grasses, *Plant Physiol.*, 80, 175, 1986.
270. Schaaf, G., ZmYS1 functions as a proton-coupled symporter for phytosiderophore- and nicotianamine-chelated metals, *J. Biol Chem.*, 279, 9091, 2004.
271. Hördt, W., Römheld, V., and Winkelmann, G., Fusarinines and dimerum acid, mono- and dihydroxamate siderophores from *Penicillium chrysogenum*, improve iron utilization by strategy I and strategy II plants, *Bio Metals*, 13, 37, 2000.
272. Yehuda, Z. et al., The role of ligand exchange in uptake of iron from microbial siderophores by graminaceous plants, *Plant Physiol.*, 111, 1273, 1996.
273. Marschner, P. and Crowley, D., Phytosiderophores decrease iron stress and pyroverdine production of *Pseudomonas fluorescens* PF-5 (PVD-INAZ), *Soil Biol. Biochem.*, 30, 1275, 1998.
274. Schaaf, G., Erenoglu, B.E., and v. Wirén, N., Physiological and biochemical characterization of metal-phytosiderophore transport in graminaceaous plant species, *Soil Sci. Plant Nutr.*, 50, 989, 2004.
275. Di Donato, R.J. et al., *Arabidopsis Yellow-Stripe-Like2* (YSL2); a metal-regulated gene encoding a plasmamembrane transporter of nicotianamine-metal complexes, *Plant J.*, 39, 2004.
276. v. Wirén, N. et al., Nicotianamine chelates both Fe^{III} and Fe^{II}: implications for metal transport in plants, *Plant Physiol.*, 119, 1107, 1999.
277. Hider, R.C. et al., Competition or complementation: the iron-chelating abilities of nicotianamine and phytosiderophores, *New Phytologist*, 264, 104, 2004.
278. Kawai, S., Tagaki., S., and Sato, Y., Mugineic acid-family phytosiderophores in root secretions of barley, corn and sorghum varieties, *J. Plant Nutr.*, 11, 633, 1988.
279. Brown, J.C., Jolley, V.D., and Lytle, C.M., Comparative evaluation of iron solubilizing substances (phytosiderophores) released by oats and corn: iron-efficient and iron-inefficient plants, *Plant Soil*, 130, 157, 1991.

280. Crowley, D.E. and Gries, D., Modeling of iron availability in plant rhizosphere, in *Biochemistry of Metal Micronutrients in the Rhizosphere*, Manthey, J.A., Crowley, D.E., and Luster, D.G., Eds., Lewis Publishers, Boca Raton, FL, 1994, p. 199.

281. Tagaki, S., Kamei, S., and Yu, M.H., Efficiency of iron extraction from soil by mugineic acid family phytosiderophores, *J. Plant Nutr.*, 11, 643, 1988.

282. Mihashi, S., Mori, S., and Nishizawa, N., Enhancement of ferric-mugineic acid uptake by iron deficient barley roots in the presence of excess free mugineic acid in the medium, *Plant Soil*, 130, 135, 1991.

283. Walter, A. et al., Effects of iron nutritional status and time of day on concentrations of phytosidero-phores and nicotianamine in different root and shoot zones of barley, *J. Plant Nutr.*, 18, 1577, 1995.

284. Tagaki, S., Mechanism of iron uptake regulation in roots and genetic differences, in *Agriculture, Soil Science and Plant Nutrition in the Northern Part of Japan*, Japanese Society of Soil Science and Plant Nutrition, 1984, p. 190.

285. Shojima, S., Nishizawa, N., and Mori, S., Establishment of a cell-free system for the biosynthesis of nicotianamine, *Plant Cell Physiol.*, 30, 673, 1989.

286. Shojima, S. et al., Biosynthesis of phytosiderophores. *In vitro* biosynthesis of 2′-deoxymugineic acid from L-methionine and nicotianamine, *Plant Physiol.*, 93, 1497, 1990.

287. Higuchi, K. et al., Purification and characterization of nicotianamine synthase from Fe-deficient barley roots, *Plant Soil*, 165, 173, 1994.

288. Kanazawa, K. et al., Inductions of two enzyme activities involved in biosynthesis of mugineic acid in Fe deficient barley roots, in *Iron Nutrition in Soils and Plants*, Abadía, J., Ed., Kluwer Academic Publishers, Dordrecht, The Netherlands, 1995, p. 37.

289. Scholz, G., Nicotianamine — a common constituent of strategy I and II of iron acquisition by plants: a review, *J. Plant Nutr.*, 15, 1647, 1992.

290. Pich, A., Scholz, G., and Stephan, U.W., Iron-dependent changes of heavy metals, nicotianamine, and citrate in different plant organs and in the xylem exudate of two tomato genotypes. Nicotianamine as possible copper translocator, *Plant Soil*, 165, 189, 1994.

291. Ma, J.F. et al., Biosynthesis of phytosiderophores, mugineic acids, associated with methionine cycling, *J. Biol. Chem.*, 270, 16549, 1995.

292. v. Wirén, N., Khodr, H., and Hider, R.C., Hydroxylated phytosiderophore species possess an enhanced chelate stability and affinity for iron (III), *Plant Physiol.*, 124, 1149, 2000.

293. Nakanishi, H. et al., Expression of a gene specific for iron deficiency (*Ids3*) in the roots of *Hordeum vulgare*, *Plant Cell Physiol.*, 34, 401, 1993.

294. Higuchi, K. et al., Cloning of nicotianamine synthase genes, novel genes involved in the biosynthesis of phytosiderophores, *Plant Physiol.*, 119, 474, 1999.

295. Takahashi, M. et al., Purification, characterization, and sequencing of nicotianamine aminotransferase (NAAT-III) expressed in Fe-deficient barley roots, in *Plant Nutrition for Sustainable Food Production and Environment*, Ando, T., Ed., Kluwer Academic Publishers, Dordrecht, The Netherlands, 1997, p. 297.

296. Takahashi, M., Enhanced tolerance of rice to low iron availability in alkaline soils using barley nicotianamine aminotransferase genes, *Nat. Biotechnol.*, 19, 466, 2001.

297. Walter, A. et al., Diurnal variations in release of phytosiderophores and in concentrations of phytosi-derophores and nicotianamine in roots and shoots of barley, *J. Plant Physiol.*, 147, 19, 1995.

298. Mori, S., Mechanisms of iron acquisition by graminaceous (strategy II) plants, in *Biochemistry of Metal Micronutrients in the Rhizosphere*, Manthey, J.A., Crowley, D.E., and Luster, D.G., Eds., Lewis Publishers, Boca Raton, FL, 1994, p. 225.

299. Cakmak, I. et al., Light-mediated release of phytosiderophores in wheat and barley under iron or zinc deficiency, *Plant Soil*, 202, 309, 1998.

300. Haase, S. et al., Iron efficiency and rhizosphere-microbial diversity of *Hordeum vulgare* as affected by atmospheric CO_2 concentration and Fe-nutritional status, *J. Environ. Qual.*, in press, 2007.

301. Nozoye, T. et al., Diurnal changes in the expression of genes that participate in phytosiderophore synthesis in rice, *Soil Sci. Plant Nutr.*, 50, 1125, 2004.

302. Negishi, T. et al., cDNA microanalysis of gene expression during Fe-deficiency stress in barley suggests that polar vesicle transport is implicated in phytosiderophore secretion of Fe-deficient barley roots, *Plant J.*, 30, 83, 2002.

303. Treeby, M., Marschner, H., and Römheld, V., Mobilization of iron and other micronutrient cations from a calcareous soil by plant-borne, microbial, and synthetic metal chelators, *Plant Soil*, 114, 217, 1989.

304. Zhang, F., Römheld, V., and Marschner, H., Release of zinc mobilizing root exudates in different plant species as affected by zinc nutritional status, *J. Plant Nutr.*, 14, 675, 1991.
305. Mench, M. and Martin, E., Mobilization of cadmium and other metals from two soils by root exudates of *Zea mays* L., *Nicotiana tabacum* L. and *Nicotiana rustica* L, *Plant Soil*, 132, 187, 1991.
306. Nomoto, K. et al., X-ray crystal structure of the copper(II)complex of mugineic acid, a naturally occuring metal chelator of graminaceous plants, *J. Chem. Soc. Chem. Commun.*, 338 1981.
307. Iwashita, T., High-resolution proton nuclear magnetic resonance analysis of solution structures and conformational properties of muguneic acids and its metal complexes, *Biochemistry*, 22, 4842, 1983.
308. Awad, F. and Römheld, V., Mobilization of heavy metals from contaminated calcareous soils by plant-borne chelators and its uptake by wheat plants, *J. Plant Nutr.*, 23, 2000.
309. Awad, F. and Römheld, V., Significance of root exudates in acquisition of heavy metals from a contaminated calcareous soil by graminaceous species, *Plant Nutr.*, 23, 1857, 2000.
310. Chaignon, V., Di Malta, D., and Hinsinger, P., Fe-deficiency increases Cu acquisition by wheat cropped in a Cu-contaminated vineyard soil, *New Phytologist*, 154, 121, 2002.
311. Shenker, M., Fan, W.M., and Crowley, D.E., Phytosiderophores influence on cadmium uptake by wheat and barley plants, *J. Environ. Qual.*, 29, 2091, 2000.
312. Hill, K.A., Lion, L.W., and Ahner, B.A., Reduced Cd accumulation in Zea mays: a protective role for phytosiderophores?, *Environ. Sci. Technol.*, 15, 5363, 2002.
313. Meharg, A.W., Arsenic in rice — understanding a new disaster in South-East Asia, *Trends Plant Sci.*, 9, 415, 2004.
314. Zhang, F., Römheld, V., and Marschner, H., Effect of zinc deficiency in wheat on release of zinc and iron mobilizing root exudates, *Z. Pflanzenernähr. Bodenk.*, 152, 205, 1989.
315. Cakmak, I. et al., Effect of zinc and iron deficiency on phytosiderophore release in wheat genotypes differing in zinc efficiency, *J. Plant Nutr.*, 17, 1, 1994.
316. Hopkins, B.G. et al., Phytosiderophore release by sorghum wheat, and corn under zinc deficiency, *J Plant Nutr.*, 21, 2623, 1998.
317. Gries, D. et al., Phytosiderophore release in relation to micronutrient metal deficiencies in barley, *Plant Soil*, 172, 299, 1995.
318. Erenoglu, B. et al., Phytosiderophore release does not relate well with Zn efficiency in different bread wheat genotypes, *J. Plant Nutr.*, 19, 1569,1996.
319. Walter, A. et al., Is the release of phytosiderophores in zinc-deficient wheat plants a response to impaired iron utilization? *Physiol Plant.*, 92, 493, 1994.
320. Rengel, Z. and Graham, D., Uptake of zinc from chelate-buffered nutrient solutions by wheat genotypes differing in zinc efficiency, *J. Exp. Bot.*, 47, 217, 1996.
321. Roberts, L.A. et al., Yellow Stripe1. Expanded roles for the maize phytosiderophore transporter, *Plant Physiol.*, 135, 112, 2004.
322. Gries, D., Klatt, S., and Runge, M., Copper-deficiency-induced phytosiderophore release in the calcicole grass Hordelymus europaeus, *New Phytologist*, 140, 95, 1998.
323. v. Wirén, N., Marschner, H., and Römheld, V., Roots of iron-efficient maize also absorb phytosiderophore-chelated zinc, *Plant Physiol.*, 111, 1119, 1996.
324. Marschner, H., Root-induced changes in the availability of micronutrients in the rhizosphere, in *Plant Roots: The Hidden Half*, Waisel, Y., Eshel, A., and Kakafi, U., Eds., Marcel Dekker, New York, 1991, p. 503.
325. Meyer, U., Gerke, J., and Römer, W., Einfluß von Citronensäure aud die Löslichkeit und die Aufnahme von Cu und Zn durch Weidelgras, *Mitt. Dtsch. Bodenkundl. Ges.*, 73, 99, 1994.
326. Krishnamurti, G.S.R. et al., Kinetics of cadmium release from soils as influenced by organic acids: implication in cadmium availability, *J. Environ. Qual.*, 26, 271, 1997.
327. Cieslinski, G. et al., Low-molecular-weight organic acids in rhizosphere soils of durum wheat and their effect on cadmium bioaccumulation, *Plant Soil*, 203, 109, 1998.
328. Wu, Q.T., Morel, J.L., and Guckert, A., Effect of nitrogen source on cadmium uptake by plants, *Compt. Rend. Acad. Sci.*, 309, 215, 1989.
329. Ebbs, S.D., Norvell, A.W.A., and Kochian, L.V., The effect of acidification and chelation agents on the solubilization of uranium from contaminated soil, *J. Environ. Qual.*, 27, 1486, 1998.
330. Wallace, A., Effect of chelating agents on uptake of trace metals when chelating agents are supplied to soil in contrast to when they are applied to solution culture, *J. Plant Nutr.*, 2, 171, 1980.

331. Kochian, L.V., Zinc absorption from hydroponic solutions by plant roots, in *Zinc in Soils and Plants*, Robson, A.D., Ed., Kluwer Academic Publishers, Dordrecht, The Netherlands, 1993, p. 45.
332. Welch, R. et al., Induction of iron (III) and copper (II) reduction in pea (*Pisum sativum* L.) roots by Fe and Cu status. Does the root-cell plasma Fe(III) reductase perform a general role in regulating cation uptake?, *Planta*, 190, 555, 1993.
333. McLaughlin, M.J., Smoders, E., and Merckx, R., Soil-root interface: physicochemical processes, in *Soil Chemistry and Ecosystem Health*, Huang, P.M., Ed., Soil Science Society of America, Madison, WI,1998, p. 233.
334. Greger, M. and Lindberg, S., Effects of Cd^{2+} and EDTA on young sugar beets (*Beta vulgaris*). I. Cd^{2+} uptake and sugar accumulation, *Physiol. Plant.*, 66, 69, 1986.
335. Godo, G.H. and Reisenauer, H.M., Plant effects on soil manganese availability, *Soil Sci. Soc. Am. J.*, 44, 993, 1980.
336. Marschner, H., Mechanisms of manganese acquisition by roots from soils, in *Manganese in Soils and Plants*, Graham, R., Hannam, R.J., and Uren, N.C., Eds., Kluwer Academic Publishers, Norwell, MA, 1988, p. 191.
337. Uren, N.C., Mucilage secretion and its interaction with soil, and contact reduction, *Plant Soil*, 155, 79, 1993.
338. Moraghan, J.T., The growth of white lupin on a Calcaquoll, *Soil Sci. Soc. Am. J.*, 55, 1353, 1991.
339. Shane, M.W. and Lambers, H., Manganese accumulation in leaves of *Hakea prostrata* (Proteaceae) and the significance of cluster roots for micronutrient uptake as dependent on phosphorus supply, *Physiol. Plant.*, 124, 441, 2005.
340. Moraghan, J.T., Manganese toxicity in flax growing on certain calcareous soils low in available iron, *Soil Sci. Soc. Am. J.*, 43, 1177, 1979.
341. Peter, M. et al., Effect of Mn deficiency and legume inoculation on rhizosphere pH in highly alkaline soils, *Plant Soil*, 262, 13, 2004.
342. Gherardi, M.J. and Rengel, Z., The effect of manganese supply on exudation of carboxylates by roots of lucerne (*Medicago sativa*), *Plant Soil*, 260, 271, 2004.
343. Römer, W., Patzke, R., and Gerke, J., Die Kupferaufnahme von Rotklee und Weidelgras aus Cu-Nitrat-, Huminstoff-Cu- und Cu-Citrat-Lösungen, in *Pflanzenernährung Wurzelleistung und Exsudation. 8. Borkheider Seminar zur Ökophysiologie des Wurzelraumes*, Merbach, W., Ed., B.G. Teubner Verlagsgesellschaft, Stuttgart, Leipzig, 1998, p. 137.
344. Yang, X., Römheld, V., and Marschner, H., Effect of bicarbonate on root growth and accumulation of organic acids in Zn-ineffcient and Zn-efficient rice varieties (*Oryza sativa* L.), *Plant Soil*, 164, 1, 1994.
345. Hajiboland, R., Zinc efficiency in rice (*Oryza sativa* L.) plants, Ph. D. thesis, Hohenheim University, Stuttgart, Verlag Grauer, 2000.
346. Lee, J.A., The calcicole-calcifuge problem revisited, *Adv. Bot. Res.*, 29, 1, 1998.
347. Hajiboland, R., Yang, X.E., and Römheld, V., Effects of bicarbonate and high pH on growth of Zn-efficient and Zn-inefficient genotypes of rice wheat and rye, *Plant Soil*, 250, 349, 2003.
348. Chvapil, M., New aspects in the biological role of zinc: a stabilizer of macromolecules membranes, *Life Sci.*, 13, 1041, 1973.
349. Bettger, W.J. and O´Dell, B.L., A critical role of zinc in the structure and function of biomembranes, *Life Sci.*, 28, 1425, 1981.
350. Fridovich, I., Biological effects of the superoxide radical, *Arch. Biochem. Biophys.*, 247, 1, 1986.
351. Wenzel, W. et al., Rhizosphere characteristics of indigenously growing nickel hyperaccumulator and excluder plants on serpentine soil, *Environ. Pollut.*, 123, 31, 2003.
352. Al-Najar, H., Schulz, R., and Römheld, V., Plant availability of thallium in the rhizosphere of hyperaccumulator plants: a key factor for assessment of phytoextraction, *Plant Soil*, 249, 97, 2003.
353. Zhao, F.J. et al., Characteristics of cadmium uptake in two contrasting ecotypes of the hyperaccumulator (*Thlaspi caerulescens*), *J. Exp. Bot.*, 53, 535, 2002.
354. Tu, S., Ma, L., and Luongo, T., Root exudates and arsenic accumulation in arsenic hyperaccumulating *Pteris vittata* and non-hyperaccumulating *Nephrolepis exaltata*, *Plant Soil*, 258, 9, 2004.
355. Kochian, V.L. et al., The physiology, genetics and molecular biology of plant aluminum resistance and toxicity, *Plant Soil.*, 274, 175, 2005.
356. Hue, N.V., Craddock, G.R., and Adams, F., Effect of organic acids on aluminum toxicity in subsoils, *Soil Sci. Soc. Am. J.*, 50, 28, 1986.

357. Bartlett, R.J. and Rigo, D.C., Effect of chelation on the toxicity of aluminium, *Plant Soil*, 37, 419, 1972.
358. Ryan, P.R., Delhaize, E., and Randall, P.J., Malate efflux from root apices and tolerance to aluminum are highly correlated in wheat, *Aust. J. Plant Physiol.*, 122, 531, 1995.
359. Zheng, S.J., Ma, J.F., and Matsumoto, H. Continuous secretion of organic acid is related to aluminum resistance in relatively long-term exposure to aluminum stress, *Physiol. Plant.*, 103, 209, 1998.
360. Ma, J.F., Zheng, S.J.Z., and Matsumoto, H., Specific secretion of citric acid induced by Al stress in *Cassia tora* L, *Plant Cell Physiol.*, 38, 1019, 1997.
361. Papernik, L.A. and Kochian, L.V., Possible involvement of Al-induced electrical signals in Al tolerance in wheat, *Plant Physiol.*, 115, 657, 1997.
362. Shen, H., Citrate secretion coupled with the modulation of soybean root tip under aluminum stress. Up-regulation of transcription, translation, and threonine-oriented phosphorylation of plasma membrane H$^+$-ATPase, *Plant Physiol.*, 138, 287, 2005.
363. Yang, Z.M., Salicylic acid-induced aluminum tolerance by modulation of citrate efflux from roots of *Cassia tora* L, *Planta*, 217, 168, 2003.
364. Li, X.F., Ma, J.F., and Matsumoto, H., Pattern of aluminum-induced secretion of organic acids differing between rye and wheat, *Plant Physiol.*, 123, 1537, 2000.
365. Tesfaye, M. et al., Overexpression of malate dehydrogenase in transgenic alfalfa enhances organic acid biosynthesis and confers tolerance to aluminum, *Plant Physiol.*, 127, 1636, 2001.
366. Pineros, M.A. et al., Aluminium resistance in maize cannot be solely explained by root organic acid exudation. A comparative physiological study, *Plant Physiol.*, 137, 231, 2005.
367. Nian, H. et al., Citrate secretion induced by aluminum stress may not be a key mechanism responsible for differential aluminum tolerance of some soybean genotypes, *J. Plant Nutr.*, 27, 907, 2004.
368. Poschenrieder, C., Tolra, R.P., and Barcelo, J., A role for cyclic hydroxamates in aluminium resistance in maize? *J. Inorg. Biochem.*, 99, 1830, 2005.
369. Heim, A. et al., Effects of aluminium treatment on Norway sprice roots: aluminium binding forms, element distribution, and release of organic substances, *Plant Soil*, 216, 103, 1999.
370. Basu, U. et al., A 43-k-Da, root exudate polypeptide co-segregates with aluminium resistance in *Triticum aestivum*, *Physiol. Plant.*, 106, 53, 1999.
371. Ligaba, A. et al., The role of phosphorus in aluminium-induced citrate and malate exudation from rape (*Brassica napus*), *Physiol. Plant.*, 120, 575, 2004.
372. Akinrinde, E. et al., Tolerance to soil acidity in cowpea genotypes is differentially affected by the P nutritional status, in *Rhizosphere 2004 — A Tribute to Lorenz Hiltner*, Hartmann, A., Schmid, M., Wenzel, W., and Hinsinger, P., Eds., GSF-Report, Munich, Neuherberg, Germany, 2005, p. 167.
373. Chang, Y.C., Ma, J.F., and Matsumoto, H., Mechanisms of Al-induced iron chlorosis in wheat (*Triticum aestivum*). Al-inhibited biosynthesis and secretion of phytosiderophores, *Physiol. Plant.*, 102, 9, 1998.
374. Jung, C. et al., Einfluß einer Kupferbehandlung auf die Exsudation von organischen Säuren bei *Helianthus annuus*, in *Pflanzenernährung Wurzelleistung und Exsudation. 8. Borkheider Seminar zur Ökophysiologie des Wurzelraumes*, Merbach, W., Ed., B.G. Teubner Verlagsgesellschaft, Stuttgart, Leipzig, 1998, p. 181.
375. Vancura, V., Root exudates of plants III. Effect of temperature and cold shock on the exudation of various compounds from seeds and seedlings of maize and cucumber, *Plant Soil*, 27, 319, 1967.
376. Rovira, A.D., Root excretions in relation to the rhizosphere effect IV. Influence of plant species, age of plant, light, temperature and calcium nutrition on exudation, *Plant Soil*, 11, 53, 1959.
377. Richter, G., *Biochemie der Pflanzen*, Thieme, Stuttgart, New York, 1996.
378. Nambiar, E.K.S., The uptake of zinc-65 by roots in relation to soil water content and root growth, *Aust. J. Soil Res.*, 14, 67, 1976.
379. Vetterlein, D. and Marschner, H., Use of a microtensiometer technique to study hydraulic lift in sandy soil planted with pearl millet (*Pennisetum americanum* L. Leeke), *Plant Soil*, 149, 275, 1993.
380. Huang, B., North, G., and Nobel, P.S., Soil sheaths, photosynthate distribution to roots and rhizosphere water relations for *Opuntia ficus-indica*, *Int. J. Plant Sci.*, 154, 425, 1993.
381. Reid, C.P.P., Assimilation, distribution and root exudation of ^{14}C by ponderosa pine seedlings under induced water stress, *Plant Physiol.*, 54, 44, 1974.
382. Reid, C.P.P. and Mexal, J.G., Water stress effects on root exudation of lodgepole pine, *Soil Biol. Biochem.*, 9, 417, 1977.

383. Davies, D.D., Anaerobic metabolism and the production of organic acids, in *The Biochemistry of Plants*, Vol. 2, Academic Press, 1980, p. 581.
384. Smucker, A.J.M. and Erickson, A.E., Anaerobic stimulation of root exudates and disease of peas, *Plant Soil*, 99, 423, 1987.
385. Rittenhouse, R.L. and Hale, M.G., Loss of organic compounds from roots. II. Effect of O_2 and CO_2 tension on release of sugars from peanut roots under axenic conditions, *Plant Soil*, 35, 311, 1971.
386. Bowes, G., Facing the inevitable: plants and increasing atmospheric CO_2, *Annu. Rev. Plant Physiol. Plant Mol. Biol.*, 44, 309, 1993.
387. Rogers, H.H., Runion, G.B., and Krupa, S.V., Plant responses to atmospheric CO_2 enrichment with emphasis on roots and the rhizosphere, *Environ. Pollut.*, 83, 155, 1994.
388. Whipps, J.M., Carbon economy, in *The Rhizosphere*, Lynch, J.M., Ed., Wiley-Interscience/John Wiley and Sons, Chichester, U.K., 1990, P. 59.
389. Hodge, A. and Millard, P., Effect of elevated CO_2 on carbon partitioning and exudate release from *Plantago lanceolata* seedlings, *Physiol. Plant.*, 103, 280, 1998.
390. Cardon, Z.G. Influence of rhizodeposition under elevated CO_2 on plant nutrition and soil organic matter, *Plant Soil*, 187, 77, 1996.
391. Barrett, D.J., Richardson, A.E., and Gifford, R.M., Elevated atmospheric CO_2 concentrations increase wheat root phosphatase activity when growth is limited by phosphorus, *J. Plant Physiol.*, 25, 87, 1998.
392. Brooks, R.R. et al., Phytomining, *Trends Plant Sci.*, 3, 359, 1998.
393. Delhaize, E. et al., Engineering high-level aluminum tolerance in barley with the ALMT1 gene, *Proc. Natl. Acad. Sci. USA*, 101, 15249, 2004.
394. Zhang, F. et al., An overview of rhizosphere processes related with plant nutrition in major cropping systems in China, *Plant Soil*, 260, 89, 2004.
395. Alvey, S. et al., Cereal/legume rotation effects in two West Africam soils under controlled conditions, *Plant Soil*, 231, 45, 2000.
396. v. Wirén, N. et al., Competition between microorganisms and roots of barley and sorghum for iron accumulated in the root apoplasm, *New Phytologist*, 130, 511, 1995.

3 Rhizodeposition and Microbial Populations

Melissa J. Brimecombe, Frans A.A.M. De Leij,
and James M. Lynch

CONTENTS

I. INTRODUCTION

In this chapter, we review the current literature available on the influence of root exudates on rhizosphere microbial populations; the effects of plant, microbial, and soil factors on the processes of rhizodeposition; and microbial colonization and activity. We first give a brief overview and definitions of some of the main concepts relating to the rhizosphere and rhizodeposition.

A. THE RHIZOSPHERE: A DEFINITION

Pioneer German agronomist and plant physiologist Lorenz Hiltner is often quoted as having said, "If plants have the tendency to attract useful bacteria by their root excretions, it would not be surprising if they would attract uninvited guests, which like the useful organisms, adapt to specific root excretions." In 1904, Hiltner attributed a "tiredness of soil" to the activities of harmful organisms in the rhizosphere of unthrifty plants and suggested that the healthy plants formed a protective bacteriorhiza.

From his early pioneering work on the *rhizosphere* (a term Hiltner used to describe specifically the interaction between bacteria and legume roots), our knowledge of the subject has greatly increased, and today perhaps a more appropriate definition of the rhizosphere is "the field of action or influence of a root" [1]. The rhizosphere is generally considered to be a narrow zone of soil subject to the influence of living roots, where root exudates stimulate or inhibit microbial populations and their activities. The rhizoplane or root surface also provides a highly favorable nutrient base for many species of bacteria and fungi. These two zones together are often referred to as the soil–plant interface. We can further define the *endorhizosphere* as the cell layers of the root itself, and the *ectorhizosphere* as the area surrounding the root (Figure 3.1).

The rhizosphere is thus an environment created by the interactions between root exudates and microorganisms, which may either utilize the organic materials, released as nutrient sources or be inhibited by them. The plant–microbe relationship is more often based on the former, where microbes

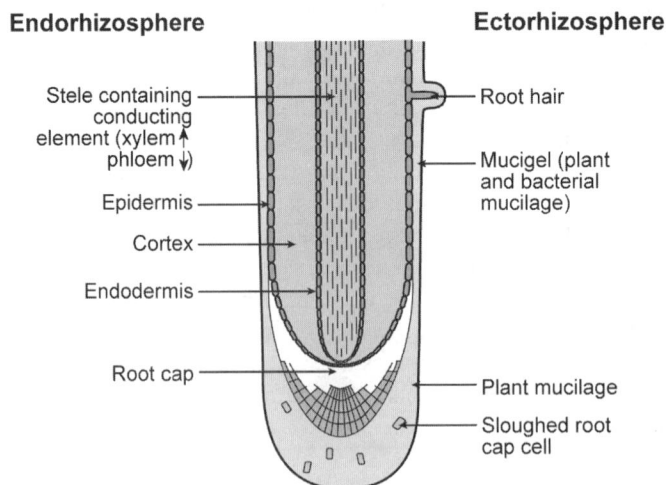

Endorhizosphere   **Ectorhizosphere**

Stele containing conducting element (xylem ↑ phloem ↓)   Root hair

Epidermis   Mucigel (plant and bacterial mucilage)

Cortex

Endodermis

Root cap   Plant mucilage

Sloughed root cap cell

FIGURE 3.1 Root region.

take advantage of nutrients provided by the plant. In return, microbes may assist the plant, for example, by making nutrients available or by producing plant growth-promoting compounds, or may cause harm to it, for example, by acting as plant pathogens. In general, the microbes that inhabit the rhizosphere serve as an intermediary between the plant, which requires soluble inorganic nutrients, and the soil, which contains the necessary nutrients but mostly in complex and inaccessible forms. Rhizosphere microorganisms thus provide a critical link between plant and soil environments [2].

B. RHIZODEPOSITION OF ORGANIC SUBSTANCES: CURRENT OPINION

Materials deposited by roots into the rhizosphere can be divided roughly into two main groups: (1) water-soluble exudates such as sugars, amino acids, organic acids, hormones, and vitamins, and (2) water-insoluble materials such as cell walls, sloughed-off materials and other root debris, and mucilage such as lysates released when cells autolyze [3,4]. In addition, carbon dioxide from root respiration often accounts for a large proportion of the carbon released from roots. Further, secretions such as polymeric carbohydrates and enzymes, depending on metabolic processes for their release, may also be regarded as root exudates. Some authors favor categorizing root exudates by their nature of release; however, as it is usually very difficult to distinguish experimentally between *true* exudates and organic compounds from other sources such as secretions and lysates, many authors use the approach of Uren and Reisenauer [5], and Rovira [6], who consider exudates to be all organic substances released by healthy and intact roots to the environment. Meharg [7] has further suggested that categorizing organic carbon compounds according to the nature of their release tends to oversimplify the interpretation of carbon budgets in the rhizosphere, as this approach does not reflect substrate availability within the rhizosphere. It was suggested that perhaps a better categorization is to classify organic carbon compounds lost from plant roots in terms of their subsequent utilization as a microbial substrate (i.e., as low-molecular-weight compounds readily assimilated by the microbial biomass; as polymeric and more complex compounds such as polysaccharides, polypeptides, nucleic acids, and pigments, etc., requiring extracellular enzymic activity to break them down before they can be assimilated; or as structural carbon sources such as cell wall materials, requiring saprophytic degradation before carbon becomes generally available to the rest of the soil biomass). This approach would appear to be more relevant to the study of rhizosphere microbial population dynamics.

Either way, microbial populations in the rhizosphere generally have access to a continuous flow of organic substrates derived from the root. Newman [8] found that these soluble and insoluble rhizodeposits ranged from 10 to 250 mg/g root produced for a number of plant species. This significant rhizodeposition has been shown to enhance microbial growth in the rhizosphere [9,10], and due to this large availability of substrate in the rhizosphere, microbial biomass and activity are generally much higher in the rhizosphere than in the bulk soil. Concentrations of microbes in the rhizosphere can reach between 10^{10} and 10^{12} cells per gram of rhizosphere soil [11], and invertebrate numbers in the rhizosphere have been shown to be at least two orders of magnitude greater than in the surrounding soil [12]. These increased population densities are largely supported by carbon input from roots. The nature of rhizodeposition is also discussed in this book (Chapter 1).

C. CARBON RHIZODEPOSITION: CURRENT ESTIMATES

The input of carbon into soil via rhizodeposition and the decay of roots has been quantified in several studies using either pulse or continuous $^{14}CO_2$-labeling techniques [4,7,13–15], and estimates of carbon rhizodeposition vary considerably. The proportion of net fixed C released from roots has been estimated to be as much as 50% in young plants [13] but less in plants grown to maturity in the field [16,17]. Lynch and Whipps [18] estimated that as much as 40% of the plant's primary C production may be lost through rhizodeposition, depending on plant species, plant age, and environmental conditions. Although many studies have attempted to quantify the amount of rhizodeposition associated with various plant species, relatively little is known about the exudation process itself. Whereas early studies of rhizodeposition assumed that once C compounds were lost from the root they were irretrievable, it has more recently been ascertained that this is an oversimplification of soil–root C fluxes. Studies by Jones and Darrah [19–21] found, for example, that the influx or resorption of soluble low-molecular-weight carbon compounds may play an important role in regulating the amount of C lost by the root. Good reviews of carbon flow in the rhizosphere and techniques used to quantify the dynamics of carbon flow include those of Meharg [7] and Grayston et al. [22]. Carbon rhizodeposition is also discussed here by Van Veen et al. (Chapter 13).

D. NITROGEN RHIZODEPOSITION: CURRENT ESTIMATES

In addition to carbon rhizodeposition, N rhizodeposition is also of considerable importance to nutrient cycling, usually as NH_4^+ [23], NO_3^- [24], amino acids [25–27], cell lysates, sloughed roots, and other root-derived debris. Despite the fact that N deposition has a significant role in N cycling and rhizosphere N dynamics, studies of N input from the root have been fewer, mainly due to methodological problems. Janzen and Bruinsma [28] estimated that for wheat, up to 50% of the assimilated N was present below ground and approximately half of this was apparently released from the root into the rhizosphere soil. In barley, 32% and 71% of the belowground N was present in rhizodeposits at sampling after 7 and 14 weeks plant growth. At maturity, the rhizodeposition of N amounted to 20% of the total plant N [29]. It is also well known that substantial amounts of N may be released from roots of legumes [23,30,31]. Jensen [29], for example, found that N deposition constituted 15% and 48% of the belowground N in pea, when determined 7 or 14 weeks after planting. At maturity, the rhizodeposition of N amounted to 7% total plant N.

E. COEVOLUTION OF PLANTS AND RHIZOBACTERIA

Whereas rhizobacteria may derive obvious benefit from the significant quantities of root exudates released into the rhizosphere, the microbes of the rhizosphere, in turn, have a significant influence on the nutrient supply to the plant by competing for inorganic nutrients and by mediating the turnover and mineralization of organic compounds. Thus the deposition of organic materials stimulates microbial growth and activity in the rhizosphere, which subsequently controls the turnover of C, N, and other nutrients [32–34]. Rhizodeposition is also considered to be of importance for soil organic matter dynamics in terms of nutrient mineralization and improvement of soil structure. These soil microbe-mediated processes of nutrient mineralization and N immobilization are strongly influenced by the presence of easily decomposable substrate in the rhizosphere — a topic covered in more detail in Chapter 4.

Whereas rhizodeposition strongly influences the size and activity of microbial populations at the soil–plant interface, the activity of these microbial populations in turn affect plant health, and thus influence both the quality and quantity of rhizodeposition. The potential for either an exudation response to bacteria or a response by bacteria to exudation suggests a certain degree of coevolution between plants and rhizobacteria [35]. As the composition and extent of exudation is largely determined genetically [36,37] and may incur a substantial metabolic cost, exudation must provide a selective advantage to plants. Indeed, Bolton et al. [37] have suggested that root exudation evolved in plants to stimulate an active rhizosphere. This is feasible if one considers that there exists a high degree of selectivity for rhizobacteria according to host plant genotype, and that certain microbial interactions in the rhizosphere have the ability to improve plant growth and plant health. Whereas components of the stimulated rhizosphere microbial community have the ability to be either beneficial or harmful to the plant, in terms of plant nutrition and plant health, there is most likely to be a balance between beneficial and detrimental organisms [38].

F. THE STUDY OF SOIL–PLANT–MICROBE INTERACTIONS

The increase in information concerning interactions at the soil–plant interface has resulted mainly from the development of new techniques to quantify microbial populations in soil, to collect and analyze root exudates, and to study microbial interactions at the root surface. The recent application of electron microscopy has further provided us with a greater understanding of the spatial distribution of microorganisms on root systems. The refinement of analytical techniques has permitted the elucidation of root exudate composition (see Chapter 1). Radioactive labeling techniques have not only permitted quantification of root exudation but have also facilitated the identification of the precise locations of exudation sites along the root. Rovira and Davey [39], for example, found that the region of meristematic cells behind the root tip is a site of major exudation of sugars and amino acids.

Besides quantification of root exudation and microbial colonization, knowledge of the growth of microorganisms in the rhizosphere in relation to the supply of organic nutrients is still an ongoing research goal. Such knowledge is necessary for evaluating the significance of microbial processes affecting plants. Several authors have assessed microbial growth in the soil and correlated the energy input provided by the addition of organic matter with the size of the observed microbial biomass. Implicit in any such calculation is a factor representing the energy requirement for maintenance of the existing population, generally expressed as maintenance coefficient (m) or specific maintenance (a). Barber and Lynch [9] investigated microbial growth in the rhizosphere of barley plants grown in solution culture either under axenic conditions or in the presence of a mixed population of microorganisms. It was found that more biomass was produced than could be accounted for by the utilization of the carbohydrates released by the roots grown in the absence of microorganisms, supporting the view that microbes stimulate the loss of soluble organic materials. It was suggested that the kinetics of growth in the rhizosphere and soil approximate most closely to those in fed batch culture. Pirt [40] had shown that a quasi steady state could be achieved in such a system, and that, therefore, the same equations could be used to describe microbial growth in the rhizosphere. Thus:

Overall rate of consumption = consumption for growth + consumption for maintenance

$$\frac{\mu x}{Y} = \frac{\mu x}{Y_G} + mx$$

where μ is specific growth rate (h^{-1}), x is the biomass (g), Y is the observed growth yield (g dry wt. g^{-1} substrate), Y_G is the true growth yield when no energy is used in maintenance (g dry wt. g^{-1} substrate) and m is the maintenance coefficient (g substrate g^{-1} dry wt. h^{-1}).

Barber and Lynch [9] used this equation to recalculate data from previous studies on microbial growth in soil, using a constant maintenance coefficient (m). They found no case where energy input exceeded the requirement for maintenance, and suggested, therefore, that apart from zones immediately around recently incorporated plant and animal residues, appreciable and continuous activity in soils can only be expected in the rhizosphere.

As an example of how this is related to microbial processes important to the plant, nitrogen fixation was considered in light of these results. Postgate [41] had shown that free-living N$_2$-fixing bacteria produced only 10 to 15 mg N g^{-1} sugar consumed, so this process was of negligible significance in the bulk soil because of the limited availability of substrate. In the rhizosphere, on the other hand, Dobereiner [42], having shown that *Azotobacter paspali* could form an association with the roots of the tropical grass *Paspalum notatum*, proposed that N$_2$-fixing processes could be of significance to the N economy of the plant. The data of Barber and Lynch [9] was used to estimate if such an association could ever be of significance to temperate cereals. Assuming the total exudation of 0.2 mg sugar mg^{-1} plant dry weight was utilized in N$_2$ fixation with the efficiency quoted by Postgate [41], only 2 to 3 μg N mg^{-1} plant dry weight would be fixed, which at most is about 15% of the total N content of the plant. The actual amount fixed would be less than this, as even when seeds are inoculated azotobacters account only for approximately 0.3% of the total rhizosphere community of the resulting plants [43]. Thus it was argued that this process could cause an appreciable increase in the N supply to temperate cereals only if larger quantities of carbohydrates were exuded by roots into the soil than were observed in previous studies [44], or that the efficiency of N_2 fixation greatly exceeded that observed by Postgate [41].

Another microbial process important to the plant is the mineralization of organic matter in soil. This process is highly dependent on the growth and activity of microbes in the rhizosphere, or those associated with organic residues present in the soil, to make mineral nitrogen available for plant uptake. Brimecombe et al. [45] found that inoculation of pea seeds with two *Pseudomonas fluorescens* strains led to increased uptake of nitrogen from ^{15}N-labeled organic residues incorporated into soil. In contrast, the mineralization of organic residues in the rhizosphere of wheat inoculated with the same strains decreased [46]. Further work in our laboratory suggests that plant-specific changes in root exudation patterns mediated by the introduced strains are responsible for the changes to soil saprophytic activities, which in turn are mediated by effects on the soil microfauna [47,48].

Stable isotope probes can be useful in studying soil–plant–microbe interactions (see Chapter 13). For example Treonis et al. [49] combined microbial community phospholipid fatty acid (PLFA) analyzes with an *in situ* stable isotope [13]CO_2-labeling approach. In the early days after label application, both fungal biomarkers and Gram-negative bacterial biomarkers showed the most ^{13}C enrichment and rapid turnover.

G. THE INFLUENCE OF PLANT AND MICROBIAL FACTORS

It has been found that many environmental factors influence the amount and composition of root exudates and hence the activity of rhizosphere microbial populations. Microbial composition and species richness at the soil–plant interface are related either directly or indirectly to root exudates and thus vary according to the same environmental factors that influence exudation. In essence, the rhizosphere can be regarded as the interaction between soil, plants, and microorganisms. Figure 3.2 shows

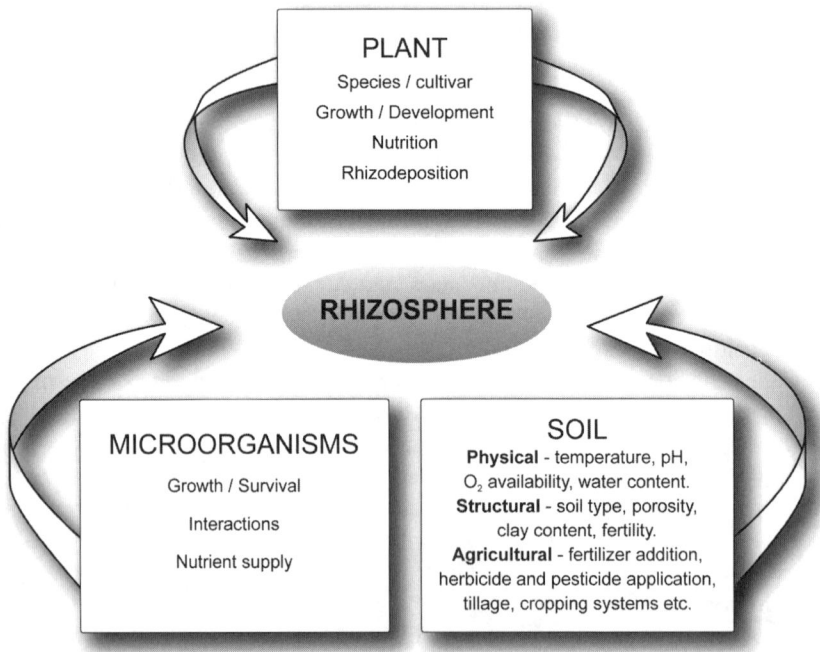

FIGURE 3.2 Factors influencing rhizosphere interactions.

some of the factors associated with these interactions, which will be discussed during the course of the chapter. Here we will mention briefly the influence of some plant and microbial factors on root exudation and rhizosphere microbial populations, while soil factors will be discussed later.

1. Plant Species

Gross differences have been observed in the amounts of fixed carbon released by annuals and perennials [50], with annuals releasing much less C than perennials. This effect may be partially due to perennials having to invest more of their assimilates to survive year-round. Between more closely related plants, several studies have reported that both the quantity and quality of root exudates vary between plant species [39,51,52]. In addition, it is also recognized that different cultivars of the same species may vary in their root exudation patterns. For example, Cieslinski et al. [53] quantified low-molecular-weight organic acids released from the roots of five cultivars of durum wheat and four cultivars of flax and found significant variation between cultivars. The quality of compounds released by plant roots appears to strongly influence the bacterial composition and activity in the rhizosphere, as shown by the preference of certain bacteria for exudates of different plant roots [54,55]. Differences in bacterial activity between cultivars of the same plant species have been shown to be related to differences in exudation spectra (i.e., subtle differences in compounds released by roots of the different cultivars) [56].

Marschner et al. [57] found that the composition of the bacterial community of rhizosphere of three plants species (chickpea, rape, and Sudan grass), as determined by PCR-DGGE of 16S rDNA, depended on the complex interaction between soil type, plant species, and root zone location. Bacterial diversity was higher in mature root zones than at the root tips in the sand and clay soils but not in the loamy sand soil. They also showed that N fertilization had no significant effect on the composition of bacterial community of the rhizosphere soil, whereas both fertilization and soil type influenced plant growth. In a further study, Marschner et al. [58] found that bacterial composition on the cluster roots and in the rhizosphere soils, determined by denaturing gradient gel electrophoresis (DGGE), differed among three species of *Banksia* (*B. attenuata* R. Brown, *B. ilicifolia* R. Brown, and *B. menziesii* R. Brown) with cluster root age (young or mature to senescing) and also between sampling times.

This suggests that it may be possible to manipulate the rhizosphere flora through genetic changes of rhizodeposition in the host plant. Of particular interest is whether different varieties, by exuding different compounds, can influence the rhizosphere flora in a way which would benefit the plant.

2. Plant Age and Stage of Development

Root exudation and microbial colonization have both been shown to change with plant age and stage of development. The quantity of both proteins [59] and carbohydrates [60] released by herbaceous plants has been shown to decrease with increasing plant age. Liljeroth and Bååth [61] found that bacterial abundance on the rhizoplane of several barley varieties significantly decreased with increasing plant age. Keith et al. [16] measured relative amounts of carbon translocated to the roots and rhizosphere during different developmental stages of wheat grown in the field. As the crop developed, it was found that proportionally less of the total photosynthesized carbon was transported below ground, with a marked decrease after flowering. Microbial numbers in the rhizosphere had previously been shown to increase over time, reaching a peak around the time of flowering and then decreasing [62]. Such changes in microbial colonization could be due to changes in total amounts of carbon exuded per unit root produced, or related to changes in the quality of exudates released. Investigating these issues at the molecular level, the rhizosphere effect measured as enrichment of bacterial population by 16S rRNA genes based on molecular analysis of DNA directly extracted from soil and rhizosphere samples was more marked for young than old roots of two maize cultivars [63].

3. Plant Growth

Prikryl and Vancura [64] found that the release of root exudates from wheat roots was positively correlated with root growth. The amounts of substances released by the roots were directly associated with root growth, and in plants where almost no root growth was observed, almost no root exudation occurred even in plants whose shoots were actively growing. This study confirmed results obtained by Prat and Retovsky [65], who found that live, intact, but nongrowing roots had lower exudation rates than the roots of plants in the period of rapid growth. A possible explanation for this may be provided by the results of Frenzel [66], who found that exudation depends considerably on the physiological state of the superficial root cells. It would appear therefore that root exudation is likely to be greatest from plants with actively growing root systems, whose superficial root cells are in an active state. This may also explain why root exudation decreases with plant age, as metabolic activity of superficial root cells decreases with plant age.

4. Presence of Microorganisms

The presence of microorganisms in the rhizosphere has been shown to increase root exudation [64,67–71]. This stimulation of exudation has been shown to occur in the presence of free-living bacteria such as *Azospirillum* spp. and *Azotobacter* spp. [72,73], and in the presence of symbiotic organisms such as mycorrhizae [74,75]. Increased root exudation has also been shown to be species specific; for example, Meharg and Killham [71] found that metabolites produced by *P. aeruginosa* stimulated a 12-fold increase in ^{14}C-labeled exudates by perennial ryegrass. However, under the same conditions, metabolites from an *Arthrobacter* species had no effect on root exudation.

II. MICROBIAL INTERACTIONS IN THE RHIZOSPHERE

During seed germination and seedling growth, the developing plant interacts with a range of microorganisms present in the surrounding soil. Plant–microbe interactions may be considered beneficial, neutral, or harmful to the plant, depending on the specific microorganisms and plants involved and on the prevailing environmental conditions.

Plant-beneficial microbial interactions can be roughly divided into three categories. First, there are those microorganisms that, in association with the plant, are responsible for its nutrition (i.e., microorganisms that can increase the supply of mineral nutrients to the plant). This group includes dinitrogen-fixing bacteria such as those involved in the symbiotic relationships with leguminous plants (e.g., *Rhizobium* and *Bradyrhizobium* species) (see Chapter 9), with monocots (e.g., *Azospirillum brasilense),* or free-living nitrogen-fixing bacteria such as *Klebsiella pneumoniae.* In addition, there are a number of microbial interactions that increase the supply of phosphorous (e.g., mycorrhizae) (see Chapter 8) and other mineral nutrients to the plant. Second, there is a group of microorganisms that stimulate plant growth indirectly by preventing the growth or activity of plant pathogens. Such organisms are often referred to as *biocontrol agents,* and they have been well documented. A third group of plant-beneficial interactions involve those organisms that are responsible for direct growth promotion, for example, by the production of phytohormones. Plant growth-promoting rhizobacteria (PGPR) or plant-beneficial microorganisms and their use to increase plant productivity have been the subject of several reviews [76–81], and examples are discussed in the following text.

Neutral interactions are found extensively in the rhizosphere of all crop plants. Saprophytic microorganisms are responsible for many vital soil processes, such as decomposition of organic residues in soil and associated soil nutrient mineralization or turnover processes. Whereas these organisms do not appear to benefit or harm the plant directly (hence the term *neutral*), their presence is obviously vital for soil nutrient dynamics, and their absence would clearly influence plant health and productivity.

Detrimental interactions within the rhizosphere include the presence and action of plant pathogens and deleterious rhizobacteria. Root exudates play a key role in determining host-specific interactions with, and the composition of, their associated rhizobacterial populations. Root exudates can attract beneficial organisms such as mycorrhizal fungi and PGPR [13,82], but they can also be equally attractive to pathogenic populations [83,84]. As mentioned earlier, it is the balance between beneficial and detrimental microorganisms that ultimately governs plant nutrition and plant health. Before discussing some of the specific interactions mentioned earlier, consideration is given to the microbial colonization of the rhizosphere.

A. COLONIZATION OF THE RHIZOSPHERE

Studies have shown that rhizobacteria are better adapted to colonization of roots than bacteria isolated from nonrhizosphere soil [85], and this is further evidence to support the theory of coevolution of plants and their associated rhizosphere microbial populations. Root colonization can be considered to involve four stages. The initial stage of root colonization is the movement of microbes to the plant root surface. Bacterial movement can be passive, via soil water fluxes, or active, via specific induction of flagellar activity by plant-released compounds (chemotaxis). The second step in colonization is *adsorption* to the root. This is a step required before anchoring and can be defined as nonspecific and based on electrostatic forces, whereas *anchoring* can be defined as the firm attachment of a bacterial cell to the root surface. Following adsorption and anchoring, specific or complex interactions between the bacterium and the host plant may ensue that lead to induction of bacterial gene expression. Table 3.1 cites examples of root exudate components involved in these processes.

TABLE 3.1
Root Colonization

Process	Exudate Component	Microbial Species	Reference
1. Movement of microbes to root surface			
A. Passive			
B. Active — induction of flagellar activity (chemotaxis)	Luteolin/phenolics	*Rhizobium meliloti*	86
	Acetosyringone	*Agrobacterium tumefaciens*	87
	Benzoate/aromatics	*Azospirillum* spp.	88
	Sugars	*Agrobacterium tumefaciens*	89
	Amino acids, nucleotides, and sugars	*Pseudomonas lachrymans*	89
	Serine	*Pseudomonas aeruginosa*	89
	Unidentified	*Pseudomonas fluorescens*	89
	Unidentified	*Pseudomonas putida*	89
	Unidentified	*Azospirrillum brasilense*	90,91
2. Adsorption			
3. Anchoring			
A. Bacterial appendages (pili, fimbriae)		*Pseudomonas fluorescens*	92
B. Agglutination	Lectins	*Pseudomonas putida*	93,94
C. Formation of cell aggregates		*Enterobacter agglomerans*	95
4. Gene expression			
A. Nodulation genes — production of nod factors	Flavonoids	*Rhizobium* spp.	96,97,98
		Bradyrhizobium spp.	
B. Virulence genes — production of virulence factors		*Agrobacterium tumefaciens*	96

Note: Table lists examples of root exudate components involved in chemotaxis, anchoring, and induction of gene expression.

B. Specific Microbial Interactions

1. Beneficial Interactions

a. Dinitrogen Fixation

Species of *Rhizobium* and *Bradyrhizobium* have long been known to induce the formation of root nodules on leguminous plants (see Chapter 9 and Chapter 11). Once formed, a differentiated form of the bacterium, the bacteroid, converts atmospheric nitrogen into ammonia, which is then used as a nitrogen source by the plant. In return, the plant provides a carbon source to the bacteroid, probably in the form of dicarboxylic acid [80]. Nodulation is host-specific, and each *(Brady)rhizobium* species can nodulate only a restricted number of legume species. Both nodulation and nitrogen fixation are complex processes, involving interactions between a number of bacterial genes and their gene products and plant products. Specific compounds released by roots of young legumes are involved in attracting these symbiotic bacteria to their roots (see Table 3.1). Following infection of the host root system, flavonoid compounds released by the root hair zone of the plant are believed to be responsible for the induction of bacterial nodulation (*nod*) genes in *Rhizobium* species [97], which results in the biosynthesis of Nod factors. Nod factors are a group of biologically active oligosaccharide signals whose effects on the host legume are similar to the early developmental symptoms of the *Rhizobium*-legume symbiosis, including root hair deformation and nodule initiation [98]. The role of organic signaling molecules between plants and microorganisms, and the biochemistry of the association between *rhizobia* and their host plants are discussed more thoroughly in Chapter 6 and Chapter 9, respectively.

In addition to N_2-fixing symbioses with legumes, associations between N_2-fixing microorganisms and the roots of monocots such as cereals and forage grasses have been reported [99]. However, the benefit of nitrogen fixation to nonlegumes was thought to be small, with studies showing that the contribution of N_2 fixation by *Azospirillum* to plant growth was minimal, with yield increases ranging from 5 to 18% [91]. It was also found that mutants unable to fix nitrogen (*Nif-*) were capable of increasing plant growth, like a wild-type N_2 fixer [100]. It was suggested that the observed beneficial effects (increased yield, increased water, and mineral uptake) of *Azospirillum* may be derived from improvements to root development as a result of the action of phytohormones, such as indoleacetic acid (IAA), gibberellins, and cytokinin-like substances, produced by the strain, rather than from its nitrogen-fixing abilities [91,100,101]. However, more recent work has initiated a reappraisal of the theory of N_2 fixation as a mechanism of plant growth promotion by *Azospirillum* spp. [91]. Most importantly perhaps, results indicate that graminaceous plants are potentially capable of establishing associations with diazotrophic bacteria in which high ammonium-secreting *Azospirillum* mutants provide the host with a source of nitrogen.

b. Mycorrhizae

The biochemistry of the association between mycorrhizae and the plant is extensively discussed in Chapter 8. Arbuscular mycorrhizal (AM) fungi interact symbiotically with approximately 80% of all plant species [102]. Mycorrhizal symbioses are present in most natural and agricultural ecosystems, where they are involved in many key processes, including nutrient cycling, maintenance of soil structure, plant health, and enhancement of nitrogen fixation by rhizobia [103]. Their primary effect on the plant is the improvement of plant growth by increasing the supply of mineral nutrients from the soil to roots. AM fungi are known to influence phosphorus uptake and plant growth in P-deficient soils [104,105]. Several mechanisms have been proposed to account for the increases in uptake of phosphorus; one of the most important being the acidification of the rhizosphere [106,107], which may explain how vesicular AM fungi increase the uptake of P as a result of plant utilization of ammonium (NH_4^+-N). Higher P uptake as a result of plant utilization of NH_4^+ occurs in both neutral and alkaline soils, and is discussed further on. More efficient utilization of ammonium by mycorrhizal than nonmycorrhizal plants has been shown by several authors [108,109] and may lead to increased H^+ secretion into the rhizosphere as a result [110,111]. Rhizosphere pH has been reported to be altered by the form of nitrogen added both in the presence and absence of mycorrhizae [108,112,113]. It has also been reported that infection by ectomycorrhizae significantly enhanced the capacity of plant roots to release H^+ into a medium, which can increase the bioavailability of compounds not readily soluble at higher pH [114]. The efflux of H^+ increases the solubility of Ca phosphates but not that of Al and Fe phosphates. If a mycorrhizal plant induces a decrease in its rhizosphere pH, this effect may contribute to more P uptake by solubilizing calcium, phosphorus, and iron and aluminum phosphates, and thus increasing P availability to both the root and hyphae. In general, the addition of nitrogen stimulates the uptake of P by the plant, especially when NH_4^+ is applied [115,116]. Ortas et al. [117] found that application of ammonium led to increased plant dry weight and P content of sorghum as compared to plants treated with nitrate. These differences were enhanced by inoculation with mycorrhizae. It has also been shown that AM fungi can take up significant quantities of Cu, Zn, and Cd [118], providing a mechanism by which plants avoid exposure to toxic quantities of these metals. The improved phosphate, nitrogen, or micronutrient uptake by plants as a result of mycorrhizal associations also has secondary effects on the uptake of other ions, such as potassium, sulfate, and nitrate. In addition, the increased uptake of phosphate indirectly stimulates nodulation and nitrogen fixation [80], as rhizobia require substantial amounts of this element for their activity. The biochemistry of the association between mycorrhizae and plants is covered in depth in Chapter 8.

c. Biocontrol

Evidence suggests that monoculture of crops eventually leads to decreased yields, as a result of an increase in plant pathogen numbers in the soil [80]. Crop rotation has previously been employed as a method to alleviate this problem, by denying pathogens a suitable host for a period of time so that their numbers in soil decline. However, it has also been noted that in some soil systems repeated monoculture eventually leads to a decrease in plant disease [80]. In these cases the disease-conducive soils are converted to disease-suppressive soils. This phenomenon has led to the isolation of a wide range of microbial biocontrol agents from disease-suppressive soils [119]. Mechanisms by which these organisms are known to antagonize plant pathogens are varied, and some of these are discussed in the following text. In some instances, control of a pathogen may result from a combination of two or more of these mechanisms.

i.   Production of Antibiotics

The production of secondary metabolites with antimicrobial properties has long been recognized as an important factor in disease suppression (see Chapter 6). Metabolites with biocontrol properties have been isolated from a large number of rhizosphere microorganisms, including the fluorescent pseudomonads (Table 3.2). To demonstrate a role for antibiotics in biocontrol, mutants lacking production of antibiotics or overproducing mutants have been used [128–130]. Furthermore, microbially produced antibiotics, such as 2,4-diacetylphloroglucinol (Phl) and phenazine-1-carboxylic acid, have been isolated from the rhizosphere of wheat following the introduction of antibiotic producing *Pseudomonas* strains into the rhizosphere of wheat [128,131–134], confirming that such antibiotics are produced *in vivo*. Furthermore, Phl production in the rhizosphere of wheat was strongly related to the density of the bacterial population present [133]. Despite the convincing evidence that antibiotics such as Phl are produced by bacteria in the rhizosphere, the mechanisms by which they enhance plant health are far from clear. For example, a colony of *P. fluorescens* F113, a Phl producing strain, will produce a large zone of inhibition when grown with other microorganism, as a result of the high concentration of Phl near the colony. Similarly, when introduced at large inoculum doses ($>10^9$ CFU/seed) the bacterium will prevent colonization of the seeds by fungi, but will also greatly inhibit the germination and seedling growth (De Leij, unpublished data). This suggests that Phl act as a true antibiotic at high concentrations. However, when seeds were coated with lower bacterial numbers (10^7 CFU/seed) and planted into soil, results were quite different. Instead of plant growth inhibition, root growth was increased by more than 50% and was longer and had more lateral roots [135]. In this situation bacterial numbers were maintained

TABLE 3.2
Antibiosis as a Mechanism of Biocontrol

Biocontrol Agent	Antibiotic Compound	Phytopathogen	Protected Host	Disease	References
Agrobacterium radiobacter	Agrocin 84	*Agrobacterium tumefaciens*	Stonefruit, roses	Crown gall	120
Pseudomonas aerofaciens Q2-87	2,4-Diacetylphloroglucinol	*Gaeumannomyces graminis* var. *tritici*	Wheat	Take-all	121
Pseudomonas fluorescens 2-79	Phenazine-1-carboxylate	*Gaeumannomyces graminis* var. *tritici*	Wheat	Take-all	122
Pseudomonas fluorescens F113	2,4-Diacetylphloroglucinol	*Pythium ultimum, Fusarium oxysporum, Phoma beta, Rhizopus stolonifera*	Sugar beet	Damping off	123
Pseudomonas fluorescens CHA0	HCN, 2,4-Diacetylphloroglucinol, pyoluteorin	*Thielviopsis basicola*	Tobacco	Black root rot	124
Pseudomonas fluorescens Pf-5	Pyrrolnitrin, pyoluteorinl	*Pythium ultimum, Rhizoctonia solani*	Cotton	Damping off	125,126
Pseudomonas fluorescens Pf-5	2,4-Diacetylphloroglucinol	*Pythium ultimum, Rhizoctonia solani*	Cotton	Damping off	127

at between 10^6 and 10^7 CFU/g root [135]. Not only did inoculation result in enhanced plant growth, but establishment of *Rhizobium* was also twofold higher compared with treatments that were inoculated with a Phl deficient strain. This suggests that Phl acts *in vivo* as a plant growth promoting substance rather than a true antibiotic. From an evolutionary point of view, this makes sense as organisms that can stimulate the plant to produce more root exudates will ensure that they benefit from a constant nutrient supply. On the other hand, organisms that invest in antibiotics that kill competitors in a situation where the nutrient source has already been depleted will gain very little. The role of microbially produced antibiotics has been the topic of some extensive reviews [136,137].

ii.   Production of Siderophores
Many plant growth-promoting bacteria, especially pseudomonads, produce high-affinity Fe^{3+} binding siderophores under conditions of low-iron concentration [138,139]. This may result in severe iron limitation in the rhizosphere, which could limit the growth of other rhizosphere bacteria and fungi. Many authors have demonstrated the importance of siderophores in the inhibition of both fungal and bacterial pathogens [140–143]. Siderophores which are produced mainly by fluorescent pseudomonads under iron-limiting conditions have a very high affinity for ferric iron. These bacterial iron chelators are thought to sequester the limited supply of iron available in the rhizosphere, making it unavailable to pathogens and thus inhibiting their growth [144,145]. Iron competition in fluorescent pseudomonads has been the subject of many studies, and the role of pyoverdine siderophores produced by these bacteria has been shown to play a major role in the control of *Pythium* and *Fusarium* species, either by comparing the effects of purified pyoverdine with synthetic iron chelators or through the use of pyoverdine minus mutants [146,147]. However, siderophores are not always implicated in disease control mediated by fluorescent pseudomonads

as several different types of siderophores are produced by these bacteria that have different properties. For example, some siderophores can only be used by the bacteria that produce them [148], whereas others can be used by many different bacteria [145]. Besides iron chelating properties, siderophores such as pyoverdine and salicylate may act as elicitors for inducing systemic resistance in plants [149,150]. This topic is discussed in more detail in Chapter 7.

iii.   Production of Volatile Compounds

Volatile compounds such as ammonia and hydrogen cyanide are produced by a number of rhizobacteria and are also believed to play a role in biocontrol. For example, *P. fluorescens* strain CHA0 can produce levels of HCN that *in vitro* are toxic to pathogenic fungi such as *Thielaviopsis basicola*, and it is possible that this is the mechanism responsible for prevention of black root rot of tobacco in the field [151].

iv.   Parasitism

Lysis by hydrolytic enzymes excreted by microorganisms is a well-known feature of mycoparasitism. Chitinase and β-1,3 glucanase (laminarase) are particularly important enzymes secreted by fungal mycoparasites capable of degrading the fungal cell wall components, chitin, and β-1,3 glucan [152–155].

Many rhizobacteria are classified as chitinolytic and, for example, *Serratia marsescens,* which excretes chitinase, was found to be an effective biocontrol agent against *Sclerotium rolfsii* [156]. Similarly, *Aeromonas caviae* was found to reduce disease caused by *Rhizoctonia solani, Fusarium oxysporum,* and *S. rolfsii* [157]. There is also evidence to support the role of β-1,3 glucanase in biocontrol of soilborne plant pathogens [158].

v.   Competition for Nutrients

Elad and Chet [159], studying biocontrol of *Pythium* damping-off by rhizobacteria, suggested that competition for available carbon and nitrogen sources may account for observed disease reduction. They found competition for nutrients between germinating oospores of *Pythium aphanidermatum* and bacteria (which was unique to isolated biocontrol strains) significantly correlated with suppression of the disease. It appeared that bacteria were competing with germinating oospores for available C and N and by eliminating these resources, the bacteria effectively reduced oospore germination.

vi.   Competition for Ecological Niche

An alternative mechanism involved in biocontrol is that of niche exclusion. For PGPR to function successfully in the field, they must inevitably be able to establish themselves effectively on plant roots. *Pseudomonas* strains, for example, are able to establish on roots from inoculated seeds relatively easily. This is a major factor contributing to the success of these strains as biological control agents. However, root colonization is not the only criterion for defining a successful PGPR strain. Most diffusible root exudates are associated with the zone of root elongation, inducing colonization by PGPR in this zone. However, some studies have indicated establishment of populations along a greater, more undefined area of the root surface, including root tips, zones of elongation, and zones of lateral root emergence.

As phytopathogenic bacteria occupy particular niches in the rhizosphere, it has been proposed that the deliberate application of a nonpathogenic mutant of the same species may prevent pathogen establishment through niche exclusion by the nonpathogenic mutant. An example of this was the deliberate release of a nonpathogenic Ice⁻ mutant of *P. syringae* to compete with plant pathogenic Ice⁺ *P. syringae* to prevent frost damage [160,161]. The Ice⁺ (pathogenic) strain of this organism causes frost damage to plants by making an ice nucleation protein that initiates ice crystallization at temperatures not normally favorable for ice formation. It was envisaged that deliberate release of the Ice⁻ mutant of *P. syringae,* from which the ice nucleation protein gene had been deleted, could prevent frost damage by colonizing niches previously occupied by the Ice⁺ strain. In the field, it was found that frost damage was indeed decreased following the application of the genetically modified Ice⁻ strain.

vii.   Induced Disease Resistance

Another mechanism responsible for the biological control of plant disease is induced systemic resistance (ISR) or induced disease resistance. ISR protects the plant systemically following induction with an inducing agent to a single part of the plant. The action of ISR is based on plant defense mechanisms that are activated by inducing agents [162]. ISR, once expressed, activates multiple potential defense mechanisms, which include increases in activity of chitinases, ß-1,3-glucanases, peroxidases, and other pathogensis-related proteins [163]; accumulation of antimicrobial low-molecular-weight substances such as phytoalexins [164]; and the formation of protective biopolymers, such as lignin, cellulose, and hydroxyproline-rich glycoproteins [165–167]. A single inducing agent can control a wide spectrum of pathogens. In cucumber, for example, treatment of the first leaf with a necrosis-forming organism protects the plant against at least 13 pathogens, including fungi, bacteria, and viruses [168]. Caruso and Kluc [169] showed systemic protection in cucumber and watermelon, resulting from induction with *Colleototrichum orbiculare* against challenge inoculum of the same pathogen. Many other cases have been reported, but widespread application has not been accomplished, as classical ISR employs pathogenic organisms as inducing agents. More recent work, however, has demonstrated that some PGPR may act as inducing agents, leading to systemic protection against pathogens. In response to earlier observations of PGPR-mediated ISR in the greenhouse, Wei et al. [170] carried out three 2-year field trials to determine the capacity of PGPR to induce systemic resistance against cucumber diseases. Results indicated that PGPR-mediated ISR was operative under field conditions, with consistent effects against challenge-inoculated angular leaf spot and naturally occurring anthracnose. Furthermore, ISR induced early-season plant growth promotion and increased yield.

In these studies, the pathogen and the resistance-inducing PGPR were applied to separate locations on the plants, excluding direct antibiosis and competition as mechanisms of disease suppression [170].

d.   Production of Plant Growth-Promoting Compounds

A large number of rhizobacteria, such as strains of *Azospirillum, Azotobacter, Pseudomonas,* and *Bacillus,* have been shown to produce plant growth-promoting compounds such as IAA, gibberellin, and cytokinin-like substances [99,101,171–173]. The presence of such compounds in the rhizosphere appears to stimulate plant growth directly.

2.   Detrimental Microbial Interactions

a.   Plant Pathogens

Interactions between microbial pathogens and plants are in general host-specific and consequently have been considered to be influenced by root exudate components. There are examples where amino acids, sugars, and other exudate components have been shown to directly stimulate pathogens [83,174]. Stimulation of chlamydospore germination of *Fusarium solani f. phaseoli* [175] and *Thielviopsis basicola* [176] was found to occur at the surface of bean seeds and roots but not in soil distant from the host. *Phytophora, Pythium, Aphanomyces,* and other examples of pathogenic oomycetes have motile zoospores and consequently tend to accumulate around roots via chemotactic or electrostatic responses to root exudates [177]. Germ tubes from spores of *Phytophora cinnamomi* have also been shown to exhibit chemotropism, becoming oriented toward the region of elongation of susceptible roots. Exudate-induced interactions unfavorable to propagule germination, growth, and colonization of roots by plant pathogens are often associated with the activity of other microorganisms in the rhizosphere. Whereas all these organisms occupy the same microhabitat, there is inevitably direct competition for nutrients and ecological niche. For example, Paulitz [178] found that hyphal growth from soil-produced sporangia of *Pythium ultimum* was stimulated by volatiles from germinating pea seeds, and this stimulation was reduced when seeds were treated with *Pseudomonas fluorescens* N1R. It was thus suggested that N1R may reduce damping-off by competing for and using volatile exudates from germinating pea seeds.

In addition, plant pathogens may be antagonized by compounds released by plant roots, plant residues, and other microorganisms present in the rhizosphere. Further, the survival and activity of specific plant pathogens may be affected by the action of antagonists present in the rhizosphere. To reiterate, it is the balance between plant pathogenic and antagonistic microorganisms that determines the effect on plant health.

b. Deleterious Rhizobacteria

Rhizobacteria that inhibit plant growth without causing disease symptoms are frequently referred to as *deleterious rhizobacteria (DRB)*, or *minor pathogens*. DRB can inhibit shoot or root growth without causing any other visual symptoms [179–181], and may be partly responsible for growth and yield reductions associated with continuous monoculture [181,182]. DRB implicated in yield decline in a number of crops have been reviewed by Nehl et al. [35]. Several mechanisms for growth inhibition by DRB have been proposed, the most likely being the production of phytotoxins such as cyanide [183,184] and other volatile and nonvolatile compounds as yet unidentified [185,186]. An alternative mechanism by which DRB inhibits plant growth may be through the production of phytohormones [181]. IAA produced by DRB has been shown to inhibit root growth in sugar beet [187] and black currant [188]. DRB may also compete with the plant and beneficial rhizobacteria for nutrients, which may also contribute to decreased plant growth and yields. Further, DRB may indirectly reduce growth by inhibiting mycorrhizal development [189] or by counteracting the effects of nitrogen-fixing rhizobacteria [35].

The pathogenicity of DRB to crop plants has been shown to be host-specific [35] and thus is conceivably linked to root exudation. Alstrom [190] found that the pathogenicity of two isolates of *Pseudomonas* was determined by the major components of the broth culture in which they were applied to bean seedlings. Both isolates were pathogenic to bean seedlings when the broth contained sucrose and peptone or sucrose and yeast extract. When the broth contained sucrose alone, one isolate was pathogenic and the other was not. Neither isolate was pathogenic when the broth contained yeast extract or peptone alone [190].

III. MICROBIAL COMPOSITION IN THE RHIZOSPHERE: METHODOLOGY

Root exudates are believed to have a major influence on the diversity of microorganisms within the rhizosphere [84,191]. Several approaches have been employed to measure the biodiversity of rhizosphere microbial communities. Most of the earliest studies relied on dilution plating procedures with selective culture media as a way to enumerate specific groups of microorganisms. Problems associated with such methods include the low culturability of soil microorganisms (1 to 10%) [192], and the difficulty in adequately defining species of different microorganisms. Subsequently, techniques of molecular biology and fatty acid analysis have increased the ease of identifying organisms without the requirement for isolation in pure culture. Community diversity in soil has been measured using a number of indices, some of which have been applied to the experimental study of rhizosphere microbial populations.

A. Microorganisms Associated with Germinating Seeds and Young Roots

Rhizosphere microorganisms may arise from seed-borne populations that survived seed storage and germination or from soilborne populations. The microbiological characteristics of the germinating seed are not often addressed but could be important for establishment of some saprophytic microbes in the endorhizosphere. It is thought that most recruitment of microorganisms for subsequent colonization of the plant root and rhizosphere soil takes place after growth stimulation of the microbes by the advancing root. The root tip may be described as a slowly advancing point source of substrate, which acts as a stimulus for resumed activity and growth by the dormant microbes in

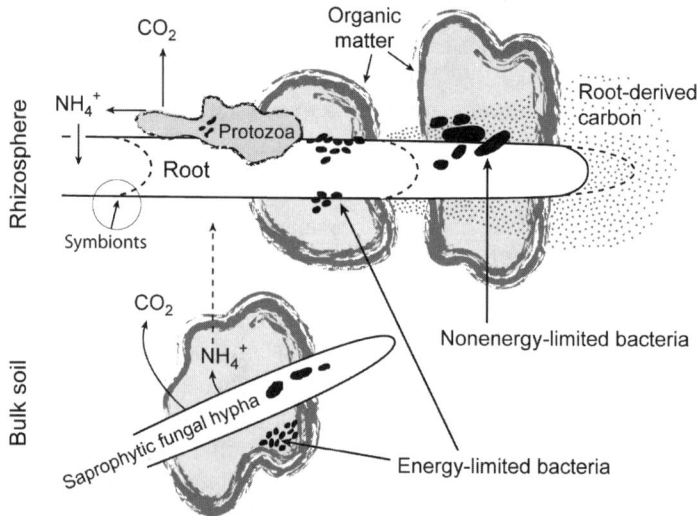

FIGURE 3.3 Model of proposed interactions in the rhizosphere and in the bulk soil.

the bulk soil. Figure 3.3 shows a model of proposed interactions within the rhizopshere in terms of microbial colonization and utilization of root exudates.

Early microbial successions have been studied by Van Vuurde and Schippers [193], who found the invasion sequence of (in order of appearance) various rhizobacteria, coryneforms, true actinomycetes, and microfungi associated with the onset of epidermal and cortical cell senescence in young roots (1 to 2 weeks old). Liljeroth et al. [194] washed 1- to 2-day-old root tips and 8- to 9-day-old root bases of wheat seedlings before maceration and plating onto 10% tryptone soya agar. Early colonizers were characterized in terms of carbon substrate utilization and clustered into 11 groups of both Gram-negative and Gram-positive bacteria. Utilization of simple sugars such as lactose, galactose, mannose, xylose, and mannitol and of organic acids such as citrate and succinate was found more often in populations from the root tips as compared to those isolated from root bases. The advancing root tip is considered to be an important site for exudation of simple sugars and organic acids, which may have stimulated the growth of rhizobacteria utilizing these compounds.

Kleeberger et al. [195] found fluorescent *Pseudomonas* spp. to be the largest group of Gram-negative bacteria and coryneforms the largest group of Gram-positive bacteria to colonize the endorhizosphere of wheat and barley when surface-sterilized root segments were plated onto plate count agar. They also suggested that the coryneforms were primarily located in the internal part of the endorhizospheres (i.e., the mucilage layer) by comparing surface-sterilized root segments with untreated roots.

B. MICROORGANISMS ASSOCIATED WITH MATURE ROOTS

Several studies have indicated that the species diversity of indigenous soil communities will influence the species composition of ectorhizosphere populations [196]. On mature roots, seasonal successions may be observed as the soil microbial activity varies with temperature, water content, nutrition, and root exudation. Acero et al. [197] found that the composition of alder (*Alnus*) rhizosphere populations alternated between one dominated by *Bacillus* spp. in autumn and winter and one dominated by *Pseudomonas* spp. in spring and summer.

Lambert et al. [198,199] investigated the total cellular protein compositions in culturable bacteria, as a way of comparing microbial diversity in crop rhizospheres. Such protein profiles, characterizing the protein composition in each clone of the culturable bacteria, were used to distinguish specific strain clusters among fast-growing rhizobacteria in the endorhizosphere of maize and sugar beet in two separate studies [198,199]. In the sugar beet study, protein profiles in whole-cell digests of organisms grown on 10% tryptone soya agar were compared after sodium dodecyl sulfate-polyacrylamide gel electrophoresis (SDS-PAGE) [199], and it was found that *P. fluorescens* or *Sphingomonas paucimobilis* were predominant on the root surface of young sugar beets until June, after which time *Xanthomonas maltophila* and *Phyllobacterium* spp. were found at increasing densities. These contributions [192,193] could be considered as pioneer work in the application of proteomics to soil.

During the plant growth season, microorganisms become more dependent on mobilization of organic matter present in the soil. Gram-positive microorganisms, including coryneforms and true actinomycetes, become increasingly more abundant in the rhizosphere of maturing plants [195]. Miller et al. [200] used 10% tryptone soya agar to obtain total bacterial counts in rhizosphere soil of 1- to 2-month-old wheat cultivars. Using two selective media for the fluorescent *Pseudomonas* spp., it was found that this group accounted for only a small percentage (~1%) of the total population. This may possibly reflect that pseudomonads are abundant only among rhizobacteria during the initial growth phases of wheat. They also found that coryneforms constituted a significant proportion of the total population. On two different wheat cultivars, coryneforms accounted for between 4 and 15% of the total populations. Coryneform numbers in the bulk soil made up to 30% of the total bacterial population, suggesting that these soilborne coryneforms colonized the rhizosphere as plants aged. In a subsequent study, Miller et al. [201] estimated the relative abundance of true actinomycetes in the rhizosphere of 3- to 10-week-old wheat cultivars. Isolation on selective chitin-oatmeal agar gave relatively stable numbers of actinomycetes over the 3- to 10-week growth period. It was concluded that root exudation did not control the establishment of true actinomycetes in the rhizosphere soil [201]. It is suggested that their high capacity for polymer degradation, resistance to desiccation, and relative insensitivity to toxic compounds gives these organisms a selective advantage when root exudates become sparser in mature plants.

De Leij et al. [202] developed a method for describing microbial community structure based on the quantification of bacterial colonies in six or seven growth classes on agar media. The idea of quantifying bacterial populations according to their growth rate is based on the ecological concept of r- and K-strategy [203]. Typical r-strategists, which do well in uncrowded, nutrient-rich environments, are inefficient in breaking down recalcitrant substrates and are also thought to be sensitive to toxins in the environment. K-strategists do relatively well in crowded environments that have reached their carrying capacity, have more efficient cell metabolism than r-strategists, and are able to utilize recalcitrant substrates. It is also thought that they are less sensitive to toxins than r-strategists. Under field conditions, it was found that as wheat root matured, the microbial community changed from one dominated by r-strategists to one distributed more toward K-strategy. De Leij et al. [204] used this method to assess the impact of field release of genetically modified *P. fluorescens* on indigenous microbial communities of wheat. The concept of r-K strategy distribution for risk assessment purposes is thought to be useful [204], as r-strategists are characteristic of environments that undergo rapid changes, whereas K-strategists dominate in stable, nonperturbed environments [203,205]. Populations of r-strategists can markedly increase when conditions are favorable, whereas, because of their poor competitive abilities and lack of long-term survival mechanisms, their numbers decline rapidly when conditions deteriorate. In contrast, the population size of typical K-strategists is more buffered against perturbation because of their slow growth and ability to form long-term survival structures. Rapidly growing bacteria and yeasts, which might all be regarded as typical r-strategists, were found to be sensitive indicators of environmental change and responded with large population decreases in the presence of wild-type and recombinant *P. fluorescens* SBW25 during early stages of wheat development [204].

IV. INFLUENCE OF SOIL FACTORS

In addition to the interactions between plants and microorganisms, a third factor, the soil, also plays a role in determining root exudation and the activity and diversity of rhizosphere microbial populations. In this section, physical and structural aspects of the soil are discussed in relation to their effects on root exudation and microbial populations. Consideration is also given to the role of agricultural management practices on rhizosphere processes. In addition, the role of other biotic factors, such as microfaunal predation, is discussed in relation to nutrient cycling in the rhizosphere.

A. PHYSICAL FACTORS

The effect of physical factors such as temperature, soil moisture content, pH, and oxygen availability on microbial survival and activity in soil are well documented (for a review see Reference 96). It is also widely acknowledged that these factors may also influence plant growth and can therefore be presumed to influence both root exudation and rhizosphere microbial populations.

1. Temperature

Extremes of temperature influence both microbial and plant growth. Cell division and expansion are temperature-dependent processes that control the rate of both root growth [206] and microbial proliferation. Any variations above and below the temperature range to which a plant is adapted, may inhibit or even irreversibly arrest root growth [207]. Similarly, above and below specific microbial adaptive temperatures, cells may become inactive or decline in number in soil. Many aspects of cell metabolism are adversely affected by temperatures that depart from the optima, including cell division, protein synthesis, respiration, and ion transportation [208]. These effects inevitably apply to both plant [209] and microbial cells. Root exudation is affected both qualitatively and quantitatively by temperature, with sudden increases or decreases in temperature stimulating exudation (for a review see Reference 22). For example, Martin and Kemp [210] found that a reduction in temperature from 15 to 10°C increased carbon loss from roots of 11 different cultivars of wheat into the rhizosphere. Qualitative differences in exudation patterns were observed when *Zea mays* seedlings, grown initially at 19°C, were subsequently grown at 5°C. This led not only to greater exudation, but also to the exudation of three new oligosaccharides and sucrose; compounds not previously observed at the higher temperature [211]. Similarly, Rovira [212] found that raising the temperature increased the amount of root exudate and the relative proportions of amino acids exuded by *Trifolium repens* and *Lycopersicon esculentum*.

Unfavorable temperatures, especially when suddenly imposed, have been shown to cause leakage of ions and metabolites from the root to the surrounding medium [208]. In addition, sudden changes of temperature cause a lowering of conductivity to water of roots and inhibition of ion transport [208]. This can indirectly affect the carbon economy of the whole plant by leading to more negative leaf water potentials, partial stomatal closure, and hence a slowing of net CO_2 assimilation.

Some roots, however, display considerable ability to acclimate to suboptimal temperatures. For example, in rye or barley acclimated by previous exposure to 8°C, net transport of K^+, Ca^{2+}, and H_2PO^{4-} into the xylem sap of detached roots was enhanced by factors of two to three as compared to controls maintained at 20°C [213,214], with a threefold increase in flux of water to the xylem. These changes almost completely compensated for the smaller root systems that developed at the lower temperature. In addition, low-temperature acclimated roots contained 9 to 15 times more soluble carbohydrates than the controls at 20°C [215]. The carbon efflux to the rhizosphere under such conditions, therefore, might be considerably enhanced if permeability to the carbohydrates was unaffected by temperature [208].

Freezing injury is another factor to be taken into account when considering the influence of low temperatures on rhizosphere dynamics. Freezing injury occurs when ice crystals penetrate the plasma membrane and cause mechanical rupture. This effect is irreversible, and upon thawing, leakage of solutes and metabolites from the cytosol is concomitant with cell death. The release of carbon compounds to the microbial community in the surrounding soil and rhizosphere of surviving roots is, therefore, likely to be considerable. In general, low temperature tends to slow root extension, rather than cell maturation; consequently, the endodermis frequently becomes suberized and thickened closer to the tip. This can lead to a block in apoplastic movement of water and solutes. The effects of suberization on the transport of carbohydrates in roots and leakage of carbon compounds from the roots to the rhizosphere have as yet received little attention.

2. Water Availability

Another major factor affecting microbial activity in soil is the availability of water, which is variable, depending on factors such as soil composition, rainfall, drainage, and plant cover. The primary importance of soil water is to provide the moisture necessary for the metabolic activities of both soil microorganisms and plants. Under conditions of water deficit, both plants and microbes may be harmed. When roots extend into dry soil, for example, the apical zone may develop a water deficit and lose turgor, so that cell expansion rates are slowed [216,217]. Osmotic adjustment takes place over a period of days during acclimation to water deficits that are gradually imposed, so that the roots and shoots can maintain low or negative water potentials and thereby maintain the flow of water from the rhizosphere to the root without the simultaneous loss of turgor [208]. In general, roots seem less sensitive to water deficit than shoots; however, growing cells in the root tip zone readily lose water to a drier environment, and — for example, in maize — irreversible damage to cells occurs when water loss exceeds 70% [218]. Desiccation causes changes to the physical and chemical properties of plant cell membranes. In mature roots of barley, for example, desiccation caused degeneration of the epidermis and outer cortex [219], and presumably resulting in the release of appreciable amounts of carbon compounds to the rhizosphere. In an earlier study, it was suggested that both C secreted as mucilage and C released by root tissue increased in zones of localized water stress, even where other parts of the root system had adequate water supply [220]. Soil water stress in tree species has been shown to both enhance and reduce root exudation [22]. Effects of water stress on root exudation are more fully discussed in Chapter 2.

3. Oxygen Deficit

Waterlogging can lead to the development of anaerobic conditions and oxygen deficit. In well-drained soils, air penetrates readily and oxygen concentrations are usually relatively high. In waterlogged soils, however, the only oxygen present is that dissolved in the soil water, and this is rapidly consumed by microbes or plant roots, leading to the development of anaerobic conditions. Under these conditions, roots and soil aerobes are in direct competition for oxygen; indeed, microorganisms have been shown to account for up to 50% of the O_2 consumption in soils densely rooted with arable crops [221]. When this oxygen has disappeared from soil pores, a combination of chemical and biological transformations takes place, resulting in the release to the soil solution of a sequence of reduced substances: NO_2^-, Mn^{2+}, Fe^{2+}, and H_2S as well as microbial end products such as acetic and butyric acid. Some of these can reach at concentrations in the rhizosphere that are damaging to roots or phytotoxic if accumulated in the leaves [222]. The rate of production of these compounds depends on the rate of supply of substrates for microbial activity and on the chemistry of the soil, including the presence of alternative electron acceptors such as NO_3^-, Fe^{3+}, Mn^{4+}, and SO_4^{2-}. However, lack of oxygen around roots alone is sufficient to induce injury [223]. Root cells quickly experience a decline in aerobic respiration if oxygen is not replenished with a concomitant decline in energy status [224,225]. Other physiological changes associated with oxygen deficiency include the inhibition of ion uptake and their transport to shoots [223], and low conductivity to water [226,227]. In addition, leakage of solutes to the soil environment can occur as a result of decreased energy status within the roots.

Oxygen availability appears to have both qualitative and quantitative effects on root exudation. Lack of oxygen has been shown to enhance root exudation [228]. Further, it has been reported, for example, that *Pisum sativum* exudes more amino nitrogen under low oxygen concentrations [229]. It has also been found that, under anaerobic conditions, *Zea mays* exudes ethanol at the expense of plant sugar content [230], and ethanol is a chemoattractant to plant pathogens. Other plant species (for example, peanut) have been shown to release more sugars under anaerobic conditions [231]. Decreasing oxygen concentrations tend to increase the permeability of the cell membrane, owing to a reduction in active transport, resulting in increased exudation [22].

4. pH and Availability of Nutrient Ions

Both pH and the availability of nutrient ions in soil play important roles in rhizosphere dynamics and are often dependent on one another. Nutrient ions move in soil toward plant roots either by mass flow with the soil water or by diffusion. Mass flow is the result of bulk convective movements of the soil solution toward roots, whereas diffusion occurs in response to a concentration gradient for a particular ion, which results from its absorption by the root and depletion from the surrounding soil. For typical agricultural soils, mass flow supplies Ca^{2+}, Mg^{2+}, Cl^-, SO_4^{2-}, and usually NO_3^-, but diffusion predominates for K^+ and $H_2PO_4^-$ because of low concentrations of these ions in free soil solution. Diffusion of K^+ and phosphate can also depend on the relative plant uptake rate, which is very high. As a consequence, their concentration is markedly reduced near the rhizoplane with formation of a steep concentration gradient, and this favors the diffusion of the two ions (discussed in Chapter 12). Where mass flow brings in ions at a rate much faster than absorption by the root surface, ions accumulate in the rhizosphere. For example, concentrations of Na^{2+} in saline soil can increase to injurious levels in the rhizosphere.

Pronounced changes in the pH of the rhizosphere by as much as one or two units often occur under agricultural conditions and especially in neutral soils where buffering capacity is least [232], due to the presence of both acid and alkaline cations. Such changes occur often because of the major influence of the form of nitrogen added on the cation or anion balance of plants. It has been demonstrated that plant uptake of nitrate occurs with the influx of two protons. This influx is also responsible for the increase of the pH value in the rhizosphere. With NH_4^+-based fertilizer, there is a marked solubilization of phosphate in calcareous soil because of the net release of H^+ by roots. In contrast, NO_3^- fertilizer leads to more alkaline conditions due to an efflux of OH^- and tends to release phosphate from the chemical forms in which it is held in an acid soil. Such differences in pH caused by nitrate or ammonium nutrition can bring about large changes in the microflora [233,234] and in the nature of microbial substrates released by the roots [38]. The rhizosphere of legumes fixing N_2 also becomes markedly more acidic. The response of root exudation to acidic conditions is important; for example, it was found that exudation from wheat was higher at pH 5.9 than at 6.4 [235]. It was postulated that the external pH altered the ionic states of compounds released and that this, in turn, affected their readsorption. The fact that external pH influences the state of released compounds (probably by modifying their charge) and affects their readsorption supports the suggestion of Jones and Darrah [19–21] that exudation is a balance reflecting both release and readsorption. Meharg and Killham [236] found that the amount of carbon lost from *Lolium perenne* increased from 12.3 to 30.6% with increasing pH from 4.3 to 6.0. It was suggested here that an increase in the microbial biomass combined with plant nitrogen limitation could have caused the observed increase in exudation with increasing pH. It has also been noted that there are often large differences in rhizosphere pH between plant species growing in the same soil. In addition, differences in pH of more than two units have been found at different points along the root of an individual plant [237]. In general, dicots tend to favor the uptake of cations over anions so that acidification of the rhizosphere is usually greater for monocots, which show a cation/anion uptake ratio close to unity.

Nutrient availability also plays a major role in exudation, with deficiencies in N, P, or K often increasing the rate of exudation [238]. It is believed that nutrient deficiency may trigger the release of substances such as organic acids or nonproteinogenic amino acids (phytosiderophores), which may enhance the acquisition of the limiting nutrient [239,240]. An example here might be the release of phenolic acids such as caffeic acid in response to iron deficiency, which results in an increase in uptake of the cation [241] (See also Chapter 2).

5. Light Intensity

Because most of the assimilated carbon in a plant is derived from photosynthesis, it follows that changes in light intensity may modify root exudation of carbon. However, Rovira [212] found that exudation of nitrogenous compounds was also affected by light intensity. In *Trifolium repens* grown under three different light intensities, the quantities of the amino acids serine, glutamic acid, and α-alanine released were considerably reduced at the lower light intensities. In *Lycopersicon esculentum,* quantities of aspartic acid, glutamic acid, phenylalanine, and leucine were significantly lower at the lower light intensities but increases in the amounts of serine and asparagine exuded were observed.

6. Carbon Dioxide Concentration

CO_2 concentrations have been rising steadily over the last 40 years and are expected to rise further though the magnitude of the increase is uncertain [242]. This could be expected to have important consequences for photosynthesis and hence exudation. However, Whipps [13] reported that the loss of assimilated carbon from *Zea mays* roots was unaffected by atmospheric CO_2 concentrations up to 1000 $\mu l \; l^{-1}$. In contrast, Norby et al. [243] found that carbon allocation to roots and root exudation increased in *Liriodendron tulipifera* grown in the presence of elevated CO_2 levels. In *Pinus echinata* seedlings, there was increased exudation under elevated CO_2 after 34 weeks but not after 41 weeks [244].

Rillig et al. [245] carried out Biolog® microplate analysis of soil C substrate utilization in the rhizosphere of *Gutierrezia sarothrae* grown in elevated atmospheric carbon dioxide. Compared to ambient CO_2 levels, they found polymers were more slowly oxidized by the microbial community, amides showed no change, and all other substrate groups were more rapidly utilized, although there was no difference in the number of viable bacteria. This change in microbial function in response to elevated carbon dioxide without any changes in total numbers of viable bacteria could have important impacts on nutrient cycling and may be due to changes in rhizodeposition under elevated CO_2. Li and Yagi [246] observed that C inputs by rice (*Oryza sativa*) grown under elevated CO_2 retarded the mineralization of organic matter in the 0 to 5 cm surface layer of paddy soil.

B. Structural Factors

Soil type and soil structure also influence the dynamics of rhizosphere microbial populations. Whether nutrients are available for bacteria in the rhizosphere often depends on the sites in the soil where nutrients are present. Organic compounds tightly bound to the soil matrix are often less available for bacteria [247], and those present in smaller pore spaces can be physically protected against mineralization. However, disturbance of the soil often causes these nutrients to become more available to soil microbes [248].

Soil textural aspects also influence bacterial survival, possibly by affecting the level of protection against predation by protozoa [96]. The presence of clay minerals such as montmorillonite or bentonite has been shown to substantially improve bacterial survival [96,249]. This appears to be true because some of the pore spaces in soil aggregates serve as protective microhabitats for soil bacteria against predation by protozoa [96], and it can thus be suggested that the greater the amount of protective pore space in heavier textured soils, such as clay soils, the greater the amount of protection against predation [250].

C. AGRICULTURAL FACTORS

1. Application of Mineral Fertilizers

There has been previous mention of the form of nitrogen fertilization and nutrient availability on rhizosphere pH and on the cation/anion balances in different plant species that affect both plant and microbial growth. However, the issue of nitrogen fertilization and its influence on root exudation and rhizosphere microbial populations is complex. Nitrogen fertilization has been shown to have variable effects on the ecology of the rhizosphere and may not always lead to predictable effects on microbial growth and activity in the rhizosphere. For example, Kolb and Martin [251] found that the application of N fertilizer stimulated root exudation in agricultural plants; and concluded that this may indirectly affect microbial growth in the rhizosphere. However, Liljeroth et al. [252] found that in wheat cultivars, stimulation of microbial growth seemed to be due to increased utilization of the root exudates themselves rather than to increased exudation rates. They also observed that N application resulted in higher bacterial abundance on seminal roots of young barley plants. In other cases, it is believed that N fertilizer may sometimes stimulate root growth at the expense of exudation, and cultivars may respond differently in terms of stimulated exudation or increased root growth [196]. On the contrary, Marschner et al. [57] found that both soil type and nitrogen fertilization affected plant growth, whereas N fertilization had no effect on bacterial community composition as determined by PCR-DGGE of 16S rDNA. In nonfertilized soil however, lower nutrient availability may limit microbial utilization of root-released carbon compounds. This is an area which clearly needs further investigation, especially under the current trend to move toward sustainable agriculture and decreasing chemical fertilizer inputs.

2. Effects of Cropping System

Alternative agricultural practices — including crop rotations, recycling of crop residues, and increased use of cover crops and green manures — contribute to high soil organic matter levels and improved soil quality, thus reducing the need for chemical fertilizers and pesticides. Agricultural management practices may be expected to have an impact on microbial diversity and thus affect soil health, crop health and yield, and ultimately sustainability. It has been hypothesized that a more diverse soil microbial community will result in greater yield stability [253].

Several studies have investigated the effects of different cropping systems and agricultural management practices on microbial biomass, activities, and diversity in soil [254–256]. However, comparatively few studies have looked at the impact of management practices on rhizosphere microbial populations and root exudation. Swinnen [257] studied carbon fluxes in the rhizospheres of barley and wheat under field conditions with conventional and integrated management using $^{14}CO_2$ pulse labeling. Compared with conventional management, integrated management leads to reduced transfer of carbon to the roots, reduced root growth, and lower total rhizodeposition.

A 15-year study in which a conventional corn–soybean rotation was compared with two low-input systems (animal manure or legumes as N-sources) was carried out by Buyer and Kaufman [258]. The effects of the three cropping systems on diversity of fast-growing aerobic culturable bacteria and fungi in the rhizosphere were studied. Following extraction and plating onto solid culture media, approximately 6000 bacteria were identified using fatty acid methyl ester analysis. Over 18,000 fungi were identified using microscopic examination of spores. However, total counts and diversity were not significantly different between the three different cropping systems.

D. BIOTIC FACTORS

In relation to nutrient cycling in the rhizosphere, another important factor to consider is the role of microfauna (protozoa, nematodes, and microarthropods), and specifically their interactions with bacterial populations. There is increasing evidence that the interactions between microflora and microfauna, especially nematodes and protozoa, are responsible for a significant portion of the mineralization of nitrogen in soil [259–264]. Extensive studies on the relationship between the soil protozoa and bacteria have revealed that the soil protozoa can feed on a wide range of bacteria, and that, indeed, bacteria are the most important food source for free-living heterotrophic protozoa [265]. Because predation by protozoa removes bacteria, it might have been expected that this would lead to a decrease in bacterial activity and consequently in the decomposition of organic matter and mineralization of nutrients. However, the opposite has been observed many times. A stimulating effect of a protozoan grazing on bacterial metabolism was demonstrated for ciliates in marine habitats and for flagellates in freshwater habitats. Hunt et al. [266] developed a simulation model for the effect of protozoan predation on bacteria in continuous culture. Results suggested that upon predation by protozoa, the growth rate of bacteria increased even though the bacterial biomass was reduced. Bacteria were thought to respond to a higher level of available carbon, nitrogen, and phosphorus upon predation.

Elliott et al. [267] presented data that suggested a significant role of soil protozoa at the soil–root interface by accelerating the mineralization of microbially immobilized nutrients. In the presence of protozoa, more mineral N was found in soil. In addition, plant shoot nitrogen concentration was higher as compared to plants grown in soils without protozoa. They hypothesised that the effect of protozoa on the mineralization of N would be greatest under the most N-limiting conditions, i.e., without the addition of mineral N fertilizer. However, it was only in the case where mineral N was added that protozoa accelerated the mineralization of microbially immobilized N.

Data have also been produced which indicate that bacteria can mineralize nitrogen from soil organic matter and that this process too can be increased by the presence of protozoa [259]. In the presence of protozoa, more nitrogen was made available to plants, and it was suggested that bacteria utilized the nitrogen from soil organic matter when supplied with a suitable energy source, i.e., root exudates. The immobilized nitrogen would then become available to plants when predators such as protozoa consumed these microorganisms and excreted excess ammonium [259]. Kuikman and Van Veen [260] investigated the impact of protozoa on the availability of this bacterially immobilized nitrogen to plants. They found that protozoa reduced bacterial numbers by a factor of eight and increased plant uptake of nitrogen by 20% in wheat. Grazing by protozoa was found to strongly stimulate the mineralization and turnover of bacterial N.

Grazing by bactivorous nematodes may also enhance the rate of N-mineralization in the rhizosphere [262, 263] by excretion of ammonium and other nitrogenous compounds [268] or indirectly by dissemination of microbial propagules through the soil [269–271] or stimulation of bacterial activity by release of growth-limiting nutrients and vitamins [272]. Furthermore, the removal of nonactive cells by microbial grazers may provide new surfaces for microbial colonization.

V. APPLICATION

Despite the fact that rhizosphere exudation patterns are governed by a large variety of factors (plant, microbial, and soil factors), it is clear that there is ample scope to modify root exudation patterns either by using agricultural practices or through plant-breeding programs. Both plant-breeding programs and agricultural practices have concentrated on reducing the negative impact of plant pathogens directly. Examples include breeding programs that render plants resistant to pathogen attack, rotation schemes that prevent excessive buildup of pathogens, or application of pesticides that protect susceptible plants against pathogenic organisms. In relation to plant nutrition, plant-breeding programs have concentrated on the development of plant varieties that give inherited high yields, provided that sufficient nutrients are available. The latter conditions are satisfied using chemical fertilizers that are directly available to plants. Even though these practices have resulted in a dramatic increase in food production during the past half century, it is questionable whether these approaches on their own are sustainable in the long term. Resistant plant varieties have limited usefulness, as plant resistance will be broken by the pathogen in the absence of measures that reduce pathogen populations. Similarly, excessive use of pesticides will inevitably lead to development of resistance in the pathogen population. Furthermore, lack of organic inputs will lead to a reduction of saprophytic biological activity in soil and, consequently, to a decline in the physical and chemical soil quality. This, in turn, leads to an increasing dependency on chemical inputs in the form of fertilizers and pesticides.

The realization that increased yield and reduced incidence of plant damage resulting from pathogens can also be achieved indirectly by programs aimed at encouraging beneficial organisms in soil and the rhizosphere opens new possibilities for plant breeding and soil management programs. Root exudates released by the plant create a rhizosphere effect, resulting in intense microbial activity in the vicinity of the roots. The influence of this microbial activity on plant health and plant nutrition depends on the net biological effect of the interactions between the rhizosphere populations, the plant, and the soil environment. With increasing knowledge of which factors influence this exudation process and their influence on rhizosphere microbial populations, it has become possible to manipulate these processes in favor of organisms that benefit the plant directly by providing biological control activity, growth stimulation, induction of resistance, or by mineralization of organic residues. Indirectly, the activity of soil microorganisms will result in improved soil quality (greater aggregate stability, improved soil structure, and better water-holding capacity), all of which benefit plant growth.

Examples of the successful stimulation of microbial populations antagonistic to pathogens such as take-all®, fusarium wilt, and cereal cyst nematodes (using continuous cropping regimes, green manures, and farmyard manures) have been well documented [273–275]. However, manipulation of rhizosphere populations using plant-breeding programs is at present not seriously pursued. The latter approach, when integrated with appropriate soil management strategies, might open new possibilities to release the full potential of microorganisms that benefit plants in a variety of ways.

A further application of the manipulation of microbial activity in the rhizosphere is their potential to remediate contaminated land. Bioremediation involves the use of microorganisms that break down contaminants. Radwan et al. [276] found that the soil associated with the roots of plants grown in soil heavily contaminated with oil in Kuwait was free of oil residues, presumably as a result of the ability of the resident rhizosphere microflora to degrade hydrocarbons. The use of plants as a means to either accumulate pollutants such as heavy metals [277] to degrade hydrocarbons and pesticides [276,278] is already widely implemented and has proven to be successful. In some cases, there is no doubt that it is the plant itself that is responsible for the removal of the contaminants. However, in most cases, it will be the interactions that take place between plant roots, the soil biota, and the soil environment that result in the desired effect. Again, more insights into the interactive processes associated with pollutant degradation will open opportunities to decontaminate land more effectively. For example, mycorrizal fungi seem to play a role in either protecting the plant against heavy metals or stimulating their uptake, but the mechanisms by which mycorrhizae affect metal uptake are far from clear, and the results are often contradictory [279]. The contribution of the rhizomicrobial population to the degradation of organic pollutants is better understood and is termed *rhizoremediation* [280,281]. Rhizoremediation can be seen as a solar driven biological pump and treatment system, where water is attracted to the root, accumulating water-soluble pollutants in the rhizosphere, where they are subsequently degraded by microorganisms [282]. Although the importance of the rhizosphere community for degradation of pollutants has been recognized, little is known on how effective rhizosphere communities are in degrading organic pollutants, and what contribution plant exudation plays in stimulating pollutant degradation. For example, Rentz et al. [283] found that phenanthrene degrading activity of *P. putida* ATCC 17484 was repressed after incubation with plant root extracts of a variety of plants. Catabolic repression as a result of the availability of alternative carbon sources, such as organic acids, carbohydrates, and amino acids, was held responsible for the apparent repression of phenantrene degradation by the bacterium. Despite this finding, many studies have shown that polycylic aromatic hydrocarbons (PAHs) are degraded more readily in vegetated soils compared to nonvegetated soils

[284–286]. The enhanced degradation of PAHs in rhizosphere soil [287] might be simply the result of the selective enrichment of PAH degrading populations in the rhizosphere [288,286] rather than induction of pollutant catabolic activity. The latter explanation was suggested by Hsu and Bartha [289], who proposed that specific root exudates could act as cometabolites for bacteria involved in the degradation of PAHs. For example, bacterial degradation of PAHs is initiated by monooxygenase or dioxygenase enzymes, which often exhibit broad substrate specificity [290]. Some aromatic root components are structural analogues of PAHs and are thought to induce the production of oxygenases involved in PAH degradation. Such exudates include l-carvone, which is produced by spearmint [291], salicylatyes that are associated with willow [292], flavones, xanthones, and other phenolic compounds associated with osage orange [293], and hydroxycinnamic acids that are produced by poplar trees [294]. The importance of such components in rhizoremediation was demonstrated by Gilbert and Crowley [291], who showed that l-carvole could induce polychlorinated biphenyl (PCB) degradation in *Arthrobacter* strain B1B. Interestingly, l-carvol is not used as a growth substrate by *Arthrobacter*.

Besides growth stimulation and induction of specific metabolic activity, some root exudates could play an important role in making PAHs more bioavailable by acting as surfactants or transporters. Crowley et al. [295] showed that degradation of PAHs was significantly enhanced in the rhizosphere of celery, and it was suggested that the production of linoleic acids by the roots led to enhanced bioavailability and subsequent degradation of PAHs in the rhizosphere. Therefore, rhizoremediation is likely to be strongly dependant on plants species, which might explain why some studies fail to show enhanced degradation of PAHs in the rhizosphere of plants [296,297].

In order to select a suitable plant for rhizoremediation a number of factors need, therefore, to be taken into account regarding root exudation. First, the exudates released into the rhizosphere should support a large microbial population [298]. Second, the exudates should contain substances that induce the production of catabolic enzymes [299]. Third, the rhizosphere should contain substrates that lead to the cometabolism of high-molecular-weight PAHs [298]. Finally, components in the exudates should improve bioavailability of pollutants such as PAHs [299]. Whereas conceptually these prerequisites for effective rhizoremediation seem sensible, we are still a long way off in proving the validity of these concepts in complex natural systems.

A critical aspect of rhizoremediation is that microorganisms need to form a very close relationship with the root (rhizosphere competence) if they are to be effective. An illustration is the use of some *Trichoderma* strains, which are very good rhizosphere competitors, in part by producing antibiotics and lytic enzymes to suppress other organisms. They can then act in concert with a wide range of plants, including trees, to catabolize a range of organics and cyanides and promote the uptake of nitrates and toxic metals [300]. This type of process has been termed *phytobial remediation*, which would appear to offer very good prospects for brownfield land remediation [301].

VI. FUTURE DIRECTIONS

There are many triggers that have driven rhizodeposition studies. Initial concerns were focused on plant nutrition. This was followed by crop protection aspects, and most recently the use of the rhizosphere in facilitating remediation of soils. However, perhaps the major environmental issue of today is climate change and the future of carbon in the environment. Rhizodeposition is the product of carbon sequestration, a key in some degree of mitigation of carbon emissions. Some of that sequestered into the asymbiotic and symbiotic root associations can make a major contribution to the soil carbon pool. Analysis of the quantities and function of this pool, such as described in Section IV.A.6 of this chapter, has never been more critical. There will be surprises in store! For example a recent report in *Nature* [302] indicates that tree roots produce the greenhouse gas methane by an abiotic process that has yet to be chemically and biochemically described. Such reports, of course, need to be thoroughly investigated but it seems very likely that rhizodeposition studies in relation to climate change will be a major research area for the future. This will need fully interdisciplinary approaches but from the soil biology angle, molecular techniques [303] will be increasingly useful.

ACKNOWLEDGMENTS

This new edition of the chapter was produced by J.M. Lynch and F.A.A.M. De Leij with the very capable and excellent secretarial support of Sue Jones. Any correspondence should be addressed to Professor Lynch at Forest Research.

REFERENCES

1. Lynch, J.M., The rhizosphere — form and function, *Appl. Soil Ecol.*, 1, 193, 1984.
2. Lynch, J.M., Introduction: some consequences of microbial rhizosphere competence for plant and soil, in *The Rhizosphere*, Lynch, J.M., Ed., John Wiley and Sons, Chichester, U.K., 1990, p. 1.
3. Rovira, A.D., Foster, R.C., and Martin, J.K., Note on terminology: origin, nature and nomenclature of organic materials in the rhizosphere, in *The Soil-Root Interface,* Harley, J.L. and Scott-Russell, R., Eds., Academic Press, London, 1979, p. 1.
4. Cheng, W. et al., *In situ* measurement of root respiration and soluble C concentrations in the rhizosphere, *Soil Biol. Biochem.*, 25, 1189, 1993.
5. Uren, N.C. and Reisenauer, H.M., The role of root exudates in nutrient acquisition, *Advances in Plant Nutrition,* Tinker, P.B. and Lauchli, A., Eds., Praeger, New York, 1988, p. 79.
6. Rovira, A.D., Plant root exudates, *Bot. Rev.,* 35, 35, 1969.
7. Meharg, A.A., A critical review of labelling techniques used to quantify rhizosphere carbon flow, *Plant Soil,* 166, 55, 1994.
8. Newman, E.I., The rhizosphere: carbon sources and microbial populations, in *Ecological Interactions in Soil,* Fitter, A.H., Ed., Blackwell Scientific Publications, Oxford, 1985, p. 107.
9. Barber, D.A. and Lynch, J.M., Microbial growth in the rhizosphere, *Soil Biol. Biochem.*, 9, 305, 1977.
10. Merckx, R. et al., Plant induced changes in the rhizosphere of maize and wheat, *Plant Soil*, 96, 85, 1986.
11. Foster, R.C., Microenvironments of soil microorganisms, *Biol. Fertil. Soils,* 6, 189, 1988.
12. Lussenhop, J. and Fogel, R., Soil invertebrates are concentrated on roots, in *The Rhizosphere and Plant Growth,* Keiser, D.L. and Cregan, P.B., Eds., Kluwer Academic Publishers, Boston, MA, 1991, p. 111.
13. Whipps, J.M., Carbon economy, in *The Rhizosphere,* Lynch, J.M., Ed., John Wiley and Sons, Chichester, U.K., 1990, p. 59.
14. Swinnen, J., Rhizodeposition and turnover of root-derived organic material under conventional and integrated management, *Agric. Ecosyst. Environ.*, 51, 115, 1994.
15. Swinnen, J., Van Veen, J.A., and Merckx, R., Root decay and turnover of rhizodeposits in field-grown winter wheat and spring barley estimated by ^{14}C pulse labelling, *Soil Biol. Biochem,* 27, 211, 1995.
16. Keith, H., Oades, J.M., and Martin, J.K., Input of carbon to soil from wheat plants, *Soil Biol. Biochem,* 18, 455, 1986.
17. Jensen, B., Rhizodeposition by ^{14}C pulse labelled spring barley grown in small field plots on sandy loam, *Soil Biol. Biochem,* 25, 1553, 1993.
18. Lynch, J.M. and Whipps, J.M., Substrate flow in the rhizosphere, *Plant Soil*, 129, 1, 1990.
19. Jones, D.L. and Darrah, P.R., Re-sorption of organic compounds by roots of *Zea mays* L. and its consequences in the rhizosphere I. Re-sorption of ^{14}C labelled glucose, mannose and citric acid, *Plant Soil,* 143, 259, 1992.
20. Jones, D.L. and Darrah, P.R., Influx and efflux of amino acids from *Zea mays* L. roots and its implications in the rhizosphere, *Plant Soil*, 163, 1, 1994.
21. Jones, D.L. and Darrah, P.R., Re-sorption of organic compounds by roots of *Zea mays* L. and its consequences in the rhizosphere III. Characteristics of sugar influx and efflux, *Plant Soil*, 178, 153, 1996.
22. Grayston, S.J., Vaughan, D., and Jones, D., Rhizosphere carbon flow in trees, in comparison with annual plants: the importance of root exudation and its impact on microbial activity and nutrient availability, *Appl. Soil Ecol.*, 5, 29, 1996.

23. Brophy, L.S. and Heichel, G.H., Nitrogen release from root of alfalfa and soybean grown in sand culture, *Plant Soil*, 116, 77, 1989.
24. Wacquant, J.P., Ouknider, N., and Jacquard, P., Evidence for a periodic excretion of nitrogen by roots of grass — legume associations, *Plant Soil*, 116, 57, 1989.
25. Rovira, A.D., Plant root excretions in relation to the rhizosphere effect I: the nature of root exudates from oats and peas, *Plant Soil*, 7, 178, 1956.
26. Boulter, A.D., Jeremy, J.J., and Wilding, M., Amino acids liberated into the culture medium by pea seedling roots, *Plant Soil*, 24, 121, 1966.
27. Hale, M.G., Moore, L.D., and Griffin, G.J., Root exudates and exudation, *Interactions between Non-Pathogenic Soil Microorganisms and Plants,* Dommergues, Y.R. and Krupa, S.V., Eds., Elsevier, Amsterdam, 1978, p. 163.
28. Janzen, H.H. and Bruinsma, Y., Methodology for the quantification of root and rhizosphere nitrogen dynamics by exposure of shoots to ^{15}N labelled ammonia, *Soil Biol. Biochem*, 21, 189, 1989.
29. Jensen, E.S., Rhizodeposition of N by pea and barley and its effect on soil N dynamics, *Soil Biol. Biochem*, 28, 65, 1996.
30. Virtanen, A.L., Von Hausen, S., and Laine, T., Investigations on the root nodule bacteria of leguminous plants XIX. Influence of various factors on the excretion of nitrogenous compounds from nodules, *J. Agric. Sci.*, 27, 332, 1937.
31. Wilson, P.W. and Wyss, O., Mixed cropping and the excretion of nitrogen by leguminous plants, *Soil Sci. Soc. Am. Proc.,* 11, 289, 1937.
32. Klemedtsson, L. et al., Microbial transformations in the root environment of barley, *Soil Biol. Biochem*, 19, 551, 1987.
33. Merckx, R. et al., Production of root-derived material and associated microbial growth in soil at different nutrient levels, *Biol. Fertil. Soils*, 5, 126, 1987.
34. Robinson, D. et al., Root-induced nitrogen mineralisation: a theoretical analysis, *Plant Soil*, 117, 185, 1989.
35. Nehl, D.B., Allen, S.J., and Brown, J.F., Deleterious rhizosphere bacteria: an integrating perspective, *Appl. Soil Ecol.*, 5, 1, 1997.
36. Hedges, R.W. and Messens, E., Genetic aspects of rhizosphere interactions, in *The Rhizosphere,* Lynch, J.M., Ed., John Wiley and Sons, Chichester, U.K., 1990, p. 129.
37. Bolton, H.J., Fredrickson, J.K., and Elliott, L.F., Microbial ecology of the rhizosphere, in *Soil Microbial Ecology,* Metting, F.B.J., Ed., Marcel Dekker, New York, 1993, p. 27.
38. Bowen, G.D. and Rovira, A.D., The rhizosphere: The hidden half of the hidden half, *Plant Roots, The Hidden Half,* Waisel, Y., Eshel, A., and Kafkafi, U., Eds., Marcel Dekker, New York, 1991, p. 641.
39. Rovira, A.D. and Davey, C.B., Biology of the rhizosphere, in *The Plant Root and Its Environment,* Carson, E.W., Ed., University Press of West Virginia, Charlottesville, VA, 1974, p. 153.
40. Pirt, S.J., *Principles of Microbe and Cell Cultivation,* Blackwell, Oxford, 1975.
41. Postgate, J.R., New advances and future potential in biological nitrogen fixation, *J. Appl. Bacteriol.,* 37, 185, 1974.
42. Dobereiner, J., Day, J.M., and Dart, P.J., Nitrogenase activity and oxygen sensitivity of the *Paspalum notatum-Azotobacter paspali* association, *J. Gen. Microbiol.*, 17, 103, 1972.
43. Patel, J.J., Microorganisms in the rhizosphere of plants inoculated with *Azotobacter chroococcum, Plant Soil,* 31, 209, 1969.
44. Barber, D.A. and Martin, J.K., The release of organic substances by cereal roots into the soil, *New Phytologist*, 76, 69, 1976.
45. Brimecombe, M.J., De Leij, F.A.A.M., and Lynch, J.M., Effect of genetically modified *Pseudomonas fluorescens* strains on the uptake of nitrogen by pea from ^{15}N enriched organic residues, *Lett. Appl. Microbiol.*, 26, 155, 1998.
46. Brimecombe, M.J., De Leij, F.A.A.M., and Lynch, J.M., Effect of introduced *Pseudomonas fluorescens* strains on the uptake of nitrogen by wheat from ^{15}N-enriched organic residues, *World J. Microbiol. Biotechnol.*, 15, 417, 1999.
47. Brimecombe, M.J., De Leij, F.A.A.M., and Lynch, J.M., Effect of introduced *Pseudomonas fluorescens* strains on soil nematode and protozoan populations in the rhizosphere of pea and wheat, *Microb. Ecol.*, 38, 387, 2000.

48. Brimecombe, M.J., De Leij, F.A.A.M., and Lynch, J.M., Use of nematode community structure as a senstive indicator of microbial perturbations induced by genetically modified *Pseudomonas fluorescens* strains, *Biol. Fertil. Soils*, 34, 270, 2001.

49. Treonis, A.M. et al., Identification of groups of metabolically-active rhizosphere microorganisms by stable isotope probing of PLFAs, *Soil Biol. Biochem.*, 36, 533, 2004.

50. Harris, W.F., Santantonio, D., and McGinty, D., The dynamic below-ground ecosystem, in *Forests: Fresh Perspectives From Ecosystem Analysis*, Waring, R.H., Ed., Oregon State University Press, Corvallis, OR, 1980, p. 119.

51. Vancura, V. and Hanzlikova, A., Root exudates of plants IV. Differences in chemical composition of seed and seedlings exudates, *Plant Soil*, 36, 271, 1972.

52. Curl, E.A. and Truelove, B., *The Rhizosphere*, Springer, New York, 1985.

53. Cieslinski, G., Van Rees, K.C.J., and Huang, P.M., Low molecular weight organic acids released from roots of durum wheat and flax into sterile nutrient solutions, *J. Plant Nutr.*, 20, 753, 1997.

54. Chan, E.C.S., Katznelson, H., and Rouatt, J.W., The influence of soil and root extracts on the associative growth of selected soil bacteria, *Can. J. Microbiol.*, 9, 187, 1962.

55. Rovira, A.D., Interactions between plant roots and soil microorganisms, *Annu. Rev. Microbiol.*, 19, 241, 1965.

56. Christensen-Weniger, C., Groneman, A.F., and Van Veen, J.A., Associative N_2 fixation and root exudation of organic acids from wheat cultivars of different aluminium tolerance, *Plant Soil*, 139, 167, 1992.

57. Marschner, P. et al., Soil plant specific effects on bacterial community composition in the rhizosphere, *Soil Biol. Biochem.*, 33, 1437, 2001.

58. Marschner, P., Grierson, P., and Rengel, Z., Microbial community composition and functioning in the rhizosphere of three species of *Banksia* species in native woodland in western Australia, *Appl. Soil Ecol.*, 28, 191, 2005.

59. Juo, P. and Stotzky, G., Electrophoretic separation of proteins from roots and root exudates, *Can. J. Bot.*, 48, 713, 1970.

60. Hamlen, R.A., Lukezic, F.L., and Bloom, J.R., Influence of age and stage of development on the neutral carbohydrate components in root exudates from alfalfa plants grown in a gnotobiotic environment, *Can. J. Plant Sci.*, 52, 633, 1972.

61. Liljeroth, E. and Bååth, E., Bacteria and fungi on roots of different barley varieties (*Hordeum vulgare* L.), *Biol. Fertil. Soils*, 7, 53, 1988.

62. Martin, J.K., Influence of plant species and plant age on the rhizosphere microflora, *Aust. J. Biol. Sci.*, 24, 1143, 1971; Schonfeld, J., Costa, R., Odnca-Hagler, L. et al., Bacterial diversity of the rhizosphere of maize (*Zea mays*) grown in tropical soil studied by temperature gradient gel eclectrophoresis, *Plant Soil*, 233, 167, 2001.

63. Gomes, N.C.H., Heuer, H., Schonfeld, J., Costa, R., Odnca-Hagler, L., and Smalla, K., Bacterial diversity of the rhizosphere of maize (*Zea mays*) grown in tropical soil studied by temperature gradient gel electrophoresis, *Plant Soil*, 233, 167, 2001.

64. Prikryl, Z. and Vancura, V., Root exudates of plants VI. Wheat root exudation as dependant on growth, concentration gradient of exudates and the presence of bacteria, *Plant Soil*, 57, 69 1980.

65. Prat, S. and Retovsky, R., Root excretion in nutrient solution, *Vestn. Kral. Ces spol. Nauk.*, 1–19, 1944.

66. Frenzel, B., Zur Atiologie der Anreicherung von Animosarum und Amiden im Wurzelraum von Helianthus annus, *Planta*, 55, 169, 1960.

67. Schonwitz, R. and Zeigler, H., Exudation of water-soluble vitamins and of some carbohydrates by intact roots of maize seedlings (*Zea mays* L.) into a mineral nutrient solution, *Z. Pflanzenphysiol.*, 107, 7, 1982.

68. Gardner, W.K. et al., The acquisition of phosphorus by *Lupinus albus* L.V. The diffusion of exudates away from roots: a computer simulation, *Plant Soil*, 72, 13, 1983.

69. Rovira, A.D., Bowen, G.D., and Foster, R.C., The significance of rhizosphere microflora and mycorrhizas on plant nutrition, in *Encyclopaedia of Plant Nutrition,* Lauchii, A. and Bielski, R.L., Eds., Springer-Verlag, Berlin, 1983, p. 61.

70. Meharg, A.A. and Killham, K., A novel method of determining root exudates in the presence of soil microflora, *Plant Soil*, 133, 111, 1991.

71. Meharg, A.A. and Killham, K., Loss of exudates from the roots of perennial ryegrass inoculated with a range of microorganisms, *Plant Soil*, 170, 345, 1995.

72. Lee, K.J. and Gaskins, M.H., Increased root exudation of ^{14}C-compounds by sorghum seedlings inoculated with nitrogen-fixing bacteria, *Plant Soil*, 69, 391, 1982.

73. Heulin, T., Gukert, A., and Balandreau, J., Stimulation of root exudation of rice seedlings by *Azospirillum* strains: carbon budget under gnotobiotic conditions, *Biol. Fertil. Soils*, 4, 9, 1987.

74. Snellgrove, R.C. et al., The distribution of carbon and the demand of the fungal symbiont in leek plants with vesicular-arbuscular mycorrhizas, *New Phytologist*, 69, 75, 1982.

75. Schwab, S.M., Leonard, R.T., and Menge, J.A., Quantitative and qualitative composition of root exudates of mycorrhizal and non-mycorrhizal plant species, *Can. J. Bot.*, 62, 1227, 1984.

76. Burr, T.J. and Caesar, A., Beneficial plant bacteria, *Crit. Rev. Plant Sci.*, 2, 1, 1984.

77. Gaskins, M.H., Rhizosphere bacteria and their use to increase plant productivity: a review, *Agric. Ecosyst. Environ.*, 12, 99, 1985.

78. Davison, J., Plant beneficial bacteria, *Biotechnology*, 6, 282, 1988.

79. Lynch, J.M., Beneficial interactions between microorganisms and roots, *Biotechnol. Advances*, 8, 335, 1990.

80. Lugtenberg, B.J.J., de Weger, L.A., and Bennett, J.W., Microbial stimulation of plant growth and protection from disease, *Curr. Opin. Biotechnol.*, 2, 457, 1991.

81. Lazarovits, G. and Nowak, J., Rhizobacteria for improvement of plant growth and establishment, *Hortic. Sci.*, 32, 188, 1997.

82. Azcon, R. and Ocampo, J.A., Factors effecting the vesicular-arbuscular infection and mycorrhizal dependency of thirteen wheat cultivars, *New Phytologist*, 87, 677, 1981.

83. Schroth, M.N. and Hildebrand, D.C., Influence of plant exudates on root-infecting fungi, *Annu. Rev. Phytopathol.*, 2, 101, 1964.

84. Hawes, M.C., Living plant cells released from the root cap: a regulator of microbial populations in the rhizosphere, *Plant Soil*, 129, 19, 1990.

85. Hozore, E. and Alexander, M., Bacterial characteristics important to rhizosphere competence, *Soil Biol. Biochem.*, 23, 717, 1991.

86. Dharmatilake, A.J. and Baker, W.D., Chemotaxis of *Rhizobium meliloti* towards nodulation gene inducing compounds from alfalfa roots, *Appl. Environ. Microbiol.*, 58, 1153, 1992.

87. Ashby, A.M., Watson, M.D., and Shaw, C.H., A Ti-plasmid determined function is responsible for chemotaxis of *Agrobacterium tumefaciens* towards the plant wound product acetosyringone, *FEMS Microbiol. Lett.*, 41, 189, 1987.

88. Lopez-de-Victoria, G. and Lovell, C.R., Chemotaxis of *Azospirillum* species to aromatic compounds, *Appl. Environ. Microbiol.*, 59, 2951, 1993.

89. Nikata, T. et al., Rapid method for determining bacterial-behavioural responses to chemical stimulii, *Appl. Environ. Microbiol.*, 58, 2250, 1992.

90. Bashan, Y. and Holguin, G., Root-to-root travel of the beneficial bacterium *Azospirillum brasilense*, *Appl. Environ. Microbiol.*, 60, 2120, 1994.

91. Bashan, Y. and Holguin, G., *Azospirillum*-plant relationships: environmental and physiological advances (1990–1996), *Can. J. Microbiol.*, 43, 103, 1997.

92. Vesper, S.J., Production of pili (fimbriae) by *Pseudomonmas fluorescens* and correlation with attachment to corn roots, *Appl. Environ. Microbiol.*, 53, 1397, 1987.

93. Anderson, A.J., Isolation from root and shoot surfaces of agglutinins that show specificity for saprophytic pseudomonads, *Can. J. Bot.*, 61, 3438, 1983.

94. Anderson, A.J., Habibzadegah-Tari, P., and Tepper, C.S., Molecular studies on the role of a root surface agglutinin in adherence and colonisation by *Pseudomonas putida*, *Appl. Environ. Microbiol.*, 54, 375, 1987.

95. Achouak, W. et al., Root colonisation by symplasmata-forming *Enterobacter agglomerans*, *FEMS Microbiol. Ecol.*, 13, 287, 1994.

96. Van Overbeek, L. and Van Elsas, J.D., Adaptation of bacteria to soil conditions: application of molecular physiology in soil microbiology, in *Modern Soil Microbiology*, Van Elsas, J.D., Trevors, J.T., and Wellington, E.M., Eds., Marcel Dekker, New York, 1997, p. 441.

97. Redmond, J.W. et al., Flavones induce expression of nodulation genes in *Rhizobium*, *Nature*, 323, 632, 1986.

98. Price, N.P.J. and Carlson, R.W., Rhizobial lipo-oligosaccharide nodulation factors: multidimensional chromatographic analysis of symbiotic signals involved in the development of legume root nodules, *Glycobiology*, 5, 233, 1995.

99. Okon, Y., *Azospirillum* as a potential inoculant for agriculture, *Trends Biotechnol.*, 3, 223, 1985.

100. Bashan, Y., Singh, M., and Levanony, H., Contribution of *Azospirillum brasilense* Cd to growth of tomato seedlings is not through nitrogen fixation, *Can. J. Bot.*, 67, 1317, 1989.
101. Elmerich, C., Molecular biology and ecology of diazotrophs associated with non-leguminous plants, *Biotechnology*, 2, 967, 1984.
102. Bonfante, P. and Perotto, S., Strategies of arbuscular mycorrhizal fungi when infecting host plants, *New Phytologist*, 130, 3, 1995.
103. Varma, A. and Hock, B., *Mycorrhiza*, Springer-Verlag, Berlin, 1995.
104. Li, X.L., George, E., and Marschner, H., Extension of the phosphorus depletion zone in VA-mycorrhizal white clover in calcareous soil, *Plant Soil*, 131, 41, 1991.
105. Li, X.L., George, E., and Marschner, H., Acquisition of phosphorus and copper by VA-mycorrhizal hyphae and root-to-shoot transport in white clover, *Plant Soil*, 135, 49, 1991.
106. Bolan, N.S., Robson, A.D., and Barrow, N.J., Effects of vesicular-arbuscular mycorrhiza on the availability of iron phosphates to plants, *Plant Soil*, 99, 401, 1987.
107. Tinker, P.B., The chemistry of phosphorus and mycorrhizal effects on plant growth, in *Endomycorrhizas*, Saunders, F.S., Mosse, B., and Tinker, P.B., Eds., Academic Press, London, 1975, p. 353.
108. Li, X.L., George, E., and Marschner, H., Phosphorus depletion and pH decrease at the root-soil and hyphae-soil interfaces of VA-mycorrhizal white clover fertilised with ammonium, *New Phytologist*, 119, 397, 1991.
109. Smith, S.E. et al., Activity of glutamine synthetase and glutamate dehydrogenase in *Trifolium subterraneum* and *Allium cepa* L., effects of mycorrhizal infection and phosphorus nutrition, *New Phytologist*, 99, 211, 1985.
110. Bolan, N.S., Hedley, M.J., and White, R.E., Processes of soil acidification during nitrogen cycling with emphasis on legume based pastures, *Plant Soil*, 134, 53, 1991.
111. Bolan, N.S., A critical review on the role of mycorrhizal fungi in the uptake of phosphorus by plants, *Plant Soil*, 134, 189, 1991.
112. Rygiewicz, P.T., Blodsoe, C.S., and Zasoski, R.J., Effects of mycorrhizae and solution pH on (^{15}N) ammonium uptake by coniferous seedlings, *Can. J. For. Res.*, 14, 885, 1984.
113. Vaast, P.H. and Zasoski, R.J., Effect of VA-mycorrhizae and nitrogen sources on rhizosphere soil characteristics, growth and nutrient acquisition of coffee seedlings (*Coffea arabica* L.), *Plant Soil*, 147, 31, 1993.
114. Rigou, L. et al., Influence of ectomycorrhizal infection on the rhizosphere pH around roots of maritime pine (*Pinus pinaster* Soland in Ait.), *New Phytologist*, 130, 141, 1995.
115. Gahoonia, T.S., Claassen, N., and Jungk, A., Mobilisation of phosphate in different soils by ryegrass supplied with ammonium or nitrate, *Plant Soil*, 140, 241, 1992.
116. Hoffmann, C. et al., Phosphorus uptake of maize as effected by ammonium or nitrate nitrogen measurements and model calculations, *Z. Pflanzenernahr. Bodenk.*, 157, 225, 1994.
117. Ortas, I., Harris, P J., and Rowell, D.L., Enhanced uptake of phosphorus by mycorrhizal sorghum plants as influenced by forms of nitrogen, *Plant Soil*, 184, 255, 1996.
118. Guo, Y., George, E., and Marschner, H., Contribution of an arbuscular mycorrhizal fungus to the uptake of cadmium and nickel in bean and maize plants, *Plant Soil*, 184, 195, 1996.
119. Hokkanen, H.M.T. and Lynch, J.M., Eds., *Biological Control: Benefits and Risks*, Cambridge University Press, Cambridge, 1995.
120. Kerr, A., Commercial release of a genetically engineered bacterium for the control of crown gall, *Agric. Sci.*, 41, November 1989.
121. Harrison, L.A. et al., Purification of an antibiotic effective against *Gaeumannomyces graminis* var. *tritici* produced by a biocontrol agent, *Pseudomonas aureofaciens*, *Soil Biol. Biochem.*, 25, 215, 1993.
122. Thomashow, L.S. and Weller, D.M., Role of phenazine antibiotic from *Pseudomonas fluorescens* in biological control of *Gaeumannomyces graminis* var. *tritici*, *J. Bacteriol.*, 170, 3499, 1988.
123. Shanahan, P. et al., Isolation of 2,4-diacetylphloroglucinol and investigation of physiological parameters influencing its production, *Appl. Environ. Microbiol.*, 58, 353, 1992.
124. Keel, C. et al., Suppression of root diseases by *Pseudomonas fluorescens* CHA0: importance of the bacterial secondary metabolite 2,4-diacetylphloroglucinol, *Mol. Plant-Microbe Interact.*, 5, 4, 1992.
125. Howell, C.R. and Stipanovic, R.D., Control of *Rhizoctonia solani* on cotton seedlings with *Pseudomonas fluorescens* and with an antibiotic produced by the bacterium, *Phytopathol.*, 69, 480, 1979.
126. Howell, C.R., and Stipanovic, R.D., Suppression of *Pythium ultimum*-induced damping-off of cotton seedlings by *Pseudomonas fluorescens* and its antibiotic pyoluteorin, *Phytopathology*, 70, 712, 1980.

127. Nowak-Thompson, B. et al., Production of 2,4-diacetylphoroglucinol by the biocontrol agent *Pseudomonas fluorescens* Pf-5, *Can. J. Microbiol.,* 40, 1064, 1994.
128. Bonsall, R.F., Weller, D.M., and Thomashow, L.S., Quantification of 2,4-diacetylphloroglucinol produced by fluorescent *Pseudomonas* spp. *in vitro* and in the rhizosphere of wheat, *Appl. Environ. Microbiol.,* 63, 951, 1997.
129. Chin-A-Woeng, T.F.C., Bloemberg, G.V., and van der Bij, A.J., Biocontrol by phenazine-l-carboxamide producing *Pseudomonas chloraphis* PCL 1391 of tomato root rot caused *by Fusarium oxysporum* f. sp. *Radicis-lycopersici, Mol. Plant-Microbe Interact.,* 11, 1069, 1998.
130. Nowak-Thompson, B. et al., Characterisation of the pyoluteorin biosynthetic gene cluster of *Pseudomonas fluorescens* Pf-5, *J. Bacteriol.,* 181, 2166, 1999.
131. Thomashow, L.S. et al., Production of the antibiotic phenazine-l-carboxylic acid by fluorescent *Pseudomonas* species in the rhizosphere of wheat, *Appl. Environ. Microbiol.,* 56, 908, 1990.
132. Fenton, A.M. et al., Exploitation of gene(s) involved in 2,4-diacetylphloroglucinol biosynthesis to confer a new biocontrol capability to a *Pseudomonas* strain, *Appl. Environ. Microbiol.,* 58, 3873, 1992.
133. Raaijmakers, J.M., Bonsall, R.F., and Weller, D.M., Effect of population density of *Pseudomonas fluorescens* on production of 2,4-diacetylphloroglucinol in the rhizosphere of wheat, *Phytopathology,* 89, 470, 1999.
134. Bainton, N.J. et al., Survival and ecological fitness of *Pseudomonas flourescens* genetically engineered with dual biocontrol mechanisms, *Microb. Ecol.,* 48, 349, 2004.
135. De Leij, F.A.A.M., Dixon-Hardy, J.E., and Lynch, J.M., Effect of 2,4-diacylphloroglucinol producing and non-producing stains of *Pseudomonas fluorescens* on root development of pea seedling in three different soil types and its effect on nodulation by *Rhizobium leguminosarium, Biol. Fertil. Soils,* 35, 114, 2002.
136. Défago, G., Keel, C., and Hass, D., Pseudomonads as biocontrol agents of diseases caused by soil-borne pathogens, in *Biological Control: Benefits and Risks,* Hokkanen, H.M.T. and Lynch, J.M., Eds., Cambridge University Press, Cambridge, 1995, p. 137.
137. Rovira, A., Ryder, M., and Harris, A., Biological control of root diseases with pseudomonads, in *Biological Control of Plant Diseases,* Tjamos, E.S., Ed., Plenum Press, New York, 1992, p. 175.
138. Kloepper, J.W. et al., *Pseudomonas* siderophores: a mechanism explaining disease-suppressive soils, *Curr. Microbiol.,* 4, 317, 1980.
139. Teintze, M. et al., Structure of ferric pseudobactin, a siderophore from a plant growth promoting *Pseudomonas, Biochemistry,* 20, 422, 1981.
140. Geels, R.P. and Schippers, B., Selection of antagonistic fluorescent *Pseudomonad* spp and their root colonisation and persistence following treatment of seed potatoes, *Phytopathology,* 108, 193, 1983.
141. Xu, G.W. and Gross, D.C., Selection of fluorescent pseudomonads antagonistic to *Erwinia carotovota* and suppressive of potato seed piece decay, *Phytopathology,* 76, 414, 1986.
142. Buyer, J.S. and Leong, J., Iron transport mediated antagonism between plant growth-promoting and plant-deleterious *Pseudomonas* strains, *J. Biol. Chem.,* 261, 791, 1986.
143. Vandenbergh, P.A. et al., Iron-chelating compounds produced by soil pseudomonads, correlation with fungal growth inhibition, *Appl. Environ. Microbiol.,* 46, 128, 1983.
144. O'Sullivan, D.J. and O'Gara, F., Traits of fluorescent *Pseudomonas* spp. Involved in suppression of plant root pathogens, *Microbiol. Rev.,* 56, 662, 1992.
145. Loper, J.E. and Henkels, M.D., Utilisation of heterologous siderophores enhances levels of iron available to *Pseudomonas putida* in the rhizosphere, *Appl. Environ. Microbiol.,* 65, 5357, 1999.
146. Loper, J.E. and Buyer, J.S., Siderophores in microbial interactions on plant surfaces, *Mol. Plant-Microbe Interact.,* 4, 5, 1991.
147. Duijf, B.J. et al., Siderophore-mediated competition for iron and induced resistance in the suppression of *Fusarium* wilt of carnation by fluorescent *Pseudomonas* spp., *Neth. J. Plant Pathol.,* 99, 277, 1993.
148. Ogena, M. et al., Protection of cucumber against *Pythium* root rot by fluorescent pseudomonads: predominant role of induced resistance over siderophores and antibiotics, *Plant Pathol.,* 48, 66, 1999.
149. Metraux, J-P. et al., Increase in salicylic acid at the onset of systemic acquired resistance in cucumber, *Science,* 250, 1004, 1990.
150. Leeman, M. et al., Iron availability affects induction of systemic resistance to *Fusarium* wilt of radish by *Pseudomonas fluorescens, Phytopathology,* 86, 149, 1996.
151. Voisard, C. et al., Cyanide production by *Pseudomonas fluorescens* helps suppress black root rot of tobacco under gnotobiotic conditions, *EMBO,* 8, 351, 1989.

152. Potgieter, H. and Alexander, M., Susceptible resistance of several fungi to microbial lysis, *Bacteriology*, 91, 1526, 1966.

153. Bartnicki-Garcia, S. and Lippman, E., Fungal cell wall composition, *Handbook of Microbiology*, Laskin, A.L. and Lechvaluer, H.L., Eds., Chemical Rubber, Cleveland, OH, 1973, p. 229.

154. Henis, Y. and Chet, I., Microbiological control of plant pathogens, *Adv. Appl. Microbiol.*, 19, 85, 1975.

155. Schroth, M.N. and Hancock, J.G., Selected topics in biological control, *Annu. Rev. Microbiol.*, 35, 453, 1981.

156. Ordentlich, A., Elad, Y., and Chet, I., The chitinase of *Serratia marcescens* for biological control of *Sclerotium rolfsii*, *Phytopathology*, 78, 84, 1988.

157. Inbar, J. and Chet, I., Evidence that chitinase produced by *Aeromonas caviae* is involved in the biological control of soil-borne plant pathogens by this bacteria, *Soil Biol. Biochem.*, 23, 973, 1991.

158. Fridlender, M., Inbar, J., and Chet, I., Biological control of soilborne plant pathogens by a β-1,3 glucanase-producing *Pseudomonas cepacia*, *Soil Biol. Biochem.*, 25, 1211, 1993.

159. Elad, Y. and Chet, I., Possible role of nutrients in biocontrol of *Pythium* Damping-off by bacteria, *Phytopathology*, 77, 190, 1987.

160. Lindow, S.E., Methods of preventing frost injury caused by epiphytic ice-nucleation-active bacteria, *Plant Dis.*, 327, March 1983.

161. Lindow, S.E. and Panopoulos, N.J., Field tests of recombinant ice- *Pseudomonas* for biological pest control in potato, in *The Release of Genetically Engineered Microorganisms*, Sussman, M., Collins, G.H., Skinner, F.A., and Stewart-Tull, D.E., Eds., *Academic Press*, London, 1988, p. 121.

162. Kloepper, J.W., Tuzun, S., and Kuc, J., Proposed definitions related to induced disease resistance, *Biocontrol Sci. Technol.*, 2, 349, 1992.

163. Lawton, M.A. and Lamb, C.J., Transcriptional activation of plant defence genes by fungal elicitor, wounding and infection, *Mol. Cell Biol.*, 7, 335, 1987.

164. Kuc, J. and Rush, J.S., Phytoalexins, *Arch. Biochem. Biophys.*, 236, 455, 1985.

165. Hammerschmidt, R. and Kuc, J., Lignification as a mechanism for induced systemic resistance in cucumber, *Physiol. Plant Pathol.*, 20, 61, 1982.

166. Hammerschmidt, R., Nuckes, E., and Kuc, J., Association of enhanced peroxidase activity with induced systemic resistance of cucumber to *Colleotrichum lagenarium*, *Physiol. Plant Pathol.*, 20, 73, 1982.

167. Hammerschmidt, R., Lamport, D.T.A., and Muldoon, E.P., Cell wall hydroxyproline enhancement and lignin decomposition as an early event in the resistance of cucumber to *Cladosporium cucumerin*, *Physiol. Plant Pathol.*, 24, 43, 1984.

168. Dean, R.A. and Kuc, J., Induced systemic protection in plants, *Trends Biotechnol.*, 3, 125, 1985.

169. Caruso, F.L. and Kuc, J., Field protection of cucumber, watermelon and muskmelon against *Colletotrichum lagenarium*, *Phytopathology*, 67, 1290, 1977.

170. Wei, G., Kloepper, J.W., and Tuzun, S., Induced systemic resistance to cucumber diseases and increased plant growth by plant growth-promoting rhizobacteria under field conditions, *Phytopathology*, 86, 221, 1996.

171. Brown, M.E., Jackson, R.M., and Burlingham, S.K., Effects produced on tomato plants by seed or root treatment with gibberellic acid and indol-3yl-acetic acid, *J. Exp. Bot.*, 19, 544, 1968.

172. Hussain, A. and Vancura, V., Formation of biologically active substances by rhizosphere bacteria and their effect on plant growth, *Folia Microbiol.*, 11, 468, 1970.

173. Eklund, E., Secondary effects of some Pseudomonads in the rhizoplane of peat grown cucumber plants, *Acta Agric. Scand. Suppl.*, 17, 1, 1970.

174. Mitchell, J.E., The effects of roots on the activity of soil-borne plant pathogens, *Physiol. Plant Pathol.*, 4, 104, 1976.

175. Schroth, M.N. and Snyder, W.C., Effect of host exudates on chlamydospore germination of the bean root rot fungus *Fusarium solani* f. *phaseoli*, *Phytopathology*, 51, 389, 1961.

176. Papavizas, G.C. and Adams, P.B., Survival of root-infecting fungi in soil XII. Germination and survival of endoconidia and chlamydospores of *Thielviopsis basicola* in fallow soil and in soil adjacent to germinating bean seed, *Phytopathology*, 59, 371, 1969.

177. Zentmyer, G.A., *Phytophora cinnamomi* and the Diseases it Causes, Monograph 10, The American Phytopathological Society, St Pauls, MN, 1980, p. 96.

178. Paulitz, T.C., Effect of *Pseudomonas putida* on the stimulation of *Pythium ultimum* by seed voltiles of pea and soybean, *Phytopathology*, 81, 1283, 1991.

179. Fredrickson, J.K. and Elliott, L.F., Effects on winter wheat seedling growth by toxin-producing rhizobacteria, *Plant Soil*, 83, 399, 1985.

180. Bakker, P.A.H.M. et al., Bioassay for studying the role of siderophores in potato growth stimulation by *Pseudomonas* spp. in short potato rotations, *Soil Biol. Biochem.*, 19, 443, 1987.

181. Schippers, A.B., Bakker, A.W., and Bakker, P.A.H.M., Interactions of deleterious and beneficial rhizosphere microorganisms and the effect of cropping practices, *Annu. Rev. Phytopathol.*, 25, 339, 1987.

182. Rovira, A.D., Elliott, L.F., and Cook, R.J., The impact of cropping systems on rhizosphere organisms affecting plant health, in *The Rhizosphere*, Lynch, J.M., Ed., John Wiley and Sons, Chichester, U.K., 1990, p. 389.

183. Bakker, A.W. and Schippers, B., Microbial cyanide production in the rhizosphere in relation to potato yield reduction and *Pseudomonas* spp-mediated plant growth-stimulation, *Soil Biol. Biochem.*, 19, 451, 1987.

184. Alstrom, S. and Burns, R.G., Cyanide production by rhizobacteria as a possible mechanism of plant growth inhibition, *Biol. Fertil. Soils*, 7, 232, 1989.

185. Bolton, H. et al., Characterisation of a toxin produced by a rhizobacterial *Pseudomonas* sp. that inhibits wheat growth, *Plant Soil*, 114, 269, 1989.

186. Astrom, B., Gustafsson, A., and Gerhardson, B., Characteristics of a plant deleterious rhizosphere pseudomonad and its inhibitory metabolites, *J. Appl. Bacteriol.*, 74, 20, 1993.

187. Loper, J.E. and Schroth, M.N., Influence of bacterial sources of indole-3-acetic acid on root elongation of sugar beet, *Phytopathology*, 76, 386, 1986.

188. Dubeikovsky, A.N. et al., Growth promotion of blackcurrant softwood cuttings by recombinant strain *Pseudomonas fluorescens* BSP53a synthesizing an increased amount of indole-3-acetic acid, *Soil Biol. Biochem.*, 25, 1277, 1993.

189. Linderman, R.G., Vesicular-arbuscular mycorrhizae and soil microbial interactions, in *Mycorrhizae in Sustainable Agriculture*, Bethlenfalvay, G.J. and Linderman, R.G., Eds., American Society of Agronomy, Madison, WI, 1992, p. 45.

190. Alstrom, S., Factors associated with detrimental effects of rhizobacteria on plant growth, *Plant Soil*, 102, 3, 1987.

191. Lemanceau, P. et al., Effect of two plant species, flax (*Linum usitatissimum* L.) and tomato (*Lycopersicon esculentum* Mill.), on the diversity of soilborne populations of fluorescent pseudomonads, *Appl. Environ. Microbiol.*, 61, 1004, 1995.

192. Torsvik, V. et al., Comparison of phenotypic diversity and DNA heterogeneity in a population of soil bacteria, *Appl. Environ. Microbiol.*, 56, 776, 1990.

193. Van Vuurde, J.W.L. and Schippers, B., Bacterial colonisation of seminal wheat roots, *Soil Biol. Biochem.*, 12, 559, 1980.

194. Liljeroth, E., Burgers, S.L.G.E., and Van Veen, J.A., Changes in bacterial-populations along roots of wheat (*Triticum aestivum*) seedlings, *Biol. Fertil. Soils*, 10, 276, 1991.

195. Kleeberger, A., Castorph, H., and Klingmuller, W., The rhizosphere microflora of wheat and barley with special reference to gram-negative bacteria, *Arch. Microbiol.*, 136, 306, 1983.

196. Sørensen, J., The rhizosphere as a habitat for soil microorganisms, *Modern Soil Microbiology*, Van Elsas, J.D., Trevors, J.T., and Wellington, E.M., Eds., Marcel Dekker, New York, 1997, p. 21.

197. Acero, N. et al., Seasonal changes in physiological groups of bacteria that participate in the nitrogen cycle in the rhizosphere of the alder, *Geomicrobiology*, 11, 133, 1994.

198. Lambert, B. et al., Rhizobacteria of maize and their antifungal activities, *Appl. Environ. Microbiol.*, 53, 1866, 1987.

199. Lambert, B. et al., Fast-growing, aerobic, heterotrophic bacteria from the rhizosphere of young sugar beet plants, *Appl. Environ. Microbiol.*, 56, 3375, 1990.

200. Miller, H.J., Henken, G., and Van Veen, J.A., Variation and composition of bacterial populations in the rhizosphere of maize, wheat and grass cultivars, *Can. J. Microbiol.*, 35, 656, 1989.

201. Miller, H.J. et al., Fluctuations in the fluorescent pseudomonad and actinomycete populations of the rhizosphere and rhizoplane during the growth of spring wheat, *Can. J. Microbiol.*, 36, 254, 1990.

202. De Leij, F.A.A.M., Whipps, J.M., and Lynch, J.M., The use of colony development for the characterization of bacterial communities in soil and on roots, *Microb. Ecol.*, 27, 81, 1993.

203. Andrews, J.H. and Harris, R.F., r- and K-selection and microbial ecology, *Adv. Microbial Ecol.*, 9, 99, 1986.

204. De Leij, F.A.A.M. et al., Impact of field release of genetically modified ?*seudomonas fluorescens* on indigenous microbial populations of wheat, *Appl. Environ. Microbiol*, 3343, 1995.

205. Panikov, N., Population dynamics of microorganisms with different life strategies, in *Environmental Gene Release: Models, Experiments and Risk Assessment,* Bazin, M.J. and Lynch, J.M., Eds., Chapman and Hall, London, 1994, p. 47.

206. Barlow, P.W., The cellular organisation of roots and its response to the physical environment, in *Root Development and Function,* Gregory, P.J., Lake, J.V., and Rose, D.A., Eds., Cambridge University Press, Cambridge, 1987, p. 1.

207. Kramer, P.J., *Water Relations of Plants*, Academic Press, Orlando, 1983, p. 489.

208. Drew, M.C., Root function, development, growth and mineral nutrition, in *The Rhizosphere,* Lynch, J.M., Ed., John Wiley and Sons, Chichester, U.K., 1990, p. 35.

209. Levitt, J., *Responses of Plants to Environmental Stresses*, Vol.1, 2nd ed., Academic Press, Orlando, FL, 1980, p. 497.

210. Martin, J.K. and Kemp, J.R., Carbon loss from roots of wheat cultivars, *Soil Biol. Biochem.*, 12, 551, 1980.

211. Vancura, V., Root exudates of plants III. Effect of temperature and 'cold shock' on the exudation of various compounds from seeds and seedlings of maize and cucumber, *Plant Soil*, 27, 319, 1967.

212. Rovira, A.D., Root excretions in relation to the rhizosphere effect. IV. Influence of plant species, age of plant, light, temperature and calcium nutrition on exudation, *Plant Soil*, 11, 53, 1959.

213. Clarkson, D.T., The influence of temperature on exudation of xylem sap from detached root systems of rye (*Secale cereale*) and barley (*Hordeum vulgare*), *Planta*, 132, 297, 1976.

214. White, P.J., Clarkson, D.T., and Earnshaw, M.J., Acclimation of potassium influx in rye (*Secale cereale*) at low root temperatures, *Planta*, 171, 377, 1987.

215. Clarkson, D.T., Shone, M.G.T., and Wood, A.V., The effect of pre-treatment temperature on the exudation of xylem sap by detached barley root systems, *Planta*, 121, 81 1974.

216. Sharp, R.E. and Davies, W.J., Solute regulation and growth by roots and shoots of water stressed maize plants, *Planta*, 147, 43, 1979.

217. Westgate, M.E. and Boyer, J.S., Transpiration and growth induced water potentials in maize, *Plant Physiol.*, 74, 882, 1984.

218. Nir, I., Klein, S., and Poljakoff-Mayber, A., Effect of moisture stress on submicroscopic structure of maize roots, *Aust. J. Biol. Sci.*, 22, 17, 1969.

219. Shone, M.G.T. and Flood, A.L., Effects of periods of localised water stress on subsequent nutrient uptake by barley roots and their adaptation to osmotic adjustment, *New Phytologist*, 94, 561, 1983.

220. Martin, J.K., Effect of soil moisture on the release of organic carbon from wheat roots, *Soil Biol. Biochem.*, 9, 303, 1977.

221. Russell, E.W., *Soil Conditions and Plant Growth*, 10th ed., Longman, London, 1973.

222. Drew, M.C. and Lynch, J.M., Soil anaerobiosis, micro-organisms and root function, *Annu. Rev. Phytopathol.*, 18, 37, 1980.

223. Trought, M.C.T. and Drew, M.C., The development of waterlogging damage in young wheat plants in anaerobic solution culture, *J. Exp. Bot.*, 31, 1573, 1980.

224. Saglio, P.H., Raymond, P., and Pradet, A., Metabolic activity and energy charge of excised maize root tips under anoxia control by soluble sugars, *Plant Physiol.*, 86, 1053, 1980.

225. Saglio, P.H., Drew, M.C., and Pradet, A., Metabolic acclimation to anoxia induced by low (2-4kPa partial pressure) oxygen pre-treatment (hypoxia) in root tips of *Zea mays, Plant Physiol.*, 86, 61, 1988.

226. Everard, J.D. and Drew, M.C., Mechanisms of inhibition of water movement in anaerobically treated roots of *Zea mays, J. Exp. Bot.*, 38, 1154, 1987.

227. Everard, J.D. and Drew, M.C., Mechanisms controlling changes in water movement through the roots of *Helianthus annuus* during continuous exposure to oxygen deficiency, *J. Exp. Bot.*, 40, 1, 1989.

228. Whipps, J.M. and Lynch, J.M., The influence of the rhizosphere on crop productivity, *Adv. Microb. Ecol.*, 6, 187, 1986.

229. Ayers, W.A. and Thornton, R.H., Exudation of amino acids by intact roots and damaged roots of wheat and pea, *Plant Soil*, 28, 193 1968.

230. Grineva, G.M., Alcohol formation and excretion by plant roots under anaerobic conditions, *Sov. Plant Physiol.*, 10, 361, 1963.

231. Rittenhouse, R.L. and Hale, M.G., Loss of organic compounds from roots II. Effect of O_2 and CO_2 tension on release of sugars from peanut roots under axenic conditions, *Plant Soil*, 35, 311, 1971.

232. Schaller, G., pH changes in the rhizosphere in relation to the pH buffering of soils, *Plant Soil*, 97, 439, 1987.

233. Smiley, R.W., Rhizosphere pH as influenced by plants, soils and nitrogen fertilisers, *Soil Sci. Soc. Am. J.*, 38, 795, 1974.

234. Smiley, R.W. and Cook, R.J., Relationship between take-all of wheat and rhizosphere pH in soils fertilised with ammonium vs. nitrate-nitrogen, *Phytopathology*, 63, 882, 1983.

235. McDougall, B.M., Movement of ^{14}C photosynthate into the roots of wheat seedlings and exudation of ^{14}C from intact roots, *New Phytologist*, 69, 999, 1970.

236. Meharg, A.A. and Killham, K., The effect of soil pH on rhizosphere carbon flow of *Lolium perenne*, *Plant Soil*, 123, 1, 1990.

237. Marschner, H. and Romheld, V., *In vivo* measurement of root-induced pH changes at the soil-root interface — effect of plant species and nitrogen source, *Z. Pflanzenphysiol.*, 111, 241, 1983.

238. Kraffczyk, I., Trolldenier, G., and Beringer, H., Soluble root exudates of maize: influence of potassium supply and rhizosphere microorganisms, *Soil Biol. Biochem.*, 16, 315, 1984.

239. Tagaki, S., Nomoto, K., and Takemoto, T., Physiological aspect of mugineic acid, a possible phyto-siderophore of graminaceous plants, *J. Plant Nutr.*, 7, 469, 1984.

240. Marschner, H., Nutrient dynamics at the soil-root interface (rhizosphere), in *Mycorrhizas in Ecosystems*, Read, D.J., Lewis, D.H., Fitter, A.H., and Alexander, I.J., Eds., CAB International, Wallingford, 1992, p. 3.

241. Jolley, V.D. and Brown, J.C., Soybean response to iron-deficiency stress as related to iron supply in the growth medium, *J. Plant Nutr.*, 10, 637, 1987.

242. King, A.W., Emanuel, W.R., and Post, W.M., Projecting future concentrations of atmospheric CO_2 with global carbon cycle models: the importance of simulating historical changes, *Environ. Manage.*, 16, 91, 1992.

243. Norby, R.J. et al., Plant Responses to Elevated Atmospheric CO_2 with Emphasis on Below-Ground Processes, ORNL/TM. 9426, Oak Ridge National Laboratory, Oak Ridge, TN, 141 1984.

244. Norby, R.J. et al., Carbon allocation, root exudation and mycorrhizal colonisation of *Pinus echinata* seedlings grown under CO_2 enrichment, *Tree Physiol.*, 3, 203, 1987.

245. Rillig, M.C. et al., Microbial carbon-substrate utilisation in the rhizosphere of *Gutierrezia sarothrae* grown in elevated atmospheric carbon dioxide, *Soil Biol. Biochem.*, 29, 1387, 1997.

246. Li, Z. and Yagi, K., Rice root-derived carbon input and its effect on decomposition of old soil carbon pool under elevated CO_2, *Soil Biol. Biochem.*, 36, 1967, 2004.

247. Knaebel, D.B. et al., Effect of mineral and organic soil constituents on microbial mineralisation of organic compounds in a natural soil, *Appl. Environ. Microbiol.*, 60, 4500, 1994.

248. Hassink, J., Effects of soil texture and structure on carbon and nitrogen mineralisation in grassland soils, *Biol. Fertil. Soils*, 14, 126, 1992.

249. Heijnen, C.E., Hok-a-Hin, C.H., and Van Elsas, J.D., Root colonisation by *Pseudomonas fluorescens* introduced into soil amended with betonite, *Soil Biol. Biochem.*, 25, 239, 1993.

250. Ladd, J.N. et al., Soil structure and biological activity, in *Soil Biochem.*, Vol. 9, Stotzky, G. and Bollag, J.M., Eds., Marcel Dekker, New York, 1996, p. 23.

251. Kolb, W. and Martin, P., Influence of nitrogen on the number of N_2-fixing and total bacteria in the rhizosphere, *Soil Biol. Biochem.*, 20, 221, 1988.

252. Liljeroth, E., Van Veen, J.A., and Miller, H.J., Assimilate translocation to the rhizosphere of two wheat lines and subsequent utilization by rhizosphere mircoorganisms at two soil nitrogen concentrations, *Soil Biol. Biochem.*, 22, 1015, 1990.

253. Cook, R.J. and Baker, K.F., *The Nature and Practice of Biological Control of Plant Pathogens*, American Phytopathology Society, St Paul, MN, 1983, 539.

254. Verstraete, W. and Voets, J.P., Soil microbial and biochemical characteristics in relation to soil management and fertility, *Soil Biol. Biochem.*, 9, 253, 1977.

255. Bolton, H. et al., Soil microbial biomass and selected soil enzyme activities: effect of fertilization and cropping practices, *Soil Biol. Biochem.*, 17, 297, 1985.

256. Harris, P.A. et al., Burning, tillage and herbicide effects on the soil microflora in a wheat-soybean double-crop system, *Soil Biol. Biochem.*, 27, 153, 1995.

257. Swinnen, J., Rhizodeposition and turnover of root-derived organic material in barley and wheat under conventional and integrated management, *Agric. Ecosyst. Environ.*, 51, 115, 1994.

258. Buyer, J.S. and Kaufman, D.D., Microbial diversity in the rhizosphere of corn grown under conventional and low input systems, *Appl. Soil Ecol.*, 5, 21, 1996.
259. Clarholm, M., Interactions of bacteria, protozoa and plants leading to mineralisation of soil nitrogen, *Soil Biol. Biochem.*, 17, 181, 1985.
260. Kuikman, P.J. and Van Veen, J.A., The impact of protozoa on the availability of bacterial nitrogen to plants, *Biol. Fertil. Soils*, 8, 13, 1989.
261. Kuikman, P.J. et al., Protozoan predation and the turnover of soil organic carbon and nitrogen in the presence of plants, *Biol. Fertil. Soils*, 10, 22, 1990.
262. Freckman, D.W., Bacterivorous nematodes and organic matter decomposition, *Agric. Ecosyst. Environ.*, 24, 195, 1988.
263. Ingham, R.E. et al., Interactions of bacteria, fungi and their nematode grazers: effects on nutrient cycling and plant growth, *Ecol. Monogr.*, 55, 110, 1985.
264. Verhoef, H.A. and Brussaard, L., Decomposition and nitrogen mineralisation in natural and agroecosystems: the contribution of soil animals, *Biogeochemistry*, 11, 175, 1990.
265. Fenchel, T., *Ecology of Protozoa*, Science Tech, Madison, WI, 1987, p. 197.
266. Hunt, H.W. et al., A simulation model for the effect of predation on bacteria in continuous culture, *Microb. Ecol.*, 3, 259, 1977.
267. Elliott, E.T., Coleman, D.C., and Cole, C.V., The influence of amoebae on the uptake of nitrogen by plants in gnotobiotic soil, *The Soil-Root Interface*, Harley, J.L. and Scott Russell, R., Eds., Academic Press, London, 1979, p. 221.
268. Wright, D.J. and Newall, D.R., Nitrogen excretion, osmotic and ionic regulation in nematodes, *The Organisation of Nematodes*, Croll, N.A., Ed., Academic Press, London, 1976, p. 163.
269. Chantano, A. and Jenson, H.J., Saprozoic nematodes as carriers and disseminators of plant pathogenic bacteria, *J. Nematology*, 1, 21, 1969.
270. Jatala, P., Jensen, H.J., and Russell, S.A., *Prisionichus iheriteri* as a carrier of *Rhizobium japonicum*, *J. Nematology*, 6, 130, 1974.
271. Bird, A.F., Adhesion of microorganisms to nematodes, *J. Nematology*, 19, 514, 1987.
272. Bouwman, L.A. et al., Short-term and long-term effects of bacterivorous nematodes and nematophagous fungi on carbon and nitrogen mineralisation in microcosms, *Biol. Fertil. Soils*, 17, 249, 1994.
273. Hams, A.F. and Wilkin, G.D., Observations on the use of predatious fungi for the control of *Heterodera spp.*, *Ann. Appl. Biol.*, 49, 515, 1961.
274. Kerry, B.R. and Crump, D.M., Two fungi parasitic on females of cyst nematodes *(Heterodera spp.)*, *Trans. Br. Mycol. Soc.*, 74, 119, 1980.
275. Kerry, B.R., Crump, D.H., and Mullen, C.A., Studies of the cereal cyst nematode *Heterodera avenae*, under continuous cereals, 1975–1978. II. Fungal parasitism of nematode females and eggs, *Ann. Appl. Biol.*, 100, 489, 1982.
276. Radwan, S.S. et al., Soil management enhancing hydrocarbon biodegradation in the polluted Kuwaiti desert, *Appl. Microbiol. Biotechnol.*, 44, 265, 1995.
277. Chaney, R.L. et al., Phytoremediation of soil metals, *Curr. Opin. Biotechnol.*, 8, 279, 1997.
278. Atlas, R.M. and Bartha, R., Hydrocarbon biodegradation and oil spill bioremediation, *Adv. Microb. Ecol.*, 12, 287, 1992.
279. Martin, F.M., Perotto, S., and Bonfante, P., Mycorrizal fungi: a fungal community at the interface between soil and roots, in *The Rhizosphere: Biochemistry and Organic Substances at the Soil-Plant Interface*, Pinton, R., Varanini, Z., and Nannipieri, P., Eds., Marcel Dekker, New York, 2001, p. 263.
280. Anderson, T.A., Guthrie, E.A., and Walton, B.T., Bioremediation in the rhizosphere, *Environ. Sci. Technol.*, 27, 2630, 1993.
281. Schwab, A.P. and Banks, M.K., Biologically mediated dissipation of polyaromatic hydrocarbons in the root zone, *Bioremediation through Rhizosphere Technology*, Anderson, T.A. and Coats, J.R., Eds., American Chemical Society, Washington, D.C., 1994, p. 132.
282. Erickson, L.E., An overview of research on the beneficial effects of vegetation in contaminated soil, *Ann. NY Acad. Sci.*, 829, 30, 1997.
283. Rentz, J.A., Alvarez, P.J.J., and Schnoor, J.L., Repression of *Pseudomonas putida* phenanthrene degradating activity by plant root extracts and exudates, *Environ. Microbiol.*, 6, 574, 2004.
284. Aprill, W. and Sims, R.C., Evaluation of the use of prairie grasses for stimulating polycyclic aromatic hydrocarbon treatment in soil, *Chemosphere*, 20, 253, 1990.

285. Liste, H.H. and Alexander, M., Plant promoted pyrene degradation in soil, *Chemosphere*, 40, 7, 2000.
286. Miya, R.K. and Firestone, M.K., Enhanced phenanthrene biodegradation in soil by slender oat root exudates and root debris, *J. Environ. Qual.*, 30, 1911, 2001.
287. Krutz, L.J. et al., Selective enrichment of a pyrene degrader population and enhanced pyrene degradation in Bermuda grass rhizosphere, *Biol. Fertil. Soils*, 41, 359, 2005.
288. Banks, M.K., Lee, E., and Schwab, A.P., Evaluation of dissipation mechanisms for benzo[a]pyrene in the rhizosphere of tall fescue, *J. Environ. Qual.*, 28, 294, 1999.
289. Hsu, T.S. and Bartha, R., Accelerated mineralization of two organophosphate insecticides in the rhizosphere, *Appl. Environ. Microbiol.*, 37, 36, 1997.
290. Cerniglia, C.E., *Microbial Transformation of Aromatic Hydrocarbons*, Petroleum Microbiology, Atlas, R.M., Ed, Macmillan Publishing, New York, 1984, p. 99.
291. Gilbert, E.S. and Crowley, D.E., Plant compounds that induce polychlorinated biphenyl biodegradation by *Arthrobacter* sp. Strain B1B, *Appl. Environ. Microbiol.*, 63, 1933, 1997.
292. Ruuhola, T.M. and Julkunen-Titto, M.R.K., Salicylates of intact Salix myrsinifolia plantlets do not undergo rapid metabolic turnover, *Plant Physiol.*, 122, 895, 2000.
293. Monache, F.D., Ferrari, F., and Pomponi, M., Flavanones and xanthones from *Maclura pomifera*, *Phytochemistry*, 23, 1489, 1984.
294. Kurkin, V.A., Braslavskii, V.B., and Zapesochnaya, G.G., High-performance liquid chromatography of extracts of propolis and poplar buds, *Russ. J. Phys. chem.*, 68, 1647, 1994.
295. Crowley, D.E. et al., Biodegradation of organic pollutants in the plant rhizosphere, in *Rhizosphere 2004*, Hartman, A., Schmid, M., Wenzel, W., and Hisinger, P.H., Eds., GSF, Neuherberg, 2005, p. 213.
296. Lalande, T.L. et al., Phytoremediation of pyrene in a Cecil soil under field conditions, *Int. J. Phytoremediation*, 5, 1, 2003.
297. Olexa, T.J. et al., Mycorrhizal colonisation and microbial community structure of annual ryegrass grown in pyrene amended soils, *Int. J. Phytoremediation*, 2, 213, 2000.
298. Gunther, T., Dornberger, U., and Fritsche, W., Effects of ryegrass on biodegradation of hydrocarbons in soil, *Chemsphere*, 33, 203, 1996.
299. Harvey, P.J. et al., Phytoremediation of polyaromatic hydrocarbons, anilines, and phenols, *Environ. Sci. Pollut. Res.*, 9, 29, 2002.
300. Harman, G.E., Lorito, M., and Lynch, J.M., *Trichoderma* spp. to alleviate soil and water pollution, *Adv. Appl. Microbiol.*, 56, 313, 2004.
301. Lynch, J.M. and Moffat, A.J., Bioremediation: prospects for the future application of innovative applied biological research, *Ann. Appl. Biol.*, 134, 217, 2005.
302. Keppler, F. et al., Methane emissions from terrestrial plants under aerobic conditions, *Nature*, 439, 187, 2006.
303. Lynch, J.M. et al., Microbial diversity in soil: ecological theories, the contribution of molecular techniques and the impact of transgenic plants and transgenic microorganisms, *Biol. Fertil. Soils*, 40, 363, 2004.

4 Nutrient Transformations in the Rhizosphere

Luigi Badalucco and Paolo Nannipieri

CONTENTS

I. INTRODUCTION

Soil processes at rhizosphere level, including nutrient transformations, are generally thought to be spatially narrow and restricted if compared to those occurring in the whole bulk soil. Coleman et al. [1] assessed that rhizosphere soil constitutes only 2 to 3% of the total soil volume. However, in spite of this low percentage of soil volume, rhizosphere processes are much more important than bulk soil processes either from qualitative and quantitative points of view. Indeed, any soil volume can be explored, sooner or later, by some roots, and be enriched by the plant rhizodeposition. Of course, the volume of soil explored by roots will be greater for plants of grasslands, which have both — the shortest life cycles and the highest root density per volume unit of soil.

It is universally recognized that one of the major problems in studying the rhizosphere is the accurate sampling of rhizosphere soil due to the difficulty of reaching inaccessible zones and to separate rhizosphere from nonrhizosphere (bulk) soil. Badalucco and Kuikman [2] reviewed soil–plant systems used to simulate the rhizosphere soil and procedures used for sampling rhizosphere soil. Usually, the soil removed by manual shaking of soil/root cores is considered the bulk soil, whereas the soil remaining attached to the roots represents the rhizosphere soil. This procedure cannot be standardized because of the root system architecture and the handling of the operator. Moreover, Badalucco and Kuikman [2] discussed that pot experiments do not allow us to assess the "rhizosphere effect" at measurable distances from the soil–root interface. On the other hand, soil of unplanted pots cannot be considered as bulk soil because it presents different chemical (ions, organic substrates, pH, etc.) and physical (redox and water potential) properties than those of the bulk soil taken from planted soils. These differences depend on the effects of processes induced by plant, such as root respiration and plant water uptake, which can affect chemical and physical properties of soil located beyond the physical boundaries of the rhizosphere effect. Consequently, both redox and water potential of soil of unplanted pots are different with respect to bulk soil of planted pots and, thus, soil microbial activity and diversity of unplanted pots may be not comparable with those of bulk soil of planted pots. In addition, different pot experiments are hardly comparable among themselves if the pot sizes are diverse, because root activity, and thus exudation, may change depending on the volume of soil to be exploited [2]. In spite of these problems, the literature on rhizosphere studies based on pot experiments is still extensive. Indeed, a literature search by Web of Science® with both the terms "rhizosphere" and "pot experiments" as requests in the title, the abstract, or simply as keywords has generated about 200 responses.

A more reliable experimental approach (than pot experiments) for studying rhizosphere effects is to use mesocosms with physical separation of the rhizosphere from bulk soil using nylon meshes for the free diffusion of gases and soluble root exudates from the rooted to the root-free soil and *vice versa*. This approach permits the study of the rhizosphere effect at measurable distances from the soil–root interface and, more importantly, without the need of using an unplanted soil as a control [3–5].

As mentioned in other chapters of this book, rhizodeposition supports microbial communities that are more active and abundant in soil around roots than in nonrhizosphere soil. Such microbial communities are chiefly crucial to the functioning of the terrestrial ecosystem, not only for their direct effects on plant growth through, for example, the release of plant promoting growth factors, but also because they affect the C-flow from plant roots to soil, thus mediating the heterotrophically driven nutrient processes in soil by rhizodeposition (Chapter 1). As such, changes in their activity or composition are likely to be reflected in the sensitivity of the overall soil functioning.

The aim of this chapter is to discuss only some aspects of some nutrient (C, N, and P) cycling occurring in the rhizosphere, because the complexity and vastness of the treated matter exceeds the limits of a single chapter. The subject is complex because it involves a series of chemical, physical, and biological interactions. An example of the latter case is the so-called "microbial loop" [6]. In the rhizosphere, the bacterial energy circulation follows a loop trajectory. Nutrients become only temporarily locked up in bacterial biomass surrounding the roots and are later released by microfaunal grazing over bacteria. Rhizodepositions trigger the microbial growth in the rhizosphere with the consequent sequestration of available plant nutrients, which would remain locked up into microbial biomass, if consumption by protozoa and nematodes would not constantly remobilize them for plant uptake [7–10]. We shall also discuss enzyme activities in the rhizosphere soil because all nutrient transformations are mediated by enzymes. Among plant nutrients we shall discuss some aspects of C, N, and P processes, because C is the major element of rhizodeposition whereas N and P availabilities are those mostly limiting both plant and microbial growth.

II. CARBON DYNAMICS IN THE RHIZOSPHERE

A. Different Components of Soil Respiration and Quantification of Rhizodepositions

Soils contain the largest active terrestrial carbon pool on Earth, and through soil respiration, their annual contribution to the flux of CO_2 to the atmosphere is 10 times greater than that from fossil fuel combustion [11]. Soil respiration is derived from both rhizosphere respiration and the microbial oxidation of stable soil organic matter by heterotrophs. Rhizosphere respiration represents the sum of root respiration and microbial respiration of labile carbon derived from live roots (rhizodeposits). Increases in rhizosphere respiration may indicate increased carbon inputs to the soil through greater photosynthesis, greater root activity or root biomass [12,13]. Increases in respiration of stable organic matter, in contrast, reduce the potential for carbon storage in the soil. If we want to predict feedbacks between global change and soil processes, we must first understand the relative contributions of rhizosphere respiration and degradation of stable organic matter to total soil respiration.

Both ^{14}C and ^{13}C tracing techniques have been used to estimate C flow in soil–plant systems, and thus also rhizosphere respiration, with a minimum of soil and root disturbance [2].

Continuous labeling with [14]CO$_2$ involves exposure of plants to an atmosphere containing a constant specific activity of [14]CO$_2$ throughout the growth of the plant. This approach allows uniform labeling of all plant C pools, that is, soluble metabolic pools, which are composed primarily of compounds derived from recently assimilated C, biochemically older C such as structural compounds, and both chemically simple and complex rhizodeposits (Chapter 13). Therefore, continuous labeling approach facilitates the quantification of total C input to soil throughout the plant growth and is potentially a powerful technique to estimate gross changes in rhizodeposition in response to increased CO$_2$ concentrations [14]. However, although continuous labeling under field conditions is possible [15], it has been generally used in controlled growth chambers, due to its high cost and problems in setting up the complex apparatus in the field.

Using continuous labeling under laboratory conditions, Liljeroth et al. [16] demonstrated that decomposition of native organic matter (evolution of nonlabeled CO$_2$) was increased from a planted soil under low nutrient conditions. They suggested that this was a consequence of the rhizosphere microbial biomass being nutrient-limited and that mineralization of native organic matter was required to support assimilation of rhizodeposition, which is typically comprised of low C-to-N ratio compounds. It is likely that this scenario will be particularly evident for ecosystems under elevated CO$_2$, where C inputs are increased, and competition for available nutrients are expected to be intensified. Therefore, at least in the short term, as soils become more depleted of plant nutrients under elevated CO$_2$ and high C-to-N inputs are released from plant (rhizodeposition, root turnover, and litterfall), net mineralization of nutrients held in organic matter may increase to sustain microbial utilization of plant inputs, thus resulting in positive feedback of CO$_2$ to the atmosphere [17]. Uniform ^{14}C labeling of soil organic matter pool is not experimentally possible, as chemically or physically protected organic materials usually have mean residence times of hundreds to thousands of years. Therefore, altered fluxes between labile and recalcitrant organic matter pools cannot be directly assessed by continuous labeling but must rely on existing soil organic matter models and use data from short-term experiments.

On the other hand, pulse chase labeling methodologies are simpler and cheaper than continuous labeling, and are readily applicable to field experiments. The partitioning of ^{14}C following pulse labeling is strongly related to the duration of the chase period prior to harvesting. Immediately following exposure to [14]CO$_2$, ^{14}C within the plant will be predominantly distributed as labile C pools (metabolic and transport fractions), and with longer chase durations, it will be increasingly present as less labile pools (structural and storage). However, due to the lack of the uniform labeling of all plant C pools, the rhizosphere respiration using the pulse chase [14]CO$_2$ labeling can be underestimated (see also Chapter 13). In addition to the advantages for field use, pulse labeling can be applied to determine the dynamics of transfer of photoassimilate through the plant, to soil and to microbial biomass [18]. Following labeling, the ^{14}C is restricted to a discrete packet of assimilated C, which can be traced through the system without interference from biochemically older C.

The use of ^{13}C analyses of organic components of terrestrial ecosystems can be useful to quantify C fluxes from either C3 or C4 plants to soil and the mineralization of native organic matter [19]. The basis of this approach is that ^{13}C-to-^{12}C ratios of C3 and C4 vegetation are distinct; C3 plants have an average $\delta^{13}C$ of −27‰ and C4 plants an average of −12‰, compared to atmospheric CO$_2$, which has a $\delta^{13}C$ value of $-7.5‰$. These differences in plant ^{13}C concentrations are a consequence of differential ^{13}C discrimination of the C3 and C4 photosynthetic pathways. Similarly, variations in soil organic matter $\delta^{13}C$ can be attributed to residue inputs by C3 or C4 plants (among plant polymers generally lignin is relatively depleted in ^{13}C), but also to the substrate-dependent discrimination during microbial mineralization of organic matter [20,21]. However, the sensitivity of natural abundance $\delta^{13}C$ techniques is lower than the sensitivity of radioisotope techniques, due to the resolution of analytical procedures.

As already mentioned, in the presence of plants, the interpretation of measurements of total soil respiration (R_t) is complicated by the CO_2 produced by rhizosphere respiration, which includes the respiration of living roots and of the microorganisms feeding on root-derived C. Only few approaches have been developed to separate R_t into its rhizosphere (R_{rh}) and soil components (R_s) under field conditions. They consist of estimating either R_{rh} or R_s and calculating the contribution of the other component by difference with R_t. The root-exclusion method (root removal, trenching, or gap formation), which is mainly used in forest ecosystems, calculates R_{rh} as the difference between CO_2 emissions rates from soil volumes in which roots are either present or excluded [22]. This technique is relatively simple and has provided realistic estimates of R_{rh} and R_s. However, soil disturbance, absence of soil–root interactions, and differences in temperature and moisture between bare soil and soil with vegetation cover may influence this determination of R_s.

The continuous ^{13}C labeling enables the calculation of respiration deriving from three distinct pools: (1) soil organic matter; (2) roots; (3) rhizosphere microorganisms [23]. This calculation is based on two assumptions: (1) the $\delta^{13}C$ value of CO_2 released as root respiration and of rhizodeposits C is the same as the $\delta^{13}C$ value of the roots [24] and (2) the $\delta^{13}C$ value of CO_2 respired by microorganisms is the $\delta^{13}C$ value of microbial biomass [25]; thus, it is only based on the determination of the $\delta^{13}C$ values of soil organic matter, roots, soil microbial biomass, and the CO_2 efflux from soil.

The flow of photosynthate into and through the soil microbial biomass has received much less attention than the partitioning of photosynthate within plants and bulk soil [26]. Because microorganisms of rhizosphere and bulk soil act as both source and sink of nutrients [27], the cycling of rhizodeposition through the microbial biomass can affect soil functioning. To assess the temporal dynamics of rhizosphere C flow through the microbial biomass, greenhouse-grown annual ryegrass plants (*Lolium multiflorum* Lam.) were labeled with $^{13}CO_2$ either during the transition between active root growth and rapid shoot growth (period 1), or 9 d later during the rapid shoot growth stage (period 2) [28]. The distribution of ^{13}C in the soil–plant system was similar between the two labeling periods, whereas microbial cycling of rhizodeposition differed between the two labeling periods. Within 24 h of labeling, approximately 12% of the total ^{13}C retained in the soil–plant system resided in the soil, most of which had already been immobilized into the microbial biomass. Average rate constants (k) and associated turnover times (the inverse of k) for the rhizosphere soils were 0.32 ± 0.07 d^{-1}, with a turnover time of 3.2 d for labeling period 1, and 0.24 ± 0.05 d^{-1} for labeling period 2, with a turnover time of 4.2 d. Bulk values were: 0.18 ± 0.02 d^{-1} with a turnover time of 5.5 d in labeling period 1, and 0.14 ± 0.07 d^{-1} and a turnover time of 7.1 d in labeling period 2. Thus turnover times were nearly twice as fast in the rhizosphere compared to the bulk soil in both labeling periods. This suggests that the microorganisms in the rhizosphere were more active than in the bulk soil, as expected. The slower turnover times observed in both the rhizosphere and bulk soils at the labeling period 2 suggest that the microbial biomass was more stable at later stage of plant growth [28].

Ostle et al. [29] suggest a 7-d turnover for soil RNA in a $^{13}CO_2$ pulse-chase experiment with grass turf. Thus, recently assimilated C moves through the soil–plant system at a very rapid pace and the use of ^{13}C pulse-chase labeling constitutes an effective approach for exploring the microbial dynamics associated with rhizosphere C cycling. Application of this methodology to a range of plant species, developmental growth stages, and environmental conditions has the potential to greatly enhance our knowledge of the dynamics of rhizosphere processes.

The measurement of C availability in the rhizosphere can be important to understand nutrient flows because it is increased by plant roots through rhizodepositions. The proportion of glucose added to soil and mineralized was shown to be increased with the rate of glucose addition whereas the proportion of substrate recovered in the chloroform-labile (microbial biomass) fraction decreased, because, at low rates of addition, the glucose was stored in the microbial biomass pool rather than used for growth [30]. These results were confirmed later by Nguyen and Guckert [31] who studied the short-term assimilation of ^{14}C-[U]glucose by soil microorganisms to assess any difference in C availability between unplanted and maize-planted soils. In unplanted soils, the kinetics of glucose uptake showed a multicomponent carrier-mediated transport; the mineralization of the substrate represented 7.8% of the ^{14}C unrecovered by 0.5-M K_2SO_4 extraction, probably because it was absorbed by soil microorganisms (97% of the added ^{14}C) after 1 h since its addition to soil at a rate of 0.07 μg C-glucose g^{-1} soil. Three days after ^{14}C-glucose addition, $^{14}CO_2$ increased to 28% of the absorbed glucose, whereas the microbial biomass ^{14}C remained constant (about 25% of the added ^{14}C). By contrast, in maize-planted soils, microorganisms mineralized a significantly higher proportion of the absorbed glucose (32%), whereas the ^{14}C activity in the chloroform-labile fraction was lower (22%).

B. DEGRADATION OF ROOT EXUDATES AND CHANGE IN THE COMPOSITION OF MICROBIAL COMMUNITIES

As discussed within Chapter 1, rhizodepositions may not occur aspecifically when they are plant responses to "sensing" the soil environment by the root system. For example, aluminum-resistant genotypes must be able to sense too-high concentrations of soluble Al^{3+} and thus to respond with the release of Al^{3+} complexing carboxylates [32]. When roots release either malate or citrate, depending on soil pH, they must be sensing directly the soil pH or a factor closely related to it [33]. Consequently, sensing soil environmental conditions is crucial to plant performance; however, any plant response by rhizodeposition can affect the rhizosphere functioning, including the effects due to signaling between the host plant and the symbiotic microorganism (see Chapter 8 and Chapter 9). The preceding considerations may support the hypothesis that plant survival mechanisms and rhizosphere microbial communities coevolve (see also Chapter 3).

Studies on the effects of root exudates on microbial activity and nutrient availability are complicated by the fact that they include a complex mixture of organic compounds [34], and it is difficult to sample soil at distinct and known distances from the soil–root interface [2]. Falchini et al. [35] tried to overcome the former problem by adding to soil model organic substrates, that is, glucose, oxalic, and glutamic acids, which are quantitatively the most representative of carbohydrates, organic acids, and amino acids, respectively, of root exudates [36], and faced the second problem by using a simple system that allows the formation of a concentration gradient with the possibility of soil sampling at various distances from a simulated rhizoplane [4]. The concentration gradient was formed in soil inside a rigid PVC cylinder by putting a cellulose disk paper wetted with the model organic substrate solution on the top of the soil core; the model organic substrate concentration decreased by increasing the distance from the cellulose paper. The oxidation of the three ^{14}C-labeled substrates to both ^{14}C- and ^{12}C-CO_2 was monitored during their diffusion throughout a sandy loam soil. The diffusion rates of the three substrates were expected to be different because of their different charge at a given soil pH, which resulted in different interactions with soil colloids, and to their different sizes and solubilities in soil solution [37,38]. After 3 and 7 d, soil was sampled from four distinct and contiguous soil layers at increasing distances from the hypothetical rhizoplane (0 to 2, 2 to 4, 4 to 6, and 6 to 14 mm) to determine residual ^{14}C in each layer [35]. The mineralization pattern of oxalic acid showed a 3-d lag phase likely due to the presence, at the early stages of exposure, of a few microorganisms able to mineralize this substrate [39]. Moreover, the diffusion of oxalic acid was limited, probably because most of the compound precipitated as Ca-oxalate, because of the presence of calcium carbonate in the soil, and remained localized in the top 0- to 2-mm soil layer. The mineralization rate of glutamate during the first 3 d was higher than that of glucose, and may indicate a preference by the soil microflora for this compound containing both C and N [35]. Composition of bacterial communities, as determined by denaturing gradient gel electrophoresis after DNA amplification by universal primers, was changed in the 0- to 2-mm layer of both oxalic and glutamic acid treated soils [35].

A full understanding of the factors affecting the composition of rhizosphere microbial community requires experimental separation of each factor (i.e., amount and composition of exudates, soil moisture, and soil nutrient status). An experimental system for the precise control of both the composition and loading rates of simulated root exudates to soil, held at constant water potential, was devised by Griffiths et al. [40]. Fructose, glucose, sucrose, succinate, malate, arginine, serine, and cysteine (compounds usually released by roots as root exudates) were continuously added at a range of concentrations (0, 188, 375, 938, 1875, 9375, and 18750 μg C d^{-1}). After 14 d, a central portion of soil, known to be influenced by the added substrate, was removed and monitored to determine changes in the overall microbial community structure by techniques such as DNA hybridization, %G + C profiling, and phospholipids–fatty acid analysis (PLFA). The trend was that microbial community structure changed consistently as the substrate loading increased, and that fungi dominated over bacteria at high substrate loading rates. Fungi might have been favored at the high substrate loadings because they are less sensitive than bacteria to high osmotic stress [41].

C. Effects of Elevated CO_2 Atmospheric Concentration on Microbial Processes of Rhizosphere Soil

Ice core data provide evidence that atmospheric CO_2 is at its highest concentration since at least 160,000 years ago. Concentrations rose from 270 ppm in the late 1800s to about 365 ppm in 1997 and may double preindustrial levels by the middle of the 21st century. Currently, atmospheric CO_2 concentration is rising 1.5 ppm/year on average [42].

Direct effects of elevated CO_2 on soil organisms are unlikely because CO_2 concentrations in soil are already 10 to 50 times higher than in the atmosphere, but increased atmospheric CO_2 can have a strong impact on terrestrial ecosystems, leading to higher C assimilation rates in plants and, hence, to greater biomass production. There are, in principle, two plant-mediated mechanisms by which increased CO_2 concentration might affect rhizosphere soil microbial communities:

1. Elevated CO_2 stimulates plant photosynthesis and, consequently, net primary production as well. The extra C fixed will be partly allocated below the ground, thus resulting in increased root biomass, root-to-shoot ratio, fine-root biomass and fine-root turnover [43,44].
2. Elevated CO_2 reduces stomatal conductance of plants, which results in higher water-use efficiency by plant with consequent decreased stand evapotranspiration and higher soil water content [45]. Matrix potential of rhizosphere soil is an important control of microbial activity, either directly through osmosis or indirectly by altering the supply of nutrients.

The increased CO_2 concentration in the atmosphere can increase photosynthetic capacity of plants, and this should be reflected in higher C turnover in the rhizosphere over extended periods, only if sink activity also increases so as to utilize the greater source pool of C [46]. Root ^{14}C-loss from pulse labeled plants has generally been found to increase at elevated CO_2 due to greater partitioning of assimilate to fine-root biomass [47].

The free air carbon dioxide enrichment (FACE) technology has been developed to study the effects of high CO_2 on intact ecosystems without the use of enclosures [48]. When the CO_2 used to fumigate during these experiments is derived from the combustion of natural gas, it contains a unique ^{13}C signature that can be followed through the experimental plots. This carbon is strongly depleted in ^{13}C and, for example, can function as a continuous stable isotopic label in an entire undisturbed forest plot.

Plants grown under elevated CO_2 conditions often exhibit increased growth, a more-than-proportional increase in C allocation to roots, and increases in other processes, such as total rhizosphere respiration and rhizodeposition [12,13]. Plant responses to elevated CO_2 may influence processes in rhizosphere soil, both as a result of changes in plant biomass allocation and by alterations in the quantity and quality of rhizodepositions [49,50]. Particularly, root growth is stimulated under elevated atmospheric CO_2. This has been attributed to nutrient limitation inducing plants to invest more carbohydrates into belowground growth and to release root exudates to utilize soil resources more effectively [51]. On the other hand, the response of soil microorganisms to elevated CO_2 seems highly variable, no matter whether activity, biomass, or effects on the N cycle were studied [51]. This variability cannot be explained by plant life forms. Studies reporting changes of soil microbial properties at elevated CO_2 often deal with soil–plant systems characterized by high belowground C inputs by plants in combination with low C content of the soil [51]. Most information on soil microbial response to elevated CO_2 originates from short-term experiments or experiments with disturbed soils. Extrapolation of these results to mature ecosystems and to longer timescales is limited [52].

Soil microorganisms hold a key position in terrestrial ecosystems as they mineralize organic matter. Therefore, any effect of elevated CO_2 on soil microorganisms may, in turn, feed back on the response of plant communities to rising CO_2 and, thus, the sequestration of extra carbon. All earlier findings confirm that rhizosphere processes play a fundamental role in C sequestration and nutrient cycling in terrestrial ecosystems [53]. The rhizosphere has been identified as one of the key fine-scale components in the overall global C cycle [54].

Assessment of denitrification rate (N_2O flux) from soil, under elevated CO_2, is of crucial importance due to the possible feedback on global warming [55]. Nitrous oxide may originate from denitrifier or nitrifier activity in soil. During denitrification, increased N_2O production may occur where labile C and available N are present under soil anoxic conditions, although whether such losses will be in the form of N_2 or N_2O requires consideration. Increased microbial activity in the rhizosphere due to the input of extra C may lead to a reduction in oxygen availability, providing an opportunity for denitrifiers to utilize rhizosphere C as matter and energy source, and soil nitrate as final acceptor of electrons [56]. The interactions between plant growth, denitrifier activity, and N_2O flux under elevated CO_2 concentrations are complex. For example, in soil microcosms planted with ryegrass at three different values of soil pH and at either ambient (450 ppm) or elevated (720 ppm) CO_2 concentration, potential denitrification rates within the rhizosphere followed a similar pattern to plant growth, suggesting that the plant growth and the size of denitrifier population within the rhizosphere are coupled [57]. On the other hand, in a 4-year old spruce–beech forest ecosystem, field-measured N_2O fluxes were not affected by atmospheric CO_2 concentration, regardless of the level of N deposition and the soil type [58]. Further work is required to elucidate the importance of soil variables (soil pH, moisture content, nutrient status, quality of rhizodeposit, etc.) on the proportion of N_2 to N_2O evolved under elevated CO_2.

The loss of N_2O from soils during the nitrification of fertilizer-derived ammonium [59] may contribute further to the accumulation of N_2O in the atmosphere. Therefore, if under elevated CO_2 concentrations, an increase in N fertilizer application occurs to sustain greater primary crop productivity, then a positive feedback effect may occur where greater quantities of N_2O released may lead to increased global warming and, in turn, to enhanced soil native organic matter mineralization.

Autotrophic nitrifiers cannot be directly affected by increased C flow under elevated CO_2 because they utilize inorganic C (CO_2-C) instead of fixed organic C as a C source even though they may be indirectly influenced, because the main source of CO_2 from soils derives from heterotrophic microbial activity, which, under elevated CO_2, is certainly stimulated by enhanced rhizodepositions. Moreover, in forest ecosystems where heterotrophic nitrification is important in supplying plant available N [60], elevated CO_2 may have an even greater impact.

III. NITROGEN DYNAMICS IN THE RHIZOSPHERE

A. NITROGEN UPTAKE BY PLANTS

Many studies on plant physiology, including those on N uptake, are based on hydroponically grown plants. However, caution is required to extrapolate results from these studies to soil–plant interactions so as to get better insights into mechanisms and processes occurring in the rhizosphere. Indeed, there are a number of ecologically crucial differences between a real soil and a nutrient solution: (1) water potential; (2) nutrient-patched vs. uniform distribution in the solution; (3) gas composition and concentration; (4) type, amount, and half-life of rhizodepositions; (5) abundance, activity, and diversity of microbial communities inhabiting the rhizosphere; and (6) symbiosis with fungi and bacteria.

Studies on plants grown in solution culture, or upon excised roots, have shown that uptake of organic N can occur at levels comparable to, or in excess of, N uptake from inorganic N sources [61]. Clearly, in solution culture, the plant uptake of organic N sources, such as amino acids may be maximized because, unlike in soil, these amino acids are readily assimilable rather than being physically or chemically bound. Moreover, competition with microorganisms for these N sources can occur in the rhizosphere soil; in addition, by using excised roots excavated from soil, uptake of, and conversion to, ammonium by rhizoplane microorganisms cannot be differentiated from those of root cells leading to large errors [62]. Significant glucosamine depletion was observed from a wheat root bathing solution (10 μM in glucosamine) in nonsterile conditions, whereas no glucosamine concentration decrease occurred at axenic conditions, thus suggesting the microbial uptake rather than root uptake occurred [63]. On the other hand, when studying the possible uptake by roots of positively charged solutes such as ammonium and some amino acids, the presence of ^{15}N in the roots does not always imply uptake because these cations can be held by the negative charges of the root cell wall without entry into the root metabolic pool. This phenomenon is well established for micronutrients and toxic metals (Al, Zn, Cd) in plants [64].

Nutrient acquisition by nonmychorrhizal roots necessitates that all ions must pass through the rhizosphere prior to uptake at the root surface, and in the rhizosphere soil, there is an intense microbial activity fuelled by the release of organic C from the root, as discussed before. This release is thought to be largely a passive flux over which the plant exerts little direct control [65], is driven by the large concentration gradient that exists between the root cytosol and the soil solution, and it is dominated by the low-molecular-weight solutes of greatest abundance in the cytoplasm, and of high membrane permeability (amino acids, sugars, organic acids) (see Chapter 12). As most of rhizosphere microorganisms possess amino acid transporters, the competition for amino N either released by root or by the mineralization of soil organic matter will be intense. However, plant root may be favored in the recapture of root exudates, such as sugars and amino acids, due to a range of H^+-ATPase-driven proton cotransporters and due to the spatial availability of these released amino acids on the rhizoplane [66]. Different is the case when a root cell dies and it is lyzed, because the competitive advantage of the root H^+-ATPase-driven proton cotransporters is eliminated; microbial activity is then stimulated because the root cell content is released into the soil immediately surrounding the root (i.e., 1-mm rhizosphere), and amino acid concentration in the soil solution can be expected to increase by three orders of magnitude from that in the bulk soil [66]. On the other hand, under circumstances such as freezing–thawing or drying–rewetting, leading to the lysis of microbial cells [67,68], but not affecting root cell metabolism, root cells may take advantage of amino acid released from microbial cells.

Most of the crop varieties grown in the developed world have been bred under conditions of high fertilizer input, approaching N soil saturation [69]. Breeding improved crop genotypes capable of more efficient N uptake and fertilizer N utilization has become an essential research topic due to either environmental and economical problems caused by the excessive application of N fertilizers in modern agriculture. The developed world has today the opportunity to use the more sustainable agriculture systems of the developing world as their cultivars have been naturally selected under low, if any, N fertilizer input [70]. The relative topic should also be considered when discussing N processes in the rhizosphere environment.

As amino acids typically build 10 to 30% of the plant dry matter, in most ecosystems they constitute the biggest input of organic N into the soil. The physiological status of the plant regulates the relative contribution of proteinic and free (monomeric) amino acids in plant tissues (usually this ratio ranges from 100:1 to 800:1) [71]. Most soil microorganisms are likely well adapted to using amino acids as C and N sources because of their ubiquity in plant residues entering the soil. Analysis of the concentration-dependent amino acid uptake kinetics indicates that microbial uptake is rapid across a wide range of soil concentrations and temperatures but remains highly dependent upon soil type [72]. Because of the similarity in amino acid transport system between plants and microorganisms, the latter have the potential to outcompete plant roots for free amino acids in the soil solution, especially in the rhizosphere, where microbial activity can be an order of magnitude higher than in the bulk soil, and the competition for labile organic N may be intense [73].

All microorganisms inhabiting either bulk and rhizosphere soil secrete substantial amounts of proteases into the extracellular soil environment in order to catalyze the hydrolysis of proteins and peptides into their component amino acids [4,74]. As discussed, dissolved organic nitrogen (DON) in the amino acid form may represent a readily available source of C and N to plant roots and soil microflora. Indeed, by using a GC–MS to measure ^{15}N-^{13}C-double labeled amino acid uptake, it was found that 31 plant species from boreal ecosystems, representing a wide variety of plant types, had the ability to take up amino acids from a mixed solution containing 15 amino acids [75].

Experiments were designed to test whether wheat roots could outcompete the rhizosphere microflora for a pulse addition of organic N in the form of three chemically contrasting amino acids (lysine, glycine, and glutamate). Amino acids were added at 100 μM, that is, a concentration typical of soil solution solutes. Both uptake and respiration of amino acids by plant and microflora were measured over a 24-h chase period [76]. The plants roots could only capture, on average, 6% of the added amino acids with the remainder captured by soil microflora, thus confirming that organic N may be of only limited importance in high-input agricultural systems, which use inorganic fertilizers. In addition to the poor competitive ability of plant roots to capture amino acids from the soil solution, the greater nitrate concentrations in agricultural soil solutions, the concentration of the various N solutes in soil solution, the exchangeable inorganic N pool, the slow movement of amino acids in soil relative to nitrate, and the rapid turnover of amino acids by soil microorganisms — all these factors concur to make amino acid N only a secondary N source for plants [76]. It is important to underline that most DON in soil solution shows a high-molecular-weight recalcitrant nature, whereas roots only have the capacity to take up low-molecular-weight DON (e.g., urea, amino acids, polyamines, small peptides) [77].

Uptake of amino acids by root is an energy-driven process whereby the outwardly directed plasma membrane H$^+$-ATPase generates the proton-motive gradient to drive inwardly directed amino acid H$^+$-cotransport [78]. Once the uptake is complete, into the root cytoplasm, amino acids are assimilated both for the production of new cell biomass and to generate energy through deamination and introduction of the keto acids into the TCA cycle. Following uptake, amino acids can also be translocated to the shoot via the xylem whereas some amino acids may also return to the root via the phloem [79].

The methodology adopted for estimating soil solution N concentrations, particularly at the low solute concentrations typical of N-poor environments, is easily liable to error [80]. As an example, significant interferences between ammonium and low-molecular-weight DON are common during analysis [81]. The major problem to be faced when assessing soil DON quantities likely concerns the extraction of the soil solution phase from soil without disrupting roots and fungal hyphae (e.g., during suctions or helped drainage). The preceding risks are more probable in forest soils, because they are usually dominated by ectomychorrhizas, and in grassland soils, because of the high density of fine roots and root hairs. Indeed, it is virtually impossible to remove roots from soil without causing the release of substantial amounts of DON, as mechanical breakage will readily facilitate cell lysis or large exudation burst leading to an overestimation of this pool size [82]. Moreover, the quick biodegradation of labile DON during the extraction procedure cannot be excluded, with a significant underestimation of the pool size. Also, the use of nondisruptive *in situ* suction samplers is prone to significant error and operational problems [83]. There is, therefore, a need to develop and validate noninvasive techniques to enable accurate measurement of DON in the field and in soil–plant mesocosms.

If we assign the value 1 to the relative soil diffusion coefficient of nitrate, the respective coefficients of ammonium, lysine, glycine, and glutamine are, respectively, 8.2×10^{-3}, 3.4×10^{-3}, 2.8×10^{-2}, 3.7×10^{-2} [76]. Because of the high diffusion, nitrate in soil is not only readily available to plant roots, but it is also easily lost from the root zone by leaching. The low diffusion coefficients of amino acids strongly limit their diffusion rate in soil (less than 1 mm d^{-1}), making their consumption by microbes more likely than their uptake by roots. Indeed, the half-life of amino acids in soil is about 4 h [70]. Thus, as discussed before, most plants may be unable to take up organic N compounds in competition with microorganisms. However, there is still a relevant controversy over the degree to which organic N is accessed by plants vs. microbes. To some extent, this may be the result of both the large microbial diversity among different soils and different experimental conditions used among different experiments. For example, environmental factors, such as soil temperature and moisture, can markedly affect microbial activity, and this makes it unrealistic to extrapolate results obtained from controlled pot experiments in growth chambers to field conditions.

Although ammonium can be taken up by plants, and in many cases it is the preferred N source, in many natural and agricultural circumstances, it may also be toxic. Cruz et al. [84] showed that ammonium inhibited the growth of 55% of a wide range of species in relation to equimolar concentration of nitrate. Although many crop plants are differently sensitive to ammonium toxicity depending on its concentration, the crucial factor seems to be the relative concentration of nitrate and ammonium [85]. Indeed plants benefit from a mixture of both nitrate and ammonium but the optimal ratio in the mixture depends on several factors, such as plant species and age, and soil pH. A continuum of plant species exist, ranging from those that prefer exclusively nitrate to those that prefer exclusively ammonium. Because urea and ammonium-based N fertilizers are commonly used, the toxicity of ammonium can have important implications in the agricultural practice. Visual symptoms of ammonium toxicity include chlorosis, growth inhibition, increased root-to-shoot ratios, and wilting (water stress). These changes are concurrent with the tissue accumulation of amino acids, lower concentrations of cations (apart from ammonium), and higher concentrations of inorganic (chloride, sulfate, and phosphate) and organic (carboxylates) anions [86]. The cation/anion imbalance, which results from switching N sources from nitrate to ammonium, is believed to be a major factor in generating toxicity and is known as "ammoniacal syndrome" [87]. Wild plants generally grow in N-limited environments and thus have been selected for optimization of N interception and assimilation. When these plants are placed in enriched N environments, an imbalance between influx, growth, and capacity storage may occur and the result can be an N efflux. Britto et al. [88] suggested that ammonium is not toxic *per se* but rather for its consequences on plant metabolism due to the high energetic costs of ammonium efflux or assimilation.

Root uptake of ammonium results in rhizosphere acidification, possibly as a means of maintaining charge balance within the plant to compensate for ammonium uptake. It has been suggested that this acidification may be a primary cause of ammonium toxicity. However, toxicity has also been observed with pH being controlled [70].

Because toxic concentrations of ammonium in most situations could be the consequence of overfertilization, it may be possible that the toxicity of ammonium depends on the fact that most plants evolved in natural environments with low ammonium concentrations. In addition, it is likely that in many plant species, efficient mechanisms for ammonium exclusion are still evolving [70].

B. THE ROLE OF MICROFAUNA IN THE N MINERALIZATION PROCESS

Populations of soil protozoa largely fluctuate through time [89], and during the decline in protozoan numbers, the respective easily degradable tissues may enter the detrital food-web. In most soils, protozoan biomass equals or exceeds that of all other soil animals taken together — with the exclusion of earthworms [90].

Many experiments using planted microcosms have shown the beneficial effects of protozoan grazing in the rhizosphere for plant growth [91]. Microfaunal stimulation of mineralization of soil organic N via the microbial loop was suggested as the main underlying mechanism [92], and this indirect stimulation by protozoa and bacterivorous nematodes may be even more important than their direct effects. Indeed, in winter wheat fields, the contribution of amoebae and nematodes to overall N mineralization was assessed as much as 18 and 5% [93], whereas their deletion from the food-web model caused reductions of 28 and 12% in N mineralization for amoebae and nematodes, respectively.

Twenty years ago, when the microbial loop concept was first proposed [6], nutrient-based models sufficiently explained the gross consequences of plant–microbes–microfauna interactions, including the C and N secretions by roots. For example, rhizodepositions can provide a very substantial N input in the legume-based grassland systems, and the amount of atmospheric-derived N in the rhizodeposits may exceed that removed by the harvested shoots [94]. Models of N transformations in the rhizosphere show that plant-derived C can support recycling of the N lost from the roots by exudation rather than the N mineralization from native soil organic N [95].

Later, further studies showed discrepancies in the microbial loop model, probably due to additional nutrient-independent effects of protozoa on plant growth [96]. Both mycorrhizal and nonmycorrhizal Norway spruce seedlings were grown in a sand culture and inoculated with naked amoebae and flagellates extracted from native forest soil, or with agar-cultured protozoa. A soil suspension after protozoa elimination was used as a control [96]. Seedlings were grown for 19 weeks in a climate chamber at 20 to 22°C. Protozoa effectively grazed bacteria extracted from the rhizoplane of both nonmycorrhizal and mycorrhizal seedlings and significantly increased seedling growth. However, concentrations of mineral nutrients did not increase in seedlings in the presence of protozoa. Thus the increased growth of seedlings was not caused by nutrients released during amoebal grazing on rhizosphere microorganisms. The protozoa presumably affected plant physiological processes and growth, either directly (via production of phytohormones), or indirectly (via modification of the structure and activity of the rhizosphere microflora) [97]. Recently, mycorrhizal colonization significantly increased the abundance of naked amoebae at the rhizoplane [96].

The exploitation of native soil organic matter by active root growth is known as "root foraging activity" [98]. To separate microfauna-mediated effects on nutrient mineralization from those due to the root foraging activity, a factorial mesocosm experiment with ryegrass grown in the presence of bacterivorous protozoa or nematodes was set up [99]. Moreover, ^{15}N-labeled plant residues were added to create hotspots of microbial activity. In the presence of protozoa the ryegrass biomass doubled, and plant N uptake and microbial incorporation of residue N increased two- and threefold, respectively. Microbial–faunal interactions were a major determinant of plant growth, because root foraging and microfauna presence accounted for 34 and 47% of plant biomass increase, respectively. It was likely that although root foraging in organic hotspots enhanced the spatial coupling of mineralization and plant uptake, microfaunal grazing increased the temporal coupling of nutrient release and plant uptake.

Protozoan grazing often stimulates nitrifying bacteria, presumably through predation on their faster-growing bacterial competitors and through the release of NH_4^+ during bacterial ingestion, resulting in high concentrations of nitrate leachate of rhizosphere soil [100]. If the increase in nitrate concentration is matched by a corresponding increase in root uptake rates, the production of significantly more roots in the presence of protozoa may enable plants to benefit from the liberated N pool [10].

The three-dimensional structure of the soil habitat favors the complex trophic interactions in the rhizosphere. Bacterial biofilms on roots and on the outer zones of soil particles may experience greater grazing pressure than bacteria hardly reachable because they are protected inside tiny crevices or because they are present in different and not communicating water films. The heterogeneity of the rhizosphere in space and time should be considered for better understanding the contribution of predator–prey interactions to the dynamics of rhizosphere processes [101,102].

Protozoan effects in the rhizosphere are likely more complex than previously assumed [103] and, recently, the first microbial loop concept has been integrated via hormonal and root growth effects [97]. Indeed, protozoa do not feed randomly but selectively, thereby stimulating nitrifiers — which take advantage from the ammonium derived from bacteria digestion — or certain bacterial strains capable of promoting plant growth by the release of hormonal substances such as indole-3-acetic acid (IAA+). Both effects may result in a greater and more branched root system, which, in turn, releases more substrates with a positive feedback on rhizosphere bacterial protozoan interactions.

IV. PHOSPHORUS DYNAMICS AND UPTAKE

Plants acquire P as phosphate (Pi) anions from the soil solution; P is likely one of the least available plant nutrients in the rhizosphere because it can be fixed by inorganic colloids, precipitates as insoluble phosphates at both acid and alkaline pH values, and is complexed by organic matter [104]. In this context, P deficiency is considered to be one of the major limitations for crop production, particularly in the tropics. Interestingly the ability of plants to acquire P increases significantly under Pi deficiency [105].

Generally, there is a great disparity in distribution of Pi between plant cells (mM) and soil solution (μM) in unmanaged agroecosystems. Extremely low levels of available P in the rhizosphere, even further reduced by the microbial competition, makes P one of the major growth-limiting factors in many natural ecosystems. The intensive agriculture has increased the total P content of soil steadily since the early 1950s due to the continued inputs of both inorganic and organic P, often in excess of crop requirements [106]; but this increased P input is mostly unavailable to plants because most of the applied P is precipitated or fixed, as discussed earlier [104,107].

Many plants species, especially those adapted to low-phosphate conditions, have developed elegant biochemical mechanisms to solubilize inorganic Pi complexes, phosphate rocks included. As discussed in Chapter 2, plants produce and secrete into the rhizosphere organic acids that, by the exchange ligand mechanism, induce the solubilization of Pi [108].

Organic P is an important source of available P if it is mineralized by phosphatases; phytates may contribute to significant portions (20 to 80%) of total organic P in soil. Phosphatases are a class of enzymes produced by both microorganisms and plants, especially those Pi-starved, as discussed in the next section. Indeed, induction of phosphatases during Pi deficiency, and their release into the rhizosphere (extracellular) or into the apoplast (intercellular) is a universal response in higher plants [109] and microorganisms [110]. Purple acid phosphatases are among the commonly observed phosphatases secreted into the rhizosphere during Pi deficiency, and they represent a distinct class of nonspecific acid phosphatases consisting of binuclear transition metal centers (Fe^{3+}-Fe^{2+}, Fe^{3+}-Mn^{2+}, Fe^{3+}-Zn^{2+}) [111].

RNases are another group of enzymes that may be involved in mobilization of Pi from organic sources during Pi deficiency [112]. Tomato cells produce extracellular cyclic nucleotide phosphodiesterases that are thought to function in concert with RNases in releasing Pi from nucleotides [113]. Diphosphohydrolases such as apyrases, capable of releasing Pi from extracellular ATP, are also induced during Pi deficiency [114].

V. ENZYME ACTIVITIES IN THE RHIZOSPHERE

A. CONCEPTS, MEANING OF MEASUREMENTS, AND ROLE OF PHOSPHATASES

As a consequence of the rhizosphere effect, the number of rhizosphere microorganisms and microbial activities, that is, essentially enzyme activities, are higher than those of the bulk soil because both are sustained by root exudates; the release of enzymes from roots is also possible [115].

Unfortunately, it is not possible to evaluate which of the enzymes differently located are responsible for the increase of the measured enzyme activity in the rhizosphere soil. Indeed, the current enzyme assays determine an overall enzyme activity, which depends on enzymes localized in root cells, root remains, microbial cells, microbial cell debris, microfaunal cells, and the relative cell debris, free in the soil solution extracellular enzymes or enzymes adsorbed or inglobed in soil organomineral particles [115]. Ultracytochemical techniques have been coupled with electron microscopy to localize enzymes in electron-transparent materials of soil such as microbial and root extracellular polysaccharides, fragments of cells walls and microbial membranes, but they cannot be applied in regions of soil with naturally electron-dense particles such as minerals [116]. Acid phosphatase has been detected in roots, mycorrhizae, soil microbial cells, and fragments of microbial membranes as small as 7×20 nm.

Plants are known to release extracellular enzymes so as to mineralize organic compounds to minerals N, P, and S [117]. The same reactions are carried out by extracellular enzymes released by soil microorganisms [115]. In addition, extracellular enzymes released from soil microorganisms initiate the degradation of high-molecular-weight substrates such as cellulose, chitin, lignin, etc. It is important to underline that enzymes attached to the outer surface of microbial cells, the ectoenzymes, can also carry out the hydrolysis of high-molecular-weight substrates [115,116]. Among the extracellular enzymes involved in the mineralization of organic to inorganic nutrients forms, the phosphatases are those more studied. As discussed earlier, acid phosphatase can be secreted, in response to P-deficiency stress, by epidermal cells of the main tip roots of white lupin and in the cell walls and intercellular spaces of lateral roots [117]. Such apoplastic phosphatase is protected against microbial degradation and cannot be adsorbed by soil colloids, but it can only be effective when soluble organophosphates, normally present in the soil solution, diffuse into the apoplastic space [117]. Both acid and alkaline phosphatase activities increased from the bulk soil to the rhizosplane of either 10-d-old clover (*Trifolium alexandrinum*) or 15-d-old wheat (*Triticum aestivum*) [118]. These increases paralleled the increase in both fungal and bacteria counts, suggesting a probable microbial origin of both enzymes in the rhizosphere soil. Both total P and organic P contents decreased in the rhizosphere soil whereas the inorganic P content increased approaching to the rhizoplane; probably, these changes in the concentrations of P forms depended on the increase in both acid and alkaline phosphatase activities of the rhizosphere soil. In addition, both phosphatase activities increased with plant age, probably as the result of the increase in microbial biomass or the increase in total root surface. It has been speculated that plants do not need to secrete phosphatase because the phosphatase activity (mostly of microbial origin) in the rhizosphere soil is generally sufficient to ensure sufficient available P [118,119]. This hypothesis seems to be confirmed by the finding that *Bacillus amyloliquefaciens* FZB45, a plant-growth-promoting rhizobacterium, stimulated growth of maize seedlings under phosphate limitation and in the presence of phytate, whereas a phytase-negative mutant strain FZB45/M2 did not stimulate plant growth [120]. Release of enzymes by plant roots has been also observed in transgenic *Nicotiana tabacum* (tobacco) or in *Arabidopsis thaliana*, modified with β-propeller phytase from *Bacillus subtilis* (*168phyA*) constitutively expressed [121] and in transgenic *Arabidopsis thaliana* modified with phytase gene (*phyA*) from *Aspergillus niger* [122]. However, transgenic *Trifolium subterraneum* L. constitutively expressing a phytase gene (*phyA*) from *Aspergillus niger* was capable of exuding phytase and taking up more P than wild-type plant when grown in agar added with phytate-P but it was not successful when growing in soil [123], probably because plant-extruded phytase was adsorbed by soil colloids or degraded by soil protease [124].

The main problems in interpreting the meaning of enzyme activities in soil are: (1) the current enzyme assays give the potential rather than the actual enzyme activity, because the conditions of incubation assays are much more ideal with respect to those *in situ*, being based on optimal pH and temperature, optimal substrate concentrations, presence of buffers, and shaking of soil slurries; and (2) as already mentioned, the present enzyme assays do not distinguish among many enzyme activities contributing to the measured enzyme activity [115,125]. It would be important to determine the intracellular enzyme activity of active microbial cells so as to get meaningful information on the microbial functional diversity [115]. Several methods have been proposed to discriminate the extracellular stabilized enzyme activity (activity due to enzyme adsorbed or inglobed in soil colloids) from intracellular enzyme activity but the methods proposed so far show several drawbacks [115,125]. The situation is made even more complex in the rhizosphere in comparison with the bulk soil by the presence of active and still intact root cells detached from the roots, mycorrhizal cells strictly linked to roots, and active bacterial, fungal, and microfaunal cells. All these cells possess a broad array of active enzymes.

B. OTHER RHIZOSPHERE ENZYME ACTIVITIES AND RELATIONSHIPS WITH THE EFFECT OF GENETICALLY MODIFIED MICROORGANISMS AND PLANTS

Bacterial and protozoan cell numbers and histidinase and casein hydrolyzing activities were monitored after 21 and 33 d of plant growth in a soil–plant (wheat) microcosm [4] as also reviewed by Badalucco and Kuikman [2] in the first edition of the book. The closer to the soil–root interface, the higher the microbial number and enzyme activities. It was hypothesized that bacteria were the main source of histidinase whereas protease activity was suggested to be indiscriminately produced by bacteria, protozoa, and root hairs.

The effect of low-molecular-weight root exudates on enzyme activities has been studied in a model rhizosphere system in which a cellulose paper wet by the solution of the root exudate simulated the rhizoplane [126]. Different root exudates were mineralized to different extents and had different stimulatory effects on microbial growth and on hydrolase activities, mostly localized in the rhizosphere zone. In particular, the rapid increase in the alkaline phosphatase activity could be considered as an indirect evidence of the important role of rhizobacteria in the synthesis of this enzyme in the rhizosphere [127].

Enzyme activities of the rhizosphere soil have been determined to study the effect due to the introduction of genetically modified microorganisms in the ecosystem [127–130]. Urease and chitobiosidase activities of the rhizosphere soil sampled at 0 to 20 cm depth from wheat, whose seeds had been inoculated with genetically modified (SBW 25 EeZY, which has the marker genes *lacZY*, kan[r], and *xylE*) *Pseudomonas fluorescence,* increased, whereas alkaline phosphatase activity was decreased [127]. If plant growth occurred in the presence of a substrate mixture composed by urea, chitin, and glycerophosphate, opposite changes in enzyme activities were observed. Arylsulfatase, phosphodiesterase, and alkaline phosphatase activities increased in the rhizosphere soil of pea inoculated with *P. fluorescens* strain F113, a wild type producing the antifungal 2,4-diacetylphloroglucinol (DAPG) and marked with *lacZY* gene cassette, with respect to the control without bacterial inoculation or to samples taken from the rhizosphere of plants inoculated with other bacteria [128]. It was hypothesized that the antifungal decreased somehow the amounts of available P and available S, which repress the synthesis of phosphatases and arylsulfatase activities, respectively.

Enzyme activities have also been measured to study the effect of transgenic plants on soil metabolism. Both dehydrogenase and alkaline phosphatase activities of soil sampled from transgenic alfalfa, regardless of association with recombinant nitrogen-fixing soil *Sinorhizobium meliloti*, were significantly lower than those of soil sampled from parental alfalfa [131].

C. LINKING ENZYME ACTIVITY TO GENE EXPRESSION AND PROTEOMICS APPROACH

Measurements of enzyme activity represent the classical biochemical determinations for bulk and rhizosphere soil [115] and only a few studies have been carried out to link these measurements with measurements of expression of genes encoding the proteins molecules with the target enzyme activity. The maximum level of transcripts of three manganese peroxidase (*mn*P) genes during the degradation of PAHs from a culture of *Phanerochaete chrysosporium* grown in presterilized soil preceded by 1 to 2 d the highest manganese peroxidase extracted from soil by the method of Bollag et al. [132], and both peaks occurred during the maximum rate of two PAHs, fluorene and chrysene, degradation [133].

However, the best attempt to cover the whole sequence of events starting from gene expression to detection of target enzyme in soil was carried by Metcalfe et al. [134]. Chitinase activity was measured by loss of chitin in buried litter-bags and by a luminescence assay by using as a substrate, 4-methylambelliferyl-$(GlcNAc)_2$, whereas the composition of community was evaluated by DNA extraction, cloning, and sequencing of PCR products by using primers for family 18 group A chitinases. The addition of sludge to the pasture soil increased the chitinase activity and the number of actinobacteria but decreased the diversity of chitinase enzymes. Unfortunately, extraction of transcripts was unsuccessful, probably due to the adsorption of mRNA by soil colloids. In addition, it would be important to extract the target enzyme proteins to complete the sequence of events.

It is well established that more protein isoforms can be synthesized by a single gene because mRNA molecules can be subjected to posttranscriptional control such as alternative splicing, polyadenylation, and mRNA editing [135–137]. The proteomics approach, thus, should complete a study on gene expression in the rhizosphere soil. However, soil proteomics is still in its infancy — mainly because it is difficult to extract the intracellular proteins from soil because there is a large background of extracellular protein N, and microbial N only accounts for 4% of the total organic N in soil [136].

VI. SOME NEW APPROACHES AND TECHNIQUES FOR STUDYING RHIZOSPHERE PROCESSES

The bulk nature of low-molecular-weight exudates is better known at present by chemically profiling the composition of root solutions in hydroponic cultures [138]. Clearly, the extrapolation of such experiments to *in situ* conditions is distorted by the absence of the soil environment, in which the physical, nutritional, chemical, and microbial factors have a determinant impact on both the amount and quality of rhizodepositions [139]. However, the tight coupling of release of C-compounds from roots to their rapid utilization by the soil microbial biomass, or their adsorption on soil surfaces, has the consequence that direct characterization of rhizodeposits from natural soils is constitutively difficult.

Recently, an attempt has been made to investigate *in situ* the nature and dynamic fluxes of root-derived neutral sugars, which are cited as the most abundant family of exudates [140] released by plant in the rhizosphere; pulse-chase isotope labeling of photoassimilated $^{13}CO_2$ has been coupled with quantitative analysis of labeled molecules by capillary gas chromatography/combustion/isotope ratio mass spectrometry (GC/C/IRMS) [141]. Seven monosaccharides were identified and quantified in the rhizosphere: glucose (37%), mannose (17%), galactose (15%), arabinose (12%), xylose (9%), rhamnose (7%), and fucose (3%) with no significant differences among sampling dates. The amount of organic ^{13}C in the rhizosphere soil, expressed as a percentage of the total photosynthetically fixed ^{13}C at the end of the labeling period, reached 16% after 1 d since the labeling and stabilized at 9% after 1 week. Glucose as moiety of polymers was the most abundant sugar in the rhizodeposits, whereas it disappeared as soluble form after 2 d. Forty percent of the root-derived C was in the form of neutral sugars, with prevalence of vegetal sugars. However, by prolonging the incubation time, the importance of microbial sugars increased, and the signature tended toward that of bulk soil organic matter [141].

The analysis of neutral sugars by GC requires a derivatization of the polar groups, thus involving the addition of several unlabeled C atoms per sugar. Different techniques have been applied to GC/C/IRMS for various monosaccharides in biogeochemistry. Macko et al. [142] used alditol acetate derivatization, whereas van Dongen [143] has adapted the methylboronic derivatization, which reduces the number of added atoms.

Models of C flows through the soil–plant system have been developed following isotope labeling of specific plant cultivars [144], but neglecting any assessment of C transfer into specific microbial taxa. Clearly, uncertainty over the role of microbial diversity in soil functioning highlights the need to assess community structure from both a taxonomic and functional perspective [145]. Traditionally, this has not been feasible due to the lack of suitable methods to identify the dominant taxa and determine functional roles, because more than 90% of soil microorganisms are unculturable [146]. Advances in molecular techniques, based on the extraction of total nucleic acids and phylogenetic analyses of amplified genes, such as 16S rRNA now allow a more accurate measure of *in situ* microbial diversity [147,148].

A major advance in linking functional activity to community structure came with the development of stable isotope probing (SIP) [149], which involves tracking of a stable isotope atom from a particular substrate into components of microbial cells that provide phylogenetic and functional information, such as lipid, DNA, or RNA. Indeed, the major advantage of the SIP technique is that ^{13}C-enriched DNA will contain the entire genome of each functionally active microbe of the community. Detailed methodology, potential, and future improvements needed for the SIP technique have been already reviewed [150–152]. Successful applications of this technique are restricted to reactions of microbial anabolism, because SIP is based on assimilatory processes [152]. Thus, nonassimilatory chemical transformations, which also occur in soil, fall outside the applicability of SIP. Even if the SIP technique can theoretically be applied to trace the assimilation of any element of biological importance that has a stable isotope, it has almost exclusively been restricted to the use of ^{13}C.

Radajewski et al. [149,153] were the first to use the ^{13}C-labeled substrates (methane and methanol) to label nucleic acids of soil. The labeled nucleic acids were separated from 'natural' (unlabeled) DNA by equilibrium density centrifugation in CsCl gradients. By amplifying functional genes involved in the oxidation of one carbon compound, it was found that not only methanol dehydrogenase and methane monooxygenase genes were involved in methanol and methane assimilation, but also species encoding ammonia monooxygenase had assimilated ^{13}C as ^{13}CO$_2$ generated by the methylotrophs [153]. Indeed, a serious drawback of the SIP approach is the possibility of secondary feeding on breakdown products of the primary substrate. Another drawback of this technique is the presence of unlabeled substrates native to the system that will compete for assimilation. This has led researchers to apply artificially high concentrations of labeled substrates into soil microcosms for extended periods of time [152] but the relevance of this approach at the real situation *in situ* has been questioned and the use of pulse ^{13}C-labeled compounds has been suggested [154,155]. Another important issue of any SIP methodology concerns the degree of labeling, which depends on number of organisms using the established substrate [152]. When the substrate is consumed by a broad diversity of organisms, the degree of labeling in any substrate-using taxa will be low, making separation by density problematic, whereas if the substrate is consumed by a small number of taxa, then the degree of labeling in the specific taxa will be high, facilitating isolation by density [152].

There are differences in using fatty acids or nucleic acids as biomarkers in the SIP technique; PLFAs are more rapidly labeled and give more quantitative information when analyzed with IRMS than nucleic acids [152]. However, extraction of PLFAs from soil is more laborious than nucleic acid extraction. In addition, the PLFA-based SIP gives an inferior phylogenetic resolution than that offered by nucleic-acid-based biomarkers and signature PLFAs have to be identified from close culturable relatives [152].

Accurate results can be obtained by applying SIP to soil if the delivery of a pulse is carefully planned by considering the ability of soil colloids to adsorb biological molecules and the solubility and volatility of the used substrates. In addition, because cell replication in soil is slow, there are limitations to the isotopic enrichment of DNA with pulse labeling, unless the duration of a pulse is extended. To solve this problem, 16S rRNA based SIP methodologies have been developed for studying microbial assimilation process in soil because rates of RNA synthesis are always higher than those of DNA due to the fact that RNA is turned over in bacteria independently of replication [152]. In spite of the fact that the use of RNA as a biomarker and the precise quantitative examination of gradient profiles has enhanced the sensitivity of SIP, the application of the technique to rhizosphere and bulk soil still presents some drawbacks such as the ability to extract clean and intact DNA or RNA from the soil or to sufficiently label nucleic acids of microorganisms involved in the metabolism of plant root exudates [152]. Grassland monoliths (400 mm diameter × 200 mm depth) were pulsed with $^{13}CO_2$ to promote the release of ^{13}C labeled root exudates into soil [29] but the analysis of 16S rRNA from root-associated soil by IRMS and equilibrium density centrifugation showed that the degree of labeling was too low to get meaningful results [152,156]. On the other hand, the PLFA–SIP analysis of soil samples derived from a $^{13}CO_2$ plant pulse showed the assimilation of root exudates by Gram-negative bacteria and fungi, but the phylogenetic resolution was low [157]. Therefore, it is problematic to use $^{13}CO_2$ to label soil microbes via root exudates because a broad range of labeled organic compounds are released as root exudates and several microbial species can use the labeled root exudates. According to Manefield et al. [152], it can be more rewarding to pulse directly, soil with labeled root exudate compounds, and monitoring microorganisms of rhizosphere soil involved in the assimilation of the target compound by the use of any SIP technique. This can be done in experimental systems simulating the delivery of root exudates into soil [2,35].

VII. CONCLUDING REMARKS

Nutrient transformations in rhizosphere are of critical economic and social importance, being central to nutrient and pest-control management strategies in agriculture and forestry, the function and maintenance of terrestrial ecosystems, the mitigation of climate, and the cleaning up of contaminated sites. Among these processes, those due to plant microorganisms and fauna interactions are the most intense and the most varied. The interplay between protozoa, bacteria, and plant roots is more complex than that previously hypothesized by the "microbial loop" [2,6] because effects of rhizobacteria on root architecture seem to be driven mainly by protozoan grazers [97]. Because the regulation of root architecture is a key determinant of nutrient- and water-use efficiency in plants, it is crucial to include soil protozoan grazers to advance our understanding of the mechanisms underlying plant–microbial interactions. Thus more research is needed to understand better the complex trophic interactions in the rhizosphere; research systems should mimic the real conditions occurring in soil because, for example, the presence of soil particles can adsorb and make ineffective molecular signals regulating the trophic interactions. Thus, plant–microorganism interactions have been extensively studied, but major questions remain still unclear because of an excessive reliance on cultivation-based techniques, and a frequent inability to clearly determine the origin and fate of mediating organic compounds from plants to microorganisms and *vice versa*. It is not always clear which microbial species carry out the investigated microbial nutrient transformations in the rhizosphere. The use of modern techniques has allowed understanding better nutrient transformation in the rhizosphere. A simultaneous measurement of C, N, and P in the rhizosphere soil solution has been conducted by Standing et al. [158] by using a tripartite reporter gene system.

The SIP methodology can allow determining the microbial species involved in the degradation of organic compounds released from roots. According to Manefield et al. [152], this should be done by pulsing directly, soil with labeled root exudate compounds, and monitoring microorganisms of rhizosphere soil involved in the assimilation of the target compound by the use of any SIP technique. Experimental systems can mimic the delivery of root exudates into soil and allow sampling soil layers at different distances from the simulated rhizoplane [2,35].

REFERENCES

1. Coleman, D.C. et al., Trophic interactions in soil as they affect energy and nutrient dynamics. I. Introduction, *Microb. Ecol.*, 4, 345, 1978.
2. Badalucco, L. and Kuikman, P.J., Mineralization and immobilization in the rhizosphere, in *The Rhizosphere — Biochemistry and Organic Substances at the Soil-Plant Interface*, Pinton, R., Varanini, Z., and Nannipieri, P., Eds., Marcel Dekker, New York, 2001, chap. 6.

3. Gahoonia, T.S. and Nielsen, N.E., A method to study rhizosphere processes in thin soil layers of different proximity to roots, *Plant Soil*, 135, 143, 1991.

4. Badalucco, L., Kuikman, P.J., and Nannipieri, P., Protease and deaminase activities in wheat rhizosphere and their relation to bacterial and protozoan populations, *Biol. Fertil. Soils*, 23, 99, 1996.

5. Yevdokimov, I. et al., Microbial immobilisation of ^{13}C rhizodeposits in rhizosphere and root-free soil under continuous ^{13}C labelling of oats, *Soil Biol. Biochem.*, 38, 1202, 2006.

6. Clarholm, M., The microbial loop in soil, in *Beyond the Biomass*, Ritz, K., Dighton, J., and Giller, K.E., Eds., Wiley-Sayce, Chichester, U.K., 1994, chap. 22.

7. Kuikman, P.J. et al., Protozoan predation and the turnover of soil organic carbon and nitrogen in the presence of plants, *Biol. Fertil. Soils*, 10, 22, 1990.

8. Griffiths, B.S. et al., Protozoa and nematodes on decomposing barley roots, *Soil Biol. Biochem.*, 25, 1293, 1993.

9. Kaye, J.P. and Hart, S.C., Competition for nitrogen between plants and soil microorganisms, *Trends Ecol. Evol.*, 12, 139, 1997.

10. Bonkowski, M. et al., Microbial-faunal interactions in the rhizosphere and effects on plant growth, *Eur. J. Soil Biol.*, 36, 135, 2000.

11. Matamala, R. et al., Impacts of fine root turnover on forest NPP and soil C sequestration potential, *Science*, 302, 1385, 2003.

12. Hungate, B.A. et al., The fate of carbon in grasslands under carbon dioxide enrichment, *Nature*, 388, 576, 1997.

13. Niklaus, P.A., Effects of six years atmospheric CO_2 enrichment on plant soil, and soil microbial C of a calcareous grassland, *Plant Soil*, 233, 189, 2001.

14. Meharg, A.A., A critical review of labelling techniques used to quantify rhizosphere carbon flow, *Plant Soil*, 122, 225, 1994.

15. Horwath, W.R., Pregitzer, K.S., and Paul, E.A., ^{14}C allocation in tree-soil systems, *Tree Physiol.*, 14, 1163, 1994.

16. Liljeroth, E., Kuikman, P.J., and Van Veen, J.A., Carbon translocation to the rhizosphere of maize and wheat and influence on the turnover of native soil organic matter at different soil nitrogen levels, *Plant Soil*, 161, 233, 1994.

17. Paterson, E., Rattray E.A.S., and Killham, K., Effect of elevated CO_2 concentration on C-partitioning and rhizosphere C-flow for tree plant species, *Soil Biol. Biochem.*, 28, 195, 1996.

18. Killham, K. and Yeomans, C., Rhizosphere carbon flow measurement and implications: from isotopes to reporter genes, *Plant Soil*, 232, 91, 2001.

19. Balesdent, J. and Balabane, M., Major contribution of roots to soil carbon storage inferred from maize cultivated soils, *Soil Biol. Biochem.*, 28, 1261, 1996.

20. Mary, B., Mariotti, A., and Morrel, J.L., Use of ^{13}C variations at natural abundance for studying the biodegradation of root mucilage, roots and glucose in soil, *Soil Biol. Biochem.*, 24, 1065, 1992.

21. Andreux, F., Humus in world soils, in *Humic Substances in Terrestrial Ecosystems*, Piccolo, A., Ed., Elsevier, Amsterdam, 1996, pp. 45–100.

22. Hanson, P.J. et al., Separating root and soil microbial contributions to soil respiration: a review of methods and observations, *Biogeochemistry*, 48, 115, 2000.

23. Rochette, P., Flanagan, L.B., and Gregorich, E.G., Separating soil respiration into plant and soil components using analyses of the natural abundance of carbon-13, *Soil Sci. Soc. Am. J.*, 63, 1207, 1999.

24. Cheng, W., Measurement of rhizosphere respiration and organic matter decomposition using natural ^{13}C, *Plant Soil*, 183, 263, 1996.

25. Santruckova, H., Bird, M.I., and Lloyd, J., Microbial processes and carbon-isotope fractionation in tropical and grassland temperate soils, *Funct. Ecol.*, 14, 108, 2000.

26. Kuziakov, Y., Ehrensberger, H., and Stahr, K., Carbon partitioning and below-ground translocation by Lolium perenne, *Soil Biol. Biochem.*, 33, 61, 2001.

27. de Neergaard, A. and Magid, J., Influence of the rhizosphere on microbial biomass and recently formed organic matter, *Eur. J. Soil Sci.*, 52, 377, 2001.

28. Butler, J.L. et al., Distribution and turnover of recently fixed photosynthate in ryegrass rhizospheres, *Soil Biol. Biochem.*, 36, 371, 2004.

29. Ostle, N. et al., Active microbial RNA turnover in a grassland soil estimated using a $^{13}CO_2$ spike, *Soil Biol. Biochem.*, 35, 887, 2003.

30. Bremer, E. and Kuikman, P.J., Microbial utilization of [14]C(U)glucose in soil is affected by the amount and timing of glucose additions, *Soil Biol. Biochem.*, 26, 511, 1994.

31. Nguyen, C. and Guckert, A., Short-term utilisation of [14]C-[U]glucose by soil microorganisms in relation to carbon availability, *Soil Biol. Biochem.*, 33, 53, 2001.

32. Kochian, L.V., Piñeros, M.A., and Owen Hoekenga, O.A., The physiology, genetics and molecular biology of plant aluminium tolerance and toxicity, *Plant Soil*, 274, 175, 2005.

33. Veneklaas, E.J. et al., Chickpea and white lupin rhizosphere carboxylates vary with soil properties and enhance phosphorus uptake, *Plant Soil*, 248, 187, 2003.

34. Grayston, S.J., Vaughan, D., and Jones, D., Rhizosphere carbon flow in trees, in comparison with annual plants: the importance of root exudation and its impact on microbial activity and nutrient availability, *Appl. Soil Ecol.*, 5, 29, 1996.

35. Falchini, L. et al., CO_2 evolution and denaturing gradient gel electrophoresis profiles of bacterial communities in soil following addition of low molecular weight substrates to simulate root exudation, *Soil Biol. Biochem.*, 36, 775, 2003.

36. Grayston, S.J. et al., Selective influence of plant species on microbial diversity in the rhizosphere, *Soil Biol. Biochem.*, 30, 369, 1998.

37. Darrah, P.R., Measuring the diffusion coefficient of rhizosphere exudates in soil. I. The diffusion of non-sorbing compounds, *J. Soil Sci.*, 42, 413, 1991.

38. Darrah, P.R., Measuring the diffusion coefficient of rhizosphere exudates in soil. II. The diffusion of sorbing compounds, *J. Soil Sci.*, 42, 421, 1991.

39. Morris, S.J. and Allen, M.F., Oxalate-metabolizing microorganisms in sagebrush steppe soil, *Biol. Fertil. Soils*, 18, 255, 1994.

40. Griffiths, B.S. et al., Soil microbial community structure: effects of substrate loading rates, *Soil Biol. Biochem.*, 31, 145, 1999.

41. Morton, J.B., Fungi, in *Principles and Applications of Soil Microbiology*, Sylvia, D.M., Fuhrmann, J.J., Hartel, P.G., and Zuberer, D.A., Eds., Prentice Hall, Upper Saddle River, NJ, 1999, p. 72.

42. IPCC Third Assessment Report-Climate Change, The Scientific Basis Technical Summary, Geneva, 2001.

43. Fitter, A.H. et al., Root production and turnover and carbon budgets of two contrasting grasslands under ambient and elevated atmospheric carbon dioxide concentrations, *New Phytologist*, 137, 247, 1997.

44. Allard, V. et al., Increased quantity and quality of coarse soil organic matter fractions at elevated CO_2 in a grazed grassland are a consequence of enhanced root growth rate and turnover, *Plant Soil*, 276, 49, 2005.

45. Körner, C., Biosphere responses to CO_2 enrichment, *Ecol. Appl.*, 10, 1590, 2001.

46. Poorter, H. and Navas, M.-L., Plant growth and competition at elevated CO_2: on winners, losers and functional groups, *New Phytologist*, 157, 175, 2003.

47. Rattray, E.A.S., Paterson, E., and Killham, K., Characterisation of the dynamics of C-partitioning within *Lolium perenne* and to the rhizosphere microbial biomass using ^{14}C pulse chase, *Biol. Fertil. Soils*, 19, 280, 1995.

48. Hendrey, G.R. et al., A free-air enrichment system for exposing tall forest vegetation to elevated atmospheric CO_2, *Glob. Change Biol.*, 5, 293, 1999.

49. Gorissen, A. and Cotrufo, M.F., Elevated carbon dioxide effects on nitrogen dynamics in grasses, with emphasis on rhizosphere processes, *Soil Sci. Soc. Am. J.*, 63, 1695, 1999.

50. Hodge, A. et al., Characterisation and microbial utilisation of exudates material from the rhizosphere of Lolium perenne grown under CO_2 enrichment, *Soil Biol. Biochem.*, 30, 1033, 1998.

51. Zak, D.R. et al., Elevated atmospheric CO_2, fine roots and the response of soil microorganisms: a review and hypothesis, *New Phytologist*, 147, 201, 2000.

52. Hu, S., Firestone, M.K., and Chapin, F.S., III, Soil microbial feedbacks to atmospheric CO_2 enrichment, *Trends Ecol. Evol.*, 14, 433, 1999.

53. Cardon, Z.G. et al., Contrasting effects of elevated CO_2 on old and new soil carbon pools, *Soil Biol. Biochem.*, 33, 365, 2001.

54. Cheng, W.X., Rhizosphere feedbacks in elevated CO_2, *Tree Physiol.*, 19, 313, 1999.

55. Baggs, E.M. and Blum, H., CH_4 oxidation and emissions of CH_4 and N_2O from Lolium perenne swards under elevated atmospheric CO_2, *Soil Biol. Biochem.*, 36, 713, 2004.

56. Nannipieri, P. and Badalucco, L., Biological processes, in *Handbook of Processes and Modeling in the Soil-Plant System*, Benbi, D.K. and Nieder, R., Eds., The Haworth Press, New York, 2003, chap. 3.
57. Hall, J.M., Paterson, E., and Killham, K., The effect of elevated CO_2 concentration and soil pH on the relationship between plant growth and rhizosphere denitrification potential, *Glob. Change Biol.*, 4, 209, 1998.
58. Hagedorn, F. et al., Responses of N fluxes and pools to elevated atmospheric CO_2 in model forest ecosystems with acidic and calcareous soils, *Plant Soil*, 224, 273, 2000.
59. Merino, P. et al., Nitrification and denitrification derived N_2O production from a grassland soil under application of DCD and Actilith F2, *Nutr. Cycl. Agroecosyst.*, 60, 9, 2001.
60. Pedersen, H., Dunkin, K.A., and Firestone, M.K., The relative importance of autotrophic and heterotrophic nitrification in a conifer forest soil as measured by ^{15}N tracer and pool dilution techniques, *Biogeochemistry*, 44, 135, 1999.
61. Raab, T.K., Lipson, D.A., and Monson, R.K., Soil amino acid utilisation among the Cyperaceee: plant and soil processes, *Ecology*, 80, 2408, 1999.
62. Neumann, G. and Römheld, V., The release of root exudates as affected by the plant's physiological status, in *The Rhizosphere — Biochemistry and Organic Substances at the Soil-Plant Interface*, Pinton, R., Varanini, Z., and Nannipieri, P., Eds., Marcel Dekker, New York, 2001, chap. 3.
63. Jones, D.L. and Darrah, P.R., Amino-acid influx at the soil-root interface of *Zea mays* L. and its implications in the rhizosphere, *Plant Soil*, 163, 1, 1994.
64. Hart, J.J. et al., Characterization of cadmium binding, uptake, and translocation in intact seedlings of bread and durum wheat cultivars, *Plant Physiol.*, 116, 1413, 1998.
65. Farrar, J.F. et al., How roots control the flux of carbon to the rhizosphere, *Ecology*, 84, 827, 2003.
66. Jones, D.L. et al., Dissolved organic nitrogen uptake by plants-an important N uptake pathway?, *Soil Biol. Biochem.*, 37, 413, 2005.
67. Fierer, N. and Schimel, J.P., Effects of drying-rewetting frequency on soil carbon and nitrogen transformations, *Soil Biol. Biochem.*, 34, 777, 2002.
68. Herrmann, A. and Witter, E., Sources of C and N contributing to the flush in mineralization upon freeze-thaw cycles in soils, *Soil Biol. Biochem.*, 34, 1495, 2002.
69. Gastal, F. and Lemaire, G., N uptake and distribution in crops: an agronomical and ecophysiological perspective, *J. Exp. Bot.*, 53, 789, 2002.
70. Miller, A.J. and Cramer, M.D., Root nitrogen acquisition and assimilation, *Plant Soil*, 274, 1, 2005.
71. Näsholm, T. et al., Accumulation of amino acids in some boreal forest plants in response to increased nitrogen availability, *New Phytologist*, 126, 137, 1994.
72. Jones, D.L. and Hodge, A., Biodegradation kinetics and sorption reactions of three differently charged amino acids in soil and their effects on plant organic nitrogen availability, *Soil Biol. Biochem.*, 31, 1331, 1999.
73. Bardgett, R.D., Streeter, T.C., and Bol, R., Soil microbes compete effectively with plants for organic-nitrogen inputs to temperate grasslands, *Ecology*, 84, 1277, 2003.
74. Kandeler, E. et al., Xylanase, invertase and protease at the soil-litter interface of a loamy sand, *Soil Biol. Biochem.*, 31, 1171, 1999.
75. Persson, J. and Näsholm, T., Amino acid uptake: a widespread ability among boreal forest plants, *Ecol. Lett.*, 4, 434, 2001.
76. Owen, A.G. and Jones, D.L., Competition for amino acids between wheat roots and rhizosphere microorganisms and the role of amino acids in plant N acquisition, *Soil Biol. Biochem.*, 33, 651, 2001.
77. Yu, Z., Contribution of amino compounds to dissolved organic nitrogen in forest soils, *Biogeochemistry*, 61, 173, 2002.
78. Fischer, W.N. et al., Amino acid transport in plants, *Trends Plant Sci.*, 3, 188, 1998.
79. Schenk, M.K., Regulation of nitrogen uptake on the whole plant level, *Plant Soil*, 181, 131, 1996.
80. Smethurst, P.J., Soil solution and other soil analyses as indicators of nutrient supply: a review, *For. Ecol. Manage.*, 138, 397, 2000.
81. Tiensing, T. et al., Soil solution extraction techniques for microbial ecotoxity testing: a comparative evaluation, *J. Environ. Monit.*, 3, 91, 2001.
82. Lorenz, S.E., Hamon, R.E., and McGrath, R.P., Differences between soil solution obtained from rhizozphere and non-rhizosphere soils by water displacement and soil centrifugation, *Eur. J. Soil Sci.*, 45, 431, 1994.
83. Wolt, J.D., *Soil Solution Chemistry: Applications to Environmental Science and Agriculture*, John Wiley and Sons, New York, 1994.

84. Cruz, C., Lips, S.H., and Martins-Loução, M.A., Interactions between nitrate and ammonium during uptake by carob seedlings and the effect and the form of earlier nitrogen nutrition, *Physiol. Plant.,* 89, 544, 1993.

85. Britto, D.T. and Kronzucker, H.J., NH_4^+ toxicity in higher plants: a critical review, *J. Plant Physiol.,* 159, 567, 2002.

86. Cramer, M.D. and Lewis, O.A.M., The influence of NO_3^- and NH_4^+ nutrition on the growth of wheat (Triticum aestivum) and maize (Zea mays) plants, *Ann. Bot.,* 72, 359, 1993.

87. Chaillou, S. and Lamaze, T., Ammoniacal nutrition of plants, in *Nitrogen Assimilation by Plants,* Morot-Gaudry, J.-F., Ed., Science Publishers, New Hampshire, 2001, p. 53.

88. Britto, D.T. et al., Futile transmembrane NH_4^+ cycling: a cellular hypothesis to explain ammonium toxicity in plants, *Proc. Natl. Acad. Sci. USA,* 98, 4255, 2001.

89. Janssen, M.P.M. and Heijmans, G.J.S.M., Dynamics and stratification of protozoa in the organic layer of a Scots pine forest, *Biol. Fertil. Soils,* 26, 285, 1998.

90. Schröter, D., Wolters, V., and De Ruiter, P.C., C and N mineralization in the decomposer food webs of a European forest transect, *Oikos,* 102, 294, 2003.

91. Ekelund, F. and Rønn, R., Notes on protozoa in agricultural soil with emphasis on heterotrophic flagellates and naked amoebae and their ecology, *FEMS Microbiol. Rev.,* 15, 321, 1994.

92. Zwart, K.B., Kuikman, P.J., and van Veen, A.J., Rhizosphere protozoa: their significance in nutrient dynamics, in *Soil Protozoa,* Darbyshire, J.F., Ed., CAB International, Wallingford, U.K., 1994, p. 93.

93. De Ruiter, P.C. et al., Simulation of nitrogen mineralization in the below-ground food webs of two winter wheat fields, *J. Appl. Ecol.,* 30, 95, 1993.

94. Høgh-Jensen, H. and Schjoerring, J.K., Rhizodeposition of nitrogen by red clover, white clover and ryegrass leys, *Soil Biol. Biochem.,* 33, 439, 2001.

95. Griffiths, B.S. and Robinson, D., Root-induced nitrogen mineralization: a nitrogen balance model, *Plant Soil,* 139, 253, 1992.

96. Jentschke, G. et al., Soil protozoa and forest tree growth: non-nutritional effects and interaction with mycorrhizas, *Biol. Fertil. Soils,* 20, 263, 1995.

97. Bonkowski, M. and Brandt, F., Do soil protozoa enhance plant growth by hormonal effects?, *Soil Biol. Biochem.,* 34, 1709, 2002.

98. Robinson, D., The response of plants to nonuniform supplies of nutrients, *New Phytologist,* 127, 635, 1994.

99. Bonkowski, M., Griffiths, B.S., and Scrimgeour, C., Substrate heterogeneity and microfauna in soil organic "hotspots" as determinants of nitrogen capture and growth of rye-grass, *Appl. Soil Ecol.,* 14, 37, 2000.

100. Verhagen, F.J.M. et al., Competition for ammonium between nitrifying bacteria and plant roots in soil in pots; effects of grazing by flagellates and fertilization, *Soil Biol. Biochem.,* 26, 89, 1994.

101. Semenov, A.M., van Bruggen, A.H.C., and Zelenev, V.V., Moving waves of bacterial populations and total organic carbon along roots of wheat, *Microb. Ecol.,* 37, 116, 1999.

102. Young, I.M. and Ritz, K., Can there be a contemporary ecological dimension to soil biology without a habitat? *Soil Biol. Biochem.,* 30, 1229, 1998.

103. Clarholm, M., Interactions of bacteria, protozoa an plants leading to mineralization of soil nitrogen, *Soil Biol. Biochem.,* 17, 181, 1985.

104. Brady, N.C. and Weil, R.R., *The Nature and Properties of Soils,* 13th ed., Prentice Hall, Upper Saddle River, NJ, 2002.

105. Schactman, D.P., Reid, R.J., and Ayling, S.M., Phosphorus uptake by plants: from soil to cell, *Plant Physiol.,* 116, 447, 1998.

106. Condron, L.M., Phosphorus — Surplus and deficiency, in *Managing Soil Quality — Challenges in Modern Agriculture,* Schjønning, P., Elmholt, S., and Christensen, B.T., Eds., CAB International, Wallingford, Oxon, U.K., 2004, chap. 5.

107. Holford, J.C.R., Soil phosphorus: its measurement, and its uptake by plants, *Aust. J. Soil Res.,* 35, 227, 1997.

108. Hinsinger, P., Bioavailability of soil organic P in the rhizosphere as affected by root–induced chemical changes: a review, *Plant Soil,* 237, 173, 2001.

109. Duff, S.M.G., Sarath, G., and Plaxton, W.C., The role of acid phosphatase in plant phosphorus metabolism, *Physiol. Plant.,* 90, 791, 1994.

110. Bosse, D. and Kock, M., Influence of phosphate starvation on phosphohydrolases during development of tomato seedlings, *Plant Cell Environ.,* 21, 325, 1998.

111. Li, D. et al., Purple acid phosphatases of Arabidopsis thaliana, *J. Biol. Chem.*, 277, 27772, 2002.

112. Green, P.J., The ribonucleases of higher plants, *Annu. Rev. Plant Physiol. Plant Mol. Biol.*, 45, 421, 1994.

113. Abel, S., Ticconi, C.A., and Delatorre, D.A., Phosphate sensing in higher plants, *Physiol. Plant.*, 115, 1, 2002.

114. Thomas, C. et al., Apyrase functions in plant phosphate nutrition and mobilizes phosphate from extracellular ATP, *Plant Physiol.*, 119, 543, 1999.

115. Nannipieri, P., Kandeler, E., and Ruggiero, P., Enzyme activities and microbiological and biochemical processes in soil, in *Enzymes in the Environment: Activity, Ecology and Applications*, Burns, R.G. and Dick, R.P., Eds., Marcel Dekker, New York, 2001, p. 1.

116. Ladd, J.N. et al., Soil structure and biological activity, in *Soil Biochemistry*, Vol. 9, Stotzky, G. and Bollag, J.-M., Eds., Marcel Dekker, New York, 1996, p. 23.

117. Wasaki, J. et al., Properties of secretory acid phosphatase from lupin roots under phosphorus-deficient conditions, in *Plant Nutrition for Sustainable Food production and Environment*, Ando, T. et al., Eds., Kluwer Academic Publisher, Dordrecht, 1997, pp. 295–300.

118. Tarafdar, C. and Jungk, A., Phosphatase activity of rhizosphere soil and its relation to the depletion of soil organic phosphorus, *Soil Biol. Biochem.*, 3, 199, 1987.

119. Tarafdar, J.C. and Marschner, H., Phosphatase activity in the rhizosphere and hyphosphere of VA mycorrhizal wheat supplied with inorganic and organic phosphorus, *Soil Biol. Biochem.*, 26, 387, 1994.

120. Idriss, E.I. et al., Extracellular phytase activity of *Bacillus amyloliquefaciens* FZB45 contributes to its plant-growth-promoting effect, *Microbiology*, 148, 2097, 2002.

121. Lung, S.-C. et al., Secretion of beta-propeller phytase from tobacco and *Arabidopsis* roots enhances phosphorus utilization, *Plant Sci.*, 169, 341, 2005.

122. Richardson, A.E., Hadobas, P.A., and Hayes, J.E., Extracellular secretion of *Aspergillus* phytase from *Arabidopsis* root enables plants to obtain phosphorus from phytate, *Plant J.*, 25, 641, 2001.

123. George, T.S. et al., Characterization of transgenic *Trifolium subterraneum* L which expresses *phyA* and release extracellular phytase: growth and P nutrition in laboratory media and soil, *Plant Cell Environ.*, 27, 1351, 2004.

124. George, T.S, Richardson, A.E., and Simpson, R.J., Behaviour of plant-derived extracellular phytase upon addition to soil, *Soil Biol. Biochem.*, 37, 977, 2005.

125. Gianfreda, L. and Ruggiero, P., Enzyme activities in soil, in *Nucleic Acids and Proteins*, Nannipieri, P. and Smalla, K., Eds., Springer, Heidelberg, 2006, p. 257.

126. Renella, G. et al., Microbial activity and hydrolase activities during decomposition of model root exudates released by a model root surface in Cd-contaminated soil, *Soil Biol. Biochem.*, 37, 133, 2005.

127. Naseby, D.C. and Lynch, J.M., Rhizosphere soil enzymes as indicators of perturbations caused by enzyme substrate addition and inoculation of a genetically modified strain of *Pseudomonas fluorescens* on wheat seed, *Soil Biol. Biochem.*, 29, 1353, 1997.

128. Naseby, D.C. and Lynch, J.M., Impact of wild-type and genetically modified *Pseudomonas fluorescens* on soil enzyme activities and microbial population structure in the rhizosphere of pea, *Mol. Ecol.*, 7, 617, 1998.

129. Naseby, D.C. and Lynch, J.M., Enzymes and microorganisms in the rhizosphere, in *Enzymes in the Environment: Activity, Ecology and Applications*, Burns, R.G. and Dick, R.P., Eds., Marcel Dekker, New York, 2002, p. 109.

130. Naseby, D.C. et al., Soil enzyme activities in the rhizosphere of field-grown sugar beet inoculated with the biocontrol agent *Pseudomonas fluorescens* F113, *Biol. Fertil. Soils*, 27, 39, 1998.

131. Donegan, K.K. et al., A field study with genetically engineered alfalfa inoculated with recombinant *Sinorhizobium meliloti*: effects on the soil ecosystem, *J. Appl. Ecol.*, 36, 920, 1999.

132. Bollag, J.-M. et al., Extraction and purification of a peroxidase from soil, *Soil Biol. Biochem.*, 19, 61, 1987.

133. Bogan, B.W. et al., Manganese peroxidase mRNA and enzyme activity levels during bioremediation of polycyclic hydrocarbon-contaminated soil with *Phanerochaete chrysosporium*, *Appl. Environ. Microbiol.*, 62, 2381, 1996.

134. Metcalfe, A.C. et al., Molecular analysis of a bacterial chitinolytic community in an upland pasture, *Appl Environ. Microbiol.*, 68, 504, 2002.

135. Graves, P.R. and Haystead, T.A.J., Molecular biologist's guide to proteomics, *Microbiol. Mol. Biol. Rev.*, 66, 39, 2002.

136. Nannipieri, P., Role of stabilised enzymes in microbial ecology and enzyme extraction from soil with potential applications in soil proteomics, in *Nucleic Acids and Proteins*, Nannipieri, P. and Smalla, K., Eds., Springer, Heidelberg, 2006, p. 75.

137. Ogunseitan, O.A., Soil proteomics: extraction and analysis of proteins from soil, in *Nucleic Acids and Proteins*, Nannipieri, P. and Smalla, K., Eds., Springer, Heidelberg, 2006, p. 95.

138. Fan, T.W. et al., Comprehensive chemical profiling of gramineous plant root exudates using high-resolution NMR and MS, *Phytochemistry*, 57, 209, 2001.

139. Toal, M.E. et al., A review of rhizosphere carbon flow modelling, *Plant Soil*, 222, 263, 2000.

140. Schulze, J. and Pöschel, G., Bacteria inoculation of maize affects carbon allocation to roots and carbon turnover in the rhizosphere, *Plant Soil*, 267, 235, 2004.

141. Derrien, D., Marol, C., and Balesdent, J., The dynamics of neutral sugars in the rhizosphere of wheat. An approach by ^{13}C pulse-labelling and GC/C/IRMS, *Plant Soil*, 267, 243, 2004.

142. Macko, S.A., Ryan, M., and Engel, M.H., Stable isotopic analysis of individual carbohydrates by gas chromatographic/combustion/isotope ratio mass spectrometry, *Chem. Geol.*, 152, 205, 1998.

143. van Dongen, B.E., Schouten, S., and Sinnoghe Damsté, J.S., Gas chromatography/combustion/isotope-ratio-monitoring mass spectrometric analysis of methylboronic derivatives of monosaccharides: a new method for determining natural ^{13}C abundance of carbohydrates, *Rapid Commun. Mass Spectrom.*, 15, 496, 2001.

144. Kuzyakov, Y., Separating microbial respiration of root exudates from root respiration in non-sterile soils: a comparison of four methods, *Soil Biol. Biochem.*, 34, 1621, 2002.

145. Griffiths, B.S. et al., An examination of the biodiversity-ecosystem function relationship in arable soil microbial communities, *Soil Biol. Biochem.*, 33, 1713, 2001.

146. Gray, N.D. and Head, I.M., Linking genetic identity and functioning communities of uncultured bacteria, *Environ. Microbiol.*, 3, 481, 2001.

147. Akkermans, A.D.L., Van Elsas, J.D., and De Bruijn, F.J., *Molecular Microbial Ecology Manual — Supplement 5*, Sec. 1, Kluwer Academic Publishers, Dordrecht, 2001.

148. Schmalenberger, A. and Tebbe, C.C., Bacterial diversity in maize rhizospheres: conclusions on the use of genetic profiles based on PCR-amplified partial small subunit rRNA genes in ecological studies, *Mol. Ecol.*, 12, 251, 2003.

149. Radajewski, S. et al., Stable-isotope probing as a tool in microbial ecology, *Nature*, 403, 646, 2000.

150. Wellington, E.M.H., Berry, A., and Krsek, M., Resolving functional diversity in relation to microbial community structure in soil: exploiting genomics and stable isotope probing, *Curr. Opin. Microbiol.*, 6, 295, 2003.

151. Radajewski, S., McDonald, I.R, and Murrell, J.C., Stable-isotope probing of nucleic acids: a window to the function of uncultured microorganisms, *Curr. Opin. Biotechnol.*, 14, 296, 2003.

152. Manefield, M. et al., Stable isotope probing: a critique of its role in linking phylogeny and function, in *Nucleic Acids and Proteins*, Nannipieri, P. and Smalla, K., Eds., Springer, Heidelberg, 2006, p. 205.

153. Radajewski, S. et al., Identification of active methylotroph populations in an acidic forest soil by stable-isotope probing, *Microbiology*, 148, 2331, 2002.

154. Jeon, C.O. et al., Discovery of a bacterium, with distinctive dioxygenase, that is responsible for *in situ* biodegradation in contaminated sediment, *Proc. Natl. Acad. Sci. USA*, 100,13591, 2003.

155. Padmanabhan, P. et al., Respiration of ^{13}C-labeled substrates added to soil in the field and subsequent 16S rRNA gene analysis of ^{13}C-labeled soil DNA, *Appl. Environ. Microbiol.*, 69, 1614, 2003.

156. Griffiths, R.I. et al., $^{13}CO_2$ pulse labelling of plants in tandem with stable isotope probing: methodological considerations for examining microbial function in the rhizosphere, *J. Microbiol. Methods*, 58, 119, 2004.

157. Treonis, A.M. et al., Identification of groups of metabolically-active rhizosphere microorganisms by stable isotope probing of PLFAs, *Soil Biol. Biochem.*, 36, 533, 2004.

158. Standing, D., Meharg, A.A., and Killham, K., A tripartite microbial reporter gene system for real-time assays of soil nutrient status, *FEMS Microbiol. Lett.*, 220, 35, 2003.

5 Nutrients as Regulators of Root Morphology and Architecture

Wolfgang Schmidt and Bettina Linke

CONTENTS

I. INTRODUCTION

Plants are sessile organisms that have to cope with and adapt to a permanently changing environment. Due to their open morphogenesis, i.e., reiterative enlargement of their body plan by new elements, developmental plasticity can be maintained during their entire life. Postembryonic development is affected by environmental cues, overriding endogenous developmental programs to anticipate forthcoming conditions. Such a plastic behavior avoids metabolic and nutritional misbalances and compensates for the restricted possibilities of plants to escape from unfavorable conditions.

Roots provide the plant with physical support and are the site of nutrient uptake, the supply of which is unevenly distributed in soils and varies greatly in time. Hence, phenotypic plasticity of roots is essential for effective soil exploration. During growth, changes in root architecture and cell fate acquisition occur according to the prevailing trophic conditions. Cells can be redifferentiated according to the plant's need. For example, the formation of root hairs is highly responsive to environmental conditions. In addition to local adjustments, nutrient acquisition is integrated with the demands of distant plant parts, a process that requires remote control of uptake systems and growth patterns by long-range signals. Responses to nutrient shortage are typical of the respective mineral. This implies that plants have evolved different sensing systems and at least partly separated signal pathways for coordinated uptake of essential nutrients. In addition, even apparently similar responses may be the result of different signals and may activate discrete downstream targets. For example, an increase in root surface area is induced by both iron and phosphate shortage, but the signals are translated by separate pathways [1].

Interspecies differences in root development are controlled by genetically inheritable traits. Ecotypic variation may cause habitat-specific responses to certain stimuli. Basic root growth parameters, such as root elongation rates and cell cycle gene activity, as well as environmental responses, may differ largely among Arabidopsis accessions, mirrored by even partly contrasting responses to nutrient deficiency [2,3]. Thus, it is not an easy task to describe a default state of the developmental changes associated with a certain environmental signal. However, despite the high variability in growth responses, some general patterns can be inferred from the available data. These general patterns, mainly studied in model plants, represent a condensation point in understanding the molecular events that underlie adaptive changes in root growth. A number of excellent recent reviews covering different aspects of root responses to the environment have been published during the last few years [4–10]. Here, we focus on the recent progress concerning signaling pathways and some of the participating regulatory cues in *trophomorphogenesis*. This term has been introduced by Forde and Lorenzo [11] to describe root growth changes induced by variations in the special or temporal bioavailability of nutrients. General aspects of root development cannot be completely covered in this review and will only be considered to understand the mechanistic or molecular basis of environmentally induced changes.

II. SIGNALS AND GROWTH RESPONSES

A. SIGNALS

Plant roots are able to respond to nutrient-rich site by enhanced growth [12]. Experimental evidence suggests that nutrients act directly as signals and not by a metabolic mechanism to induce such growth changes. A decrease in the shoot-to-root ratio, a characteristic response to nitrate shortage, was observed in the wild type and also in transgenic plants with low nitrate reductase activity [13]. Furthermore, localized lateral root proliferation in response to spots with high nitrate availability was increased both in wild-type plants and in mutants defective in the assimilation of nitrate [14]. Similarly, the phosphate analog phosphite is not metabolized in plants but is able to suppress phosphate deficiency responses [15 and references therein]. Rapid induction of regulatory genes in response to phosphorus, potassium, and iron deficiency further underlines the assumption that nutrients can be sensed directly and do not act indirectly by improving the metabolic state of the plants [16].

In addition to cell-autonomous responses, nutrient homeostasis is controlled systemically at the whole-plant level [17–20]. Thus, formally a remote, shoot-borne signal has to be assumed that communicates the shoot's nutrient status to the roots. This information is transmitted from its arrival in the stele of the roots to the peripheral cells in which nutrient uptake takes place. It can therefore be proposed that two signals approach peripheral root cells: a systemic one communicating the overall nutrient status of the plant and a local signal that monitors nutrient availability either apoplasmatically in the immediate surroundings of the rhizodermal cells or symplasmatically in a locally restricted cell population. The two incoming signals are then integrated and uptake rates and cell differentiation are adjusted accordingly.

Mineral nutrients are dissolved in water, and their bioavailability is dependent on the presence of water in the rhizosphere. Recently it has been demonstrated that osmotic potential can affect root architecture [21]. Lack of water availability reversibly reduces root proliferation, but lateral root primordia are maintained to allow a fast response when water becomes more available.

B. GROWTH RESPONSES

1. Lateral Roots

Lateral roots are initiated from pericycle cells adjacent to the protoxylem poles in the differentiation zone of the parent root in an acropetal sequence. Depending on their position behind the root tip, pericycle cells primed for the formation of lateral roots either remain meristematic after displacement from the root apical meristem and continue to cycle or may dedifferentiate and reenter the cell cycle at the G2 phase [22,23]. Mutants defective in the production of the nuclear-localized protein ALF4 show a premature loss of meristem-like properties of the xylem-adjacent pericycle and a reduced formation of lateral roots. It is suggested that ALF4 participate in maintenance of pericycle cells in the mitotically competent state [24]. The first events in lateral root primordia initiation are asymmetric transverse cell divisions of pericycle founder cells adjacent to the protoxylem poles. Sequential periclinal divisions led to the formation of a domelike four-layered structure. Emergence of the primordium from the parent root is mainly driven by cell expansion rather than cell division (Figure 5.1). Continued growth is achieved by activation of the lateral root meristem [25].

In a series of classical experiments, Drew [12] showed that length and number of lateral roots are stimulated when plants are exposed to local patches of high phosphate, nitrate, or ammonium concentrations. Nutrient-induced changes in root architecture have been studied in detail in response to the presence or absence of either phosphate or nitrate. Limited P availability generally favors lateral root development and decreases primary root growth by reduced cell division and elongation [26–29], allowing the plant to acquire phosphate from horizons with higher phosphate levels. This response has been referred to as *topsoil foraging* [26]. For example, the number of lateral roots was found to be fivefold higher in low-phosphate medium compared to plants grown with optimal levels of phosphate [28]. Inhibition of primary root growth is caused by induction of a determinate developmental program that causes a reduction of cell elongation followed by cessation of cell proliferation and premature cell differentiation processes in former meristematic cells [30]. Lower mitotic activity under low-phosphate conditions, evidenced by kinetic analysis of expression patters of the cell cycle marker *CycB*, was not restricted to primary roots and was also observed in mature laterals leading to a more branched root system.

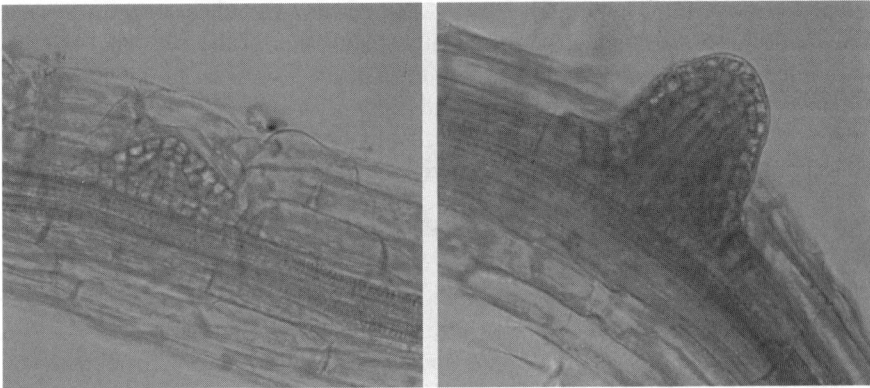

FIGURE 5.1 Development of lateral roots. Primordia of lateral roots reveal a dome-shaped structure developing from the inner pericycle layer. Left: initial stage; right: outgrowth of lateral root meristem.

An extreme case of determinate growth is represented by the development of cluster (proteoid) roots (see Chapter 2). Cluster roots are closely spaced laterals that emerge synchronously on the same plant, leading to a bottlebrush-like appearance of the root. The number of cluster roots is greatly enhanced in response to phosphate shortage. In some species, a similar response is induced by suboptimal iron concentrations [31,32]. This coincidence suggests that phosphate and iron trigger similar developmental programs that aid in the acquisition of sparingly soluble nutrients. Whether similar downstream components are involved in transducing the signals or whether separate courses lead to the formation of clusters under phosphate- and iron-deficient conditions has not yet been determined.

Unlike what happens upon exposure to phosphate shortage, a decreasing nitrate concentration results in enhanced primary root growth [14]. Lateral root elongation is repressed by a uniform supply of high nitrate levels. In contrast, local zones of high nitrate availability have a stimulatory effect if the overall nitrate status of the plant is low, suggesting that integration of both local (promotive) and systemic (repressive) signals defines the fate of lateral root primordia and allows for efficient nitrogen foraging [14]. The differences in the responses to phosphate and nitrate availability may be related to the contrasting solubility or distribution of these two nutrients in soils. Enhanced growth of the primary root may confer a competitive advantage by exploring nitrogen resources in lower soil horizons that are not available to competing neighbors, whereas phosphate levels are generally higher in upper soil layers because the relative immobility of phosphate restricts its movement to lower soil profiles.

2. Root Hairs

The root epidermis of arabidopsis provides a simple model for studying both intrinsic developmental programs and environmentally induced response pathways. Epidermal cells consist of only two cell types: hair cells and nonhair cells that adopt their fate by a combination of apoplastic and symplastic signals. Epidermal cells that are in contact with two of the eight underlying cortical cells develop into hair cell and those that touch only one cortical cell develop into a nonhair cell. Laser ablation experiments that cause invading of an epidermal cell in a new position result in a change in cell fate [34], indicating that the epidermal pattern is biased by positional information. The molecular nature of the positional signal has not yet been determined.

Differentiation of epidermal cells in both leaves and roots is controlled by a network of interacting factors, the core of which is a complex of MYB and bHLH transcription factors associated with a WD40 repeat protein [35–37]. In roots, association of the R-like bHLH proteins GL3 and EGL3 with the MYB-type transcription factor WER and the WD40 protein TTG in nonhair cells promotes the expression of the single-repeat MYB protein CPC and of the homeodomaine leucine zipper protein GL2 (Figure 5.2). GL2 acts as a positive regulator of the nonhair cell fate. The WER/GL3/EGL3/TTG complex further inhibits expression of the *GL3* and *EGL3* genes. In cells that develop into hairs, CPC (and possibly other related small MYB proteins such as TRY and ETC1) suppress the expression of *WER,* and a complex composed of CPC/GL3/EGL3 and TTG is formed. This complex blocks the expression of *CPC* and *GL2* in future hair cells. WER controls the transcription of *CPC* and *GL2* directly by binding to their promoter regions [38]. Specification of the epidermal cell types is achieved by a bidirectional signaling mechanism, in which CPC moves from nonhair cells into hair cells, suppressing the binding of WER to the GL3/TTG complex, and GL3/EGL3 moves from hair cells to nonhair cells to form the WER/GL3/EGL3/TTG complex. Negative autoregulation of *GL3* and *EGL3* is dependent on bidirectional signaling between H and N cells [39]. Upstream of this transcription factor cascade, *SCRAMBLED (SCM)*, a newly identified gene encoding a putative plasma-membrane-bound, leucine-rich, receptor-like protein kinase (LRR-RLK), transduces the (unknown) positional signal and controls expression of the *GL2, CPC, WER,* and *EGL3* genes [40].

In aging roots, patterning of epidermal cells becomes susceptible to environmental cues. The regular pattern of seedlings, in which all cells in the permissive position develop into root hairs,

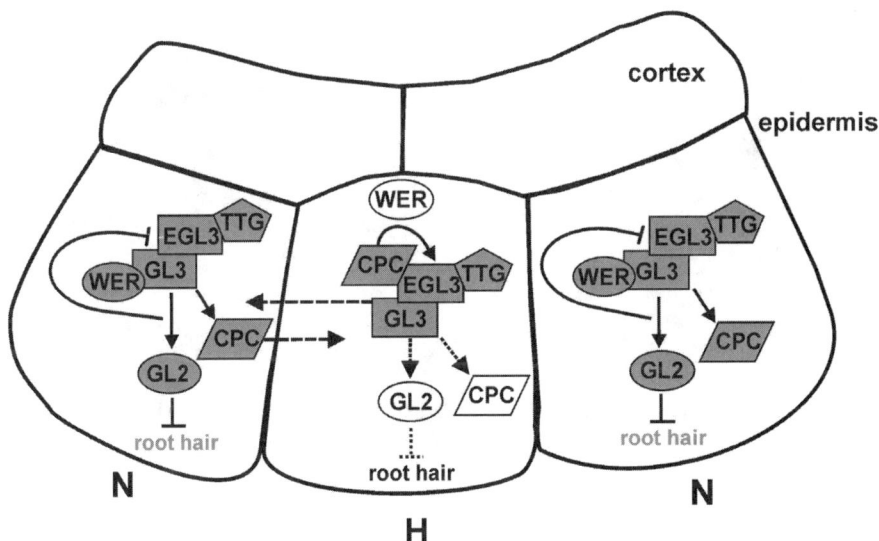

FIGURE 5.2 Model of epidermal cell specification (adapted from Bernhardt, C. et al., *Development*, 130, 6431, 2003; Bernhardt, C. et al., *Development*, 132, 291, 2005). In nonhair cells (N), a high abundance of WER induces the formation of a complex consisting of WER, GL3/EGL3, and TTG to activate expression of GL2, which specifies the nonhair fate (by repressing hair formation). Additionally, the WER/GL3/EGL3/TTG complex stimulates expression of CPC and limits the expression of GL3/EGL3 by an autoregulatory feedback loop. In future hair (H) cells, a high level of CPC contributes to the formation of an inactive CPC/EGL3/GL3/TTG complex that prevents activation of GL2 or CPC, resulting in the positive progression of root hair specification. The presence of CPC in H cells further activates expression of GL3/EGL3. Protein movement from N to H cells (CPC) and *vice versa* (Gl3/EGL3) links the regulatory network between the two cell types. Low protein abundance is indicated by white-colored symbols, broken lines demonstrate protein movement, unbroken lines show transcriptional regulation, and dotted lines mark a minor regulatory activity on gene expression (further information in the text).

is markedly reduced in adult plants [41]. The pattern of *GL2* expression, however, remains unchanged. Root hair length and frequency is further dependent on the composition of the growth medium [42] and is particularly affected by the availability of sparingly soluble nutrients such as phosphate and iron [43,44]. Root hairs of phosphate-deficient plants are longer and denser compared to plants treated with sufficient phosphate concentrations. The increased surface area confers a competitive advantage; Arabidopsis accessions with more and longer root hairs have high-phosphate acquisition efficiency [45]. The higher frequency of root hairs in phosphate-deficient plants is due to an increased number of epidermal cells developing into hairs both in the normal location and in positions normally occupied by nonhair cells [1,43]. Under iron deficiency, Arabidopsis uses a strategy different from that observed under phosphate-deficient conditions to increase the root surface area. The number of root hairs per unit root length is not significantly enhanced, but a high percentage of hairs with bifurcated tips is formed [41,44]. Similar to the aging root, under control conditions, the spatial expression of GL2 is unchanged by Fe deficiency (Figure 5.3).

Although the molecular nature of nutrient-induced changes in epidermal patterning has not yet been elucidated, experiments with mutants harboring defects in cell specification genes strongly suggest that hairs produced in each growth type are dependent on a distinct set of genes that differ from these resulting in hairs formed in normal or ectopic positions . In a survey of mutants defective at various stages of root hair development, divergence in root hair patterning was most pronounced in mutants with defects in genes that affect the first stages of differentiation, suggesting that nutritional signals affect cell specification genes [41].

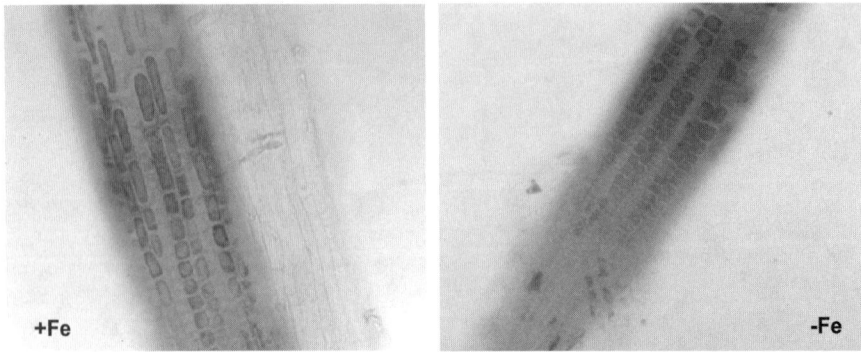

FIGURE 5.3 *In situ* hybridization of Fe-sufficient and Fe-deficient *Arabidopsis* roots (Fe+/Fe−) with *GLABRA2 (GL2)*. *GL2* transcripts reveal comparable expression patterns (B. Linke, unpublished).

3. Transfer Cells

Besides developing into root hairs, root epidermal cells can undergo ultrastructural changes that are presumably important for nutrient acquisition. Rhizodermal cells in the zone of metaxylem differentiation may form polarized wall ingrowths, which are primarily located on outer tangential walls. In the root epidermis, these transfer cell-like structures are formed in response to various nutrient stresses such as salt, exposure to cadmium, or shortage in phosphate and iron, and are not observed under control conditions (see Offler et al., 46, for a review). Both hair and nonhair cells can differentiate into transfer cells. Transfer cells are associated with high proton fluxes; the density of P-type H+-ATPase molecules was found to be significantly higher in cells that have adopted the transfer cell fate [47].

Transfer cells are not formed in the rhizodermis of Arabidopsis [44], although this cell type is found elsewhere in this species [48]. An interesting parallel is the induction of transfer cells by invasion of symbiotic or parasitic organisms [49], and nematode attack [50]. In both the cases of cyst nematode infection and iron deficiency, structures such as plasmalemmasomes and paramural bodies were found in addition to the general features of transfer cells, suggesting that abiotic and biotic stress can induce similar developmental programs [50,51]. In roots affected by cyst or root-knot nematodes, wall ingrowth formation is independent of the formation of giant cells or syncytia, the feeding structures induced by the nematodes [52].

III. SENSING

Sensors involved in nutrient homeostasis may be located either in the plasma membrane, monitoring the external concentration of mineral ions, or may measure their intracellular level after absorption from the soil solution. Most of the nutrients that affect root development appear to trigger both cell-autonomous decisions and systemic signals, ultimately leading to adaptive changes of developmental programs. This dual regulation has been well described for phosphate and nitrate [53,54]; the molecular nature of the sensors is, however, still unknown. Local sensors of nutrient concentrations may be located in the quiescent center (QC), a mitotically less active spot behind the root tip. Phosphate deficiency is associated with a high frequency of periclinal divisions of QC cells, reflected by an increased number of cells expressing a QC marker [30]. These changes preceded other morphological responses to phosphate deficiency such as premature cell differentiation.

Forward genetic approaches have revealed promising candidates for the interpretation of local nutritional signals. The *phosphate deficiency response (pdr)* mutation causes an increase in sensitivity and amplitude of phosphate starvation responses [55]. The *prd2* mutant can be rescued by the nonmetabolizable phosphate analog phosphite, suggesting that local rather than systemic signaling is affected. PDR2 may be involved in surveillance of phosphate availability and in transmitting this information to meristematic cells. Interestingly, *pdr2* mutants show a biphasic dose–response curve when primary root extension is plotted vs. phosphate concentration, indicating that different sensing systems are active at different phosphate concentrations [55].

In yeast, sensing of external phosphate is achieved by low-affinity orthophosphate transporters, all of which carry an SPX domain (named after the proteins SYG1 and PHO81 of yeast and the human XPR1 protein) that has suggested to be involved in G-protein-associated signaling [56]. An SPX domain has recently been reported for the Arabidopsis protein PHO1, initially identified as a transporter involved in xylem loading [57]. Although neither the function of the *PHO1* gene family nor of the SPX domain has been precisely defined, this coincidence suggests that PHO1-like proteins are possibly involved in phosphate sensing [58].

A plausible candidate for sensing nitrogen levels is the high-affinity nitrate transporter NRT2.1. *NRT2.1* is allelic to *LIN1* and has been isolated by a screen for mutants that can initiate lateral roots under high sucrose to nitrogen ratios; conditions that almost completely inhibit the formation of lateral roots in the wild type [59,60]. Reduced nitrate uptake rates that are expected to decrease rather than increase the number of lateral roots are displayed by *lin1* mutants. This may be interpreted as an involvement of the NRT2.1 transporter in nitrate surveillance. Interestingly, *NRT2.1* is strongly expressed by phosphate starvation.

His–Asp phosphorelay systems (two-component systems) are implicated in a wide variety of cellular responses to environmental stimuli and are involved in nutrient sensing in yeast. They consist of a sensor histidine kinase and a cognate response regulator whose activity is modulated by the sensor. Similar systems have been described in plants. Some response regulators are upregulated by nitrate-starvation in Arabidopsis and may play a role in nitrate sensing. However, the proposed mechanisms involve an indirect sensing of nitrate mediated by cytokinin, leaving the question of the upstream components open [61,62].

In mammalian cells, the stability or translatability of mRNAs of iron-responsive genes is affected by iron regulatory proteins (IRPs) that bind to a specific stem–loop sequence in the untranslated region of the mRNAs that is known as the iron-responsive element (IRE) [63]. One of two IRPs in mammals is cytosolic aconitase, which can be converted from aconitase function to a posttranscriptional regulator with RNA binding properties. Aconitase function is dependent on the presence of a [4Fe–4S] cluster. Oxidation of the cluster by NO leads to cluster disassembly and to the loss of aconitase activity. In plants, regulation of ferritin via a NO-mediated pathway has recently been demonstrated [64]. Repression of ferritin synthesis under low iron supply occurs at the level of transcription and thus differs from the animal IRE/IRP system. Three out of four ferritin genes in arabidopsis, *AtFer1*, *AtFer3*, and *AtFer4,* are responsive to both iron [65] and NO [66]. It is thus tempting to speculate that a similar system is active in plants.

IV. SIGNALING

A. AUXIN

The plant hormone auxin is involved in almost all root developmental processes [67–69]. The development of lateral roots is no exception; most mutants with defects in lateral root formation can be related to defects in the transport of or sensitivity to auxin [4]. Exogenously applied auxin rescues the phenotype of several root hair defective mutants [70], causes an increase in root-hair length and density of the wild type [43,71,72], and induces rhizodermal transfer cells under nonpermissive conditions [73].

The role of auxin in transducing environmental cues into morphological changes is less clear. Exogenously applied auxin induces the formation of cluster roots, and auxin inhibitors reduce phosphate stress-induced development of cluster roots in white lupin and lateral root formation in

Arabidopsis [28,74]. In addition, transcripts of the auxin-inducible gene *HRGP* accumulate in roots of phosphate-deficient Arabidopsis plants [29]. It is thus tempting to speculate that auxin is involved in phosphate starvation responses. However, several auxin-related mutants show wild-type responses to phosphate and nitrate availability, i.e., a decrease in primary root growth under low-phosphate conditions and an increase in lateral root elongation induced by nitrate-rich patches [27,28,33]. The short-root phenotype of phosphate-deficient plants cannot be rescued by auxin [75]. Moreover, phosphate-deficient plants exhibited decreased rather than increased expression of the synthetic auxin-responsive promoter DR5. Analysis of *low phosphate-resistant root* (*lpr1*) mutant lines, unable to respond to phosphate starvation by forming more lateral roots, further negates an involvement of auxin in architectural alterations in phosphate-deficient plants. The *lpr1* mutation is allelic to *BIG*, which codes for a protein that is required for polar auxin transport [76]. Because *lpr1* mutants responded to low phosphate by inhibited primary root growth and formation of additional root hairs, the authors suggested that two different pathways are involved in root architectural responses to phosphate limitation. Reduction in primary root growth and increase in root hair density during P shortage is thought to be independent of BIG and auxin transport. In the auxin-dependent pathway, BIG is necessary for the formation of lateral root primordia. Maturation of primordia into fast-growing lateral roots due to low phosphate availability may be a subsequent auxin and BIG-independent step in the second pathway [75].

Hence, at least for the case of phosphate, most experimental evidence argues against an essential role for auxin in signal transduction of trophomorphogenetic responses. It appears that this held also true for the phosphate-deficiency-induced formation of extra root hairs. Whereas auxin signaling is essential for the increase in root surface area in iron-deficient roots, no such requirement was observed for induction of the phenotype typically of phosphate-deficient plants [1]. This effect is most striking in the *trh1* mutant that forms root hairs neither under control conditions [77] nor in response to iron shortage but does so under low phosphate conditions [41]. *TRH1* was shown to encode a potassium transporter, but high concentrations of external potassium do not induce root hair formation in *trh1* mutants. Exogenously applied auxin rescues morphological *trh1* phenotypes, providing evidence for a requirement of TRH1 for auxin transport [78]. Apparently, a functional *TRH1* product and, hence, auxin transport is not necessary for the induction of the phosphate-deficient phenotype.

The situation might be different for the development of lateral roots under sulfur-deficient conditions. Growth of Arabidopsis plants starved upon sulfate results in an accelerated development and a higher frequency of lateral roots, and leads to transcriptional activation of *NITRILASE 3* (*NIT3*, [79]). *NIT3* promoter activity was found to be highest in root conductive tissue and lateral root primordia. *NIT3p:uidA* plants strongly upregulate promoter activity upon addition of the cysteine precursor *O*-acetyserine to the growth medium. *O*-acetyserine accumulates under sulfur deficiency and may communicate sulfate starvation to *NIT3*. A similar role for *O*-acetyserine has been reported for enzymes of sulfate assimilation. Sulfur deprivation further leads to an intensified turnover of the indole-3-acetonitrile (IAN) precursor glucobrassicin. A regulatory loop was proposed in which nitrilase converts IAN into IAA that then may prime pericycle cells for the formation of lateral root primordia [79].

B. CYTOKININS

Less detailed information is available on the role of hormones other than auxin in the translation of environmental signals into root developmental changes. A number of studies have associated cytokinins with nutrient starvation responses, and some experimental evidence suggests an involvement of cytokinins in trophomorphogenesis. For instance, cytokinins have been shown to antagonistically affect the formation of cluster roots [80]. Other evidence comes from transcriptional profiling studies in white lupin. In cluster roots, cytokinin oxidase was found to be differentially expressed during various stages of development [81] that is consistent with a role of cytokinin in the regulation of this response.

Cytokinins have been suggested to play a role in integrating root and shoot development by transmitting information on the root's nutritional status directed to the shoot via a His–Asp phosphotransfer system. Such a system has been proposed for interorgan communication of the nitrogen status [82] and for nitrogen-dependent root development [83]. A two-component signaling circuit typical for cytokinin responses has also been implicated in the regulating phosphate and sulfate homeostasis [84,85]. A functional context of cytokinin signaling in trophomorphogenesis remains, however, to be established. The lack of major differences in the phosphate deficiency responses among wild type and double mutants defective in the cytokinin receptors CYTOKININ RESPONSE 1 (CRE1) and ARABIDOPSIS HISTIDINE KINASE 3 (AHK3) grown with part of the roots in P-free medium negates a prominent role for cytokinin in phosphate signaling [86]; see following text for a discussion of long-range signaling.

C. Ethylene, Abcisic Acid, and Brassinosteroids

Ethylene has been implicated in many trophomorphogenetic responses including root elongation, lateral root growth, and root hair formation. Phosphate deficiency was suggested to alter ethylene responsiveness [28,43], but ethylene does not appear to play a major role in lateral root formation or in the development of extra root hairs in response to phosphate deficiency, because ethylene-related mutants do not differ much in their responses from wild-type plants [1,28]. This contrasts with effects on the formation of bifurcated root hairs in response to iron deficiency. Similar to what has been observed for auxin, the development of branched hairs is inhibited both by antagonists of ethylene synthesis or action as well as in ethylene mutants [1]. Application of the ethylene precursor 1-aminocyclopropane-1-carboxylic acid (ACC) inhibited lateral root formation independent of the phosphate status of the plants [28], taking the argument further against an involvement of ethylene in stress-induced development of lateral roots. Ethylene inhibitors restored decreased primary root growth caused by low water potential in plants treated with fluridone that imposes ABA deficiency; thus, it is suggested that accumulation of ABA at low water potential serves to prevent excess ethylene production [87].

Several ABA-insensitive mutants do not show the typical inhibitory effect of high nitrate concentrations on the formation of lateral root primordia, implying a role of ABA in nitrate signaling [88]. Suboptimal nitrate supply mimics in some respects water shortage. It has been suggested that the typical soil-drying response is due to a limitation in nitrate supply. ABA may integrate nitrate and osmotic sensing [89]. Both ABA-deficient mutants and mutants in *LRD2*, a newly identified gene presumed to be involved in osmotic stress response, show an enlarged root system both under control conditions and under mild osmotic stress. This implies that both ABA and *LRD2* are necessary to repress the outgrowth of lateral root primordia [21].

Other potential players in the translation of environmental cues into morphogenetic responses are brassinosteroids (BRs). Recent studies have shown that BRs positively and autonomously affect root growth. A BR-deficient mutant displays differential expression of genes involved in lateral root growth and root hair formation [90]. An involvement of BRs in trophomorphogenesis has, however, not yet been demonstrated.

V. REGULATION

A. Transcription Factors

Although the mechanism underlying the perception of available nutrients remain still obscure, some downstream components have been identified during the last few years. PHR1 is a conserved MYB transcription factor related to the *PSR1* gene from *Chlamydomonas reinhardtii* that was identified in genetic screen for plants with reduced responses to phosphate starvation [91]. PHR1 binds as a dimer to an imperfect palindromic sequence in the promoter of P starvation-responsive genes, indicating that PHR1 is a downstream component of the phosphate signaling pathway. Interestingly, a similar consensus sequence was found in the promoters of the white lupin phosphate transporter *LaPT1* and secreted acid phosphatase *LaSAP1* genes, suggesting that *cis* acting promoter elements in phosphate starvation signaling are highly conserved [92].

The MADS-box transcription factor ANR1 was isolated in a screen for nitrate-regulated genes in Arabidopsis roots. Lateral roots of *ANR1*-antisense plants do not respond to external nitrate, indicating that ANR1 is part of the signaling pathway that integrates the availability of nitrate with the development of lateral roots [4].

A novel class of transcription factors regulating root growth has recently been identified in Arabidopsis. *BREVIS RADIX (BRX)* controls the rate of cell proliferation and elongation and is thus a potential candidate for translating environmental cues into changes in root growth [93].

B. MicroRNAs

During recent years, small RNA molecules (microRNAs or miRNAs) have been identified as novel regulators of posttranscriptional gene expression. MicroRNAs are ~22 nucleotides in length; their maturation from imperfect stem loop precursors is mediated by different regulators, such as ARGO-NAUTE (AGO, [94], see following text). In plants, miRNAs can act to control cell fate determination by regulating the mRNA abundance of transcription factors or other key players involved in developmental processes [94]. The F-box protein NAC1, which transduces auxin signals for lateral root emergence, is repressed in mRNA abundance by miR164. Thus, miR164 indirectly downregulates auxin signals for lateral root initiation. Increase of miR164 levels by auxin is dependent on AXR1, AXR2, and TIR3, which are required for auxin-mediated lateral root development [95]. *TIR1,* recently identified as an auxin receptor [96,97], promotes auxin-induced lateral root growth [98]. Transcripts of *TIR1* are targets of miR393, indicating that miRNAs regulate auxin homeostasis at various points [99].

This coincides with findings that several members of the ARF (auxin response factor) family are subjected to microRNA-mediated regulation [100]. ARF17 is targeted by miR160; plants carrying a miR160-resistant version of *ARF17* display reduced root length and decreased root branching [100]. ARF17 regulates expression of *GH3*-like genes, which are likely to play an important role in auxin responsiveness. Repressing *GH3* genes mediated by *ARF17* negatively regulates the formation of adventitious roots in *ago1* mutants [101]. As mentioned earlier, presence of AGO1 is required in the miRNA biogenesis pathway [94], indicating a correlation between formation of adventitious roots and miRNA-regulated steps of auxin responses. Because *ago1-3* mutants are not affected in lateral root development, it is suggested that the formation of lateral roots and adventitious roots is controlled by different regulatory pathways.

In addition to genes involved in developmental processes, various stress-responsive genes have been identified as immediate targets of miRNAs [102,103]. Expression of miR395, a sulfurylase-targeting miRNA, increases upon sulfate starvation, confirming that miRNAs can be induced by nutritional stress. Thus, it can be assumed that miRNAs or other small regulatory might participate in regulatory control during transition of environmental stimuli.

C. Chromatin Remodeling and Protein–Protein Interactions

Adaptation to environmental stimuli also includes regulatory mechanisms at the DNA level [104]. In eukaryotic cells, accessibility of chromatin is crucial for cell differentiation. Chromatin remodeling has been demonstrated to be important in responses to environmental stimuli such as nitrogen starvation in yeast [105]. In Arabidopsis, mutations in histone acetylation show reduced root growth and disturbed responses to abiotic stress [106]. The *bru1* mutation affects the stability of heterochromatin organization but does not interfere with genomic DNA methylation processes; *bru1-1* mutants exhibit a retarded growth of primary roots [107]. This may point to an involvement of chromatin modification in the adaptation to changing environmental conditions. Interestingly, histone acetylation has recently been shown to affect the position-dependent expression of patterning genes in the root epidermis of Arabidopsis [108].

A novel control mechanism of protein modification or turnover by sumoylation has been shown to participate in morphological phosphate deficiency responses [109]. Sumoylation is a posttranslational modification that involves the covalent attachment of a SUMO (small ubiquitin-related modifier), an ubiquitin-like modifier protein to substrate proteins. SUMO proteins are present in yeast, mammals, and plants and are highly conserved among eukaryotic kingdoms. Conjugation of SUMO to a target protein affects its subcellular localization and enhances its stability. Mutants defective in a SUMO-conjugating enzyme (SUMO E3 ligase SIZ) show exaggerated responses to phosphate starvation, including reduction in primary root elongation, enhanced lateral root development, increased root to shoot mass ratio, and enhanced root hair number [109]. In Arabidopsis, the SUMO E3 ligase AtSIZ acts as a repressor of the low phosphate-induced responses. The transcription factor PHR1 was found to be sumoylated by an AtSIZ1-dependent process. These results suggest that protein modification by sumoylation is important in phosphate homeostasis.

D. Interorgan Communication

A variety of experimental evidences points to an interorgan regulation of nutrient acquisition mechanisms by means of long-distance signaling [17–20]. In contrast, most morphological responses to nutrient availability appear to be rather controlled by the local availability of nutrients than by remote signals from the shoot. This held true for lateral root development in response to phosphate shortage [33] and for the formation of transfer cells and root hairs in phosphate and iron-deficient plants [28,110,111], as well as for altered root growth angle induced by phosphate deficiency [112]. However, long-distance signal pathways may intersect with cell local sensing of nutrients. Although the formation of root hairs in response to phosphate and Fe deficiency is dependent on local signals, divided root studies and mutant analyses revealed that shoot-borne signals can influence the fate of rhizodermal cells [113]. Dual regulation by local and systemic signals was first described for the development of lateral roots that can be induced by high local nitrate concentrations and inhibited by a high overall nitrogen status of the plant [14]. The mechanisms underlying shoot control of root development and the nature of internal long-range signals have not yet been identified. Long-range movement via the phloem has been shown for proteins as well as for RNA [114,115]. The transport of these molecules is passive by bulk flow. Recently destination-selective trafficking of phloem proteins has been demonstrated, indicating an active control of long-distance transport of macromolecules partly regulated by protein–protein interactions [116].

A putative serine or threonine receptor kinase that is required for shoot-controlled regulation of root growth has been isolated in *Lotus japonicus* [117]. Mutants defective in the *HYPERNOD-ULATION ABERRANT ROOT1 (HAR1)* locus show reduced growth of the primary root and excessive growth of lateral roots. Grafting of wild-type shoots onto *har1-3* roots rescues the mutant. HAR1 is required for nodule organogenesis and nitrate sensitivity and shows high sequence similarity to the *CLAVATA1* receptor kinase gene of Arabidopsis. A model has been suggested in which autoregulation of nodulation is achieved by a root-derived signal communicating nodule number that is sent to and integrated in the shoot. Further nodulation is then inhibited by interaction of HAR1 with a shoot-localized ligand [117,118]. A similar mechanism could account for nutrient homeostasis in nonsymbiotic root development.

Not only shoot-to-root, but also root-to-shoot signaling appears to be of importance in integrating environmental signals into developmental programs. A functional product of the Arabidopsis gene *BYPASS1 (BPS1)* is required to negatively regulate a root-derived signal that modulate both root and shoot architecture [119]. BPS1 is expressed in all root tissues and developmental stages [120]. In analogy to P-deficient plants, *bps1* mutants produce long root hairs and show reduced expression of the synthetic auxin-responsive promoter DR5 [119], suggesting a possible involvement of BPS1 in the responses to phosphate shortage.

VI. CONCLUSIONS

Within inherent genetic boundaries, plant roots display a wide range of phenotypical plasticity to cope with suboptimal availability of nutrients. Morphological adaptations necessitate local sensing of nutrients, integration of whole-plant demands by feedback regulation and redifferentiation of target tissues. At the cellular level, a number of players in cell fate decisions have been identified during the last few years, but still some unexplored levels remain. It has been estimated that 2 to 4% of a plant's genome is involved in the control of nutrient homeostasis [121] that mirrors the complexity of concerted responses. Deciphering the mechanisms involved in integration of genetic, epigenetic, and remote control trophomorphogenesis will be a major challenge for the coming years.

REFERENCES

1. Schmidt, W. and Schikora, A., Different pathways are involved in phosphate and iron stress-induced alterations of root epidermal cell development, *Plant Physiol.*, 125, 2078, 2001.
2. Beemster, G.T.S. et al., Variation in growth rate between Arabidopsis ecotypes is correlated with cell division and A-type cyclin-dependent kinase activity, *Plant Physiol.*, 129, 854, 2002.
3. Chevalier, F. et al., Effects of phosphate availability on the root system architecture: large-scale analysis of the natural variation between *Arabidopsis* accessions, *Plant Cell Environ.*, 26, 1839, 2003.
4. Casimiro, I. et al., Dissecting *Arabidopsis* lateral root development, *Trends Plant Sci.*, 8, 165, 2003.
5. Casson, S.A. and Lindsey, K., Genes and signalling in root development, *New Phytologist*, 158, 11, 2003.
6. López-Bucio, J., Cruz-Ramírez, A.C., and Herrera-Estrella, L., The role of nutrient availability in regulating root architecture, *Curr. Opin. Plant Biol.*, 6, 280, 2003.
7. Casal, J.J. et al., Signalling for developmental plasticity, *Trends Plant Sci.,* 9, 309, 2004.
8. Malamy, J.E., Intrinsic and environmental response pathways that regulate root system architecture, *Plant Cell Environ.*, 28, 67, 2005.
9. Walter, A. and Schurr, U., Dynamics of leaf and root growth: endogenous control versus environmental impact, *Ann. Bot.*, 95, 891, 2005.
10. Ueda, M., Koshino-Kimura, Y., and Okada, K., Stepwise understanding of root development, *Curr. Opin. Plant Biol.*, 8, 71, 2005.
11. Forde, B.G. and Lorenzo, H., The nutritional control of root development, *Plant Soil*, 232, 51, 2001.
12. Drew, M.C., Comparison of the effects of a localized supply of phosphate, nitrate, ammonium and potassium, *New Phytologist*, 75, 479, 1975.
13. Scheible, W.R. et al., Accumulation of nitrate in the shoot acts as a signal to regulate shoot-root allocation in tobacco, *Plant J.*, 11, 671, 1997.
14. Zhang, H. and Forde, B.G., An *Arabidopsis* MADS box gene that controls nutrient-induced changes in root architecture, *Science*, 279, 407, 1998.
15. Ticconi, C.A. and Abel, S., Short on phosphate: plant surveillance and countermeasures, *Trends Plant Sci.*, 9, 548, 2004.
16. Wang, Y.H., Garvin, D.F., and Kochian, L.V., Rapid induction of regulatory and transporter genes in response to phosphorus, potassium, and iron deficiencies in tomato roots. Evidence for cross talk and root/rhizosphere-mediated signals, *Plant Physiol.*, 130, 1361, 2002.
17. Grusak, M.A. and Pezeshgi, S., Shoot-to-root signal transmission regulates root Fe(III) reductase activity in the *dgl* mutant of pea, *Plant Physiol.*, 110, 329, 1996.
18. Liu, C., Muchhal, U.S., and Raghothama, K.G., Differential expression of TPSI1, a phosphate star-vation-induced gene in tomato, *Plant Mol. Biol.*, 33, 867, 1997.
19. Gansel, X. et al., Differential regulation of the NO_3^-- and NH_4^+-transporter genes *AtNrt2.1* and *AtAmt1.1* in Arabidopsis: relation with long-distance and local controls by N status of the plant, *Plant J.*, 26, 143, 2001.
20. Shane, M.W. et al., Shoot P status regulates cluster-root growth and citrate exudation in *Lupinus albus* grown with a divided root system, *Plant Cell Environ.*, 265, 2003.
21. Deak, K.I. and Malamy, J., Osmotic regulation of root system architecture, *Plant J.*, 43, 17, 2005.
22. Dubrovsky, J.G. et al., Pericycle cell proliferation and lateral root initiation in Arabidopsis, *Plant Physiol.*, 124, 1648, 2000.

23. Beeckman, T., Burssens, S., and Inzé, D., The peri-*cell*-cycle in *Arabidopsis, J. Exp. Bot.*, 52, 403, 2001.

24. DiDonato, R.J. et al., Arabidopsis ALF4 encodes a nuclear-localized protein required for lateral root formation, *Plant J.*, 37, 340, 2004.

25. Malamy, J.E. and Benfey, P.N., Down and out in *Arabidopsis*: the formation of lateral roots, *Trends Plant Sci.*, 390, 1997.

26. Lynch, J.P. and Brown, K.M., Topsoil foraging — an architectural adaptation to low phosphorus availability, *Plant Soil*, 237, 225, 2001.

27. Williamson, L.C. et al., Phosphate availability regulates root system architecture in Arabidopsis, *Plant Physiol.*, 126, 875, 2001.

28. López-Bucio, J. et al., Phosphate availability alters architecture and causes changes in hormone sensitivity in the Arabidopsis root system, *Plant Physiol.*, 129, 244, 2002.

29. Al-Ghazi, Y. et al., Temporal responses of *Arabidopsis* root architecture to phosphate starvation: evidence for the involvement of auxin signaling, *Plant Cell Environ.*, 26, 1053, 2003.

30. Sánchez-Calderón, L. et al., Phosphate starvation induces a determinate developmental program in the roots of *Arabidopsis thaliana, Plant Cell Physiol.*, 46, 174, 2005.

31. Skene, K.R., Cluster roots: model experimental tools for key biological problems, *J. Exp. Bot.*, 52, 479, 2000.

32. Neumann, G. and Martinoia, E., Cluster roots — an underground adaptation for survival in extreme environments, *Trends Plant Sci.*, 7, 162, 2002.

33. Linkohr, B.I. et al., Nitrate and phosphate availability and distribution have different effects on root system architecture of *Arabidopsis, Plant J.*, 29, 751, 2002.

34. Berger, F. et al., Positional information in root epidermis is defined during embryogenesis and acts in domains with strict boundaries, *Curr. Biol.*, 8, 421, 1998.

35. Schellmann, S. et al., *TRIPTYCHON* and *CAPRICE* mediate lateral inhibition during trichome and root hair patterning in *Arabidopsis. EMBO J.*, 21, 5036, 2002.

36. Larkin, J.C., Brown, M.L., and Schiefelbein, J., How do cells know what they want to be when they grow up? Lessons from epidermal patterning in *Arabidopsis, Ann. Rev. Plant Biol.*, 54, 403, 2003.

37. Bernhardt, C. et al., The bHLH genes *GLABRA3* (*GL3*) and *ENHANCER OF GLABRA3* (*EGL3*) specify epidermal cell fate in the *Arabidopsis* root, *Development*, 130, 6431, 2003.

38. Koshino-Kimura, Y. et al., Regulation of *CAPRICE* transcription by MYB proteins for root epidermis differentiation in *Arabidopsis, Plant Cell Physiol.*, 46, 817, 2005.

39. Bernhardt, C. et al., The bHLH genes *GL3* and *EGL3* participate in an intercellular regulatory circuit that controls cell patterning in the *Arabidopsis* root epidermis, *Development*, 132, 291, 2005.

40. Kwak, S.H., Shen, R., and Schiefelbein, J., Positional signaling mediated by a receptor-like kinase in *Arabidopsis, Science*, 307, 1111, 2005.

41. Müller, M. and Schmidt, W., Environmentally induced plasticity of root hair development in *Arabidopsis, Plant Physiol.*, 134, 409, 2004.

42. Wubben, M.J.E., II, Rodermel, S.R., and Baum, T.J., Mutation of a UDP-glucose-4-epimerase alters nematode susceptibility and ethylene responses in *Arabidopsis* roots, *Plant J.*, 40, 712, 2004.

43. Ma, Z. et al., Regulation of root elongation under phosphorus stress involves changes in ethylene responsiveness, *Plant Physiol.*, 131, 1381, 2003.

44. Schmidt, W., Tittel, J., and Schikora, A., Role of hormones in the induction of Fe deficiency responses in *Arabidopsis* roots, *Plant Physiol.*, 122, 1109, 2000.

45. Narang, R.A., Bruene, A., and Altmann, T., Analysis of phosphate acquisition efficiency in different *Arabidopsis* accessions, *Plant Physiol.*, 124, 1786, 2000.

46. Offler, C.E. et al., Transfer cells: cells specialized for a special purpose, *Ann. Rev. Plant Biol.*, 54, 431, 2003.

47. Schmidt, W., Michalke, W., and Schikora, A., Proton pumping by tomato roots. Effect of Fe deficiency and hormones on the activity and distribution of plasma membrane H^+-ATPase in rhizodermal cells, *Plant Cell Environ.*, 26, 361, 2003.

48. Haritatos, E., Medville, R., and Turgeon, R., Minor vein structure and sugar transport in *Arabidopsis thaliana, Planta*, 211, 105, 2000.

49. Berry, A.M., McIntyre, L., and McCully, M.E., Fine structure of root hair infection leading to nodulation in the *Frankia-Alnus* symbiosis, *Can. J. Bot.*, 64, 292, 1986.

50. Golinowski, W., Grundler, F.M.W., and Sobczak, M., Changes in the structure of *Arabidopsis thaliana* during female development of the plant-parasitic nematode *Heterodera schachtii*, *Protoplasma*, 194, 103, 1996.

51. Schmidt, W. and Bartels, M., Formation of root epidermal transfer cells in *Plantago*, *Plant Physiol.*, 110, 217, 1996.

52. Jones, M.G.K. and Northcote, D.H., Multinucleate transfer cells induced in *Coleus* roots by the root-knot nematode, *Meloidogyne arenaria*, *Protoplasma*, 75, 381, 1972.

53. Abel, S., Ticconi, C.A., and Delatorre, C.A., Phosphate sensing in higher plants, *Physiol. Plant.*, 115, 1, 2002.

54. Forde, B.G., Local and long-range signaling pathways regulating plant responses to nitrate, *Ann. Rev. Plant Biol.*, 53, 203, 2002.

55. Ticconi, C.A. et al., *Arabidopsis pdr2* reveals a phosphate-sensitive checkpoint in root development, *Plant J.*, 37, 801, 2004.

56. Pinson, B. et al., Low affinity orthophosphate carriers regulate *PHO* gene expression independently of internal orthophosphate concentration in *Saccharomyces cerevisiae*, *J. Biol. Chem.*, 279, 35273, 2004.

57. Poirier, Y. et al., A mutant of *Arabidopsis* deficient in xylem loading of phosphate, *Plant Physiol.*, 97, 1087, 1991.

58. Wang, Y. et al., Structure and expression profile of the *Arabidopsis PHO1* gene family indicates a broad role in inorganic phosphate homeostasis, *Plant Physiol.*, 135, 400, 2004.

59. Malamy, J.E. and Ryan, K.S., Environmental regulation of lateral root initiation in *Arabidopsis*, *Plant Physiol.*, 127, 899, 2001.

60. Rao, H., Little, D., and Malamy, J., A. high affinity nitrate transporter regulates lateral root initiation, *Proc. 14th Int. Conf. Arabidopsis Res.*, abstract number 250, University of Wisconsin, Madison, WI, 2005.

61. Sakakibara, H., Taniguchi, M., and Sugiyama, T., His-Asp phosphorelay signaling: a communication avenue between plants and their environment, *Plant Mol. Biol.*, 42, 273, 2000.

62. Sakakibara, H.J., Nitrate-specific and cytokinin-mediated nitrogen signaling pathways in plants, *J. Plant Res.*, 116, 253, 2003.

63. Eisenstein, R.S., Iron regulatory proteins and the molecular control of mammalian iron metabolism, *Ann. Rev. Nutr.*, 20, 627, 2000.

64. Murgia, I., Delledonne, M., and Soave, C., Nitric oxide mediates iron induction of ferritin accumulation in *Arabidopsis*, *Plant J.*, 30, 521, 2002.

65. Petit, J.M., Briat, J.F., and Lobreaux, S., Structure and differential expression of the four members of the *Arabidopsis thaliana* ferritin gene family, *Biochem. J.*, 359, 575, 2001.

66. Parani, M. et al. Microarray analysis of nitric oxide responsive transcripts in *Arabidopsis*, *Plant Biotechnol. J.*, 2, 359, 2004.

67. Leyser, O., Molecular genetics of auxin signaling, *Annu. Rev. Plant Biol.*, 53, 377, 2002.

68. Berleth, T., Krogan, N.T., and Scarpella, E., Auxin signals — turning genes on and turning cells around, *Curr. Opin. Plant Biol.*, 7, 553, 2004.

69. Teale, W.D. et al., Auxin and the developing root of *Arabidopsis thaliana*, *Physiol. Plant.* 123, 130, 2005.

70. Masucci, J.D. and Schiefelbein, J.W., Hormones act downstream of *TTG* and *GL2* to promote root hair outgrowth during epidermis development in the *Arabidopsis* root, *Plant Cell*, 8, 1505, 1996.

71. Bates, T.R. and Lynch, J.P., Stimulation of root hair elongation in *Arabidopsis thaliana* by low phosphorus availability, *Plant Cell Environ.*, 19, 529, 1996.

72. Pitts, R.J., Cernac, A., and Estelle, M., Auxin and ethylene promote root hair elongation in *Arabidopsis*, *Plant J.*, 16, 553, 1998.

73. Schikora, A. and Schmidt, W., Iron stress-induced epidermal cell fate is regulated independently from physiological acclimations to low iron availability, *Plant Physiol.*, 125, 1679, 2001.

74. Gilbert, G.A. et al., Proteoid root development of phosphorus deficient lupin is mimicked by auxin and phosphonate, *Ann Bot.*, 85, 921, 2000.

75. López-Bucio, J. et al., An auxin transport independent pathway is involved in phosphate stress-induced root architectural alterations in *Arabidopsis*. Identification of *BIG* as a mediator of auxin in pericycle cell activation, *Plant Physiol.*, 137, 681, 2005.

76. Gil, P. et al., BIG: a calossin-like protein required for polar auxin transport in *Arabidopsis*, *Genes Dev.*, 15, 1985, 2001.

77. Rigas, S. et al., *TRH1* encodes a potassium transporter required for tip growth in *Arabidopsis* root hairs, *Plant Cell*, 13, 139, 2001.

78. Vincence-Agullo, F. et al., Potassium carrier TRH1 is required for auxin transport in *Arabidopsis* roots, *Plant J.*, 40, 523, 2004.

79. Kutz, A. et al., A role for nitrilase 3 in the regulation of root morphology in sulphur-starving *Arabidopsis thaliana*, *Plant J.*, 30, 95, 2002.

80. Neumann, G. et al., Physiological aspects of cluster root function and development in phosphorus-deficient white lupin (*Lupinus albus* L.), *Ann. Bot.*, 85, 909, 2000.

81. Uhde-Stone, C. et al., Nylon filter arrays reveal differential gene expression in proteoid roots of white lupin in response to phosphorus deficiency, *Plant Physiol.*, 131, 1064, 2003.

82. Takei, K. et al., Multiple routes communicating nitrogen availability from roots to shoots: a signal transduction pathway mediated by cytokinin, *J. Exp. Bot.*, 53, 971, 2002.

83. Schmülling, T., New insights into the function of cytokinins inplant development, *J. Plant Growth Reg.*, 21, 40, 2002.

84. Maruyama-Nakashita, A. et al., A novel regulatory pathway of sulfate uptake in *Arabidopsis* roots: implication of CRE1/WOL/AHK4-mediated cytokinin-dependent regulation, *Plant J.*, 38, 779, 2004.

85. Franco-Zorrilla, J.M. et al., The transcriptional control of plant responses to phosphate limitation, *J. Exp. Bot.*, 55, 285, 2004.

86. Franco-Zorrilla, J.M. et al., Interaction between phosphate-starvation, sugar, and cytokinin signaling in *Arabidopsis* and the roles of cytokinin receptors CRE1/AHK4 and AHK3, *Plant Physiol.* 138, 847, 2005.

87. Spollen, W.G. et al., Abscisic acid accumulation maintains maize primary root elongation at low water potentials by restricting ethylene production, *Plant Physiol.*, 122, 967, 2000.

88. Signora, L. et al., ABA plays a central role in mediating the regulatory effects of nitrate on root branching in *Arabidopsis*, *Plant J.*, 28, 655, 2001.

89. Wilkinson, S. and Davies, W.J., ABA-based chemical signalling: the co-ordination of responses to stress in plant, *Plant Cell Environ.*, 25, 195, 2002.

90. Müssig, C., Shin, G.H., and Altmann, T., Brassinosteroids promote root growth in *Arabidopsis*, *Plant Physiol.*, 133, 1261, 2003.

91. Rubio, V. et al., A conserved MYB transcription factor involved in phosphate starvation signaling both in vascular plants and in unicellular algae, *Genes Dev.*, 15, 2122, 2001.

92. Liu, J. et al., Signaling of phosphorus deficiency-induced gene expression in white lupin requires sugar and phloem transport, *Plant J.*, 41, 257, 2005.

93. Mouchel, C.F., Briggs, G.C., and Hardtke, C.S., Natural genetic variation in *Arabidopsis* identifies BREVIS RADIX, a novel regulator of cell proliferation and elongation in the root, *Genes Dev.*, 18, 700, 2004.

94. Bartel, D.P., MicroRNAs: genomics, biogenesis, mechanism, and function, *Cell*, 116, 281, 2004.

95. Guo, H.S. et al., Micro RNA164 directs *NAC1* mRNA cleavage to downregulate auxin signals for lateral root development, *Plant Cell*, 17, 1376–1386, 2005.

96. Kepinski, S. and Leyser, O., The Arabidopsis F-box protein TIR1 is an auxin receptor, *Nature*, 435, 446, 2005.

97. Dharmasiri, N., Dharmasiri, S., and Estelle, M., The F-box protein TIR1 is an auxin receptor, *Nature*, 435, 441, 2005.

98. Xie, Q. et al., *Arabidopsis NAC1* transduces auxin signal downstream of *TIR1* to promote lateral root development, *Genes Dev.*, 14, 3024, 2000.

99. Jones-Rhoades, M.W. and Bartel, D.P., Computational identification of plant microRNAs and their targets, including a stress induced miRNA, *Mol. Cell*, 14, 787, 2004.

100. Mallory, A.C., Bartel, D.P., and Bartel, B., MicroRNA-directed regulation of *Arabidopsis auxin RESPONSE FACTOR17* is essential for proper development and modulates expression of early auxin response genes, *Plant Cell*, 17, 1360, 2005.

101. Sorin, C. et al., Auxin and light control of adventitious rooting in *Arabidopsis* require argonaute1, *Plant Cell* 17, 1343, 2005.

102. Sunkar, R. and Zhu, J.K., Novel and stress-regulated microRNAs and other small RNAs from *Arabidopsis*, *Plant Cell*, 16, 2001, 2004.

103. Zhang, B.H. et al., Identification and characterization of new plant microRNAs using EST analysis, *Cell Res.*, 15, 336, 2005.

104. Arnholdt-Schmitt, B., Efficient cell reprogramming as a target for functional-marker strategies? Towards new perspectives in applied plant-nutrition research, *J. Plant Nutr. Soil Sci.*, 168, 617, 2005.

105. Mizuno, K. et al., Counteracting regulation of chromatin remodeling at a fission yeast cAMP responsive element-related recombination hotspot by stress-activated protein kinase, cAMP-dependent kinase and meiosis regulators, *Genetics*, 159, 1467, 2001.

106. Nelissen, H. et al., The *elongata* mutants identify a functional Elongator complex in plants with a role in cell proliferation during organ growth, *Proc. Natl. Acad. Sci. USA*, 102, 7754, 2005.

107. Takeda, S. et al., BRU1, a novel link between responses to DNA damage and epigenetic gene silencing in *Arabidopsis, Genes Dev.*, 18, 782, 2004.

108. Xu, C.R. et al., Histone acetylation affects expression of cellular patterning genes in the *Arabidopsis* root epidermis, *Proc. Natl. Acad. Sci. USA*, 102, 14469, 2005.

109. Miura, K. and Hasegawa, P.M., The *Arabidopsis* SUMO E3 ligase SIZ1 controls phosphate deficiency responses, *Proc. Natl. Acad. Sci. USA*, 102, 7760, 2005.

110. Martin, A.C. et al., Influence of cytokinins on the expression of phosphate starvation genes in *Arabidopsis, Plant J.*, 24, 559, 2000.

111. Schikora, A. and Schmidt, W. Acclimative changes in root epidermal cell fate in response to Fe and P deficiency. A specific role for auxin?, *Protoplasma*, 218, 67, 2001.

112. Bonser, A.M., Lynch, J., and Snapp, S., Effect of phosphorus deficiency on growth angle of basal roots in *Phaseolus vulgaris*, *New Phytologist*, 132, 281, 1996.

113. Müller, M. et al., Root hair development in adaptation to Fe and P deficiency. Paper T02-109, *Proc. Int. Conf. Arabidopsis Res.*, Berlin, 2004.

114. Gomez, G., Torres, H., and Pallas, V., Identification of translocatable RNA-binding phloem proteins from melon, potential components of the long-distance RNA transport system, *Plant J.*, 41,107, 2005.

115. Haywood, V. et al., Phloem long-distance trafficking of gibberellic acid-insensitive RNA regulates leaf development, *Plant J.*, 42, 49, 2005.

116. Aoki, K. et al., Destination-selective long-distance movement of phloem proteins, *Plant Cell,* 17, 1801, 2005.

117. Krusell, L. et al., Shoot control of root development and nodulation is mediated by a receptor-like kinase, *Nature,* 420, 422, 2002.

118. Nishimura, R. et al., HAR1 mediates systemic regulation of symbiotic organ development, *Nature*, 420, 426, 2002.

119. Van Norman, J.M., Frederick, R.L., and Sieburth, L.E., BYPASS1 negatively regulates a root-derived signal that controls plant architecture, *Curr. Biol.*, 14, 1739, 2004.

120. Birnbaum, K. et al., A gene expression map of the *Arabidopsis* root, *Science*, 302, 1956, 2003.

121. Lahner, B. et al., Genomic scale profiling of nutrient and trace elements in *Arabidopsis thaliana*, *Nat. Biotechnol.*, 21, 1215, 2003.

6 Root Membrane Activities Relevant to Nutrient Acquisition at the Plant–Soil Interface

Zeno Varanini and Roberto Pinton

CONTENTS

I. INTRODUCTION

Roots play an important role in the relationship between plant and soil, acting as anchoring organs and allowing the acquisition of water and nutrients. These functions are complicated by the nature of soil, which erects barriers of a mechanical, physical, chemical, and biological nature against the activities and development of the root system. These conditions determine numerous forms of anatomical, physiological, biochemical, and molecular adaptations by which the plant can modify, to its own advantage, the characteristics of the soil; it has been found that there are profound modifications — physical, chemical, and biological [1,2] — in the area immediately around the roots (rhizosphere).

In this interplay, the plasma membrane (PM) of the root cells, which is the main barrier between rhizosphere soil and the cytoplasm, plays a prominent role. This organelle is characterized by selective permeability that allows both the entrance of essential ions and metabolites into the cell and the extrusion of a variety of exudates (H$^+$, electrons, organic acids etc.), which can alter the conditions at the rhizosphere. The biochemical mechanisms that regulate the interactions between root and soil must be both sensitive and reactive to the metabolic status of the plant and the conditions at the rhizosphere; this implies the need to possess, at the PM, sensors of the external environment, and the possibility of efficient regulation of enzymatic activities and transport.

Considering the structure of the PM, it is clear that in the interaction between the soil and the root cells, which involves processes of nutrient acquisition, the activity and the function of the proteins associated with this organelle is of great importance, and, in particular, H^+ pumps (H^+-ATPase), carriers, and protein channels. Despite the enormous quantity of information of a molecular nature, few works connect the activity of these structures with the conditions at the rhizosphere, in spite of the evidence, recently clearer than ever, that multiple isoforms are present to which it has rarely been possible to assign a precise physiological role. The aim of this chapter is to review our knowledge on the activities taking place at the plasma membrane that appear to be involved in the soil–root interaction, considering when possible, the regulative and functional aspects in relation to the conditions present at the rhizosphere.

II. THE PLASMA MEMBRANE (PM) H^+-ATPase

The PM H^+-ATPase is an electrogenic enzyme that can couple the chemical energy released from the hydrolysis of ATP to the transport of H^+ from the cytoplasm into the apoplast; this activity creates a transmembrane gradient of electric potential ($-100 \div -200$ mV, negative inside) and pH (about 2 units, more acidic outside the cell), which can be exploited in a variety of physiological processes, such as the transport of nutrients and metabolites, preservation of intra- and extracellular pH, cell turgor, and related processes [3]. In the soil–plant relationships, the main roles of the PM H^+-ATPase concern energization of nutrient transport, acquisition of nutrients by rhizosphere acidification, and the response to abiotic stress and rhizosphere signals.

A. STRUCTURE

The enzyme able to catalyze the transport of little more than 10^2 H^+ ions per second consists of a catalytic polypeptide with an approximate molecular weight of 100 kDa and is one of the most abundant proteins on the plasma membrane with an average value of about 1% [4]. Investigations on the topology of PM H^+-ATPases present in higher plants (Figure 6.1) reveal that the enzyme is made up of a hydrophobic part consisting of 10 membrane-spanning regions and 4 cytoplasm domains (N-terminal region, small cytoplasmic loop, large cytoplasmic loop, and C-terminal region) [5]. Among the transmembrane regions of particular significance are number 4, 5, and 6 because of the role they play in the transport of H^+ from the cytoplasm into the apoplast. The specific function of the N-terminal region is still unknown. The small cytoplasmic loop is thought to be involved in the changes in conformation that occur during the catalytic cycle and its function seems to be linked to the coupling between ATP hydrolysis and H^+ transport. The large cytoplasmic loop contains the aspartate residue that is phosphorylated during the catalytic cycle and is the site of the ATP link. The C-terminal region exerts a self-inhibiting function on the enzyme activity; the inhibition is removed when, after phosphorylation, proteins of the 14-3-3 family link to this domain.

Studies with solubilized enzymes show that the minimal functional unit of the PM H^+-ATPase is a monomer; however radiation–inactivation experiments indicate that the enzyme exists *in vivo* as a dimer [6]. Recently it has been shown that the PM H^+-ATPase isoform PMA2 of *Nicotiana plumbaginifolia*, present as a dimer, can be converted into a hexameric form upon phosphorylation and interaction with the 14-3-3 regulatory proteins [7].

In several plant species, it has been found that the PM H^+-ATPase is encoded by a multigene family (about 10 genes) belonging to 5 subfamilies; of these, only 2 appear to be abundantly expressed in the tissues [8]. It has been suggested that the heterogeneity of isoforms can be linked to the multicellular nature of plants and the need for a fine regulation of the enzymatic activity.

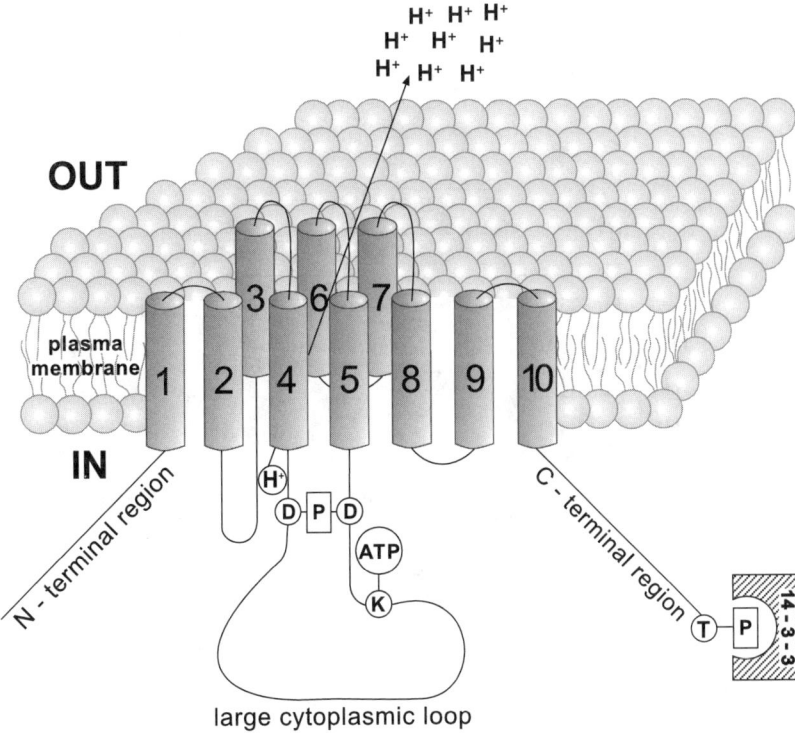

FIGURE 6.1 Predicted topology of the plant plasma membrane H^+-ATPase. Binding sites involved in catalytic and regulatory functions are also shown.

B. Regulation of the PM H^+-ATPase as a Response to Variations in the Conditions at the Rhizosphere

Even though the structural and kinetic characteristics of the PM H^+-ATPase and its properties of enzyme regulation have been widely studied, little is known about how they affect behaviors that in nature ensure the performance of functions at the level of root cells.

1. Energization of Nutrient Transport

The movement of nutrients through carriers or channels is made possible by the pH gradient (more acidic outside the cell) and the electric potential (negative inside the cell) generated by the activity of the PM H^+-ATPase (Figure 6.2).

The concentration of the nutrients in the soil solution varies, though to different extents, depending on their dynamics in the soil; furthermore, in this environment nutrients are distributed in a nonhomogeneous manner [9]. The fact that the cytoplasm concentration, though variable to a lesser degree, is often very far from (and usually higher) that of the soil solution [10], indicates the intervention of mechanisms of selective acquisition and the dependence of the process on the metabolic energy available (Table 6.1).

While the uptake of most cations is only dependent on metabolic energy to the extent that ATP is required to maintain the electrogenic component of the electrical potential difference between the symplast and the apoplast (rhizosphere) of the root cortex, metabolic energy in the form of ATP is needed to energize the thermodynamically uphill movement of anions into the root cells (see following text). Recently, a series of transporters for oxoanion forms of essential nutrients (N,P,S) that operate in symport with H^+ have been discovered and cloned. In this context, the

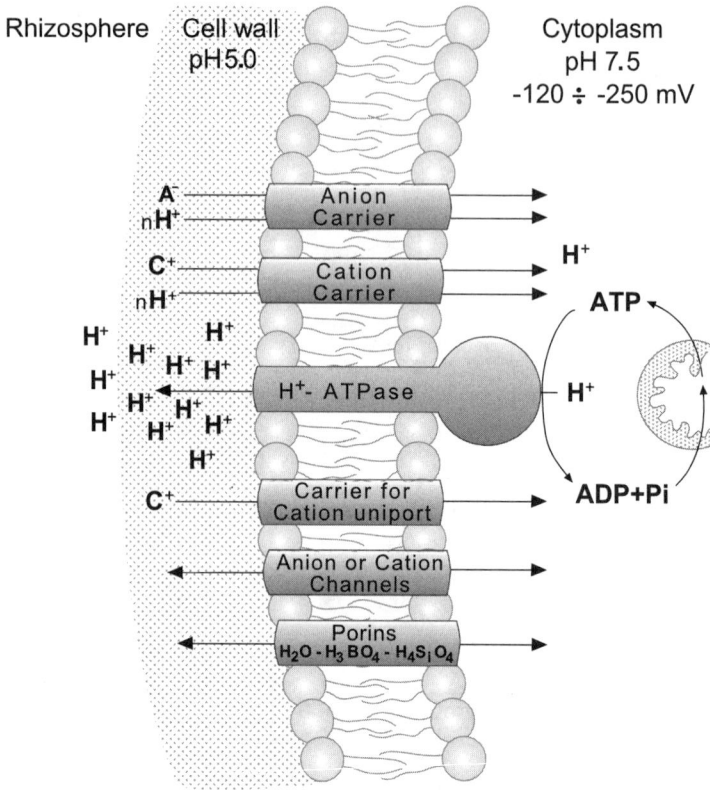

FIGURE 6.2 Primary and secondary transport systems operating at the plant plasma membrane.

TABLE 6.1
Examples of Average Concentration of Mineral Nutrients in the Soil Solution and in the Cytoplasm of Root Cells

Nutrient	Soil Solution (mM)	Root Cytoplasm (mM)
Nitrate	1.31	15
Ammonium	0.20	5
Phosphate	0.001	10
Sulfate	0.55	10
Potassium	1.28	150
Calcium	1.87	0.001
Magnesium	3.00	3

Note: Concentrations in the soil solution can vary considerably depending on the soil type and environmental conditions, whereas those within the plant change as a consequence of different external concentrations and physiological status.

Source: Modified from Cacco, G. and Varanini, Z., *Biochimica agraria*, Scarponi, L., Ed., Patron, Bologna, 2003, p. 837.

fundamental role that the PM H$^+$-ATPase can play in the process of nutrient acquisition is evident. However, so far there is little experimental evidence of the specific involvement of the PM H$^+$ pump in the uptake of various nutrients at the root cell–rhizosphere interface. Immunodetection and gene expression analysis revealed that the PM H$^+$-ATPase is expressed at high levels in tissues or cells involved in active ion transport [3]. Jahn et al. [11] have shown that in the epidermal and root cortex cells, the PM H$^+$-ATPase is unevenly spread through the plasmalemma in relation to the direction of the nutrient flow.

The close relationship between the activity of the PM H$^+$ pump of the root cells and transmembrane transport of anionic nutrients has been demonstrated in the case of NO$_3^-$. In fact, it has been shown that following an increase in NO$_3^-$ uptake rates brought about by root exposure to the ion (induction, see following text) for 24 h, and thus a greater demand for H$^+$ driving force, there is a simultaneous increase in enzyme activity, for the most part due to the increased levels of protein at the PM [12]. Interestingly, immunodetection experiments have shown that after 24 h of NO$_3^-$ induction plants had higher levels of PM H$^+$-ATPase in the epidermis and cortex cells compared to roots of plants that had not been treated with the anion (De Marco et al., unpublished). Such behavior could be explained in terms of the greater number of cells involved in the influx of the anion due to induction and the experimental system used (hydroponic culture). More recently, time-course studies on the induction of higher uptake rates have revealed a close parallelism between NO$_3^-$ transport rate and the activity and quantity of the enzyme [13]. This behavior is supported by the preferential expression of the genes (*MHA3* and *MHA4*, belonging to subfamily II of the PM H$^+$-ATPase) of two of the five isoforms so far found in maize. In particular, the isoform MHA4 was found to be more sensitive than MHA3 to NO$_3^-$ treatment, with a greater up- or downregulation. These results suggest that within the multigene family of the PM H$^+$-ATPase, there exists a specialization of functions with a greater, if not exclusive, involvement in the transport of nutrients of some isoforms in respect to others.

Information regarding variations in the levels of PM H$^+$-ATPase activity in relation to fluctuations in transport rate of other macronutrient anions is not available in the literature.

In principle, the electrical component of the H$^+$ driving force should be sufficient to guarantee the inflow of cations; however, for some of them (e.g., NH$_4^+$ and K$^+$), especially when present at the rhizosphere at very low concentrations and at much higher levels in the cytosol of the root cells (see Table 6.1), it becomes necessary to energize cation transport by both electrical and chemical (pH gradient) components of the H$^+$-driving force.

In regard to the cationic form of nitrogen (NH$_4^+$), it has been shown that fluctuations in the availability of NH$_4^+$ determine opposite variations in the uptake rate and the activity of the PM H$^+$-ATPase in roots of sugar-beet [14], suggesting that, in this case, PM H$^+$-ATPase activity is used to preserve cytoplasm pH homeostasis and transmembrane electric potential rather than directly energize the nutrient transport process [15]. It has also been shown that PM H$^+$-ATPase activity and levels were insensitive to variations in K$^+$ supply [16], suggesting that the H$^+$ gradient needed to sustain the uptake of the cation at a low external concentration (i.e., that requiring a H$^+$-driven cotransport mechanism) was maintained close to the plasma membrane because of the activity of the PM H$^+$-ATPase extruding H$^+$ into the confined space of the apoplast [17].

2. Nutrient Acquisition through Rhizosphere Acidification

The activity of the PM H$^+$-ATPase, which extrudes H$^+$ into the apoplast and from here into the rhizosphere soil, causes an acidification of the latter [18]. This phenomenon is important because it favors the release of cations from the exchange sites of the cell wall and the soil colloids, allowing them to pass in solution and reach the binding sites of the transport proteins on the external surface of the PM. The acidifying effect of the H$^+$ pump can be modified by secondary transport processes in function of the cation–anion absorption ratio [19]. Typically, the uptake of nitrogen as NH$_4^+$ or NO$_3^-$, respectively, can increase or decrease acidification [20].

Rhizosphere acidification by the PM H$^+$-ATPase is very important for the solubilization of scarcely available essential nutrients. Lowering the rhizosphere pH by an amount of up to 2 units

[21], is an aggression of the soil by the roots, aimed at obtaining nutrients in a soluble form (see also Chapter 2). The most widely studied case is that of Fe. This element is extremely insoluble at neutral and alkaline pH values. Plants (dicots and nongraminaceous monocots) are endowed with mechanisms that can oppose the increasing insolubility of Fe at neutral and alkaline pH levels. Among these mechanisms, the ability to acidify the rhizosphere can determine an increase in the amount of Fe^{3+} in solution, in fact by lowering the pH by one unit there is an increase equal to 1000 times the concentration of Fe^{3+} in the soil solution. It has been shown that, at least for the species characterized by a greater capacity for acidification, under conditions of Fe deficiency, the activity of the PM H^+-ATPase increases [22]. A similar increase is observed in the quantity of the enzyme, which appears to be concentrated particularly in the rhizodermal and root-hair cells in the subapical area of the roots [23]. Moreover, transfer cells developing in the roots of tomato plants, after Fe deprivation, reveal an increased PM H^+-ATPase protein density associated with a greater capacity to acidify the external surroundings [24]. Recently, it has been found that the roots of Fe-deficient cucumber plants accumulate transcripts of specific forms of the enzyme. In particular, the expression of the *CsHA2* gene, found both in roots and leaves, was not influenced by the Fe nutritional status of the plant, while the gene *CsHA1*, expressed exclusively in the roots was upregulated by Fe deprivation [25].

3. Response to Abiotic Stress

The activity of the PM H^+-ATPase is also modulated by different anomalous conditions that occur at the rhizosphere. A particular case is that of a lack of P. Though this nutrient may be present in the soil in high quantities, its availability to the plant is limited by the presence of scarcely soluble forms such as phytin and Ca, Fe, and Al phosphates. Plants able to respond efficiently to this situation have developed mechanisms that enable them to acquire this essential element in adequate quantities to satisfy their nutritional needs. One of the mechanisms involved in this active response to P-deprivation is the release of large amounts of negatively charged organic acids, such as citrate and malate (see Chapter 2). In *Lupinus albus* — a plant adapted to acid soil containing scarce amounts of P — it was determined that the release of citrate is favored by a contemporaneous efflux of H^+ [26]. Modifications of the activity and quantity of the PM H^+-ATPase suggest that the plasmalemma H^+ pump can provide and maintain the flow of H^+ needed for organic anion release [27,28]. Recently, it has been shown that the greater release of carboxylates by the proteoid roots of *L. albus* grown under conditions of limited P availability is closely linked to an activation of the PM H^+-ATPase (Tomasi et al., unpublished).

Likewise, also the presence of toxic amounts of Al at the rhizosphere, which often occurs in acidic soils, can induce, in resistant plants, an abundant release of organic acid anions from the subapical regions of the roots (see Chapter 2). Also, in this case, it has been shown that the PM H^+-ATPase carries out an important role. In fact, its activation is higher in resistant cultivars than in those sensitive to Al [29]. The changes in enzyme activity have been ascribed to transcriptional and posttranscriptional regulation of the protein. In regard to this latter aspect, the activation of the enzyme due to the presence of Al is, at least, partly caused by increased phosphorylation of a threonine residue localized in the autoinhibitory (C-terminal) domain of the PM H^+-ATPase. This suggests that 14-3-3 proteins are involved in its regulatory mechanism. Increases in enzyme activity have also been observed in the roots of plants adapted to acidic soils, implying that the PM H^+-ATPase helps preserve cytoplasm pH homeostasis [30].

One of the most common stresses is salt stress caused by an excessive amount of Na in the soil solution. The majority of plants adapted to salinity are able to maintain a relatively low concentration of Na in the cytoplasm by the active exclusion of Na^+ into the vacuole or the apoplast. It seems evident how the extrusion of a greater quantity of H^+ can contribute the energization of the Na^+/H^+ antiport system that operates at the tonoplast and the plasma membrane. In regard to this, it has been shown that PM H^+-ATPase gene expression in specific tissues was enhanced by NaCl or salt stress [31]; in tomato, this response is ascribed to the increased expression of a single isoform [32]. It has also been suggested that in a situation of osmotic shock, a posttranslational regulation can take place that, through the interaction with 14-3-3 proteins, increases the H^+/ATP coupling ratio [33].

4. Response to Rhizosphere Signals

a. Humic Substances

Humic substances are the result of biological and chemical transformations of plant, animal, and microbial residues carried out by soil microorganisms. In the soil, humic substances are linked to mineral components through various chemical–physical interactions. Humic substances have also been found in the soil solution, with a highly variable concentration — ranging from 1 to 400 mg/l — that depends on soil type, and it is plausible that these compounds may interact with the root cells [34]. Humic substances extracted from the soil can, in fact, influence plant metabolism affecting various physiological and biochemical mechanisms, stimulating growth and increasing the amount of nutrients taken up by the plant [35]. Stimulation of active H^+ extrusion from the roots [36] and transmembrane potential hyperpolarization [37] indicated the involvement of the PM H^+-ATPase in the increased nutrient uptake observed in the presence of humic substances. Direct proof of an interaction between humic molecules and the PM H^+-ATPase has been obtained by Varanini et al. [38] who demonstrated that low-molecular-weight (<5 kDa) humic molecules at concentrations compatible with those present at the rhizosphere can stimulate the phosphohydrolytic activity of this enzyme in isolated PM vesicles. Further proof of the action of humic molecules on PM H^+-ATPase activity and on nutrient uptake mechanisms was obtained when studying the effect of these molecules on NO_3^- uptake. As already stated (see preceding text), transport of this nutrient is a substrate-inducible process and involves H^+ cotransport. At higher uptake rates, the levels and activity of root PM H^+-ATPase were observed to increase [12]. The short-term (4-h) contact of roots with a low-molecular-weight water-extractable fraction of humic substances (WEHS), in the presence or absence of NO_3^-, caused a more rapid development of the NO_3^- uptake capacity and a further increase in PM H^+-ATPase activity measured in PM vesicles isolated from maize roots [39]. Because no increase in protein amount was observed, this effect was attributed to a posttranslational regulation of the PM H^+-ATPase. On the other hand, a prolonged treatment with high-molecular-weight humic acids isolated from earthworm compost determined a promoting effect on activity and amount of the PM H^+-ATPase [40], which was attributed to the presence of auxin bound in an exchangeable form to the humic molecules. More recently, an increase in transcript levels of the PM H^+-ATPase isoform MHA2 in maize roots, treated for 48 h with an earthworm low-molecular-weight humic fraction, endowed with auxin, was observed [41]. The action of humic molecules on the PM H^+-ATPase can also positively affect the acquisition of sparingly soluble nutrients, such as Fe [42]. Increased PM H^+-ATPase activity can contribute to Fe nutrition in several ways (see the following text and Chapter 2): (1) by solubilizing Fe in the apoplast and the rhizosphere, (2) by maintaining favorable conditions for the activity of the Fe^{3+}-chelate reductase (low apoplast pH and transmembrane electrical potential homeostasis), and (3) by favoring uptake of free Fe^{2+} or Fe^{3+} complexes (e.g., Fe–phytosiderophores).

Taken together, these results strongly support the view that the PM H^+-ATPase in root cells plays a primary role in the interactions between roots and soil components, such as humic substances, which can be present at the rhizosphere.

b. Plant Growth Promoting Rhizobacteria

It has been shown that plant growth promoting rhizobacteria (PGPR, see Chapter 3) can affect root development and nutrient acquisition by plants. The positive effect of PGPR on plant growth is generally correlated to remarkable changes in root morphology and architecture [43]. It is assumed that this developmental response is triggered by phytohormones (e.g., auxin) produced by the bacteria. On the other hand, an increased uptake of different nutrients has been reported, which could not be explained simply by the increased root surface [44]. In fact, inoculation of oilseed rape with an *Achromobacter* strain caused a more rapid development of the NO_3^- uptake capacity and an increase in the net anion uptake [45]. In addition, the uptake of K^+ and H^+ efflux were also stimulated by the treatment, suggesting that rhizobacteria can affect mineral nutrient acquisition via a stimulation of the PM H^+-ATPase.

III. ION CARRIERS

It is generally assumed that the transport of ions into cells is facilitated by carrier proteins in the PM, which provide a gated hydrophilic channel through the hydrophobic lipid bilayer. These membrane proteins can move solutes either up or down gradients at rates of 10^2 to 10^4 molecules per second. Among the carrier proteins there are symporters, antiporters, and uniporters. Symporters and antiporters can transport a substance against its electrochemical gradient by contemporaneously transporting a second substance (usually, H^+ in plants), the movement of which occurs down the electrical or chemical gradient. Uniporters move substances down a concentration gradient. Knowledge of the mechanisms involved in the movement of ions through the root cell PM is essential to understand the complex phenomena that regulate the acquisition of nutrients (and the tolerance to toxic elements) at the root–soil interface. These aspects are particularly important to improve the use efficiency of the nutrient pool and the plant's adaptability to adverse conditions.

A. Anion Carriers

In the rhizosphere–root interaction, the activity of the transporters involved in the movement of oxoanions (NO_3^-, Pi and SO_4^{2-}) is particularly important. These proteins appear to belong to the major facilitator superfamily (MFS); members of this family are single polypeptide secondary carriers capable of transporting small solutes utilizing chemiosmotic ion gradients. MFS is a divergent group of proteins that are typically 500 to 600 amino acids in length, and with a membrane topology in which two sets of six transmembrane helices are connected by a cytosolic loop [46]. In addition, the presence, on the cytoplasm domains, of motifs that suggest the possibility of posttranslational regulation via phosphorylation and dephosphorylation has been evidenced. In recent years, numerous genes have been discovered that encode for oxoanion transporters in different types of plants (e.g., in *Arabidopsis*: 11 for NO_3^- [47], 14 for SO_4^{2-} [48], and 13 for Pi [49]); however, of these, only a limited number is believed to be responsible for the uptake of these nutrients from the soil. The remainder probably encodes transporters that take part in internal redistributions.

As regards NO_3^-, which is the most important source of mineral N for plants growing in agricultural aerated soils, it has been found that plants acquire it from the soil via the combined activity of low- and high-affinity systems encoded by different genes (*NRT1* and *NRT2*, respectively), whose products act as nH^+/anion cotransporters. The Km values estimated for the high affinity NO_3^- transport are in the range 7 to 110 μM, whereas those for low-affinity transport system are in the range 170 to 25000 μM. So far, seven members of the subfamily *NRT2* (belonging to the NNP family, NO_3^--NO_2^- porters) have been found in *Arabidopsis*. For two of these (*AtNRT2.1* and *AtNRT2.2*), which are mainly expressed in roots and are induced by root contact with NO_3^- (IHATS, inducible high affinity transport system), a close correlation has been observed between the levels of gene expression and influx of ions in conditions of low external concentration [50]. The fact that the products of these two genes play a fundamental role in the acquisition of NO_3^- at the soil–root interface is demonstrated in experiments with T-DNA *Arabidopsis* mutants disrupted in the *AtNRT2.1* and *AtNRT2.2* genes, which exhibited severe, specific impairments in their IHATS function [51,52]. High-affinity constitutive transporters (CHATS, possibly *AtNRT2.4*, *AtNRT2.5*, *AtNRT2.6*) operating at low capacity facilitate the entry of NO_3^- into roots allowing for the induction of high-affinity and high-capacity transporters (IHATS). Recently [53], it has been shown that the functional activity of a barley NO_3^- transporter (*HvNRT2.1*) in oocytes required the presence of a much smaller protein with only one transmembrane domain encoded by the gene *HvNAR2.3*. Confirmation of the role of the *NAR*-like genes has been obtained using two T-DNA *Arabidopsis* mutants disrupted in the *NAR* (renamed by the authors *AtNRT3.1*) promoter and encoding regions, respectively [54]. In wild-type plants exposed to NO_3^-, the transcript levels of *AtNRT3.1* followed an expression pattern (upregulation followed by downregulation) similar to that of *AtNRT2.1*, whereas in *AtNRT3.1* mutants the expression actually declined. In both mutants, high-affinity NO_3^- influx was reduced by more than 90%. Although the mechanisms that render the coexpression of the two genes indispensable for the transport of NO_3^- are still unclear, these results have important implications for the genetic manipulation of membrane transport processes aimed at improving NO_3^- uptake. It should be remembered, for example, that the overexpression of a IHATS transporter in tobacco (*NpNRT2.1*) failed to yield any increase in NO_3^- uptake [55]. These results could depend on the missing coexpression of a NAR-like partner gene, but it is also possible that posttranscriptional regulations intervene to modulate anion uptake. By transcript analysis and using inhibitors of the reductive assimilation pathway while measuring NO_3^- flux, possible posttranscriptional regulation mechanisms of the proteins involved in anion transport were identified [56]. It is interesting to note that in the structure of the transporters belonging to subfamily NRT2, there are a number of conserved protein kinase C recognition motifs [57]. The existence of these motifs could indicate that phosphorylation and dephosphorylation reactions play a part in the regulation of the activity of the NRT2.

Recently, it has been suggested that *NRT2.1* could play a role not only in mediating high-affinity NO_3^- uptake but also in modulating changes in root structure [58]. In particular, *NRT2.1* could be a key factor in coordinating root development both indirectly, through its role as a major NO_3^- uptake system that determines the nitrogen uptake-dependent responses of the root architecture, and directly through a specific action on the initiation of lateral roots under nitrogen-limited conditions [59].

The transport of NO_3^- into plant cells from a solution containing a high concentration of the anion (≥ 1 mM) is usually attributed to low-affinity transporters (LATS) belonging to the subfamily NRT1. In turn, the NRT1 belong to the family of POT or PTR (polypeptide transporters) that is widely distributed in both prokaryotes and eukaryotes, most members of which function as H^+/oligopeptide cotransporters in the plasma membrane [60]. In *Arabidopsis*, four *NRT1* genes have been identified among which *NRT1.1* and, to a lesser extent, *NRT1.2*, are expressed depending on NO_3^- concentration. The LATS function of NRT1.1 was revealed in heterologous expression investigations in oocytes, which showed that the products of these genes cotransport H^+ and NO_3^- with a Km between 4 and 8.5 mM [61,62].

AtNRT1-deletion mutants revealed that *NRT1.1* could also play a role in the high-affinity NO_3^- uptake in plants previously grown at elevated NH_4^+ concentrations [63]. This result suggests that *NRT1.1* might operate as a dual affinity NO_3^- transporter. The switch between the two modes of action is regulated by phosphorylation at the threonine residue 101; when phosphorylated, *NRT1.1* functions as a high-affinity NO_3^- transporter, whereas when dephosphorylated, it operates as a low-affinity NO_3^- transporter [64].

Another interesting property of the *Arabidopsis NRT1.1* is its ability to be induced, in the absence of external NO_3^-, by a rapid decrease in the external pH [65], a condition that is often found in the apoplast of root cells because of the activity of nutrient uptake and the release of exudates from the roots.

As reported earlier, genes of the POT family appear to mediate the H^+-coupled transport of NO_3^-, amino acids, and oligopeptides. For example, in *Brassica napus*, the *BnNRT1.2* gene product can transport L-hystidine as well as NO_3^-. Interestingly, the optimum pH for the two substrates is quite different: hystidine transport is favored at an alkaline pH and NO_3^- transport at an acidic one. The ecological significance of this double function is still unclear, though it should be remembered that amino acids can be present at the rhizosphere as a result of microbial and root activity.

To understand the role of the *NRT1* and *NRT2* gene products in the root–soil interaction that takes place at the rhizosphere, it is essential to define the spatial and developmental regulation of the transporters. Consistent with their postulated role in high-affinity NO_3^- uptake, evidence indicates that *NRT2* genes are generally expressed more strongly in roots than in aerial tissues [66,67]. Moreover, *AtNRT2.1* has been shown to be expressed in epidermal cells of older tissues and in root hairs, but in very low amounts at the root tip [50,68]. In contrast, *AtNRT1.1* appears to be primarily expressed in the epidermal cells closer to the root tip. In older roots, the gene is expressed in cells deeper in the cortex, including the endodermis. The constitutive *AtNRT1.2* gene was primarily expressed in root hairs and the epidermis in both young (tips) and mature regions of roots.

Considering the concentrations of NO_3^- normally encountered in cultivated soil (see Table 6.1), the question arises concerning the relative importance of low- and high-affinity transport systems. On the basis of models of the movement of NO_3^- in the soil solution, it was calculated that a concentration equal to 300 to 400 μM of this nutrient in the external solution is sufficient to maintain the high-affinity uptake system saturated [69]. Besides, at a concentration of NO_3^- in the soil solution in the millimolar range, we could assume that downregulation of HATS due to the effect of glutamine accumulated in the tissue should occur [56]; this would imply that under these conditions, the contribution of this transport system to NO_3^- nutrition is quite limited. On the other hand, it is well-known that cereals and other crops respond to applications of N fertilizers far beyond those needed to produce very high NO_3^- concentrations in the soil solution. Furthermore, although NO_3^- is the predominant form of N in well-aerated and pH-balanced soils, NH_4^+ is also present and can inhibit NO_3^- uptake by roots both directly [70] and through the action of its assimilation product, glutamine, on transporter gene expression [56]. Given the spatial and temporal distribution of *AtNRT2.1* and *AtNRT1.1* genes, it has been hypothesized that young roots may derive most of their absorbed NO_3^- via low-affinity transporters. This activity could reduce the amount of NO_3^- present around the older parts of the roots, thus allowing the expression of the IHATS genes which provide for the acquisition of the remaining NO_3^- [71].

However, it must be remembered that the concentrations of NO_3^- are extremely variable depending on environmental (rainfall, temperature) and human factors (fertilizing and irrigation) and subject to the equilibrium that is established between the different forms of nitrogen in the soil, as the distribution is not uniform [72]. In addition, it must be remembered that other biotic factors present in the rhizosphere, such as PGPR, can alter NO_3^- acquisition, root development, and the expression of the NO_3^- transporters [73].

Recent observations [Giorgio et al., unpublished] seem to indicate that the expression of the genes involved in the induction of NO_3^- uptake [13] is influenced by the contemporaneous presence of SO_4^{2-}, because induction is limited when the external solution is S-depleted. The rate at which the phenomenon occurs confirms the idea that the uptake of the two anions is coregulated.

A further complication emerges from recent studies that show that organic nitrogen compounds (amino acids) present at the rhizosphere can be taken up by the roots and thus contribute to the nitrogen nutrition of the plant [74,75] (see also Chapter 4).

Sulfur is present in the soil solution mainly as SO_4^{2-}, and in this form, it is preferentially taken up by the plants [76]. Physiological studies have shown that its transport across the PM of root cells is a process that requires energy in the form of H^+, so a system of cotransport nH^+/SO_4^{2-} [77] is formed. Unlike NO_3^-, SO_4^{2-} uptake rates increase under conditions of S deprivation; a process that is then reverted by S supply [78]. In higher plants, there are numerous genes encoding for SO_4^{2-} transporters; 14 have been identified in *Arabidopsis*, and a similar number is probably present in other species [48]. On the ground of their amino acid sequences, SO_4^{2-} transporters in *Arabidopsis* have been grouped into five subfamilies (from *AtSULTR1* to *AtSULTR5*) with different catalytic properties. The members of the *AtSULTR1* family encode for high-affinity transporters, among these *AtSULTR1.1* and *AtSULTR1.2* are mainly localized in root epidermis cells and are thus believed to be involved in SO_4^{2-} uptake from the rhizosphere [79]. *AtSULTR1.1* is strongly expressed under conditions of S depletion, whereas the expression of *AtSULTR1.2* only increases weakly when S is lacking; it seems plausible that *AtSULTR1.2* is involved in constitutive uptake, whereas *AtSULTR1.1* is responsible for absorption when sulfur is scarcely available. The high-affinity SO_4^{2-} uptake system is mainly regulated at a transcriptional level in response to the nutritional status of the plant: higher rates are observed when the intracellular concentration of SO_4^{2-}, cysteine, and glutathione decreases [80]. When sulfur supply is adequate, cysteine and glutathione in particular would repress the genes encoding for the transporters, whereas the accumulation of cysteine precursor, *O*-acetylserine, would exert a positive regulation. However, it must be noted that the levels of gene expression and SO_4^{2-} uptake rates do not go hand in hand always [81,82]. Investigations on the structure of high-affinity SO_4^{2-} transporters have revealed the presence of a potentially phosphorylable serine residue in the C-terminal domain, that could be involved in posttranscriptional regulation.

The kinetic characteristics of the high-affinity SO_4^{2-} transport system observed both *in vivo* (Km $\cong$ 10 μM) and in heterologous systems (Km $\cong$ 1.5 to 6.9 μM) [83] indicate that its activity is compatible with a situation where sulfur availability at the rhizosphere is very low or when a great demand for this nutrient, the movement of which in the soil solution mainly occurs by diffusion, determines a drop in its concentration at the root surface (rhizoplane), to such a level as to require the intervention of these mechanisms.

Phosphorus is taken up by plants in the orthophosphate (Pi) forms $H_2PO_4^-$ and HPO_4^{2-}. At pH levels ranging from 4.5 to 5.0, phosphorus is mainly taken up as $H_2PO_4^-$ [84]. In most soils, the concentration of Pi in the solution is several orders of magnitude lower than that in plant tissues (see Table 6.1). Accordingly, an energy-mediated cotransport process driven by a H^+ gradient, generated by the PM H^+-ATPase has been proposed for Pi uptake in plants [85]. Like SO_4^{2-}, the deprivation of P determines the derepression of the genes encoding for high-affinity Pi transporters with a consequent increase in the uptake rates of the anion. Biomolecular research has revealed that Pi uptake from the soil is mediated by proteins encoded by genes belonging to the Pht1 family. To date, nine genes encoding for high-affinity Pi transporters have been identified in *Arabidopsis,* and a few less in other plant species such as tomato, potato, *Medicago* and white lupine [49]. All the members of the plant Pht1 family exhibit high sequence similarity with each other and with fungal Pi transporters. Detailed studies on the two tomato Pi transporter genes *LePT1* and *LePT2* [84] show that *LePT1* mRNA was detectable in both root and shoot tissues, whereas *LePT2* mRNA was only found in the roots. Cytolocalization investigations indicate a preferential expression of *LePT1* in rhizodermal cells, outer layers of the cortex, root cap and root hairs, and of *LePT2* in root epidermal cells. Experiments with cultured tobacco cells expressing the *Arabidopsis PHT1* gene revealed a Km value of 3.1 μM [86], which is in agreement with data obtained from physiological studies, showing Km ranging from 2 to 10 μM [87]. Taken together, these data support a role for some members of the Pht1 family in the acquisition of Pi from the soil. Transcript levels of *LePT1* and *LePT2* increased upon Pi starvation and decreased after resupplying the anion. The use of LePT1-specific antibodies revealed that changes in *LePT1* transcript amounts, induced by fluctuations in Pi availability, were accompanied by modifications in the levels of the transport protein present at the plasma membrane of root cells. These data indicate a preferential transcriptional regulation of Pi transporters. The presence in the sequence of the Pi transporter of potential phosphorylation or glycosylation sites, suggests that posttranslational regulation may also occur.

An important factor modulating the expression of high-affinity Pi transporters appears to be the availability of other nutrients at the rhizosphere. Low levels of SO_4^{2-} or NO_3^- may limit the derepression of Pi transporters, which occurs under conditions of Pi-starvation [88]. On the other hand, Zn deficiency causes an overexpression of certain transporters of the Pht1 family, even if external concentrations of the anion are not limiting, suggesting a specific role of Zn in the signal transduction pathway that regulates the expression of these genes [89].

Most plant species can form symbiotic associations with mycorrhizal fungi and draw benefits in terms of greater access to soil Pi (see Chapter 8). In mycorrhized plants, Pi is taken up by the fungal hyphae from areas distant from the root surface and is then transferred to the roots. In two mycorrhizal VA fungi (*Glomus versiforme* and *Glomus intraradices*), Pi transporters have been identified that have characteristics and functions similar to those of the high-affinity transporters of the Pht1 family, present in plants [90,91]. The translocation of Pi inside the hyphae involves processes of mass flow and cytoplasmic streaming. In the VA mycorrhizae, the Pi transported inside the fungal hyphae must be released into the apoplastic space that separates the PM of the root cells from the fungal membrane to be transferred into the root cells. This type of mycorrhizae are known to form branched structures (arbuscules) inside the cortical cells. This phenomenon causes a modification of the PM of the host cell with the formation of a symbiotic interface characterized by a large surface area (periarbuscular membrane) separated from the wall of the fungus and its membrane by a thin apoplastic space (interface matrix).

The molecular components involved in the release of the Pi from the hyphal interface matrix have not yet been identified. On the other hand, uptake into the root cells via the periarbuscolar membrane seems to involve a high-affinity Pi transporter having a similar sequence to that of plant and fungal Pht1 transporters [92]. The functioning of the transporter depends on the availability of H^+-driving force supplied by the activity of an H^+-ATPase localized on the membrane itself [93].

B. Cation Carriers

Despite the extensive amount of investigations carried out over the last decade and aimed at shedding light on the molecular bases of K^+ transport at the soil–root interface, it is still unclear what the specific role, function, and importance of each carrier identified are [94]. The pioneering work of Epstein et al. [95] helped establish that at external concentrations of $K^+ < 1$ mM, the uptake of this cation at the PM of root cells is mediated by a saturable, high-affinity transport system (Km $\cong 20$ μM), whereas a second system operates at higher concentrations of K^+. Detailed studies in *Arabidopsis* roots have shown that at low external concentrations of K^+ (< 0.2 mM), to reach the very high concentrations observed in the cytosol ($\cong 150$ mM), the uptake of the cation needs to be energized presumably via a H^+/K^+ symport [96]; at higher external concentrations, K^+ uptake seems to be mediated by an inward K^+-selective channel (see the following text). High-affinity transporters that mediate K^+ uptake have been cloned in barley and *Arabidopsis* and named *HAK, KT,* or *KUP* [94]. These transporters possess 12 putative membrane-spanning regions and are members of a large, conserved family of transporters that are highly K^+-selective and probably function as H^+-coupled systems [97]. In barley, a transporter belonging to this family (*HvHAK1*) has been found to be exclusively expressed in the roots, and its expression is induced by K^+ deprivation [97]. This result is in accordance with induction of a high-affinity K^+ uptake observed in K^+-starved intact plants. More recent studies reveal that in barley, the genes encoding for the HvHAK1A and HvHAK1B transporters were strongly upregulated by K^+ deprivation, whereas *HvHAK2* and *HvHAK3* genes were only weakly influenced by K^+ deprivation [98]. Moreover, in *Arabidopsis* roots subjected to K^+ deprivation, an upregulation of the *AtHAK5* gene (homologue of *HvHAK1*) takes place with a maximum expression in the epidermal tissues of the main and lateral roots and the stele of the main root [99]. The use of T-DNA insertion mutants has helped establish the principal role of this transporter in the high affinity K^+ uptake system induced by the deprivation of this nutrient in *Arabidopsis* roots. Thirteen genes belonging to the *AtHAK/KT/KUPs* family have been identified in *Arabidopsis*, ten of which are expressed in root hairs while only five in the root apex [100]; this suggests that the root hairs play an important role in the uptake of K^+ from the soil solution.

As for K^+, two systems have been described for NH_4^+, one displaying high affinity (saturable and with Km in the range 17 to 188 μM) and one having low affinity, and apparently not saturable [101,102]. The activity of the latter is evident at NH_4^+ concentrations ≥ 1 mM and probably depends on the activity of channels (see following text).

It has been estimated that at an external concentration ≤ 10 μM, to preserve a cytosol concentration of 1mM or more, a high-affinity transport system coupled with the availability of energy, probably in the form of H^+, is needed. Numerous high-affinity NH_4^+ transporters have been cloned in different plant species [103]; these transporters, belonging to the AMT subfamily, have 11 membrane-spanning regions with an extra cytosolic N-terminus and a cytosolic C-terminal domain [104]. In *Arabidopsis*, six AMT genes have been identified, two of which (*AMT1.1* and *AMT1.3*) play a role in NH_4^+ uptake at root level. Only AMT1.1 is upregulated by nitrogen deprivation; both transporters appear to be expressed in cortex and rhizodermal cells including root hairs under conditions of nitrogen deficiency. Their contribution to NH_4^+ uptake by the roots has been estimated to be about 30%, indicating the need to find a role for other transporters of this nutrient. An unexpected characteristic of a root hair NH_4^+ transporter in tomato (*LeAMT1.1*) has emerged when expressed in *Xenopus* [105] oocytes: this investigation revealed that NH_4^+ transport into the oocytes was independent of the external concentration of H^+, suggesting that the NH_4^+ uptake process depends on the transmembrane electric potential and concentration gradient of the cation. These results seem to suggest a uniport mechanism with a Km ($10\ \mu M$) similar to that observed in intact tomato roots, but this hypothesis does not agree with the data of numerous researchers concerning the cytosol concentration of NH_4^+ [69].

Recently, a transporter has been found in the mychorrizal fungus *Glomus intraradices* (*GintAMT1*) that could be involved in NH_4^+ uptake, when the external solution contains micromolar concentrations of the cation [106].

Numerous transporters able to mediate the absorption of micronutrient cations from the soil solution into the root cells have been found using different molecular methods [107]. Members of the ZIP (zinc-regulated transporters, iron regulated transporter-like protein) gene family can transport a variety of bivalent cations. One of the best studied transport system is that of Fe; the acquisition of this micronutrient depends on a series of coordinated systems that differ according to the plant species considered (see Chapter 2) and are defined as strategy I (all plants except grasses) and strategy II (grasses). In strategy I plants, Fe is mainly taken up from the soil solution as Fe^{2+} by the IRT1 transporter [108]. This protein consists of eight membrane-spanning domains and belongs to the ZIP family [109]; it has a Km of about $6\ \mu M$ for Fe^{2+} and is also able to transport Mn^{2+}, Zn^{2+}, Cd^{2+} and Co^{2+}. The use of site-directed mutagenesis has revealed that the substrate specificity of IRT1 is only defined by a few amino acid residues [110]. In *Arabidopsis*, transcript and protein levels greatly increase in the roots of Fe-deficient plants; because the IRT1 gene is exclusively expressed in the root epidermis and the protein appears to be localized at the PM, it seems evident that IRT1 plays a fundamental role in the uptake of Fe^{2+} from the rhizosphere solution [111]. The expression of the transporter goes hand in hand with that of the *FRO2* gene that encodes for the protein responsible for Fe^{3+} reduction and the consequent splitting of the Fe^{III}-chelate bond [112]. By means of split-root experiments it was possible to show that the expression of *IRT1* and *FRO2* is controlled both at local level by the Fe pool present at the roots and at systemic level by a still unknown signal from the shoot [113]. It has also been demonstrated that the activity IRT1 and that of FRO2 are subject to posttranscriptional regulation [114,115].

Hortologues of *IRT1* have been identified in different plant species, such as pea [116] and tomato [117]. In both tomato and *Arabidopsis* other *IRT* (*IRT2*) genes have been found that, though not greatly upregulated by Fe deficiency, could carry out a role in the acquisition of Fe^{2+} from the soil solution.

Members of the ZIP family are also involved in Zn^{2+} transport. In *Arabidopsis*, *ZIP1* and *ZIP3* genes are expressed in the roots in response to Zn^{2+} deficiency, indicating their role in the transport of this micronutrient from the soil solution into the root cells [109].

Another class of transporters that could have a role in the uptake and homeostasis of various divalent cations, including Fe^{2+}, is the NRAMP (natural resistance-associated macrophage proteins) family; their involvement in metal transport has been demonstrated in *Arabidopsis* [118]. However, it remains to be clarified whether these transporters, which are generally scarcely specific for single cations, are involved in uptake processes from the soil or only play a limited role in the distribution within the plant of the cations they transport.

A particular case is that of Cu^{2+}. Five transporters belonging to the Ctr family have recently been identified in *Arabidopsis* and named COPT1-5 [119]. Ctr proteins contain three predicted membrane-spanning regions and a variable number of metal-binding motifs at their extracellular N-terminal domain. COPT1 is expressed in the peripheral cells of a limited apical zone of the root and a reduction in its expression causes a significant decrease in the rate of Cu^{2+} transport [120]. These two observations seem to confirm that COPT1 plays an important role in the uptake of Cu^{2+} from the soil.

Cationic micronutrients can also be taken up by the roots as complexes with low-molecular-weight compounds. Plants (grasses) that use strategy II to acquire Fe are the best studied. In these plants, the micronutrient is taken up from the soil solution as a ferric complex formed with phytosiderophores (PS), low-molecular-weight compounds released in large amounts from the roots when the availability of Fe is limited. PS, characterized by a high stability constant for Fe^{3+}, can solubilize the micronutrient and form complexes (Fe^{III}–PS) with it, which are then taken up by a transporter (YS1) present in the PM of the root cells (see Chapter 2). The *ZmYS1* gene identified using the yellow stripe 1 maize mutant [121] encodes for a protein of 682 amino acid residues with 12 putative transmembrane domains and belongs to the OPT family of oligopeptide transporters [122]. ZmYS1 is expressed both in the roots and the leaves and is overexpressed in both tissues under conditions of Fe deficiency; this protein has a Km $\cong$ 10 μM and its activity seems to be linked to membrane potential and H^+ symport [123]. ZmYS1 is able to transport complexes between PS and various metals, such as Fe^{3+}, Zn^{2+}, Cu^{2+}, and Ni^{2+}, as well nicotianamine (NA) complexed with Ni^{2+}, Fe^{2+}, and Fe^{3+}. This latter function seems to be linked to the transport of the metals inside the plants rather than their acquisition from the rhizosphere solution. On the other hand, in barley, a plant that is particularly efficient at acquiring Fe in calcareous soils, a transporter has been identified, the expression of which depends on the scarcity of Fe. It is highly specific both for the metal and the ligand and can thus only transport the Fe^{III}–PS complex [124]. This transporter, in contrast to ZmYS1, is exclusively expressed in roots.

Eighteen putative *OsYSL* (*Oryza sativa YS1*-like) genes have been identified in rice [125]. However, in this graminaceous plant that is cultivated in flooded soils with a high availability of Fe^{2+}, a homologue of the *Arabidopsis IRT1* gene, *OsIRT1* has been isolated [126], indicating how the expression of the molecular components typical of a certain strategy for acquiring iron can also be present in plants of a different strategy group, as an adaptive response to specific growth conditions [127].

YS-like genes are also found in plants that use strategy I to acquire Fe [128]. These transporters may play a role in preserving the homeostasis of Fe and other metals in the plant. A putative transporter for oligopeptides (*AtOPT3*), with a sequence similar to *ZmYS1* and *AtYSL* genes, has been found to be strongly upregulated in the root tissues of Mn-deficient *Arabidopsis* plants [129].

IV. CHANNELS

Protein channels facilitate the diffusion of water and ions down energetically favorable gradients. These proteins can open and close to form pores through which organic and inorganic ions and water molecules can pass at very rapid rates: from 10^6 to 10^7 ion/molecule/second/channel.

As regards the root–soil relationship and nutrient acquisition, a fundamental part is played both by the structures involved in cation (K^+, NH_4^+, and Ca^{2+}) transport and the efflux of anions (NO_3^-, Cl^-, and organic anions). Moreover, channels specialized in water transport (aquaporins) are fundamental, as well as similar structures ascribed to the transport of nutrients that are absorbed by the roots mainly in undissociated forms (e.g., Si and B).

A. CATION CHANNELS

The uptake of K^+ from the soil solution appears to be mediated, besides by carriers (see the preceding text) also by channels, among which the main role is played by AKT1, a shaker-type channel found in several species including *Arabidopsis* [130]. This type of channel is made up of four subunits arranged around a central pore. The hydrophobic domain of each subunit consists of six transmembrane segments, the fourth of which has a repeated series of basic residues that function as voltage sensors. Within the structure there is a loop able to confer selectivity for K^+ and a regulatory C-terminal cytosolic domain with the presumed linking site for cyclic nucleotides. AKT1 mediates the flow of K^+ into the root peripheral cells and appears to be activated by the hyperpolarization of the transmembrane potential, a condition induced by limited K^+ availability. This phenomenon probably depends on the activation of the PM H^+-ATPase, in turn determined by the acidification of the cytoplasm, because of the lowered K^+ concentration in the cytoplasm [131]. Though it is generally recognized that the channels are responsible for the uptake of K^+ from an external solution with a high K^+ concentration, it has been shown that AKT1 can mediate the transport of this cation at an external concentration in the μM range if the transmembrane potential is sufficiently negative [132]. It has been suggested that the activity of AKT1 could be regulated by a physical interaction with the α subunit of AtKC1, another shaker-type of potassium channel expressed in the root hairs and the endodermis [133]. AKT1 could also act as a low-affinity NH_4^+ transporter [69].

There is little information on the molecular nature of the proteins involved in the transport of Ca^{2+} and Mg^{2+} from the rhizosphere solution. For both cations, genes have been identified that could carry out this function. As regards Ca^{2+}, for example, the *Arabidopsis AtTPC1* gene appears to encode for a channel that favors the influx of Ca^{2+} [134]; moreover, in this species, there is a family of ten *MRS2* (mitochondrial RNA splicing2)-like genes that could mediate the transport of Mg^{2+} into the plant [135].

B. ANION CHANNELS

Numerous types of anion-transporting channels have been found at the plasma membrane of root cells with different physical properties and presumably physiological functions. These can be grouped into two categories depending on whether they transport inorganic or organic anions.

In regard to the influx of inorganic ions, channels have been identified that could mediate the transport of NO_3^- and Cl^- [136]. However, considering the differences in the concentration of the two anions that are normally present in the soil solution and the cytosol, and the transmembrane electric potential, their contribution to the uptake of these nutrients seems improbable. However in saline soil, where there is an elevated influx of Na^+, it is presumably accompanied by an influx of Cl^- so as to prevent an excessive depolarization of the membrane potential induced by the cation.

Channels for the efflux of inorganic anions have been identified and well characterized at the plasma membrane of epidermis cells and root hairs of *Arabidopsis* [137]. These channels can transport SO_4^{2-}, Cl^-, and NO_3^- from the cytoplasm into the apoplast of the root cells [138]. Their presence raises the problem of the physiological significance that the efflux of anions could have on plant nutrition.

The best-studied case is that of NO_3^- and how its efflux can affect the use efficiency of this nutrient. It has been suggested that the channels mediating NO_3^- transport could act as overspill mechanisms when there is an unbalance between the uptake of the anion and the metabolic demand for the nutrient [69].

The efflux mechanism, on the other hand, could be instrumental for controlling the accumulation, at cytoplasm level, of anions such as Cl^- (e.g., under saline conditions) and SO_4^{2-} (for which there is a limited capacity for accumulation in root-cell vacuoles).

Anion channels in the root periphery could play an important role in resistance to excessive concentrations of B at the rhizosphere [139], favoring the efflux of borate from the cytosol through a channel localized in the PM of the root cells.

C. PORINLIKE CHANNELS

At the PM of root cells there are particular channels involved in water transport, called aquaporins [140]. When open, aquaporins facilitate the passive movement of water molecules down a water potential gradient. In *Arabidopsis*, thirty genes have been found that encode for aquaporin homologues.

Some of these genes encode for highly abundant, constitutively expressed proteins and some are known to be temporally and spatially regulated during plant development and in response to stress. Some of the aquaporins that are regulated by a process of phosphorylation and dephosphorylation appear to be involved in cytosol osmoregulation but also for the bulk flow of water in plants.

Channels similar to aquaporins can mediate the transmembrane transport of nutrients in the form of undissociated acids, and this is the case of B and Si. The uptake of B as H_3BO_3 takes place through PIP1, PIP1a, PIP2a, and PIP2b-type channels [141]. Recently a molecular mechanism (*Ls1* gene) controlling Si accumulation has been identified in rice [142]. The *Ls1* gene, belonging to the aquaporin family, is constitutively expressed in roots and localized on the plasma membrane of the exodermis and endodermis. When expressed in *Xenopus* oocytes *Ls1* only transported Si, in the form of monomeric H_4SiO_4.

D. ORGANIC ANION CHANNELS

The release of low-molecular-weight organic compounds from the roots can alter the chemical characteristics (see Chapter 2) and microbial composition of the rhizosphere (see Chapter 3). This phenomenon is particularly important in the case of nutritional deficiencies (Pi in particular) or in the presence of toxic elements such as Al.

One of the mechanisms by which the plant can increase the availability of Pi at the rhizosphere is the release of negatively-charged organic acids, such as citrate, malate, tartrate, and succinate. This type of activity has been studied in *Arabidopsis* and *L. albus* using an electrophysiological approach. In *Arabidopsis*, the transport of organic acids seems to be mediated by a Pi-regulated root anion channel that is only activated by P-deficiency [137]. It has been hypothesized that the activity of this channel could be regulated by a still unknown cytosolic factor. In the case of *L. albus*, the channel, which has electrophysiological characteristics different from those found in *Arabidopsis*, mediates the efflux of organic acids also in P-sufficient plants but its activity increases significantly following P starvation [143].

Tolerance to high concentrations of Al^{3+} present at the rhizosphere is associated with the activity of channels that are highly permeable to malate or citrate [144]. The activation of these channels depends on the presence of extra cellular Al^{3+} and could involve a mechanism of phosphorylation. Recently, in wheat, a gene (*ALMT1*) that encodes for a membrane protein constitutively expressed in the root apex of a wheat line tolerant to Al^{3+} has been cloned [145]. The expression of *ALMT1* in *Xenopus* oocytes and in cultured cells of rice and tobacco can determine an efflux of malate activated by the presence of Al^{3+}.

V. CONCLUDING REMARKS AND FUTURE PERSPECTIVES

The huge amount of information accumulated during the last decade, using a molecular approach, has shown the existence of an abundance of genes that encode for primary and secondary transport proteins. Presumably, these are responsible for the acquisition of nutrients, and their distribution within the plant. Many of them are situated in the PM, but only for a few have a nutrient transport function from the rhizosphere into the root cells been clearly assigned. This indicates that, at least in some cases, each isoform has a specialized function, linked perhaps to the need to respond to specific signals and situations that could occur in the distinctive, changeable environment at the rhizosphere. It is evident that the activity of these proteins can respond to local or systemic signals and is subject to different types of regulation, whether transcriptional or posttranscriptional. Concerning the latter, in only a few cases has the mechanism involved been identified and there is very little information on the modulating factors. The elucidation of these phenomena is the object of future research.

Some proteins can carry out secondary transport, with a varying affinity for the ion to be transported, which could be explained by the need to adapt to rapid changes in ion concentration of the soil solution.

The principal way to understand the mechanisms of nutrient acquisition operating in the root–soil system is to evaluate their activity and regulation under conditions as close as possible to the true ones. Because of the inevitable simplification of experimental designs used up to now, the results obtained do not always explain which could be the specific function, regulation, and contribution of each transporter in relation to the concentration and distribution of the nutrients at the rhizosphere. These aspects are also linked to the often unnatural way the plants have been grown (e.g., the use of C sources, such as sucrose, which are themselves able to modulate transport activities or concentrations of nutrients that are not usually encountered in nature) and the use of model plants (e.g., *Arabidopsis*), which do not always represent the features of crop species [94].

Other complications are the complexity of the ion composition in the rhizosphere solution and the presence in this environment of organic molecules able to modulate the activity of the ion transport mechanisms [34]. In regard to the former aspect, it now appears that there is a mutual relationship between the phenomena induced by variations in single nutrients that suggest the existence of one or more systems for the coordination and coregulation of the uptake of nutrients that are essential for the plant [146]. It is reasonable to think that further advances in the understanding of nutrient transport from the rhizosphere will only be possible using a molecular approach in soil-grown plants which takes into account the different factors that interact at the rhizosphere and make it a unique environment.

REFERENCES

1. Pinton, R., Varanini, Z., and Nannipieri, P., *The Rhizosphere: Biochemistry and Organic Substances at the Soil-Plant Interface*, Marcel Dekker, New York, 2001.
2. Waisel, Y., Eshel, E., and Kafkafi, U., *Plant Roots: The Hidden Half*, 3rd ed., Marcel Dekker, New York, 2002.
3. Sondergaard, T., Schulz, A., and Palmgren, M.G., Energization of transport processes in plants. Roles of the plasma membrane H^+-ATPase, *Plant Physiol.*, 136, 2475, 2004.
4. Sussman, M.R., Molecular analysis of proteins in the plasma membrane, *Annu. Rev. Plant Phys. Mol. Biol.*, 45, 211, 1994.
5. Morsomme, P. and Boutry, M., The plant plasma membrane H^+-ATPase: structure, function and regulation, *Biochim. Biophys. Acta,* 1465, 1, 2000.
6. Palmgren, M.G., Plant plasma membrane H^+-ATPases: powerhouses for nutrient uptake, *Annu. Rev. Plant Phys. Mol. Biol.*, 52, 817, 2001.
7. Kanczewska, J. et al., Activation of the plant plasma membrane H^+-ATPase by phosphorilation and binding of 14-3-3 proteins converts a dimer into a hexamer, *Proc. Natl. Acad. Sci. USA*, 102, 11675, 2005.
8. Arango, M. et al., The plasma membrane proton pump ATPase: the significance of gene subfamilies, *Planta*, 216, 355, 2003.
9. Barber, S.A., *Soil Nutrient Bioavailability: A Mechanistic Approach*, John Wiley and Sons, New York, 1995.
10. Cacco, G. and Varanini, Z., Lo zolfo nel sistema suolo pianta, in *Biochimica agraria*, Scarponi, L., Ed., Patron, Bologna, 2003, p. 837.
11. Jahn, T. et al., Plasma membrane H^+-ATPase in the root apex: evidence for strong expression in xylem parenchyma and asymmetric localization within cortical and epidermal cells, *Physiol. Plant*, 104, 311, 1998.
12. Santi, S. et al., Plasma membrane H^+-ATPase in maize roots induced for NO_3^- uptake, *Plant Physiol.*, 109, 1277, 1995.
13. Santi, S. et al., Induction of nitrate uptake in maize roots: expression of a putative high-affinity nitrate transporter and plasma membrane H^+-ATPase isoforms, *J. Exp. Bot.*, 54, 1851, 2003.
14. Monte, R., Meccanismi di assorbimento di forme inorganiche dell'azoto in mais e barbabietola da zucchero: caratterizzazione fisiologica e biochimica, Ph. D. thesis, University of Udine, Italy, 2004.
15. Yamashita, K. et al., Stimulation of H^+ extrusion and plasma membrane H^+-ATPase activity of barley roots by ammonium-treatment, *Soil Sci. Plant Nutr.*, 41, 133, 1995.

16. Samuels, A.L., Fernando, M., and Glass, A.D.M., Immunofluorescent localization of plasma membrane H^+-ATPase in barley roots and effect of K nutrition, *Plant Physiol.*, 99, 1509, 1992.

17. Grignon, C. and Sentenac, H., pH and ionic conditions in the apoplast, *Annu. Rev. Plant Phys. Mol. Biol.*, 42, 103, 1991.

18. Mengel, K. and Schubert, S., Active extrusion of protons into deionized water by roots of intact maize plants, *Plant Physiol.*, 71, 618, 1985.

19. Hisinger, P. et al., Origins of root-mediated pH changes in the rhizosphere and their responses to environmental constraints: a review, *Plant Soil*, 248, 43, 2003.

20. Schubert, S. and Yan, F., Nitrate and ammonium nutrition of plants: effects on acid/base balance and adaptation of root cell plasmalemma H^+ ATPase, *Z. Pflanzenernähr. Bodenk.*, 160, 275, 1997.

21. Marschner, H., *Mineral Nutrition of Higher Plants*, 2nd ed., Academic Press, London, 1997.

22. Dell'Orto, M.D. et al., Fe-deficiency responses in cucumber (*Cucumis sativus* L.) roots: involvement of plasma membrane H^+-ATPase activity, *J. Exp. Bot.*, 51, 695, 2000.

23. Dell'Orto, M.D. et al., Localization of the plasma membrane H^+-ATPase in Fe-deficient cucumber roots by immunodetection, *Plant Soil*, 241, 11, 2002.

24. Schmidt, W., Michalke, W., and Schikora, A., Proton pumping by tomato roots. Effect of Fe deficiency and hormones on the activity and distribution of plasma membrane H^+-ATPase in rhizodermal cells, *Plant Cell Environ.*, 26, 361, 2003.

25. Santi, S. et al., Two plasma membrane H^+-ATPase genes are differentially expressed in iron-deficient cucumber plants, *Plant Physiol. Biochem.*, 43, 287, 2005.

26. Kania, A. et al., Use of plasma membrane vesicles for examination of phosphorus deficiency-induced root excretion of citrate in cluster roots of white lupin, in *Plant Nutrition — Food Security and Sustainability of Agro-Ecosystems*, Horst, W.J. et al., Eds., Kluwer Acad. Publ., The Netherlands, 2001, p. 546.

27. Ligaba, A. et al., Phosphorous deficiency enhances plasma membrane H^+-ATPase activity and citrate exudation in greater purple lupin (*Lupinus pilosus*), *Funct. Plant Biol.*, 31, 1075, 2004.

28. Yan, F. et al., Adaptation of H^+-pumping and plasma membrane H^+-ATPase activity in proteoid roots of white lupin under phosphate deficiency, *Plant Physiol.*, 129, 50, 2002.

29. Shen, H. et al., Citrate secretion coupled with the modulation of soybean root tip under aluminum stress. Up-regulation of transcription, translation, and threonin-oriented phosphorilation of plasma membrane H^+-ATPase, *Plant Physiol.*, 138, 287, 2005.

30. Yan, F. et al., Adaptation of active proton pumping and plasmalemma ATPase activity of corn roots to low root medium pH, *Plant Physiol.*, 117, 311, 1998.

31. Niu, X. et al., NaCl regulation of plasma membrane H^+-ATPase gene expression in a glycophyte and halophyte, *Plant Physiol.*, 103, 713, 1993.

32. Kalampanayl, B.D. and Wimmers, L.E., Identification and characterization of salt-stress-induced plasma membrane H^+-ATPase in tomato, *Plant Cell Environ.*, 24, 999, 2001.

33. Kerkeb, L. et al., Enhanced H^+/ATP coupling ratio of H^+-ATPase and increased 14-3-3 protein content in plasma membrane of tomato cells upon osmotic shock, *Physiol. Plant.*, 116, 37, 2002.

34. Varanini, Z. and Pinton, R., Direct versus indirect effects of soil humic substances on plant growth and nutrition, in *The Rhizosphere: Biochemistry and Organic Substances at the Soil-Plant Interface*, Pinton, R., Varanini, Z., and Nannipieri, P., Eds., Marcel Dekker, New York, 2001, p. 141.

35. Varanini, Z. and Pinton, R., Humic substances and plant nutrition, in *Progress in Botany*, 56 Lüttge, U., Ed., Springer Verlag, Berlin,1995, p. 97.

36. Pinton, R. et al., Soil humic substances stimulate proton release by intact oat seedling roots, *J. Plant Nutr.*, 20, 857, 1997.

37. Slesak, E. and Jurek, J., Effects of potassium humate on electric potentials of wheat roots, *Acta Univ. Wratislav.*, 37, 13, 1988.

38. Varanini, Z. et al., Low molecular weight humic substances stimulate H^+-ATPase activity of plasma membrane vesicles isolated from oat (*Avena sativa* L.) roots, *Plant Soil*, 153, 61, 1993.

39. Pinton, R. et al., Modulation of NO_3^- uptake by water-extractable humic substances: involvement of root plasma membrane H^+-ATPase, *Plant Soil*, 215, 155, 1999.

40. Canellas, L.P. et al., Humic acids isolated from earthworm compost enhance root elongation, lateral root emergence, and plasma membrane H^+-ATPase activity in maize roots, *Plant Physiol.*, 130, 1951, 2002.

41. Quaggiotti, S. et al., Effect of low molecular size humic substance on nitrate uptake and expression of genes involved in nitrate transport in maize (*Zea mays* L.), *J. Exp. Bot.*, 55, 803, 2004.

42. Varanini, Z. and Pinton, R., Plant-soil relationship: role of humic substances in iron nutrition, in *Iron Nutrition in Plants and Rhizospheric Microorganisms*, Barton, L.L. and Abadia, J., Eds., Springer, Heidelberg, 2006, p.153.
43. Bloemberg, G.V. and Lugtenberg, B.J.J., Molecular bases of plant growth promotion and biocontrol by rhizobia, *Curr. Opin.Plant Biol.*, 4, 343, 2001.
44. Mantelin, S. and Touraine, B., Plant growth-promoting bacteria and nitrate availability: impacts on rhizobacteria root development and nitrate uptake, *J. Exp. Bot.*, 55, 27, 2004.
45. Bertrand, H. et al., Stimulation of the ionic transport system in *Brassica napus* by a plant growth-promoting rhizobacteria (*Achromobacter* sp.), *Can. J. Microbiol.*, 46, 229, 2000.
46. Pao, S.S., Paulsen, I.T., and Saier, M.H., Major facilitator superfamily, *Microbiol. Mol. Biol. Rev.*, 62, 1, 1998.
47. Glass, A.D.M. et al., Nitrogen transport in plants, with an emphasis on the regulation of fluxes to match plant demand, *J. Plant Nutr. Soil Sci.*, 164, 199, 2001.
48. Hawkesford, M.J., Transporter gene families in plants: the sulphate transporter gene family — redundancy or specialization?, *Physiol. Plant.*, 117, 155, 2003.
49. Rausch, C. and Bucher, M., Molecular mechanisms of phosphate transport in plants, *Planta*, 216, 23, 2002.
50. Okamoto, M., Vidmar, J.J., and Glass, A.D.M., Regulation of NRT1 and NRT2 gene families of *Arabidopsis thaliana*: responses to nitrate provision, *Plant Cell Physiol.*, 44, 304, 2003.
51. Cerezo, M. et al., Major alterations of the regulation of root NO_3^- uptake are associated with the mutation of *Nrt2.1* and *Nrt2.2* genes in *Arabidopsis, Plant Physiol.*, 127, 262, 2001.
52. Filleur, S. et al., An *Arabidopsis* T-DNA mutant affected in *Nrt2* genes is impaired in nitrate uptake, *FEBS Lett.*, 489, 220, 2001.
53. Tong, Y. et al., A two component high-affinity nitrate uptake system in barley, *Plant J.*, 41, 442, 2005.
54. Okamoto, M. et al., High-affinity nitrate transport in roots of *Arabidopsis* depends on expression of the *NAR2*-like gene *AtNRT3.1*, *Plant Physiol.*, 140, 1036, 2006.
55. Fraisier, V. et al., Constitutive expression of a putative high-affinity nitrate transporter in *Nicotiana plumbaginifolia*: evidence for post-transcriptional regulation by a reduced nitrogen source, *Plant J.*, 23, 489, 2000.
56. Vidmar, J.J. et al., Regulation of high-affinity nitrate transporter genes and high-affinity nitrate influx by nitrogen pools in roots of barley, *Plant Physiol.*, 123, 307, 2000.
57. Forde, B.G., Nitrate transporters in plants: structure, function and regulation, *Biochim. Biophys. Acta*, 1465, 219, 2000.
58. Little, D.Y. et al., The putative high-affinity nitrate transporter NRT2.1 represses lateral root initiation in response to nutritional cues, *Proc. Natl. Acad. Sci. USA*, 102, 13693, 2005.
59. Remans, T. et al., A central role for the nitrate transporter NRT2.1 in the integrated morphological and physiological responses of the root system to nitrogen limitation in *Arabidopsis, Plant Physiol.*, 140, 909, 2006.
60. Steiner, H.Y., Naider, F., and Becker, J.M., The PTR family: a new group of peptide transporters, *Mol. Microbiol.*, 16, 825, 1995.
61. Huang, N.C. et al., CHL1 Encodes a component of the low-affinity nitrate uptake system in *Arabidopsis* and shows cell type-specific expression in roots, *Plant Cell*, 8, 2183, 1996.
62. Zhou, J.J. et al., Cloning and functional characterization of a *Brassica napus* transporter that is able to transport nitrate and histidine, *J. Biol. Chem.*, 273, 12017, 1998.
63. Wang, R., Liu, D., and Crawford, N.M., The *Arabidopsis* CHL1 protein plays a major role in high-affinity nitrate uptake, *Proc. Natl. Acad. Sci. USA*, 95, 15134, 1998.
64. Liu, K.-H. and Tsay, Y.-F., Switching between the two action modes of the dual-affinity nitrate transporter CHL1 by phosphorylation, *EMBO J.*, 22, 1005, 2003.
65. Tsay, Y.-F. et al., The herbicide sensitivity gene CHL1 of *Arabidopsis* encodes a nitrate-inducible nitrate transporter, *Cell*, 72, 705, 1993.
66. Quesada, A. et al., PCR-identification of a *Nicotiana plumbaginifolia* cDNA homologous to the high-affinity nitrate transporters of the crnA family, *Plant Mol. Biol.*, 34, 265, 1997.
67. Krapp, A. et al., Expression studies of *Nrt2:1Np*, a putative high-affinity nitrate transporter: evidence for its role in nitrate uptake, *Plant J.*, 14, 723, 1998.
68. Nazoa, P. et al., Regulation of the nitrate transporter gene *ATNRT2.1* in *Arabidopsis thaliana*: responses to nitrate, amino acids and developmental stage, *Plant Mol. Biol.*, 52, 683, 2003.

69. Forde, B. and Clarkson, D.T., Nitrate and ammonium nutrition of plants: physiological and molecular perspectives, *Adv. Bot. Res.*, 30, 1, 1999.
70. Lee, R.B. and Drew, M.C., Rapid, reversible inhibition of nitrate influx in barley by ammonium, *J. Exp. Bot.*, 40, 741, 1989.
71. Glass, A.D.M., Nitrogen use efficiency of crop plants: physiological constraints upon nitrogen absorption, *Crit. Rev. Plant Sci.*, 22, 453, 2003.
72. Hodge, A., The plastic plant: root responses to heterogeneous supplies of nutrients, *New Phytologist*, 162, 9, 2004.
73. Mantelin, S. et al., Nitrate-dependent control of root architecture and N nutrition are altered by a plant growth-promoting *Phyllobacterium* sp., *Planta*, 223, 591, 2006.
74. Thornton, B. and Robinson, D., Uptake and assimilation of nitrogen from solutions containing multiple N sources, *Plant Cell Environ.*, 28, 813, 2005.
75. Okamoto, M. and Okada, K., Differential responses of growth and nitrogen uptake to organic nitrogen in four gramineous crops, *J. Exp. Bot.*, 55, 1577, 2004.
76. Barrow, N.J., Effects of adsorption of sulfate by soils on the amount of sulfate present and its availability to plants, *Soil Sci.*, 108, 193, 1969.
77. Hawkesford, M.J., Davidian, J.-C., and Grignon, C., Sulfate proton cotransport in plasma membrane vesicles isolated from roots of *Brassica napus* L. Increased transport in membrane isolated from sulfur-starved plants, *Planta*, 190, 297, 1993.
78. Lee, R.B., Selectivity and kinetics of ion uptake by barley plants following nutrient deficiency, *Ann. Bot.*, 50, 429, 1982.
79. Yoshimoto, N. et al., Two distinct high-affinity sulfate transporters with different inducibilities mediate uptake of sulfate in *Arabidopsis* roots, *Plant J.*, 29, 465, 2002.
80. Hawkesford, M.J. and Wray, J.L., Molecular genetics of sulphate assimilation, *Adv. Bot. Res.*, 33, 159, 2000.
81. Smith, F.W. et al., Regulation of expression of a cDNA from barley roots encoding a high affinity sulphate transporter, *Plant J.*, 12, 875, 1997.
82. Hopkins, L. et al., O-acetylserine and regulation of expression of genes encoding components for sulfate uptake and assimilation in potato, *Plant Phys.*, 138, 433, 2005.
83. Smith, F.W. et al., Plant members of a family of sulfate transporters reveal functional subtypes, *Proc. Nat. Acad. Sci. USA*, 92, 9373, 1995.
84. Raghothama, K.G., Phosphate acquisition, *Annu. Rev. Plant Physiol. Mol. Biol.*, 50, 665, 1999.
85. Ullrich-Eberius, C.I. et al., Relationship between energy-dependent phosphate uptake and the electrical membrane potential in *Lemna gibba* G1, *Plant Physiol.*, 67, 797, 1981.
86. Mitsukawa, N. et al., Overexpression of an Arabidopsis thaliana high-affinity phosphate transporter gene in tobacco cultured cells enhances cell growth under phosphate-limited conditions, *Proc. Natl. Acad. Sci. USA*, 94, 7098, 1997.
87. Mimura, T., Regulation of phosphate transport and homeostasis in plant cells, *Int. Rev. Cytol.*, 191, 149, 1999.
88. Smith, F.W. et al., Phosphate transport in plants, *Plant Soil*, 248, 71, 2003.
89. Huang, C. et al., Zinc deficiency up-regulates expression of high-affinity phosphate transporter genes in both phosphate-sufficient and -deficient barley roots, *Plant Phys.*, 124, 415, 2000.
90. Harrison, M.J. and van Buuren, M.L., A phosphate transporter from mycorrhizal fungus *Glomus versiforme*, *Nature*, 378, 626, 1995.
91. Maldonado-Mendoza, I.E., Dewbre, G.R., and Harrison, M.J., A phosphate transporter gene from the extraradical mycelium of an arbuscular mycorrizal fungus *Glomus intraradices* is regulated in response to phosphate in the environment, *Mol. Plant-Microbe Interact.*, 14, 1140, 2001.
92. Rausch, C. et al., A phosphate transporter expressed in arbuscule-containig cells in potato, *Nature*, 414, 462, 2001.
93. Gianninazzi-Pearson, V. et al., Differential activation of H^+-ATPase genes by an arbuscular mycorrhizal fungus in root cells of transgenic tobacco, *Planta*, 211, 609, 2000.
94. Mäser, P., Gierth, M., and Schroeder, J.I., Molecular mechanisms of potassium and sodium uptake in plants, *Plant Soil*, 247, 43, 2002.
95. Epstein, E., Rains, D.W., and Elzam, O.E., Resolution of dual mechanisms of potassium absorption by barley roots, *Proc. Natl. Acad. Sci. USA*, 49, 684, 1963.

96. Maathuis, F.J. and Sanders, D., Mechanism of high-affinity potassium uptake in roots of *Arabidopsis thaliana*, *Proc. Natl. Acad. Sci. USA*, 91, 9272, 1994.

97. Santa-Maria, G.E. et al., The *HAK1* gene of barley belongs to a large gene family and encodes a high-affinity potassium transorter, *Plant Cell*, 9, 2281, 1997.

98. Vallejo, A.J., Peralta, M.L., and Santa-Maria, G.E., Expression of potassium-transporter coding genes, and kinetics of rubidium uptake, along a longitudinal root axis, *Plant Cell Environ.*, 28, 850, 2005.

99. Gierth, M., Mäser, P., and Schroeder, J.I., The potassium transporter *AtHAK5* functions in K$^+$ deprivation-induced high-affinity K$^+$ uptake and *AKT1* K$^+$ channel contribution to K$^+$ uptake kinetics in Arabidopsis roots, *Plant Physiol.*, 137, 1105, 2005.

100. Ahn, S.J., Shin, R., and Schachtman, D.P., Expression of *KT/KUP* genes in Arabidopsis and the role of root hairs in K$^+$ uptake, *Plant Physiol.*, 134, 1135, 2004.

101. Wang, M. et al., Ammonium uptake by rice roots. II. Kinetics of $^{13}NH_4^+$ influx across the plasmalemma, *Plant Physiol.*, 103, 1259, 1993.

102. Kronzuzucker, H.J., Siddiqi, M.J., and Glass, A.D.M., Kinetics of NH_4^+ influx in spruce, *Plant Physiol.*, 110, 773, 1996.

103. Loqué, D. and von Wiren, N., Regulatory levels for the transport of ammonium in plant roots, *J. Exp. Bot.*, 55, 1293, 2004.

104. Schwacke, R. et al., ARAMEMNON, a novel database for Arabidopsis integral membrane protein, *Plant Physiol.*, 131, 16, 2003.

105. Ludewig, U., von Wiren, N., and Frommer, W.B., Uniport of NH_4^+ by root hair plasma membrane ammonium transporter LeAMT1;1, *J. Biol. Chem.*, 277, 13548, 2002.

106. López-Pedrosa, A. et al., *GintAMT1* encodes a functional high-affinity ammonium transporter that is expressed in the extraradical mycelium of *Glomus intraradices*, *Fungal Genet. Biol.*, 43, 102, 2006.

107. Fox, C.T. and Guerinot, M.L., Molecular biology of cation transport in plants, *Annu. Rev. Plant Physiol. Mol. Biol.*, 49, 669, 1998.

108. Curie, C. and Briat, J.-F., Iron transport and signaling in plants, *Annu. Rev. Plant Biol.*, 54, 183, 2003.

109. Guerinot, M.L., The ZIP family of metal transporters, *Biochim. Biophys. Acta*, 1465, 190, 2000.

110. Rogers, E.E., Eide, D.J., and Guerinot, M.L., Altered selectivity in an Arabidopsis metal transporter, *Proc. Natl. Acad. Sci. USA*, 97, 12356, 2000.

111. Vert, G. et al., IRT1, an Arabidopsis transporter essential for iron uptake from the soil and for plant growth, *Plant Cell*, 14, 1223, 2002.

112. Robinson, N.J. et al., A ferric-chelate reductase for iron uptake from soils, *Nature*, 397, 694, 1999.

113. Vert, G.A., Briat, J.-F., and Curie, C., Dual regulation of the Arabidopsis high-affinity root iron uptake system by local and long-distance signals, *Plant Physiol.*, 132, 796, 2003.

114. Connolly, E.L., Fett, J.P., and Guerinot, M.L., Expression of the IRT1 metal transporter is controlled by metals at the levels of transcript and protein accumulation, *Plant Cell*, 14, 1347, 2002.

115. Connolly, E.L. et al., Overexpression of the FRO2 ferric chelate reductase confers tolerance to growth on low iron and uncovers posttranscriptional control, *Plant Physiol.*, 133, 1102, 2003.

116. Cohen, C.K. et al., The role of iron-deficiency stress responses in stimulating heavy-metal transport in plants, *Plant Physiol.*, 116, 1063, 1998.

117. Eckhardt, U., Mas Marques, A., and Buckhout, T.J., Two iron-regulated cation transporters from tomato complement metal uptake-deficient yeast mutants, *Plant Mol. Biol.*, 45, 437, 2001.

118. Thomine, S. et al., Cadmium and iron transport by members of a plant metal transporter family in Arabidopsis with homology to Nramp genes, *Proc. Natl. Acad. Sci. USA*, 97, 4991, 2000.

119. Sancenón, V. et al., Identification of a copper transporter family in *Arabidopsis thaliana*, *Plant Mol. Biol.*, 51, 577, 2003.

120. Sancenón, V. et al., The Arabidopsis copper transporter COPT1 functions in root elongation and pollen development, *J. Biol. Chem.*, 279, 15348, 2004.

121. Curie, C. et al., Maize yellow stripe 1 encodes a membrane protein directly involved in Fe(III) uptake, *Nature*, 409, 346, 2001.

122. Yen, M.R., Tseng, Y.H., and Saier, M.H., Jr., Maize Yellow Stripe1, an iron phytosiderophore uptake transporter, is a member of the oligopeptide transporter (OPT) family, *Microbiology*, 147, 2881, 2001.

123. Schaaf, G. et al., ZmYS1 functions as a proton-coupled symporter for phytosiderophore- and nicotianamine-chelated metals, *J. Biol. Chem.*, 279, 9091, 2004.

124. Murata, Y. et al., A specific transporter for iron(III)–phytosiderophore in barley roots, *Plant J.*, 46, 563, 2006.

125. Kobayashi, T., Nishizawa, N.K., and Mori, S., Molecular analysis of iron-deficient graminaceous plants, in *Iron Nutrition in Plants and Rhizospheric Microorganisms*, Barton, L.L. and Abadia, J., Eds., Springer, Heidelberg, 2006, p. 395.

126. Bughio, N. et al., Cloning an iron-regulated metal transporter from rice, *J. Exp. Bot.*, 53, 1677, 2002.

127. Ishimaru, Y. et al., Rice plants take up iron as an Fe^{3+}-phytosiderophore and as Fe^{2+}, *Plant J.*, 45, 335, 2006.

128. Roberts, I.A. et al., Yellow Stripe 1. Expanded roles for the maize iron-phytosiderophore transporter, *Plant Physiol.*, 135, 112, 2004.

129. Wintz, H. et al., Expression profiles of *Arabidopsis thaliana* in mineral deficiencies reveal novel transporters involved in metal homeostasis, *J. Biol. Chem.*, 278, 47644, 2003.

130. Véry, A.-A. and Sentenac, H., Molecular mechanisms and regulation of K^+ transport in higher plants, *Annu. Rev. Plant Biol.*, 54, 575, 2003.

131. Walker, D.J., Black, C.R., and Miller, A.J., The role of cytosolic potassium and pH in the growth of barley roots, *Plant Physiol.,* 118, 957, 1998.

132. Hirsch, R.E. et al., A role for the AKT1 potassium channel in plant nutrition, *Science*, 280, 918, 1998.

133. Reintanz, B. et al., ATKC1, a silent Arabidopsis potassium channel α-subunit modulates root hair K^+ influx, *Proc. Natl. Acad. Sci. USA*, 99, 4079, 2002.

134. Véry, A.-A. and Sentenac, H., Cation channels in the *Arabidopsis* plasma membrane, *Trends Plant Sci.*, 7, 168, 2002.

135. Gardner, R.C., Genes for magnesium transport, *Curr. Opin. Plant Biol.*, 6, 263, 2003.

136. Skerret, M. and Tyerman, S.D., A channel that allow inwardly directed fluxes of anions in protoplast derived from wheat roots, *Planta*, 192, 295, 1994.

137. Diatloff, E. et al., Characterization of anion channels in the plasma membrane of Arabidopsis epidermal root cells and the identification of a citrate-permeable channel induced by phosphate starvation, *Plant Physiol.*, 136, 4136, 2004.

138. Roberts, S.K., Plasma membrane anion channels in higher plants and their putative functions in roots, *New Phytologist*, 169, 647, 2006.

139. Hayes, J.E. and Reid, R.J., Boron tolerance in barley is mediated by efflux of boron from the roots, *Plant Physiol.*, 136, 3376, 2004.

140. Luu, D.-T. and Maurel, C., Aquaporins in a challenging environment: molecular gears for adjusting plant water status, *Plant Cell Environ.*, 28, 85, 2005.

141. Brown, P.H. et al., Boron in plant biology, *Plant Biol.*, 4, 205, 2002.

142. Ma, J.F. et al., A silicon transporter in rice, *Nature*, 440, 688, 2006.

143. Zhang, W.H., Ryan, P.R., and Tyerman, S.D., Citrate-permeable channels in the plasma membrane of cluster roots from white lupin, *Plant Physiol.*, 136, 3771, 2004.

144. Ryan, P.R. and Delhaize, E., Function and mechanism of organic anion exudation from plant roots, *Annu. Rev. Plant Physiol. Mol. Biol.*, 52, 527, 2001.

145. Sasaki, T. et al., A wheat gene encoding an aluminium-activated malate transporter, *Plant J.*, 37, 645, 2004.

146. Wang, Y.-H., Garvin, D.F., and Kochian, L.V., Rapid induction of regulatory and transporter genes in response to phosphorus, potassium, and iron deficiencies in tomato roots. Evidence for cross talk and root/rhizosphere-mediated signals, *Plant Physiol.*, 130, 1361, 2002.

7 Function of Siderophores in the Plant Rhizosphere

David E. Crowley and Stephan M. Kraemer

CONTENTS

I. INTRODUCTION

Siderophores are iron chelating agents that are secreted by microorganisms and graminaceous plants in response to iron deficiency. Given the essential requirement for iron by almost all living organisms, these compounds are important not only for iron nutrition but are also speculated to function in the ecology of microorganisms in the plant rhizosphere [1]. Siderophores have been studied for their importance in plant disease suppression by mediating nutritional competition for iron [2,3], and they contribute directly to the rhizosphere competence of root-colonizing bacteria [4,5]. In research on plant ecology, siderophores have been investigated in relation to calcicole and calcifuge plant species and as one of the factors that may explain the distribution of various plant species in different soils [6]. Still other research has focused on plant microbe systems for phytoremediation of heavy metals, in which siderophores and phytosiderophores facilitate heavy metal uptake and food chain transfer of metals [7–11].

Despite the wealth of information on siderophores, it is still debated as to how they function in the plant rhizosphere and the degree to which they accumulate in soils. Much of this debate has been due to inadequate methodology for detecting siderophores at microsite locations in the rhizosphere, and the lack of analytical methods for *in situ* study of the interaction of siderophores and other iron-mobilizing substances. Using simplified systems in the laboratory, it is possible to examine many different scenarios as to how siderophores might function. Yet, for the most part, there is still almost no information on what siderophores actually occur in the rhizosphere, which compounds are the predominant iron sources, and how this varies at different root locations. As highlighted in this chapter, new molecular techniques for the fingerprinting of microbial communities and identification of the predominant microbial species in the rhizosphere will enable us to better evaluate the role of siderophores in microbial ecology.

One ecological aspect that must be considered in research on microbial iron nutrition in the rhizosphere is the influence of plant root growth on competition for iron. Because new elongating plant roots have the first access to iron that is mobilized at the root tips, it is likely that the roots take up much of the readily available iron at the same time that new microbial growth begins to occur in the zone of elongation. Thus, plant iron demand combined with rhizodeposition of carbon may be the driving force behind competition for iron during primary colonization of the plant roots [1]. The degree to which root-colonizing microorganisms compete for iron depends on where they are located in the rhizosphere, the pH and redox conditions, and the efficacy of plant iron stress responses such as rhizosphere acidification, release of reductants, and secretion of iron mobilizing root exudates that increase the bioavailability of iron.

A major focus of this chapter is the review of plant microbial interactions that regulate sidero-phore production and their role in mediating competition for iron in the plant rhizosphere. In understanding the factors that control siderophore production, it is important to consider that iron-mobilizing substances in the rhizosphere include not only siderophores, which are produced by virtually all microorganisms, but also a milieu of other compounds that chelate iron. New research on the interactions of siderophores and plant-produced organic acids shows that mixtures of these substances may act synergistically to solubilize iron from iron hydroxides. Understanding how siderophores function in microbial competition thus requires consideration of kinetic factors that control the dissolution of iron from minerals, the mobilization of iron bound to organic substances, and the exchange of iron between different siderophores. Other topics that are considered include the way plants use microbial siderophores, the role of siderophores in rhizosphere competence of plant growth, promoting and disease suppressive bacteria, and the relevance of siderophores to nitrogen fixation.

II. SIDEROPHORES AND IRON OXIDE DISSOLUTION

Under aerated conditions at neutral to alkaline pH, soluble iron concentrations are limited at extremely low levels by iron oxides, hydroxides, and oxohydroxides [12]. Their low solubility and slow dissolution rates impose severe limitations on iron availability to plants and microorganisms. The solubility of iron oxides is minimum in the near neutral pH range that occurs in calcareous soils. Coincidentally, the rates of dissolution are also at a minimum in this pH range. A number of strategies can be employed that increase iron solubility and accelerate the dissolution of the mineral phases by specific mechanisms, which include lowering of the redox potential (reductive dissolution mechanism), lowering of the pH (proton-promoted dissolution mechanism), or the exudation of organic ligands (ligand-controlled dissolution mechanism). For example, elevated phytosiderophore concentrations in the soil solution increase the solubility of iron [13], and phytosiderophores adsorbed to iron oxide surfaces accelerate iron dissolution by a ligand-controlled dissolution mechanism [11]. It is important to note that a high affinity of the ligand for iron complexation is not necessarily paired with a strong accelerating effect on mineral dissolution rates and *vice versa*. For example, the model bacterial siderophores desferrioxamine-B and desferrioxamine-D form extremely stable iron complexes and have a strong effect on iron solubility, but their low propensity for adsorption limits their effect on dissolution rates. In contrast, low-molecular-weight organic acids, such as oxalate, malonate, or fumarate, have a low affinity for iron compared to bacterial or plant siderophores but can strongly accelerate dissolution. Consequently, synergistic effects on iron oxide dissolution have been observed in mixtures of siderophores and low-molecular-weight organic acids with complementary properties [11,14]. Some plant species are known to exude phytosiderophores in a diurnal pattern. The efficiency of this process is enhanced by the continuous kinetic destabilization of iron at the mineral surface by low-molecular-weight organic acids and the rapid release of the accumulated labile iron during the time of maximum phytosiderophore release [11].

Large differences exist between the solubility and dissolution rates of various iron oxide minerals that coexist in soils in thermodynamic disequilibrium [15]. Modeling the dissolution of iron from these minerals is difficult because solubility and dissolution rates depend on the size, specific surface area, and crystallinity of the minerals, as well as the local chemical environment that affects the weathering of the iron oxides [15,16]. Published solubility constants, rates, or rate coefficients have been measured over a wide range of conditions in the laboratory but can diverge from field observations by orders of magnitude [15]. In soils containing mixtures of iron oxides, it is likely that the least stable minerals are the most important for plant nutrition [14]. Nonetheless, the plant iron stress response is adaptive and permits use of the highly crystalline iron oxihydroxide goethite (α-FeOOH), which has been shown to serve as iron source for barley in hydroponic culture [11].

III. IRON SPECIATION AND LIGAND EXCHANGE REACTIONS IN SOLUTION

Due to low concentrations of hydrolysis species in equilibrium with iron oxides in the near neutral pH range, the acquisition of iron from soluble organic complexes is an important iron acquisition strategy. This includes iron bound to humic and fulvic substances, siderophores, or other low-molecular-weight organic ligands contained in root or microbial exudates. Low-molecular-weight root exudates include organic acids that are secreted by plant roots as a specific response to iron deficiency [17], or that are released constitutively at low levels in root exudates and lysates [18]. Further adding to this complexity, humic and fulvic acids, as well as microbial produced chelators contained in compost, can also serve as sources of mobilizable iron for uptake by plants and microorganisms [19–21].

In the rhizosphere of graminaceous plants, the iron stress response manifests as an increase in the production and release of organic acids after which the plant increases its production of highly efficient chelators, phytosiderophores, that are co-exuded in localized zones behind the root tips [22]. Comprehensive analysis of root exudate composition by ^{2}H and ^{13}C multidimensional nuclear magnetic resonance (NMR) and by silylation gas chromatography-mass spectrometry (GC-MS) has proven to be particularly useful for identifying the constituents of root exudates and how they change in relation to iron deficiency [23]. As shown for barley that has been subjected to different levels of iron stress, many organic and amino acids, as well as several derivatives of mugineic acid phytosiderophores, can be identified; the major one being 3-epihydroxymugineic acid (Table 7.1). Quantification of all major components during changes in the plant iron nutritional status revealed a sevenfold increase in total exudation under moderate iron deficiency with 3-epihydroxymugineic acid comprising approximately 22% of the exudate mixture during early stages of iron deficiency. As iron deficiency increased, total quantities of exudate per gram of root remained unchanged, but the relative quantity of carbon allocated to phytosiderophore increased to approximately 50% of the total carbon released by the roots. The role of phytosiderophores in iron acquisition is unequivocal. However, considering that mixtures of organic ligands can have synergistic effects on iron oxide dissolution rates as discussed earlier, the complex exudation pattern may have a distinct function in the iron acquisition process as well.

Given the inherent complexity of root exudates and mixtures of other organic substances in the rhizosphere, as well as theoretical consideration of all of the factors that govern production, diffusion, degradation, chelation chemistry, and mineral surface chemistry of these compounds,

TABLE 7.1
Effect of Iron Stress on Chemical Composition of Root Exudates Collected from Barley Plants as Determined by Combined NMR and GC–MS Analysis

	pFe[a]			
Compound	16.5	17.0	17.5	18.0
Lactate	1.1	4.1	5.0	4.8
Alanine	0.3	4.4	1.5	0.7
Succinate	0.1	1.0	0.2	0.1
g-Aminobutyrate	0.0	2.5	0.4	0.3
Fumarate	0.1	0.3	0.2	0.2
Malate	0.0	2.8	0.2	0.2
Glycine betaine	0.5	5.9	1.7	2.7
Acetate	1.6	4.7	1.7	2.3
3-Epi-OH-MA	1.1	8.1	16.2	17.4
Total	5.0	37.0	29.0	32.0

[a] Barley plants grown in chelators-buffered hydroponic solutions with free activity of iron at given pFe values ranging from 16.5 (iron sufficient) to 18.0 (severe iron deficiency).

soil chemists have increasingly turned to computer models to describe the chemistry of the soil solution. These include chemical equilibrium models that describe the solubility, adsorption, and speciation of metal ions in the presence of calcium and other minerals at a given pH and redox potential [24] and kinetic models that describe the rates of slow reactions including dissolution and ligand exchange reactions [11], as well as conceptual and mathematical models of factors that influence siderophore production and accumulation in the rhizosphere [1,25]. A current objective is the validation of these models using data obtained from microsite analysis of the rhizosphere soil solution or with gene reporter systems that are calibrated to measure the potential production and accumulation of siderophores in the rhizosphere [26,27].

The distribution of soluble iron among hydrolysis species and organic complexes can be predicted from chemical equilibrium equations using numerical models such as GEOCHEM [24], as demonstrated in Figure 7.1. In general, these models are appealing and highly useful tools for the designing of hydroponic media to examine iron stress responses and to measure uptake rates by plants and microorganisms. However, the few empirical data that are available indicate that the situation is much more complex in soils. In soils, differences in metal dissolution rates seldom correspond to the stability constants of different chelators and complexing agents but are instead controlled by the differing abilities of organic acids and chelators to attack mineral surfaces [13,28,29] and by kinetic limitations of subsequent ligand or metal exchange reactions.

Models that can predict rates of exchange are critical for understanding competition for iron based on preferential utilization of different siderophore types. The exchange of metals between tetradentate carboxylate siderophores and phytosiderophores has been proposed as a mechanism for plant use of microbial siderophores [29–31]. However, the exchange of iron between hexadentate siderophores such as desferrioxamine-B and other ligands can be extremely slow [32]. The exchange can be considerably accelerated by microbial degradation of bacterial siderophores [33]. Conversely, it has also been shown that microbial siderophores may strip iron from phytosiderophores [34]. For this reason, there has been some debate as to how plants and microorganisms interact in their mutual problem of acquiring iron, and the mechanisms that are used for iron acquisition from different iron sources.

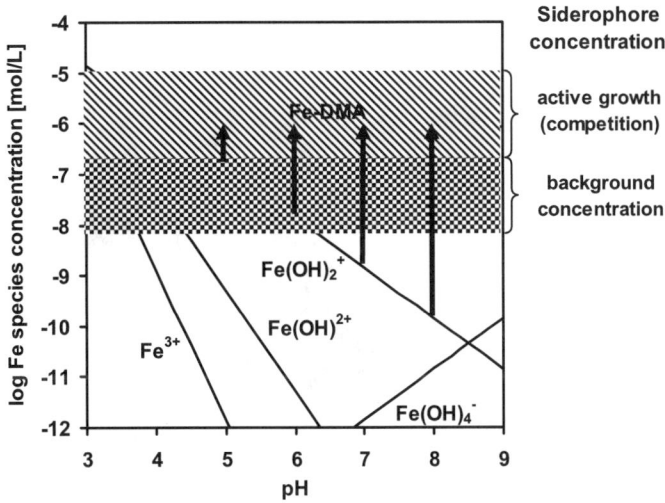

FIGURE 7.1 Calculated iron speciation in equilibrium with soil iron oxide [139], and 1 μM dissolved deoxymugineic acid as a function of pH. (Equilibrium constants from Martell et al., *Coord. Chem. Rev.*, 133, 39, 1994; Murakami, T. et al., *Chem. Lett.*, 12, 2137, 1989. With permission.) Arrows indicate the increase in iron solubility due to the presence of the phytosiderophore.

The major driving factors for kinetically slow processes such as ligand exchange and dissolution reactions are the disequilibria introduced by siderophore release and iron uptake. This process is described in Figure 7.1. In the presence of deferrated siderophores, the solution concentrations of the inorganic iron hydrolysis species are depleted to very low concentrations below equilibrium levels maintained by the mineral phase. With strong chelators such as microbial siderophores, even a small amount of excess demetallated chelate can suppress inorganic metal concentrations down to levels of only a few atoms per liter [12,24]. Consequently, the mineral undergoes further dissolution to reestablish equilibrium concentrations. In the meantime, the only available iron is that provided by metal chelators and complexes. The maintenance of disequilibria is, therefore, critical to drive processes that increase the bioavailability of iron.

Plant and microbial competition for iron involves complex interactions that are influenced by a number of factors (Figure 7.2). These include differences in the level of siderophore production by all of the competing microorganisms, the chemical stabilities of various siderophores and other chelators with iron, their resistance to degradation, and the ability of different chelators to solubilize iron by attack of mineral surfaces. Moreover, all of the different siderophores in the soil solution may interact through ligand exchange. At the level of the individual microorganism, competition is further mediated by the ability to use different siderophore types and the transport kinetics for these compounds. To date, there has been very little research on the interactions of complex mixtures of chelators. The best data so far are those generated with computer models for synthetic chelators in simple systems with at the most two chelators in a chemically defined media at a specific pH.

To get around the problem of predicting dissolution and exchange rates, there is an increasing emphasis on using empirical data based on soil extractions with different chelators. However, a problem that has arisen in this methodology is the introduction of artifacts caused by using soil slurries in shaking flasks as opposed to percolation of intact soil columns. The former technique has been criticized for exposing new mineral surfaces by disruption of soil pore spaces [35]. Soil pores, which contain most of the soil water, are exposed to continuous redox fluctuations as they undergo wetting and drying, and are likely to have different mineral types, crystallinities, and surface properties as compared to the internal matrix of soil peds that are exposed during slurry extractions.

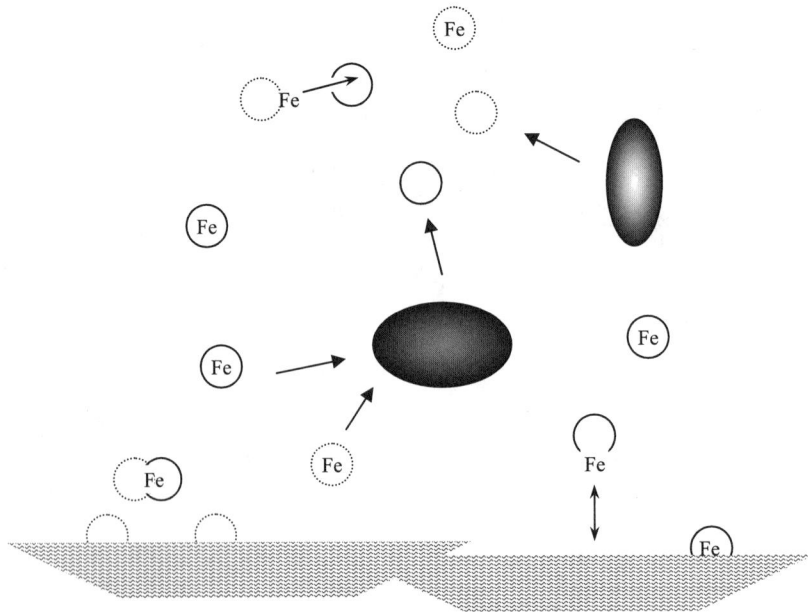

Competition for Iron

Chelation kinetics	Quantity produced
Stability constants	Transport specificity
Ligand exchange	Degradation rate

FIGURE 7.2 Conceptual model showing competition for iron between two microorganisms producing different siderophore types. Much of the siderophore present in the soil solution during active growth is in the deferrated form. The deferrated chelate mobilizes iron by two mechanisms: (1) sequestration of dissolved iron present at very low concentrations that causes further dissolution of the solid phase, and (2) by direct attack of mineral surfaces. Once in solution, the chelated iron can donate iron to other chelators by ligand exchange, or by releasing inorganic iron into the solution in accordance with chemical equilibria-governing chelate stability. Competition for iron is based on the quantity of chelator, its efficacy in mobilizing iron, stability with iron under given pH, redox conditions, and its rate of degradation by other microbes that use the chelator as a carbon source for growth. At the organism level, competition is further mediated by transport specificity and the kinetics of iron uptake from different chelators.

Upon formation of a metal chelate or complex, the next rate-limiting step in delivering iron to the cell is the diffusion of iron complexes through the soil in response to diffusion gradients. In the vicinity of plant roots, metal chelates and complexes may also move by bulk flow in the transpiration stream. However, depending on their charge characteristics and hydrophobicity, metal chelators and complexes can become adsorbed to oxide minerals, clays, and organic matter that may then decrease their mobility and bioavailability to plants and microorganisms [36–38]. This can greatly complicate the estimation of siderophore and metal chelator quantities that are actually functional in metal transport, especially if metal chelate concentrations are determined by soil extraction.

Generally, concentrations of a chelator are estimated in relation to soil water content at an appropriate value, e.g., 10% soil moisture, which may vary depending on the soil texture. Repeated extractions of soil with water or buffered salt solutions or the use of large extraction solution volumes can result in over estimation of effective concentrations for chelators that are normally sorbed to soils. Alternatively, concentrations of ionic, charged chelators, and metal complexes that accumulate at soil microsites may be underestimated when they are diluted and become bound to clay and organic matter during the extraction process. For example, when the model microbial siderophore desferrioxamine-B is added to soil, on an average only 3% can be recovered even after repeated extraction [39]. The recovery is dependent on the clay and organic matter content. This latter problem may account for the difficulty in detecting siderophores in soil extracts, as well as the general finding that microbial siderophore concentrations averaged over the entire rhizosphere are only present in nanomolar concentrations [28,40]. These concentrations are orders of magnitude lower than would be needed to be physiologically relevant for iron uptake by either plants or microorganisms.

IV. IRON UPTAKE MECHANISMS FROM SIDEROPHORES AND PHYTOSIDEROPHORES

A. PLANT IRON UPTAKE MECHANISMS

Metal chelates and complexes deliver iron to cells by diffusion through plant and microbial cell walls, where they are presented to highly selective metal ion or metal-chelate transport systems that are deployed in the cell membranes of plants, bacteria, and fungi (see reviews, Reference 40 to Reference 45). Charge characteristics of the wall itself, along with the accumulation of various ions and organic materials in the wall matrix, and the size of the molecule in relation to the wall matrix — all can potentially influence the diffusion rates of metal chelators across the cell wall. The operation of reductase enzymes present in cell walls can also strip iron from the chelate, resulting in the precipitation of amorphous iron oxides or sorption of the metal ion to the wall matrix [46]. In order to be taken up by the cells, these metal ions must then be remobilized or else reside in the wall until they are released by decomposition by saprophytic microorganisms. This metal deposition process may occur primarily with low stability metal complexes, but it also has been shown to occur with synthetic chelates such as ethylenediaminetetraacetic acid (EDTA) and phytosiderophores produced by grasses [47]. In this case, localized secretion of phytosiderophores and microbial siderophores can remobilize precipitated iron. Other naturally occurring compounds, such as humic and fulvic acid or heme compounds, can also function in iron nutrition by donating iron to transport proteins or reductases at the cell surface. Recently, it has been shown that iron containing low-molecular-weight humic molecules can be reduced by both intact roots and isolated plasma membrane vesicles [19,48]. Iron reduction may occur by means of an nicotinamide adenine dinucleotide (NADH) dependent, iron-stress-regulated reductase [49] that releases Fe^{2+} from Fe^{3+}-specific chelates and complexes and by biosynthetic reducing agents such as riboflavin [50] and caffeic acid [51], which have been proposed to enhance mobilization of iron that has been precipitated in the cell wall.

In plants and microorganisms that are well adapted to iron-limiting growth conditions, the responses to iron stress are quite diverse and are highly effective. Iron efficient dicotyledonous plants, termed *Strategy I* [52], produce organic acids that are released into the soil in the vicinity of the growing root tips to solubilize iron. They also increase their root surface area, numbers of root tips, and increase the activity of a proton ATPase that acidifies the rhizosphere and helps to increase iron solubility. Certain plants such as lupine release extremely large quantities of citrate in response to phosphorus deficiency, but concurrently take up greater quantities of metals as a result of this response [53]. As mentioned earlier, plants may also increase the activity of a plasma membrane bound reductase in the rhizodermal cells that reduces chelated ferric iron to ferrous iron. Under iron deficiency, the kinetics of metal transport by the carrier protein are altered to permit a much higher Vmax, suggesting increased induction of carrier protein synthesis or an alteration in the gate channel that regulates iron transport [54,55].

The use of microbial siderophores by dicotyledonous plants appears to involve uptake of the entire metallated chelate [56–58] or an indirect process in which the siderophore undergoes degradation to release iron [31,59]. As demonstrated in initial studies examining this question, there was concern that iron uptake from microbial siderophores may be an artifact in which radiolabeled iron is accumulated by microorganisms [60]. Consequently, evidence for direct uptake of iron from microbial siderophores has required the use of axenic plants. In experiments with cucumber, it was shown that the microbial siderophore ferrioxamine-B could be used as an iron source at concentrations as low as 5 μM and that the siderophore itself entered the plant [58].

The role of membrane bound chelate reductases for iron uptake from microbial siderophores has been examined for several plant species [49,61,62]. With certain microbial siderophores, such as rhizoferrin and rhodotorulic acid, the reductase may easily cleave iron from the siderophore to allow subsequent uptake by the ferrous iron transporter. However, with the hydroxamate siderophore, ferrioxamine-B, which is produced by actinomycetes and used by diverse bacteria and fungi, it has been shown that the iron-stress-regulated reductase is not capable of cleaving iron at rates that would be relevant for plant nutrition. This suggests that ferrioxamine-B may be utilized by another mechanism involving uptake of the siderophore. In studies with sterile onion plants, the rate of iron uptake and translocation by plants supplied with ^{55}Fe-ferrioxamine-B was identical to that for ferrated ^{14}C-ferrioxamine-D that was shown to accumulate in low concentrations in the root and shoot tissues [57]. In the case of coprogen, a fungal siderophore, the hydrolysis products of the parent compound were shown to be an excellent source of iron for both strategy I and strategy II plants [33]. The hydrolysis products, which are presumably the same as those generated during decomposition by microbial hydrolases, were stable with iron but were much more easily reduced by the strategy I chelate reductase; the chelated iron was more easily exchanged with phytosiderophores. This suggests that the decomposition and turnover of microbial siderophores in the rhizosphere could provide an important pool of bioavailable iron to generalists that lack specific transport systems for iron.

In the case of graminaceous plants, phytosiderophores are secreted to mobilize iron and are purportedly reabsorbed by means of a highly specific transport protein. These compounds are secreted at the root tips in short pulses of 2 to 4 h duration, shortly after the onset of daylight with both their synthesis and release being regulated by diurnal light and temperature rhythms [22,63]. Differences in phytosiderophore production have been correlated with iron efficiency [17,64], but several other factors also have been shown to contribute to the relative iron efficiency of various monocot grasses. These include variations in the minimum iron requirement of the plant tissues for different species and cultivars, differences in the ability to respond by early and rapid production of phytosiderophores prior to the onset of deficiency symptoms, and alterations in the relative growth rate to prevent deficiency from becoming severe [65]. Alterations in root to shoot ratio and increases in specific root length and numbers of root tips also occur during the onset of iron stress. Iron uptake rates from synthetic chelators, microbial siderophores, and phytosiderophores may also be enhanced to different degrees depending on the plant cultivar [40].

In addition to iron, phytosiderophores can also function for the uptake of zinc, copper, and other trace metals that are fortuitously mobilized under iron deficiency conditions. Specific release of phytosiderophores in response to zinc deficiency has been shown for certain wheat species and wild grasses [66,67] but was not observed in barley [68]. In contrast to the iron stress response, in which phytosiderophore release is correlated with iron efficiency, differences in zinc uptake efficiency among wheat cultivars do not consistently correspond to differences in phytosiderophore release rates [66].

Nonetheless, both zinc and iron deficiencies have been shown to increase the uptake of copper in wheat (*Triticum aestivum*) [69]. A comparison of two cultivars showed that cv. Aroona was induced to take up copper by both zinc and iron deficiency, whereas a second cultivar, cv. Songlen, accumulated greater amounts of copper only when induced by iron deficiency. These results suggest that there is differential induction of phytosiderophore release in some cultivars in response to different trace metal deficiencies and demonstrate that phytosiderophores can mobilize other trace metals. In the case of copper, most of the metal was retained in the root system, suggesting that it was removed from the phytosiderophore and bound to another metal-binding substance, such as a phytochelatin, in the cells. In addition to copper, zinc, and iron, iron deficiency can cause phyto-siderophore-mediated uptake of cadmium by wheat and barley plants [70]. As in the case of copper, the majority of cadmium is retained in the plant roots.

The mechanisms by which grasses utilize iron from rhizoferrin, ferrioxamine-B, EDTA, and deoxymugineic acid have been studied for barley and corn plants grown in nutrient solution [29]. By supplying iron to the plants during the morning or evening, it was possible to examine the effect of concurrent phytosiderophore release on iron uptake rates from the different iron sources. Uptake and translocation rates from Fe chelates paralleled the diurnal rhythm of phytosiderophore release except in corn, in which a similar uptake and translocation rates were observed both in the morning and in the evening. Later, it was shown that for corn there was a constant rate of phytosiderophore release during the 14-h light period rather than the normal diurnal pattern observed in other plants. The results strongly suggested that iron supplied by rhizoferrin is taken up by Strategy II plants by an indirect mechanism that involves ligand exchange between the ferrated microbial siderophore and phytosiderophores. This hypothesis was further verified by *in vitro* ligand-exchange experiments showing the exchange of iron between rhizoferrin and phytosiderophore.

B. Microbial Iron Uptake Mechanisms

In the rhizosphere, microorganisms either utilize organic acids or phytosiderophores to trans-port iron, or produce siderophores. There are a wide variety of siderophores in nature, and some of them have now been identified and chemically purified [71]. Presently, three general mechanisms are recognized for utilization of these compounds by microorganisms. These include a shuttle mechanism in which chelators deliver iron to a reductase on the cell surface, direct uptake of metallated siderophores with destructive hydrolysis of the chelator inside the cell, and direct uptake followed by reductive removal of iron and resecretion of the chelator (see review [72]).

Depending on the ability of specific transport systems to utilize the predominant metal chelates present in the soil solution, competition may occur between plants and microorganisms, and between different types of microorganisms for available iron. This has been particularly well studied for *Pseudomonas* spp., which not only produce highly unique iron chelators that are utilized in a strain-specific manner but also retain the ability to use more generic siderophores produced by other microorganisms [72,73]. In a pure culture iron-limiting media, pseudomonads can produce extremely large quantities of siderophore that are inhibitory to iron uptake by other bacteria and fungi, and which can also cause iron deprivation to plants [74,75].

V. PLANT MICROBIAL INTERACTIONS

Attempts to unravel complex interactions in iron competition or cooperation between microorgan-isms and plants have generally involved pure culture systems in agar or liquid media, or the use of hydroponic culture systems with plants. These studies have revealed basic mechanisms by which various plants and microorganisms respond to iron deficiency, and their ability to use different metal chelators at defined concentrations. However, there has frequently been a tendency to extrap-olate too far from these studies and overlook the differences that occur between soil and hydroponic systems. Hydroponic studies on plant use of microbial siderophores and phytosiderophores immerse the entire root system in uniform concentrations of the chelate. In actuality, phytosiderophores are produced behind the root tips [76], where they accumulate to high concentrations in localized zones of the rhizosphere [77]. Similarly, microbial siderophores, if produced in significant quantities in rhizosphere, are most likely secreted at sites of high microbial activity behind the root apices and at the sites of lateral root emergence [1,75].

One particular controversy derived from hydroponic studies was the observation that organic acids and phytosiderophores are subject to rapid degradation [78] and that they would thus be ineffective for mobilizing iron in soils. This concern originated from hydroponic studies in which siderophore degrading bacteria increased to such high population densities that phytosiderophores could not even be collected from the culture medium or were rendered ineffective for mobilization of iron from solid phase minerals [75,79]. It has since been shown that degradation is a localized phenomenon in sites of high microbial activity that can be ameliorated by pulsed release of phyto-siderophore at the root tips [1,80].

In a carefully conducted study that examined the effect of bacterial distribution in the rhizo-sphere on phytosiderophore degradation, maize plants were grown axenically or inoculated with a mixture of microorganisms in a limestone substrate supplemented with iron oxide. Those that were grown axenically developed well and released phytosiderophores; whereas, inoculated plants showed severe symptoms of Fe deficiency, suggesting microbial degradation of phytosiderophores in the apical root zones [81]. By manipulating the watering pattern with a wick system, the plants were able to secrete enough phytosiderophore in localized zones to facilitate iron uptake. Similar results were obtained in a silt loam soil, where short-term periodic flooding resulted in a uniform distribution pattern of rhizosphere microorganisms and iron chlorosis. In contrast, nonflooded plants had lower numbers of microorganisms in the apical root zones and were not affected by iron chlorosis.

Differences in the utilization of microbial siderophores by dicotyledonous plants have been correlated with the iron efficiency stress reaction [82]. In an experiment with iron efficient and inefficient varieties of hydroponically grown carnation, the microbial siderophore, pseudobactin, was shown to serve as an iron source for the more efficient cultivar but not for the inefficient cultivar. Interestingly, the synthetic chelate Fe-EDDHA was equally well used by both cultivars, suggesting a unique mechanism for use of the microbial siderophore that only occurred in the iron-efficient cultivar. In later researches on monocots by these investigators, plant use of pseudobactin, the synthetic chelator ethylenediaminedi (*o*-hydroxyphenylacetic acid) EDDHA, and phytosidero-phore was compared using barley plants grown under axenic and nonsterile conditions [83]. Results showed that phytosiderophore was the preferred iron source, but pseudobactin at 4 μM enhanced iron uptake by barley under nonsterile conditions in which the phytosiderophore was subject to rapid degradation. In comparison, research by other investigators examining short-term Fe uptake from ^{59}Fe-pyoverdine as compared with ^{59}FeEDTA by cucumber and ^{59}Fe-phytosiderophores by maize showed that ferrated pyoverdine was a relatively poor Fe source for either maize or cucumber. Iron uptake from pyoverdine by cucumber was 40 times slower than from EDTA. For maize plants, iron uptake from pyoverdine was 665 times lower than phytosiderophores [75].

VI. SIDEROPHORE PRODUCTION IN THE RHIZOSPHERE

A. TYPES OF SIDEROPHORES PRODUCED

The production and accumulation of different types of siderophores in the rhizosphere is largely unknown except for a few specific compounds that have been estimated using bioassays or antibiotic detection procedures [84,85]. Using chemical detection procedures, only one study has reported the direct extraction of siderophores in soil. This particular study examined schizokinen [86] that was produced in flooded soil under rice culture. After extracting the compound from soil with water, the siderophore was sorbed on to a cation exchange medium, purified, and analyzed by high performance liquid chromatography. More commonly used methods for detection of siderophores have involved bioassays. In general, these bioassays use siderophore auxotrophic strains, which require siderophores in order to grow on agar media or to cause turbidity during growth, so that it can be used to estimate siderophore concentrations in a variation of the most probable number assay. *Arthrobacter flavescens* JG-9 has been one of the most widely used auxotrophic strains and is used to detect hydroxamate-type siderophores. Because *Arthrobacter* responds to a variety of hydroxamate siderophores, the bioassay is usually calibrated to desferrioxamine-B, which is available commercially (Figure 7.3).

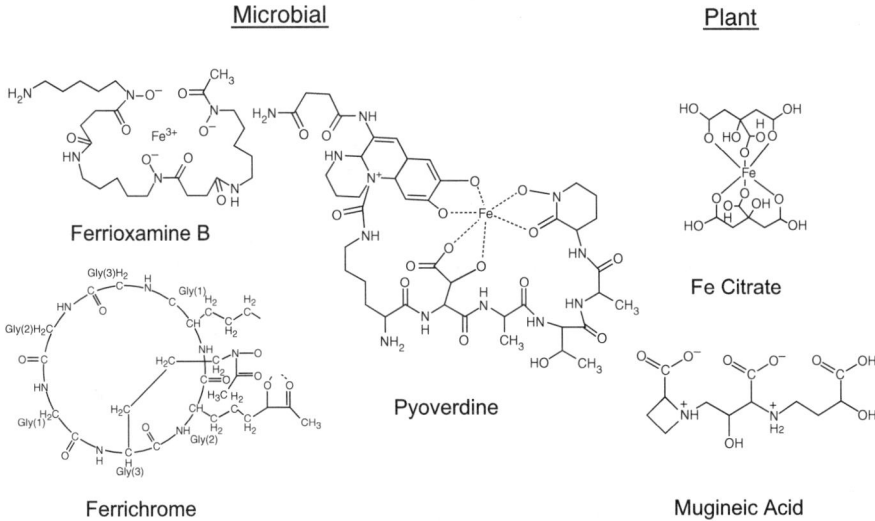

FIGURE 7.3 Chemical structures of representative microbial siderophores, phytosiderophore, and organic acid iron complexes produced by microorganisms and plants.

Another assay has used *Escherichia coli* mutants to detect hydroxamate and catecholate siderophores [85]. Presently, several surveys have shown that hydroxamate siderophores commonly occur in soil at nanomolar concentrations [85,87,88], which can be increased by amendment with various carbon substrates used to simulate rhizosphere conditions at nutrient-rich sites [87].

In early studies on the rhizosphere, it was presumed that pseudomonads and various Gram-positive or Gram-variable bacteria, such as *Bacillus* and *Arthrobacter*, were the predominant bacteria that grew on plant roots. There has also been some investigation of siderophore production by nitrogen-fixing bacteria, mycorrhizal fungi, and selected fungi such as rhizopus. However, to date, most of our knowledge of rhizosphere communities and the siderophores that they produce has been obtained in studies that focus on culturable microorganisms, particularly root colonizing pseudomonads. Early work by Vancura [89] and Kleeberger et al. [90], suggested that pseudomonads were the predominant bacteria in the rhizosphere, comprising over 50 to 70% of the culturable bacterial species. Later, a comprehensive study of culturable bacteria associated with sugar beet (*Beta vulgaris*) revealed 102 species from 40 genera, which were identified from among 556 isolates. The ten most common genera were *Bacillus* (14%), *Arthrobacter* (12%), *Pseudomonas* (11%), *Aureobacterium* (9%), *Micrococcus* (6%), *Xanthomonas* (5%), *Alcaligenes* (4%), *Flavobacterium* (3%), *Agrobacterium* (3%), and *Microbacterium* (3%), all of which accounted for 70% of isolates. Since then, some 775 genera of bacteria have been isolated from the environment and studies using genetic methods suggest that there may be as many as 4000 species per gram of soil [89]. Culturable microorganisms thus may represent only 1 to 5% of the total bacteria that occur in nature; more than 60% of the bacterial species identified from a 16S rRNA gene library of rhizosphere and bulk soils from the southwestern U.S. are completely unknown species [91]. Many of these bacteria appeared to belong to a major bacterial lineage that has never been identified before, except for one single cultured representative, and almost none of the bacteria that were isolated were from genera that are typically isolated on agar media. Based on these data, it has become increasingly apparent that we do not even know what siderophores to look for yet, much less how they function in rhizosphere ecology and plant nutrition.

Siderophore production by rhizosphere bacteria that are culturable on agar media has been estimated by plating out colonies on an indicator agar medium containing chrome azurol, a weak iron-chelating complex that changes color upon deferration [92]. Among 240 isolates that were screened, 80% failed to grow on the assay medium, highlighting the problem of medium selectivity in working with soil bacteria. Bacteria that grew on the assay medium produced siderophores *in vitro* at concentrations ranging from 100 to 230 μM, a concentration range that would be effective for providing plants with chelated iron in case they have a mechanism for use of the chelator. A more fundamental question is whether these concentrations ever exist in the rhizosphere, or instead are concentrations that might occur only temporarily at specific microsites of high microbial activity or under carbon-rich conditions in iron-deficient media. Because most bacteria and fungi have siderophore transport systems that are designed for much lower concentrations, which are suppressed by iron-chelate concentrations above 5 to 10 μM, it is unlikely that ferrated siderophore concentrations above these values would ever accumulate in the rhizosphere. Hence, if large concentrations of siderophore are produced at microsites, it is likely that they would be in the deferrated form.

The potential ability of pseudomonads to produce very high concentrations of siderophores at localized sites of high microbial activity in the rhizosphere is particularly well shown by the use of root fingerprints on iron limiting agar media. As shown in Figure 7.4 for barley roots pressed onto an agar plate containing iron-deficient medium selective for *Pseudomonas* spp., culturable cells of root colonizing pseudomonads are located primarily in the zone of elongation behind the root tips. After growth of the bacteria on the agar plate, the siderophore produced by the root-associated bacteria is visualized by exposure of the plate to ultraviolet light, which causes the iron-chelating fluorescent siderophore to give off a green color that can be photographed. The green halo associated with the pseudomonad colonies is highly localized at these same locations, i.e., at the root tips and sites of lateral root emergence. These sites correspond to those that are visualized with autophotography using a bioluminescent strain of pseudomonas (Figure 7.5A and 7.5B) and

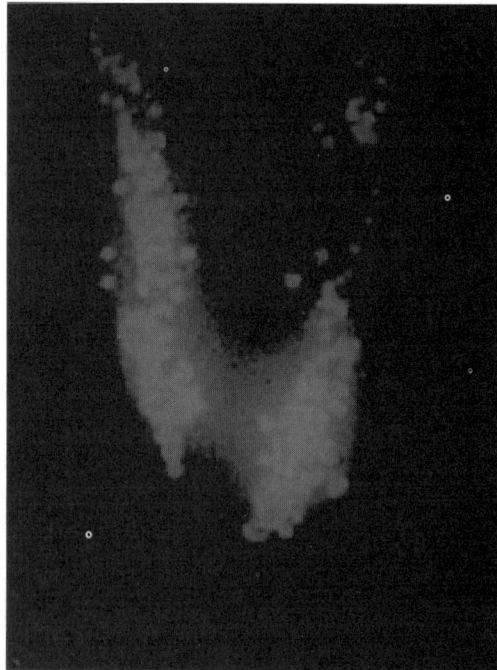

FIGURE 7.4 Root fingerprints of *Pseudomonas* spp. associated with barley seedlings, showing the production of siderophore by actively growing bacteria located in the zone of elongation behind the root tips. Roots were pressed on to an iron-deficient minimal medium selective for pseudomonas. After growth of the colonies, the production of siderophore was visualized by exposure of the agar plate to ultraviolet light, which causes the siderophore to fluoresce.

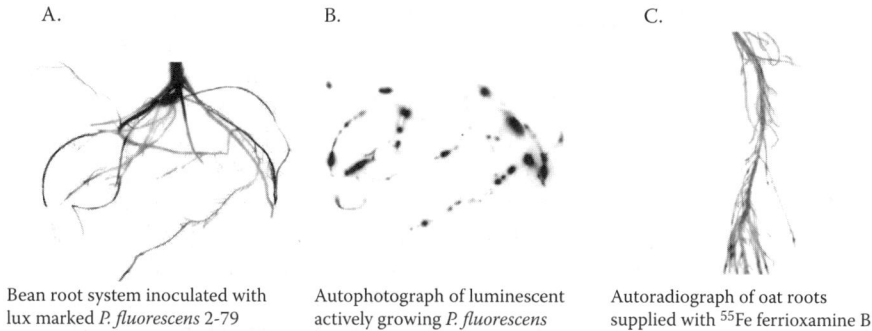

A.

B.

C.

Bean root system inoculated with
lux marked *P. fluorescens* 2-79

Autophotograph of luminescent
actively growing *P. fluorescens*

Autoradiograph of oat roots
supplied with ^{55}Fe ferrioxamine B

FIGURE 7.5 Sites of high microbial activity on plant roots as depicted by autophotography using a bioluminescent pseudomonad *P. fluorescens* 2-79RL that produces light during active growth. (A) Corn root system
inoculated with the lux-marked bacterium; (B) autophotograph generated by placing film over the root system,
which was then exposed to light produced by bioluminescent bacteria (autophotograph shows that microbial
activity is greatest in the zone of elongation at the root tips and at sites of lateral root emergence); and (C)
autoradiograph of a nonsterile, iron-stressed oat plant showing iron accumulation from ^{55}Fe-siderophore
desferrioxamine-B. Dark areas caused by exposure of film to the radioactive iron are located in the zones of
elongation and at sites of lateral root emergence, corresponding to the areas of microbial growth in Figure 7.5B.

to those labeled and visualized with autoradiography following exposure of roots to radiolabeled-
iron chelates (Figure 7.5C).

Iron uptake by bacteria at sites of lateral root emergence has been further confirmed by using
another technique employing 7-nitrobenz-2-oxa-1,3-diazole-desferrioxamine-B, which is a deriva-
tized siderophore, that becomes fluorescent after it is deferrated [93]. In this case, iron uptake from
the siderophore ferroxamine-B was associated primarily with microbially colonized roots, but both
plant and iron uptake from this chelate occurred in the regions just behind the root tips.

Several studies have shown that ferrated-pyoverdine-type siderophores can be used as iron
sources for plants when added to soils [94,95]. This may involve decomposition of the siderophore
after addition to the soil. However, to date, almost all attempts to supply iron to plants by inoculation
of hydroponic solutions with siderophore-producing bacteria or by inoculating soils with
pseudomonads have been unsuccessful [61,81,96]. In experiments with cucumber, inoculation of a
hydroponic medium with *P. putida* or with soil microorganisms and amendment with autoclaved
soil as an iron source failed to provide iron for plant growth [75]. In other related work, the extent
to which plants benefit from root colonization by siderophore-producing bacteria was examined for
iron efficient and inefficient oat cultivars, inoculated with six strains of bacteria that produced high
concentrations of siderophores, and grown in a calcareous soil [91]. Three species of *Pseudomonas*
colonized the roots of both cultivars in numbers greater than 10^6 cells g^{-1} root, but they had no
significant effect on Fe acquisition by the plants. Instead, plants fertilized with 5 μM Fe were larger
and supported greater numbers of rhizosphere bacteria per gram of root than plants that were not
supplied with iron and inoculated with bacteria.

B. Microsite Differences in Siderophore Production

Differences in siderophore production in the rhizosphere are likely to occur depending on the root
location, the plant growth stage, its mineral nutritional status, the presence or absence of mycorrhizae,
and on the soil's chemical properties. Because siderophores are produced only in response to iron
stress, which is a function of the relative growth rate, it cannot be presumed that similar types of
siderophores will occur at specific root locations or that siderophores will occur in all root zones.
Microbial growth rates are controlled by carbon availability, which varies along the root axis, and

it has long been established that there is a successional development of microbial communities as roots elongate and mature [97,98]. New root tips that are elongating and which secrete root exudates in the zone of elongation are sites of high microbial activity, and they are presumably colonized by rapid growing, opportunistic microorganisms such as *Pseudomonas* spp. [99]. Older root zones, on the other hand, are crowded environments that are relatively oligotrophic once the roots become suberized. Nonelongating secondary root tips and the sites of lateral root emergence are still other microsites that may be presumed to have unique microbial communities. Thus, it may not be possible to make general conclusions regarding which siderophores are predominant in the rhizosphere, and whether or not any particular siderophore type accumulates at a concentration that would be relevant to plant nutrition when it is only produced at a microsite location. Similarly, siderophores may have different functions depending on the root zone. For example, it can be speculated that pyoverdines may contribute to rhizosphere competence of pseudomonads at the root tips, but they may be irrelevant in the older, carbon-limited root zones where iron is either not limiting or is chelated with various, more generic siderophores.

Recent studies have further examined the iron stress response of pseudomonads using an iron-regulated, ice nucleation gene reporter (inaZ) for induction of the iron stress response [26,27,100]. This particular reporter system was developed by Loper and Lindow [101] for study of microbial iron stress on plant surfaces but was later employed in soil assays. In initial studies, cells of *P. fluorescens* and *P. syringae* that contained the pvd-inaZ fusion were shown to express iron-responsive, ice nucleation activity in the bean rhizosphere and phyllosphere. Addition of iron to leaves or soil reduced the apparent transcription of the pvd-inaZ reporter gene, as shown by a reduction in the number of ice nuclei produced.

This reporter system has been used in a number of studies examining regulation of the iron stress response in soils and in short-term experiments after inoculation of plant roots [26]. Ice nucleation activity expressed by rhizosphere populations of *P. fluorescens* Pf-5 was at a maximum within 12 to 24 h following inoculation of the bacterium onto bean roots and typically decreased gradually during the following 4 d. The finding that iron stress decreased 1 to 2 d after the bacterium was inoculated onto root surfaces, suggested that iron became more available to rhizosphere populations of Pf-5, once they were established in the rhizosphere. This led to the speculation that iron acquisition systems of plants and other rhizosphere organisms may provide available sources of iron to established rhizosphere populations of *P. fluorescens*.

In related researches involving a longer term of study, the *P. fluorescens* Pf-5 strain was inoculated into soil that was used to grow lupine (*Lupinus albus*) and barley (*Hordeum vulgare* L.) over several weeks [27]. It was shown that iron stress was greatest in the zone behind the root tips as compared to older root zones and in the bulk soil. Calibration of the ice nucleation reporter activity to siderophore production *in vitro* allowed estimation of potential siderophore production on a per cell basis *in vivo*. By using the regression between ice nucleation activity and pyoverdine production, and assuming a *P. fluorescens* population density of 10^8 CFU g^{-1} root, the maximum possible pyoverdine concentration was estimated to be 0.5 and 0.8 nmol g^{-1} root for lupine and barley, respectively. The low-ice-nucleation activity measured in the rhizosphere suggested that nutritional competition for iron in the rhizosphere may not be a major factor influencing root colonization by *P. fluorescens* Pf-5 (pvd-inaZ). These data also suggest that concentrations of siderophore in the rhizosphere would be too low for plant nutrition.

In another experiment with rice and barley, it was further shown that the iron stress response of strain Pf-5 could be shut down by production of phytosiderophores after induction of the iron stress response by plant roots [100]. When barley and rice seedlings were treated with foliar applications of ferric citrate, the iron stress response of the plant was suppressed such that phyto-siderophore production decreased and caused increased iron stress in the root-colonizing pseudomonads. This study also tested the ability of this pseudomonad to use phytosiderophore and ferric citrate as iron sources for growth *in vitro*, and showed that both were effective in supplying iron and in suppressing expression of the iron-regulated, ice nucleation reporter.

Altogether, these studies strongly suggest that siderophore production by root-colonizing micro-organisms is only induced when it becomes necessary to supplement an inadequate level provided by phytosiderophores and organic acids due to plant iron stress response. Thus, the plant iron stress response may control iron availability to microorganisms in the rhizosphere with plant use of microbial siderophores, causing some feedback to the plant that may partially mediate the plant response.

VII. EFFECT OF PLANT IRON STRESS ON MICROBIAL COMMUNITY STRUCTURE

A. DENATURING GRADIENT GEL ELECTROPHORESIS FINGERPRINTING OF MICROBIAL COMMUNITIES

A now common genetic technique that has been developed for characterization and comparison of microbial communities from different environmental samples is the use of denaturing gradient gel electrophoresis (DGGE) [102]. Using the 16S rRNA gene database, portions of the 1500 base pair sequence encoding 16S rRNA can be analyzed to provide identifications of bacteria [103]. DGGE is considered to be a low-resolution method in that it provides a *fingerprint*, or DNA-banding pattern, that can be used to compare the similarities of microbial communities. This is in contrast to high-resolution methods in which the predominant bacterial species that are present are identified by construction of a clone library that carries all of the sequences contained in the community. Individual clones can then be analyzed to determine their 16S rRNA gene sequences and similarity to described taxa of bacteria and archaea. High-resolution methods are still expensive to carry out for large numbers of samples due to the cost of sequencing hundreds or thousands of clones from individual samples. The advantage of DGGE is the relatively low cost and ability to compare community structures of bacteria, including the noncultured bacteria that comprise the majority of most environmental samples. The end product of DGGE is a broad snapshot of the community structure that is analogous to a bar code pattern in which each DNA band represents a different group or species of bacteria. The predominant bands can also be used for identification and monitoring of specific microbial populations in environmental samples [104–106].

To generate microbial community fingerprints, the total bacterial DNA is extracted from the sample, after which all of the 16S rRNA gene sequences from the community are amplified simultaneously using polymerase chain reaction (PCR). DGGE is then used to separate the sequences based on differences in their guanosine cytosine (GC) content and thermal melting points as the DNA migrates through the gradient gel. This results in discrete bands for each species, or group of bacteria, having sequences with similar melting points. Once the DNA patterns are generated, image analysis is used to compare fingerprints from samples that have been subjected to different experimental treatments. Caution must be used in interpreting DGGE data because bands having the same position can represent completely different species. Individual DNA bands may also be comprised of sequences from many different unrelated species of bacteria. Thus statistical analyses comparing similarity may be confounded by the low resolution. This can be improved by using genus or phylum-specific primer sets. Nonetheless, DGGE should be used as a preliminary screening method to determine possible community similarities or dissimilarities that should then be evaluated using high-resolution methods by construction of 16S rRNA gene libraries and sequencing. DGGE data are commonly used to estimate differences in species richness and diversity for communities from different samples, but such data should again be interpreted with caution since the method provides only very low resolution that may obscure true differences.

B. ANALYSIS OF FE STRESS EFFECTS ON MICROBIAL COMMUNITY STRUCTURES IN THE RHIZOSPHERE OF BARLEY

In studies examining the effect of plant iron stress on microbial communities in the rhizosphere, experiments were conducted with barley plants grown in replicate root box microcosms in an iron-limiting soil [107]. Half of the plants were treated with foliar-applied iron to alleviate plant iron stress and shut down phytosiderophore production at the root tips. Previously, this treatment had been shown to cause an increase in iron stress for cells of *P. fluorescens* located at the root tips [100], due to increased reliance on endogenous siderophore production. To examine differences in community composition at root microsites, 1-cm root segments were harvested from different locations on Fe stressed and nonstressed plants and subjected to PCR-DGGE using universal primers for bacteria. Comparison of the DGGE profiles for bacterial communities associated with similar locations on replicate plants showed that the rhizosphere community associated with specific locations was highly structured and reproducible for replicate root segments obtained from the same plant or from replicate plants grown in different microcosms. Differences in microbial communities associated with different root parts are shown in Figure 7.6 [107].

Statistical analysis of the microbial communities that were associated with the root tips of iron stressed and nonstressed plants revealed the formation of distinct communities in response to plant iron nutritional status. The appearance of a few predominant bands at the root tips is indicative of rapid growth of a few bacterial groups during primary colonization. As the roots matured, the number of bands and evenness (peak height) increased and reflected a crowded community with few predominant species. The effect of plant iron status was observed primarily at the root tips where phytosiderophores are released. Communities associated with the older root parts clustered similarly, whereas sites of lateral root emergence and nongrowing root tips differentiated to a lesser degree than the communities associated with the rapidly growing primary root tips. This type of analysis clearly demonstrates the impact of plant iron nutritional status on rhizosphere microbial community composition and how this varies for different root zones.

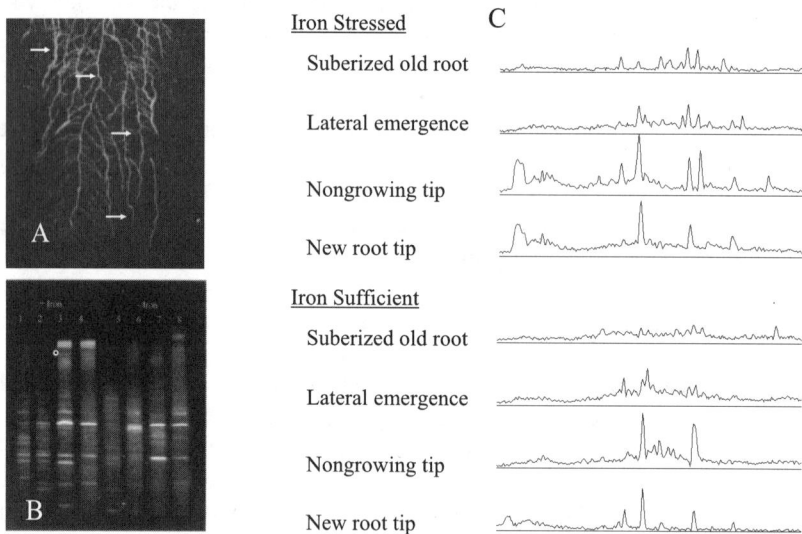

FIGURE 7.6 Genetic fingerprints of microbial communities associated with the roots of iron sufficient and iron-stressed barley grown in an iron-limiting soil medium. Fingerprints were generated using PCR primers for 16S rRNA genes, which were then separated on a polyacrylamide gel using denaturing gradient gel electrophoresis (DGGE). Each band corresponds to an individual species or group of bacteria having similar 16S rRNA gene sequences. The advantage of this technique is that it provides a culture-independent method for analyzing microbial community structure and species composition. (A) Root locations selected for microbial community analysis; (B) 16S rRNA gene band patterns generated by PCR-DGGE; and (C) line profiles generated by image analysis of DNA banding patterns shown in Figure 7.6B. (From Yang, C.-H. and Crowley, D.E., *Appl. Environ. Microbiol.*, 66, 345, 2000. With permission.)

VIII. FUNCTIONS OF SIDEROPHORES
IN RHIZOSPHERE COMPETENCE

One of the driving forces behind research on siderophores and their importance in rhizosphere colonization has been the finding that various bacteria can have a wide range of effects on plant growth. Certain bacteria, termed *plant growth promoting*, are thought to produce hormones, vitamins, and growth factors that enhance plant growth; whereas, other bacteria produce compounds that are deleterious to plants. The ability to manipulate rhizosphere populations via inoculation with siderophore-producing bacteria or by using soil amendments to alter iron availability has thus been a matter of practical concern. In one such study, over 100 bacterial strains were tested for their plant growth promoting and plant deleterious effects [108]. These strains belonged to the genera azotobacter, azospirillum, bacillus, enterobacter, and pseudomonas and were all isolated from the rhizosphere of various crops. *In vitro* inoculation of maize seedlings with these strains resulted in a 50% decrease to a 70% increase in plant growth as compared to noninoculated controls. Subsequent research has examined the relationship between siderophore production and competitive fitness of these strains, but it was unable to demonstrate any relationship between the ability to use pseudobactin type siderophores and rhizosphere competence. On the other hand, a contrasting study by other researchers showed that a mutant of *P. fluorescens* that was selected for overproduction of siderophore was shown to have an improved ability to control plant disease and to significantly enhance plant growth as compared to the parent strain [109].

Genetic studies have shown that nucleotide sequence of the *Pseudomonas* sp. strain M114 PbuA gene, encoding the outer membrane receptor for ferric pseudobactin M114, displays characteristics in common with other outer membrane proteins [110]. This sequence also has strong homology with the TonB boxes of both *E. coli* and pseudomonas receptors. More extensive homologies were found with the PupA receptor of *P. putida* WCS358 and the FhuE and BtuB receptors of *E. coli*. Based on these findings, it was suggested that a PbuA-like receptor may be widely distributed among pseudomonas rhizosphere isolates. Those areas of the PbuA receptor exhibiting the least homology between different pseudomonads may represent ferric siderophore-specific recognition sites for different types of pseudobactin.

The specificity of certain pseudomonads for strain-specific siderophores has been demonstrated for *P. putida* WCS358, which can be recovered efficiently on a medium amended with 300 μM pseudobactin 358 [111]. Low-population densities of indigenous pseudomonads ($\leq 10^3$ g^{-1} of soil or root) recovered on the pseudobactin 358 revealed that natural occurrence of the pipA gene, required for use of this siderophore, was limited to a very small number of indigenous *Pseudomonas* spp. that are very closely related to *P. putida* WCS358. In a subsequent experiment [5], the potential of different pseudomonas strains that are to utilize heterologous siderophores was compared with their competitiveness in the rhizosphere of radish. Interactions were examined for microbial populations of *P. putida* WCS358 and *P. fluorescens* WCS374 and between strain WCS358 and eight indigenous *Pseudomonas* strains capable of utilizing pseudobactin 358. During four successive plant growth cycles of radish, strain WCS358 significantly reduced rhizosphere population densities of the wild-type strain WCS374 by up to 30 times. In comparison, a derivative strain WCS374 (pMR), harboring the siderophore receptor PupA for ferric pseudobactin 358, was able to maintain its population density. Whereas this suggests that the ability to use pseudobactin was directly involved in rhizosphere competence, parallel studies on interactions between strain WCS358 and the eight different indigenous *Pseudomonas* strains showed that population densities of these bacteria were all reduced by up to 20-fold by strain WCS358, despite the ability to use pseudobactin 358. The authors concluded that siderophore-mediated competition for iron is a major determinant in interactions between some strains of bacteria but that other traits also contribute to the rhizosphere competence of fluorescent pseudomonads.

The ability of pyoverdine-type siderophores produced by pseudomonads to suppress the growth of other bacterial genera is now being more closely scrutinized; in fact, these compounds have been shown in certain cases to actually promote the growth of different bacteria. In a study conducted with mixed batch cultures, *P. cepacia* stimulated the growth of *Bacillus polymyxa* in low iron medium, and this effect could be replicated in pure culture by addition of pyoverdine produced by either *P. fluorescens* or *P. cepacia* [112]. In another study, two strains of *Bradyrhizobium japonicum* that normally utilize ferric citrate and the hydroxamate siderophores, ferrichrome and rhodotorulic acid, were also shown to use the pyoverdine-type siderophore pseudobactin St3 [113]. Crossfeeding of bacteria with heterologous siderophore transport systems thus appears to be a somewhat common mechanism for iron acquisition in many different bacterial species and can completely alter the outcome of competition for iron.

IX. ROLE OF SIDEROPHORES IN BIOCONTROL OF PLANT PATHOGENS

Unraveling the role of siderophores in disease suppression has been complicated and is one of the best examples of the problems that may be encountered by extrapolating to broad conclusions from simple experimental systems. Early research frequently employed screening assays using coculture of pseudomonads and plant pathogens on agar plate media to demonstrate that siderophores may be responsible for suppressing the growth of various fungal pathogens under iron-limiting conditions. More recently, it has been recognized that this is a narrow approach, and that these assays may still have utility for ruling out whether or not siderophores have an effect on the pathogen [114]. In these plate assays, the siderophore sequesters all of the available iron to the detriment of the pathogen. Competition for iron is due to the superior ability of the *Pseudomonas* sp. siderophores to chelate iron, which is based on their high stability constant for iron as compared to the siderophores produced by most fungal pathogens. Siderophores have also been added to soil in spent media to achieve biological control and in some instances are as effective as adding the bacteria themselves [3]. However, much larger quantities of siderophores are produced during growth in nutrient-rich, iron-limiting media than are detected in soils, and the relative importance of siderophores in disease suppression in comparison to antibiotic production is controversial.

An increase in the number of siderophore-producing organisms in the rhizosphere is associated with increased disease suppression and can be achieved by amending soils with compost [115]. However, it appears that although many disease-suppressive bacteria produce large quantities of siderophores when screened *in vitro*, the function of siderophores in directly antagonizing pathogens in the rhizosphere is questionable. Suppression of fungal pathogens in the rhizosphere by root-colonizing pseudomonads involves several mechanisms that may operate individually or collectively to antagonize the pathogen. These mechanisms involve nutritional competition, production of antibiotics [116], release of cyanide [117], and production of siderophores. Some pseudomonads also cause disease suppression by causing a systemic resistance response that is induced by the host plant after colonization by pseudomonads that produce pyoverdine and other salicylate-based siderophores [118].

One of the most powerful methods for studying the function of siderophores for biological control of root-disease-causing microorganisms has involved the generation of mutant strains that are defective in producing antibiotics or siderophores. The bacteria are then inoculated into soils to assay their efficacy in disease suppression (see review [119]). Examples where siderophores cause direct antagonistic effects on pathogens are rare and appear to operate in a strain-dependent fashion in very specific disease interactions. For example, the sole mechanism involved in the suppression of fusarium wilt of radish by pseudomonas strain WCS358 is siderophore-mediated competition for iron, whereas strain WCS374 suppresses this disease by induction of systemic resistance [120]. Other examples where siderophores have been shown to have no discernable function in disease suppression include studies with *Enterobacter cloacae* CT-501, which suppresses pythium damping-off of cucumber and other plant hosts. This biocontrol bacterium produces the hydroxamate siderophore aerobactin and a catechol siderophore tentatively identified as enterobactin. After generating mutants defective in either aerobactin or enterobactin synthesis, or double mutants defective in both of these siderophores, it was shown that neither of these siderophores contributed to the ability of *E. cloacae* to suppress pythium damping-off of cotton or cucumber [121]. Similarly, the antagonism of five bacterial isolates, including *Acinetobacter* sp., *B. polymyxa*, *B. subtilis*, *P. cepacia*, and *P. putida*, against the pathogens *Sclerotinia sclerotiorum*, *S. minor*, and *Gaeumannomyces graminis* was shown to be due to antibiotics rather than siderophore production [122].

Interactions between the induction of systemic acquired resistance in plants and the involvement of siderophores in direct antagonism of root pathogens have turned out to be subtle and difficult to discriminate. In one such study [2], the influence of iron availability on induction of systemic resistance in radish (*Raphanus sativus* L.) against fusarium wilt as mediated by *P. fluorescens* was examined. In the actual experiment, the pathogen (*Fusarium oxysporum* sp. raphani) and a strain of *Pseudomonas*, salicylic acid (SA), or a pseudobactin were applied at separate locations on the plant root. Strain WCS374 and its pseudobactin-minus Tn5 mutant gave greater disease control in the induced systemic resistance bioassay when iron availability in the radish nutrient solution was low rather than when it was high. Mutants of *P. fluorescens* strains WCS374 and WCS417, lacking the O-antigenic side chain of the lipopolysaccharide, induced resistance at low, but not at high, iron concentrations. Interestingly, the purified pseudobactin of strain WCS374, but not the pseudobactins of strains WCS358 and WCS417, induced resistance to *Fusarium*. It was subsequently shown by gas chromatography that strains WCS374 produced high concentrations of SA under conditions of low iron availability and that SA also induced systemic resistance to root disease. This led to the conclusion that SA produced by selected *P. fluorescens* strains, as well as pseudobactin produced by WCS374, may both be involved in causing induced systemic resistance.

Related research examining the effects of SA in causing induced systemic resistance under iron limiting conditions suggests that SA itself is a siderophore as it chelates iron and is released in response to iron stress [2]. The plant growth-promoting rhizobacterium *Pseudomonas aeruginosa* 7NSK2 produces three siderophores when it is iron limited. These include pyoverdine, the salicylate derivative of pyoverdine, called *pyochelin*, and SA. The role of pyoverdine and pyochelin in the suppression of *Pythium splendens* was investigated by using various siderophore-deficient mutants derived from *P. aeruginosa* 7NSK2 in a bioassay with tomato (*Lycopersicon esculentum*). Production of either pyoverdine or pyochelin proved to be necessary to achieve wild-type levels of protection against Pythium-induced damping-off. Because pyoverdin and pyochelin are both siderophores, siderophore-mediated iron competition could explain the observed antagonism, but the authors could not exclude the possibility that the siderophores acted in an indirect way.

The relative importance of antibiotic production vs. production of siderophores for disease suppression has recently been studied by Mulya and coworkers [123]. In experiments with *P. fluorescens* strain PfG32 isolated from the rhizosphere of the onion, it was shown that the bacterium actively suppressed the occurrence of bacterial wilt disease of tomato and produced both antibiotics and siderophore in pure culture. After isolating several mutants defective in antibiotic substances or siderophore production, the suppression of bacterial wilt by the different strains was then compared. Mutants that did not produce antibiotics but produced siderophores were less effective for disease biocontrol than were strains that produced both substances, or that produced antibiotic but no siderophores. This suggests that, for this particular bacterium, the antibiotic was more important than the siderophore and that both substances contributed to the ability of the bacterium to cause disease suppression.

Whether antibiotics or siderophores or both are involved in disease suppression, an obvious role of siderophores is the enhanced competitiveness that may be associated with the ability to produce and use siderophores for iron acquisition under iron-limiting, high-carbon conditions in the rhizosphere. Generally, pseudomonad population densities of 10^5 CFU g^{-1} soil are associated with disease suppressiveness. Disease suppression in soils providing natural biocontrol of take-all decline of wheat is associated with high numbers of pseudomonads that produce the antibiotic 2,4-diacetylphloroglucinol [124]. Below a cell density of 10^5 CFU g^{-1} soil, disease suppression rapidly declines; whereas, above this threshold density there appears to be no additional benefit for further enhancing the disease suppressive effects. The importance of maintaining a critical threshold density for suppression of *Fusarium* wilt of radish was also shown in soil inoculated with *P. putida*. In this study, population densities of 10^5 CFU g^{-1} of root were required to provide biocontrol by either strain WCS358 that produces siderophores, or strain WCS374 that causes systemic resistance. Although it may be speculated that the ability to utilize additional siderophores may increase the fitness and population density of pseudomonads used for biocontrol, it appears that crossfeeding among pseudomonads does not necessarily result in enhanced numbers of pseudomonads in the rhizosphere [4]. Transfer of a plasmid conferring the ability to utilize additional pseudobactins had no effect on increasing the population size of pseudomonad inoculants in soil.

X. FUNCTION OF SIDEROPHORES IN NITROGEN FIXATION

Iron is required in large quantities for nitrogen-fixing bacteria, and the importance of siderophores in the iron stress response of these microorganisms has been well established [125–127]. Iron is an essential component of the nitrogenase enzyme complex and is required for respiration, oxygen binding and delivery by leghemoglobin, and for synthesis of ferredoxin, which delivers electrons to nitrogenase. Siderophores may also be important for temporary storage of iron in the peribacteroid space of the symbiosome of rhizobia nodules [128]. Siderophores found in the peribacteroid space are different for various species of *Rhizobium*. The prototypical siderophore produced by *Rhizobium* is termed *rhizobactin* and is structurally characterized as a derivative of citrate in which the distal carboxyl groups of the citrate molecule are joined by an amide bond with two side chains [129]. Other studies have revealed that not all *Bradyrhizobium* and *Rhizobium* strains produce sidero-phores, and that the types of siderophores produced include both hydroxamate and catecholate type siderophores [130]. In screening assays of 31 strains, it was shown that siderophore production was correlated with nitrogen-fixing efficiency. Other studies suggests that this may not always be the case [131,132].

Differences in siderophore production have been observed for the free-living forms of the bacterial symbionts [133]. However, it has not been determined whether the production of high concentrations of siderophores corresponds with an increase in nitrogen fixation efficiency or adaptation of *Rhizobium* strains to iron-limiting soils. In one study examining various siderophore mutants of *R. meliloti*, it was shown that the high-affinity iron acquisition system expressed by the free-living form of this bacterium is not essential for nitrogen fixation; although it can affect the early events of nodulation [134]. In contrast, results of another study suggest that the ability to produce siderophores is essential for nitrogen fixation to occur. In this particular case, mutants of *R. meliloti* strain 1021 that were defective in rhizobactin synthesis produced nodules on alfalfa but fixed insignificant amounts of dinitrogen [135]. Mutants of *R. meliloti* constitutive for rhizobactin synthesis also produced nodules, but nitrogen fixation was low. These contrasting data suggest that there is a need for siderophore production during symbiosis, but that the exact role of siderophores in nodule formation is still unclear.

One explanation for the disagreement over the relative importance of siderophore production by *Rhizobium* may be the plant–host related effects that influence iron availability to the symbiont, and the ability of free-living rhizobium to obtain iron from sources other than its own siderophore. Jadhav and Desai [136] showed that whereas cowpea *Rhizobium* GN1 (peanut isolate) produced siderophore under iron-starved conditions, other nutritional and environmental factors also affected siderophore production and iron assimilation by this microorganism. Maximum siderophore pro-duction was obtained with maltose and urea as carbon and nitrogen source, respectively. With citrate as a sole source of carbon there was complete repression of siderophore production without any effect on growth. The involvement of citrate in iron transport was confirmed by ^{55}Fe-citrate uptake studies. This suggests that plant species that produce large quantities of citrate in response to iron stress may provide iron for rhizobium in the rhizosphere and during nodulation.

Positive interactions between plant growth-promoting rhizobacteria and rhizobium have been reported, but to date, there is little evidence that siderophores produced by pseudomonads are beneficial for promoting nodulation and nitrogen fixation. In experiments examining the role of siderophore production on nodulated clover plants, siderophore defective mutants were shown to stimulate growth of nodulated clover plants similarly to the siderophore-producing parent strain [137].

XI. SUMMARY

The primary function of plant and microbial siderophores in the rhizosphere is iron acquisition under iron-limiting conditions. Almost all microorganisms have been found to produce siderophores and can potentially compete with each other for iron, depending on their ability to utilize different siderophore types or based on the uptake kinetics of their siderophore transport systems. Similarly, plants and microorganisms can compete for iron under certain conditions; although the extent to which this influences plant ecology in nature can be questioned. Since the discovery of microbial siderophores by Lankford in the 1950s and phytosiderophores by Takagi in the 1970s, there has been extensive research on the characterization of siderophores, the proteins that function for the uptake of iron, and the molecular regulation of high-affinity iron transport systems that synthesize and transport these compounds. In comparison, much less is known about the function of sidero-phores in microbial ecology.

It is evident from studies on the rhizosphere that the plant iron stress response is the dominant factor in determining whether microorganisms are subjected to iron stress and are thereby induced to secrete siderophores. In response to iron deficiency, dicotyledonous plants release increased quantities of organic acids, acidify the rhizosphere, and secrete reductants — all of which increase iron availability to the plant and the root associated microflora. Grasses that produce phytosidero-phores in response to iron deficiency also can increase iron availability to microorganisms that utilize these iron-chelating substances. Many microorganisms have iron transport systems that function with organic acids such as citrate, and both organic acids and phytosiderophores have been shown to serve as effective iron sources for pseudomonads that colonize plant roots. For this reason, it can be assumed that siderophores are produced by microorganisms only in quantities necessary to augment iron that is not provided by the plant's iron stress response.

Initial studies examining the function of siderophores in the rhizosphere have focused on practical problems related to agricultural biotechnology. Early research suggested the involvement of siderophores in plant disease suppression by certain root-colonizing pseudomonads. Since then, it has become apparent that siderophore-mediated interactions are only one of several factors that influence rhizosphere competence during the colonization of plant roots. Many factors influence competition in soils, including the fitness of microorganisms in a given niche, the ability to use different carbon substrates, and the growth strategy of microorganisms that are differentially adapted to copiotrophic or oligotrophic growth conditions. Keeping in mind Liebig's law of the minimum, competition for iron will occur only when all other growth factors have been optimized. In the rhizosphere, easily utilizable carbon substrates are provided as root exudates and lysates but are localized in specific root zones that comprise a relatively small fraction of the total root surface area. Production of siderophores is most likely to occur in spot locations during intense competition for easily used carbon substrates. In the oligotrophic environment of the older root zones, which comprises most of the rhizosphere, carbon is available only in the form of much more recalcitrant substrates such as cellulose. On the older root parts, it is unlikely that competition for iron will ever occur as sufficient iron may be provided by turnover of plant tissues, microbial cells, back-ground siderophore concentrations, and organic matter iron complexes. The production and release of siderophores is very expensive in terms of carbon allocation for cells. In constructing a carbon budget for the oligotrophic zones of the rhizosphere, it should be evident that any microorganism releasing large quantities of siderophore would be at a disadvantage and would simply be providing a labile carbon substrate for other bacteria and fungi.

The extent to which siderophores shape the ecology of the rhizosphere is an interesting question that remains to be investigated. New tools in molecular ecology are now providing insight into microbial community structure and species composition in different root zones. It has been shown that the plant iron stress response can alter the structure of the microbial community at the root tips but that there is little effect of plant iron nutritional status on the microbial community composition of the older root zones. An interesting question that remains to be investigated is whether competition for iron that occurs during primary colonization of plants also has an effect on microbial community development in older root zones as these communities mature and undergo succession into a crowded, oligotrophic community.

Plants, like microorganisms, appear to have the ability to use a variety of iron sources for iron nutrition. Studies in hydroponic culture have suggested that siderophores can function similarly to synthetic chelates in providing iron that is taken up by a variety of mechanisms. Depending on the chelator, these uptake mechanisms can include reductive release of iron, degradative release of iron during decomposition of labile siderophores, uptake of the intact iron chelate, and ligand exchange with phytosiderophores and organic acids. However, the extent to which plants rely on microbial siderophores for iron nutrition is questionable, because siderophores do not ever occur uniformly at high concentrations throughout the rhizosphere, and when they do, they are primarily in the deferrated form, which is inhibitory to plant iron uptake. Background levels of siderophores detected in soil extracts are generally very low [88,138], being measured at nanomolar concentrations that may be of some use for oligotrophic microorganisms but are too low to be of physiological relevance for plant nutrition. Depending on their charge characteristics, siderophores can also be strongly adsorbed to clay and organic matter, and at low concentrations may not diffuse readily through the soil matrix [39,87].

The importance of siderophore production in rhizosphere competence of bacteria and fungi very likely will depend on the growth strategy of the microorganism. Most studies on rhizosphere microorganisms have focused on pseudomonads, which are generally characterized as opportunistic bacteria that grow rapidly on root exudates, and thus are primary colonizers of new root tissues. Certainly siderophores play a role in rhizosphere competence of these bacteria, but as much of the root system is relatively oligotrophic, it would not be surprising to find that the predominant microorganisms that occupy most of the root surface are little influenced by siderophores. Nonetheless, the fact that most culturable bacteria and fungi maintain high-affinity iron transport systems suggests that there is selection pressure for maintenance of these systems for occasional growth under conditions that require the synthesis of siderophores for iron acquisition.

Given the complexity of modeling soil solutions and in predicting the chemistry of siderophores in complex mixtures with different solid phase minerals and constantly fluctuating ionic strength and redox conditions, it is a considerable challenge to unravel how siderophores mediate competition for iron in the rhizosphere. An immediate need for scientists studying the plant rhizosphere will be to develop conceptual understanding of how siderophores can potentially function in different root zones and under different soil conditions. One of the most important avenues of research for future studies on siderophores will be the investigation of how these compounds influence heavy metal transport and bioavailability in soils, and the possible role of siderophores in groundwater contamination and food chain transfer. Understanding of these siderophores and their transport systems may also have application in the development of plant microbial systems for phytoremediation, and the use of siderophores as iron fertilizers.

REFERENCES

1. Crowley, D.E. and Gries, D., Modeling of iron availability in the plant rhizosphere, in *Biochemistry of Metal Micronutrients in the Plant Rhizosphere*, Manthey, J.A., Crowley, D.E., and Luster, D.G., Eds., Lewis Publishers, Ann Arbor, 1994, p. 199.
2. Buysens, S. et al., Involvement of pyochelin and pyoverdin in suppression of *Phythium*-induced damping-off of tomato by *Pseudomonas aeruginosa* 7NSK2, *Appl. Environ. Microbiol.*, 62, 865, 1996.

3. Neilands, J.B. and Leong, S.A., Siderophores in relation to plant-growth and disease, *Ann. Rev. Plant Physiol.*, 37, 187, 1986.
4. Moenne-Loccoz, Y. et al., Rhizosphere competence of fluorescent *Pseudomonas* sp B24 genetically modified to utilise additional ferric siderophores, *FEMS Microbiol. Ecol.*, 19, 215, 1996.
5. Raaijmakers, J.M. et al., Dose-response relationships in biological control of fusarium wilt of radish by *Pseudomonas* spp., *Phytopathology*, 85, 1075, 1995.
6. Gries, D. and Runge, M., The ecological significance of iron mobilization in wild grasses, *J. Plant Nutr.*, 15, 1727, 1992.
7. Chen, Y. et al., Stability constants of pseudobactin complexes with transition-metals, *Soil Sci. Soc. Am. J.*, 58, 390, 1994.
8. John, S.G. et al., Siderophore mediated plutonium accumulation by Microbacterium flavescens (JG-9), *Environ. Sci. Technol.*, 35, 2942, 2001.
9. Mench, M.J. and Fargues, S., Metal uptake by iron-efficient and inefficient oats, *Plant Soil*, 165, 227, 1994.
10. Burd, G.I., Dixon, D.G., and Glick, B.R., Plant growth-promoting bacteria that decrease heavy metal toxicity in plants, *Can. J. Microbiol.*, 46, 237, 2000.
11. Reichard, P.U. et al., Goethite dissolution in the presence of phytosiderophores: rates, mechanisms, and the synergistic effect of oxalate, *Plant Soil*, 276, 115, 2005.
12. Lindsay, W.L. and Schwab, A.P., The chemistry of iron in soils and its availability to plants, *J. Plant Nutr.*, 5, 821, 1982.
13. Kraemer, S.M., Iron oxide dissolution and solubility in the presence of siderophores, *Aquat. Sci.*, 66, 3, 2004.
14. Cheah, S.F. et al., Steady-state dissolution kinetics of goethite in the presence of desferrioxamine B and oxalate ligands: implications for the microbial acquisition of iron, *Chem. Geol.*, 198, 63, 2003.
15. Schwertmann, U., Solubility and dissolution of iron oxides, in *Iron Nutrition and Interactions in Plants*, Chen, Y. and Hadar, Y., Eds., Kluwer Acad. Pub., Boston, MA, 1990, p. 3.
16. Cornell, R.M. and Schwertmann, U., *The Iron Oxides*, Wiley-VCH, Weinheim, 2003, p. 664.
17. Römheld, V., Existence of two different strategies for the acquisition of iron in higher plants, in *Iron Transport in Animals, Plants, and Microorganisms*, Winkelmann, G., Van der Helm, D., and Neilands, J.B., Eds., VCH Chemie, Weinheim, Germany, 1987, p. 353.
18. Jones, D.L., Darrah, P.R., and Kochian, L.V., Critical evaluation of organic acid mediated iron dissolution in the rhizosphere and its potential role in root iron uptake, *Plant Soil*, 180, 57, 1996.
19. Cesco, S. et al., Uptake of Fe-59 from soluble Fe-59-humate complexes by cucumber and barley plants, *Plant Soil*, 241, 121, 2002.
20. Nikolic, M. et al., Uptake of iron (Fe-59) complexed to water-extractable humic substances by sunflower leaves, *J. Plant Nutr.*, 26, 2243, 2003.
21. Chen, L.M. et al., Fe chelates from compost microorganisms improve Fe nutrition of soybean and oat, *Plant Soil*, 200, 139, 1998.
22. Ma, J.F. and Nomoto, K., Effective regulation of iron acquisition in graminaceous plants. The role of mugineic acids as phytosiderophores, *Physiol. Plant.*, 97, 609, 1996.
23. Fan, T.W.M. et al., Comprehensive analysis of organic ligands in whole root exudates using nuclear magnetic resonance and gas chromatography mass spectrometry, *Anal. Biochem.*, 251, 57, 1997.
24. Parker, D.R., Chaney, R.L., and Norvell, W.A., Chemical equilibrium models: applications to plant nutrition research, in *Chemical Equilibrium and Reaction Models*, Soil Sci. Soc. Am. Spec. Pub. 42, Madison, WI, 1995, p. 163.
25. Darrah, P.R., The rhizosphere and plant nutrition: a quantitative approach, *Plant Soil*, 155, 1, 1993.
26. Loper, J.E. and Henkels, M.D., Availability of iron to *Pseudomonas fluorescens* in rhizosphere and bulk soil evaluated with an ice nucleation reporter gene, *Appl. Environ. Microbiol.*, 63, 99, 1997.
27. Marschner, P. and Crowley, D.E., Iron stress and pyoverdin production by a fluorescent pseudomonad in the rhizosphere of white lupine (*Lupinus albus* L) and barley (*Hordeum vulgare* L), *Appl. Environ. Microbiol.*, 63, 277, 1997.
28. Treeby, M., Marschner, H., and Romheld, V., Mobilization of iron and other micronutrient cations from a calcareous soil by plant-borne, microbial, and synthetic metal chelators, *Plant Soil*, 114, 217, 1989.

29. Yehuda, Z. et al., The role of ligand exchange in the uptake of iron from microbial siderophores by gramineous plants, *Plant Physiol.*, 112, 1273, 1996.
30. Shenker, M., Hadar, Y., and Chen, Y., Kinetics of iron complexing and metal exchange in solutions by rhizoferrin, a fungal siderophore, *Soil Sci. Soc. Am. J.*, 63, 1681, 1999.
31. Hordt, W., Romheld, V., and Winkelmann, G., Fusarinines and dimerum acid, mono- and dihydroxamate siderophores from *Penicillium chrysogenum*, improve iron utilization by strategy I and strategy II plants, *Biometals*, 13, 37, 2000.
32. Tufano, T.P. and Raymond, K.N., Coordination chemistry of microbial iron transport compounds 21 kinetics and mechanism of iron exchange in hydroxamate siderophore complexes, *J. Am. Chem. Soc.*, 103, 6617, 1981.
33. Hoerdt, W., Roemheld, V., and Winkelmann, G., Fusarinines and dimerum acid, mono- and dihydroxamate siderophores from *Penicillium chrysogenum*, improve iron utilization by strategy I and strategy II plants, *Biometals*, 13, 37, 2000.
34. Jurkevitch, E. et al., Indirect utilization of the phytosiderophore mugineic acid as an iron source to rhizosphere fluorescent *Pseudomonas*, *Biometals*, 6, 119, 1993.
35. Matschonat, G. and Vogt, R., Equilibrium solution composition and exchange properties of disturbed and undisturbed soil samples from an acid forest soil, *Plant Soil*, 183, 171, 1996.
36. Lavie, S. and Stotzky, G., Interactions between clay-minerals and siderophores affect the respiration of *Histoplasma capsulatum*, *Appl. Environ. Microbiol.*, 51, 74, 1986.
37. O'Connor, G.A., Lindsay, W.L., and Olsen, S.R., Iron diffusion to plant roots, *Soil Sci.*, 119, 285, 1975.
38. Powell, P.E. et al., Hydroxamate siderophores in the iron nutrition of plants, *J. Plant Nutr.*, 5, 653, 1982.
39. Powell, P.E. et al., Occurrence of hydroxamate siderophore iron chelators in soils, *Nature*, 287, 833, 1980.
40. Crowley, D.E. et al., Mechanisms of iron acquisition from siderophores by microorganisms and plants, *Plant Soil*, 130, 179, 1991.
41. Boukhalfa, H. and Crumbliss, A.L., Chemical aspects of siderophore mediated iron transport, *Biometals*, 15, 325, 2002.
42. Braun, V. and Braun, M., Iron transport and signaling in *Escherichia coli*, *FEBS Lett.*, 529, 78, 2002.
43. Johnson, G.V., Lopez, A., and La Valle Foster, N., Reduction and transport of Fe from siderophores, *Plant Soil*, 241, 27, 2002.
44. Stintzi, A. et al., Microbial iron transport via a siderophore shuttle: a membrane ion transport paradigm, *Proc. Nat. Acad. Sci.*, 97, 10691, 2000.
45. Schmidt, W., Mechanisms and regulation of reduction-based iron uptake in plants, *New Phytologist*, 141, 1, 1999.
46. Bienfait, H.F., Vandenbriel, W., and Meslandmul, N.T., Free space iron pools in roots — generation and mobilization, *Plant Physiol.*, 78, 596, 1985.
47. Mihashi, S., Mori, S., and Nishizawa, N., Enhancement of ferric-mugineic acid uptake by iron deficient barley roots in the presence of excess free mugineic acid in the medium, *Plant Soil*, 130, 135, 1991.
48. Pinton, R. et al., Water-extractable humic substances enhance iron deficiency responses by Fe-deficient cucumber plants, *Plant Soil*, 210, 145, 1999.
49. Bienfait, H.F., Regulated redox processes at the plasmalemma of plant-root cells and their function in iron uptake, *J. Bioenerg. Biomembr.*, 17, 73, 1985.
50. Susin, S. et al., Flavin excretion from roots of iron-deficient sugar-beet (*Beta vulgaris* L), *Planta*, 193, 514, 1994.
51. Solinas, V. et al., Reduction of the FeIII-desferrioxamine-B complexes by caffeic acid: a reduction mechanism of biochemical significance, *Soil Biol. Biochem.*, 28, 649, 1996.
52. Marschner, H. and Romheld, V., Strategies of plants for acquisition of iron, *Plant Soil*, 165, 261, 1994.
53. Dinkelaker, B., Hengeler, C., and Marschner, H., Distribution and function of proteoid roots and other root clusters, *Bot. Acta*, 108, 183, 1995.
54. Holden, M.J. et al., Fe^{3+}-chelate reductase-activity of plasma-membranes isolated from tomato (*Lycopersicon esculentum* Mill) roots — comparison of enzymes from Fe-deficient and Fe-sufficient roots, *Plant Physiol.*, 97, 537, 1991.
55. Kochian, L.V., Mechanisms of micronutrient uptake, translocation, and interactions in plants, in *Micronutrients in Agriculture*, Mortvedt, J.J., Cox, F.R., Shjuman, L., and Welch, R.M., Eds., Soil Sci. Soc. Am., Madison, WI, 1991, p. 229.

56. Crowley, D.E., Reid, C.P.P., and Szaniszlo, P.J., Utilization of microbial siderophores in iron acquisition by oat, *Plant Physiol.*, 87, 680, 1988.
57. Manthey, J.A., Tisserat, B., and Crowley, D.E., Root responses of sterile-grown onion plants to iron deficiency, *J. Plant Nutr.*, 19, 145, 1996.
58. Wang, Y. et al., Evidence for direct utilization of a siderophore, ferrioxamine-B, in axenically grown cucumber, *Plant Cell Environ.*, 16, 579, 1993.
59. Barness, E. et al., Short term effects of rhizosphere microorganisms on Fe uptake from microbial siderophores by maize and oat, *Plant Physiol.*, 100, 451, 1992.
60. Crowley, D.E. et al., Root-microbial effects on plant iron uptake from siderophores and phytosiderophores, *Plant Soil*, 142, 1, 1992.
61. Romheld, V. and Marschner, H., Mechanism of iron uptake by peanut plants.1. FeIII reduction, chelate splitting, and release of phenolics, *Plant Physiol,*, 71, 949, 1983.
62. Voelker, C. and Wolf-Gladrow, D.A., Physical limits on iron uptake mediated by siderophores or surface reductases, *Marine Chem.*, 65, 227, 1999.
63. Mori, S., Mechanisms of iron acquisition by graminaceous (strategy II) plants, in *Biochemistry of Metal Micronutrients in the Rhizosphere*, Manthey, J.A., Crowley, D., and Luster, D.G., Eds., Lewis Publishers, London, 1994, p. 225.
64. Kawai, S., Takagi, S.I., and Sato, Y., Mugineic acid family phytosiderophores in root secretions of barley, corn and sorghum varieties, *J. Plant Nutr.*, 11, 633, 1988.
65. Crowley, D.E. et al., Quantitative traits associated with adaptation of three barley (*Hordeum vulgare* L) cultivars to suboptimal iron supply, *Plant Soil*, 241, 57, 2002.
66. Cakmak, I. et al., Zinc efficient wild grasses enhance release of phytosiderophores under zinc deficiency, *J. Plant Nutr.*, 19, 551, 1996.
67. Zhang, F.S., Romheld, V., and Marschner, H., Diurnal rhythm of release of phytosiderophores and uptake rate of zinc in iron-deficient wheat, *Soil Sci. Plant Nutr.*, 37, 671, 1991.
68. Gries, D. et al., Phytosiderophore release in relation to micronutrient metal deficiencies in barley, *Plant Soil*, 172, 299, 1995.
69. Chaignon, V., Di Malta, D., and Hinsinger, P., Fe-deficiency increases Cu acquisition by wheat cropped in a Cu-contaminated vineyard soil, *New Phytologist*, 154, 121, 2002.
70. Shenker, M., Fan, T.W.M., and Crowley, D.E., Phytosiderophores influence on cadmium mobilization and uptake by wheat and barley plants, *J. Environ. Qual.*, 30, 2091, 2001.
71. Winkelmann, G., Specificity of iron transport in bacteria and fungi, in *CRC Handbook of Microbial Iron Chelates*, Winkelmann, G., Ed., CRC Press, Boca Raton, FL, 1991, p. 65.
72. Winkelmann, G., Microbial siderophore-mediated transport, *Biochem. Soc. Trans.*, 30, 691, 2002.
73. Raaijmakers, J.M. et al., Utilization of heterologous siderophores and rhizosphere competence of fluorescent *Pseudomonas* spp., *Can. J. Microbiol.*, 41, 126, 1995.
74. Becker, J.O., Hedges, R.W., and Messens, E., Inhibitory effect of pseudobactin on the uptake of iron by higher plants, *Appl. Environ. Microbiol.*, 49, 1090, 1985.
75. Walter, A. et al., Iron nutrition of cucumber and maize: effect of *Pseudomonas putida* YC 3 and its siderophore, *Soil Biol. Biochem.*, 26, 1023, 1994.
76. Marschner, H., Romheld, V., and Kissel, M., Localization of phytosiderophore release and iron uptake along intact roots of barley, *Physiol. Plant.*, 71, 157, 1987.
77. Romheld, V., The role of phytosiderophores in acquisition of iron and other micronutrients in gramineous species — an ecological approach, *Plant Soil*, 130, 127, 1991.
78. Watanabe, M. and Wada, H., Mugineic acid decomposing bacteria isolated from the rhizoplane of iron-deficient barley, *Jpn. J. Soil. Sci. Plant Nutr.*, 60, 413, 1989.
79. Clark, R.B., Romheld, V., and Marschner, H., Iron uptake and phytosiderophore release by roots of sorghum genotypes, *J. Plant Nutr.*, 11, 663, 1988.
80. Von Wiren, N. et al., Competition between microorganisms and roots of barley and sorghum for iron accumulated in the root apoplasm, *New Phytologist*, 130, 511, 1995.
81. Von Wiren, N. et al., Influence of microorganisms on iron acquisition in maize, *Soil Biol. Biochem.*, 25, 371, 1993.
82. Duijff, B.J., Bakker, P.A.H.M., and Schippers, B., Ferric pseudobactin 358 as an iron source for carnation, *J. Plant Nutr.*, 17, 2069, 1994.

83. Duijff, B.J. et al., Influence of pseudobactin 358 on the iron nutrition of barley, *Soil Biol. Biochem.*, 26, 1681, 1994.

84. Buyer, J.S., Kratzke, M.G., and Sikora, L.J., A method for detection of pseudobactin, the siderophore produced by a plant-growth-promoting Pseudomonas strain, in the barley rhizosphere, *Appl. Environ. Microbiol.*, 59, 677, 1993.

85. Nelson, M. et al., An *Escherichia coli* bioassay of individual siderophores in soil, *J. Plant Nutr.*, 11, 915, 1988.

86. Akers, H.A., Isolation of the siderophore schizokinen from soil of rice fields, *Appl. Environ. Microbiol.*, 45, 1704, 1983.

87. Bossier, P. and Verstraete, W., Ecology of *Arthrobacter* JG-9 detectable hydroxamate siderophores in soils, *Soil Biol. Biochem.*, 18, 487, 1986.

88. Reid, R.K. et al., Comparison of siderophore concentration in aqueous extracts of rhizosphere and adjacent bulk soils, *Pedobiologia*, 26, 263, 1984.

89. Vancura, V., Fluorescent pseudomonads in the rhizosphere of plants and their relation to root exudates, *Folia Microbiol.*, 25, 168, 1980.

90. Kleeberger, A., Castorph, H., and Klingmuller, W., The rhizosphere microflora of wheat and barley with special reference to gram-negative bacteria, *Arch. Microbiol.*, 136, 306, 1983.

91. Kuske, C.R., Barns, S.M., and Busch, J.D., Diverse uncultivated bacterial groups from soils of arid southwestern United States that are present in many geographic regions, *Appl. Environ. Microbiol.*, 63, 3614, 1997.

92. Alexander, D.B. and Zuberer, D.A., Use of chrome azurol S reagents to evaluate siderophore production by rhizosphere bacteria, *Biol. Fertil. Soils*, 12, 39, 1991.

93. Bar Ness, E. et al., Iron uptake by plants from microbial siderophores — a study with 7-nitrobenz-2 oxa-1,3-diazole-desferrioxamine as fluorescent ferrioxamine B-analog, *Plant Physiol.*, 99, 1329, 1992.

94. Bar Ness, E. et al., Siderophores of Pseudomonas putida as an iron source for dicot and monocot plants, in *Iron Nutrition and Interactions in Plants*, Chen, Y. and Hadar, Y., Eds., Kluwer Acad. Publishers, Boston, MA, 1992, p. 271.

95. Jurkevitch, E., Hadar, Y., and Chen, Y., Involvement of bacterial siderophores in the remedy of iron induced chlorosis in peanut, *Soil Sci. Soc. Am. J.*, 52, 1032, 1988.

96. Alexander, D.B. and Zuberer, D.A., Responses by iron-efficient and inefficient oat cultivars to inoculation with siderophore-producing bacteria in a calcareous soil, *Biol. Fertil. Soils*, 16, 118, 1993.

97. Barber, D.A. and Lynch, J.M., Microbial growth in the plant rhizosphere, *Soil Biol. Biochem.*, 9, 305, 1976.

98. Bowen, G.D. and Rovira, A.D., Microbial colonization of plant roots, *Ann. Rev. Phytopath.*, 14, 121, 1976.

99. Sorensen, J., Jensen, L.E., and Nybroe, O., Soil and rhizosphere as habitats for *Pseudomonas* inoculants: new knowledge on distribution, activity and physiological state derived from micro-scale and single-cell studies, *Plant Soil*, 232, 97, 2001.

100. Marschner, P. and Crowley, D.E., Phytosiderophores decrease iron stress and pyoverdine production of *Pseudomonas fluorescens* Pf-5 (PVD-INA Z), *Soil Biol. Biochem.*, 30, 1275, 1998.

101. Loper, J.E. and Lindow, S.E., A biological sensor for iron available to bacteria in their habitats on plant surfaces, *Appl. Environ. Microbiol.*, 60, 1934, 1994.

102. Muyzer, G. and Smalla, K., Application of denaturing gradient gel electrophoresis (DGGE) and temperature gradient gel electrophoresis (TGGE) in microbial ecology, *Ant. von Leeuwenhoek*, 73, 127, 1998.

103. Woese, C.R., Bacterial evolution, *Microbiol. Rev.*, 51, 181, 1987.

104. Jensen, S. et al., Diversity in methane enrichments from an agricultural soil revealed by DGGE separation of PCR amplified 16S rDNA fragments, *FEMS Microbiol. Ecol.*, 26, 17, 1998.

105. Stephen, J.R. et al., Analysis of beta subgroup proteobacterial ammonia oxidizer populations in soil by denaturing gradient gel analysis and hierarchical phylogenetic probing, *Appl. Environ. Microbiol.*, 64, 2958, 1998.

106. Vallaeys, T. et al., Evaluation of denaturing gradient gel electrophoresis in the detection of 16S rDNA sequence variation in rhizobia and methanotrophs, *FEMS Microbiol. Ecol.*, 24, 279, 1997.

107. Yang, C.-H. and Crowley, D.E., Rhizosphere microbial community structure in relation to root location and plant iron nutritional status, *Appl. Environ. Microbiol.*, 66, 345, 2000.

108. Forlani, G. et al., Root colonization efficiency, plant-growth-promoting activity and potentially related properties in plant-associated bacteria, *J. Genet. Breed.*, 49, 343, 1995.

109. Lim, H.S., Lee, J.M., and Kim, S.D., A plant growth-promoting *Pseudomonas fluorescens* GL20: mechanism for disease suppression, outer membrane receptors for ferric siderophore, and genetic improvement for increased biocontrol efficacy, *J. Microbiol. Biotechnol.*, 12, 249, 2002.

110. Morris, J. et al., Nucleotide sequence analysis and potential environmental distribution of a ferric pseudobactin receptor gene of *Pseudomonas* sp strain M114, *Mol. Gen. Genet.*, 242, 9, 1994.

111. Raaijmakers, J.M. et al., Siderophore receptor PupA as a marker to monitor wild-type *Pseudomonas putida* WCS358 in natural environments, *Appl. Environ. Microbiol.*, 60, 1184, 1994.

112. Chiarini, L., Tabacchioni, S., and Bevivino, A., Interactions between rhizosphere microorganisms under iron limitation, *Arch. Microbiol.*, 160, 68, 1993.

113. Plessner, O., Klapatch, T., and Guerinot, M.L., Siderophore utilization by *Bradyrhizobium japonicum*, *Appl. Environ. Microbiol.*, 59, 1688, 1993.

114. Omar, S.A. and Abd-Alla, M.H., Biocontrol of fungal root rot diseases of crop plants by the use of rhizobia and bradyrhizobia, *Folia Microbiol.*, 43, 431, 1998.

115. De Brito Alvarez, M.A., Gagne, S., and Antoun, H., Effect of compost on rhizosphere microflora of the tomato and on the incidence of plant growth-promoting rhizobacteria, *Appl. Environ. Microbiol.*, 61, 194, 1995.

116. Cook, R.J. et al., Molecular mechanisms of defense by rhizobacteria against root disease, *Proc. Natl. Acad. Sci.*, 92, 4197, 1995.

117. Laville, J. et al., Characterization of the hcnABC gene cluster encoding hydrogen cyanide synthase and anaerobic regulation by ANR in the strictly aerobic biocontrol agent *Pseudomonas fluorescens* CHAO, *Phytopathology*, 180, 3187, 1998.

118. Maurhofer, M. et al., Induction of systemic resistance of tobacco necrosis virus by the root colonizing *Pseudomonas fluorescens* strain CHAO: influence of the gacA gene and of pyorverdine production, *Phytopathology*, 84, 139, 1994.

119. Loper, J.E., Contributions of molecular biology towards understanding mechanisms by which rhizo-sphere pseudomonads effect biological control, in *Improving Plant Productivity with Rhizosphere Bacteria, 3rd Int. Workshop on Plant Growth Promoting Rhizobacteria*, Ryder, M.H., Stephens, P.M., and Bowen, G.D., Eds., CSIRO Publications, East Melbourne, Australia, 1994, p. 89.

120. Costa, J.M. and Loper, J.E., Characterization of siderophore production by the biological control agent *Enterobacter cloacae*, *Mol. Plant Microbe Interact.*, 7, 440, 1994.

121. Oedjijono, M.A.L. and Dragar, C., Isolation of bacteria antagonistic to a range of plant pathogenic fungi, *Soil Biol. Biochem.*, 25, 247, 1993.

122. Leeman, M. et al., Iron availability affects induction of systemic resistance to fusarium wilt of radish by *Pseudomonas fluorescens*, *Phytopathology*, 86, 149, 1996.

123. Mulya, K. et al., Suppression of bacterial wilt disease in tomato by root-dipping with *Pseudomonas fluorescens* PfG32: the role of antibiotic substances and siderophore production, *Ann. Phytopath. Soc. Jpn.*, 62, 134, 1996.

124. Raaijmakers, J. and Weller, D.M., Natural plant protection by 2,4-diacetylphloroglucinol producing *Pseudomonas* spp. in take-all decline soils, *Mol. Plant Microbe Interact.*, 11, 144, 1988.

125. Boyer, G.L., Iron uptake and siderophore formation in the actinorhizal symbiont *Frankia*, in *Biochemistry of Metal Micronutrients in the Rhizosphere*, Manthey, J.A., Crowley, D.E., and Luster, D.G., Eds., Lewis Publishers, Boca Raton, FL, 1994.

126. Fabiano, E. et al., Extent of high-affinity iron transport systems in field isolates of rhizobia, *Plant Soil*, 164, 177, 1994.

127. Neilands, J.B., Overview of bacterial iron transport and siderophore systems in rhizobia, in *Iron Chelation in Plants and Soil Microorganisms*, Barton, L.L. and Hemming, B.C., Eds., Academic Press, London, 1993, p. 179.

128. Wittenberg, J.B. et al., Siderophore-bound iron in the peribacterial space of soybean root nodules, *Plant Soil*, 178, 161, 1996.

129. Persmark, M. et al., Isolation and structure of rhizobactin 1021, a siderophore from the alfalfa symbiont *Rhizobium meliloti* 1021, *J. Am. Chem. Soc.*, 115, 3950, 1993.

130. Duhan, J.S. and Dudeja, S.S., Effect of exogenous iron, synthetic chelator and rhizobial siderophores on iron acquisition by pigeonpea host in pigeonpea-Rhizobium symbiosis, *Microbiol. Res.*, 153, 37, 1998.

131. Duhan, J.S., Dudeja, S.S., and Khurana, A.L., Siderophore production in relation to N_2 fixation and iron uptake in pigeon pea-*Rhizobium* symbiosis, *Folia Microbiol.*, 43, 421, 1998.

132. Fabiano, E. et al., Siderophore-mediated iron acquisition mutants in *Rhizobium meliloti* 242 and its effect on the nodulation kinetic of alfalfa nodules, *Symbiosis*, 19, 197, 1995.

133. Barton, L.L. et al., Siderophore-mediated iron metabolism in growth and nitrogen fixation by alfalfa nodulated with *Rhizobium meliloti*, *J. Plant Nutr.*, 19, 1201, 1996.

134. Jadhav, R.S. and Anjana, D., Effect of nutritional and environmental conditions on siderophore production by cowpea *Rhizobium* GN1 (peanut isolate), *Indian J. Exp. Biol.*, 34, 436, 1996.

135. Dhul, M., Suneja, S., and Dadarwal, K.R., Role of siderophores in chickpea (*Cicer arietinum* L)-*Rhizobium* symbiosis, *Microbiol. Res.*, 153, 47, 1998.

136. Jadhav, R.S. and Desai, A., Role of siderophore in iron uptake in cowpea *Rhizobium* GN1 (peanut isolate): possible involvement of iron repressible outer membrane proteins, *FEMS Microbiol. Lett.*, 115, 185, 1994.

137. Marek-Kozaczuk, M., Deryto, M., and Skorupska, A., Tn5 insertion mutants of *Pseudomonas* sp 267 defective siderophore production and their effect on clover (*Trifolium pratense*) nodulated with *Rhizobium leguminosarum* bv trifolii, *Plant Soil*, 179, 269, 1996.

138. Bossier, P. and Verstraete, W., Detection of siderophores in soil by a direct bioassay, *Soil Biol. Biochem.*, 18, 481, 1986.

139. Lindsay, W.L., *Chemical Equilibria in Soils*, Blackburn Press, Caldwell, NJ, 2001, p. 449.

140. Martell, A.E., Hancock, R.D., and Motekaitis, R.J., Factors affecting stabilities of chelate, macrocyclic and macrobicyclic complexes in solution, *Coord. Chem. Rev.*, 133, 39, 1994.

141. Murakami, T. et al., Stabilities of metal-complexes of mugineic acids and their specific affinities for iron(III), *Chem. Lett.*, 12, 2137, 1989.

8 Mycorrhizal Fungi: A Fungal Community at the Interface between Soil and Roots

Francis M. Martin, Silvia Perotto, and Paola Bonfante

CONTENTS

I. INTRODUCTION

The rhizosphere is a dynamic environment in which bacteria, viruses, fungi, and microfauna, including arthropods and nematodes, develop, interact with each other, and take advantage of the organic matter released by the root [1]. A substantial consequence of this richness, in comparison with the bulk soil, is an intense microbial activity with feedback effects on root development and the growth of the whole plant. The diversity of microorganisms resident in the rhizosphere and the complexity of interactions occurring among them and with the plant have been the subject of several studies that are summarized in some recent reviews [2–5].

Mycorrhizal symbiosis, a mutualistic plant–fungus association, is an essential feature of the biology and ecology of most terrestrial plants; thanks to nutrient exchanges, the plant receives mineral nutrients and improves its vegetative growth, whereas the fungus obtains carbohydrates and accomplishes its life cycle [6]. The concept that many events essential for a successful infection occur in the rhizosphere is a more recent acquisition [1], and detailed investigations of the occurrence and role of mycorrhizal fungi in the rhizosphere began more recently [5]. Leake et al. [7] stated that the extraradical mycelia of mycorrhizal fungi represent a network of power and influence, because they control biogeochemical cycling, plant community composition, and agroecosystem functioning. Mycorrhizal fungi reside in the rhizosphere as spores, hyphae, and propagules, and occupy the rhizoplane (the root surface) during their interaction with the root [8]. The more specific terms, *mycorrhizosphere* and *mycorrhizoplane* — that is, the surroundings and the surface of mycorrhizal roots [9] — are accompanied by the *hyphosphere*, namely the region not directly influenced by the root where mycorrhizal hyphae and soil particles interact. As significant components of the rhizosphere populations, mycorrhizal fungi interact with other microorganisms producing beneficial effects on plant nutrition and health, and on soil stability.

Mycorrhizal research is currently part of mainstream biology, thanks to DNA technologies and genomics, which provide us with new tools to discover symbiont diversity and to reveal the contribution of symbiotic partners to ecosystem functioning. Another leading concept developed during this last decade and pertinent to the rhizosphere is the evolutionary and molecular similarities existing between the legumes/rhizobia symbiosis and the symbiosis of legumes with a specific group of mycorrhizal fungi, the arbuscular mycorrhizal (AM) fungi. As a consequence, the recent years have witnessed an extensive blooming of literature on the molecular, cellular, and physiological aspects of plant–fungus communication in mycorrhizal roots in general, [10–13] and, more specifically, in arbuscular mycorrhizas [14–18].

This chapter offers a synoptic overview of the literature on the interactions between mycorrhizal fungi and their rhizospheric environment. Here, we first discuss recent advances in molecular tools used to estimate the diversity of mycorrhizal species. We then describe the roles of mycorrhizal fungi in the establishment of a bridge between the soil and the root, their crucial role in the acquisition and assimilation of nutrients and water, metal detoxification, stabilization of the soil, and colonization of neighboring plant roots. Finally, the major insights derived from cellular, biochemical, and molecular studies of mycorrhiza development at the soil–root interface will be summarized and the gaps in our current knowledge will be highlighted.

II. MYCORRHIZAL FUNGI ARE GENETICALLY DIVERSE

Mycorrhizal fungi are major components of the microbial soil community, mediating soil-to-plant transfer of nutrients. They are a heterogeneous group of soil fungi, which colonize the roots of about 240,000 plant species in nearly all terrestrial ecosystems. Despite their impressive genetic diversity, all mycorrhizal fungi complete their life cycle in close association with the roots through the establishment of a symbiosis. The taxonomic position of the plant and fungal partners defines the type of mycorrhiza [19], each association being distinguished by specific anatomical and physiological features (Table 8.1). They are divided into two main categories: endomycorrhiza (arbuscular, ericoid, and orchid mycorrhiza) and ectomycorrhiza. Because mycorrhizal plants occur in a wide range of habitats and soil conditions, a high degree of diversity should be expected in the genetic and physiological abilities of the fungal endophytes [1]. In addition to the Glomeromycota, the new taxon proposed to classify AM fungi [20], thousands of Asco- and Basidiomycotina species have been recorded as being mycorrhizal. Determination of their inter- and intraspecific genetic diversity has now been made possible by the extensive use of molecular techniques. Molecular techniques developed for these symbiotic species provide new tools for answering a

TABLE 8.1
Characteristics of Mycorrhizae Symbiont Range

Mycorrhizae	Host Plants	Fungal Symbionts	Fungal Structures
Ectomycorrhizas	Many trees and shrubs, especially of temperate regions	Many fungi, including species from 25 families of Basidiomycotina, 7 families of Ascomycotina, and 1 genus of Zygomycotina (Endogone)	Bundles, fungal mantle, Hartig net
Arbuscular mycorrhizas	Many plant species, including representatives of bryophytes, gymnosperms, and many angiosperms	Glomeromycota	Appressoria, inter- and intracellular hyphae, coils, arbuscules, vesicles, spores
Orchid mycorrhizas	All members of the Orchidaceae	Many isolates from sterile mycelia referable to form genus *Rhizoctonia*, induced to form sexual stages, referable to about 8 genera of Basidiomycotina including some pathogens	Coils
Ericoid mycorrhizas	Members of the Ericales with fine hair roots, especially Ericoideae, Vaccinoideae, Rhododendroideae, Epacridaceae, and Empetraceae	*Hymenoscyphus* isolates, *Oidiodendrum griseum* sterile isolate	Coils
Arbutoid mycorrhizas	Members of the Ericales with sturdier roots including *Arbutus*, *Arctostaphylos*, and Pyrolaceae	Ectomycorrhizal fungi on other types of plants	Fungal mantle, Hartig net, and coils
Monotropoid mycorrhizas	Achlorophyllous members of Ericales such as *Monotropa*, *Sarcodes*, *Plerospora*	Ectomycorrhizal fungi on other types of plants	Fungal mantle, penetration peg

range of questions, from their spatiotemporal dynamics in ecosystems to the mechanisms of phenotypic plasticity.

In most ecosystems, roots are exposed to several mycorrhizal fungal species, each represented by a large population whose individuals almost invariably display some genetic diversity [21]. This diversity has significant ecological consequences. Individual fungal populations vary in their potential range of host species, ability to colonize different host genotypes and promote plant growth, and adaptation to abiotic factors (e.g., soil pH, toxic levels of heavy metals, and nutrient shortage) that are likely to affect both the establishment and progress of a beneficial symbiosis and their dissemination in the ecosystem. The physiological status of the root is highly dependent on the creation and efficient functioning of the symbiosis. This, in turn, has a clear impact on the rhizospheric environment and the microorganisms involved through the secretion of carbohydrates, amino acids, secondary metabolites, and various ions.

The origin and maintenance of genetic diversity in fungal communities and populations critically determine their temporal and spatial distribution during the development of an ecosystem; understanding of these issues will be needed to provide a mechanistic basis for these biological processes taking place in the soil–rhizosphere environment.

Over the past decade, a variety of polymerase chain reaction (PCR)-based methods have been developed that allow direct surveys and descriptions of AM and ectomycorrhizal (ECM) fungal species in their native habitats [13,22–24]. PCR amplification of targeted genomic sequences followed by terminal restriction fragment length polymorphism (T-RFLP), allele-specific hybridization, direct sequencing, or single-strand conformation polymorphisms have been largely used to detect ECM [25,26], AM [27,28], ericoid [29–32], and orchid [33–36] fungi in natural ecosystems. PCR primers based on highly conserved regions of nuclear and mitochondrial ribosomal DNA have been designed [23,25,29,37,38] to amplify two polymorphic noncoding regions, namely the internal transcribed spacers (ITS) and the intergenic spacers (IGS). Ribosomal DNA sequencing alone has identified hundreds of mycorrhizal species. Existing public sequence databases are insufficient for ecological purposes. The development of sequence databases with environmental specifications will build knowledge of where, when, and under what conditions sequences of mycorrhizal fungi were retrieved. Such databases can provide mechanisms for data exchange within the research community and provide a data catalog that can be mined. The UNITE database is such an rDNA sequence database focused on ECM asco- and basidiomycetes from boreal forests [39]. The database currently holds 1100 ITS sequences from 630 species. The sequences are generated from fruit bodies, collected and identified by specialists, and deposited in public herbaria.

Microsatellite analysis, microsatellite-primed PCR, RAPD, and repeated DNA probes are highly efficient approaches for identification of the genotypes of mycorrhizal fungi and have been employed to determine the genetic structure of their populations and assess gene flux between introduced and indigenous strains [40–43]. DNA microarrays and DNA bar coding have considerable potential for the high-throughput identification of many mycorrhizal species at the level of population and communities [113]. Metagenomics, based on production-scale genome sequencing and bioinformatics analysis, will take this direct observation of native mycorrhizal assemblages to the next level of complexity.

The application of these molecular methods has provided detailed insights into the complexity of ECM fungal communities and offers exciting prospects to elucidate processes that structure ECM fungal communities [13,22,24,44]. They will improve our understanding of plant ecology, such as plant interactions and ecosystem processes. About 50 such ECM community studies have been published over the past 5 years. They have shown the following:

1. Any single mycorrhiza can potentially be identified to species either by PCR-RFLP of the nuclear ribosomal DNA internal transcribed spacers or by DNA sequencing.
2. Sporocarp production is unlikely to reflect belowground symbiont communities. Not all ectomycorrhizal fungi produce conspicuous epigeous sporocarps and of those fungi that do produce conspicuous sporocarps, a species' sporocarp production does not necessarily reflect its belowground abundance.
3. A few fungal ECM taxa account for most of the mycorrhizal abundance and are widely spread, whereas the majority of species are only rarely encountered.
4. The spatial variation of ECM fungi is very high and most species show aggregated distributions. As stressed by Dahlberg [44], little is known about the relative importance of vegetative spread and longevity of genotypes vs. novel colonization from meiospores for any ECM fungal species: this deserves attention if we want to understand the dynamics and structure of ECM communities/populations [13,43].

Similar advances are being made in our understanding of AM communities. Morphological and molecular methods to identify spores demonstrated that a single patch of habitat may support 30 to 40 AM fungal species [45] and a considerable diversity can be detected in a single root system [27]. This high diversity suggests that in natural conditions selectivity is the norm and that AM fungi show a functional specificity. Some of them might be more effective at P transport, whereas others at pathogen defense or at drought resistance [45]. These features (high number of distinct fungal types and selectivity) suggest that AM fungal taxa are more than the about 150 to 200 fungal species which have been described so far (http://www.tu-darmstadt.de/fb/bio/bot/schuessler/amphylo/amphylogeny.html). However, the diversity revealed by molecular analysis does not map onto the conventional morphological taxonomy [45]. Differently from ECMs where systematics are better grounded, AM taxonomy strongly suffers from our ignorance on the genetics of these apparently haploid and asexual organisms. Notwithstanding their clonal reproduction, they possess a high genetic heterogeneity and such a variability is shown by variants in the sequences of ribosomal and constitutive genes, or in the patterns of amplified fragments from the same spore and from spores of the same isolate, respectively [21,23,46–48]. This is probably guaranteed by the presence of genetically different nuclei in a single AM spore. According to this view, AM fungi can be seen as multigenomic organisms, and the variation in rDNA sequences within individuals may have relevant consequences for understanding their evolutionary ecology and functioning [49].

The use of molecular methods has also deeply modified our view on the specificity of mycorrhizal fungi toward their host plants. Ericoid mycorrhizas, originally thought to be a very specific association between the single fungal species *Rhizoscyphus* (*Hymenoscyphus*) *ericae* and a single plant tribe (Ericoideae), offer an excellent example: on one hand, mycorrhizal plants have been found to interact with a wide range of fungal symbionts, each colonizing a different part of the root system and creating a mosaic of populations [29,32,50]. On the other hand, the typical ericoid mycorrhizal fungus *R. ericae* was found to colonize the rhizoids of the leafy liverwort *Cephaloziella exiliflora* [51], thus confirming previous observations by Duckett and Read [52]. Other observations indicate that the host range of ericoid mycorrhizal fungi may be extended also to ECM plants in nature [53–55]. Another fascinating example of how molecular tools have provided new cues to understand plant ecology is the identification of the endomycorrhizal symbionts of forest orchids. *Rhizoctonia* spp. were reported until recently as the dominant mycorrhizal symbionts of orchids, but direct amplification of fungal DNA from mycorrhizal roots of achlorophyllous and green forest orchids demonstrated that the main symbionts are unculturable fungi that belong to known ECM taxa [33–36]. The ecological implications of these mycorrhizal connections is described in more detail in Section V.D.

DNA tools are essential for our understanding of the distribution and interactions of mycorrhiza fungi in space and time, and for predicting how populations and communities respond to changes in their environment. At the same time, molecular biologists are making important advances in understanding the molecular and cellular processes required for symbiosis development and functioning. Understanding the interplay between these molecular mechanisms with ecosystem biology is now crucial. In the following text, we highlight some of the recent studies in ecologically relevant traits.

III. SIGNALING MOLECULES AT THE PLANT–FUNGUS INTERFACE

A. RHIZOSPHERIC SIGNALS

Signaling molecules produced early in the interaction between mycorrhizal fungi and their host plants elicit discrete responses in the partners as the initial step in the cascade of events leading to contact at the host surface and eventual symbiosis [56,57]. Interplay of signals probably coordinates and organizes the responses of the symbiotic cells and modulates their respective biochemical and molecular differentiation. Identifying the potential signaling molecules, which are active in the rhizosphere and regulating the information flow between mycorrhizal fungi and host roots is currently an area of intense research [12,15,58]. Fungal spore germination, chemoattraction of the hyphae by the root cells, adhesion to root surface, root penetration, and development of symbiotic structures in the root probably depend on precisely tuned host-derived signaling molecules, whereas molecules secreted by the hyphae (e.g., phytohormones) generate drastic morphological and molecular changes in the host roots.

At least as far as concerns AM fungi, little is known about the molecular mechanisms that govern signaling and recognition between these symbiotic fungi and their hosts. It is well known that root exudates cause a specific phenotype in AM fungi, that usually show an extensive branching in the vicinity of host roots, prior to the formation of the appressorium [59,60]. Hyperbranching can be also

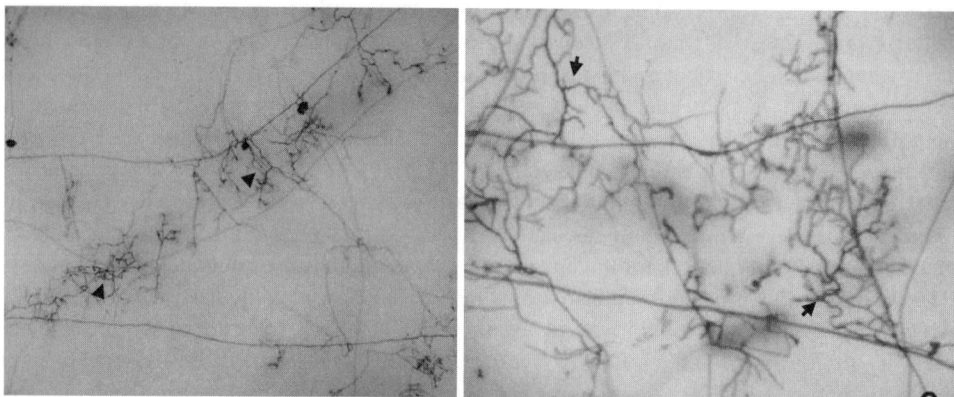

FIGURE 8.1 Mycelium of *Gigaspora margarita* growing in the presence of the host plant but physically separated from the root by a nitrocellulose membrane with pore diameter of 45 *μm*. The hyphal branching (arrows) demonstrate that the fungus has perceived the root exudates.

observed when the partners are physically separated: in this case, the fungus perceives the plant presence but symbiotic structures are not developed (Figure 8.1). By using this phenotype as a bioassay, a "branching factor" has been recently isolated from the roots exudates of *Lotus japonicus* and identified as a strigolactone, 5-deoxy-strigol [58]. Interestingly, this sesquiterpene, which was first characterized in the parasitic weeds *Striga* and *Orobanche*, induces an extensive branching in *Gigaspora* at very low concentrations (Figure 8.2). The branching factor leads to the activation of specific fungal genes, like those involved in respiration and oxidative burst [61,62]. It would be very interesting to investigate whether *Lotus* mutants affected in their symbiotic capabilities (see Section IV.B) produce these active compounds. The AM fungal partners, on the other hand, may release diffusible molecules (described as a potential Myc factor) that are perceived by host roots in the absence of direct physical contact. Using a novel *in vitro* culture technique Kosuta et al. [63] demonstrated that the AM fungal factor released by *Gigaspora rosea*, that may represent one or several molecules, induces the activation of the nodulation-inducible gene *ENOD 11* gene in transformed roots of *Medicago truncatula*. All these analogies between rhizobium/legumes and AM symbioses [64] have led to the hypothesis that Myc factors akin to Nod factors may be produced by AM fungi [65].

B. CHEMODIFFERENTIATION AT THE ROOT SURFACE

During their initial contact with the host surface, ECM and AM fungal hyphae undergo a morphogenetic switch and produce repeated apical branching that result in a labyrinthine growth form on

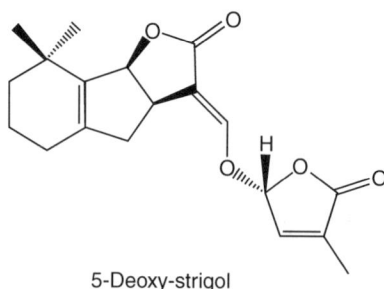

5-Deoxy-strigol

FIGURE 8.2 Chemical structure of the branching factor isolated from *Lotus japonicus*. (Akiyama, K. et al., *Nature*, 435, 824, 2005.)

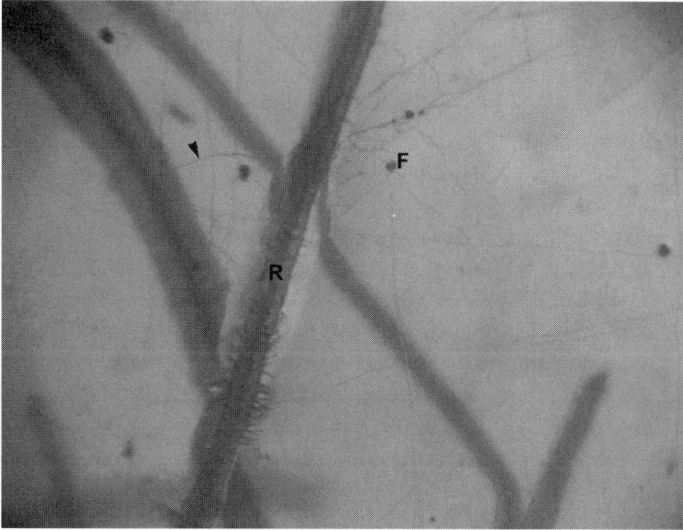

FIGURE 8.3 Extraradical mycelium produced by *Gigaspora margarita*, an arbuscular mycorrhizal fungus, in the presence of a root. A network of larger (arrows) and thinner hyphae are formed as well as groups of auxiliary cells (F) (× 100).

this surface (Figure 8.3) [66,59]. A concentration gradient of signaling molecules diffusing through the rhizosphere probably drives the tip toward the root surface and only induces this switch at high concentrations on the surface itself [56,57]. Hyperbranching ensures intimate contact with the root surface and appears to be a prerequisite for subsequent penetration and further differentiation of the specialized symbiotic structures (e.g., AM appressorium [Figure 8.4], ECM mantle). Whereas

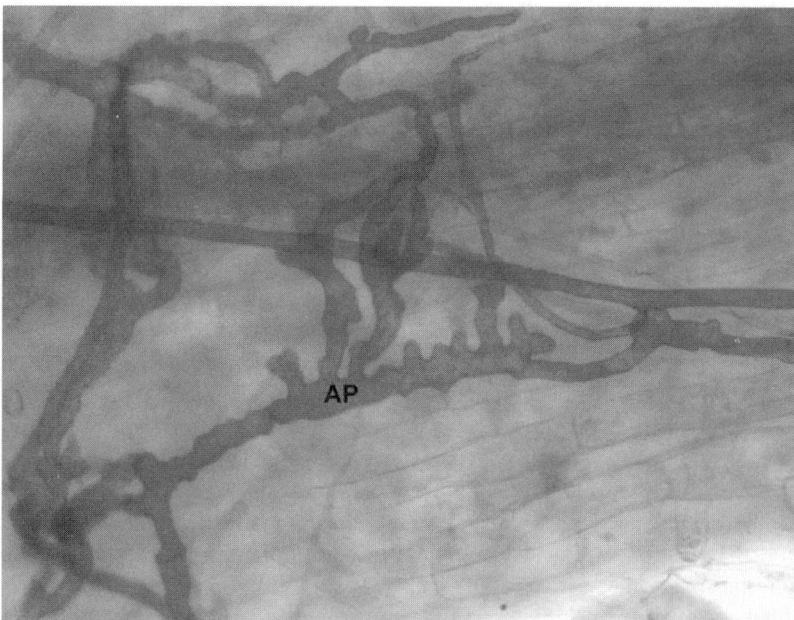

FIGURE 8.4 When the fungus *Gigaspora margarita*, makes contact with the root surface, a specialized swollen and branched structure is formed, called appressorium (AP) (× 200).

at least one group of active molecules has been identified in plants establishing AM symbiosis (Section III.A), identification of these signaling molecules remains a major challenge for ectomycorrhiza research. Flavonoids, such as quercetin, present in eucalyptus root exudates, induce rapid and striking changes in *Pisolithus microcarpus* hyphal morphology, leading to hyperbranching and an increased proportion of hydrophilic hyphae [67]. As growing hyphae of filamentous fungi contain a tip-high Ca^{2+} gradient thought to be vital for establishing and maintaining apical organization, morphogenesis, and growth [68], it is tempting to speculate that the signaling molecules involved in this hyperbranching alter the Ca^{2+} homeostasis by regulating or interacting with ion transport systems (e.g., Ca^{2+}-dependent ATPases) and transduction pathways. This contention is supported by the observation that reduced levels of calcineurin, a Ca^{2+}/calmodulin-regulated serine/threonine phosphoprotein phosphatase, cause growth arrest preceded by hyperbranching in *Neurospora crassa* [69]. This hyperbranching may be caused by alterations in the synthesis and rigidity of the cell wall, resulting in turgor pressure. Hyperbranching requires new membrane and cell wall synthesis; interestingly in the ECM truffle *Tuber borchii*, a secreted and surface-associated phospholipase A2 (TbSP1) was found to be strongly expressed on the branched mantle and Hartig net [70]. In addition to the low-molecular-weight diffusible molecules released by both plant roots and AM hyphae in the rhizosphere (see Section III.A), contact events between partners are also crucial for the signaling events. They are located in the rhizoplane, the root surface, and involve epidermal cells and extraradical hyphae.

The use of plant mutants impaired in AM symbiosis (Section IV.B) led to the identification of genetically defined steps in the development of the symbiotic interaction. Results obtained on *Lotus japonicus* have demonstrated that colonization is a multistep, genetically regulated process under the control of specific loci, that involves significant changes in the viability of root epidermal cells and in the cytoskeletal organization [71]. Molecular analyses performed on *Medicago truncatula* convincingly demonstrate that the plant perceives the fungal contact during appressoria formation, and differentially expresses genes closely involved in the signal transduction pattern [72]. On the other hand, cellular analyses performed by using *M. truncatula* plants transformed with different GFP constructs revealed that — prior to infection — the epidermal cells assemble a transient intracellular structure with a novel cytoskeletal organization [73]. All these findings demonstrated how the root surface is another "hot spot" for the generation of signaling events.

C. Fungal Phytohormones Are Produced in the Rhizosphere

Comparisons of root morphology and branching following auxin application or fungal colonization indicate that fungal auxins play a key role in ectomycorrhiza formation [74]. Further evidence in support of this view comes from the observation that overproduction of indole-3-acetic acid (IAA) by a fluoroindole-resistant mutant of *Hebeloma cylindrosporum* induces abnormal proliferation of the intercellular network of hyphae [24,75], whereas the stimulation of ethylene production by *P. microcarpus* during early ectomycorrhiza formation is presumably triggered by the production of IAA [76]. Auxins and ethylene would thus appear to be chemical signals that regulate fungal penetration and several anatomical features of an ectomycorrhiza. This morphogenetic role may imply targeted secretion of auxins and their conjugates, and local regulation of their metabolism. Other fungal indolic compounds, such as hypaphorine, a tryptophan betaine of *P. microcarpus* [77], are also involved in ECM development. Evidence of competitive antagonism between IAA and hypaphorine includes organ development, gene expression, or molecule–molecule interaction levels. This hypaphorine/IAA competition likely involves extracellular and intracellular signaling pathways. Hypaphorine with other active indole alkaloids should be regarded as a new class of IAA antagonists finely regulating specific steps of plant growth or development [78,79]. Apart from these, clear indications that fungal auxins and their derivatives modulate root morphogenesis during symbiosis development; however, their precise role in the control of the cascade of molecular events leading to the mature ectomycorrhiza is still uncertain [79].

In conclusion, there is increasing evidence that there is an intense signal interplay prior to physical interactions between partners [18]; plants perceive signals produced by their fungal symbionts, whereas root exudates affect fungal phenotype. The next task will be to characterize the chemical nature of these cross-talk factors, and to identify their role in the establishment of the symbiotic phase. We can hypothesize that more than a single molecule, such as strigol [58], controls the widespread colonization by AM fungi and their conserved behavior.

IV. MYCORRHIZAL FUNGI DEVELOP STRUCTURES OUTSIDE AND INSIDE THE ROOTS

According to their ability to colonize the root cells, mycorrhizal fungi are divided into two main categories: endomycorrhizal (arbuscular, ericoid, and orchid) and ECM fungi (Table 8.1).

Endomycorrhizal hyphae adopt a variety of colonization patterns in their penetration of the host root cells. Glomalean fungi are highly dependent on their host and cannot survive for long in its absence. Their hyphae form appressoria on the epidermal cells, penetrate the cortical tissue and, eventually, form highly branched structures called arbuscules (Figure 8.5 to Figure 8.7) [66].

Colonization of the Ericales takes place via simple hyphal structures whose organization is not deeply modified during the presymbiotic and the intraradical symbiotic phase. Hyphae of ericoid fungi may form loops and bundles outside the host plant, ill-defined appressoria and coils inside the epidermal root cells (Figure 8.8) [80]. Basidiomycetous fungi produce specialized coils (Figure 8.9) in the orchid cells during both the protocorm and the mature stages [81]. During the symbiotic phase, ectomycorrhizal fungi form a mantle sheath that surrounds the root, and progress into the apoplastic space of the rhizodermic (Angiosperms) and cortical cells (Gymnosperms), producing the Hartig net, an intercellular hyphal network inside the root tissues (Figure 8.10, Figure 8.11).

A gallery of these mycorrhizal types is available in many Web sites (http://mycorrhiza.ag.utk.edu/mimag.htm; http://www.ffp.csiro.au/research/mycorrhiza/; http://invam.caf.wvu.edu/), whereas updated descriptions are found in Reference 82 and Reference 83. A detailed structural definition of mycorrhiza is out of the main aim of this review chapter, and major attention will be given to the interface involved in the nutrient exchange between the symbionts, as well as to the molecular and cellular bases of the symbiosis development.

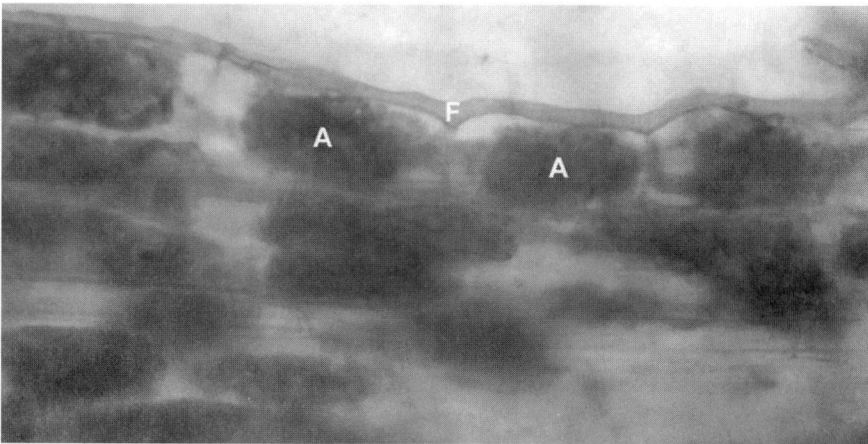

FIGURE 8.5 When *Gigaspora margarita*, an AM fungus (F), penetrates the cortical roots cells, highly branched, intracellular structures are produced — the arbuscules (A) ($\times$ 200).

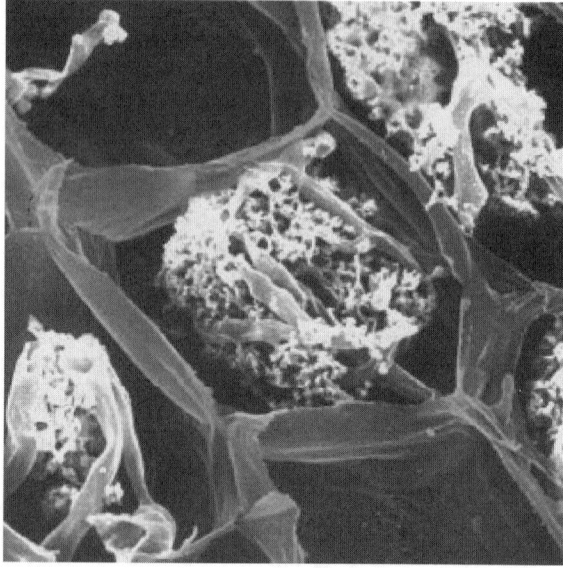

FIGURE 8.6 Scanning electron micrography of a cortical cell of *Ginkgo biloba* colonized by a *Glomus* strain. The symbiont produces a conspicuous arbuscule (× 1500).

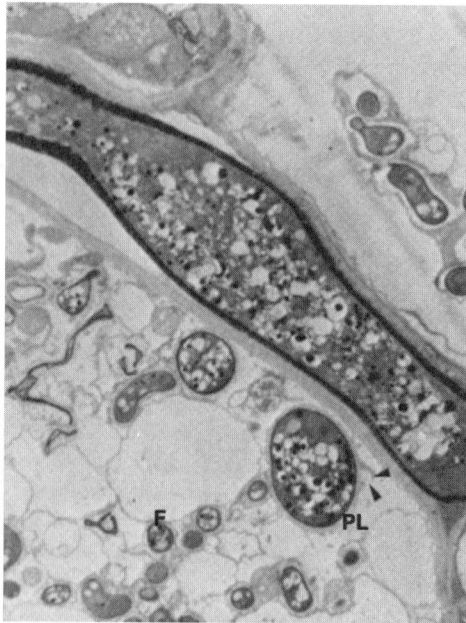

FIGURE 8.7 Transmission electron micrography of a cortical cell of *Ginkgo biloba* colonized by a *Glomus* strain. A large intracellular hypha is seen surrounded by smaller branches (F). Each hypha is surrounded by the invaginated host membrane (arrows) (× 5800).

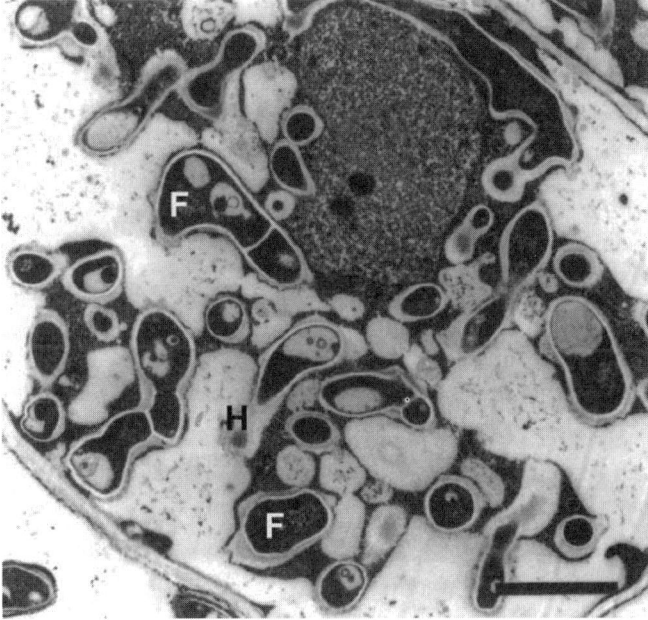

FIGURE 8.8 Hair root of *Calluna vulgaris* colonized by an ericoid mycorrhizal strain. The ascomycetous fungus is a dark sterile mycelium and produces an intercellular coil, which is surrounded by the host membrane ($\times$ 15000).

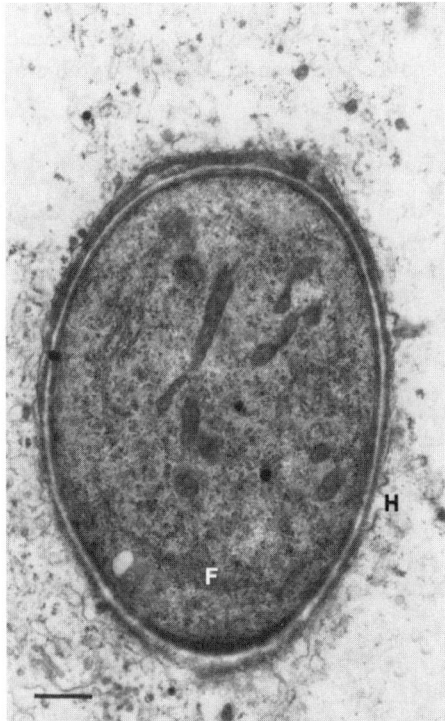

FIGURE 8.9 Detail of an orchid root cell colonized by an orchid symbiont. The basidiomycetous fungus (F) has a thick wall and is surrounded by the host (H) membrane ($\times$ 21000).

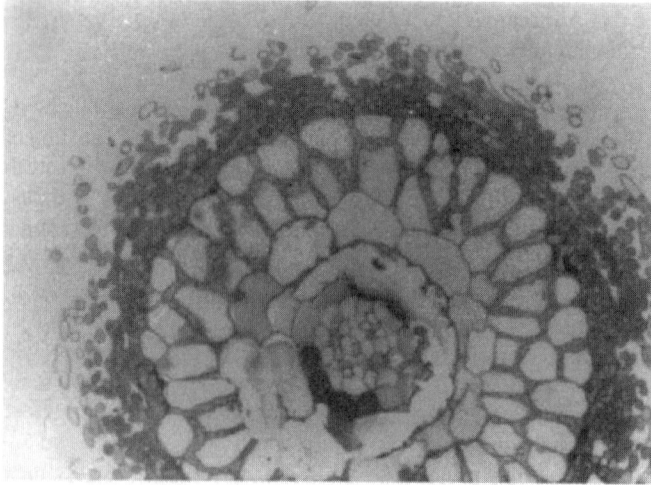

FIGURE 8.10 Ectomycorrhizal root of *Quercus suber.* The fungus produces a well-developed mantle and a Hartig net involving the outer root layers.

Irrespective of their taxonomy, mycorrhizal fungi can be described as a living interface located between the plant and its soil environment. In their extraradical phase, they enlarge the nutrient depletion zone around the root, increasing therefore the plant–soil nutrients interface [84]. During their intraradical growth, mycorrhizal fungi develop an extended contact area with the root cell, which changes structurally depending on the intercellular or intracellular location of the fungus.

We can therefore identify two interfaces, or exchange surfaces, at both ends of the fungal biomass; an outer interface, between extraradical hyphae and soil, and an inner interface, between intraradical fungal structures and the host plant cells. These two biological surfaces have profound

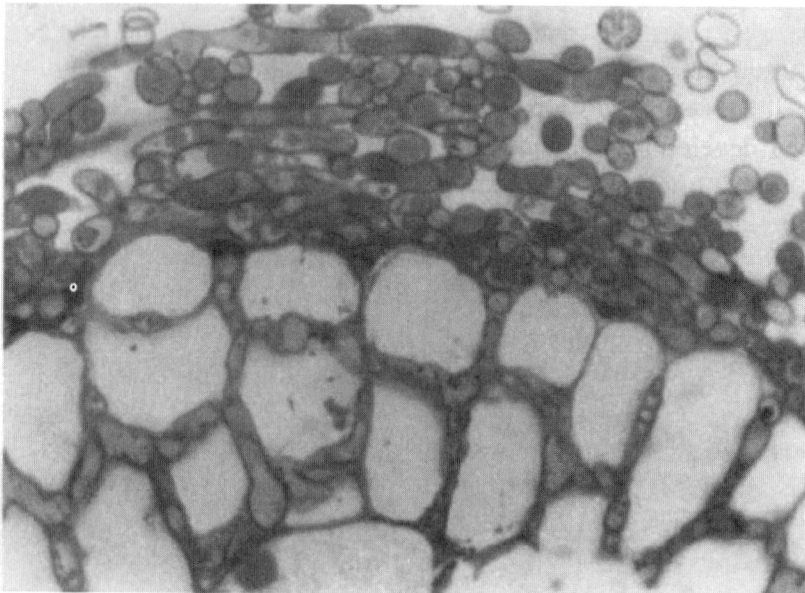

FIGURE 8.11 Ectomycorrhizal root of *Pisolithus tinctorius.* The fungus produces a well-developed mantle and a Hartig net involving the outer root layers.

morphological and functional differences. On the one hand, the role of external hyphae is to explore the neighboring soil through maximal extension, actively acquiring nutrients and water from the environment for feeding both the fungus and the plant (Section V). This interface has been described as a continuously extending surface [85], with unidirectional transport capabilities (at least as far as AM symbiosis is concerned). The inner interface, by contrast, first requires the building up of a new compartment in contact with the host plant, and in most mycorrhizal associations, is the site of bidirectional exchange.

A. AT THE PLANT–FUNGUS INTERFACE: STRUCTURE, ROLE, AND BIOGENESIS OF A NEW COMPARTMENT

Early interactions between the cell walls of the host plant and the colonizing fungus and changes in their composition are essential morphogenetic events in the constitution of a functioning mycorrhiza. The result of such structural interactions is a specialized interface, which is essential for nutritional exchanges [66,86]. During endosymbiosis, irrespective of the engaged partners, the microorganism is engulfed by a plant-derived membrane, which is part of a complex developmental program leading to the intracellular accommodation of microbes by plants [14,87]. In AM, the new compartment is known as the "interfacial compartment" [66,86,88] and consists of the invaginated host membrane, of cell wall-like material, and of the fungal wall and plasma membrane. Cellular and molecular approaches have provided many insights into the structure, the function, and the biogenesis of this complex and ever-changing compartment.

Bonfante et al. [89] were among the first to use monoclonal antibodies and enzyme–gold complexes to investigate the nature of the molecules at the interface between the fungal wall and the host plasma membrane in AM roots. They first revealed pectins and cellulose (Figure 8.12), and many additional wall components, among which were glucans, hydroxyprolin-rich glycoproteins,

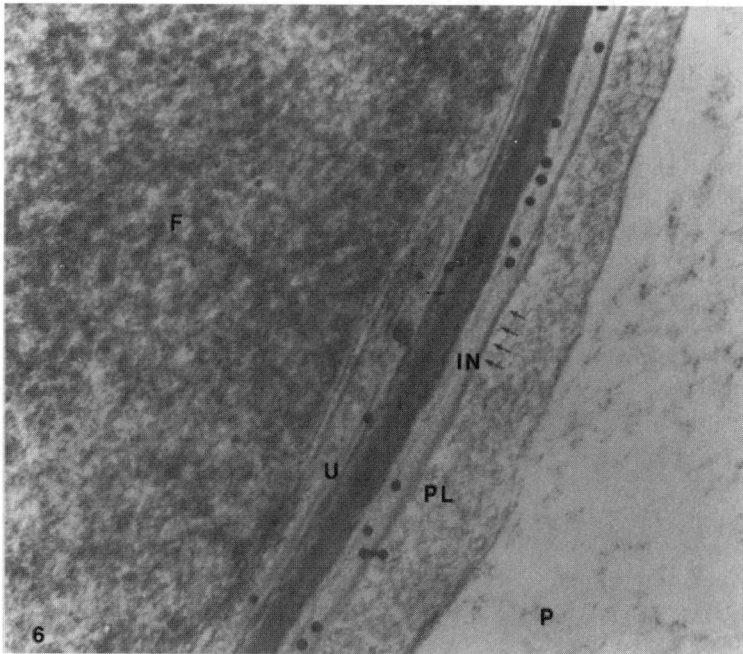

FIGURE 8.12 Detail of the interface space (IN) between the fungal wall (U) of *Glomus versiforme* and the host membrane of leek (PL), as seen using electron microscopy. Xyloglucan molecules are revealed by using a specific antibody and colloidal gold granules (arrow) (× 30000).

and more recently expansins [66,90]. Expansins, which increase plant cell wall extensibility *in vitro*, seem to be upregulated during mycorrhization [90] and could play a crucial role in the plant accommodation process by acting as wall loosening factors. These studies indicate that the AM interface is an apoplastic space of high molecular complexity, where the boundaries of the partners, however, cannot be easily defined.

The examination of other endomycorrhizal systems has demonstrated that their interface is morphologically similar, but different in composition. Cellulose and pectins are only present at the interface in orchids when the endophyte is collapsing [91] and absent in ericoid mycorrhizae [80]. It is clear, therefore, that invagination of the host plasma membrane following fungal penetration is a common biotrophic interaction, whereas the nature of the molecules forming the interface is closely dependent on the nutritional capabilities of the fungus and its relationship with the host plant.

The intraradical colonization patterns are a consequence of events taking place at the root surface. In one of the many studies of this subject, Gollotte et al. [92] showed that AM fungi fail to develop regular structures in pea mutants. The first sign of this failure was the deposition of callose on the surface of the epidermal cells. Similarly, in ECM early contacts between the partners are followed by alterations in cell wall ultrastructure and composition leading to the formation of a new interface [66,82,93]. Plant and fungal walls are always in direct contact, as shown in a detailed analysis of *Corylus avellana* and *Tuber magnatum* [94], whose map of the symbiont cell wall components also shows that the fungus causes subtle changes in the host walls. The results suggest that a cementing material embedding the mantle hyphae and obscuring the plant–fungal contacts partly corresponds to an upregulated fungal cell wall component.

Identification of the cell wall and surface proteins involved in fungal attachment and penetration is crucial toward an understanding of how hyphae establish early interactions with their host [95,96]. The protein composition of cell walls of *P. microcarpus* is strikingly altered by the symbiotic interaction [93]. Many 31- and 32 kDa, symbiosis-regulated acidic polypeptides (SRAPs), are found in this cellular compartment [97]. The central part of the SRAP sequence contains the Arg-Gly-Asp (RGD) motif, a cell adhesion sequence found in several extracellular matrix adhesins, for example, vitronectins and fibronectins. These ligands could interact with integrins, which mediate cell adhesion and signal transduction in mammal, yeast, and plant cells. Mycorrhiza development induces the accumulation of SRAP transcripts and the corresponding proteins thus gather in the cell wall when the ECM sheath is aggregating around the colonized roots and the infecting hyphae penetrate between the epidermal cells. Upregulation of the synthesis of cell wall proteins in ectomycorrhiza is not limited to SRAPs. Transcripts of three hydrophobins, another class of fungal-secreted cell wall proteins, similarly accumulate severalfold in *P. microcarpus* hyphae colonizing the *Eucalyptus* root surface [98].

Interestingly, enzymatic proteins may also be located at the fungal–plant interface. TbSP1 (Section III.B) is a secreted and surface-associated phospholipase A2 previously found to be upregulated in C- or N-deprived free-living mycelia from the ECM ascomycete *Tuber borchii* [99]. An *in vitro* symbiotic system between *Cistus incanus* and *T. borchii* allowed to demonstrate a substantially enhanced TbSP1 mRNA expression compared to nutrient-limited, free-living mycelia, and a similar expression trend was revealed by the immunolocalization experiments, which located TbSP1 on the fungal wall, mostly on the branched Hartig net hyphae [70].

Changes in cell wall protein composition may regulate the molecular architecture of protein networks in a manner that allows new developmental outcomes for both fungal cell adhesion and root colonization. Incompatibility between ECM hyphae and the host roots detected during the initial contacts is generally expressed in the form of polyphenol accumulation in host tissue and thickening of host cell walls abutting the incompatible isolate [100,101]. These complex interactions are evidently the result of early events in the mycorrhizoplane.

The development of refined molecular tools has led to new questions concerning the genesis of the interfacial material and the activation of genes involved in cell wall and membrane synthesis [88]. Deposition of cell wall material requires, in fact, the combined activities of both polysaccharide-synthase and lytic enzymes. Two xyloglucan endo-transglycosilases (XETs) genes have been isolated from *M. truncatula* [102], one being only expressed in mycorrhizal roots. The authors suggest that the gene product may be involved either in facilitating hyphae penetration by allowing localized cell wall loosening or in modifying the structure of xyloglucans in the interface compartment. Lytic and transglycosilation events during cell wall deposition may be facilitated by the interface pH that is becoming more acidic owing to the activity of an H^+-ATPase [103]. The transcript profile of *M. truncatula* roots during the AM symbiosis with *Glomus versiforme* has been the object of extensive investigations [104,105]. By using a cDNA arrays approach, a gene (*MtCel1*) induced specifically during the symbiosis was predicted to be involved in cell wall modifications. In mycorrhizal roots, *MtCel1* expression is associated specifically with cells that contain arbuscules and, considering the membrane domain, MtCel1 was suggested to be located in the periarbuscular membrane and involved in the assembly of the cellulose/hemicellulose matrix at the interface [104].

New techniques based on genetic transformation, which allow *in vivo* observations, will be very useful to study the building up of the interface compartment. By using transformed roots of *Medicago sativa* that express different GFP-conjugated proteins labeling specific intracellular structures (Section III.A), Genre et al. [73] demonstrate the presence of a so far undescribed prepenetration cell rearrangement in root epidermis. This complex structure involves cytoskeleton, membranes, and nucleus and may represent an early step of the interface buildup.

B. THE GENETIC BASIS OF FUNGAL COLONIZATION

Formation of functional mycorrhizas requires finely coordinated regulation of plant and fungal gene expression. Activation of specific genes is likely to be the consequence of perception of primary signals at the cell periphery and subsequent transduction of the signal into a cascade of events, eventually reaching the nucleus of the cell to be colonized [16]. Such hypothesis can be currently checked only for AMs, thanks to the availability of plant mutants unable to form symbiosis: these mutant lines provide, in fact, a powerful tool to identify genetically defined steps in the development of the symbiotic interaction. The genetic dissection of AM development has been pioneered by the isolation of pea mutants impaired in AM symbiosis. These mutants were initially identified through their altered root nodule symbiosis with rhizobium. Subsequently, it was found that a subset of the nodulation mutants was also affected in the AM symbiosis [106]. This finding demonstrated an overlap in the genetic programs for the two diverse symbiotic interactions [106–108]. Unfortunately, the isolation of the affected genes from pea is hampered by its large genome size. *Lotus japonicus* and *Medicago truncatula* represent more amenable legumes to isolate symbiotic mutants with the final objective of cloning and functionally characterizing the responsible genes. A significant discovery has been made with the identification of the so called "sym" genes, which are essential for the establishment of the microbial symbiosis [15]. Analysis of mutants defective in nodule and AM formation in legumes allowed the identification of a number of components of the signal perception and transduction pathway shared by both symbioses: (1) a leucine-rich-repeat receptor-like kinase (LRR-RLK); (2) the DMI1 protein, a putative cation channel; (3) DMI3, a calcium- and calmodulin-dependent protein kinase [15 and references there in, 109] and, (4) a putative K channel, surprisingly located in the plastids [110]. The corresponding genes are constitutively expressed in plant roots. Identification of such genes suggests a sequence of events leading to the establishment of both symbioses (Figure 8.13). In this model, LRR-RLK binds signals through its extracellular domain and transduces them to the intracellular kinase domain. The activity of LRR-RLK and DMI1 then leads to the so-called electrochemical prelude: membrane depolarization, calcium influx, and calcium spiking, consisting in prolonged oscillations of intracellular calcium concentration and indicating a very quick change in gene expression. The DMI3 protein acts downstream of the calcium spiking. Molecular events are mirrored by morphological changes, which indicate the presence of "check points" required for the formation of a functional AM. For example, LjSym4 [110] is required for the initiation or coordinated expression of the host plant

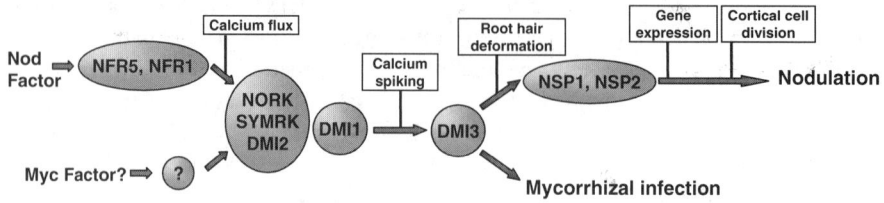

FIGURE 8.13 Sym genes involved in the Nod factor and mycorrhizal signaling pathways. This signaling pathway has been defined through genetics in the legumes *Lotus japonicus* and *Medicago sativa*. The genes identified are defined in round boxes, whereas physiological or morphological landmarks of the Nod factor signaling pathway are indicated in boxes. Mutations in all these genes have been characterized for calcium spiking except SYMRK. In addition, it has been shown that mutations in NFR5 and NFR1 lack the calcium flux, whereas mutations in DMI1 and DMI2 show the first phase of the flux response. Components of the Nod factor signaling pathway are conserved with mycorrhizal signaling. (Modified from Oldroyd, G. et al., *Plant Physiol.*, 137, 1205, 2005.)

cell's accommodation program and allows the passage of both microsymbionts through the epidermis layer [111,112].

C. ECTOMYCORRHIZA DEVELOPMENT: A COORDINATED EXPRESSION OF GENE NETWORKS

To gain a predictive understanding of the complex biological systems that evolve from mycorrhizal interactions, a surge of studies based on functional genomics (large-scale EST sequencing, cDNA array analysis of gene expression, proteomics) has allowed an assessment of the development and functioning of AM and ECM symbioses on a larger scale [104,105,113,114]. Global gene expression analyses [115–118] added new information to existing models of ECM development. Expression profiling showed that developmental reprogramming takes place in host roots and colonizing hyphae. A marked change in the gene expression in *Eucalyptus/Pisolithus* and *Betula/Paxillus* symbiotic tissues was observed at multiple levels: (1) a general activation of the fungal protein synthesis machinery and primary carbon metabolism probably supporting an intense cell division/proliferation, (2) the increased accumulation of transcripts coding for cell surface proteins in fungal hyphae (e.g., hydrophobins) probably involved in the mantle and symbiotic interface formation, and (3) the upregulation of defense reactions and hormone metabolism in colonized roots (Figure 8.14). Changes in transcript levels for symbiosis-regulated (SR)-genes found in cDNA array studies [117,118] were confirmed by digital northern [119] and northern RNA blotting [for example, see Reference 97 and Reference 98]. The fact that several of these cellular functions are regulated in both the *Betula/Paxillus and Eucalyptus/Pisolithus* symbioses suggests the induction of common genetic programs in various ECM systems. At the different developmental stages studied, development of *Eucalyptus/Pisolithus* and *Betula/Paxillus* symbioses does not induce the expression of ECM-specific genes [115,116,118]. The apparent lack of ECM-specific genes is striking and suggests that ontogenic and metabolic programs leading to the symbiosis development and functioning are driven by changes in the organization of gene networks (e.g., differential arrays of preexisting transcriptional factors or transduction pathways), rather than the specific expression of symbiosis-specific transcriptional factors or signaling components. A more complete analysis of this key question will await the completion of larger sets of ECM expression profiles on a wider range of associations using genome-wide microarrays [120,121].

Changes in morphology associated with mycorrhizal development are thus accompanied by changes in transcript patterns and these changes commenced at the time of contact between the two partners long before the formation of functional ECM. These gene profiling experiments have

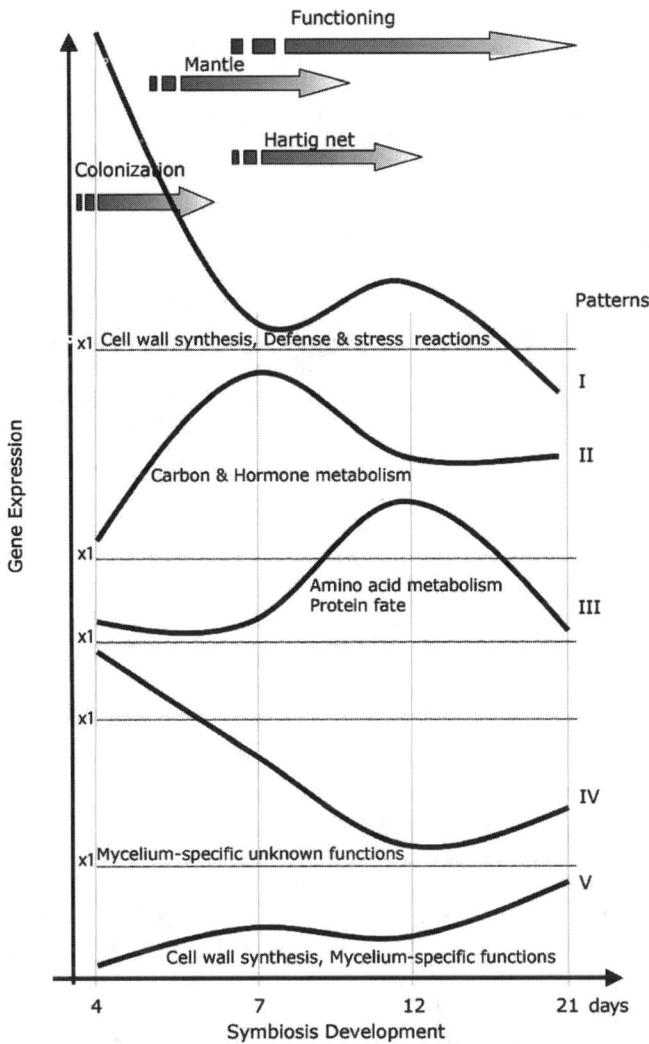

FIGURE 8.14 Schematic drawing describing the five major expression patterns of plant and fungal genes during the development of the *Eucalyptus*/*Pisolithus* mycorrhiza. (Adapted from Duplessis, S. et al., *New Phytologist*, 165, 599, 2005.)

stressed the importance of coordination between development of mycorrhiza and the differential gene expression in both partners. Understanding the synchronization of these events is essential for understanding the determinants of symbiosis compatibility and mycorrhiza ontogenesis. Further studies are now needed to investigate the expression of the identified SR-genes in environmental samples.

If there is a basic repertoire of fungal symbiotic genes, it can be accessed only by comparing whole genomes of saprobic (e.g., *Coprinus cinereus*, *Phanerochaete chrysosporium*) and pathogenic (e.g., *Magnaporthe grisea*, *Ustilago maydis*) species with mycorrhizal genomes. The availability of genome sequences from ecologically and taxonomically diverse fungi will not only allow ongoing research on those species, but will enhance the value of other sequences through comparative studies of gene evolution, genome structure, metabolic and regulatory pathways, and symbiosis/pathogenesis. One of the major strengths of rhizosphere studies for addressing these issues is that realistic ecological interactions can be investigated in a restricted micro- or mesocosm under environmentally controlled conditions with organisms whose genomes have been completely defined [121] or genetically modified. In the next few years, numerous fungal genomes are scheduled to be sequenced, owing largely to the fungal genome initiatives at the Broad Institute (Cambridge, MA, U.S. — http://www-genome.wi.mit.edu/annotation/fungi/fgi/) and the Joint Genome Institute (Walnut Creek, CA, U.S. — http://www.jgi.doe.gov/). The ongoing sequencing programes include well-studied models important to human health, plant pathogens, as well as mycorrhizal species (e.g., the ectomycorrhizal *Laccaria bicolor* and *Glomus intraradices*) [121].

Whole genome availability will certainly allow an in-depth analysis of rapidly evolving genes that may code for specific functions, such as symbiosis. Analysis of this wealth of information is certain to provide breakthroughs in understanding of the molecular and cellular mechanisms involved in the development and biochemical pathways in symbiotic partners. In addition, it will allow us to answer fundamental questions about whether parasitic and symbiotic habits evolved through gene acquisition and loss, or gene regulation. The promoter analysis of the current compendium of mycorrhiza-regulated genes will provide the basis for a more precise molecular dissection of the complex genetic networks that control symbiosis development and function. Determination of entire genome sequences, however, is only the first step in understanding the inner workings of an organism. The next critical step is to elucidate the functions of these sequences and give biochemical, physiological, and ecological meaning to this information.

V. THE ROLE OF MYCORRHIZAL FUNGI IN NUTRIENT CYCLING AT THE SOIL–ROOT INTERFACE

Despite the fact that mycorrhizal fungi play an important role in N, P, and C cycling in ecosystems in decomposing organic materials, the detailed function of fungi in nutrient dynamics *in situ* is still unknown. Mycorrhizal fungi differ in their functional abilities and the different mycorrhizas they establish thus offer distinct benefits to the host plant. Some fungi may be particularly effective in scavenging organic N and may associate with plants for which acquisition of N is crucial [122]; others may be more effective at P uptake and transport. An important goal is therefore to develop approaches by which the functional abilities of the symbiotic guilds are assessed in the field. In any case, it is necessary to measure microbial activities below the ground to assess what is really happening [123]. Analyses of ^{13}C and ^{15}N isotopic signatures have a significant potential to provide information in this area [124], although further investigation is required to understand the isotopic enrichment phenomenon [125]. Combined community/population structure and function studies applying genomics may, in the future, significantly promote our understanding of the interactions between mycorrhizal fungal species with their hosts, and with their biotic and abiotic environments [113]. A first step toward this type of environmental genomics is to explore fungal community-functioning under simulated forest conditions using microcosm systems [126]. In these systems, intact mycorrhizal root systems comprising individual species (e.g., abundant taxa likely to be functionally important) or natural communities can be manipulated and analyzed to determine, for example, C and N relations and host root–fungus metabolic activities that contribute to plant growth or plant community productivity. Using cDNA array profiling, Morel et al. [127] compared the levels of expression of approximately 1200 fungal genes in the ECM root tips and the connected extraradical mycelium (EM) for the *Paxillus involutus-Betula pendula* ectomycorrhizal association grown on peat in a microcosm system. Their results suggest that (1) there is a spatial difference in the patterns of fungal gene expression between ECM and EM, (2) urea and polyamine transporters could facilitate the translocation of nitrogen compounds within the EM network, and (3) changes in lipid metabolism may contribute to membrane remodeling during ectomycorrhiza formation.

Symbiotic roots provide a niche for mycorrhizal fungi. To bring about a symbiosis, the host plant must trade the fungus demand for carbon for respiration and growth, which is met primarily by glycolytic and anaplerotic processes requiring carbon sources, against its provision of extra nitrogen, phosphate, and minerals [128–130]. Hyphae prospecting the soil absorb nutrients by active metabolism and transport ions and assimilated metabolites to the host root via their strands and rhizomorphs. This mechanism is crucial for the absorption of nutrients that are poorly mobile, such as inorganic phosphate (Pi) and K^+, or bound to soil particulates (NH_4^+). Because ions rapidly absorbed by nonmycorrhizal plant roots become scarce in the rhizosphere, a zone of deficiency forms and the root's absorption rate mainly depends upon their diffusion rate rather than its own activity. Mycorrhizal hyphae counteract this deficiency, because nutrients translocate through the fungal cells to any sink, such as the root cells, more quickly than they diffuse in the soil [6]. This faster translocation rate is sufficient to explain the enhanced absorption rates of symbiotic roots. In exchange, the fungus receives their carbon compounds. These two-way flows of nutrients and other metabolites take place when physiologically active cells of both partners are in intimate contact [103].

A. Nitrogen Acquisition in Mycorrhizal Fungi

The morphological and physiological dissimilarities between mycorrhizal symbioses probably determine their success and their distinct patterns in different ecosystems [6]. Nitrogen (N) available to both AM and ECM plants should not be regarded a single pool open to free competition. Specialization of its acquisition and utilization in a given habitat is an important feature of plant and microbial community structure, whereas the fact that the ability to exploit its sources (and those of other limited nutrients) is not the same in all species may result in niche differentiation. If habitat specialization is a reflection of differences between mycorrhizal types, ECM and AM species could co-occur, because they exploit different niches in the same ecosystem.

In the forest ecosystems of Eurasia and North America, where ECM associations are dominant, there are often wide variations in the environmental concentration of N and its forms, and its limited availability to plants is due to N microbial transformation in soil [131]. The leaf litter produced by most tree species is relatively slow to decompose and thus forms a distinct layer of acidic, organically enriched material. Acidity, a high C:N ratio, seasons marked by low temperatures, and surface drying are major obstacles to nitrification and ammonification. Mineralization of N is so slow in many forests that available N becomes the main growth limiting factor [6,131,132]. Most trees have therefore elaborated mycorrhizal associations (Figure 8.15) and a wide range of alternative trophic adaptations (e.g., N_2-fixing symbioses, cluster roots) to be able to compete for the limited resources of specific nutrients [133]. The beneficial effect of AM colonization on N

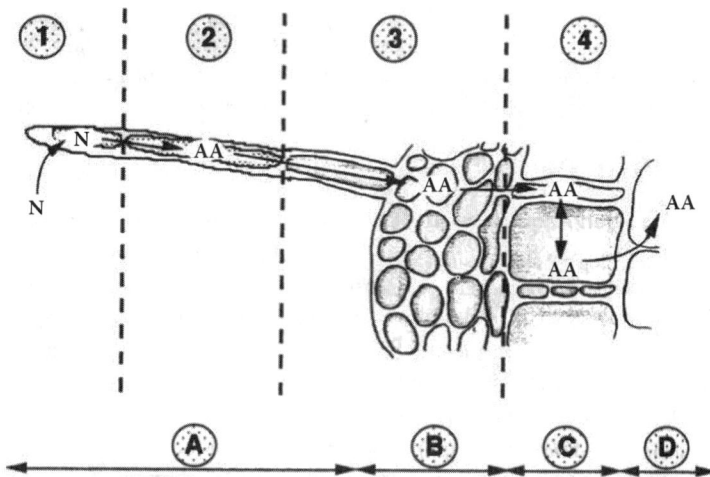

FIGURE 8.15 The different steps of nitrogen metabolism in the extraradical hyphae, ectomycorrhizal roots, and roots of the host plant.

acquisition has been overlooked [6]. Its significance in plant-N uptake in both agricultural and natural ecosystems, however, is now becoming increasingly clear [134,135].

1. Utilization of Organic N

The low temperatures and low soil pH that usually prevail at higher altitudes and latitudes (e.g., heathlands) restrain nitrification and (to a lesser extent) ammonification [136]. Studies of N relations in temperate and boreal ecosystems have demonstrated the importance of its organic forms for plant nutrition [136]. Some ECM and ericoid fungi use complex organic N, such as proteins, and their host plants have access to peptides and proteins. Soluble amino acids are also a substantial source for all types of mycorrhizal associations in these ecosystems [129,136].

The extensive extramatrical mycelium of the ECM fungi is ideally placed for nutrient acquisition in the top 10 cm of soil, where most of the local pools is sequestered in organic form [136]. A major contributing factor in N acquisition by ECM trees is the continuous growth of this mycelium into the patchily distributed soil resources, which it absorbs and transports for storage by the colonized roots. Degradation of organic N residues gives rise to free amino acids and (through microbial processes) to inorganic N forms, that is, NH_4^+, NO_3^-. ECM fungi contribute to N nutrition of the host by conversion of litter and complex soil N into forms more readily utilized by either the fungus or the host, and by absorption, assimilation, and translocation of inorganic N compounds from the soil to its roots. Most ECM fungi readily take up amino acids, such as glutamine, glutamate, and alanine, which predominate in soil solution, and peptides released by protein degradation [129,136]. This ability is retained in the symbiosis state and supplies the host with organic N [136]. Absorption of soil proteins requires their enzymatic degradation to peptides and then amino acids. Both ericoid and ECM fungi secrete a wide range of proteinases when grown on animal proteins (casein, gelatin, albumin) and protein fractions from beech forest litter as the substrate [129]. Marked inter- and intraspecific variations in the ability to use protein N probably express genetic differences between fungal strains and in the availability of host-derived carbon compounds. Competition for protein N between saprotrophic and ECM fungi in forest soils is presumably very tight, but symbiotic fungi are favored by the host's continuous provision of the carbon compounds needed to capture this N form through the synthesis of proteolytic enzymes.

AM fungi also increase decomposition and subsequent capture of inorganic N from organic materials [135]; they show therefore a kind of response which (for long times) has been considered characteristic of ECM fungi [7]. There is clearly a need to understand the mechanisms involved in such organic N mobilization by AM fungi and to detect the molecular basis of such events. A putative aminoacid permease, like those characterized in yeast, has been recently detected in *Glomus mosseae* [137]; the analysis of its expression, exclusively located in the extraradical hyphae and N dependent, might provide some insights into this still "hidden" capacity of AM fungi.

2. Uptake of NO_3^- and NH_4^+

AM and ECM mycelia are extremely active scavengers of inorganic forms of N, such as either NH_4^+ or NO_3^- [134,138]. Experiments performed on *Pisolithus* demonstrate that the fungus has access to NH_4^+ and Ca^{2+} ions trapped in between the vermiculite 2:1 layer. As vermiculite samples were separated from the mycelium by cellophane films, soluble fungal exudates may be considered responsible for phyllosilicate weathering. In the experiment, NH_4^+ ions, usually considered as retrograded, were mobilized from the interlayer spaces and replaced by Mg^{2+}, Al^{3+}, Ca^{2+}, and Na^+ ions [139].

NH_4^+ absorbed by mycelia, or derived from NO_3^- reduction, is rapidly assimilated into glutamate and glutamine, which are then used to synthesize other amino acids, such as alanine and γ-aminobutyrate, within the foraging hyphae. Next, assimilated N is either incorporated into mycelial proteins or translocated to the host, glutamine being regarded as the main translocation form [140]. NH_4^+ and NO_3^- are apparently assimilated a long way from the mycorrhizal roots when the hyphal web is permeating the rhizosphere and various soil horizons. In natural ecosystems, therefore, primary NH_4^+ assimilation is carried out by the fungus, then conversion to glutamine and its transfer to the host occur [129]. In addition, both AM and ECM fungi mediate N-transfer between plants through mycelial links [141]. In grassland and forest ecosystems, where plants are grown in very close association, these webs may be crucial in between-plant N cycling.

The molecular bases of NO_3^- and NH_4^+ uptake have been deeply investigated in many ECM [142] and AM fungi. Transporters and assimilating enzymes have been characterized in *Pisolithus, Laccaria*, and *Tuber*; in this last ECM, glutamine synthase represents one of the most expressed genes both during fruit body ripening [143] and N starvation [144]. The most detailed analysis has been carried out in the ectomycorrhizal *Hebeloma cylindrosporum* [142]. This fungus contains at least three ammonium transporters (Amts). The *AMT1* cDNA encodes a 477-amino-acid protein (50.9 kDa). Analysis of predicted sequences from cDNA showed 67.8 and 46.7% identity with Amt2 and Amt3, respectively. The expression of *AMT1* and *AMT2* only in ammonium-limiting conditions [142] is consistent with a role for the high-affinity ammonium transporter in scavenging low concentrations of ammonium, whereas the low-affinity ammonium transporter Amt3 would be required for growth in ammonium-sufficient conditions. Similar conclusions were drawn for MepA and MeaA of *A. nidulans* [142]. The control of AMT1 and AMT2 mRNA levels by glutamine, which was deduced from experiments with glutamine synthetase inhibitors and with glutamine as sole nitrogen source, is similar to that observed in *Saccharomyces cerevisiae*. The control of *AMT1* and *AMT2* mRNA levels by the intracellular level of glutamine in *H. cylindrosporum* certainly requires complex regulatory mechanisms, similar to those observed in other fungi. In *S. cerevisiae*, nitrogen catabolic repression (NCR) is the mechanism designed to prevent or reduce the unnecessary diversion of the cell's synthetic capacity to the formation of enzymes and permeases for the utilization of compounds that are nonpreferred N sources when a preferred N source is available.

B. UTILIZATION OF SOIL CARBON COMPOUNDS BY MYCORRHIZAL FUNGI

As described in (Section IV), mycorrhizas differ in their morphology. Some have abundant external hyphae with high metabolic activity, whereas others are smooth ECMs with little or no exploratory mycelium. These features reflect differences in the need for carbon and the way in which it is shared between the symbionts. Carbon metabolism provides the mycelia and host cells of all types of mycorrhizas with energy and reducing power, and the skeletons required for the synthesis of various metabolites (e.g., amino acids). Pathways of hexose metabolism have been investigated in symbiotic and in free-living ECM fungi in axenic cultures [130]. Carbon is acquired by the fungus via: (1) host photosynthesis and translocation, (2) carbon dioxide fixation in hyphal and root cells, and (3) assimilation following the degradation of soil carbon polymers. This section will focus on assimilation, which takes place in both the rhizosphere and the hyphosphere.

ECM and ericoid fungi typically inhabit organic soil horizons containing high levels of lignins, soluble phenolic acids, and polyphenolics derived from plant litter. Most ECM fungi so far investigated have limited polyphenol-degrading activities, whereas those of ericoid fungi are well developed. A few ECM fungi (e.g., *Hebeloma crustuliniforme*) have well-developed ligninolytic abilities [145] and their use of lignin as a carbon source reduces the amount of C needed from the host plant. A full complement of lignin-degrading enzymes has been identified in the recently sequenced genome of *Laccaria bicolor* (http://genome.jgi-psf.org/Lacbi1/Lacbi1.home.html). Whether these genes are expressed remains to be determined. The ecological significance of this ligninolysis of ECM, however, is unknown.

Ericoid fungi, on the other hand, produce an array of hydrolytic enzymes during their extraradical phase and can thus exploit both simple and complex organic matter in the soil [146,147]. *Hymenoscyphus ericae*, the strain that has been best characterized, is able to grow on a variety of complex organic substrates (see Reference 146) and seems well equipped to degrade most of the polymeric components of plant and fungal cell walls included in the organic matter. Polysaccharides, such as carboxymethyl cellulose, tylose, laminarin [148], and xylans [147] are utilized through the secretion of hydrolytic enzymes. Chitin, the structural polysaccharide of the fungal wall, is degraded, as well as proteins and even lignin (see Reference 146). Pectin is another important component of the plant cell wall debris and polygalacturonase, an enzyme involved in its degradation, is produced by a wide range of ericoid fungi [149]. Several polygalacturonase isoforms are produced by ericoid isolates [149], depending on the species involved. These biochemical data may be of great ecological significance together with the observation of multiple root occupancy by genetic analysis of ericoid fungi [29]. Ericaceous plants may thus enhance their exploitation of complex soil substrates by widening their metabolic capabilities through an association with several fungi endowed with different functional enzymes.

C. Utilization of Phosphorus by Mycorrhizal Fungi

Plants have developed different strategies to ensure and enhance Pi acquisition; modifying root architecture and extension to explore larger portions of soil; and secreting organic acids or phosphatases that allow the release of bound Pi (150, see Chapter 2). As an alternative, they can establish symbiotic association with soil microorganisms, in particular with mycorrhizal fungi. The major aspects of this topic which has been extensively discussed in many recent reviews (for example, see Reference 16) will be discussed here from a "rhizosphere" perspective.

1. Pi Uptake by ECM Fungi

The growth of ECM trees is frequently improved by their increased phosphorus (Pi) accumulation [6] and this, in turn, is related to the intensity of the mycorrhizal infection. ECM fungi solubilize insoluble forms of Al and Ca phosphates, as well as inositol hexaphosphates, through secreted enzymes though a wide interstrain variability has been recorded [151]. Pi in soil solutions is easily taken up by ECM hyphae and then translocated to the host roots. Its absorption and efflux are probably regulated by intracellular Pi and inorganic polyphosphates (PolyP) pools. Excess intracellular Pi is stored as PolyP by most ECM fungi. Most of these PolyP are oligophosphates with an average chain length of 10 phosphate residues [152]. NMR comparisons of PolyP *in vivo* with those in model solutions suggest that they are low-soluble aggregates in ECM fungi [152]. PolyP are the only macromolecular anions in the fungal vacuole [152], and their roles in basic amino acid and cation retention and osmoregulation have been demonstrated both *in vivo* and *in vitro*. Several low- and high-affinity Pi transporters have been identified in the genome sequence of *Laccaria bicolor* [F. Martin, unpublished results] and other ECM fungi (e.g., *Hebeloma cylindrosporum*) (Section 5.3.2), but the molecular processes controlling Pi uptake in ECM fungi are so far unknown.

2. Pi Uptake by AM Fungi

Orthophosphate uptake is greatly enhanced during an AM association [153,154]. There are many possible explanations for this increased efficiency [6]. An AM fungal mycelium explores the soil more efficiently than the root itself and spreads beyond the phosphate depletion zone [150]. It takes phosphate from the soil and transfers it to the plant root [153] by using sources of Pi which might not be available to roots [6]. The kinetics of Pi uptake into hyphae may differ from that of roots, leading to a more effective absorption. For all these reasons AM fungi are considered powerful tools for low-input agricultural practices. Mycorrhizal plants therefore can acquire Pi either directly from the soil through plant specific phosphate transporters (PT), or through uptake and transport systems of the fungal symbiont. It has been demonstrated that both systems can work simultaneously, but there is a preferential uptake *via* fungal hyphae. This seems to occur independently from nutrients availability or growth effect, thus indicating that Pi transport in the AM symbiosis plays a role less obvious than previously expected [16].

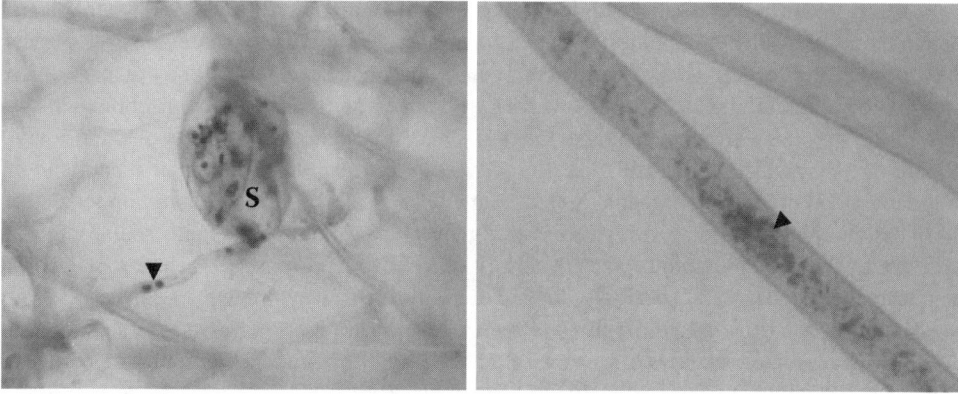

FIGURE 8.16 Mycelium of *Glomus mosseae* stained with Toluidine blue O for the histochemical detection of polyphosphate (S: spore, arrowheads: polyphosphate granules).

In the fungus-mediated uptake, Pi is absorbed into extraradical mycelium by means of a high-affinity phosphate transporter [155–157] and accumulated in the vacuoles of extraradical hyphae (Figure 8.16) in the form of polyP [158]. PolyP chains are supposed to be transferred by means of a motile tubular vacuolar network [159] in the intraradical compartment, where Pi ions resulting from polyP hydrolysis are assumed to be released by membrane passive carriers into the periarbuscular space [160]. Mycorrhiza-specific PT, possibly responsible for plant Pi uptake in arbuscule-containing cells, has recently been characterized in potato, barley, and *M. truncatula* [161–164].

PT genes isolated from the AM fungi *Glomus versiforme* (*GvPT*), *G. intraradices* (*GiPT*), and *G. mosseae* (*GmPT*) [155–157] encode for high-affinity proton-coupled transporters. They share structural and sequence similarity with other plant and fungal high-affinity PT. The apparent K_m of GvPT, evaluated in a heterologous system, is in the micromolar range, a value comparable to free Pi concentration generally found in soil solution. The fungal *PT* transcripts are predominantly detected in extraradical mycelium, thus indicating a role in Pi acquisition from the soil. It has been demonstrated that *PT* expression responds to external Pi concentrations and also to overall mycorrhiza Pi content [156,157]. Unlike *GiPT*, *GmPT* shows significant expression also in the mycorrhizal roots, opening new questions about the role and the functioning of the high-affinity PTs in AM fungi.

Taken in their whole, these data provide a rather satisfying picture of the molecular mechanisms that operate along the P fungus/plant pathway to guarantee the fungus–plant phosphate exchange.

D. THE WOOD-WIDE WEB CONCEPT

The hyphal organization of mycorrhizal fungi, their ability to grow in the soil, and the possibility to establish symbiosis with the roots of one or more individual plants, leads to the formation of a subterranean hyphal network that connects distinct plants within the community. Depending on the specificity of the plant–fungus interactions, such hyphal connections, named by Simard et al. [165] the "wood-wide web," may involve plants of the same or different species. Although mycorrhizal networks can be formed by both ECM and endomycorrhizal fungi, the functional and ecological roles of the hyphal networks have been investigated mostly for fungi involved in ectomycorrhiza. In particular, an exciting discovery was that, among nutrients, organic carbon is transferred from one plant to another along the hyphae of the mycorrhizal fungal web, following the general rule of a source-to-sink movement [165].

The wood-wide web represents a highway of horizontal nutrient movement and a pool of organic carbon in the ECM plant communities; it is therefore not surprising that some organisms have learned to exploit it. In the recent years, it has become clear that many achlorophyllous plants, which are heterotrophic because of the lack of photosynthesis, obtain organic carbon by connecting to the wood-wide web through their mycorrhizal endophytes [7]. The mycorrhizal symbionts of achlorophyllous plants are usually recalcitrant to isolation and growth in axenic culture, and their identification has been greatly aided by molecular methods (for example, see Reference 33). A common feature of the achlorophyllous plants so far investigated is that they all associate with fungi capable of forming mycorrhiza on surrounding autotrophic species. Another feature common to achlorophyllous plants is the unusually high degree of specificity toward their mycorrhizal symbionts. Exclusive associations with a single (or a narrow range of) fungal species have been reported for several achlorophyllous angiosperms forming associations as diverse as orchid, monotropoid, and arbuscular mycorrhiza (for example, see Reference 33, Reference 36, and Reference 166), as well as for liverworts [167].

Studies on the biodiversity of mycorrhizal symbionts of achlorophyllous species have mostly focused on orchids and monotropoids. Direct amplification and sequencing of fungal DNA from roots have identified, as endomycorrhizal endophytes, fungi known to form ECM on tree species. It is intriguing that the same fungus can display a completely different morphogenetic program depending on the host plant, and form either intracellular symbioses or typical ECM. Tuber offers a nice example of this strategy: it is an ECM fungus, but it has been recently detected in orchid roots [34]. From an ecological point of view, the occurrence of a common symbiont in achloro-phyllous and photoautotrophic host provides cues to understand the nutritional strategy of these nonphotosynthetic species. A direct transfer of radiolabeled carbon from autotrophic plants to achlorophyllous species has actually been demonstrated for the orchid *Corallorhiza trifida* [168], as well as for the nonphotosynthetic liverworts *Cryptothallus mirabilis* [167]. Indirect evidence of plant nutrition through the associated mycorrhizal fungus, a strategy named *mycoheterotrophy* by Leake [169], also comes from the analysis of stable C and N isotopes [7].

Recent papers strongly suggest that understorey plants containing chlorophyll, but with ineffi-cient photosynthesis due to environmental limitation, may adopt the same mycoheterotrophic strategy as fully achlorophyllous species [35,36,170]. The wood-wide web appears to be an essential component for the survival of these endangered plant species.

VI. DO MYCORRHIZAL FUNGI PROTECT THEIR HOST FROM ELEMENTAL POLLUTANTS?

The inflow of high amounts of heavy metals, to land and water ecosystems, is currently regarded as one of the most significant human factors affecting forest health and productivity. Because heavy metals cannot be degraded, *in situ* bioremediation is currently based on the use of plants and microorganisms to either decontaminate the soil from heavy metals, or to stabilize these compounds in the soil to reduce leakage of metal ions.

The ability of mycorrhizal fungi to grow in polluted soils and withstand high heavy-metal concentrations has been reported for both endo- and ECM fungi [171–174]. There are two major strategies organisms can adopt to protect themselves against heavy-metal toxicity. Avoidance restricts entry of metal ions into the cytoplasm, and relies on decreased uptake or increased efflux of metal ions, or by their immobilization outside the cell. When present, the cell wall is an important site of metal immobilization. Sequestration occurs to reduce cytoplasmic concentration of free metal ions, either through the synthesis of chelating compounds or by compartmentalization into the vacuole [175]. The cellular and molecular mechanisms of heavy metal tolerance in mycorrhizal fungi are not fully understood, but the general mechanisms observed in fungi [176] have also been found in mycorrhizal fungi [177,178]. More detailed studies have been carried out for some specific contam-inants. For example, the mechanism of arsenic tolerance in ericoid mycorrhizal fungi has been investigated by Sharples et al. [174]. This element enters the cell through the phosphate transporters, causing mycorrhizal fungi to enhance both phosphate and arsenate uptake. Sharples et al. [174] found that active and specific efflux mechanisms are adopted by ericoid mycorrhizal fungi from polluted sites, so as to decrease cellular concentrations of arsenic while retaining phosphate.

Metal-specific tolerance mechanisms have also been demonstrated for ECM fungi [179–181]. Increasing concentrations of toxic aluminum (Al) are being reported in the acidic soils of temperate forests. ECM fungi on tree roots modify the compartmentalization of absorbed Al and protect their host against its toxic effects [179,181], even if contrasting results have been reported [182]. Al has been detected in P-rich granules in the vacuoles [180] and in cells walls [179] by energy-dispersive X-ray microanalysis and electron energy loss imaging [179,181,183], suggesting that PolyP sequester heavy metal ions and Al is part of a fungus's detoxification mechanism. ^{27}Al NMR has shown that mixed solvation complexes of Al-PolyP occur in the vacuoles of fungal cells [152].

Phytoextraction leads to the removal of toxic metals from the soil, thus recuperating land to agriculture, whereas phytostabilization makes use of plants and microorganisms to immobilize metals into insoluble or complexed forms, thus decreasing their toxicity and avoiding contamination of the water table [186]. In both strategies, the first goal that must be achieved is the revegetation of contaminated sites with plants and microorganisms able to tolerate toxic concentrations of heavy metals. Because mycorrhizal fungi mediate the uptake and transfer of elements from the soil particles to the roots of mycorrhizal plants, several studies have focused on their role on heavy-metal uptake and transfer to the plant [5,171,177,178,185]. Increased tolerance of both endo- and ectomycorrhizal plants in metal contaminated sites is well documented [5,171,177,181,186,187,188], By contrast, the influence of mycorrhizal fungi on root metal uptake and transfer to the leaves (phytoextraction) appears to be quite variable and specific of metal and plant [187,189].

Although this paragraph is focused on inorganic pollutants, the possible exploitation of rhizosphere interactions for bioremediation of organic soil pollution should be mentioned. Saprotrophic fungi as well as rhizobacteria seems to be playing a major role in rhizodegradation, but data on the degradation of organic pollutants in soil planted with mycorrhizal plants suggest interesting developments in this area [5,190,191].

VII. BACTERIA AND MYCORRHIZAL FUNGI IN THE RHIZOSPHERE

Mycorrhizas are often described as tripartite interactions, because in natural conditions, bacteria are associated to AM and ECM fungi as microbes that colonize the extraradical hyphae or as endobacteria living in the cytoplasm of at least some fungal taxa.

Understanding of the interactions between the microorganisms found routinely in the rhizosphere is an essential prelude to describe the nature of the soil–plant interface. The interaction of mycorrhizal hyphae with other microorganisms either directly, or indirectly by modifying host physiology and the pattern of root exudation, was demonstrated long ago [192], and represent an extensively investigated topic [193]. Associations between mycorrhizal fungi and soil bacteria, particularly plant growth-promoting rhizobacteria (PGPR) such as rhizobia, pseudomonads, and *Azospirillum*, have profound effects on plant health through beneficial synergisms. PGPR interact with both mycorrhizal fungi and the plant root in a wide variety of ways.

Many experiments have demonstrated that bacteria stimulate mycorrhiza formation. Garbaye [194] has elegantly shown how bacterial strains increase a root's ability to establish an ectomycorrhizal symbiosis. He has also proposed a new bacterial category to describe this effect: the mycorrhization helper bacteria (MHB). Similar effects have been reported for AM fungi, where several rhizosphere bacteria appear to stimulate the growth and development of endomycorrhizal fungi or increase root mycorrhization, as well as root architecture [5,9,195]. Several mechanisms have been proposed to explain these effects — for example, the production of vitamins, amino acids, phytohormones, or cell wall hydrolytic enzymes. Some of these effects may directly influence the germination and growth rate of fungal structures, whereas others may act on root development and susceptibility to infection.

Better use of mycorrhizal fungi as biocontrol and biofertilizer agents requires a more mechanistic understanding of the biological balance between them and rhizosphere bacteria. Because bacteria form biofilms around the hyphae of ecto- and endomycorrhizal fungi [196,197], Perotto and Bonfante [198] have suggested that their attachment to the root and fungal surface is a crucial moment of the interaction. Investigation of the mechanisms by which this attachment takes place has shown that several molecular components, including adhesive proteins, flagella, and extracellular polysaccharides, are involved [199]. Several bacteria described as good root colonizers also adhere to hyphae, indicating that similar attachment mechanisms are adopted. For example, PGPR isolated as strong root colonizers, such as strain WCS 365 of *Pseudomonas fluorescens*, form a coat around the hyphae of AM fungi, as do strains of *Rhizobium leguminosarum* [196]. The significance of this bacterial attachment to a solid surface is not clear. It may be a way to avoid dispersion by percolating soil water. In addition, according to Boddey et al. [200], rhizobacteria interacting with mycorrhizal fungi may use hyphae as a vehicle for their distribution or to enhance root colonization. In view of the widespread distribution of mycorrhizal fungi and the ability of AM fungi to colonize most plants, this strategy would offer these bacteria a good chance of finding new niches.

Mycorrhizal fungi may have an impact on genotypic and functional diversity of bacterial communities, because some of them seem to be specific to mycorrhizal plants [201,202]. This impact may be related to the plant root; mycorrhizal establishment changes the chemical composition of root exudates [5,201] and these are often a food source for soil bacteria. It may also be related to fungal metabolism [203]. Similarly, fruit bodies of ECMs are specific niches for bacterial communities. In truffle ascomata, α-Proteobacteria showing significant similarity values with members of the *Sinorhizobium/Ensifer* Group, *Rhizobium* and *Bradyrhizobium* spp. have been detected as well as members of the β- and γ-Proteobacteria [204]. In *Cantharellus* sporocarps, the prominent culturable bacterial population is *P. fluorescens* biovar I, an user of trehalose, the most abundant storage carbohydrate in fruiting bodies [205]. Trehalose accumulated in ECM mycelium may play a role in the structuration of ECM-associated bacterial communities as specific *Pseudomonas* strains are able to use this disaccharide [203].

Andrade et al. [206] found that the composition, but not the size, of bacterial communities in the rhizosphere and hyphosphere of *Sorghum*, in the presence and absence of mycorrhizal fungi, varied both within and among AM fungal treatments, suggesting that fungal taxa may have a qualitative effect on bacterial diversity. In addition, a *Burkholderia cepacia* strain was always present in the rhizosphere [206]. This result is of particular interest because *Burkholderia*-related isolates have been described as endosymbionts living in the cytoplasm of spores of *Gigaspora margarita*, an AM fungus [207]. Bacteria-like organisms (BLOs) in the cytoplasm of AM fungi were first observed ultrastructurally in the early 1970s [208], but confirmation of their prokaryotic nature was prevented by their refusal to grow on cell-free media. A combined morphological and molecular approach has now shown that the cytoplasm of *G. margarita* spores harbors a homogeneous population of bacteria identified from the sequence of the 16S ribosomal RNA gene as *Candidatus* Glomeribacter gigasporarum [209]. PCR assays with oligonucleotides specific for this sequence have revealed these bacteria in all stages of the fungal life cycle (spores and symbiotic mycelia). In addition, isolates of different origin from the three AM fungal families (Glomaceae, Gigasporaceae, and Acaulosporaceae) display bacteria when observed by confocal microscopy using a fluorescent dye specific for bacterial staining. However, PCR amplification performed with universal eubacterial primers and primers specifically designed for the endobacteria of *G. margarita* on DNA preparations from AM spores demonstrated that the endobacteria were widespread among AM fungi, whereas *Candidatus* G. gigasporarum is limited to some Gigasporaceae. Thus, it seems that these intracellular bacteria may be a general feature of AM spores and not merely sporadic components [210].

Associations between endosymbiotic bacteria and the Homoptera, Blattaria, and Coleoptera are common. One of the best known is that between Buchnera and the aphids [211]. Both partners are obligate and mutualistic symbionts and the aphids cannot survive without the bacteria [211]. Buchnera, in fact, possesses genes involved in DNA metabolism, protein secretion, carbohydrate degradation, and ATP generation, as well as synthesis of essential amino acid; one of its functions is the biosynthesis of essential amino acids such as tryptophan, cysteine, and methionine for the aphid [211]. Similarly, the results so far available for *Candidatus* G. gigasporarum strongly suggest that this bacterium is also an obligate endosymbiont at least in some AM fungi. *Candidatus* G. gigasporarum does not grow in pure culture, but the setting up of an isolation protocol has allowed us to define some of its physiological features; it is morphologically described as a Gram⁻ microbe — it keeps its viability for some days but does not grow in culture, and its genome is extremely reduced, consisting of a chromosome and a plasmid for a total size of 1.4 Mb [212]. Bacteria move along the fungal generations, following a vertical transmission mechanism [213].

The functional significance of endosymbiotic bacteria in AM fungi is not clear. Using the *Buchnera*–aphid system as an analogy, we are investigating this issue. The finding that a genomic library developed from *G. margarita* spores is also representative of *Candidatus* genome [214] will help to establish the features of this rhizospheric bacterial population, which has so far remained hidden inside its fungal hosts.

The occurrence of intrafungal bacteria, in both live and dead cells, has been recently observed in pure culture and natural mycelial samples of the ECM fungus *L. bicolor* [215,216]. However, contrary to the bacterial endosymbionts of the Glomeromycota, the endobacteria of *L. bicolor* are not systematically present and homogeneously distributed in the mycelium. Moreover, they belong to different bacterial groups, depending on the origin of the fungal samples; α-proteobacteria were predominant in *L. bicolor* mycelium collected in artificial substrates and natural soils from nurseries or forest, whereas *Paenibacillus* predominated in pure cultures of the fungus. These results suggest an environmental acquisition of the endobacteria by the fungus from the reservoir of the soil bacterial communities. So far, the role of these endobacteria is completely unknown, contrary to some of the extracellular bacterial communities that inhabit the *L. bicolor* ectomycorhizosphere. Indeed, the analysis of the functional diversity of the culturable *Pseudomonas fluorescens* population associated to *L. bicolor* ectomycorrhizas revealed that this population presents a high potential for phosphorus and iron mobilization, as well as against fungal phytopathogens, that would be beneficial to plant nutrition and health [202].

In conclusion, the analysis of the multiple interactions established by mycorrhizal fungi with plant and bacterial cells offers new cues for understanding the complexity of mycorrhizal symbiosis as a tripartite multitrophic association. This aspect will lead to the definition of new parameters in the design of mixed inocula containing mycorrhizal fungi and associated bacteria and opens up new strategies for the practical use of mycorrhizal fungi in the control of soilborne pathogens, the improvement of plant nutrition and growth, or generally speaking, in low-input agriculture and sustainable forestry [201]. To select new beneficial strains of fungal-associated bacteria in the future, screening strategies will rely on data from genome-wide transcriptomics [218].

VIII. CONCLUSIONS

Mycorrhizas are usually described as the result of a symbiotic interaction between plant and fungal genomes. There is increasing evidence that many of the complex events described inside the root (morphogenetic changes, regulated development, nutritional exchanges) are the consequence of extraradical events, some of which have been highlighted here — genetic and functional diversity, development of fungal structures outside the roots, role of mycorrhizal fungi in nutrient cycling, plant protection against pollutants, and living substrate for bacterial populations.

Whereas there have been substantial advances in recent years in our understanding of developmental processes leading to the formation of mycorrhizal symbiosis, many questions concerning the morphogenesis of fungal structures in the rhizosphere remain unanswered. A major theme in future studies in this research area will be the molecular signaling involved in host recognition and the coordinated development of specialized symbiotic structures. The signals identified might be analogous to those that regulate formation of other symbiotic tissues and may well offer insights into the processes that regulate organogenesis in fungi and plants.

Mycorrhizal fungi are a crucial component of the rhizosphere, where they work at the soil–plant interface and also free from the influence of the root by interacting with a number of other organisms to create a living underground web [103]. The other topic of great interest will be, therefore, the interaction between mycorrhizal fungi and rhizosphere microbes. A good knowledge of these multiple interactions will offer new tools for the development of "biofertilizers" to be used in the frame of sustainable agriculture. Lastly, it will be important to explore the potential of mycorrhizal fungi as bioremediation agents in polluted soils. There is a strong research need in this field.

Molecular approaches based on DNA technology, genomics as well as the emerging fields of the soil metagenomics [218] will be surely crucial to clarify at least some of these problems; an improved knowledge will provide new stimuli for the better dissection of the role and functioning of mycorrhizal fungi in the rhizosphere. Identifying genes of ecological and evolutionary relevance is of high priority [219].

ACKNOWLEDGMENTS

The investigations carried out in Francis Martin's laboratory were supported by grants from the INRA (programs *Microbiology, Sequencing Symbiont & Pathogen Genomes, Lignome, and ECOGER*) and the Région de Lorraine. Research in Paola Bonfante's laboratory was supported by the Consiglio Nazionale delle Ricerche, by the Ministero della Ricerca Scientifica (MURST, Prin Projects, FIRB), by CEBIOVEM, by the European projects (GENOMYCA, INTEGRAL), by CRT and Compagnia San Paolo. The authors thank Dr A.M. Cantisani and Dr M. Novero for their help in the preparation of the reference list and illustrations. They would like to acknowledge the useful discussions with Pascale Frey-Klett, Jean Garbaye, Sébastien Duplessis and Michel Chalot (UMR IaM, INRA-Nancy).

REFERENCES

1. O'Gara, F., Dowling, D.N., and Boesten, B., *Molecular Ecology of Rhizosphere Microorganisms,* VCH, Weinheim, 1994.
2. Bais, H.P. et al., How plants communicate using the underground information superhighway, *Trends Plant Sci.,* 9, 26, 2004.
3. Morrissey, J.P. et al., Are microbes at the root of a solution to world food production?, *EMBO Rep.,* 5, 922, 2004.
4. Singh, B.K. et al., Unravelling rhizosphere–microbial interactions: opportunities and limitations, *Trends Microbiol.,* 12, 386, 2004.
5. Barea, J.M. et al., Microbial co-operation in the rhizosphere, *J. Exp. Bot.,* 56, 1761, 2005.
6. Smith, S.E. and Read, D.J., *Mycorrhizal Symbiosis,* Academic Press, San Diego, CA, 1997.
7. Leake, J.R. et al., Networks of power and influence: the role of mycorrhizal mycelium in controlling plant communities and agroecosystem functioning, *Can. J. Bot.,* 82, 1016, 2004.
8. Bianciotto, V. and Bonfante, P., Presymbiotic versus symbiotic phase in arbuscular endomycorrhizal fungi: morphology and cytology, in *Mycorrhizas: Structure, Function, Molecular Biology and Biotechnology,* Varma, A. and Hoch, B., Eds., Springer-Verlag, Berlin Heidelberg, 1998, p. 229.
9. Lynch, J.M., *The Rhizosphere,* John Wiley and Sons, 1990.
10. Gianinazzi, S. et al., *Mycorrhizal Technology in Agriculture: From Genes to Bioproducts,* Birkhauser Verlag, Basel, 2002.
11. Bonfante, P. and Perotto, S., Strategies of arbuscular mycorrhizal fungi when infecting host plants, *New Phytologist,* 130, 3, 1995.

12. Martin, F. et al., Developmental cross talking in the ectomycorrhizal symbiosis: signals and communication genes, *New Phytologist,* 151, 145, 2001.

13. Marmeisse, R. et al., *Hebeloma cylindrosporum* — a model species to study ectomycorrhizal symbiosis from gene to ecosystem, *New Phytologist,* 163, 481, 2004.

14. Kistner, C. and Parniske, M., Evolution of signal transduction in intracellular symbiosis, *Trends Plant Sci.,* 7, 511, 2002.

15. Parniske, M., Molecular genetics of the arbuscular mycorrhizal symbiosis, *Curr. Opin. Plant Biol.,* 7, 414, 2004.

16. Karandashov, V. and Bucher, M., Symbiotic phosphate transport in arbuscular mycorrhizas, *Trends Plant Sci.,* 10, 22, 2005.

17. Hause, B. and Fester, T., Molecular and cell biology of arbuscular mycorrhizal symbiosis, *Planta,* 221, 184, 2005.

18. Harrison, M.J., Signalling in the arbuscular mycorrhizal symbiosis, *Annu. Rev. Microbiol.,* 59, 19, 2005.

19. Brundrett, M.C., Coevolution of roots and mycorrhizas of land plants, *New Phytologist,* 154, 275, 2002.

20. Schüßler, A., Schwarzott, D., and Walker, C., A new fungal phylum, the Glomeromycota: phylogeny and evolution, *Mycol. Res.,* 105, 1413, 2001.

21. Rosendhal, S. and Taylor, J.W., Development of multiple genetic markers for studies of genetic variation in arbuscular mycorrhizal fungi using AFLP, *Mol. Ecol.,* 6, 821, 1997.

22. Horton, T.R. and Bruns, T.D., The molecular revolution in ectomycorrhizal ecology: peeking into the black-box, *Mol. Ecol.,* 10, 1855, 2001.

23. Sanders, I.R., Clapp, J.P., and Wiemken, A., The genetic diversity of arbuscular mycorrhizal fungi in natural ecosystems: a key to understanding the ecology and functioning of the mycorrhizal symbiosis, *New Phytologist,* 133, 123, 1996.

24. Lilleskov, E.A. et al., Detection of forest stand-level spatial structure in ectomycorrhizal fungal communities, *FEMS Microbiol. Ecol.,* 49, 319, 2004.

25. Gardes, M., Identification of indigenous and introduced symbiotic fungi in ectomycorrhizae by amplification of nuclear and mitochondrial ribosomal DNA, *Can. J. Bot.,* 69, 180, 1991.

26. Dickie, I.A., Xu, B., and Koide, R.T., Vertical niche differentiation of ectomycorrhizal hyphae in soil as shown by T-RFLP analysis, *New Phytologist,* 156, 527, 2002.

27. Husband, R. et al., Molecular diversity of arbuscular mycorrhizal fungi and patterns of host association over time and space in a tropical forest, *Mol. Ecol.,* 11, 2669, 2002.

28. Zèzè, A. et al., Intersporal genetic variation of *Gigaspora margarita,* a vesicular arbuscular mycorrhizal fungus, revealed by M13 minisatellite-primed PCR, *Appl. Environ. Microbiol.,* 63, 676, 1997.

29. Perotto, S. et al., Molecular diversity of fungi from ericoid mycorrhizal roots, *Mol. Ecol.,* 5, 123, 1996.

30. Monreal, M., Berch, S.M., and Berbee, M., Molecular diversity of ericoid mycorrhizal fungi, *Can. J. Bot.,* 77, 1580, 1999.

31. Sharples, J.M. et al., Genetic diversity of root associated fungal endophytes from *Calluna vulgaris* at contrasting field sites, *New Phytologist,* 148, 153, 2000.

32. Allen, T.R. et al., Culturing and direct DNA extraction find different fungi from the same ericoid mycorrhizal roots, *New Phytologist,* 160, 255, 2003.

33. Taylor, D.L., Mycorrhizal specificity and function in myco-heterotrophic plants, in *Ecological Studies, 157: Mycorrhizal Ecology,* Van der Heijden, M.G.A. and Sanders, I., Eds., Springer, Berlin, 2002, p. 375.

34. Selosse, M.A. et al., Chlorophyllous and achlorophyllous specimens of *Epipactis microphylla* (Neottieae, Orchidaceae) are associated with ectomycorrhizal septomycetes, including truffles, *Microb. Ecol.,* 47, 416, 2004.

35. McCormick, M.K., Whigham, D.F., and O'Neill, J., Mycorrhizal diversity in photosynthetic terrestrial orchids, *New Phytologist,* 163, 425, 2004.

36. Bidartondo, M.I. et al., Changing partners in the dark: isotopic and molecular evidence of ectomycorrhizal liaisons between forest orchids and trees, *Proc. R. Soc. Lond. B,* 271, 1799, 2004.

37. Simon, L., Lalonde, M., and Bruns, T.D., Specific amplification of 18S fungal ribosomal genes from vesicular-arbuscular endomycorrhizal fungi colonizing roots, *Appl. Environ. Microbiol.,* 58, 291, 1992.

38. Gardes, M. and Bruns, T.D., ITS primers with enhanced specificity for Basidiomycetes: application to identification of mycorrhizae and rusts, *Mol. Ecol.,* 2, 113, 1993.

39. Kõljalg, U. et al., UNITE: a database providing web-based methods for the molecular identification of ectomycorrhizal fungi, *New Phytologist,* 166, 1063, 2005.
40. Longato, S. and Bonfante, P., Molecular identification of mycorrhizal fungi by direct amplification of microsatellite regions, *Mycol. Res.,* 101, 425, 1997.
41. Martin, F. et al., Genomic fingerprinting of ectomycorrhizal fungi by microsatellite-primed PCR, in *Mycorrhiza Manual,* Varma, A. and Hock, B., Eds., Springer Lab Manual, Berlin, 1998, p. 463.
42. Gryta, H. et al., Fine-scale structure of populations of the ectomycorrhizal fungus *Hebeloma cylindrosporum* in coastal sand dune forest ecosystems, *Mol. Ecol.,* 6, 353, 1997.
43. Selosse, M.A. et al., Spatio-temporal persistence and distribution of an American inoculant strain of the ectomycorrhizal basidiomycete *Laccaria bicolor* in a French forest plantation, *Mol. Ecol.,* 7, 561, 1998.
44. Dahlberg, A., Community ecology of ectomycorrhizal fungi: an advancing interdisciplinary field, *New Phytologist,* 150, 555, 2001.
45. Fitter, A.H., Global environmental change and the biology of arbuscular mycorrhizas: gaps and challenges, *Can. J. Bot.,* 82, 1133, 2004.
46. Sanders, I.R. et al., Identification of ribosomal DNA polymorphisms among and within spores of the Glomales: application to studies on the genetic diversity of arbuscular mycorrhizal fungal communities, *New Phytologist,* 130, 419, 1995.
47. Zèzè, A. et al., Intersporal genetic variation of *Gigaspora margarita,* a vesicular arbuscular mycorrhizal fungus, revealed by M13 minisatellite-primed PCR, *Appl. Environ. Microbiol.,* 63, 676, 1997.
48. Lanfranco, L., Delpero, M., and Bonfante, P., Intrasporal variability of ribosomal sequences in the endomycorrhizal fungus *Gigaspora margarita, Mol. Ecol.,* 8, 37, 1999.
49. Hijri, M. and Sanders, I.R., Low gene copy number shows that arbuscular mycorrhizal fungi inherit genetically different nuclei, *Nature,* 433, 160–163, 2005.
50. Bougoure, D.S. and Cairney, J.W.G., Assemblages of ericoid mycorrhizal and other root-associated fungi from *Epacris pulchella* (Ericaceae) as determined by culturing and direct DNA extraction from roots, *Environ. Microbiol.,* 7, 819, 2005.
51. Chambers, S.M. et al., Molecular identification of *Hymenoscyphus* sp. from rhizoids of the leafy liverwort *Cephaloziella exiliflora* (Tayl.) Steph. in Australia and Antarctica, *Mycol. Res.,* 103, 286, 1999.
52. Duckett, J.G. and Read, D.J., Ericoid mycorrhizas and rhizoid-ascomycete associations in liverworts share the same mycobiont isolation of the partners and resynthesis of the associations *in vitro, New Phytologist,* 129, 439, 1995.
53. Bergero, R. et al., Ericoid mycorrhizal fungi are common root associates of a Mediterranean ectomycorrhizal plant (*Quercus ilex*), *Mol. Ecol.,* 9, 1639, 2000.
54. Vrålstad, T., Fossheim, T., and Schumacher, T., *Piceirhiza bicolorata* — the ectomycorrhizal expression of the *Hymenoscyphus ericae* aggregate?, *New Phytologist,* 145, 549, 2000.
55. Villarreal-Ruiz, L., Anderson, I.C., and Alexander, I.J., The interaction between an isolate from the *Hymenoscyphus ericae* aggregate and roots of *Pinus and Vaccinium, New Phytologist,* 164,183, 2004.
56. Horan, D.P. and Chilvers, G.A., Chemotropism: the key to ectomycorrhizal formation, *New Phytologist,* 116, 297, 1990.
57. Giovannetti, M.C. et al., Analysis of factors involved in fungal recognition responses to host-derived signals by arbuscular mycorrhizal fungi, *New Phytologist,* 133, 65, 1996.
58. Akiyama, K., Matsuzaki, K., and Hayashi, H., Plant sesquiterpenes induce hyphal branching in arbuscular mycorrhizal fungi, *Nature,* 435, 824, 2005.
59. Giovannetti, M. et al., Differential hyphal morphogenesis in arbuscular mycorrhizal fungi during pre-infection stages, *New Phytologist,* 125, 587, 1993.
60. Buee, M. et al., The pre-symbiotic growth of arbuscular mycorrhizal fungi is induced by a branching factor partially purified from plant root exudates, *Mol. Plant Microbe Interact.,* 13, 6, 693, 2000.
61. Tamasloukht, M. et al., Root factors induce mitochondrial-related gene expression and fungal respiration during the developmental switch from asymbiosis to presymbiosis in the arbuscular mycorrhizal fungus *Gigaspora rosea, Plant Physiol.,* 131, 1468, 2003.
62. Lanfranco, L., Novero, M., and Bonfante, P., The mycorrhizal fungus *Gigaspora margarita* possesses a CuZn superoxide dismutase that is up-regulated during symbiosis with legume hosts, *Plant Physiol.,*137, 319, 2005.
63. Kosuta, S., A diffusible factor from arbuscular mycorrhizal fungi induces symbiosis-specific MtENOD11 expression in roots of *Medicago truncatula, Plant Physiol.,* 131, 952, 2003.

64. LaRue, T.A. and Weeden, N.F., The symbiosis genes of the host in *Proc. 1st Eur. Nitrogen Fixation Conf.,* Kiss, G.B. and Endre, G., Eds., Officina Press, Szeged, Hungary, 1994, p. 147.

65. Becard, G. et al., Partner communication in the arbuscular mycorrhizal interaction, *Can. J. Bot.,* 82, 1186, 2004.

66. Bonfante, P., At the interface between mycorrhizal fungi and plants: the structural organization of cell wall, plasma membrane and cytoskeleton, in *Mycota, IX: Fungal Associations,* Hock, B., Ed., Springer Verlag, Berlin, 2001, p. 45.

67. Lagrange, H., Jay-Allemand, C., and Lapeyrie, F., Rutin, the phenolglycoside from eucalyptus root exudates, stimulates *Pisolithus* hyphal growth at picomolar concentrations, *New Phytologist,* 149, 349, 2001.

68. Heath, I.B., Integration and regulation of hyphal tip growth, *Can. J. Bot.,* 73, 131, 1995.

69. Prokisch, H. et al., Impairment of calcineurin function in *Neurospora crassa* reveals its essential role in hyphal growth, morphology and maintenance of the apical Ca^{2+} gradient, *Mol. Gen. Genet.,* 256, 104, 1997.

70. Miozzi, L. et al., Phospholipase A(2) up-regulation during mycorrhiza formation in *Tuber borchii, New Phytologist,* 167, 229, 2005.

71. Genre, A. and Bonfante, P., Epidermal cells of a symbiosis-defective mutant of *Lotus japonicus* show altered cytoskeleton organisation in the presence of a mycorrhizal fungus, *Protoplasma,* 219, 43, 2002.

72. Weidmann, S. et al., Fungal elicitation of signal transduction-related plant genes precedes mycorrhiza establishment and requires the dmi3 gene in *Medicago truncatula, Mol. Plant-Microbe Interact.,* 17, 1385, 2004.

73. Genre, A. et al., Arbuscular mycorrhizal fungi elicit a novel intracellular apparatus in *Medicago truncatula* root epidermal cells before infection, *Plant Cell,* 17, 3489, 2005.

74. Beyrle, H., The role of phytohormones in the function and biology of mycorrhizas, in *Mycorrhiza: Structure, Molecular Biology and Function,* Varma, A.K. and Hock, B., Eds., Springer, Berlin Heidelberg, 1995, p. 365.

75. Gea, L. et al., Structural aspects of ectomycorrhiza of *Pinus pinaster* (Ait.) Sol. formed by an IAA-overproducer mutant of *Hebeloma cylindrosporum* (Romagnesi), *New Phytologist,* 128, 659, 1994.

76. Rupp, L.A., Mudge, K.W., and Negm, F.B., Involvement of ethylene in ectomycorrhiza formation and dichotomous branching of roots of mugo pine seedlings, *Can. J. Bot.,* 67, 477, 1989.

77. Beguiristain, T. et al., Hypaphorine accumulation in hyphae of the ectomycorrhizal fungus *Pisolithus tinctorius, Phytochemistry,* 40, 1089, 1995.

78. Beguiristain, T. and Lapeyrie, F., Host plant stimulates hypaphorine accumulation in Pisolithus tinctorius hyphae during ectomycorrhizal infection while excreted fungal hypaphorine controls root hair development, *New Phytologist,* 136, 525, 1997.

79. Jambois, A. et al., Competitive antagonism between IAA and indole alkaloid hypaphorine must contribute to regulate ontogenesis, *Physiol. Plant.,* 123, 120, 2005.

80. Perotto, S. et al., Ericoid mycorrhizal fungi: cellular and molecular bases of their interactions with the host plant, *Can. J. Bot.,* 73, 557, 1995.

81. Peterson, R.L. and Uetake, Y., Orchid symbiosis: the fungi involved and cellular events during their establishment, *Symbiosis,* 25, 29, 1998.

82. Peterson, R.L. and Massicotte, H.B., Exploring structural definitions of mycorrhizas, with emphasis on nutrient-exchange interfaces, *Can. J. Bot.,* 82, 1074, 2004.

83. Peterson, R.L., Massicotte, H.B, and Melville L.H., *Mycorrhizas: Anatomy and Cell Biology,* NRC Research press, Ottawa, 2004.

84. Leake, J.R., Myco-heterotroph/epiparasitic plant interactions with ectomycorrhizal and arbuscular mycorrhizal fungi, *Curr. Opin. Plant Biol.,* 7, 422, 2004.

85. Smith, S.E, Smith, F.A., and Jakobsen, I., Mycorrhizal fungi can dominate phosphate supply to plants irrespective of growth responses, *Plant Physiol.,* 133, 16, 2003.

86. Bonfante, P. and Perotto, S., Strategies of arbuscular mycorrhizal fungi when infecting host plants, *New Phytologist,* 130, 3, 1995.

87. Parniske, M., Intracellular accommodation of microbes by plants: a common developmental program for symbiosis and disease?, *Curr. Opin. Plant Biol.,* 3, 320, 2000.

88. Balestrini, R. and Bonfante, P., The interface compartment in arbuscular mycorrhizae: a special type of plant cell wall?, *Plant Biosyst.,* 139, 1, 2005.

89. Bonfante, P. et al., Cellulose and pectin localization in roots of mycorhizal *Allium porrum:* labelling continuity between host cell wall and interfacial material, *Planta,* 180, 537, 1990.

90. Balestrini, R., Cosgrove, D.J., and Bonfante, P., Differential location of alpha-expansin proteins during the accommodation of root cells to an arbuscular mycorrhizal fungus, *Planta*, 220, 889, 2005.

91. Peterson, R.L. et al., The interface between fungal hyphae and orchid protocorm cells, *Can. J. Bot.*, 74, 1861, 1996.

92. Gollotte, A. et al., Cellular localization and cytochemical probing of resistance reactions to arbuscular mycorrhizal fungi in a 'locus a' myc- mutant of *Pisum sativum* L., *Planta*, 191, 112, 1993.

93. Tagu, D. and Martin, F., Molecular analysis of cell wall proteins expressed during the early steps of ectomycorrhiza development, *New Phytologist*, 133, 73, 1996.

94. Balestrini, R., Hahn, M.G., and Bonfante, P., Location of cell-wall components in ectomycorrhizae of *Corylus avellana* and *Tuber magnatum*, *Protoplasma*, 191, 55, 1996.

95. Jones, E.B.G., Fungal adhesion, *Mycol. Res.*, 98, 981, 1994.

96. Martin, F. et al., Cell wall proteins of the ectomycorrhizal basidiomycete *Pisolithus tinctorius*: identification, function, and expression in symbiosis, *Fungal Genet. Biol.*, 27, 161, 1999.

97. Laurent, P. et al., A novel class of cell wall polypeptides in *Pisolithus tinctorius* contain a cell-adhesion RGD motif and are up-regulated during the development of *Eucalyptus globulus* ectomycorrhiza, *Mol. Plant-Microbe Interact.*, 12, 862, 1999.

98. Tagu, D., Nasse, B., and Martin, F., Cloning and characterization of hydrophobins-encoding cDNAs from the ectomycorrhizal basidiomycete *Pisolithus tinctorius*, *Gene*, 168, 93, 1996.

99. Soragni, E. et al., A nutrient-regulated, dual localization phospholipase A(2) in the symbiotic fungus *Tuber borchii*, *EMBO J.*, 20, 5079, 2001.

100. Lei, J. et al., Infectivity of pine and eucalypt isolates of *Pisolithus tinctorius* (Pers.) Coker and Couch on roots of *Eucalyptus urophylla* S. T. Blake *in vitro*. II. Ultrastructural and biochemical changes at the early stage of mycorrhiza formation, *New Phytologist*, 116, 115, 1990.

101. Bonfante, P. et al., Morphological analysis of early contacts between pine roots and two ectomycorrhizal *Suillus* strains, *Mycorrhiza*, 8, 1, 1998.

102. Maldonado-Mendoza, I.E. et al., Expression of a xyloglucan endotransglucosylase/hydrolase gene, Mt-XTH1, from *Medicago truncatula* is induced systemically in mycorrhizal roots, *Gene*, 345, 191, 2005.

103. Smith, S.E. and Smith, F.A., Structure and function of the interfaces in biotrophic symbiosis as they relate to nutrient transport, *New Phytologist*, 114, 1, 1990.

104. Liu, J. et al., Transcript profiling coupled with spatial expression analyses reveals genes involved in distinct developmental stages of an arbuscular mycorrhizal symbiosis, *Plant Cell*, 15, 2106, 2003.

105. Journet, E.P. et al., Exploring root symbiotic programs in the model legume *Medicago truncatula* using EST analysis, *Nucleic Acids Res.*, 30, 5579, 2002.

106. Duc, G. et al., First report of non-mycorrhizal plant mutants (Myc–) obtained in pea (*Pisum sativum* L.) and fababean (*Vicia faba* L.), *Plant Sci.*, 60, 215, 1989.

107. Hirsch, A.M. and Kapulnik, Y., Signal transduction pathways in mycorrhizal associations: comparisons with the Rhizobium-legume symbiosis, *Fungal Genet. Biol.*, 23, 205, 1998.

108. Albrecht, C., Geurts, R., and Bisseling, T., Legume nodulation and mycorrhizae formation; two extremes in host specificity meet, *EMBO J.*, 18, 281, 1999.

109. Oldroyd, G., Harrison, M., and Udvardi, M.K., Peace talks and trade deals: keys to long-term harmony in legume-microbe symbioses, *Plant Physiol.*, 137, 1205, 2005.

110. Imaizumi-Anraku, H. et al., Plastid proteins crucial for symbiotic fungal and bacterial entry into plant roots, *Nature*, 433, 527, 2005.

111. Bonfante, P. et al., The *Lotus japonicus* LjSym4 gene is required for the successful symbiotic infection of root epidermal cells, *Mol. Plant-Microbe Interact.*, 13, 1109, 2000.

112. Novero, M. et al., Dual requirement of the *LiSym4* gene for mycorrhizal development in epidermal and cortical cells of *Lotus japonicus* roots, *New Phytologist*, 154, 741, 2002.

113. Martin, F., Frontiers in molecular mycorrhizal research — genes, loci, dots and spins, *New Phytologist*, 150, 499, 2001.

114. Wiemken, V. and Boller, T., Ectomycorrhiza: gene expression, metabolism and the wood-wide web, *Curr. Opin. Plant Biol.*, 5, 355, 2002.

115. Voiblet, C. et al., Identification of symbiosis-regulated genes in *Eucalyptus globulus-Pisolithus tinctorius* ectomycorrhiza by differential hybridization of arrayed cDNAs, *Plant J.*, 25, 181, 2001.

116. Johansson, T. et al., Transcriptional responses of *Paxillus involutus* and *Betula pendula* during formation of ectomycorrhizal root tissue, *Mol. Plant-Microbe Interact.*, 17, 202, 2004.

117. Duplessis, S. et al., Transcript patterns associated with ectomycorrhiza development in *Eucalyptus globulus* and *Pisolithus microcarpus, New Phytologist,* 165, 599, 2005.

118. Le Quere, A. et al., Global patterns of gene regulation associated with the development of ectomycorrhiza between birch *Betula pendula (*Roth9 and *Paxillus involutus* (Batsch) Fr., *Mol. Plant-Microbe Interact.,* 18, 659, 2005.

119. Peter, M. et al., Analysis of expressed sequence tags from the ectomycorrhizal basidiomycetes *Laccaria bicolor* and *Pisolithus microcarpus, New Phytologist,* 159, 117, 2003.

120. Martin, F., Exploring the transcriptome of the ectomycorrhizal symbiosis, in *Molecular Genetics and Breeding of Forest Trees,* Kumar, S. and Fladung, M., Eds., Haworth's Food Products Press, New York, 2004, p. 81.

121. Martin, F. et al., Symbiotic sequencing for the *Populus* mesocosm, *New Phytologist,* 161, 330, 2004.

122. Peter, M., Ayer, F., and Egli, S., Nitrogen addition in a Norway spruce stand altered macromycete sporocarp production and below-ground ectomycorrhizal species composition, *New Phytologist,* 149, 311, 2001.

123. Lilleskov, E.A. and Bruns, T.D., Nitrogen and ectomycorrhizal fungal communities: what we know, what we need to know, *New Phytologist,* 149, 156, 2001.

124. Hobbie, E.A. et al., Foliar nitrogen concentrations and natural abundance of (^{15})N suggest nitrogen allocation patterns of Douglas-fir and mycorrhizal fungi during development in elevated carbon dioxide concentration and temperature, *Tree Physiol.,* 21, 1113, 2001.

125. Henn, M.R. and Chapela, I.H., Differential C isotope discrimination by fungi during decomposition of C(3)- and C(4)-derived sucrose, *Appl. Environ. Microbiol.,* 66, 4180, 2000.

126. Timonen, S., Tammi, H., and Sen, R., Characterisation of the host genotype and fungal diversity in Scots pine ectomycorrhiza from natural humus microcosms using isozyme and PCR-RFLP analyses, *New Phytologist,*135, 313, 1997.

127. Morel, M. et al., Identification of genes differentially expressed in extraradical mycelium and ectomycorrhizal roots during *Paxillus involutus-Betula pendula* ectomycorrhizal symbiosis, *Appl. Environ. Microbiol.,* 71, 382, 2005.

128. Jakobsen, I., Carbon metabolism in mycorrhiza, *Methods Microbiol.,* 23, 149, 1991.

129. Botton, B. and Chalot, M., Nitrogen assimilation: enzymology in ectomycorrhizas, in *Mycorrhiza: Structure, Molecular Biology and Function,* Varma, A. and Hoch, B., Eds., Springer, Berlin Heidelberg, New York, 1995, p. 325.

130. Hampp, R. and Schaeffer, C., Mycorrhiza — carbohydrates and energy metabolism, in *Mycorrhiza: Structure, Molecular Biology and Function,* Varma, A. and Hoch, B., Eds., Springer Verlag, Berlin Heidelberg, 1995, p. 267.

131. Attiwill, P.M. and Adams, M.A., Nutrient cycling in forests, *New Phytologist,* 124, 561, 1993.

132. Francis, R. and Read, D.J., The contributions of mycorrhizal fungi to the determination of plant community structure, in *Management of Mycorrhizas in Agriculture, Horticulture and Forestry,* Robson, A.D., Abbott, L.K., and Malajczuk, N., Eds., Kluwer Academic Publishers, Dordrecht, 1994, p. 11.

133. Pate, J.S, The mycorrhizal association: just one of many nutrient acquiring specializations in natural ecosystems, in *Management of Mycorrhizas in Agriculture, Horticulture and Forestry,* Robson, A.D., Abbott, L.K., and Malajczuk, N., Eds., Kluwer Academic Publishers, Dordrecht, 1994, p. 1.

134. Johansen, A., Finlay, R.D., and Olsson, P.A., Nitrogen metabolism of external hyphae of the arbuscular mycorrhizal fungus *Glomus intraradices, New Phytologist,* 133, 705, 1996.

135. Hodge, A., Campbell, C.D., and Fitter, A.H., An arbuscular mycorrhizal fungus accelerates decomposition and acquires nitrogen directly from organic material, *Nature,* 413, 297, 2001.

136. Read, D.J., Mycorrhizas in ecosystems, *Experientia,* 47, 376, 1991.

137. Cappellazzo, G. et al., unpublished results.

138. Finlay, R.D. et al., Mycelial uptake, translocation and assimilation of nitrogen from ^{15}N-labelled ammonium by *Pinus sylvestris* plants infected with four different ectomycorrhizal fungi, *New Phytologist,*110, 59,1988.

139. Paris, F. et al., Weathering of ammonium -or calcium-saturated 2:1 phyllosilicates by ectomycorrhizal fungi *in vitro, Soil Biol. Biochem.,* 27, 1237, 1995.

140. Martin, F. and Botton, B., Nitrogen metabolism of ectomycorrhizal fungi and ectomycorrhiza, *Adv. Plant Pathol.,* 9, 83, 1993.

141. Read, D.J., Mycorrhizal fungi: the ties that bind, *Nature,* 388, 517, 1997.

142. Javelle, A. et al., High-affinity ammonium transporters and nitrogen sensing in mycorrhizas, *Trends Microbiol.,* 11, 53, 2003.

143. Lacourt, I. et al., Isolation and characterization of differentially expressed genes in the mycelium and fruit body of *Tuber borchii*, *Appl. Environ. Microbiol.*, 689, 4574, 2002.

144. Montanini, B. et al., Distinctive properties and expression profiles of glutamine synthetase from a plant symbiotic fungus, *Biochem. J.*, 373, 357, 2003.

145. Bending, G.D. and Read, D.J., Lignin and soluble phenolic degradation by ectomycorrhizal and ericoid fungi, *Mycol. Res.*, 101, 1348, 1997.

146. Leake, J.R. and Read, D.J., Experiments with ericoid mycorrhiza, in *Methods in Microbiology*, Vol. 23, Norris, J.R., Read, D.J., and Varma, A.K., Eds., Academic Press, London, 1991, p. 435.

147. Cairney, J.W. and Burke, R.M., Plant cell wall-degrading enzymes in ericoid and ectomycorrhizal fungi, in *Mycorrhizas in Integrated Systems: From Genes to Plant Development*, Azcon-Aguilar, C. and Barea, J.M., Eds., Official Publications of the European Community, Luxemburg, 1996, p. 218.

148. Varma, A.K. and Bonfante, P., Utilization of cell-wall related carbohydrates by ericoid mycorrhizal endophytes, *Symbiosis*, 16, 301, 1994.

149. Perotto, S. et al., Production of pectin-degrading enzymes by ericoid mycorrhizal fungi, *New Phytologist*, 135, 151, 1997.

150. Marshner, H., *Mineral Nutrition of Higher Plants*, 2nd ed., Academic Press, London, 1995.

151. Lapeyrie, F., Ranger, J., and Vairelles, D., Phosphate-solubilizing activity of ectomycorrhizal fungi *in vitro*, *Can. J. Bot.*, 69, 342, 1991.

152. Martin, F. et al., Aluminium polyphosphate complexes in the mycorrhizal basidiomycete *Laccaria bicolor*: a ^{27}Al NMR study, *Planta*, 194, 241, 1994.

153. Smith, S.E. and Gianinazzi-Pearson, V., Physiological interactions between symbionts in vesicular arbuscular mycorrhizal plants, *Annu. Rev. Plant Physiol. Plant Mol. Biol.*, 39, 221, 1988.

154. Marschner, G.E. and Jakobsen, I., Role of AM fungi in uptake of phosphorus and nitrogen from soil, *Crit. Rev. Biotechnol.*, 15, 257, 1995.

155. Harrison, M.J. and Van Buuren, M.L., A phosphate transporter from the mycorrhizal fungus *Glomus versiforme*, *Nature*, 378, 626, 1995.

156. Maldonado-Mendoza, I.E., Dewbre, G.R., and Harrison, M.J., A phosphate transporter gene from the extra-radical mycelium of an arbuscular mycorrhizal fungus *Glomus intraradices* is regulated in response to phosphate in the environment, *Mol. Plant-Microbe Interact.*, 14, 1140, 2001.

157. Benedetto, A. et al., Expression profiles of a phosphate transporter gene (*GmosPT*) from the endomycorrhizal fungus *Glomus mosseae*, *Mycorrhiza*, 15, 620, 2005.

158. Ezawa, T. et al., Rapid accumulation of polyphosphate in extraradical hyphae of an arbuscular mycorrhizal fungus as revealed by histochemistry and a polyphosphate kinase/luciferase system, *New Phytologist*, 161, 387, 2003.

159. Uetake, Y. et al., Extensive tubular vacuole system in an arbuscular mycorrhizal fungus, *Gigaspora margarita*, *New Phytologist*, 154, 761, 2002.

160. Ezawa, T., Smith, S.E., and Smith, F.A., P metabolism and transport in AM fungi, *Plant and Soil*, 244, 221, 2002.

161. Harrison, M.J., Dewbre, G.R., and Liu, J., A phosphate transporter from *Medicago truncatula* involved in the acquisition of phosphate released by arbuscular mycorrhizal fungi, *Plant Cell*, 14, 2413, 2002.

162. Karandashov, V. et al., Evolutionary conservation of a phosphate transporter in the arbuscular mycorrhizal symbiosis, *Proc. Natl. Acad. Sci. USA*, 101, 6285, 2004.

163. Paszkowski, U. et al., Rice phosphate transporters include an evolutionarily divergent gene specifically activated in arbuscular mycorrhizal symbiosis, *Proc. Natl. Acad. Sci. USA.*, 99, 13324, 2002.

164. Rausch, C. et al., A phosphate transporter expressed in arbuscule-containing cells in potato, *Nature*, 414, 462, 2001.

165. Simard, S.W. et al., Net transfer of carbon between ectomycorrhizal tree species in the field, *Nature*, 388, 579, 1997.

166. Bidartondo, M.I. and Bruns, T.D., Fine-level mycorrhizal specificity in the Monotropoideae (Ericaceae): specificity for fungal species groups, *Mol. Ecol.*, 11, 557, 2002.

167. Bidartondo, M.I. et al., Specialized cheating of the ectomycorrhizal symbiosis by an epiparasitic liverwort, *Proc. R. Soc. Lond. B.*, 270, 835, 2003.

168. McKendrick, S.L. et al., Symbiotic germination and development of myco-heterotropicplants in nature: ontogeny of *Corrallorhiza trifida* and characterization of its mycorrhizal fungi, *New Phytologist*, 145, 523, 2000.

169. Leake, J.R., The biology of mycoheterotrophic ('saprophytic') plants, *New Phytologist,* 127, 171, 1994.
170. Julou, T. et al., Mixotrophy in orchids: insights from a comparative study of green individuals and nonphotosynthetic individuals of *Cephalanthera damasonium, New Phytologist,* 166, 639, 2005.
171. Turnau, K. et al., Role of arbuscular mycorrhiza and associated micro-organisms in phytoremediation of heavy metal polluted sites, in *Trace Elements in the Environment Biogeochemistry, Biotechnology and Bioremediation,* Prasad, M.N.V., Sajwan, D., and Ravi, S., Eds., CRC Press/Lewis Publishers, Boca Raton, FL, 2005.
172. Colpaert, J.V. et al., Genetic variation and heavy metal tolerance in the ectomycorrhizal basidiomycete *Suillus luteus, New Phytologist,* 147, 367, 2000.
173. Martino, E. et al., Ericoid mycorrhizal fungi from heavy metal polluted soils: their identification and growth in the presence of heavy metals, *Mycol. Res.,* 104, 338, 2000.
174. Sharples, J.M. et al., Symbiotic solution to arsenic contamination, *Nature,* 404, 951, 2000.
175. Gadd, G.M., Microbial influence on metal mobility and application for bioremediation, *Geoderma,* 122, 109, 2004.
176. Gadd, G.M., Bioremedial potential of microbial mechanisms of metal mobilization and immobilization, *Curr. Opin. Biotech.,* 11, 271, 2000.
177. Leyval, C., Turnau, K., and Haselwandter, K., Effect of heavy metal pollution on mycorrhizal colonization and function: physiological, ecological and applied aspects, *Mycorrhiza,* 7, 139, 1997.
178. Perotto, S. and Martino, E., Molecular and cellular mechanisms of heavy metal tolerance in mycorrhizal fungi: what perspectives for bioremediation?, *Minerva Biotecnol.,* 13, 55, 2001.
179. Väre, H., Aluminum polyphosphate in the ectomycorrhizal fungus *Suillus variegatus* (Fr.) O. Kuntze as revealed by energy dispersive spectrometry, *New Phytologist,* 116, 663, 1990.
180. Kottke, I. and Martin, F., Demonstration of aluminium in polyphosphate of *Laccaria amethystea* (Bolt. ex Hooker) Murr. by means of electron energy loss spectroscopy, *J. Microsc.,* 174, 225, 1994.
181. Wilkins, D.A., The influence of sheathing (ecto-) mycorrhizas of trees on the uptake and toxicity of metals, *Agric. Ecosyst. Environ.,* 35, 245, 1991.
182. Jentschke, G.D., Godbold, L., and Hüttermann, A., Culture of mycorrhizal tree seedlings under controlled conditions: effects of nitrogen and aluminium, *Physiol. Plant.,* 81, 408, 1991.
183. Kottke, I., Electron energy loss spectroscopy and imaging techniques for subcellular localization of elements in mycorrhizas, *Methods Microbiol.,* 23, 369, 1991.
184. Meagher, R.B., Phytoremediation of toxic elemental and organic pollutants, *Curr. Opin. Plant Biol.,* 3, 153, 2000.
185. Kahn, A.G. et al., Role of plants, mycorrhizae and phytochelators in heavy metal contaminated land remediation, *Chemosphere,* 41, 197, 2000.
186. Bradley, R.A., Burt, J., and Read, D.J., Mycorrhizal infection and resistance to heavy metal toxicity in *Calluna vulgaris, Nature,* 292, 335, 1981.
187. Hall, J.L., Cellular mechanisms for heavy metal detoxification and tolerance, *J. Exp. Bot.,* 53, 1, 2002.
188. Adriaensen, K. et al., A zinc-adapted fungus protects pines from zinc stress, *New Phytologist,* 161, 549, 2004.
189. Lasat, M.M., Phytoextraction of toxic metals: a review of biological mechanisms, *J. Environ. Qual.,* 31, 109, 2002.
190. Meharg, A.A. and Cairney, J.W.G., Ectomycorrhizas — extending the capabilities of rhizosphere remediation?, *Soil Biol. Biochem.,* 32, 1475, 2000.
191. Joner, E.J. and Leyval, C., Phytoremediation of organic pollutants using mycorrhizal plants: a new aspect of rhizosphere interactions, *Agronomie,* 23, 495, 2003.
192. Bowen, G.D. and Theodorou, C., Interactions between bacterial and ectomycorrhizal fungi, *Soil Biol. Biochem.,* 11, 119, 1979.
193. De Boer, W. et al., Living in a fungal world: impact of fungi on soil bacterial niche development. *FEMS Microbiol. Rev.,* 29, 795, 2005.
194. Garbaye, J., Helper bacteria: a new dimension to the mycorrhizal symbiosis, *New Phytologist,* 128, 197, 1994.
195. Gamalero, E. et al., Impact of two fluorescent pseudomonads and an arbuscular mycorrhizal fungus on tomato plant growth, root architecture and P acquisition, *Mycorrhiza,* 14, 185, 2004.
196. Bianciotto, V. et al., Cellular interactions between arbuscular mycorrhizal fungi and rhizosphere bacteria, *Protoplasma,* 193, 123, 1996.

197. Nurmiaho-Lassila, E.L. et al., Bacterial colonization patterns of intact *Pinus sylvestris* mycorrhizospheres in dry pine forest soil: an electron microscopy study, *Can. J. Microbiol.,* 43, 1017, 1997.
198. Perotto, S. and Bonfante, P., Bacterial associations with mycorrhizal fungi: close and distant friends in the rhizosphere, *Trends Microbiol.,* 5, 496, 1997.
199. Chin-a-woeng, T.F.C. et al., Description of the colonization of a gnotobiotic tomato rhizosphere by *Pseudomonas fluorescens* biocontrol strain WCS365, using scanning electron microscopy, *Mol. Plant-Microbe Interact.,* 10, 79, 1997.
200. Boddey, R.M. et al., Biological nitrogen fixation associated with sugar cane, *Plant Soil,* 137, 111, 1991.
201. Johansson, J.F., Paul, L.R., and Finlay, R.D., Microbial interactions in the mycorrhizosphere and their significance for sustainable agriculture, *FEMS Microbiol. Ecol.,* 48, 1, 2004.
202. Frey-Klett, P. et al., Ectomycorrhizal symbiosis affects functional diversity of rhizosphere fluorescent pseudomonads, *New Phytologist,* 165, 317, 2005.
203. Frey, P. et al., Metabolic and genotypic fingerprinting of fluorescent pseudomonads associated with the Douglas fir-*Laccaria bicolor* mycorrhizosphere, *Appl. Environ. Microbiol.,* 63, 1852, 1997.
204. Barbieri, E. et al., New evidence for bacterial diversity in the ascoma of the ectomycorrhizal fungus *Tuber borchii* Vittad, *FEMS Microbiol. Lett.,* 247, 23, 2005.
205. Danell, E., Alström, S., and Ternström, A., *Pseudomonas fluorescens* in association with fruit bodies of the ectomycorrhizal mushroom *Cantharellus cibarius, Mycol. Res.,* 97, 1148, 1993.
206. Andrade, G. et al., Bacteria from rhizosphere and hyphosphere soils of different arbuscular-mycorrhizal fungi, *Plant Soil,* 192, 71, 1997.
207. Bianciotto, V. et al., An obligately endosymbiotic fungus itself harbors obligately intracellular bacteria, *Appl. Environ. Microbiol.,* 62, 3005, 1996.
208. Scannerini, S. and Bonfante, P., Bacteria and bacteria like objects in endomycorrhizal fungi (Glomaceae), in *Symbiosis as Source of Evolutionary Innovation: Speciation and Morfogenesis,* Margulis, L. and Fester, R., Eds., The MIT Press, Cambridge, MA, 1991, p. 273.
209. Bianciotto, V. et al., '*Candidatus Glomeribacter gigasporarum*', an endosymbiont of arbuscular mycorrhizal fungi, *Int. J. Syst. Evol. Microbiol.,* 53, 121, 2003.
210. Bianciotto, V. et al., Detection and identification of bacterial endosymbionts in arbuscular mycorrhizal fungi belonging to Gigasporaceae, *Appl. Environ. Microbiol.,* 66, 4503, 2002.
211. Baumann, P., Genetics, physiology and evolutionary relationships of the genus *Buchnera*: intracellular symbionts of aphids, *Annu. Rev. Microbiol.,* 49, 55, 1995.
212. Jargeat, P. et al., Isolation, free-living capacities, and genome structure of "*Candidatus Glomeribacter gigasporarum,*" the endo cellular bacterium of the mycorrhizal fungus *Gigaspora margarita, J. Bacteriol.,* 186, 6876, 2004.
213. Bianciotto, V. et al., Vertical transmission of endobacteria in the arbuscular mycorrhizal fungus *Gigaspora margarita* through generation of vegetative spores, *Appl. Environ. Microbiol.,* 70, 3600, 2004.
214. van Buuren, M. et al., Construction and characterization of genomic libraries of two endomycorrhizal fungi: *Glomus versiforme* and *Gigaspora margarita, Mycol. Res.,* 103, 955, 1999.
215. Bertaux, J. et al., *In situ* identification of intracellular bacteria related to *Paenibacillus spp.* in the mycelium of the ectomycorrhizal fungus *Laccaria bicolor* S238N, *Appl. Environ. Microbiol.,* 69, 4243, 2003.
216. Bertaux, J. et al., Occurrence and distribution of endobacteria in the plant-associated mycelium of the ectomycorrhizal fungus *Laccaria bicolor* S238N, *Environ. Microbiol.,* 7, 1786, 2005.
217. Frey-Klett, P. and Garbaye, J., Mycorrhiza helper bacteria: a promising model for the genomic analysis of fungal-bacterial interactions, *New Phytologist,* 168, 4, 2005.
218. Daniel, R., The metagenomics of soil, *Nat. Rev. Microbiol.,* 3, 470, 2005.
219. Jackson, R.B. et al., Linking molecular insight and ecological research, *Trends Ecol. Evol.,* 17, 409, 2002.

9 Molecular Biology and Ecology of the *Rhizobia*–Legume Symbiosis

Dietrich Werner

CONTENTS

I. INTRODUCTION

The progress in understanding the symbiosis between the various genera of rhizobia and their legume host plants was significant during the past few years. It was promoted by several genome projects on the bacterial side (*Rhizobium* NGR 234, *Sinorhizobium meliloti*, *Mesorhizobium loti*, *Bradryrhizobium japonicum*), as well as on the plant side (*Medicago sativa*, *Lotus japonicus*) by proteomics and by new methods in molecular cell biology. In the new seven volume series on *"Nitrogen Fixation: Origins, Applications and Research Progress,"* this symbiosis plays a major role in three volumes [1–3]. From all procaryotic genomes so far sequenced, three rhizobia species have the largest genomes, with 9.1 Mb for *Bradyrhizobium japonicum*, 7.6 Mb for *Mesorhizobium loti,* and 6.7 Mb for *Sinorhizobium meliloti*. On the other side, some animal pathogenic bacteria have the smallest genome, such as *Borrellia burgdorferi* with 1.2 Mb and *Chlamydia trachomatis* with 1.0 Mb [4].

This chapter covers the symbiosis in the following order:

- The ecology of rhizobia and legumes
- Signal exchange between host plants and rhizobia in the rhizosphere
- Nod factor perception and related processes
- Nodule and compartmentation development
- Soil stress factors affecting the symbiosis
- The ecological and agricultural impact of the symbiosis for the nitrogen cycle

The rationale to combine molecular biology and the ecology of the symbiosis is that understanding the functions of genes and the signaling between the two symbiotic partners is the basis to understand the ecological success of the symbiosis. The ecological and agricultural importance of legumes [5], on the other side, is a major argument, to study the molecular biology of the partners with all available methods.

II. THE ECOLOGY OF RHIZOBIA AND LEGUMES

The following genera of rhizobia are established with a still open number of species:

* The genus *Rhizobium (Allorhizobium)*
* The genus *Sinorhizobium*
* The genus *Mesorhizobium*
* The genus *Bradyrhizobium*
* The genus *Azorhizobium*

It is now confirmed that species from other genera can form nitrogen-fixing nodules on specific legumes such as *Mimosa* sp. and *Crotolaria* sp:

* The genus *Ralstonia* (Wautersia) [6]
* The genus *Burkholderia* [7]
* The genus *Methylobacterium*[8]

The species concept of rhizobia has critically been discussed by Pablo Vinuesa [9], Peter Young [10], Kristina Lindström [11] and Esperanza Martinez-Romero [12]. The basic physiological characteristics of rhizobia as widely distributed soil bacteria have been described in detail by Werner [13] and Bottomley [14]. New results have been found in the detailed studies on quorum sensing in rhizobia, where the bacteria communicate with each other, before they reach the critical population density to communicate and invade their host plants [15]. In *Rhizobium leguminosarum* bv *viciae,* four quorum sensing systems have been identified, three located on the symbiotic plasmid: *rai, rhi,* and *tra,* whereas the *cinRI* system genes are located on the chromosome. The rhi system comprises *rhiR*, which is a *luxR* homolog, *rhiI* (a *luxI* homolog), and the *rhiABC* operon, which is controlled by Rhi R, whose expression is controlled by flavonoids (Figure 9.1). The gene *rhiA* is well expressed in free-living cells in the rhizosphere, but repressed in bacteroids. Signal molecules in quorum sensing are acylated homoserine lactones (AHLs). The gene *rhiI* is responsible for the synthesis of short chain AHLs such as C_6-HSL and C_8-HSL [16]. Also, long-chain HSLs have been identified such as 3-OH-$C_{14:1}$-HSL, encoded by the cinRI locus [17]. The gene for the AHL synthase (*cinI*) is regulated by the long-chain AHL. Mutations in the three quorum sensing genes *rai, tra,* and *cin* do not affect nodulation [18]; however, mutations in the rhi system affect nodulation efficiency. *Sinorhizobium meliloti* has two characterized quorum sensing systems. The genes *sinR* and *sinI* encode the production of rather large AHLs with up to C_{18}-HSL [19]. Both genes are also required for exopolysaccharide (EPS)-synthesis regulation and may affect nodulation efficiency via *expR* [20]. A completely different mechanism is apparently used by *Bradyrhizobium japonicum* for quorum sensing. The population density is here mediated by bradyoxetin: 2-(4-((4-(3-aminooxetan-2-yl)phenyl)(imino)methyl(phenyl l) oxetan-3-ylamine [21]. The response regulator nwsb for quorum sensing acts together with nolA to repress nod genes. The connection to nutrient supply for the bacteria is given by the observation that synthesis of bradyoxetin is enhanced under Fe-limitation. Twenty years ago, the hypothesis was already published that *Bradyrhizobium* cells invading root hairs of their host plant soybeans find in the first place there is a high Fe (and Co, Ca, and Mo) concentration and, only later, the N/C exchange dominates the mutualistic interaction [22,23].

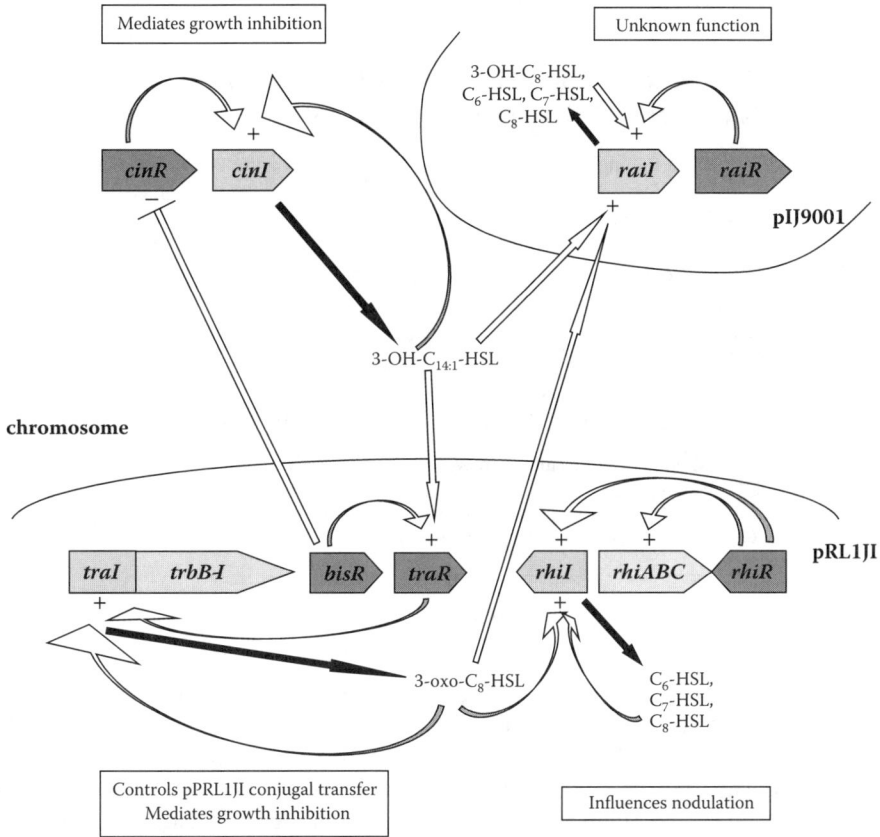

FIGURE 9.1 *Rhizobium leguminosarum* bv. *viciae* quorum-sensing network [15].

Biofilms composed of a large number of bacterial species are another fascinating new area of microbial ecology. Here, the prokaryotic cells differentiate in layers and niches similar to multicellular eukaryotic species, communicating with each other with a very complex set of signal molecules [24,25]. The spatial arrangement of microcolonies, with clusters of different species, competes for nutrients at, probably, rather low concentrations. The genes involved in the transition of planktonic life to biofilm life are of special interest, for they have been overlooked in decades of laboratory cultures, using only liquid cultures and fermenter techniques. Biofilms on plant surfaces are even more complicated, because here the signaling from and to the plant is also involved [26]. The composition of a biofilm from water tanks was a big surprise: members of Rhizobiales were the most abundant group, not typical species from the aquatic environment (Table 9.1).

Molecular techniques (including those for metagenomics) also permit the study of the large number of unculturable microorganisms from soils as well as from many other habitats with their genes, their proteins, and also with their metabolites [29,30]. By using DNA microchips from biofilms, a large number of clones can be studied at the same time [31]. For the proteomic approach, microarray techniques allow the screening of hundreds of different proteins, including enzymes [32]. The ecological importance of legumes can be hardly overestimated. With more than 16000 species, this order is present worldwide in all ecosystems, from the arctic tundra with lupins to tropical forests with a large number of legume tree species. The agricultural production of the most important grain legumes is presented in Table 9.2.

TABLE 9.1
Composition of Bacterial Families from a Biofilm
Taken from a Drinking-Water Reservoir

Sequences Belonging to the Families	Percentage of the Population
Rhizobiales	32
Pseudomonadales	15
Enterobacteriales	11
Burkholderiales	7
Actinomycetales	6
Xanthomonadales	3
Alteromonadales	2
Other groups	19

Note: Modified from results from W. Streits laboratory.

Source: Schmeisser, C. et al., *Appl. Environ. Microbiol.*, 69, 7298, 2003;
Steele, H.L. and Streit, W.R., *FEMS Microbiol. Lett.*, 247, 105, 2005.

Soybeans are by far the most important legume production crop, with about 189 million tons per year harvested and cultivated on about 80 million hectares. This is a larger area than the combined land area of Italy and Germany. The U.S., Brazil, Argentina, China, and India are the largest producers. This crop is covered in several chapters in [1]. For many other countries, especially in the tropics and subtropics, several other grain legumes listed in Table 9.2 are of equal importance.

The land area covered with fodder and pasture legumes may be even larger than that used for grain legumes production. When we assume that, on average, 10% of all pasture areas are covered with legumes such as *Trifolium* sp., *Medicago* sp., or *Stylosanthes* sp., more than 300 million hectares are covered with legumes [13]. Tree legumes, such as the (more than 1200) Acacia species [39] may cover an even larger area but with a low density in many regions.

TABLE 9.2
World Production of Grain Legumes

Legume	Million Tons per Year
Soybeans	189
Groundnuts	36
Dry beans (*Phaseolus* and *Vigna*)	19
Dry peas	11
Chickpeas	6
Dry faba beans	4
Lentils	3
Green beans	5
Green peas	7

Source: FAO Statistical Yearbook, FAO, Rome, 2004.

III. SIGNAL EXCHANGE BETWEEN HOST PLANTS
AND RHIZOBIA IN THE RHIZOSPHERE

A. FLAVONOID-DEPENDENT SIGNALING

After the publication of the first edition of this volume in 2001, with the chapter from the author on "Organic Signals between Plants and Microorganisms," a number of reviews have been published on signal exchange and biochemical modification [33–35]. The most comprehensive contribution came from Jim Cooper [36], covering the synthesis, release, biochemical modifications, gene-inducing and regulating activities of flavonoids. The signal exchange starts with release of small concentrations (micro- to nanomolar) of flavonoids or isoflavonoids from specific root zones. Some of the compounds act also as chemoattractants from rhizobia. A selected list of nod-gene-inducing compounds from legumes is summarized in Table 9.3.

The largest set of different nod-gene-inducing compounds has been identified in the root exudate of *Phaseolus vulgaris*, *Glycine max*, and *Medicago sativa*. These are also the most prominent research species of legumes, and they are very important crops in agriculture. The best-studied nod gene inducers are luteolin (a flavone), daidzein (an isoflavone), naringenin (a flavanone), and isoliquiritigenin (a chalcone). Completely different structures are present in the betains stachydrine and trigonellin found in *Medicago sativa* exudates [40] and in aldonic acids found in *Lupinus albus* [41]. The microsymbionts are able to degrade the nod-gene-inducing flavonoids and isoflavonoids. As shown for luteolin degradation by *Sinorhizobium meliloti* and for daidzein by *Bradyrhizobium japonicum* in Figure 9.2, caffeic acid, protocatechuic acid, coumaric acid, and phenylacetic acid are major products of the degradation. These compounds can effect growth of several rhizospere microorganisms and, therefore, also the survival and competition of rhizobia.

The major function of the compounds in Table 9.2 is certainly the triggering of the early nodulation events (Figure 9.3).

In nano- to micromolar concentrations, they bind to the Nod D receptor in the rhizobia-cell surface, triggering the induction of all other nod genes involved in the production of the nodulation factors, called also nod factors (NF) or lipochitooligosaccharides (LCOs). Combinations of two or three different flavonoids are, in some cases, more effective than single compounds, as shown for the combination of daidzein and isoloquiritigenin in *Rhizobium leguminosarum* bv *phaseoli* [42]. The nod genes, the species in which they have been identified, and their functions are summarized in Table 9.4.

Nod A, B, and C, responsible for the synthesis of the LCO backbone, and nod I and J, involved in the final secretion of the LCOs, are common in all rhizobia; other nod genes involved in the regulation of nod gene expression and in substitution reactions are only present in some rhizobial species and absent in others. The general structure of NF is shown in Figure 9.4.

Recently, in a specific strain of *Rhizobium etli*, a nod factor with six glucosamine rings has been identified as the major LCO [43].

The induction of genes responsible for type III secretion systems (TTSS) by flavonoids is only recently established for several *Rhizobium* and *Bradyrhizobium* strains, such as *Rhizobium* NGR 234 [44,45], *Bradyrhizobium japonicum* [46] and *Sinorhizobium fredii* [47]. The TTSS of the last two species have only little sequence homology. On the other side, TTSS genes of some rhizobia have a high degree of homology to genes from the pathogen Yersinia [48]. It is also very interesting that in the genome of *Sinorhizobium meliloti* [49], no TTSS homologues have been found. Proteins secreted by the TTSS apparatus are called Nops (nodulation outer proteins), equivalent to Yops (Yersinia outer proteins) [50]. At least five different Nops have been identified, with three of them associated with pili, other Nops may be involved in the MAP–kinase pathway [51] (see next paragraph). The function of Nops for the development of the symbiosis and nodule formation are not yet understood; mutations in TTSS genes can in some cases increase nodule number, in other cases decrease the number. In one case, where the wild-type cannot nodulate a specific cultivar of

TABLE 9.3
Nod-Gene-Inducing Compounds from the Root Exudate of Legumes

Host Legume	Compound
Phaseolus vulgaris	Delphinidin (3,5,7,3',4',5'-hexahydroxyflavylium)
	Kaempferol (3,5,7,4'-tetrahydroxyflavonol)
	Malvidin (3,5,7,4'-pentahydroxy-3',5'-dimethoxyflavylium)
	Myricetin (3,5,7,3',4',5'-hexahydroxyflavone)
	Petunidin (3,5,7,4',5'-pentahydroxy-3'methoxyflavylium)
	Quercetin (3,5,7,3',4'-pentahydroxyflavonol)
	Eriodictyol
	Genistein
	Naringenin (5,7,4'-trihydroxyflavonone)
	Daidzein
	Liquiritigenin
	Isoliquiritigenin
	Coumestrol
Glycine max	Daidzein (7,4'-dihydroxyisoflavone)
	Genistein (5,7,4'-trihydroxyisoflavone)
	Coumestrol (3,9-dihydroxycoumestan)
	Isoliquiritigenin (4,2',4'-triihydroxychalcone)
	Genistein-7-*O*-glucoside
	Genistein-7-*O*-(6'-*O*-malonylglucoside)
	Daidzein-7-*O*-(6'-*O*-malonylglucoside)
Medicago sativa	Luteolin (5,7,3'4'-tetrahydroxyflavone)
	Chrysoeriol (3'-methoxy-5,7,4'-trihydroxyflavone)
	Liquiritigenin (7,4'-Dihydroxyflavonone)
	7,4'-Dihydroxyflavone
	Methoxychalcone (4,4'-dihydroxy-2'-methoxychalcone)
	Stachydrine (betaine)
	Trigonelline (betaine)
Robinia pseudoacacia	Isoliquiritigenin
	Naringenin
	Apigenin
	4,7-Dihydroxyflavone
	Chrysoeryol
Vicia sativa	3,5,7,3'-Tetrahydroxy-4'-methoxyflavonone
	7,3'-Dihydroxy-4'-methoxyflavonone
	Four more partially characterized flavonones
Trifolium repens	7,4'-Dihydroxyflavone
	Geraldone (7,4'-dihydroxy-3'-methoxyflavone)
	4'-Hydroxy-7-methoxyflavone
Cowpea	Daidzein
	Genistein
	Coumestrol
Pisum sativum	Apigenin-7-*O*-glucoside (5,7,4'-trihydroxyflavone-7-*O*-glucoside)
	Eriodictyol (5,3,3',4'-tetrahydroxyflavanone)
Lupinus albus	Erythronic acid (aldonic acid)
	Tetronic acid (aldonic acid)
Sesbania rostrata	Liquiritigenin

Source: Modified from Cooper, J.E., *Advances in Botanical Research*, Vol. 41, *Incorporating Advances in Plant Pathology*, 2004, p. 1; Werner, D., *Fast Growing Trees and Nitrogen Fixing Trees*, Werner, D. and Müller, P., Eds., G. Fischer, Stuttgart, 1990, p. 3; Scheidemann, P. and Wetzel, A., *Trees*, 11, 316, 1997; Sprent, J.I., *Agriculture, Forestry, Ecology and the Environment*, Werner, D. and Newton, W.E., Eds., Kluwer Academic, Dordrecht, 2005, p. 113.

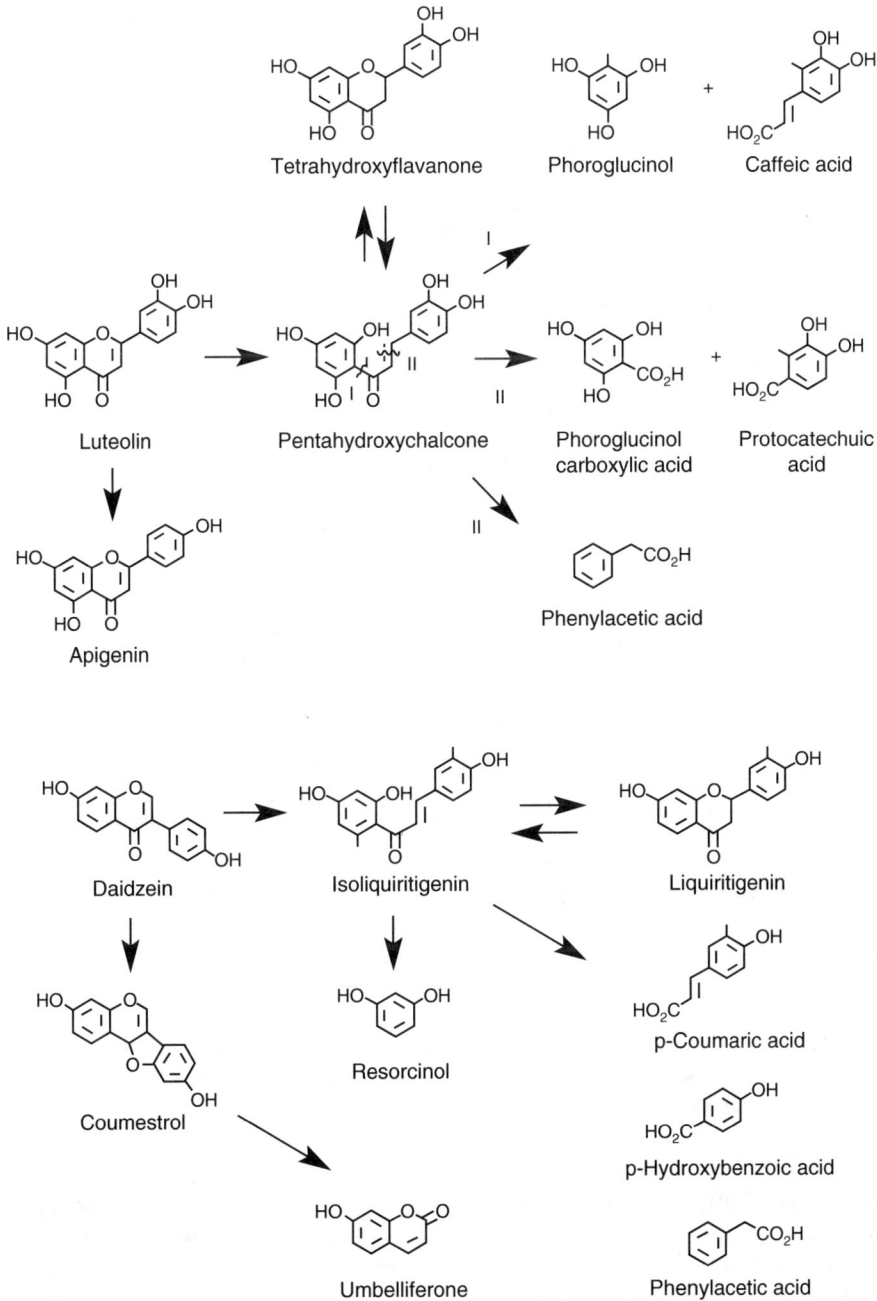

FIGURE 9.2 Proposed degradation pathway for luteolin by *Sinorhizobium meliloti* and for daidzein by *Bradyrhizobium japonicum* [33].

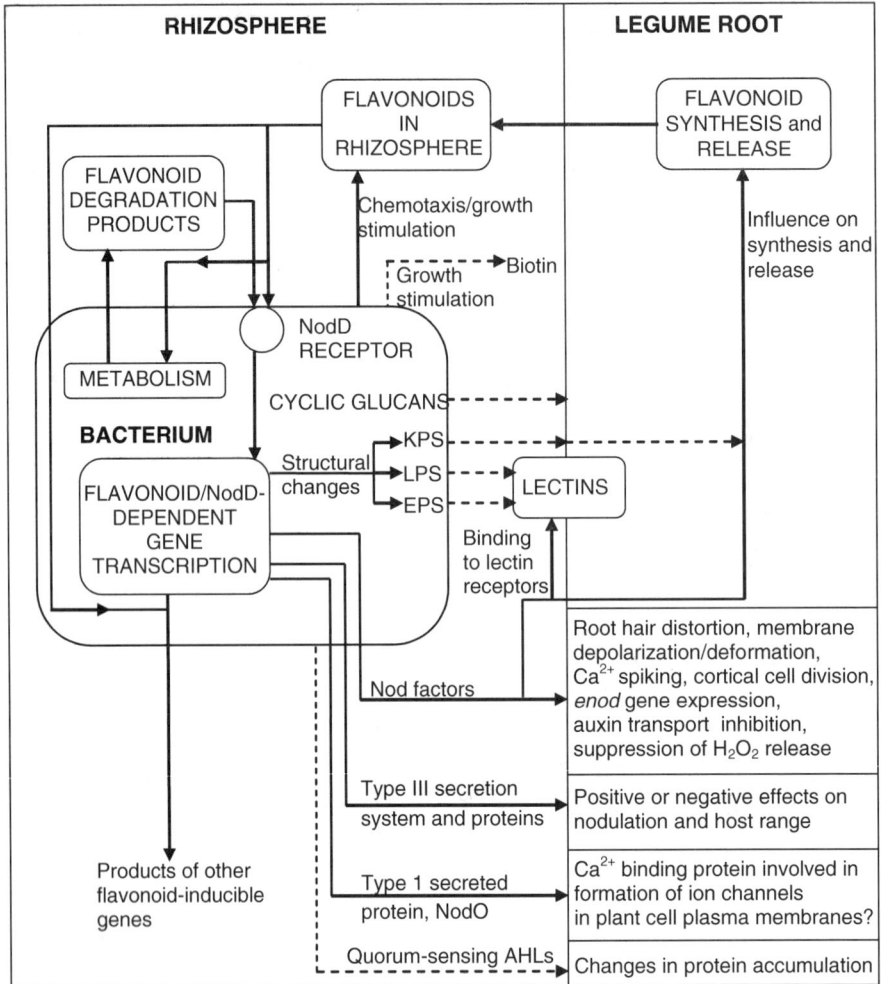

FIGURE 9.3 Early signaling events in legume–rhizobia symbioses. Some pathways are not operative in all symbioses. EPS — extracellular polysaccharide; KPS — capsular polysaccharide; LPS — lipopolysaccharide; AHL — *N*-acyl homoserine lactone [36].

Glycine max (Mc Call), a mutant (in the *rhe* gene) can form nodules [47]. This indicates that the so-called host specificity of rhizobia is in no way a sufficient parameter for systematics and taxonomy, as used in the manuals of bacteriology for a long time. For type I secreted proteins, which is flavonoid inducible (Figure 9.3), there is only one example: Nod O, which may be involved in ion channels formation [52] and infection thread formation in root hairs [53].

B. OTHER COMPONENTS OF COMMUNICATION: CYCLIC GLUCANS, EPS, AND LPS

In rhizobia, the cyclic glucans are polymers of 17 to 40 units of glucose, linked by 1,2 or 1,3, or 1,6 glycosidic bonds and substituted by *sn*-1-phosphoglycerole or phosphocholine [54,55]. Besides the major function in the free-living state in the soil against osmotic changes, they may have a specific function in the communication with the host plant in suppression of the defence response [56]. Also, an increased solubility and transport of flavonoids and NF mediated by cyclic glucans are discussed [57].

TABLE 9.4
Biochemical Function of Nodulation Gene Products Involved in Nod Factor Synthesis and Transport

Nod Gene	Species	Gene Product Function
Regulation of Nod Gene Expression		
nodD	*Common*	LysR-type regulator
nodW	*Bj*	Two-component family sensor
nodW	*Bj*	Two-component family regulator
nolA	*Bj*	MerR-type regulator
nolR	*Rm*	LysR-type regulator
syrM	*Rm*	LysR-type regulator
Synthesis of the Chito Oligosaccharide Backbone		
nodM	*Rm,Rlv,Rlt*	D-glucosamine synthase
nodC	*Common*	UDP-GLcNAc transferase
nodB	*Common*	De-*N*-acetylase
N-Substitutions at Nonreducing End		
nodE	*Rm,Rlv,Rlt*	β-ketoacyl synthase
nodF	*Rm,Rlv,RM,Ml*	Acyl carrier protein
nodA	*common*	*N*-acyltransferase
nodS	*Rn,Rt,Ac,Bj,Rf*	*S*-adenosyl methionine methyl transferase
O-Substitutions at Nonreducing End		
nodL	*Rm,Rlv,Rlt*	6-*O*-acetyltransferase
nodU	*Rn,Rt,Ac,Bj,Rf*	6-*O*-carbamoyltransferase
O-Substitutions at Reducing End_		
nodP	*Rm,Rt*	ATP sulfurylase
nodQ	*Rm,Rt*	ATP sulfurylase, APS kinase
nodH	*Rm,Rt*	Sulfotransferase
nodZ	*Bj,Rn*	Fucosyl transferase
nodZ	*Ac*	Glycosyl transferase
nolK	*Ac*	Sugar epimerase
nodX	*Rlv TOM*	Acetyl transferase
O-Substitutions at Nonterminal Residue		
nod	*Rg*	3-*O*-acetyltransferase
Secretion of NF		
nodI	*Common*	ATP-binding protein
nodJ	*Common*	Membrane protein
nodT	*Rlv,Rlt*	Outer-membrane protein
nodFGHI	*Rm*	Membrane proteins

Note: Abbreviations: Ac = *Azorhizobium caulinodans*; Be = *Bradyrhizobium elkani*; Bj = *B. japonicum*; Re = *Rhizobium etli*; Rf = *R. fredii*; Rg = *R.* sp. GRH2; Ml = *R. loti*; Rlt = *R.* l. bv *trifolii*; Rlv = *R. leguminosarum* bv *viciae*; Rm = *R. meliloti*, Rn = *R.* sp. NGR 234; Rt = *R. tropici*; Rg = *R. galegae.*

Source: From Werner, D. and Müller, P., *Environmental Signal Processing and Adaptation*, Heldmaier, G. and Werner, D., Eds., Springer, Berlin, 2003, p. 9.

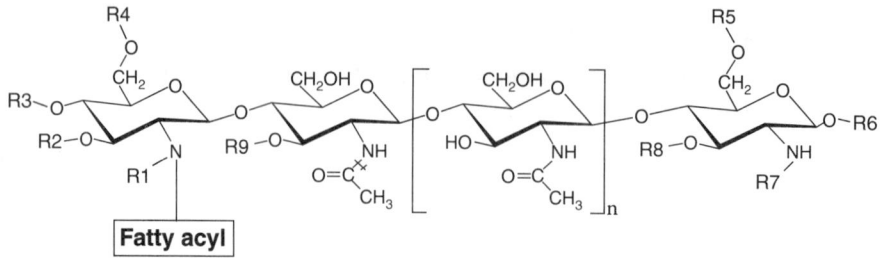

FIGURE 9.4 General structures of the Nod factors produced by rhizobia. The presence of substituents numbered R1 to R9 is variable within various strains of rhizobia. In the absence of specific substituents, the R groups stand for hydrogen (*R1*), hydroxy (*R2, R3, R4, R5, R6, R8,* and *R9*), and acetyl (*R7*) [33].

The EPS genes in rhizobia are well studied. Figure 9.5 summarizes the data for *Sinorhizobium meliloti*. Two types of EPS have been separated from *Sinorhizobium meliloti*, EPS I as a succino-glucan and EPS as a galactoglucan, with size classes ranging from 8 to 40 saccharides up to several thousand units [56]. EPS mutants in *Bradyrhizobium japonicum* elicit a pronounced defence response [58], indicating a role of EPS in suppression of host defence response. Symbiosis and parasitism are also in many other regulatory networks closely related [59,60]. The exo P cluster of *Rhizobium* NGR 234, which has a very broad host range, infecting more than 112 legume genera, differs from the cluster in *Sinorhizobium meliloti*, with a very narrow host range, by the lack of exoH and exo TWV [61,62].

The typical structure of lipopolysaccharides (LPS) is composed of three major components: the core chain, the lipid A, and the O-antigen chain [63,64]. The genes for the synthesis of the core chain and the O-antigen chain have localized in *Rhizobium etli* on a plasmid [65]. The LPS in rhizobia is more heterogeneous than the EPS. Three different components of rhizobial lipid are shown in Figure 9.6.

A glucose-amino-disaccharide backbone carries one very long (C 27 or C 28) hydroxy-fatty acid and four other shorter fatty acids. In different species of *Rhizobium,* the disaccharide can be phosphorylated, galacturonated, or modified to deoxygluconate [66]. It is not unexpected that the LPS from *Agrobacterium tumefaciens* is very similar to the structures found in *Rhizobium* species [67]. The repression of host plant defense responses by complete LPS structures has already been assumed some time ago [68]. More recently, it has been shown that the LPS from *Sinorhizobium meliloti* can suppress the defense response in *Medicago sativa* cell suspensions after addition of a yeast extract elicitor [69,70]. For the early stages of infection, LPS are not essential; however, a mutation in the C27/C29 fatty acid had a phenotype of delayed nodulation [71]. To understand this effect, it is very important to know that LPS can move in the form of vesicles from the surface of the microsymbiont to, or even into, the plant surface layers [72]. The next stages of nodule development are definitely affected by the LPS structures. LPS mutants in the core or the O-antigen formed only pseudonodules or ineffective nodules in *Phaseolus vulgaris* and *Sesbania rostrata* [73,74]. During the infection process and the differentiation of bacteroids, the LPS may be changed. Specific flavonoids can affect the microsymbiont to produce a more hydrophobic LPS by integrating more 2-*O*-methyl-fucose instead of rhamnose [75,76]. A modification of the LPS to a reduced rhamnose content resulted also in ineffective (not nitrogen fixing) nodules. Studies on the signaling sequence indicate that NF and LPS are consecutive signals in the symbiosis development [77].

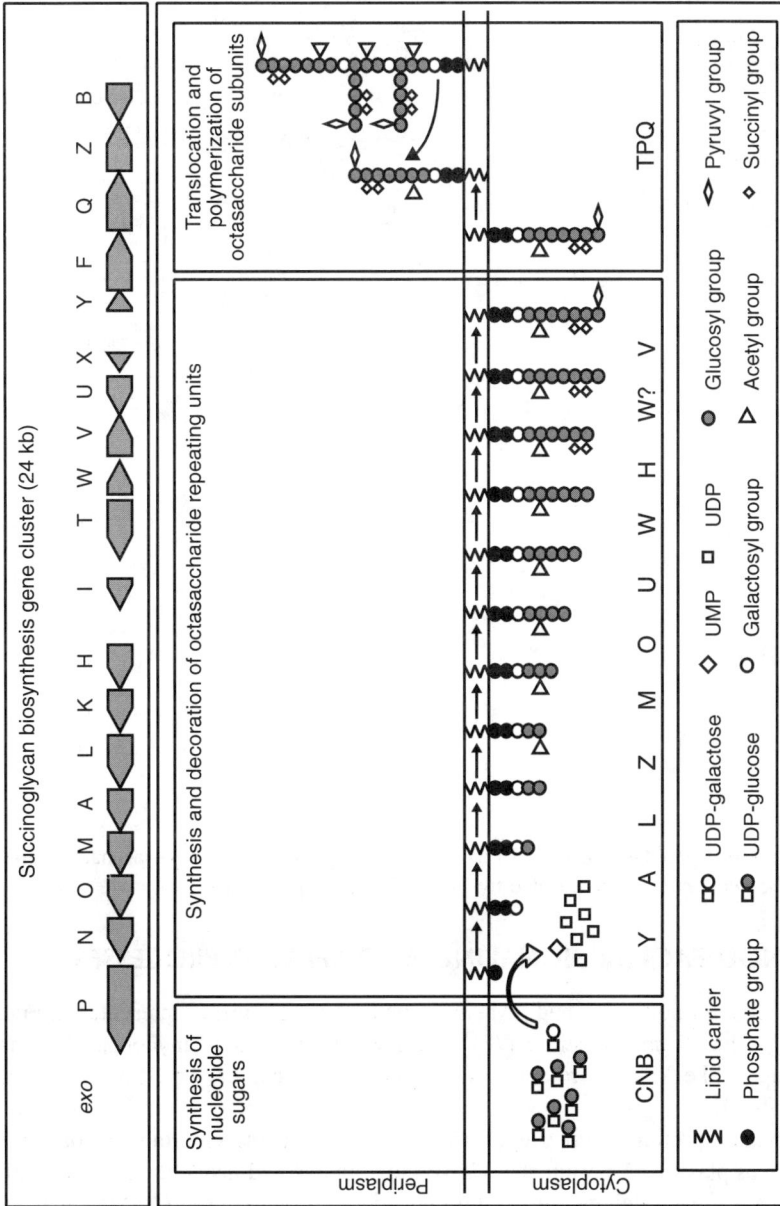

FIGURE 9.5 Structure and gene cluster of exopolysaccharide (EPS)–succinoglycan biosynthesis in *Sinorhizobium meliloti* (communicated by Anke Becker, Bielefeld).

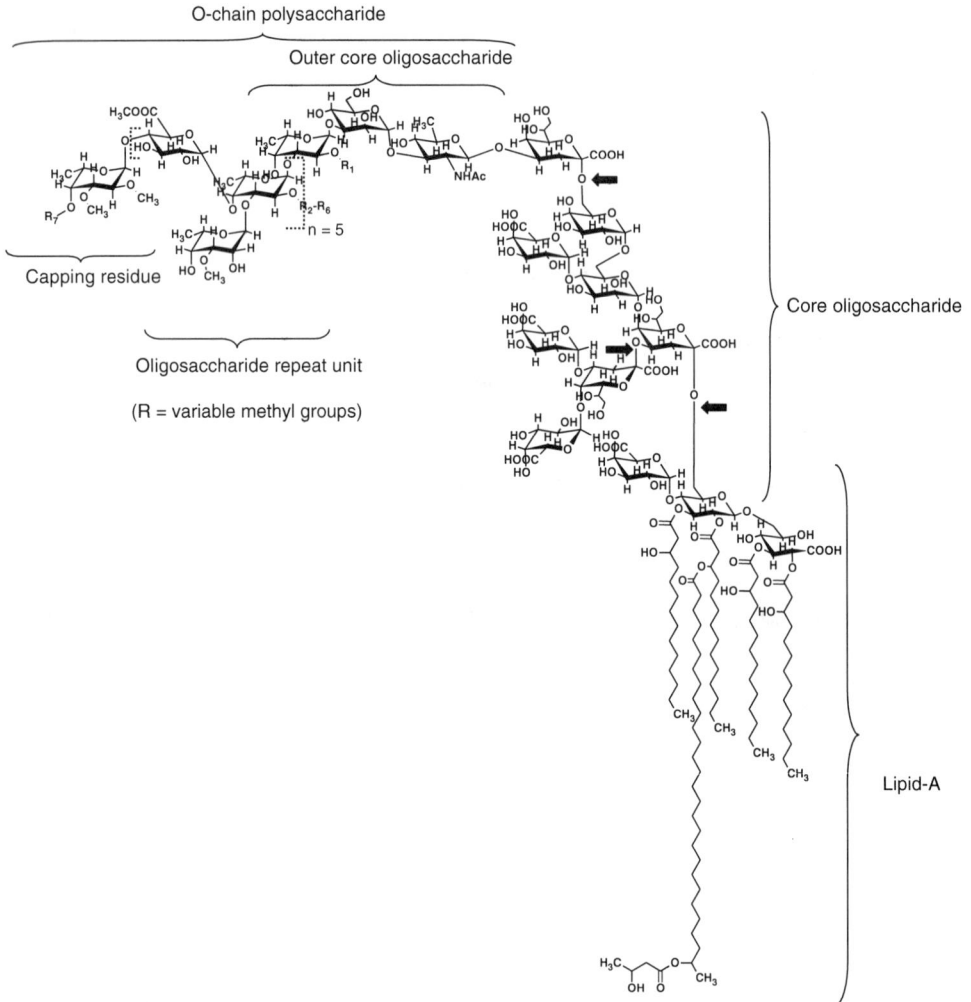

FIGURE 9.6 Structure of lipopolysaccharide (LPS) with core oligosaccharide, oligosaccharide repeat unit, and lipid A from *Rhizobium etli* (communicated by Elmar Kannenberg, Tübingen).

IV. NOD FACTOR SIGNALING AND RELATED PROCESSES

A summarizing scheme for nod factor and AM signal pathways in legumes is presented in Figure 9.7 from a review by Udvardi and Scheible [78]. The NF and equivalent mycorrhizal signals are perceived by receptor-like kinase proteins NFR (nod-factor receptor) or SYMRK (symbiosis-receptor kinase) [79,80].

The phosphorylated proteins then trigger a Ca^{2+} influx from the environment or from other compartments such as plastids. A perinuclear Ca-spiking outside and inside the nucleus of the host plants can activate a nuclear calcium/calmodulin-dependent protein kinase (CcaMK) that activates two transcriptional activators NSP1 and NSP 2 (GRAS-type regulators), which trigger the transcription of early root hair nodulins and other early nodulins (enods). Nodulins are symbiosis-specific proteins in nodules. The calcium/calmodulin-dependent protein kinase gene from *Medicago truncatula* has been designated as dmi3 and sequenced [81]. A mutant in this gene can still bind nod factor, but cannot transduce the signal downstream of the Ca-spiking. A very similar publication with the same host plant was published by other authors including statement for AM-symbiosis

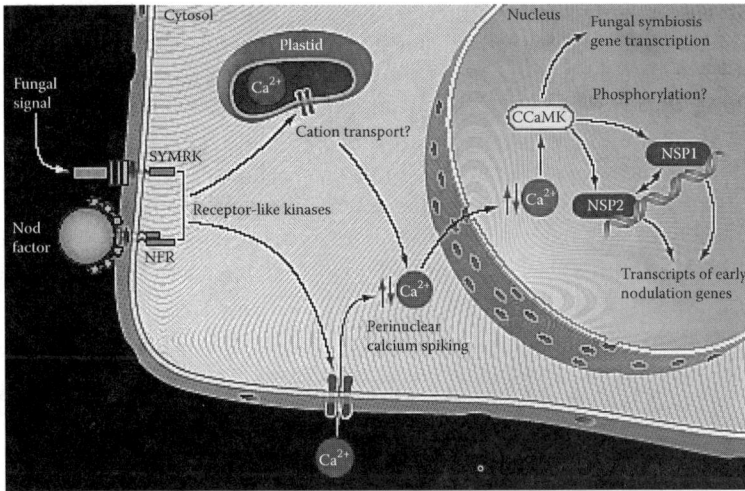

FIGURE 9.7 Possible signaling pathway triggered by bacterial nod factor in plant root cells. Nod factors are perceived at the cell surface by receptor-like kinase proteins (NFR — Nod factor receptor complex, and SYMRK — symbiosis receptor kinase). Protein phosphorylation may then lead to influx of extracellular calcium (Ca^{2+}) or release of intracellular calcium stores by channel proteins into the cytoplasm. Rhythmic calcium spiking, especially around and possibly within the nucleus, may activate a nuclear calcium/calmodulin-dependent protein kinase (CCaMK), which in turn could activate the GRAS-type transcriptional regulators NSP1 and NSP2 by phosphorylation. NSP1 and NSP2 then induce transcription of early nodulation genes, leading to root hair deformation and formation of nodules. CCaMK probably also activates expression of genes required for fungal (mycorrhizal) [78].

[82], but both groups do not recognize the work with *Lotus japonicus* on this signaling pathway. The same situation can be found in competing groups working with different species of rhizobia; some laboratories working with *Sinorhizobium meliloti* neglect work done with *Bradyrhizobium japonicum* and *vice versa*. The function of the dmi2 gene was also studied in *Medicago truncatula*, but with reference to other systems [83]. This gene is not induced by NF; however, it is very significantly induced in the primordia of nodules and it is suggested that it plays a role in the continuing dialogue between the plant and the microsymbionts. The expression of *dmi2* was not effected by mutations in other symbiotic genes such as *dmi1*, *dmi3*, *nsp1*, and *nsp2*. Mutants in *dmi2* can grow normal with a mineral nitrogen source. It is very interesting that a dmi2 mutant is also more sensitive to touching of leaves than the wild-type, indicating a role also in other plant organs than nodules and roots [84]. The NSP 1 and NSP 2 proteins belong to the GRAS protein family, present only in plants and apparently not in prokaryotes and in animals [85]. The name GRAS is derived from the first identified proteins (GAI, RGA, SCR). So far, 33 different GRAS proteins have been identified in *Arabidopsis*. They are composed of 400 to 770 amino acids. Their major sequence character is a VHIID motif, flanked by two leucine-rich areas. The C terminal of the proteins is highly conserved, the N-terminal rather variable. After the second leucine-rich domain, a tyrosine phosphorylation site is also widespread in these proteins. GRAS proteins are involved in such different areas of plants development as the following:

- Root and nodule development
- Meristem development in general
- Shoot meristem development
- Phytochrome signaling
- Gibberellin signal transduction
- Meiosis development

They are probably essential regulatory proteins for the communication of all plant organs, which have to communicate during growth and development. It has been shown that GRAS proteins have transactivation domains and can thereby act as activators of transcription [86,87]. A phylogenetic tree of the GRAS protein family is given in Figure 9.8.

The GRAS protein NSP2 is localized in the endoplasmic reticulum (ER) and relocalized to the nucleus after nod-factor application [88]. In this respect, it is very important to remember that the ER plays also a decisive role in the development of the symbiosome (peribacteroid) membrane [89,90]. Therefore, it is very likely that GRAS proteins may play also a role in the differentiation of the symbiosome membrane, which is in addition under the control of several genes from the

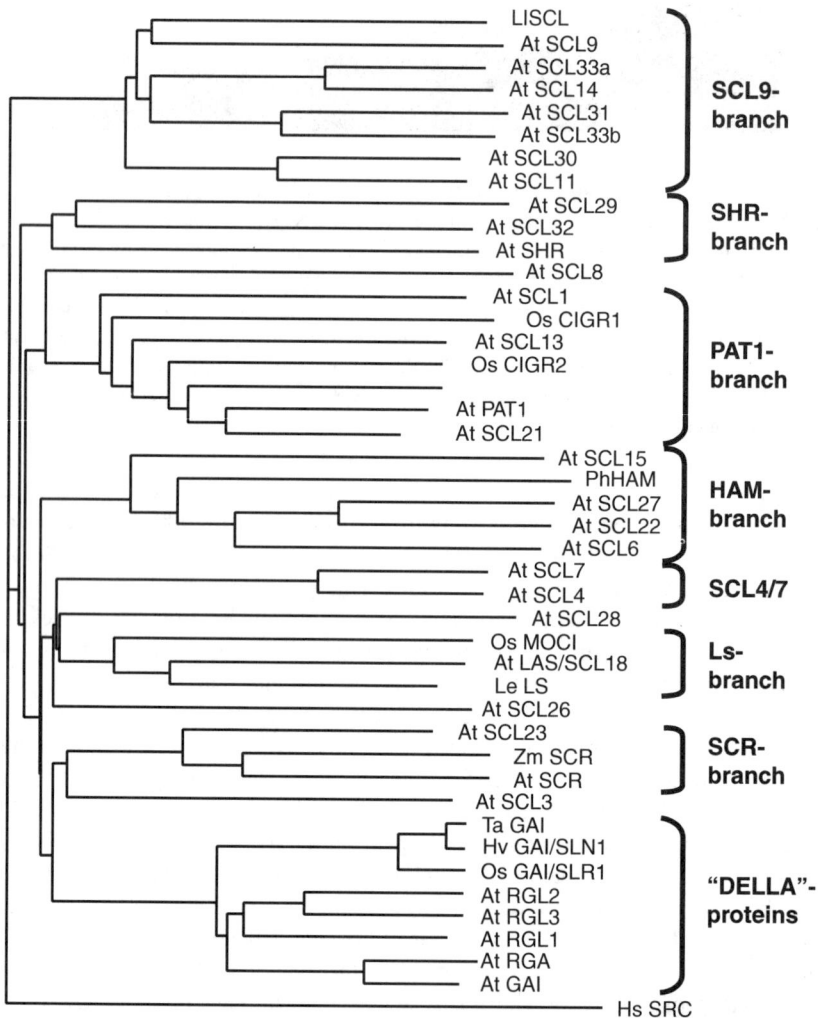

FIGURE 9.8 Phylogenetic tree of GRAS proteins. The rooted tree summarizes the evolutionary relationship among the 33 members of the *Arabidopsis thaliana* GRAS protein family (*At*) including several GRAS proteins from *Petunia x hybrida* (petunia: *Ph*), *Lycopersicon esculentum* (tomato: *Lelium longiflorum* (lily: *Ll*), *Oryza sativa* (rice: *Os*), *Hordeum vulgare* (barley: *Hv*) and *Zea mays* (maize: *Zm*). The program used for sequence aligment was CLUSTAL W. Phylogenetic analysis was done using the DAMBE4 Program (Xia and Xie 2001) with the neighbor-joining method using a human (Homo sapiens) STAT protein (Hs SRC; NP004374) as an outgroup. AtSCL 33 was regarded as two independent proteins, AtSCL 33a and b [85].

rhizobia, as shown for signal peptidase genes in the symbiosis of *Glycine max* and *Bradyrhizobium japonicum* [91,92]. NSP 1 is constitutively expressed, preferentially in roots [93]. Besides plant-specific gene families such as GRAS, gene families, active for example, in mammalian cells during membrane differentiation such as the GRIP-proteins [94], may be of interest for symbiosis development.

The infection process during symbiosis starts in most cases in root hair cells. The method to isolate root hair cells in large quantities with a freeze fracture method was developed by Röhm and Werner [95], and the first root hair specific proteins were identified with this method. After nod-factor treatment, a more than 100-fold increase in early nodulin protein with a molecular weight of 11 kDa (ENOD 11) was found in *Medicago truncatula* root hairs and an increase by a factor of 30 for ENOD 40 [96]. Inhibition of the phospholipase D and phospholipase C blocked the ENOD 11 activation [97]. In *Lotus japonicus*, two ENOD 40 proteins were expressed with a different pattern after infection with the bacteria [98]. The regulation of the ENOD 40 promoter is species-specific in *Medicago sativa* and *Trifolium*. The cytokinin BAP induces the expression, whereas in *Lotus japonicus* no such effect was observed [99]. Root hairs themselves are highly differentiated plant surface cells with functions in nutrient uptake, water uptake, and root exudation. Their growth rate and also their microtubule dynamics is reduced by addition of NF (LCOs) [100].

Inside the root hairs as receiving cells, the communication with the plastids is also essential. Two new genes, called Castor and Pollux, have been characterized in *Lotus japonicus*, involved in ion fluxes between cytosol and plastids (Figure 9.9). Mutations in either gene have a Nod⁻ and Myc⁻ phenotype. Homologues of this gene exists not only in other legumes such as *Pisum sativum*, *Glycine max*, and *Medicago truncatula*, but also in *Oryza sativa* (rice) and *Arabidopsis thaliana*. This indicates, that a widespread gene has developed in legumes to a symbiosis-specific gene, localized in plastids [101,102].

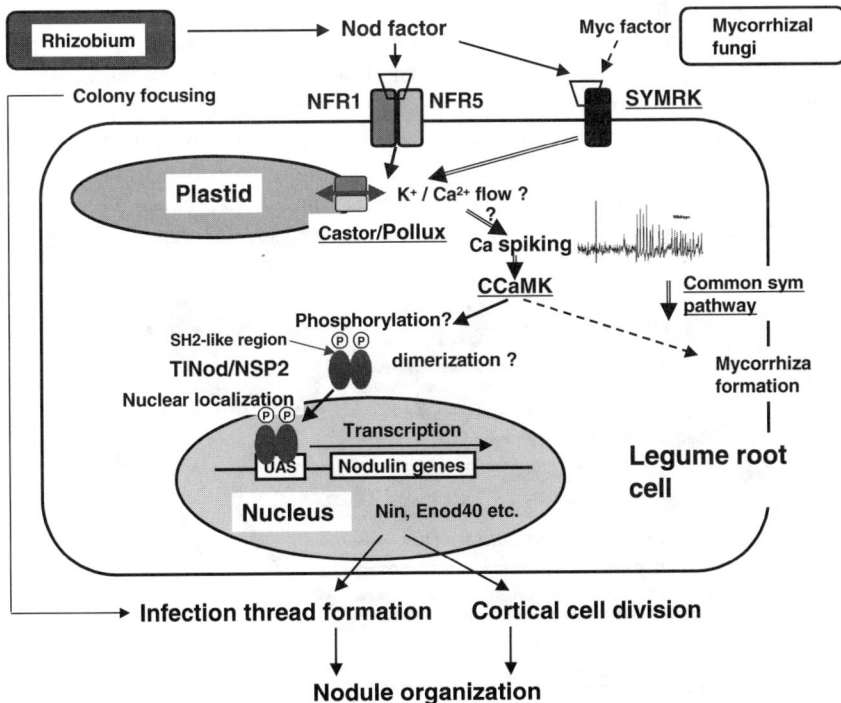

FIGURE 9.9 Lotus component corresponding to the NSP2 of *Medicago truncatula* [101,102].

Using C-DNA techniques, huge numbers of genes overexpressed or repressed during the symbiosis development have been detected, but the real functions remain open. More than 100 overexpressed genes have been found in *Lotus japonicus* after infection by *Mesorhizobium loti* [103]. These genes include those for 32 nodulins, and proteins involved in (1) defense responses, (2) phytohormone synthesis, (3) cell wall synthesis, (4) membrane transport, (5) signal transduction, and (6) transcriptional regulation.

In *Medicago truncatula*, the other model legume, 46 sequences were differentially expressed after inoculation with *Sinorhizobium meliloti* [104], functioning in (1) ribosome biogenesis, (2) defense response, (3) stress response, (4) signal transduction, and (5) transcriptional regulation.

As already pointed out for related signal pathways between the rhizobia–legumes symbiosis and the arbuscular mycorrhiza–legume symbiosis (Figure 9.7 and Figure 9.9 [105]), the *Frankia*-host plant and the rhizobia–legume symbiosis also share some transcriptional regulations, as indicated for the subtilisin-like Serine protease gene cg 12 [106]. Even nematodes can elicit the same signal transduction as NF, as shown with *Lotus japonicus* [107].

V. INFECTION AND NODULE COMPARTMENTATION DEVELOPMENT

The further steps of symbiosis development become more and more complex, because the number of microsymbionts per host plant cell increases and all other cell organelles have to respond to the new organelle, the symbiosome. But before this step, the invasion through the infection thread is an essential process, as shown in Figure 9.10 [108].

The infection thread is a cylinder, growing inside root hairs toward the root cortex, surrounded by a probably modified host plasma membrane and a probably modified cell wall. The infection threads grow with a speed of around 10 μm/h in *Medicago* [109]. Inside this compartment, extensins (glycoproteins) are continuously changing their cross-linking, triggered by H_2O_2, producing a more solidified matrix from a fluid matrix [110]. In this solidified matrix the rhizobia stop to divide, which means that according to this model, only around 25 to 50 bacteria at the tip of the infection threads are actively dividing in the infection thread [108]. The matrix material consists of glycoproteins with dominating hydroxyproline residues and arabinogalactan glycosylations.

The next steps of nodule initiation and development are closely linked to the cell cycle regulation. In the nodule development of *Medicago truncatula,* more than 10 layers of root cortex cells are invaded by the infection threads and the bacteria continue to produce NF, which send out a mitogenic signal; the root cortex cells do not divide but undergo several rounds of endoreduplication,

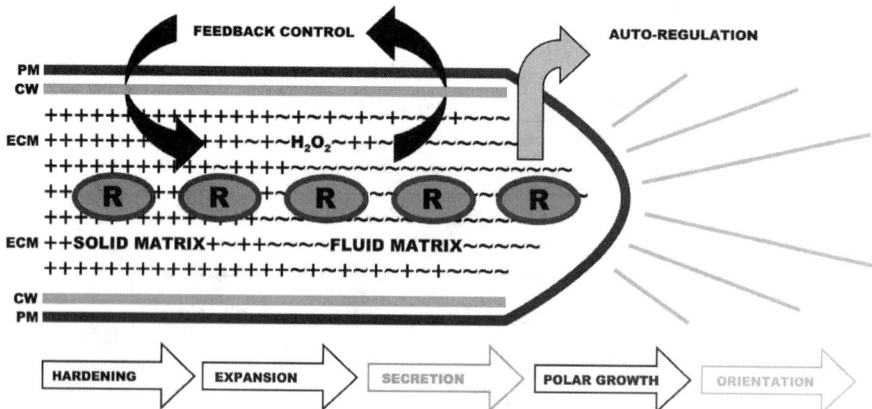

FIGURE 9.10 Model for the growth of an infection thread [110].

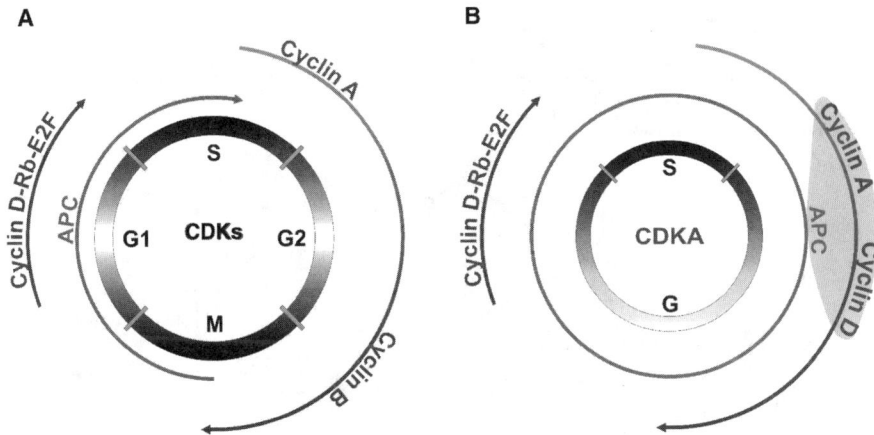

FIGURE 9.11 Mitotic cell cycle and endoreduplication cycle in nodule development (Figure supplied by Adam and Eva Kondorosi).

ending in polyploid and significantly enlarged root cells [111]. The DNA content of the cells can increase from 2 to 64 C, after five rounds of endoreduplication [112].

The basic mitotic cell cycle is controlled by cyclin-dependent cyclin-kinases, where the transition of G1-S in affected by cyclin D, the transition of S-G 2 by cyclin A, and the G2-M transition by cyclin B (Figure 9.11). The NF (LCOs) reactivate cells arrested at the GO phase by inducing the cyclin A2. The promoter of the *cycA2* gene has a number of auxin-responsive elements. Application of auxins led to a strong expression of *cycA2* in front of the xylem poles, the site of nodule initiation [113]. The switch from the mitotic cycle to the endoreduplication cycle is mediated by the switch gene *ccs52A*. The gene product provokes a degradation of cyclin B converting the mitotic cell cycle to the endoreduplication cycle. In ccs52A antisense plants, the nodules were much smaller, the degree of polyploidy was significantly reduced, and the cells included no nitrogen-fixing bacteroids [114]. The gene *ccs52A* is expressed in determinate as well as in indeterminate nodules.

The next level of signaling determines the number of nodules per plant. It is evident here that the shoot-to-root communication should play an important role, as the number of nodules determines (with a large variation of the efficiency of the single nodules) how much nitrogen per plant is fixed, and the whole plant must control how much nitrogen it needs during the various growth phases. The latest model of autoregulation of nodulation, based on the long time work of the laboratory of Peter Gresshoff is presented in Figure 9.12.

Nod factors perception is attenuated by a signal coming from the shoot/leaves, mediated by NARK (nodule autoregulation receptor kinase), which also receives a signal from the roots [115]. With grafting experiments it was shown that the genotype of the shoot controls the number of nodules on the root system [116]. The NARK receptor kinase consists of a signal peptide, leucine-rich repeats, one transmembrane domain, and the kinase sequence [117]. The similar har 1 gene was cloned from *Lotus japonicus* [116,118], with the largest level of identity to the CLAVATA1 receptor kinase of *Arabidopsis*, which is involved in the negative regulation of apical meristems of shoots. The phytohormone jasmonate suppresses nodule formation and could be part of the signaling pathway [119].

Compartmentation in the infected nodule cells with the development of stable symbiosomes are the next even more complex steps in the symbiosis development. A fully infected large host cell of *Glycine max* nodules contains more than 10000 symbiosomes, functioning for more than two weeks in a largely unexplored cooperation with all other cell organelles such as mitochondria,

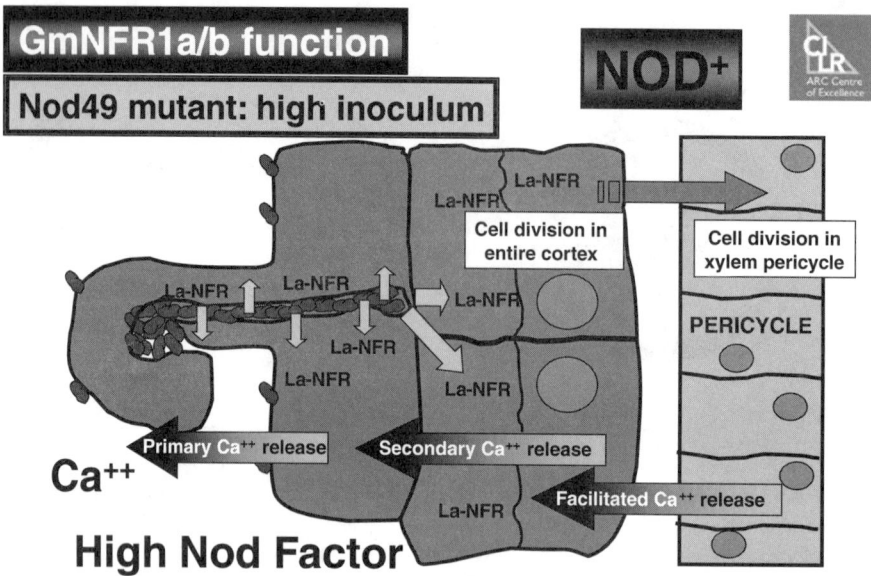

FIGURE 9.12 Autoregulation of nodulation circuit in legumes [115,116].

amyloplasts, ER–Golgi complex, and the cell nucleus (Figure 9.13). There is no other example in cell biology of such a massive invasion of other cells and genomes into a eukaryotic host cell leading to a stable symbiosis [13]. For comparison, the number of plastids per eukaryotic cell is in the range between 1 and around 400. With chemically induced mutants of *Bradyrhizobium japonicum* already in the 1980s, a number of unstable symbiosome membranes and host-cell compartmentations have been found as phenotypes of the symbiosis [120–123]. With TnphoA

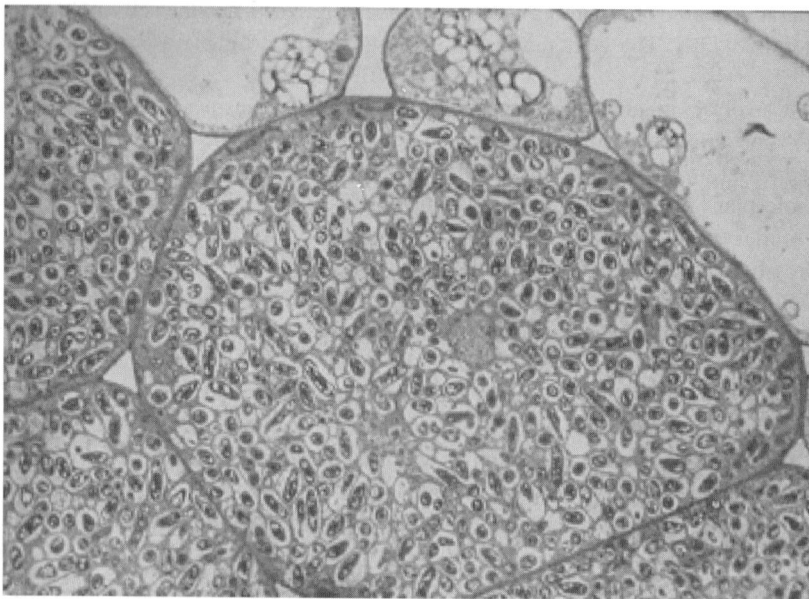

FIGURE 9.13 Optimal infected nodule cells with several thousand symbiosomes per host cell in nodules of *Glycine max*.

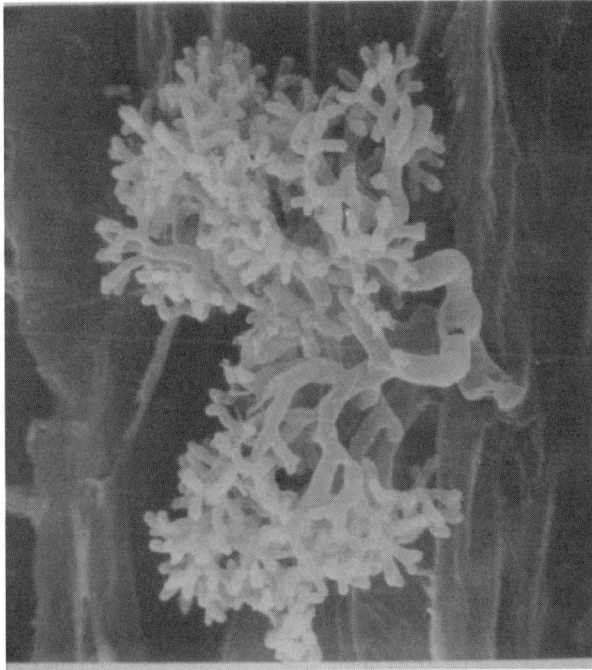

FIGURE 9.14 Optimal branched arbuscule from an AM fungus inside the host cell.

insertion mutagenesis, new genes relevant for the later stages of symbiosis and beyond the known *har, nod, nol, noe,* and *nif* genes could be identified with the *sip* (signal-peptidase) genes. A specific symbiotic phenotype could be produced by mutation in the *sipS* gene, because *Bradyrhizobium japonicum* contains, with sip F, a second functional signal peptidase, which allows an undisturbed growth in the free-living stage [91,92,124]. The release from the infection thread and the stability of the bacteroides in the symbiosomes were affected in sip S mutants. The genome regions around sip F were studied further by the use of multipurpose transposon Tn KPK2, producing ineffective (not nitrogen fixing) nodules [125]. Also, outside the symbiotic region of the *Bradyrhizobium japonicum* USDA 110 genome, the new genes *srrB* and *srrC* were found, producing a phenotype with a very low number of symbiosomes per infected cell [126]. The sum of all symbiosomes in the infected nodule cell is functionally comparable to the arbuscule in the arbuscular mycorrhiza symbiosis (Figure 9.14). However, we still know much less about this compartment, because it is not yet possible, to isolate arbuscules in a similar manner as symbiosomes for functional and biochemical studies. On the other side, the molecular genetics of this symbiosis has made significant progress in the past few years [79,127].

VI. SOIL STRESS FACTORS AFFECTING THE SYMBIOSIS

Our knowledge about the symbiosis development described so far is to a large extent based on laboratory experiments and observations. The real world of legumes and rhizobia are, of course, the various ecosystems such as arable agricultural land, pastures, savannahs and forests, where the standard development is affected by a large number of environmental factors such as:

- Soil water stress (osmotic stress)
- Nutrient stress
- pH stress and competition stress

The best-studied system for osmotic stress in rhizobia is *Sinorhizobium meliloti*, which accumulates very different compatible solutes such as glutamic acid, glycine betaine, proline betaine, trehalose, and *N*-acetylglutaminylglutamine amide [128,129]. The basic understanding of osmoregulation is even more advanced in *Escherichia coli* and *Bacillus subtilis* [130,131]. In *Bacillus subtilis,* there are two classes of osmotically regulated genes:

- Genes that are permanently switched on at high osmolarity are the Opu Transporters and genes for the proline biosynthesis. Altogether more than 100 genes belong to this group, most of them have unknown functions. A central role has the two-component regulatory system DegS/DegU [132].
- Genes that are only transiently upregulated, such as the *yhaSTU* genes, responsible for K^+/Na^+ export. Mechanosensitive channel proteins hereby play an important role.

Besides osmotic adaptation, the tolerance to desiccation is another important factor to water supply and shortage. But the genetics and biochemistry of desiccation in rhizobia is poorly studied [133]. A decrease in unsaturated fatty acids under these conditions has been reported [134]. When we include the plant partner it is remarkable that N_2 fixation is more sensitive to desiccation than photosynthesis and nitrate assimilation [135]. Nutrient stress and limitation has a fundamental impact on growth and development of the host plants as well as the microsymbionts. The limitation of a single essential element of the around 20 essential soil elements also finally limits the symbiosis. In the keynote lecture of the International Nitrogen Fixation Conference in Peking [23], it was remembered that already 20 years ago [136] it was demonstrated, that root hairs of soybeans contain almost 10 times more calcium, cobalt, and iron than soybean roots and 3 to 8 times more than wheat root hairs. Soybean roots contain also about 10 times more molybdenum than wheat roots (Table 9.5). For all these essential elements, rhizobia have an unusual high requirement for growth and differentiation. The central hypothesis from these (limiting) data is that the symbiotic interaction does not start with a carbon/nitrogen exchange, which takes place only in already developed nodules, but with the supply of essential trace elements by the host plants to the rhizobia in a competition-limiting environment [136]. In the past few years it has been established that calcium spiking in root hairs of legumes plays a central role in the signaling pathway for NF reception (see Figure 9.7) [137]. In many soils, the limitation of available phosphate is a major constraint, and also for this element, the symbiotic bacteria and the legume host plants compete. The phosphorus concentration in fix$^+$ nodules is about 60% higher than in root tissue, but the same as in fix$^-$ nodules [138]. On the other side, the chloride concentration in both types of nodules is reduced by about 80% compared to roots (both 20 d old).

The acquisition of phosphate is, in many cases, a limiting factor of legume development [139]. Therefore, the tripartite symbiosis of legumes with AM fungi and rhizobia is of special interest, because the AM fungi reduce the phosphate limitation and the rhizobia the nitrogen limitation [140].

TABLE 9.5

Calcium, Iron, and Cobalt Accumulation in Root Hairs of Soybean (*Glycine max*)

Element	Ppm (Dry Matter)			
	Soybean Root Hair	Soybean Root	Wheat Root Hair	Wheat Root Hair
K	11740 ± 2450	12840 ± 2640	4670 ± 1010	4780 ± 990
S	530 ± 165	560 ± 170	180 ± 55	190 ± 60
Fe	**414 ± 138**	**31 ± 5**	**120 ± 35**	**44 ± 26**
Co	**7.9 ± 3.8**	**0.88 ± 0.4**	**2.6 ± 0.8**	**1.3 ± 1.1**
Ca	**22000 ± 460**	**287 ± 70**	**246 ± 60**	**288 ± 70**
Mo	3.1 ± 0.5	5.4 ± 0.7	0.6 ± 0.12	0.5 ± 0.3

Source: From Werner, D. et al., *Z. Naturforsch.*, 40c, 912, 1985.

Besides the nutrient acquisition, the AM-fungal symbiosis also protects the host plants against drought stress [141] and also salinity stress [142]. Also PGPR (plant growth promoting rhizobacteria) can assist AM fungi in solubilizing phosphate for the host plants [143,144]. The complexity of the interactions has further increased by the observations that endosymbiotic bacteria are present within certain species of AM fungi, such as *Gigaspora* and have been identified as the new species *Candidatus Glomeribacter gigasporum* [145]. The bacteria are stable cytoplasm components, are transmitted over vegetative spore generations [146], and are considered as obligate endosymbionts [147]. Bacterial responses to pH, in general, have been summarized in detail by R.K. Poole [148]. The sensor/regulator pair ActS/ActR has been studied in most detail in *Sinorhizobium meliloti* [149]. The most interesting special feature of rhizobia compared to other rhizosphere bacteria is that they inhabit in both parts of their life cycle an acid environment — as free-living bacteria in the rhizosphere before infection and as bacteroides in the symbiosome. This has been used to identify additional new genes responsible for the later stages of nodule colonization, by producing acid-sensitive mutants in the free-living stage and test them in competition experiments during nodule compartmentation. With the acid-tolerant strain *Rhizobium* CIAT 899, the gene *atvA* ("acid tolerance and virulence") was identified, with a strong homology to the *acvB* gene in *Agrobacterium tumefaciens* [150]. A serine/alanine exchange in the lipase motive of the protein leads to an acid-sensitive phenotype with a significantly reduced competitiveness for nodule occupancy. Mutation in the upstream localized *lpiA* gene also leads to much reduced competitiveness. Also, mutants with an increased acid tolerance have been created. A leucine biosynthesis mutant was able to survive at pH 3.5, which the wild-type strain could not [151]. The wild-type could survive and grow down to pH 4.0. The mutant was able to raise the pH of the medium from 3.5 to 3.8 as a leucine auxotroph, where the wild-type did not raise the pH of the medium. It is known that catabolization of amino acids leads to an alkalinization of the medium. This mechanism may explain that in the rhizosphere of many legumes amino acid auxotrophs can be found. Two other genes involved in acid tolerance have been most recently also identified in *Rhizobium tropici* — the gene *sycA* has strong identity with a CIC chloride channel and the second gene in the same operon *olsC* is homologous to aspartyl-asparginyl-ß-hydroxylase [152]. This hydroxylase modifies two ornithine-containing membrane lipids of the microsymbionts. Both genes are involved in acid tolerance in the free-living state as well as in colonization of the infected cells by affecting competitiveness. The organization of these genes is shown in Figure 9.15.

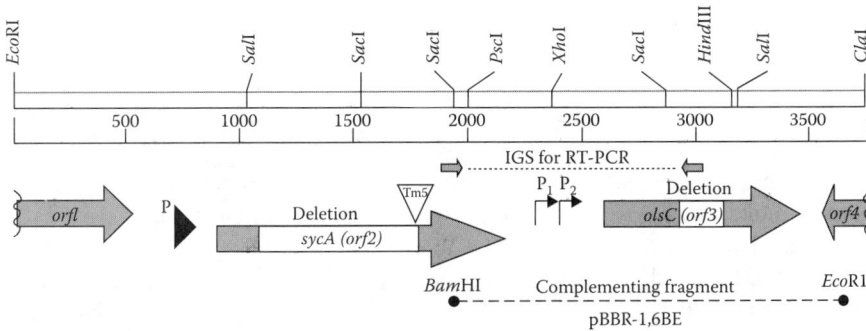

FIGURE 9.15 Genetic and physical maps of the 3,761-bp *Eco*RI-*Cla*I region from *Rhizobium tropici* CIAT899 analyzed in this study (accession number AY954450). Selected restriction sites are shown. Four open reading frames (ORF) (represented by arrows) were detected. The site of the Tn5 insertion located in *sycA* between nucleotides C1763 and T1764 is indicated by an open triangle. Nonpolar deletion mutants lacking the regions shown in white were generated in *sycA* and *olsC*. Predicted promoters are shown as thin arrows. The dotted line represents the intergenic spacer between *sycA* and *olsC* subjected to reverse transcriptase–polymerase chain reaction (RT-PCR) analysis (shown in D). The dashed line shows the location of the 1.66-kb *Bam*HI-*Eco*RI fragment cloned into pBBR-MCS5 and used to complement strain 899-olsCΔ1 [152].

The two soybean-nodulating rhizobia also differ in their acid tolerance, *Bradyrhizobium japonicum* can still grow at pH 4.5 but not under alkaline conditions of pH 9.0, whereas *Sinorhizobium fredii* will not grow at pH 4.5 but at 9.0 [153]. From genistoid legumes, the new acid-tolerant species *Bradyrhizobium canariense* has been isolated and characterized [154], which can nodulate *Chamaecytisus proliferus*, *Teline stenopetala,* and *Lupinus luteus*, but not *Glycine max* and *Glycine soja*. This new species has been characterized in relation to population genetics and the phylogeny with all other *Bradyrhizobium* species: *Bradyrhizobium japonicum*, *Bradyrhizobium elkanii*, *Bradyrhizobium liaoningense*, *Bradyrhizobium yuanningese*, and *Bradyrhizobium betae* [155].

VII. THE ECOLOGICAL AND AGRICULTURAL IMPACT OF THE SYMBIOSIS FOR THE NITROGEN CYCLE

A detailed description of the agricultural relevant nitrogen fixation of legumes has been published in the volume *Nitrogen Fixation in Agriculture, Forestry, Ecology and the Environment* [156–160]. The two other major processes in the nitrogen cycle have been covered in the same volume: nitrification by Fiencke et al. [161] and denitrification by van Spanning et al. [162]. Increased nitrogen fixation for legume crop production will most likely come from traditional techniques rather than from genetic engineering in the next few years. Too many promises have been made during the last ten years from these techniques, which are extremely useful for studying the molecular mechanism of symbiosis development, but have almost nothing contributed to increased legume production so far. The one classical technique is the use of improved inoculum production and application of competitive and stress-tolerant strains of rhizobia selected for very different soil types, climates, and host cultivars [163]. The other technique comes from modern plant breeding with the following elements, as summarized by Peter Graham et al. [164]:

- Use of genetic variation with traits associated with nitrogen fixation and symbiosis
- Use of progress in understanding pathogen resistance in legumes
- Understanding trait genetics
- Development of field-based and inexpensive methods for progeny selection
- Better use of new insights in source/sink relationships and organ development

VIII. CONCLUSIONS

The symbiosis between different species of rhizobia with their limited or very large number of legume host plants species has reached the stage, where we know a large number of genes involved and the signal molecules exchanged in the chemical dialogue, the symbiotic partners speak to each other. Involved in the symbiosis are the development of an organelle-like structure, the "bacteroid" and the development of a new and unique plant organ, the "nodule." Therefore, it is also a perfect model system of evolution of organelles and organs. The symbiosis attracts many researchers, because the bacteria as well as the plants involved have been used for genome projects and also because of their agricultural and ecological importance for the nitrogen nutrition of plants. Nitrogen fertilizers present the largest cost factor in agriculture besides labor. Also fascinating is the aspect that the arbuscular mycorrhiza, the other ecologically very important root symbiosis of higher plants (in phosphate mobilization), uses some similar signaling pathways in addition to others.

REFERENCES

1. Werner, D. and Newton, W.E., Eds., *Nitrogen Fixation in Agriculture, Forestry, Ecology and Environment*, Vol. 4, Book series *Nitrogen Fixation: Origins, Applications, and Research Progress*, Springer, Dordrecht, 2005.
2. Palacios, R. and Newton, W.E., Eds., *Genomes and Genomics of Nitrogen-Fixing Organisms*, Vol. 3, Book series *Nitrogen Fixation: Origins, Applications, and Research Progress*, Springer, Dordrecht, 2005.
3. Sprent, J., Dilworth, M., and Newton, W.E., Eds., *Nitrogen-Fixing Leguminous Symbioses*, Vol. 7, Book series *Nitrogen Fixation: Origins, Applications, and Research Progress*, Springer, Dordrecht, 2006.
4. Varma, A., Abbott, L., Werner, D., and Hampp, R., The state of the art, in *Plant Surface Microbiology*, Varma, A., Abbott, L., Werner, D., and Hampp, R., Eds., Springer, Heidelberg, 2004, p. 1.
5. FAO Statistical Yearbook, FAO, Rome, 2004.
6. Chen, W.-M. et al., Nodulation of *Mimosa* spp. by the beta-proteobacterium *Ralstonia taiwanensis*, *Mol. Plant-Microbe Interact.*, 16, 1051, 2003.
7. James, E.K. et al., Novel *Mimosa*-nodulating strains of *Burkholderia* from South America, in *Biological Nitrogen Fixation, Sustainable Agriculture and the Environment: Proceedings of the 14th International Nitrogen Fixation Congress*, Wang, Y.-P., Lin, M., Tian, Z.X., Elmerich, C., and Newton, W.E., Eds., *Current Plant Science and Biotechnology in Agriculture*, Vol. 41, Springer, Dordrecht, 2005, p. 391.
8. Young, J.M. et al., A revision of *Rhizobium* Frank 1889, with an emended description of the genus, and the inclusion of all species of *Agrobacterium* Conn 1942 and *Allorhizobium undicola* de Lajudie et al. 1998 as new combinations: *rhizobium radiobacter, R. rhizogenes, R. rubi, R. undicola* and *R. vitis, Int. J. Syst. Evol. Microbiol.*, 51, 89, 2001.
9. Vinuesa, P. and Silva, C., Species delineation and biography of symbiotic bacteria associated with cultivated and wild legumes, in *Biological Resources and Migration*, Werner, D., Ed., Springer, Heidelberg, 2004, p. 143.
10. Mutch, L.A. and Young, J.P.W., Diversity and specificity of *Rhizobium leguminosarum* biovar viciae on wild and cultivated legumes, *Mol. Ecol.*, 13, 2004.
11. Lindström, K. et al., The species paradigm in bacteriology: from a cross-disciplinary species concept to a new prokaryotic species definition (with emphasis on rhizobia), in *Biological Nitrogen Fixation, Sustainable Agriculture and the Environment: Proceedings of the 14th International Nitrogen Fixation Congress*, Wang, Y.-P., Lin, M., Tian, Z.X., Elmerich, C., and Newton, W.E., Eds., *Current Plant Science and Biotechnology in Agriculture*, Vol. 41, Springer, Dordrecht, 2005, p. 373.
12. Martinez-Romero, E. and Caballero-Mellado, J., *Rhizobium* phylogenies and bacterial genetic diversity, *Crit. Rev. Plant Sci.*, 113, 1996.
13. Werner, D., *Symbiosis of Plants and Microbes*, 2nd ed., Chapman and Hall, London, 1992.
14. Bottomley, P.J., Ecology of *Bradyrhizobium* and *Rhizobium*, in *Biological Nitrogen Fixation*, Stacey, G., Burris, R., and Evans, J.H., Eds., Chapman and Hall, New York, 1992, p. 293.
15. González, J.E. and Marketon, M.M., Quorum sensing in nitrogen-fixing rhizobia, *Microbiol. Mol. Biol. Rev.*, 67, 574, 2003.
16. Rodelas, B. et al., Analysis of quorum-sensing-dependent control of rhizosphere-expressed (*rhi*) genes in *Rhizobium leguminosarum* bv. *Viciae, J. Bacteriol.*, 181, 3816, 1999.
17. Lithgow, J.K. et al., The regulatory locus *cinRI* in *Rhizobium leguminosarum* controls a network of quorum-sensing loci, *Mol. Microbiol.*, 37, 81, 2000.
18. Wisniewski-Dye, F. et al., *raiR* genes are part of a quorum-sensing network controlled by *cinI* and *cinR* in *Rhizobium leguminosarum, J. Bacteriol.*, 184, 1597, 2002.
19. Marketon, M.M. et al., Characterization of the *Sinorhizobium meliloti sinR/sinI* locus and the production of novel N-acyl homoserine lactones, *J. Bacteriol.*, 184, 5686, 2002.
20. Marketon, M.M. et al., Quorum sensing controls exopolysaccharide production in *Sinorhizobium meliloti, J. Bacteriol.*, 185, 325, 2003.
21. Loh, J. et al., Bradyoxetin, a unique chemical signal involved in symbiotic gene regulation, *Proc. Natl. Acad. Sci. USA*, 99, 14446, 2002.
22. Werner, D. et al., Calcium, iron and cobalt accumulation in root hairs of soybean (*Glycine max*). *Z. Naturforsch.*, 40c, 912, 1985.

23. Werner, D. et al., Sustainable agriculture/forestry and biological nitrogen fixation, in *Biological Nitrogen Fixation, Sustainable Agriculture and the Environment: Proceedings of the 14th International Nitrogen Fixation Congress*, Wang, Y.-P., Lin, M., Tian, Z.X., Elmerich, C., and Newton, W.E., Eds., *Current Plant Science and Biotechnology in Agriculture*, Vol. 41, Springer, Dordrecht, 2005, p. 13.

24. Davey, M.E. and O'Toole, A., Microbial films: from ecology to molecular genetics, *Microbiol. Mol. Biol. Rev.*, 64, 847, 2000.

25. Stoodley, P. and Sauer, K., Biofilms as complex differentiated communities, *Annu. Rev. Microbiol.*, 56, 187, 2002.

26. Ramey, B.E. and Koutsoudis, M., Biofilm formation in plant-microbe associations, *Curr. Opin. Microbiol.*, 7, 602, 2004.

27. Schmeisser, C. et al., Metagenome survey of biofilms in drinking-water networks, *Appl. Environ. Microbiol.*, 69, 7298, 2003.

28. Steele, H.L. and Streit, W.R., Metagenomics: advances in ecology and biotechnology, *FEMS Microbiol. Lett.*, 247, 105, 2005.

29. Schloss, P.D., Larget, B.R., and Handelsman, J., Integration of microbial ecology and statistics: a test to compare gene libraries, *Appl. Environ. Microbiol.*, 70, 5485, 2004.

30. Schloss, P.D. and Handelsman, J., Introducing DOTUR, a computer programme defining operational taxonomic units and estimating species richness, *Appl. Environ. Microbiol.*, 71, 1501, 2005.

31. Sebat, J.L., Colwell, F.S., and Crawford, R.L., Metagenomic profiling: microarray analysis of an environmental genomic library, *Appl. Environ. Microbiol.*, 69, 4927, 2003.

32. Angenendt, P. et al., Subnanolitre enzymatic assays on microarrays, *Proteomics*, 5, 420, 2005.

33. Werner, D. and Müller, P., Communication and efficiency in the symbiotic signal exchange, in *Environmental Signal Processing and Adaptation*, Heldmaier, G. and Werner, D., Eds., Springer, Berlin, 2003, p. 9.

34. Werner, D., Signalling in the rhizobia-legumes symbiosis, in *Plant Surface Microbiology*, Varma, A., Abbott, L., Werner D., and Hampp, R., Eds., Springer, Heidelberg, 2004, pp. 99–119.

35. Holsters, M. et al., Signalling for nodulation in a water-tolerant legume, in *Biological Nitrogen Fixation, Sustainable Agriculture and the Environment: Proceedings of the 14th International Nitrogen Fixation Congress*, Wang, Y.-P., Lin, M., Tian, Z.X., Elmerich, C., and Newton, W.E., Eds., *Current Plant Science and Biotechnology in Agriculture*, Vol. 41, Springer, Dordrecht, 2005, p. 161.

36. Cooper, J.E., Multiple responses of rhizobia to flavonoids during legume root infection, in *Advances in Botanical Research*, Vol. 41, *Incorporating Advances in Plant Pathology*, 2004, p. 1.

37. Werner, D., Forests — trees — cells, in *Fast Growing Trees and Nitrogen Fixing Trees*, Werner, D. and Müller, P., Eds., G. Fischer, Stuttgart, 1990, p. 3.

38. Scheidemann, P. and Wetzel, A., Identification and characterization of flavonoids in the root exudate of *Robinia pseudoacacia*, *Trees*, 11, 316, 1997.

39. Sprent, J.I., Nodulated legume trees, in *Agriculture, Forestry, Ecology and the Environment*, Werner, D. and Newton, W.E., Eds., Springer, Dordrecht, 2005, p. 113.

40. Phillips, D.A. and Streit, W.R., Rhizosphere signals and ecochemistry, in *Environmental Signal Processing and Adaptation*, Heldmaier, G. and Werner, D., Eds., Springer, Berlin, 2003, p. 39.

41. Gagnon, H. and Ibrahim, R.K., Aldonic acids: a novel family of nod gene inducers of *Mesorhizobium loti*, *Rhizobium lupine*, and *Sinorhizobium meliloti*, *Mol. Plant-Microbe Interact.*, 11, 988, 1998.

42. Bolaños-Vasquez, M.C. and Werner, D., Effect of *Rhizobium tropici*, *R. etli*, and *R. leguminosarum* bv. *phaseoli* on *nod* gene-inducing flavonoids in root exudates of *Phaseolus vulgaris*, *Mol. Plant-Microbe Interact.*, 10, 339, 1997.

43. Pacios-Bras, C. et al., Novel lipochitin oligosaccharide structures produced by *Rhizobium etli* KIM5s, *Carbohydr. Res.*, 337, 1193, 2002.

44. Freiberg, C. et al., Molecular basis of symbiosis between *Rhizobium* and legumes, *Nature*, 387, 394, 1997.

45. Marie, C. et al., Characterization of Nops, nodulation outer proteins, secreted via the type III secretion system of NGR234, *Mol. Plant-Microbe Interact.*, 16, 743, 2003.

46. Göttfert, M. et al., Potential symbiosis-specific genes uncovered by sequencing a 410 kilobase DNA region of the *Bradyrhizobium japonicum* chromosome, *J. Bacteriol.*, 183, 1405, 2001.

47. Krishnan, H.B. et al., Extracellular proteins involved in soybean cultivar-specific nodulation are associated with pilus-like surface appendages and exported by a type III protein secretion system in *Sinorhizobium fredii* USDA257, *Mol. Plant-Microbe Interact.*, 16, 617, 2003.

48. Viprey, V. et al., Symbiotic implications of type III protein secretion machinery in *Rhizobium, Mol. Microbiol.*, 28, 1381, 1998.
49. Galibert, F. et al., The composite genome of the legume symbiont *Sinorhizobium meliloti, Science,* 293, 668, 2001.
50. Marie, C., Broughton, W.J., and Deakin, W.J., *Rhizobium* type III secretion systems: legume charmers or alarmers?, *Curr. Opin. Plant Biol.*, 4, 336, 2001.
51. Bartsev, A.V. et al., Purification and phosphorylation of the effector protein NopL from *Rhizobium* sp. NGR234, *FEBS Lett.*, 554, 271, 2003.
52. Sutton, J.M., Lea, E.J.A., and Downie, J.A., The nodulation signaling protein NodO from *Rhizobium leguminosarum* biovar *viciae* forms ion channels in membranes, *Proc. Natl. Acad. Sci. USA*, 91, 9990, 1994.
53. Walker, S.A. and Downie, J.A., Entry of *Rhizobium leguminosarum* bv. *viciae* into root hairs requires minimal Nod factor specificity, but subsequent infection thread growth requires nodO or nodE, *Mol. Plant-Microbe Interact.*, 13, 754, 2000.
54. Breedveld, M.W. and Miller, J.K., Cell-surface -glucans, in *The Rhizobiaceae: Molecular Biology of Model Plant-Associated Bacteria*, Spaink, H.P., Kondorosi, A., and Hooykaas, P.J.J., Eds., Kluwer, Dordrecht, 1998, p. 81.
55. Rolin, D.B. et al., Structural studies of a phosphocholine substituted β-(1,3); (1,6) macrocyclic glucan from *Bradyrhizobium japonicum* USDA 110, *Biochim. Biophys. Acta*, 1116, 215, 1992.
56. Werner, D., Organic signals between plants and microoganisms, in *The Rhizosphere*, Pinton, R., Varanini, Z., and Nannipieri, P., Eds., Marcel Dekker, New York, 2001, p. 197.
57. Schlaman, H.R.M. et al., Chitin oligosaccharides can induce cortical cell division in roots of *Vicia sativa* when delivered by ballistic microtargeting, *Development,* 124, 4887, 1997.
58. Parniske, M. et al., Plant defense responses of host plants with determinate nodules induced by EPS-defective *exoB* mutants of *Bradyrhizobium japonicum, Mol. Plant Microbe Interact.*, 7, 631, 1994.
59. Werner, D. et al., Symbiosis and defence in the interaction of plants with microorganisms, *Symbiosis*, 32, 83, 2002.
60. Parniske, M., Intracellular accommodation of microbes by plants: a common developmental program for symbiosis and disease?, *Curr. Opin. Plant Biol.*, 3, 320, 2000.
61. Streit, W.R. et al., An evolutionary hot spot: the pNGR234b replicon of *Rhizobium* sp. Strain NGR234, *J. Bacteriol.*, 186, 535, 2004.
62. Staehelin, C. et al., Characterization of mutants from *Rhizobium* sp. NGR234 with defective exopolysaccharide synthesis, in *Biological Nitrogen Fixation, Sustainable Agriculture and the Environment: Proceedings of the. 14th International Nitrogen Fixation Congress,* Wang, Y.-P., Lin, M., Tian, Z.X., Elmerich, C., and Newton, W.E., Eds., *Current Plant Science and Biotechnology in Agriculture*, Vol. 41, Springer, Dordrecht, 2005, p. 213.
63. Raetz, C.R. and Whietfield, C., Lipopolysaccharide endotoxins, *Annu. Rev. Biochem.*, 71, 635, 2002.
64. Vedam, V. et al., The pea-nodule environment restores the ability of a *Rhizobium leguminosarum* lipopolysaccharide ACPXL mutant to add 27-hydroxyoctacosanoic acid to its lipid-A, *J. Bacteriol.*, 188, 2126, 2006.
65. Vinuesa, P. et al., Identification of a plasmid-borne locus in *Rhizobium. etli* KIM5s involved in lipopolysaccharide O-chain synthesis and nodulation of *Phaseolus vulgaris, J. Bacteriol.*, 181, 5606, 1999.
66. Becker, A., Fraysse, N., and Sharypova, L., Recent advances in studies on structure and symbiosis-related function of rhizobial K-antigens and lipopolysaccharides, *Mol. Plant-Microbe Interact.*, 18, 899, 2005.
67. Silipo, A. et al., Full structural characterization of the lipid A components from the *Agrobacterium tumefaciens* strain C58 lipopolysaccharide fraction, *Glycobiology*, 14, 805, 2004.
68. Schonejans, E., Expert, D., and Toussaint, A., Characterization and virulence properties of *Erwinia chrysanthemi* lipopolysaccharide-defective øEC2-resistant mutants, *J. Bacteriol.*, 169, 4011, 1987.
69. Albus, U. et al., Suppression of an elicitor-induced oxidative burst in *Medicago sativa* cell-cultures, by *Sinorhizobium meliloti* lipopolysaccharides, *New Phytologist*, 151, 597, 2001.
70. Scheidle, H., Gross, A., and Niehaus, K., The lipid A substructure of the *Sinorhizobium meliloti* lipopolysaccharides is sufficient to suppress the oxidative burst in host plants, *New Phytololgist*, 165, 559, 2005.
71. Sharypova, L.A. et al., *Sinorhizobium meliloti* AcpXL mutant lacks the C28 hydroxylated fatty acid moiety of lipid A and does not express a slow migrating form of lipopolysaccharide, *J. Biol. Chem.*, 278, 12946, 2003.

72. Beveridge, T.J., Structures of gram-negative cell walls and their derived membrane vesicles, *J. Bacteriol.*, 181, 4725, 1999.
73. Noel, K.D., Forsberg, L.S., and Carlson, R.W., Varying the abundance of O antigen in *Rhizobium etli* and its effect on symbiosis with *Phaseolus vulgaris*, *J. Bacteriol.*, 182, 5317, 2000.
74. Gao, M. et al., Knockout of an Azorhizobial dTDP-L-Rhamnose synthase affects lipopolysaccharide and extracellular polysaccharide production and disables symbiosis with *Sesbania rostrata*, *Mol. Plant-Microbe Interact.*, 14, 857, 2001.
75. Fraysse, N. et al., Symbiotic conditions induce structural modification of *Sinorhizobium* sp. NGR234 surface polysaccharides, *Glycobiology*, 12, 741, 2002.
76. Noel, K.D., Box, J.M., and Bonne, V.J., 2-O-methylation of fucosyl residues of a rhizobial lipopolysaccharide is increased in response to host exudates and is eliminated in a symbiotically defective mutant, *Appl. Environ. Microbiol.*, 70, 1537, 2004.
77. Mathis, R. et al., Lipopolysaccharides as a communication signal for progression of legume endosymbiosis, *Proc. Natl. Acad. Sci. USA*, 102, 2655, 2005.
78. Udvardi, M.K. and Scheible, W.-R., GRAS genes and the symbiotic green revolution, *Science*, 308, 1749, 2005.
79. Stracke, S. et al., A plant receptor-like kinase required for both bacterial and fungal symbiosis, *Nature*, 417, 959, 2002.
80. Endre, G. et al., A receptor kinase gene regulating symbiotic nodule development, *Nature*, 417, 962, 2002.
81. Mitra, R.M. et al., A Ca²⁺/calmodulin-dependent protein kinase required for symbiotic nodule development: gene identification by transcript-based cloning, *Proc. Natl. Acad. Sci. USA*, 101, 4701, 2004.
82. Lévy, J. et al., A putative Ca²⁺ and calmodulin-dependent protein kinase required for bacterial and fungal symbioses, *Science*, 303, 1361, 2004.
83. Bersoult, A. et al., Expression of the *Medicago truncatula DMI2* gene suggests roles of the symbiotic nodulation receptor kinase in nodules and during early nodule development, *Mol. Plant-Microbe Interact.*, 18, 869, 2005.
84. Esseling, J.J., Lhuissier, F.G.P., and Emons, A.M.C., A nonsymbiotic root hair tip growth phenotype in NORK-mutated legumes: implications for nodulation factor-induced signalling and formation of a multifaceted root hair pocket for bacteria, *Plant Cell*, 16, 933, 2004.
85. Bolle, C., The role of GRAS proteins in plant signal transduction and development, *Planta*, 218, 683, 2004.
86. Itoh, H. et al., The gibberellin signalling pathway is regulated by the appearance and disappearance of SLENDER RICE1 in nuclei, *Plant Cell*, 14, 57, 2002.
87. Morohashi, K. et al., Isolation and characterization of a novel *GRAS* gene that regulates meiosis-associated gene expression, *J. Biol. Chem.*, 278, 20865, 2003.
88. Kaló, P. et al., Nodulation signalling in legumes requires NSP2, a member of the GRAS family of transcriptional regulators, *Science*, 308, 1786, 2005.
89. Mellor, R.B. et al., Phospholipid transfer from ER to the peribacteroid membrane in soybean nodules, *Z. Naturforsch.*, 40c, 73, 1985.
90. Mellor, R.B. and Werner, D., Peribacteroid membrane biogenesis in mature legume root nodules, *Symbiosis*, 3, 75, 1987.
91. Müller, P. et al., A TnphoA insertion within the *Bradyrhizobium japonicum sipS* gene, homologous to prokaryotic signal peptidases, results in extensive changes in the expression of PBM-specific nodulins of infected soybean (*Glycine max*) cells, *Mol. Microbiol.* 18, 831, 1995.
92. Bairl, A. and Müller, P., A second gene for Type I signal peptidase in *Bradyrhizobium japonicum*, *sipF*, is located near genes involved in RNA processing and cell division, *Mol. Gen. Genet.*, 260, 346, 1998.
93. Smit, P. et al., NSP1 of the GRASS protein family is essential for rhizobial Nod factor-induced transcription, *Science*, 308, 1789, 2005.
94. Derby, M.C. et al., Mammalian GRIP proteins differ in their membrane binding properties and are received to distinct domains of the TGN, *J. Cell Sci.*, 117, 5865, 2004.
95. Röhm, M. and Werner, D., Isolation of root hairs from seedlings of *Pisum sativum*. Identification of root hair specific proteins by *in situ* labelling, *Physiol. Plant.*, 69, 129, 1987.
96. Sauviac, L. et al., Transcript enrichment of Nod factor-elicited early nodulin genes in purified root hair fractions of the model legume *Medicago truncatulata, J. Exp. Bot.*, 56, 2507, 2005.

97. Charron, D. et al., Pharmacological evidence that multiple phospholipids signalling pathways link *Rhizobium* nodulation factor perception in *Medicago truncatula* root hairs to intracellular responses, including Ca^{2+} spiking and specific ENOD gene expression, *Plant Physiol.*, 136, 3582, 2004.

98. Takeda, N. et al., Expression of LjENOD40 genes in response to symbiotic and non-symbiotic signals: LjENOD40-I and LjENOD40-2 are differentially regulated in *Lotus japonicus, Plant Cell Physiol.*, 46, 1291, 2005.

99. Grønlund, M. et al., Analysis of promoter activity of the early nodulin *Enod40* in *Lotus japonicus, Mol. Plant-Microbe Interact.*, 18, 414, 2005.

100. Vassileva, V.N., Kouchi, H., and Ridge, R.W., Microtubule dynamics in living root hairs: transient slowing by lipochitin oligosaccharide nodulation signals, *Plant Cell*, 17, 1777, 2005.

101. Imaizumi-Anraku, H. et al., Plastid proteins crucial for symbiotic fungal and bacterial entry into plant roots, *Nature*, 433, 527, 2005.

102. Imaizumi-Anraku, H. et al., *Castor* and *Pollux*, the twin genes that are responsible for endosymbioses in *Lotus japonicus*, in *Biological Nitrogen Fixation, Sustainable Agriculture and the Environment: Proceedings of the 14th International Nitrogen Fixation Congress*, Wang, Y.-P., Lin, M., Tian, Z.X., Elmerich, C., and Newton, W.E., Eds., *Current Plant Science and Biotechnology in Agriculture*, Vol. 41, Springer, Dordrecht, 2005, p. 195.

103. Kouchi, H. et al., Large-scale analysis of gene expression profiles during early stages of root nodule formation in a model legume, *Lotus japonicus, DNA Res.*, 31, 263, 2004.

104. Mitra, R.M., Shaw, S.L., and Long, S.R., Six nonnodulating plant mutants defective for Nod factor-induced transcriptional changes associated with the legume-rhizobia symbiosis, *Proc. Natl. Acad. Sci. USA*, 101, 10217, 2004.

105. Ane, J.M. et al., *Medicago truncatula* DMII required for bacterial and fungal symbioses in legumes, *Science*, 303, 1364, 2004.

106. Svistoonoff, S. et al., Infection-related activation of the cg12 promoter is conserved between actinorhizal and legume-rhizobia root nodule symbiosis, *Plant Physiol.*, 136, 3191, 2004.

107. Weerasinghe, R.R., Bird, D.M., and Allen, N.S., Root-knot nematodes and bacterial Nod factors elicit common signal transduction events in *Lotus japonicus, Proc. Natl. Acad. Sci. USA*, 102, 3147, 2005.

108. Rathbun, E.A. and Brewin, N.J., Root nodule extensions in infection thread development, in *Biological Nitrogen Fixation, Sustainable Agriculture and the Environment: Proceedings of the 14th International Nitrogen Fixation Congress*, Wang, Y.-P., Lin, M., Tian, Z.X., Elmerich, C., and Newton, W.E., Eds., *Current Plant Science and Biotechnology in Agriculture*, Vol. 41, Springer, Dordrecht, 2005, p. 193.

109. Gage, D.J., Analysis of infection thread development using Gfp- and DsRed-Expressing *Sinorhizobium meliloti, J. Bacteriol.*, 184, 7042, 2002.

110. Brewin, J.N., Plant cell wall remodelling in the Rhizobium-legume symbiosis, *Crit. Rev. Plant Sci.*, 23, 293, 2004.

111. Foucher, F. and Kondorosi, E., Cell cycle regulation in the course of nodule organogenesis in Medicago, *Plant Mol. Biol.*, 43, 773, 2000.

112. Kondorosi, A. et al., Cell cycle and symbiosis, in *Biological Nitrogen Fixation, Sustainable Agriculture and the Environment: Proceedings of the 14th International Nitrogen Fixation Congress*, Wang, Y.-P., Lin, M., Tian, Z.X., Elmerich, C., and Newton, W.E., Eds., *Current Plant Science and Biotechnology in Agriculture*, Vol. 41, Springer, Dordrecht, 2005, p. 147.

113. Roudier, F. et al., The *Medicago* species A2-type cyclin is auxin regulated and involved in meristem formation but dispensable for endoreduplication-associated developmental programs, *Plant Physiol.*, 131, 1091, 2003.

114. Vinardell, J.M. et al., Endoreduplication mediated by the anaphase-promoting complex activator CCS52A is required for symbiotic cell differentiation in *Medicago truncatula* nodules, *Plant Cell*, 15, 2093, 2003.

115. Gresshoff, P.M. et al., Functional genomics of the regulation of nodule number in legumes, in *Biological Nitrogen Fixation, Sustainable Agriculture and the Environment: Proceedings of the 14th International Nitrogen Fixation Congress*, Wang, Y.-P., Lin, M., Tian, Z.X., Elmerich, C., and Newton, W.E., Eds., *Current Plant Science and Biotechnology in Agriculture*, Vol. 41, Springer, Dordrecht, 2005, p. 173.

116. Krusell, L. et al., Shoot control of root development and nodulation is mediated by a receptor-like kinase, *Nature*, 422, 2002.

117. Searle, I.R. et al., Long-distance signaling in nodulation detected by a CLAVATA1-like receptor kinase, *Science*, 299, 109, 2003.

118. Nishimura, R. et al., HAR1 mediates systemic regulation of symbiotic organ development, *Nature*, 420, 426, 2002.

119. Kawaguchi, M., "Activator" and "inhibitor" leading to generation and stabilization of symbiotic organ development in legume, in *Biological Nitrogen Fixation, Sustainable Agriculture and the Environment: Proceedings of the 14th International Nitrogen Fixation Congress*, Wang, Y.-P., Lin, M., Tian, Z.X., Elmerich, C., and Newton, W.E., Eds., *Current Plant Science and Biotechnology in Agriculture*, Vol. 41, Springer, Dordrecht, 2005, p. 179.

120. Werner, D. et al., Lysis of bacteroids in the vicinity of the host cell nucleus in an ineffective (fix-) root nodule of soybean (*Glycine max*), *Planta*, 162, 8, 1984.

121. Mellor, R.B., Christensen, T.M.I.E., and Werner, D., Choline kinase II is present only in nodules that synthesize stable peribacteroid membranes, *Proc. Natl. Acad. Sci. USA*, 83, 659, 1986.

122. Werner, D. et al., Particle density and protein composition of the peribacteroid membrane from soybean root nodules is affected by mutation in the microsymbiont *Bradyrhizobium japonicum*, *Planta*, 174, 263, 1988.

123. Werner, D., Physiology of nitrogen fixing legume nodules — compartments and functions, in *Biological Nitrogen Fixation*, Stacey, G., Burris, R., and Evans, J.H., Eds., Chapman and Hall, New York, 1992, p. 399.

124. Müller, P., Klaucke, A., and Wegel, E., Tn*phoA*-induced symbiotic mutants of *Bradyrhizobium japonicum* that impair cell and tissue differentiation in *Glycine max* nodules, *Planta*, 197, 163, 1995.

125. Müller, P., Use of the multipurpose transposon Tn *KPK2* for the mutational analysis of chromosomal regions upstream and downstream of the sipF gene in *Bradyrhizobium japonicum*, *Mol. Gen. Genomics*, 271, 359, 2004.

126. Becker, B.U. et al., A novel genetic locus outside the symbiotic island is required for effective symbiosis of *Bradyrhizobium japonicum* with soybean *Glycine max*, *Res. Microbiol.*, 155, 770, 2004.

127. Yoshida, S. and Parniske, M., Regulatory mechanisms of SYMRK kinase activity, in *Biological Nitrogen Fixation, Sustainable Agriculture and the Environment: Proceedings of the 14th International Nitrogen Fixation Congress*, Wang, Y.-P., Lin, M., Tian, Z.X., Elmerich, C., and Newton, W.E., Eds., *Current Plant Science and Biotechnology in Agriculture*, Vol. 41, Springer, Dordrecht, 2005, p. 183.

128. Botsford, J.L., Osmoregulation in *R.meliloti*: inhibition of growth by salts, *Arch. Microbiol.*, 137, 124, 1984.

129. Boscari, A. et al., BetS is a major glycine betaine/proline betaine transporter required for early osmotic adjustment in *Sinorhizobium meliloti*, *J. Bacteriol.*, 184, 2654, 2002.

130. Schiefner, A. et al., Cation-phi interactions as determinants for binding of the compatible solutes glycine betaine and proline betaine by the periplasmic ligand-binding protein ProX from *Escherichia coli*, *J. Biol. Chem.*, 279, 5588, 2004.

131. Holtmann, G. and Bremer, E., Thermoprotection of Bacillus subtilis by glycine betaine and structurally related compatible solutes: involvement of Opu-transporters, *J. Bacteriol.*, 186, 1683, 2004.

132. Steil, L. et al., Genome-wide transcriptional profiling analysis of adaptation of *Bacillus subtilis* to high salinity, *J. Bacteriol.*, 185, 6358, 2003.

133. Sadowsky, M.J., Stress factors influencing symbiotic nitrogen fixation, in *Agriculture, Forestry, Ecology and the Environment*, Werner, D. and Newton, W.E., Eds., Kluwer Academic, Dordrecht, 2005, p. 89.

134. Boumahdi, M., Mary, P., and Hornez, J.P., Changes in fatty acid composition and degree of unsaturation of (brady)rhizobia as a response to phases of growth, reduced water activities and mild desiccation, *Antonie Van Leeuwenhoek*, 79, 73, 2001.

135. Purcell, L.C. and King, C.A., Drought and nitrogen source effects on nitrogen nutrition, seed, growth, and yield in soybean, *J. Plant Nutrit.*, 19, 969, 1996.

136. Werner, D. et al., Chair's comments: the nitrogen cycle, in *Nitrogen Fixation: Global Perspectives*, Finan, T.M., O'Brian, M.R., Layzell, D.B., Vessey, J.K., and Newton, W., Eds., CABI, Oxon, New York, 2002, p. 297.

137. Wais, R.J., Keating, D.H., and Long, S.R., Structure-function analysis of nod factor-induced root hair calcium spiking in *Rhizobium*-legume symbiosis, *Plant Physiol.*, 129, 211, 2002.

138. Kuhlmann, K.-P. et al., Mineral composition of effective and ineffective nodules of *Glycine max* in comparison to roots: characterization of developmental stages by differences in nitrogen, hydrogen, sulfur, molybdenum, potassium and calcium content, *Angew. Bot.*, 56, 315, 1982.

139. Vance, C.P., Symbiotic nitrogen fixation and phosphorus acquisition. Plant nutrition in a world of declining renewable resources, *Plant Physiol.*, 127, 390, 2001.

140. Barea, J.M. et al., Interactions of arbuscular mycorrhiza and nitrogen fixing symbiosis in sustainable agriculture, in *Agriculture, Forestry, Ecology and the Environment*, Werner, D. and Newton, W.E., Eds., Kluwer Academic, Dordrecht, 2005, p. 199.

141. Ruiz-Lozano, J.M. et al., Arbuscular mycorrhizal symbiosis can alleviate drought-induced nodule senescence in soybean plants, *New Phytolologist*, 151, 493, 2001.

142. Augé, R.M., Water relations, drought and vesicular-arbuscular mycorrhizal symbiosis, *Mycorrhiza*, 11, 3, 2001.

143. Barea, J.M., Azcón, R., and Azcón-Aguilar, C., Mycorrhizosphere interactions to improve plant fitness and soil quality, Antonie van Leeuwenhoek, 81, 343, 2002.

144. Barea, J.M. et al., The rhizosphere of mycorrizal plants, in *Mycorrhiza Technology in Agriculture, from Genes to Bioproduction*, Gianinazzi, S., Schüepp, H., Barea, J.M., and Haselwandter, K., Eds., Birkhäuser, Basel, 2002, p. 1.

145. Bianciotto, V. et al., '*Candidatus Glomeribacter gigasporum*', an endosymbiont of arbuscular mycorrhizal fungi, *Int. J. Syst. Evol. Microbiol.*, 53, 121,2002.

146. Bianciotto, V. et al., Vertical transmission of endobacteria in the arbuscular mycorrhizal fungus *Gigaspora margarita* through vegetative spore generations, *Appl. Environ. Microbiol.*, 70, 3600, 2004.

147. Bonfante, P., The rhizosphere and rhizoplane continuum: where plants, fungi and bacteria meet, in Rhizosphere 2004 — Perspectives and Challenges — A Tribute to Lorenz Hiltner, GSF-Bericht 05/05, Neuherberg, 2005, p. 99.

148. Poole, R.K., Summary, in *Bacterial Responses to pH*, Chadvick, J. and Cardew, G., Eds., John Wiley and Sons, Chichester, U.K., 1999, p. 1.

149. Glenn, A.R. et al., Acid tolerance in root nodule bacteria, in *Bacterial Responses to pH*, Chadvick, J. and Cardew, G., Eds., John Wiley and Sons, Chichester, U.K., 1999, p. 112.

150. Vinuesa, P. et al., Genetic analysis of a pH-regulated operon from *Rhizobium tropici* CIAT899 involved in acid tolerance and nodulation competitiveness, *Mol. Plant-Microbe Interact.*, 16, 159, 2003.

151. Steele, H.L., Vinuesa, P., and Werner, D., A *Rhizobium tropici* CIAT899 leucine biosynthesis mutant which can survive and grow at pH 3.5, *Biol. Fertil. Soils*, 38, 84, 2003.

152. Rojas-Jimenez, K. et al., A CIC chloride chanel homolog and ornithine-containing membrane lipids of *Rhizobium tropici* CIAT899 are involved in symbiotic efficiency and acid tolerance, *Mol. Plant-Microbe Interact.*, 11, 1175, 2005.

153. Sadowsky, M.J., Keyser, H.H., and Bohlool, B.B., Biochemical characterization of fast- and slow-growing rhizobia that nodulate soybeans, *Int. J. Syst. Bacteriol.*, 33, 716, 1983.

154. Vinuesa, P. et al., *Bradyrhizobium canariense* sp. nov., an acid-tolerant endosymbiont that nodulates endemic genistoid legumes (Papilionoideae:Genisteae) from the Canary Islands, along with *Bradyrhizobium japonicum* bv. Genistearum, *Bradyrhizobium* genospecies α and *Bradyrhizobium* genospecies β, *Int. J. Syst. Evol. Microbiol.*, 55, 569, 2005.

155. Vinuesa, P. et al., Population genetics and phytogenetic inference in bacterial molecular systematics: the role of migration and recombination in *Bradyrhizobium* species cohesion and delineation, *Mol. Phylogenet. Evol.*, 34, 29, 2005.

156. Werner, D., Production and biological nitrogen fixation of tropical legumes, in *Agriculture, Forestry, Ecology and the Environment*, Werner, D. and Newton, W.E., Eds., Kluwer Academic, Dordrecht, 2005, p. 1.

157. Pueppke, S.G., Nitrogen fixation in soybean in North America, in *Agriculture, Forestry, Ecology and the Environment*, Werner, D. and Newton, W.E., Eds., Kluwer Academic, Dordrecht, 2005, p. 15.

158. Hungria, M. et al., The importance of nitrogen fixation to the soybean cropping in South America, in *Agriculture, Forestry, Ecology and the Environment*, Werner, D. and Newton, W.E., Eds., Kluwer Academic, Dordrecht, 2005, p. 25.

159. Mahna, S.K., Production, regional distribution of cultivars and agricultural aspects of soybean in India, in *Agriculture, Forestry, Ecology and the Environment*, Werner, D. and Newton, W.E., Eds., Kluwer Academic, Dordrecht, 2005, p. 43.

160. Ruiz Sainz, J.E. et al., Soybean cultivation and BNF in China, in *Agriculture, Forestry, Ecology and the Environment*, Werner, D. and Newton, W.E., Eds., Kluwer Academic, Dordrecht, 2005, p. 67.

161. Fiencke, C., Spieck, E., and Bock, E., Nitrifying bacteria, 255 in *Agriculture, Forestry, Ecology and the Environment*, Werner, D. and Newton, W.E., Eds., Kluwer Academic, Dordrecht, 2005, p. 255.

162. Spanning, R., van Delgado, M.J., and Richardson, D.J., The nitrogen cycle: denitrification and its relationship to N_2 fixation, in *Agriculture, Forestry, Ecology and the Environment*, Werner, D. and Newton, W.E., Eds., Kluwer Academic, Dordrecht, 2005, p. 277.
163. Hungria, M. et al., Inocula preparation, production and application, in *Agriculture, Forestry, Ecology and the Environment*, Werner, D. and Newton, W.E., Eds., Kluwer Academic, Dordrecht, 2005, p. 223.
164. Graham, P., Hungria, M., and Tlusty, B., Breeding for better nitrogen fixation in grain legumes: where do the rhizobia fit in?, *Crop Managem.*, 10, 1094, 2004.

10 Biocontrol of Plant Pathogens: Principles, Promises, and Pitfalls

Ben Lugtenberg and Johan Leveau

CONTENTS

I. INTRODUCTION TO BIOCONTROL-RELATED PHENOMENA IN THE RHIZOSPHERE

In the past decades, the severity of several plant diseases has been decreased by the use of (partially) resistant crops or by the use of chemical pesticides. Even with these measures in place, plant diseases can be responsible for a loss of 20 to 30% of the crop yield. It should be noted that not all pathogens can be controlled by resistance or by chemicals. Chemical control is increasingly banned because of negative effects of some chemicals on human beings and the environment. The major alternative for chemical control is the use of resistant plants. However, resistance comes at a cost of an average of 3% of crop yield. Moreover, genetically engineered resistant plants are not accepted by many countries, including those of the EU.

Biological control is an environmentally friendly alternative for chemical control [1]. Biological control of soilborne plant diseases is defined as the exploitation of the natural ability of soils to suppress disease [2–9]. Alternatives to soils, for example, rockwool, can also become disease-suppressive; rockwool on which cucumber plants have been grown can suppress cucumber root and crown rot caused by *Pythium aphinidermatum* [10,10a]. Soils can develop resistance to disease naturally or after inoculation with a biocontrol agent (BCA). Several hundreds of such BCAs are presently already on the market [11]. Presently, biocontrol is less consistent than chemical control. To make biocontrol more reproducible, it is important to know the mechanisms used by BCAs as well as the influence of abiotic, biotic, and ecological factors, which influence biocontrol efficacy [12,13].

Many aspects of the rhizosphere will be treated in other chapters of this book. Therefore, we will focus here only on topics relevant to understanding biocontrol. In this chapter, we will focus on mechanisms of biocontrol, on nutrients available in the rhizosphere for pathogens and biocontrol agents, on chemical signaling between microbes with each other and with the plant in the rhizosphere, and on microbial biosensors that can be used to improve our knowledge of the rhizosphere environment. The work in Leiden focuses mainly on TFRR (tomato foot and root rot) as a model disease. It is caused by the pathogenic soilborne fungus *Fusarium oxysporum* f. sp. *radicis-lycopersici*, a major pathogen [14]. Many examples presented in this review have been taken from this model system.

Plants usually grow in the very complex environment of the soil, but some commercial crops are grown also on rockwool, perlite, or tuff. One gram of soil contains approximately 10^{10} bacteria of which only 0.1 to 10% can be cultured [1a]. Moreover, it contains 1 to 5 km of hyphae, 30,000 protozoa, and 50 to 100 nematodes. Biocontrol of soilborne diseases takes mainly place in the rhizosphere [1,15,16]. Rhizosphere organisms not only interact with the root but also with each other. For example, bacteria form biofilms on hyphae. In the past decade, chemical signaling between rhizosphere organisms has been shown to play a major role in their activities. Furthermore, predation of bacteria by protozoans, called bacterial grazing, is believed to be the major factor whereby the biomass produced by bacteria reenters the food web [16a,17]. Apart from grazing, the level of exudate nutrients is another control factor that determines the bacterial load on roots.

The most important target soilborne pathogens are the bacteria *Erwinia* and *Streptomyces*, the fungi Fusarium [14], *Rhizoctonia* and *Verticillium*, and the funguslike oomycetes *Phytophtora* and *Pythium* [10]. It is interesting to note that a *Pseudomonas* strain exists, which is pathogenic for both animals and plants and uses common bacterial virulence factors to attack both types of organisms [18,19]. This suggests that the virulence factors have evolved early in evolution.

The best known BCAs [11] are the bacterial species *Bacillus* [20–22], *Burkholderia, Pseudomonas*, and *Streptomyces*, and the fungal species *Gliocladium* [23,24] and *Trichoderma* [25].

In the past two decades, many new techniques have become available for the study of biocontrol. To track biocontrol microbes they can be marked, for example, with different autofluorescent proteins, such as green fluorescent protein (GFP) and subsequently visualized by fluorescence microscopy or confocal laser scanning microscopy [26,27]. The use of reporter genes has made it possible to measure the activity of selected genes in biocontrol agents, and to assess their individual contribution to the biocontrol phenotype [28]. Furthermore, bioreporter strains for habitat exploration [29] are being used to understand the performance of the biocontrol agent in the context of the chemical, physical, and biological conditions that these microorganisms encounter in the rhizosphere. Genetic techniques also play a major role in the study of rhizosphere microbiology. They can be used to monitor the influence of the introduction of BCAs on populations of culturable and unculturable microbes [30]. Many of these techniques are based on the fact that 16S ribosomal RNA (rRNA) genes contain conservative as well as variable regions. Sequencing the 16S rDNA provides the most complete information. Several other methods based on the sequence also provide useful information. Denaturing gradient gel electrophoresis (DGGE) [30] is a culturing-independent method that can be used to obtain population profiles based on interspecies differences in 16S rDNA sequences. Amplified ribosomal DNA restriction analysis (ARDRA) is a fast way to identify strain differences by generating profiles of their rDNA restriction fragments. FISH (fluorescence *in situ* hybridization) is used to identify microbes with microscopy or flow cytometry using rRNA-targeted probes that may range from species-specific to kingdom-specific.

Genomics is going to play an increasingly important role in understanding and monitoring biocontrol because sequences of pathogens and biocontrol agents are rapidly becoming available. High-throughput methods such as microarray technology will become increasingly important for appreciating the identity and activity of the vast amount of microorganisms that inhabit the rhizosphere and for assessing the effect of BCAs on the activity of target organisms.

II. RHIZOSPHERE COMPETENCE

Competitive root tip colonization plays an important role in various mechanisms of biocontrol (see Section III) [31–36]. It is crucial for a biocontrol microbe that it is able to establish itself in the rhizosphere and to compete with the indigenous microflora for nutrients and niches on the root. The Leiden laboratory has especially been involved in dissecting the process of competitive root tip colonization by identifying genes and traits required for tomato rhizosphere colonization by *P. fluorescens* biocontrol strain WCS365. A review covering the majority of the results [35] has been published a few years ago and therefore the topic will only be treated briefly here. The newer data will be discussed more extensively.

Many studies on the mechanism of action of competitive rhizosphere colonization have initially been carried out in a gnotobiotic system with sterile quartz sand as the substrate and plant nutrient solution without a carbon source as growth medium. Therefore, seed and root exudates served as the carbon source. Two different microbes, for example a wild type and a putative colonization mutant, are coated on seeds or seedlings (one of several possible ways to apply them) and, after growth of the plant, the root tip was inspected for its microbial population using selective medium to discriminate the two populations [37]. In this way, it is possible to identify competitive colonization mutants or to test which one of two wild-type strains is the best colonizer. Although the gnotobiotic system is much simpler than realistic production systems, it appeared to have a highly predictive value. All important conclusions from the gnotobiotic system were verified in potting soil and, practically, all of them also appeared to hold in this system [38,39].

It appeared that biocontrol pseudomonads are not plant-specific in their competitive colonization ability although some quantitative differences in colonization ability for roots of different plants exist. Similarly, competitive colonization traits on one plant also appeared to be colonization traits on most tested other plants, although to a different extent. Pseudomonads are excellent colonizers whereas bacilli are poor colonizers.

Analysis of the composition of tomato seed and root exudates revealed that organic acids are better represented than sugars and amino acids [16]. Citric acid, malic acid, and lactic acid are the major organic acids [16]; glucose, xylose, and fructose are the major sugars [13,40]; and aspartic and glutamic acid are the major amino acids [41]. Moreover, it appeared that putrescine is the major nitrogen-containing tomato exudate compound [42]. The important role of exudates for rhizosphere competence was shown by Kuiper et al. [43] and Kamilova et al. [13], who showed that selection for enhanced colonizers selects for strains that function well in the rhizosphere [44] and efficiently use exudate components for growth in the rhizosphere. Consistent with the finding that organic acids are better represented in exudates than sugars is the finding that a mutant that is not able to utilize sugars colonizes the root as efficiently as the wild type [40], whereas a mutant, which is a poor utilizer of organic acids is unable to compete for the root tip [45]. Based on these results, it was concluded that utilization of organic acids is the nutritional basis for rhizosphere colonization [33,35,44,45].

Motility is a major competitive colonization trait [46]. Recently, we could show that chemotaxis rather than nondirectional motility is required [47,48]. The major chemoattractants for

Pseudomonas fluorescens WCS365 toward tomato exudates appeared to be the organic acids, malic acid and citric acid, and the amino acid L-isoleucine, whereas exudate sugars, representing a significant part of the total exudates [40], are inactive as chemoattractants [47].

Several other cell surface components are also involved in competitive colonization, such as pili [49], part of the O-antigen of the lipopolysaccharide [39,50] and cellulose [35].

Being able to synthesize crucial small molecules, including those for building its own macromolecules, is important for rhizosphere competence because mutants unable to synthesize amino acids, vitamins, and uracil are poor colonizers. Interestingly, putrescine was identified as a major exudate component based on the observation that one of the competitive colonization mutants was impaired in maintaining optimal intracellular levels of putrescine. Subsequent studies showed the presence of putrescine but not of the other polyamines in tomato exudate and that high intracellular putrescine concentrations are bacteriostatic for biocontrol strain WCS365 [42].

The molecular analysis of one colonization mutant indicated that Sss, a member of the lambda integrase family of site-specific recombinases, is involved in rhizosphere competence [38]. The genes *xerC* of *E. coli* and *sss* of *P. aeruginosa* promote reciprocal recombination between two small DNA fragments, which can result in inversion or excision of the DNA fragment situated between the two small recognition sites. We favor the hypothesis of Dybvig [51] that subpopulations generated by DNA rearrangements enable a bacterial population to respond adequately to environmental changes. According to this idea, the *Sss* mutant is locked in a nonrhizosphere competent genetic configuration [38]. These DNA rearrangements often result in a change in colony morphology designated as colony phase variation. Later experiments by Rivilla's group supported this notion and extended it by the observation that in addition to Sss, another site-specific recombinase, XerD, is also involved in phase variation [52].

More detailed experiments revealed that colony phase variation in *Pseudomonas* can affect expression of genes involved in the production of a lipopeptide biosurfactant, and of the exoenzymes chitinase, lipase, and protease [53,54]. The lipopeptide was shown to be involved in biocontrol whereas the other factors might also be involved in biocontrol [55].

Genes involved in phase variation of pseudomonads in the rhizosphere include, next to the site-specific recombinases encoded by *sss* [38] and *xerD* [52], the *gac* two-component system *gacA-gacS*, *rpoS*, and *mutS* [56,57]. Mutations in the *gac* system prevent synthesis of the lipopeptide and exoenzymes mentioned previously. Because these mutants have a growth advantage, it can be expected that the biocontrol-positive phase has advantages in the presence of the pathogen, whereas the biocontrol-negative phase has competitive advantages in the absence of the pathogen. It is likely that the biocontrol-positive phase has similar competitive root tip colonization disadvantages as an *sss* mutant [38]. Colonization strategies of pseudomonads have been correlated with differential expression of flagellin [58,59] and of indole-3-acetic acid (IAA) [58].

Protein secretion plays an important role in competitive root tip colonization; mutants in *secB* and in the type three secretion system (TTSS) [35,60] are impaired in competitive root tip colonization. Whereas the interpretation for the general secretion component *secB* is difficult, we favor the following explanation for the role of the TTSS in competitive colonization. Based on sequence similarities and on microscopic data, it is generally accepted that the TTSS and the flagellar apparatus have evolved after duplication of genes for the flagellar apparatus [61]. We hypothesize that prior to the evolution of the fine-tuning of the secretion of effector protein, the needle of the TTSS was open and was used by the microbe to get access to the juices of the eukaryotic host as a source of nutrients [45]. This conclusion is consistent with the fact that a TTSS is present in many plant-colonizing pseudomonads [62,63].

Endophytic colonization is getting more and more attention [64,65]. The cell wall degrading enzymes endoglucanase and endopolygalacturonase of *Burkholderia* sp. PsJN have been suggested to play a role in gaining entry into root internal tissues of grapevine [66]. Ethylene suppresses endophytic colonization of *Klebsiella pneumonia.* Moreover, plant defense responses seem to limit endophytic colonization.

The search for genes that contribute to rhizosphere competence has been facilitated by the ability to create random or targeted mutations in genes of interest, and comparing the behavior of wild-type and mutant strains in a rhizosphere setting. This has led to the identification of loci that are believed to contribute significantly to the ability of rhizomicroorganisms to establish themselves in the rhizosphere. A major drawback of the mutation approach is that only genes with a substantial contribution to rhizosphere competence are identified by this method. To circumvent this problem, other approaches have been developed and tried that ask the question what genes are expressed during rhizosphere colonization of competent microorganisms. The idea behind this strategy is that at least a subset of these genes should contribute incrementally to rhizosphere competence. One early example is the work by van Overbeek and van Elsas [67], who inserted a promoterless *lacZ* gene into random chromosomal locations of *P. fluorescens* R2f, and tested the resulting strains for expression of β-galactosidase in response to wheat root exudates. One exudate-inducible strain was challenged with a suite of individual compounds, and proline was the only one that invoked expression of the *lacZ* fusion, suggesting (1) that proline is present in root exudates, and (2) that the ability to utilize this amino acid may be a determinant of rhizosphere competence. More recently, *in vivo* expression technology (IVET) has been used to identify genes that are specifically induced by *Pseudomonas* strains during root colonization [68,69]. IVET [71] is often based on the *in vivo* survival of auxotrophic mutants of a wild-type bacterium as a result of, in this case, rhizosphere-inducible promoter sequences upstream from a promoterless gene that complements the phenotype of the auxotroph. Analysis of these rhizosphere-induced (*rhi*) genes [68] or genes with root-adapted promoters (*rap*) [69] revealed various genes involved in functions such as nutrient acquisition, stress response, secretion, chemotaxis, and motility. Most surprisingly, in both studies many genes with as-of-yet unknown function were identified, suggesting that our understanding of genes that contribute to rhizosphere competence is far from complete.

III. MECHANISMS OF BIOCONTROL

Biocontrol traits of microbes have been attributed to a variety of mechanisms [60,71–73].

A. ANTIBIOSIS

This is the process of the production and delivery of molecules that kill or decrease the growth of the target pathogen, and is the best-known mechanism by which microbes can control plant diseases [74,75]. The reason why this mechanism is best known presumably is that it is easier to do a preselection for microbes that inhibit the target pathogen *in vitro* and subsequently test the antibiosis-positive strains in a biocontrol test than to do biocontrol tests on all isolates.

The best-known antibiotics produced by Gram-negative bacteria are phenazines, 2,4-diacetylphloroglucinol [76–79], pyrrolnitrin [80–82], pyoluteorin [83–86], oomycin A [87], ammonia [88] and hydrogen cyanide [89]. Some biocontrol bacilli produce the antibiotics zwittermycin A and kanosamine [90]. *Trichoderma* and *Gliocladium* can produce antimicrobial compounds such as gliovirin and gliotoxin [24].

The structures and mode of action of many antimicrobial compounds have recently been extensively reviewed [60,69]. The role of the antibiotic in the biocontrol by such strains has been elucidated by testing the biocontrol ability of mutants in one of the structural genes for antibiotic production [71]. In the case of the PCN (phenazine-1-carboxamide) producer *P. chlororaphis* PCL1391, it was shown that colonization of the root surface by the bacterium is required to control the disease. This result was interpreted in the sense that PCN should be delivered along the whole root surface to protect the plant against the *Fusarium* pathogen, which can be present deep in the soil [32]. It is likely that in most cases, when antibiosis is the mechanism, root colonization is required as the delivery system of the antibiotic. In this respect it should be noted that some bacilli also produce antibiotics, whereas bacilli are extremely poor colonizers compared to pseudomonads.

Because the antibiotic 2,4-diacetylphloroglucinol also appeared to act as an inducer of systemic resistance [91], it may be that biocontrol bacilli rather use the latter mechanism than antibiosis.

More recently, biosurfactants have received considerable attention as antimicrobial compounds [92–95]. Because pathogens often form a biofilm on the root surface, it is interesting to note that some biosurfactants prevent biofilm formation and even degrade existing biofilms [92,96]. Rhamnolipid biosurfactants appear to be important for the development of mushroomlike structures in biofilms [97]. Cyclic lipopeptides surfactants such as viscosinamide [93] and tensin [94] are produced by *P. fluorescens* and have antifungal activity against *Rhizoctonia solani* and *Pythium ultimum* [95]. More recently, it was shown that a cyclic lipopeptide of *P. fluorescens* is involved in reducing hyacinth root rot and acts by causing lysis of zoospores of *Pythium intermedium* [98]. Moreover, the cyclic lipopeptide amphisin was shown to play a role in the colonization of sugar beet seeds [98a].

B. DEGRADATION OF AHLs

The synthesis of many antibiotics is dependent on the density of the producing bacterial population [99–101]. This phenomenon is designated as quorum sensing, involving extracellular molecules called acyl-homoserine lactones (AHLs) and is treated in detail elsewhere in this book (see Chapter 11). It is interesting to note that biocontrol pseudomonads form biofilms on the root which are covered by a mucoid layer [26,102]. In the biofilm, the bacteria will easily reach the required population density, whereas the mucoid layer may decrease or even prevent the diffusion of the N-acyl-homoserine lactones required for quorum sensing [102].

N-acyl-homoserine lactones are required for the synthesis of several antifungal metabolites and exoenzymes, factors that can play a role in biocontrol. Bacteria have been isolated which degrade N-acyl-homoserine lactones [103,104] using N-acyl-homoserine lactinase, encoded by the gene *aiiA*, which opens the lactone bond of the AHL ring [103] and therefore can interfere with biocontrol. This form of biocontrol is promising for protecting plants against maceration by *Erwinia carotovora* and *Erwinia amylovora,* which produce, in a quorum-sensing-dependent way, macerating enzymes as virulence factors [105]. Analogues of N-acyl-homoserine lactones have been synthesized, which antagonize bacterial quorum sensing [105].

C. PREDATION AND PARASITISM

The best example of predation and parasitism is biocontrol by *Trichoderma* fungi. Recently an excellent review appeared [106]. In addition to being parasites of other fungi, *Trichoderma* induces resistance responses, enhances root growth and development, crop productivity, resistance to stresses, and uptake and use of nutrients [106].

D. COMPETITION FOR FE³⁺ IONS

Competition for Fe^{3+} ions was one of the first described mechanism of biocontrol (Chapter 7 and Reference 107 to Reference 114). Iron is an essential cofactor for growth of all organisms. However, the amount of solubilized iron available to organisms in soil is low at neutral and alkaline pH values. Fluorescent *Pseudomonas* species, growing under iron limitation, produce large amounts of siderophores, which are Fe^{3+}–chelators, which bind Fe^{3+} with a very high affinity. The Fe^{3+}-siderophore complexes are bound to protein receptors at the bacterial surface, which are also induced under iron-limiting conditions. Some strains in addition have the ability to take up siderophores produced by other organisms [115,116]. When the iron uptake system of the pathogen is less efficient than that of the BCA, the pathogen will be impaired in its growth.

E. INDUCTION OF SYSTEMIC RESISTANCE

After contact with a necrotizing pathogen or a nonpathogenic biocontrol bacterium, a state of physiological immunity can be induced in plants, which protects the plant against subsequent viral, bacterial, or fungal attacks [117,118]. This phenomenon of systemic resistance is characterized by remote action, long-lasting resistance, and protection against a large number of other pathogens. The immunity caused by infection with a necrotizing pathogen is known as *systemic acquired resistance* (SAR). At the physiological level, salicylic acid accumulates and the production of pathogenesis-related (PR) proteins is induced. Certain nonpathogenic root-colonizing bacteria and fungi are able to induce resistance toward a variety of diseases in several plants. This phenomenon is designated as induced systemic resistance (ISR) [119–121]. Systemic resistance can be shown when the biocontrol strain is physically separated from the pathogen. When the biocontrol strain is present on the root, the pathogen is either applied on leaves [122–124] or, using a split root system, on the other part of the root [13]. Most resistance-inducing microbes described so far are Gram-negative bacteria, especially *Pseudomonas* and *Serratia* strains [120]. However, a number of Gram-positive bacteria also can induce resistance [125].

Systemic resistance cannot only be activated by living cells but also by a collection of chemically diverse cellular components such as lipopolysaccharide [126,127], flagella [120,126], salicylic acid [128], the siderophores pyochelin and pyocyanin [124], the cyclic peptide syringolin [129], 2,4 diacetylphloroglucinol [91], an *N*-trialkylated benzylamine derivative produced by *P. putida* [130], and the cyclic lipopeptide massetolide A, which induces systemic resistance in tomato [131]. Recently, it was shown that also *N*-acyl-homoserine lactones of *Serratia liquefaciens* induce resistance toward *Alternaria alternate* in tomato [131a]. Volatiles from *Bacillus* spp. have also been reported to induce ISR [131b]. In contrast to BCAs, in which the action is based on antibiosis (see Section III.A), biocontrol based on ISR does not require extensive colonization of the whole root system [132].

The response of the plant toward interaction with chitin fragments [133], LPS [127], and flagella or their flg22 peptide (which represents the elicitor-active epitope of flagellin) has been studied. Flagellin perception by *Arabidopsis thaliana* triggers resistance to pathogenic bacteria by restricting bacterial invasion and induces the expression of defense-related genes [134,135]. Flg22 seems to belong to a group of elicitors corresponding to what is called pathogen-associated molecular patterns in human pathogenic bacteria and fungi, to which also bacterial LPS and the fungal components chitin and ergosterol belong. In mammals, these are recognized by so-called Toll-like receptors [136]. Resistance proteins in plants have homology with these Toll-like receptors in mammals, for example, in *A. thaliana* fla22 is recognized by *FLS2*. Toll-like receptors play a role in innate immunity of mammals [134]. The induction of systemic resistance in the plant and innate immunity in mammals are evolutionary-related processes [137,138]. The fact that systemic resistance in the plant can be induced by the large number of chemically unrelated bacterial components mentioned earlier was initially surprising because it was difficult to incorporate in a simple model. However, because there are many Toll-like receptors which have a certain specificity for certain elicitors, it is tempting to speculate that the induction of systemic resistance by different bacterial components is mediated by Toll-like receptors (see Figure 10.1). In this model, efficient systemic resistance would depend on the right combination of elicitors, perhaps even in the right concentrations.

A very thorough study on the mechanism of action of ISR was done with *P. fluorescens* biocontrol strain WCS417r. Verhagen et al. surveyed the transcriptional analysis of 8,000 *Arabidopsis* genes after treatment of the root with the biocontrol strain. Whereas they observed a substantial change in the expression of 97 genes in the root, none of the tested genes was changed in the leaves. However, after challenge inoculation of the WCS417r-treated plants with the leaf pathogen *P. syringae* pv. tomato DC3000, 81 genes showed augmented expression in ISR-expressing leaves, suggesting that these genes were primed to respond faster or more strongly upon pathogen attack. The majority of the primed genes were predicted to be regulated by jasmonic acid or ethylene signaling [139]. One of these genes encodes the MYB72 transcription factor in the root. Upon challenge with strain WCS417r, a knockout mutant in this gene did not show ISR anymore. Evidence was obtained that MYB72 is an intrinsic part of local, ethylene-dependent signaling events that eventually leads to systemic expression of ISR in the leaves [140].

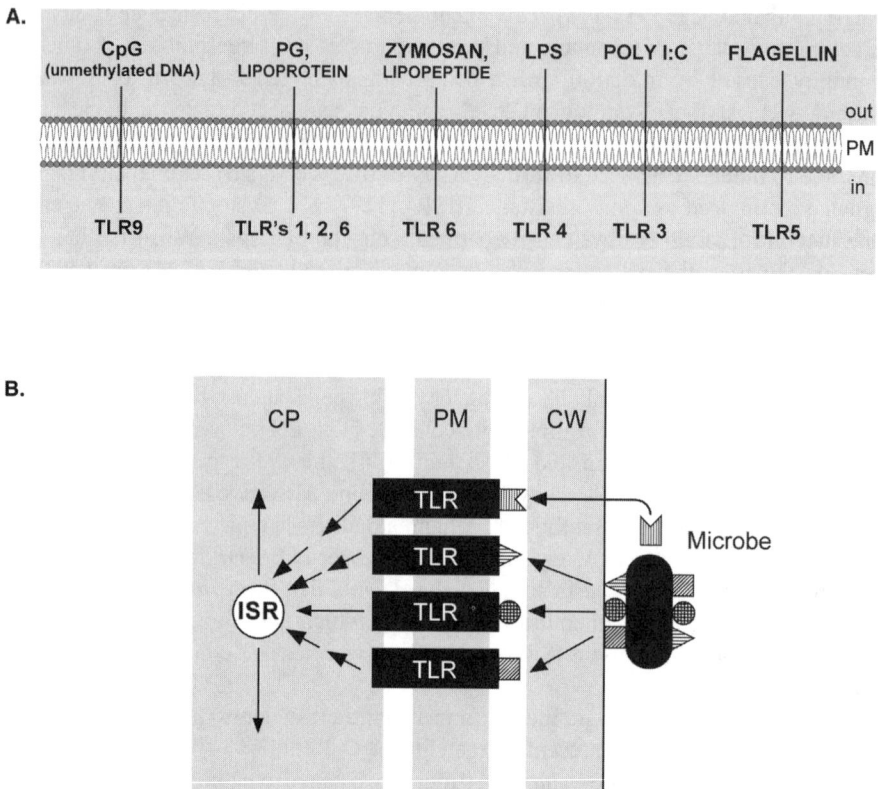

FIGURE 10.1 Are innate immunity in animals and systemic resistance in plants related? A. Innate immunity in animals can be induced by a variety of molecules (or parts thereof) from microbes. These molecular structures can be of bacterial, fungal, or viral origin. They are common for many different microbes, so that infection, in principle, by any microbe can be detected. The structures include lipopolysacharide, flagellin, zymosan, and nucleic acids and are not specific for pathogens although they are often called PAMPs (pathogen associated molecular patterns). These structures are recognized by so-called TLRs (Toll-like receptors) in the host membrane and the recognition eventually leads to inflammation and clearing of the infection. This form of pathogen control is designated as innate immunity. Ten TLRs are known in humans. Unmethylated DNA (CpG) binds to TRL 9, lipopeptide to TRL 1, peptidoglycan to TRL 2, double stranded DNA to TRL 3, flagellin to TRL 5, zymosan and lipopeptide to TLR 6, and bacterial endotoxin (or lipopolysaccharide; LPS) to TRL 4. (Figure after Clark, W. ASM News, 70, 317, 2004.) B. Plants recognize microbes in a similar way as animals (see Section III.A). Recognition of a wide range of structures, produced by beneficial microbes, by plant resistance proteins, which share homology with TLRs, results in systemic resistance as outlined in Section III.A.

F. Competition for Nutrients and Niches

Competition for nutrients and niches has been claimed for a long time as a mechanism of biocontrol. Although the claim sounds logical, experimental support for this claim hardly exists. We reasoned that, if this is indeed a mechanism of biocontrol, the procedure described by Kuiper et al. [43] for the enrichment of enhanced root tip colonizing bacteria from the total rhizosphere microbial population can be used to isolate such strains. Kamilova et al. [13] showed that this is indeed the case. These authors used total rhizosphere microbes to inoculate seeds and, after growth of the seedling, selected those microbes that reach the root tip. After repeating this procedure twice, they could show that among five isolates from the root tip, four showed some form of biocontrol.

One strain, *P. fluorescens* PCL1751, was selected for further studies, which suggested that the strain indeed uses the mechanism "competition for nutrients and niches" [13]. This method for the isolation of biocontrol bacteria has the advantages that (1) it is a method by which biocontrol strains can be selected (instead of screened for), and (2) the selected strains usually do not produce antibiotics. Antibiotic production is a disadvantage for registration as a BCA.

G. OTHER POSSIBLE MECHANISMS

It has repeatedly been published that fungal hyphae can be colonized by bacteria *in vitro* [141–147]. Bolwerk et al. [148] showed that also in the rhizosphere bacteria can form biofilms on hyphae. Nelson et al. [149] produced evidence that the ability of *Enterobacter cloacae* to function as a BCA on seed surfaces is related to its ability to bind to hyphae of *Pythium* and to inhibit further hyphal infection and development at the seed surface. Support for this interpretation was recently obtained from an observation of Kamilova et al. [149a], who incubated the biocontrol bacterium *P. fluorescens* WCS365 and hyphae of *Fusarium* in exudate, both individually and in combination. Compared to incubation of each of the microbes alone, incubation of the mixture resulted in a ten-fold decrease of viable fungi. This strongly suggests that the bacteria kill the fungi. Therefore, colonization of hyphae and subsequent steps may be a trait contributing to biocontrol [150].

Bolwerk et al. observed that *Trichoderma* strains inhibit *Fusarium* spore germination. Mutant studies indicated that endo- and exochitinases are involved in this process [151]. Therefore, inhibition of the pathogen's spore germination may be a mechanism involved in biocontrol.

H. OTHER REMARKS ON THE MECHANISMS OF BIOCONTROL

1. It should be realized that the results of biocontrol experiments are not very accurate, in the sense that, if the results show one major mechanism of biocontrol, they certainly do not exclude other mechanisms. The poor accuracy only makes it difficult to prove that the strain also uses a second mechanism.

2. Care should be taken with the interpretation that a biocontrol strain is always beneficial. Examples are known of biocontrol strains that suppress disease in one plant but can be growth-inhibiting for another plant. For example, Slininger and Shea-Andersh [152] have shown that phloroglucinol production by certain *P. fluorescens* strains applied on wheat seeds reduced germination up to 40 to 50%. However, growth conditions could be found to optimize survival of the strain and to minimize antibiotic production, thereby optimizing conditions for seed coating and subsequent biocontrol [152].

3. Biocontrol products do not always cause consistent results. We believe that an important reason is that tests for antibiosis are an easy way to preselect potential biocontrol strains, and that, therefore, there exists a bias toward BCAs that act through antibiosis. The production of antifungal factors is subject to a large number of factors, such as the following:

   a. Syntheses of the most-often found antifungal metabolites phloroglucinol and phenazine are subject to regulation by a variety of environmental factors [153], such as pH, ions [153,154], and composition of the growth medium [12,155,156], all of which can play a role under the changing conditions in the rhizosphere.

   b. Syntheses of the most-often found antifungal metabolites phloroglucinol [1,37,157] and phenazine [8,99,101] are subject to a variety of genetic factors [3,158–161].

   c. Synthesis of phenazine requires quorum sensing through *N*-acyl-homoserine lactones [99]. Rhizobacteria have been found which degrade *N*-acyl-homoserine lactones [103,104].

  d. Various organisms produce substances that inhibit synthesis of antifungal metabolites. For instance, the pathogenic fungus *Fusarium* produces fusaric acid, a phytotoxin that inhibits the syntheses of the antifungal factors phloroglucinol [162] and phenazine [163]. The molecular explanation is that fusaric acid is a quorum-sensing inhibitor because it inhibits *N*-acyl-homoserine lactone synthesis.
  e. Many pseudomonads show colony phase variation. It was observed that the syntheses of antifungal metabolites are subject to phase variation [55,164]. Because this phenomenon also plays a role in the rhizosphere [141], it can have a negative influence on biocontrol.
  f. Also metabolites in the rhizosphere can influence antifungal factor synthesis. The antifungal factor 2,4-diacetylphloroglucinol induces its own synthesis, whereas its synthesis is suppressed by the bacterial metabolites salicylate and pyoluteorin. In the biocontrol strain Pf-5 the antifungal factors pyoluteorin and 2,4-diacetylphloroglucinol mutually inhibit one another's production [37].

4. Other reasons why biocontrol is not always working optimally are that at least three organisms, the substrate, as well as biotic and abiotic factors influencing the interactions between organisms and substrate are involved. Considering this complexity, one might say that it is a miracle that biocontrol often works!
5. Goodman's group [165–167] has found that biocontrol efficacy is influenced by the plant cultivar.

IV. BIOREPORTERS AND BIOCONTROL

Bioreporters [29,168,169] have contributed considerably to our understanding of biocontrol. With bioreporter technology it has become possible to answer questions that prior were amenable to speculation only. Are BCAs delivered along the root in a manner that maximizes their effect on the target pathogen? Once they are in place, do they find themselves in an environment that allows them to survive and carry out their appointed tasks, and if so, is the biocontrol phenotype expressed at the right time and with the desired effect?

In the context of this chapter, it is most convenient to classify bioreporters into three categories. The first are those that allow for visualization of the BCA *in vivo*. The most commonly used reporter gene for this purpose is *gfp*, as it does not require substrate or cofactors and because it allows the visualization of individual BCA cells. In combination with a constitutive promoter, the *gfp* gene is expressed independently of environmental factors, and accumulation of its product (GFP) renders the cells green fluorescent and readily detectable by fluorescence microscopy or confocal laser-scanning microscopy. This has been exploited to study rhizomicroorganisms at the highest most relevant resolution, that is, at the micrometer scale of individual cells, to reveal, for example, preferred sites of colonization and clustering into microcolonies. A detailed discussion of insights obtained by bioreporters follows in Section VIII. Other types of GFP-like fluorescent proteins are also available, and this has allowed for the simultaneous observation of more than one microorganism in the rhizosphere [27,170] (also see Section VIII). The *lux* gene is another reporter gene that has been used for detection of microorganisms in the rhizosphere, but its use seems to be restricted to low-resolution applications, for example, for the detection of BCA populations along entire root systems [171–175].

The second class of bioreporters allows monitoring the general well-being or activity of BCAs in the rhizosphere. The rationale behind this is that presence and correct localization alone are no guarantee for proper functioning of the BCA; it should also be fit to perform. There are several dyes available that can be used to determine whether bacteria or fungi are dead or alive, and for bacteria the metabolic activity can be estimated from their ribosome content as determined by FISH. However, both methods are rather invasive (i.e., requiring the addition of dyes and FISH-probes, respectively) so that their applicability in *in vivo* situations is rather limited. Bioreporter technology offers several alternatives to these methods. One is based on constitutive expression of the *lux* genes, which only results in bioluminescence when the cells are metabolically active, because of the high energy requirement of light generation. Several groups [176–179] have used this bioreporter system to show that, in general, the metabolic activity of *Pseudomonas* inoculants is higher in rhizosphere soil than in bulk soil, and in the rhizosphere, metabolic activity decreases over time after introduction of the inoculated BCA [28]. Another bioreporter system for assessing metabolic activity is based on the expression of *lux* genes [172,180] or unstable GFP [181,182] from a growth-rate-dependent ribosomal promoter. This approach has been used in the rhizosphere of barley seedlings [182] to reveal that the activity of *P. putida* in natural rhizosphere settings was highly variable among individual cells and that the majority of cells seemed to be in a starved state.

A third group of bioreporters is aimed at providing information on specific activities of BCAs in the rhizosphere. Besides *gfp* and *lux*, other reporter genes can be used for this, including *lacZ*, *xylE*, *gusA*, and *inaZ*. Although these reporters do not allow direct assessment of reporter gene activity in individual cells, each of them have their particular advantages [183]. For example, InaZ offers the greatest sensitivity and widest range of detection [183a], LacZ measurements can be performed with fluorogenic substrates to exploit the low detection limit of fluorescence [183b], and XylE has the advantage of a low background activity in most microbial habitats including the rhizosphere [183c]. A newer type of reporter system, based on recombination *in vivo* expression technology or RIVET [184] has been described and tested in the rhizosphere of barley by Casavant et al. [170,185]. It is based on the conditional expression of *gfp* by a cascade of events that involves activation of the site-specific recombination machinery of bacteriophage P22, consequent excision of the *cI* repressor gene, resulting in derepression of *gfp* expression from a *cI*-regulated promoter.

In bioreporters of specific activity, the expression of the reporter gene is driven by promoter sequences that are selected based on their biological function. For example, to test whether a particular gene that is suspected to be involved in BCA performance is indeed expressed in the rhizosphere, it is possible to fuse the promoter of that gene to a reporter gene, introduce this fusion into the BCA and compare reporter gene activity in the rhizosphere or in the presence of rhizosphere factors (such as exudates or rhizomicroorganisms) to that *in vitro* [44,186–196]. Another possibility is to choose a promoter that is known to be responsive to a certain component or condition with the intention to assess the rhizosphere for the presence of that component or condition. For example, several bioreporters are available that have been designed for the detection of specific sugars or amino acids and they have been used to demonstrate the presence of, for example, sucrose [197], galactose, melibiose, raffinose [198], arabinose [170], proline [67], and tryptophan [197] in the rhizosphere of various plants. Other examples of bioreporters that have been used in the rhizosphere are responsive to the bioavailability of carbon [175,199], nitrogen [199, 200], phosphate [173,178,199,201,202], oxygen [203], and iron [204–207]. Several bioreporters have been developed for the specific detection of pollutants in the rhizosphere such as (chloro) biphenyl [181,208], naphthalene [44], and toluene [185], and of heavy metals such as copper [209], mercury, and arsenite [210].

Many factors should be taken into consideration when interpreting the output of bioreporters [29]. For the purpose of this chapter, we would like to stress two such factors. First of all, a considerable number of the bioreporters that have been used in the rhizosphere and that are described here are based on the insertion of a promoterless reporter gene in a chromosomal locus with a rhizosphere-inducible phenotype. It is crucial that the gene activity of such reporters is understood in light of the possibility that inactivation of the inserted gene may have an effect on the bioreporter's rhizosphere competence, on its ability to report properly, or on both. In this respect, it would be perhaps more appropriate to work with bioreporters that carry a promoter-gene fusion on a chromosomally neutral location or on a plasmid. The second point of caution concerns the observation that many of the bioreporters for the availability of, for example, nitrogen, phosphate, and iron are based on what are essentially stress-related promoters. For example, most of the bioreporters for iron are based on genes that produce siderophores, the synthesis of which is induced under conditions of iron limitation. In other words, when iron is abundant, these bioreporters will not show reporter gene activity, whereas when iron becomes limited they do. Given the observation that a significant population of bioreporter cells in the rhizosphere is dead or in a state of viable-but-not-culturable (VBNC) [28], and assuming that dead or VBNC bioreporters probably exhibit no reporter activity independent of iron availability, it would actually be very difficult to discriminate between dead or VBNC bioreporters and bioreporters that experience no iron limitation. In such a case, a correct interpretation of reporter data would require additional information, such as whether or not cells are dead, VBNC, or alive and active. As has been suggested before [29], the activity of a bioreporter should always be interpreted within the context of its biology.

V. NATURAL ROLES OF SMALL MOLECULES IN THE RHIZOSPHERE

Rhizodeposition is the process of excretion of plant compounds into the rhizosphere. This deposition can be quite substantial: for example, plants may excrete as much as 40% of their photosynthate into the rhizosphere [211]. Low-molecular-weight exudate components, such as organic acids [16,45], sugars [40], and amino acids [41], as well as the polyamine putrescine [42] are the major known nutrient sources for biocontrol agents and pathogens. Some exudates may act not only as food source but also as attractors. It was shown that the biocontrol agent *P. fluorescens* WCS365 moves toward the root because it shows a chemotactic response toward some organic acids and toward amino acids whereas no chemotaxis was observed toward sugars [47]. Chemotaxis toward fungal hyphae is thought to be the first step in the colonization of hyphae by biocontrol bacteria. Fusaric acid appears to be the major chemoattractant secreted by *Fusarium* for chemotaxis of *P. fluorescens* WCS365 [48].

Our understanding of both the quality and the quantity of plant compounds in the rhizosphere has grown with the publication of several bioreporter studies. The availability of nitrogen in the barley rhizosphere, as recorded by a *P. fluorescens* bioreporter of nitrogen, was found to be higher in sterilized than in unsterilized soils [202]. This implies that competition for nitrogen in the natural rhizosphere of barley is quite likely to occur. *P. fluorescens* DF57 reported phosphate starvation in a gnotobiotic but not a natural root system of barley [178], suggesting that the presence of an indigenous microbial population prevents phosphate starvation by strain DF57. Several factors influence root exudate composition, including plant nutrition and stress. For example, high nitrate concentrations decrease carbon flow from the roots of common barley, probably due to changes in root architecture, that is, shorter root length and reduced number of root tips [201]. When exposed to pollutant stress, *Plantago lanceolata* roots exuded increased amounts of carbon, as measured by a *lux*-based bioreporter [179]. In turn, differences or changes in root exudation may influence gene expression in the microbial population and thus affect biocontrol efficacy. The production of 2,4-diacetyl-phloroglucinol (DAPG) by *P. fluorescens* CHA0, as measured by a bioreporter strain, is greater in the rhizosphere of maize and wheat than in bean and cucumber [193]. Also, plant age has a profound effect on the effect of *phlA* expression in the rhizosphere, as does infection with the target pathogen [193]. Expression of the phenazine antibiotic locus in *P. aureofaciens* PGS12 differs on germinating seeds from sugar beet, radish, or wheat [190]. Other factors in the rhizosphere may also affect the availability of nutrients. A good example is the observation that the availability of nitrogen to *P. fluorescens* in soil amended with straw is decreased by the presence of *Trichoderma harzianum* [200]. This has been explained by the production of cellulases by the fungus, subsequent mobilization of carbon from the straw, and an increase in the demand for nitrogen [200].

Bioreporter studies also have shown that plant exudates are not distributed evenly across the root surface. On the roots of alfalfa, galactosides such as galactose occurred patchily around zones of lateral root initiation and around root hairs, but not around root tips [198]. Sucrose and tryptophan showed very different distribution patterns on roots of the annual grass *Avena barbata*, with sucrose most abundant at the root tip and tryptophan higher up in the root system [197]. With common barley (*Hordeum vulgare*), arabinose could not be detected at the root tips, but instead near the root–seed junction and on the seminal roots [170]. Surprisingly, and in contrast to this result, exudation of carbon from barley roots as determined by another bioreporter strain was greater at the tip than on other regions of the root [202]. Perhaps such contradictory results are due to differences between plant cultivars in terms of rhizodeposition or to differences in experimental parameters. Intriguingly, the differences may also be explained as a difference in how two very different types of bioreporters, one a *Sinorhizobium meliloti* [170], the other a *P. fluorescens* [202], experience the barley rhizosphere.

A *P. syringae* GFP bioreporter revealed substantial heterogeneity in the availability of iron in the rhizosphere of bean plants [204], suggesting that competition for iron may achieve different levels in different parts of the root system. This study also demonstrates the added value of using GFP-based reporters. Earlier work [205] had suggested that the rhizosphere is generally not an environment that is limiting in iron availability, but this conclusion was based on *inaZ* bioreporters that could not be interrogated individually but only as a population. Thus, while the average cell is not limited in iron, the heterogeneity observed with the GFP-based bioreporter indicates that there are subpopulations, some of which are and some of which are not iron-limited. Most interestingly, similar types of bioreporters could be used to show that *Pseudomonas* bacteria can use the siderophores of other bacteria [206] or even of plants [206,207] to sequester iron.

The microscale heterogeneity in the chemical composition of the rhizosphere (which is in large part determined by rhizodeposition), together with the observation that the expression of many biocontrol genes is influenced by the chemical composition of the rhizosphere, seems to also suggest that efficacy of biocontrol can be interpreted on a micrometer scale. Hence, consideration for the microscale heterogeneity may be warranted in understanding some of those cases in which the biocontrol does not seem to work as would be expected from a macroscale point of view.

N-acyl-homoserine lactones [212] are produced by several rhizobacteria [102,103,213–217]. They play a role in many rhizosphere processes, such conjugation between microbes [218] and the syntheses of several exoenzymes [219], antibiotics [101,219] and biosurfactants [220]. They influence both beneficial [99,221,222] as well as pathogenic [223, 224] traits. The biosurfactant rhamnolipid produced by *P. aeruginosa* increases the solubility and thereby the bioactivity of the cell-to-cell signal molecule *Pseudomonas* quinolone signals [225]. Production of the phenazine antibiotic produced by the biocontrol microbe *P. aureofaciens* 30 to 84 is positively affected by a rhizosphere subpopulation that secretes AHLs [226] but is negatively affected by a second subpopulation. The signal responsible for the latter effect was not identified and is not extractable with ethyl acetate [100].

Interference with quorum sensing by other organisms is widespread in nature [157,227]. Several rhizosphere bacteria disrupt the quorum-sensing process by degrading the *N*-acyl-homoserine lactone signal [105]. Moreover, a halogenated furanone compound produced by the marine alga *Delisea pulchra* [228] as well as garlic extract and 4-nitro-pyridine-*N*-oxide are active as quorum sensing quenchers [229]. Also the inhibition of phenazine-1-carboxamide synthesis by the *Fusarium* phytotoxin fusaric acid interferes at or before the level of inhibition of quorum sensing [156].

Also, plants seem to have the potential to interfere with bacterial communication through the production of molecules that mimic AHLs. All three *E. coli lux* bioreporters recognizing different types of AHLs became bioluminescent when inoculated onto seedlings of pea (*Pisum sativum*) [196], suggesting that the plants are able to produce AHL-like substances. Interestingly, the reporter signal intensity differed for different sections of the root, which may indicate that the production or secretion of such mimics by the plant can vary locally. It was also shown that the plant can interfere with the perception of AHL by bioreporter strains [196], by as-of-yet unknown mechanisms. It is presently unclear if and how plant-produced AHL-mimics influence biocontrol.

IAA is a plant growth hormone produced by various microorganisms associated with plants roots. In fact, many BCAs are capable of producing IAA or auxin-like compounds [230–242]. Several of these BCAs have a plant growth promoting effect, probably by the production of IAA at stimulatory concentrations. In most cases, it is not clear whether this effect is responsible for, or linked in any way to, biocontrol. In at least two instances, it has been recorded that the ability to produce IAA did not seem to contribute to biocontrol efficacy [234,241]. However, at least in one case, IAA was shown to have an antifungal activity against the dry rot causative pathogen *Gibberella pulicaris* and to suppress dry rot infection of wounded potatoes [232]. Interestingly, introduction of the gene for ACC (1-aminocyclopropane-1-carboxylic acid) deaminase into BCA *P. fluorescens* CHA0 improves its ability to protect cucumber against *Pythium* damping-off, and potato tubers against *Erwinia* soft rot, but not tomato against *Fusarium* crown and root rot [243]. ACC deaminase degrades a precursor of the plant hormone ethylene, and this promotes root elongation of plant seedlings due to the reduced ability to synthesize ethylene. Thus, plant hormone manipulation may be one of the mechanisms contributing to some of the biocontrol phenotype of *P. fluorescens* CHA0.

VI. ENHANCING OF BIOCONTROL EFFICACY BY MANIPULATION

Interference with *N*-acyl-homoserine lactone-mediated communication can be predicted to have positive as well as negative effects, depending on whether the manipulation influences beneficial or pathogenic traits [244]. The natural communication processes mentioned in Section V can be engineered for biocontrol purposes. Tobacco plants genetically modified to produce an *N*-acyl-homoserine lactone are able to complement *N*-acyl-homoserine lactone mutants of *P. aureofaciens* and *Erwinia carotovora* in the rhizosphere [245]. Similarly, modified potato plants have become more susceptible to infection by *E. carotovora* [219]. In contrast, Mae et al. [246] found that incorporation of the *N*-acyl-homoserine lactone biosynthetic gene in plants results in enhanced resistance toward the pathogenic bacterium *Erwinia carotovora*. It is clear that this type of genetic engineering is still far away from being applicable.

The level of antifungal metabolites can be manipulated by genetic engineering. This can be illustrated by the following examples:

1. Interference with phase variation by introducing *sss* genes [38] in multiple copies can increase biocontrol efficacy of some strains [132].
2. Transformation of the 2,4-diacetylphloroglucinol producing *P. fluorescens* Q8r1-96 with genes encoding phenazine-1-carboxylic acid biosynthesis enhanced the biocontrol properties of the strain to the extent that a one to two orders of magnitude lower dose of cells is required for biocontrol of *Rhizoctonia* root rot [247].
3. Incorporation of the *phzH* gene in strains that already are able to synthesize phenazine-1 carboxylic acid enable the modified strains to control tomato foot and root rot because the new constructs can synthesize phenazine-1-carboxamide, which is more effective than phenazine-1-carboxylic acid in biocontrol of tomato foot and root rot [248].
4. By bringing the genes for pyrrolnitrin synthesis under the control of p.*tac*, the pyrrolnitrin level, and therefore the biocontrol efficacy of the BCA could be enhanced [249].
5. Exu- and rhizosphere-induced promoters have been identified [67–69]. They can be handy if one wants to express a gene in the plant environment.
6. Rhizosphere microbes can substantially influence one another's functioning. Strains of the fungus *Aspergillus*, isolated from the rhizosphere, were tested for their ability to influence the production of the nematocidal compound 2,4-diacetylphloroglucinol and the biocontrol performance of *P. fluorescens* biocontrol strain CHA0. It appeared that *A. niger* enhanced the Phl production and biocontrol activity whereas *A. qudrilineatus* repressed such activities [250].

Cocktails of BCAs have been tried by several groups with the idea to combine different mechanisms of biocontrol in one product. For example, biocontrol of the root knot nematode

Meloidogyne javanica by *P. fluorescens* CHA0 is improved by the presence of *Trichoderma harziarum*. The effect is explained by showing that the culture filtrate of *Trichoderma* enhances the expression of the genes encoding the synthesis of 2,4-diacetylphloroglucinol [195]. The Leiden group has tested a variety of cocktails and the result was that, at best, a level of biocontrol could be reached that was the same as that of individual strains.

Survival and efficacy of a biocontrol product can be severely influenced by additives. Chemically characterized compounds, such as a saponin from pepper (*Capsicum frutescens* L), benzaldehyde, chitosan [251], and 2-deoxy-D-glucose are being studied as natural fungicides [252]. Moreover, bacteria can be selected or constructed that are able to utilize a nutrient that most other rhizosphere bacteria cannot use [253–256]. This has also been referred to as creating a "biased rhizosphere" [257]. Finally, nutrients such as amino acids, gelatin, glucose, lactose, wheat bran, and maize cobs can affect activity of biocontrol organisms by providing a food base to aid proliferation [258].

VII. WAR IN THE RHIZOSPHERE: ATTACK AND DEFENSE

Until a few years ago, the pathogenic fungus was considered as a simple victim of the biocontrol process. In particular, Brian Duffy's and Jos Raaymakers' groups have put the role of the fungus in a different perspective [259]. Several fungi have developed ingenious strategies to defend themselves. These strategies can be different in strains of the same species [260]:

1. Repression of the synthesis of biocontrol traits is a widely distributed strategy. Expression of the chitinase genes *ech*42 and *nag*1 of the biocontrol fungus *Trichoderma atroviride* strain P1, which contribute to biocontrol activity, is inhibited by the *Fusarium* mycotoxin deoxynivalenol [261]. Similarly, the phytotoxin fusaric acid produced by many Fusarium strains inhibits syntheses of the antimicrobial factors 2,4-diacetyl phloroglucinol [37] and phenazine-1-carboxamide [12,163] in different *Pseudomonas* biocontrol strains. For the phytopathogenic fungus *Pythium ultimum*, it has been shown that it is able to downregulate the expression of ribosomal RNA in BCA *P. fluorescens* F113 [262]. It has been suggested that this reduces the ability of the bacterium to respond adequately to conditions that would otherwise support rapid growth. Interestingly, mutants of strain F113 affected in rRNA expression were not different from the wild type in controlling the fungus [263], indicating that downregulation by the fungus is not occurring during biocontrol. Fungal stimulation instead of repression of bacterial gene expression has also been observed: root infection of tomato and cucumber by *P. ultimum* stimulated expression of the *phlA* gene for DAPG production in *P. fluorescens* CHA0 [193].
2. Detoxification of antifungal metabolites is another strategy [259]. For example, many *Fusarium oxysporum* strains tolerant to the antifungal metabolite 2,4-diacetyl phloroglucinol deacetylate the latter compound to the less fungitoxic derivative monoacetyl phloroglucinol [260].

VIII. VISUALIZATION OF BIOCONTROL

Rhizomicroorganisms constitutively expressing GFP can be thought of as bioreporters of microlocation (see Section IV). They have proven to be instrumental in understanding the rhizosphere at the micrometer scale, as they have shown how microorganisms are distributed along the root surface and in relation to root features and other root microorganisms.

Bacteria on the root occur hardly as individual cells but are mostly arranged as biofilms, covering only a small part of the root [102,264–268]. Some areas, including the root tip, are almost devoid of microbes [102,264–268]. Pseudomonads are mainly present at junctions between root epithelial cells and at sites where side roots emerge [36,266]. They are often covered by a mucoid layer [102], most likely consisting of root material [271], which may facilitate the quorum-sensing-dependent conjugation process [102,272].

The process of colonization of tomato roots upon bacterization of seeds with the *P. fluorescens* BCA WCS365 has been studied in a gnotobiotic system [41]. Bacteria multiply fast on the seed and colonize the growing roots but the number of colony-forming units on the lower parts of the root can be four orders of magnitude lower than that at the root base [41,102]. When bacterial numbers become higher, the colonizing bacteria start to form biofilms or microcolonies consisting of hundreds or even thousands of cells, especially along the junctions between epidermal cells as shown by scanning electron microscopy [102]. Distribution patterns of other *Pseudomonas* biocontrol strains and of *P. mendicina*, *Acidovorax facilis* and *Xanthomonas oryzae* were indistinguishable from that of *P. fluorescens* WCS365 [102]. In contrast, *Acinetobacter radioresistens* and *Rhizobium* strains appeared to be poor colonizers of the tomato root; they colonize the root in lower numbers and do not form a biofilm. Rhizobia show no preference for the junctions between epidermis cells like pseudomonads do [102].

Infection of the tomato root by the pathogenic fungus *Fusarium oxysporum* f. sp. *radicis-lycopersici* was studied using confocal laser electron microscopy (CLSM) and a GFP-labeled fungus [273]. This marker appeared to be stable and did not influence pathogenicity. The fluorescent signal was clearly visible in the hyphae as well as in the chlamydiospores and conidia [273]. The processes of attachment, colonization, infection, and disease development on tomato roots were visualized at the cellular level. For technical reasons, visualization using CLSM requires the use of a simple biocontrol system. Attachment starts at the root hairs. Interestingly, like for biocontrol pseudomonads, the junctions between the epidermal cells are the preferred colonization sites [273].

Agents that control TFRR should interfere in these steps. How the BCAs *Pseudomonas fluorescens* WCS365, *P.chlororaphis* PCL1391 [148], the nonpathogenic strain *Fusarium oxysporum* F047 [148] *T. atroviride* P1, and *T. hazarzianum* [151] colonize the rhizosphere and how they control TFRR has recently been visualized. In this gnotobiotic biocontrol assay, the seeds or seedlings are inoculated by the *Pseudomonas* bacteria, whereas spores of the pathogen *Fusarium oxysporum* f. sp. *radicis-lycopersici* were mixed through the sterile quartz sand. After attachment to the root hairs, hyphae from the pathogen colonize the intercellular junctions on the root surface. When added alone, cells of *Pseudomonas fluorescens* WCS365 and *P. chlororaphis* PCL1391 colonize the same sites and subsequently form biofilms on the root [26,27]. The fact that totally different microbes occupy the same sites on the root can be explained by assuming that these are the sites where nutrients are exuded from the root. When both pathogen and BCA were present, the BCA reaches the root first and severely delays and diminishes the hyphal biomass that can reach the root and that can infect the root. The BCAs not only form biofilms on the root but also on the hyphae. It has been suggested that colonization of hyphae contributes to biocontrol [148,150]. Analysis of viable counts of BCA and pathogen after incubation in exudates has shown that the bacterium increases in numbers whereas the number of hyphae decreases [13].

For control of TFRR by the nonpathogenic Fusarium F047 (F047) an excess of at least 50-fold of F047 over Forl is required. Alone, F047 hyphae attached earlier to the root than Forl hyphae. When both microbes are present, root colonization by the pathogen was reduced and arrested at the stage of initial attachment to the root. Furthermore, results indicated that prior to competition for the root surface, a process takes place that is negative for the pathogen. It appeared that the percentage of F047 spores that germinates in exudates is higher than that of the pathogen Forl [274].

Control of TFRR by *Trichoderma* strains *T. atroviride* P1 and *T. harzianum* T22 showed competition for sites on the root hairs. Both trichodermas are poor colonizers of the main root. Moreover, it was shown that the culture supernatant of *Trichoderma* spp. reduces germination of Forl spores and that an endochitinase and an exochitinase are involved in both reduction of spore germination as well as in disease suppression [151].

Visualization of the biocontrol process is not limited to localizing and following BCAs on root surfaces. With bioreporter strains carrying fusions of reporter genes to inducible promoters (see Section IV), it is also possible to follow general or more specific activities of BCAs or other rhizomicroorganisms as they colonize the rhizosphere. The picture that emerges is generally one of heterogeneity; only a fraction of the total BCA or rhizomicroorganism population in the rhizosphere scores positive for the activity that is being assessed. Sometimes, but not always, this activity can be correlated to the microlocation of the BCA. Some examples are given here. In the rhizosphere of barley, *P. putida* CRR3000 cells were found to colonize root sites such as hairs and tips, being most abundant in the crevices between neighboring plant root epidermal cells [182]. Yet, only a fraction of all CRR3000 cells was actively growing, namely those that were associated with the sloughing root sheath cells, and only in the first two days after inoculation [182]. Boldt et al. [181] were able to see actively growing *Pseudomonas* cells only at the root tips and sites of lateral root emergence of alfalfa. In *P. fluorescens* DR54, genes for motility were expressed at the base of the barley root system, in single cells or in microcolonies along the root cells, root hairs, or mucus layers, whereas *P. fluorescens* DF57 expressed genes for nitrogen starvation patchily along the roots of barley seedlings [200].

It is important to realize the implication of this observed apparent heterogeneity to BCA activity in the rhizosphere. If we see, for example, in the same rhizosphere setting, heterogeneity in activity A with one bioreporter and also heterogeneity in activity B with another bioreporter, does that mean that we are possibly dealing with four subpopulations of BCA cells, namely those that are active for A and B, those that are inactive for A but active for B, those that are active for A but not B, and those that are not active for A and not for B? Which one of these has the biocontrol activity? Or is activity A always correlated (positively or negatively) with activity B? For example, does biocontrol activity require high or low carbon, actively growing or resting cells, siderophore production or not? These are still unanswered questions with great implications for the understanding of and ability to improve on biocontrol strategies. There are currently no rhizosphere bioreporters that report on more than one activity on a single-cell basis, so it is, as of yet, practically impossible to answer some of these questions. This is challenge for the future.

IX. MONITORING THE BIOCONTROL PROCESS

Monitoring the biocontrol process allows for the identification of possible causes of BCA failure or for the improvement of existing biocontrol strategies. First, it is important to know the enemy, that is, to characterize the target of the BCA. Its presence and abundance in natural rhizosphere settings can be determined in many different ways, including plating and molecular techniques [275]. Most common is the use of rRNA specific primers for the semiquantitative detection of the pathogen. Rhizosphere-specific behavior of the pathogenic target can be studied using bioreporter technology, for which excellent examples are available [27]. A similar set of tools can be used to assess the presence, abundance, and persistence of the BCA after introduction into the rhizosphere [276,277]. Several so-called biological containment systems have been developed that ensure survival of the BCA only in the presence of the plant (for example, Reference 278), and validation of such systems relies heavily on very sensitive methods for detection of the BCA. Bioreporter technology can be and has already been used to see whether the BCA actually comes into close proximity to the target, whether it is active, and whether it expresses the genes that are essential for biocontrol activity (Section IV). Success rate of the biocontrol process is usually assessed by scoring the incidence or severity of disease symptoms. By relating this information to abundance and activity of both pathogen and BCA throughout the process, the contribution of individual steps in the biocontrol process can be assessed and possibly refined.

X. COMMERCIALIZATION OF BIOCONTROL

The road from a good laboratory biocontrol result to a commercial product is a long one [279]. The best biocontrol strain, preferentially active on several plants and against several pathogens under realistic practical conditions, has to survive and be active under a variety of conditions.

Production in bulk amounts has to be inexpensive [150]. The cells should survive formulation, which usually involves drying [280], and have a shelf life of approximately 14 months to enable the grower to use material left over from the previous year. The biocontrol microbe should be formulated in such a way that its application is compatible with agricultural or horticultural practice. Formulation procedures are often company secrets and very few informative publications on this topic exist [281,282].

Registration is a point that needs attention from the beginning. Major aspects are taxonomy and antibiotic production. Relationship with a pathogen or growth at the temperature of the human body can be a problem. Also, the production of antibiotics that cause cross resistance with applied antibiotics is problematic. Moreover, the antibiotic should not cause environmental damage. For example, it should not be harmful toward beneficial organisms. It is advisable to become aware of the registration procedure in the countries where the product should be applied in a very early stage. Once a product is on the market, it is important to be able to monitor the effect on the plant as well as the viability and activity of the biocontrol agent.

ACKNOWLEDGMENT

We thank Dr. Faina Kamilova for critical reading of the manuscript and for her help during its preparation.

REFERENCES

1. Haas, D. and Defago, G., Biological control of soil-borne pathogens by fluorescent pseudomonads, *Nat. Rev. Microbiol.*, 3, 307, 2005.
1a. Torsvik, V.L., Sorheim, R., and Goksoyr, J., Total bacterial diversity in soil and sediment communities-a review, *J. Ind. Microbiol.*, 17, 170, 1996.
2. Schroth, M.N. and Hancock, J.G., Disease suppressive soil and root colonizing bacteria, *Science*, 216, 1376, 1981.
3. Alabouvette, C., Fusarium wilt suppressive soils from the Chateaurenard region: reviews of a 10 year study, *Agronomie*, 6, 273, 1986.
4. Scher, F.M. and Baker, R., Mechanism of biological control in a Fusarium-suppressive soil, *Phytopathology*, 72, 1567, 1980.
5. Schippers, B., Lugtenberg, B.J.J., and Weisbeek, P.J., Plant growth control by fluorescent pseudomonads, in *Innovative Approaches to Plant Disease Control*, Chet, I., Ed., Wiley, New York, 1987, p. 19.
6. Weller, D.M., Biological control of soilborne plant pathogens in the rhizosphere with bacteria, *Annu. Rev. Phytopathol.*, 26, 379, 1988.
7. Bloemberg, G.V. et al., Rhizosphere colonisation by biocontrol Pseudomonas spp. in *Proceedings of the 5th International PGPR Workshop*, Loper, J. et al., Eds., 2000, p. 1.
8. Chin-A-Woeng, T.F.C., Lugtenberg, B.J.J., and Bloemberg, G.V., Mechanisms of biological control of phytopathogenic fungi by Pseudomonas spp., in *Plant-Microbe Interactions,* Vol 6., Stacey, G. and Keen, N.T., Eds., The American Phytopathologial Society, St. Paul, MN, 2003, p. 173.
9. Handelsman, J. and Stabb, E.V., Biocontrol of soilborne plant pathogens, *Plant Cell*, 8, 1855, 1996.
10. Postma, J., Willemsen-de Klein, M.J.E.I.M., and van Elsas, J.D., Effect of the indigenous microflora on the development of root and crown rot caused by *Pythium aphanidermatum* in cucumber grown on rockwool, *Phytopathology*, 90, 125, 2000.
10a. Postma, J. et al., Characterization of the microbial community involved in the suppression of *Pythium aphanidermatum* in cucumber grown on rockwool, *Phytopathology*, 95, 808, 2005.
11. Copping, L.G., Ed., *The Manual of Biocontrol Agents; Third edition of the BioPesticide Manual*, British Crop Protection Council, Alton, 2004
12. Bloemberg, G.V. et al., Visualisation of microbes and their interactions in the rhizosphere using auto fluorescent proteins as markers, in *Molecular Microbial Ecology Manual*, Kowalchuk, G.A. et al., Eds., Springer, Berlin, Germany, 2004, p. 1257.

13. Kamilova, F. et al., Enrichment for enhanced competitive plant root tip colonizers selects for a new class of biocontrol bacteria, *Env. Microbiol.*, 7, 1809, 2005.

14. Roberts, P.D., McGovern, R.J., and Datnoff, L.E., http://edis.ifas.ufl.edu/PG082, 2000.

15. Lugtenberg, B.J.J., Chin-A-Woeng, T.F.C., and Bloemberg, G.V., Microbe-plant interactions: principles and mechanisms, *Antonie Van Leeuwenhoek*, 81, 373, 2002.

16. Lugtenberg, B.J.J. and Bloemberg, G.V., Life in the rhizosphere, in *Pseudomonas*, Vol. 1, Ramos, J.L., Ed., Kluwer Academic/Plenum Publishers, New York, 2004, p. 403.

16a. Clarholm, M., Interactions of bacteria protozoa and plants leading to mineralization of soil nitrogen, *Soil Biol. Biochem.*, 17, 181, 1985.

17. Azam, F. et al., The ecological role of water-column microbes in the sea, *Mar. Ecol. Prog.*, 10, 257, 1983.

18. He, J. et al., The broad host range pathogen *Pseudomonas aeruginosa* strain PA14 carries two pathogenicity islands harbouring plant and animal virulence genes, *Proc. Natl. Acad. Sci. USA*, 101, 2530, 2004.

19. Rahme, L.G. et al., Plants and animals share functionally common bacterial virulence factors, *Proc. Natl. Acad. Sci. USA*, 16, 8815, 2000.

20. Emmert, E.A.B. and Handelsman, J., Biocontrol of plant disease: a (Gram-) positive perspective, *FEMS Microbiol. Lett.*, 171, 1, 1999.

21. Handelsman, J. et al., Biological control of damping-off of alfalfa seedlings with *Bacillus cereus* UW85, *Appl. Environ. Microbiol.*, 56, 713, 1999.

22. Pusey, P.L., Use of *Bacillus subtillis* and related organisms as biofungicides, *Pest. Sci.*, 27, 133, 1999.

23. Di Pietro, A. et al., Endochitinase from *Gliocladium virens*: isolation, characterisation and synergistic antifungal activity in combination with gliotoxin, *Phytopathology*, 83, 308, 1993.

24. Howell, C.R., Stipanovic, R.D., and Lumsden, R.D. Antibiotic production by strains of *Gliocladium virens* and its relation to the biocontrol of cotton seeding diseases, *Biocontrol Sci. Technol.*, 3, 435, 1993.

25. Lorito, M. et al., Synergistic interaction between fungal cell wall degrading enzymes and different antifungal compounds enhances inhibition of spore germination, *Microbiology*, 140, 623, 1994.

26. Bloemberg, G.V. et al., Green fluorescent protein as a marker for *Pseudomonas* spp., *Appl. Environ. Microbiol.*, 63, 4543, 1997.

27. Bloemberg, G.V. et al., Simultaneous imaging of *Pseudomonas fluorescens* WCS365 populations expressing three different autofluorescent proteins in the rhizosphere: new perspectives for studying microbial communities, *Mol. Plant-Microbe Interact.*, 13, 1170, 2000.

28. Sørensen, J., Jensen, L.E., and Nybroe, O., Soil and rhizosphere as habitats for *Pseudomonas* inoculants: new knowledge on distribution, activity and physiological state derived from micro-scale and single-cell studies, *Plant Soil*, 232, 97, 2001.

29. Leveau, J.H.J. and Lindow, S.E., Bioreporters in microbial ecology, *Curr. Opin. Microbiol.*, 5, 259, 2002.

30. Handelsman, J. and Smalla, K., Conversations with the silent majority, *Curr. Opin. Microbiol.*, 6, 271, 2003.

31. Bull, C.T. et al., Relationship between root colonization and suppression of Gaeumannomyces graminis var. tritici by *Pseudomonas fluorescens* strain 2–79, *Phytopathology*, 81, 954, 1991.

32. Chin-A-Woeng, T.F.C. et al., Root colonization by phenazine-1-carboxamide-producing bacterium *Pseudomonas chlororaphis* PCL1391 is essential for biocontrol of tomato foot and root rot, *Mol. Plant Microbe Interact.*, 13, 1340, 2000.

33. Lugtenberg, B.J.J. and Dekkers, L.C., What makes *Pseudomonas* bacteria rhizosphere competent?, *Environ. Microbiol.*, 1, 9, 1999.

34. Lugtenberg, B.J.J. et al., *Pseudomonas* genes and traits involved in tomato root colonization, in *IC-MPMI Congress Proceedings: Biology of Plant-Microbe Interactions*, Vol. 2, International Society for Molecular Plant-Microbe Interactions, St. Paul, MN, de Wit, P.J. G.M., Bisseling, T. and Stiekema, W.J., Eds., 1999, p. 324.

35. Lugtenberg, B.J.J., Dekkers, L.C., and Bloemberg, G.V., Molecular determinants of rhizosphere colonization by *Pseudomonas, Annu. Rev. Phytopathol.*, 39, 461, 2001.

36. Chin-A-Woeng, T.F.C., Bloemberg, G.V., and Lugtenberg, B.J.J., Root colonisation following seed inoculation, in *Plant Surface Microbiology*, Varma, A. K. et al., Eds., Springer, Berlin, Germany, 2004, p. 13.

37. Schnider-Keel, U. et al., Autoinduction of 2,4-diacetylphloroglucinol biosynthesis in the biocontrol agent *Pseudomonas fluorescens* CHA0 and repression by the bacterial metabolites salicylate and pyoluteorin, *J. Bacteriol.*, 182, 1215, 2000.

38. Dekkers, L.C. et al., A site-specific recombinase is required for competitive root colonization by *Pseudomonas fluorescens* WCS365, *Proc. Natl. Acad. Sci. USA*, 95, 7051, 1998.

39. Dekkers, L.C. et al., Role of the O-antigen of lipopolysaccharide, and possible roles of growth rate and of NADH: ubiquinone oxidoreductase (nuo) in competitive tomato root-tip colonization by *Pseudomonas fluorescens* WCS365. *Mol.Plant-Microbe Interact.*, 11, 763, 1998.

40. Lugtenberg, B.J.J., Kravchenko, L.V., and Simons, M., Tomato seed and exudate sugars: composition, utilization by *Pseudomonas* biocontrol strains and role in rhizosphere colonization, *Environ. Microbiol.*, 1, 439, 1999.

41. Simons, M. et al., Gnotobiotic system for studying rhizosphere colonization by plant growth-promoting *Pseudomonas* bacteria, *Mol. Plant-Microbe Interact.*, 9, 600, 1996.

42. Kuiper, I. et al., Increased uptake of putrescine in the rhizosphere inhibits competitive root colonization by *Pseudomonas fluorescens* strain WCS365, *Mol. Plant Microbe Interact.*, 14, 1096, 2001.

43. Kuiper, I., Bloemberg, G.V., and Lugtenberg, B.J.J., Selection of a plant-bacterium pair as a novel tool for rhizostimulation of polycyclic aromatic hydrocarbon-degrading bacteria, *Mol. Plant Microbe Interact.*, 14, 1197, 2001.

44. Kuiper, I. et al., *Pseudomonas putida* strain PCL1444, selected for efficient root colonization and naphtalene degradation, effectively utilizes root exudate components, *Mol. Plant-Microbe Interact.*, 15, 734, 2002.

45. De Weert, S. et al., Role of competitive root tip colonization in the biological control of tomato foot and root rot, in *Biological Control of Plant Diseases*, Chincolcar, S.B. and Mukerji, K.G., Eds., Haworth Press, 2007, in press.

46. De Weger, L.A. et al., Flagella of a plant growth stimulating *Pseudomonas fluorescens* strain are required for colonization of potato roots, *J. Bacteriol.*, 169, 2769, 1987.

47. De Weert, S. et al., Flagella-driven chemotaxis towards exudate components is an important trait for tomato root colonization by *Pseudomonas fluorescens*, *Mol. Plant Microbe. Interact.*, 15, 1173, 2002.

48. De Weert, S. et al., Role of chemotaxis toward fusaric acid in colonization of hyphae of Fusarium oxysporum f.sp. radicis-lycopersici by *Pseudomonas* WCS365, *Mol. Plant-Microbe Interact.*, 16, 1185, 2004.

49. Camacho, M.M., Molecular Characterization of Type 4 pili, NDHI and PyrR in Rhizosphere Colonization of Pseudomonas Fluorescens WCS365, Ph.D. thesis, Leiden University, 2001.

50. De Weger, L.A. et al., *Pseudomonas* spp. with mutational changes in the O-antigenic side chain of their lipopolysaccharide are affected in their ability to colonize potato roots, in *Signal Molecules in Plants and Plant-Microbe Interactions*, NATO ASI Series H, Lugtenberg, B.J.J., Ed., 1989, p. 197.

51. Dybvig, K., DNA rearrangements and phenotypic switching in prokaryotes, *Mol. Microbiol.*, 10, 465, 1993.

52. Martínez-Granero, F. et al., Two site-specific recombinases are implicated in phenotypic variation and competitive rhizosphere colonization in *Pseudomonas fluorescens*, *Microbiology*, 151, 975, 2005.

53. Achouak, W. et al., Phenotypic variation of *Pseudomonas brassicacearum* as a plant root-colonization strategy, *Mol. Plant-Microbe Interact.*, 17, 872, 2004.

54. Chabeaud, P. et al., Phase-variable expression of an operon encoding extracellular alkaline protease, serine protease homologue and lipase in *Pseudomonas brassicacearum*, *J. Bacteriol.*,183, 2117, 2001.

55. Van den Broek, D. et al., Biocontrol traits of *Pseudomonas* spp. are regulated by phase variation, *Mol. Plant-Microbe Interact.*, 16, 1003, 2003.

56. Van den Broek, D. et al., Molecular nature of spontaneous modifications in gacS which cause colony phase variation in *Pseudomonas* sp. strain PCL1171, *Microbiology*, 151, 1403, 2005.

57. Blumer, C. et al., Global GacA-steered control of cyanide and exoprotease production in *Pseudomonas fluorescens* involves specific ribosome binding sites, *Proc. Natl Acad Sci. USA*, 96, 14073, 1999.

58. Achouak, W. et al., Phase variable effects of *Pseudomonas brassicacearum* on Arabidopsis thaliana root architecture, in *Biology of Molecular Plant-Microbe Interactions*, Vol. 4, Proceedings of the 2003 Symposium of xxxxxx, St. Petersburg, Russia, Lugtenberg, B., Tikhonovich, I., and Provorov, N., Eds., International Society for Plant-Microbe Interactions, St Paul, MN, 2004, p. 440.

59. Sánchez-Contreras, M. et al., Phenotypic selection and phase variation occur during Alfalfa root colonization by *Pseudomonas fluorescens* F113, *J. Bacteriol.*, 184, 1587, 2002.

60. Chin-A-Woeng, T.F.C., Lugtenberg, B.J.J., and Bloemberg, G.V., Mechanisms of biocontrol of phytopathogenic fungi by *Pseudomonas* spp., in *Molecular Plant Microbe Interactions*, Vol. 6, Stacey, G. and Keen, N., Eds., 2003, p. 173.

61. Sheng, Y.H., Type III protein secretion systems in plant and animal pathogenic bacteria, *Annu. Rev. Phytopathol.*, 36, 363, 1998.

62. Preston, G.M., Bertrand, N., and Rainey, P.B., Type III secretion in plant growth-promoting *Pseudomonas fluorescens* SBW25, *Molec. Microbiol.*, 41, 999, 2001.

63. Mazurier, S. et al., Distribution and diversity of type III secretion system-like genes in saprophytic and phytopathogenic fluorescent pseudomonads, *FEMS Microbiol. Ecol.*, 49, 455, 2004.

64. Hallmann, J. et al., Endophytic bacteria in agricultural crops, *Can. J. Microbiol.*, 43, 895, 1997.

65. Krechel, A. et al., Potato-associated bacteria and their antagonistic potential towards plant-pathogenic fungi and the plant-parasitic nematode *Meloidogyne incognita* (Kofoid and White) Chitwood, *Can. J. Microbiol.*, 48, 772, 2002.

66. Compant, S. et al., Endophytic colonization of *Vitis vinifera* L. by plant growth-promoting bacterium Burkholderia sp. strain PsJN, *Appl. Environ. Microbiol.*, 2005, 1685, 2005.

67. Van Overbeek, L.S. and van Elsas, J.D., Root exudate-induced promoter activity in *Pseudomonas fluorescens* mutants in the wheat rhizosphere, *Appl. Environ. Microbiol.*, 61, 890, 1995.

68. Rainey, P.B., Adaptation of *Pseudomonas fluorescens* to the plant rhizosphere, *Environ. Microbiol.*, 1, 243, 1999.

69. Ramos-Gonzalez, M.I., Campos, M.J., and Ramos, J.L., Analysis of *Pseudomonas putida* KT2440 gene expression in the maize rhizosphere: *in vitro* expression technology capture and identification of root-activated promoters, *J. Bacteriol.*, 187, 4033, 2005.

70. Angelichio, M.J. and Camilli, A., *In vivo* expression technology, *Infect. Immun.*, 70, 6518, 2002.

71. Thomashow, L.S. and Weller, D.M., Current concepts in the use of introduced bacteria for biological disease control: mechanisms and antifungal metabolites, in *Plant-Microbe Interact.*, Vol. 1., Stacey, G. and Keen, N.T., Eds., 1996, p. 187.

72. Cook, R.J. et al., Molecular mechanisms of defense by rhizobacteria against root disease, *Proc. Natl. Acad. Sci. USA,* 92, 4197, 1995

73. O'Sullivan, D.J. and O'Gara, F., Traits of fluorescent *Pseudomonas* spp. involved in suppression of plant root pathogens, *Microbiol. Rev.*, 56, 662, 1992.

74. Dowling, D.N. and O'Gara, F., Metabolites of *Pseudomonas* involved in the biocontrol of plant disease, *TIBTECH*, 12, 133, 1994.

75. Fravel, D.R., Role of antibiosis in the biocontrol of plant diseases, *Annu. Rev. Phytopathol.*, 26, 75, 1988.

76. Bangera, M.G. and Thomashow, L.S., Identification and characterization of a gene cluster for synthesis of the polyketide antibiotic 2, 4-diacetylphloroglucinol from *Pseudomonas fluorescens* Q2-87, *J. Bacteriol.*, 181, 3155, 1999.

77. Delany, I. et al., Regulation of production of the antifungal metabolite 2,4-diacetylphloroglucinol in *Pseudomonas fluorescens* F113: genetic analysis of *phlF* as a transcriptional repressor, *Microbiology*, 146, 537, 2000.

78. Fenton, A. et al., Exploitation of gene(s) involved in 2,4-diacetylphloroglucinol biosynthesis to confer a new biocontrol capability to a *Pseudomonas* strain, *Appl. Environ. Microbiol.*, 58, 3873, 1992.

79. Keel, C. et al., Suppression of root diseases by *Pseudomonas fluorescens* CHA0: importance of the bacterial secondary metabolite 2,4-diacetylphloroglucinol, *Mol. Plant-Microbe Interact.*, 5, 4, 1992.

80. Arima, K. et al., Pyrrolnitrin, a new antibiotic substance, produced by *Pseudomonas, Agric. Biol. Chem.*, 28, 575, 1964.

81. Habte, M. and Alexander, M., Further evidence for the regulation of bacterial populations in soil by protozoa, *Arch. Microbiol.*, 113, 181, 1977.

82. Pfender, W.F., Kraus, J., and Loper, J.E., A genomic region from *Pseudomonas fluorescens* Pf-5 required for pyrrolnitrin production and inhibition of Pyrenophora tritici-repentis in wheat straw, *Phytopathology*, 83, 1223, 1993.

83. Howell, C.R. and Stipanovic, R.D., Suppression of *Pythium ultimum*-induced damping-off of cotton seedlings by *Pseudomonas fluorescens* and its antibiotic, pyoluteorin, *Phytopathology*, 70, 712, 1980.

84. Kraus, J. and Loper, J.E., Characterization of a genomic region required for production of the antibiotic pyoluteorin by the biological control agent *Pseudomonas fluorescens* Pf-5, *Appl. Environ. Microbiol.*, 61, 849, 1995.

85. Maurhofer, M., Keel, C., and Défago, G., Pyoluteorin production by *Pseudomonas fluorescens* strain CHA0 is involved in the suppression of Pythium damping-off of cress but not of cucumber, *Eur. J. Plant Pathol.*, 100, 221, 1994.

86. Brodhagen, M. et al., Positive autoregulation and signaling properties of pyoluteorin, an antibiotic produced by the biological control organism *Pseudomonas fluorescens* Pf-5, *Appl. Environ. Microbiol.*, 70, 1758, 2004.

87. Gutterson, N., Microbial fungicides: recent approaches to elucidating mechanisms, *Crit. Rev. Biotechnol.*, 10, 69, 1990.

88. Howell, C.R., Beier, R.C., and Stipanovic, R.D., Production of ammonia by Enterobacter cloacae and its role in the biological control of *Pythium* preemergence damping-off by the bacterium, *Phytopathology*, 78, 1075, 1988.

89. Voisard, C. et al., Cyanide production by *Pseudomonas fluorescens* helps suppress black root rot of tobacco under gnotobiotic conditions, *EMBO J.*, 8, 351, 1989.

90. Silo-suh, L.A. et al., Biological activities of two fungistatic antibiotics produced by *Bacillus cereus* UW85, *Appl. Environ. Microbiol.*, 60, 2023, 1994.

91. Iavicoli, A. et al., Induced systemic resistance in *Arabidopsis thaliana* in response to root inoculation with *Pseudomonas fluorescens* CHAO, *Mol. Plant Microbe Interact.*, 16, 851, 2003.

92. Bais, H.P., Fall, R., and Vivanco, J.M., Biocontrol of *Bacillus subtilis* against infection of *Arabidopsis* roots by *Pseudomonas syringae* is facilitated by biofilm formation and surfactin production, *Plant Physiol.*, 134, 307, 2004.

93. Nielsen, T.H. et al., Viscosinamide, a new cyclic depsipeptide with surfactant and antifungal properties produced by *Pseudomonas fluorescens* DR54, *J. Appl. Microbiol.* 87, 80, 1999.

94. Nielsen, T.H. et al., Structure, production characteristics and fungal antagonism of tensin — a new antifungal cyclic lipopeptide from *Pseudomonas fluorescens* strain 96.578, *J. Appl. Microbiol.*, 89, 992, 2000.

95. Thrane, C. et al., Viscosinamide-producing *Pseudomonas fluorescens* DR54 exerts a biocontrol effect on Pythium ultimum in sugar beet rhizosphere, *FEMS Microbiol. Ecol.*, 33, 139, 2000.

96. Kuiper, I. et al., Characterization of two *Pseudomonas putida* lipopeptide biosurfactants, putisolvin I and II, which inhibit biofilm formation and break down existing biofilms, *Mol. Microbiol.*, 51, 97, 2004.

97. Lequette, Y. and Greenberg, E.P., Timing and localization of rhamnolipid synthesis gene expresiion in *Pseudomonas aeruginosa* biofilms, *J. Bacteriol.*, 187, 37, 2005.

98. De Souza, J.T. et al., Biochemical, genetic and zoosporicidal properties of cyclic lipopeptide surfactants produced by *Pseudomonas fluorescens*, *Appl. Environ. Microbiol.*, 69, 7161, 2003.

98a. Nielsen, T.H. et al., Genes involved in lipopeptide production are important for seed and staw colonization by *Pseudomonas* sp. strain DSS73, *Appl. Environ. Microbiol.*, 71, 4112, 2005.

99. Chin-A-Woeng, T.F.C. et al., Phenazine-1-carboxamide production in the biocontrol strain *Pseudomonas chlororaphis* PCL1391 is regulated by multiple factors secreted into the growth medium, *Mol. Plant-Microbe Interact.*, 14, 969, 2001.

100. Morello, J.E., Pierson, E.A., and Pierson, L.S., III, Negative Cross-communication among wheat rhizosphere bacteria: effect on antibiotic production by the biological control bacterium *Pseudomonas aureofaciens* 30-84, *Appl. Environ. Microbiol.*, 70, 3103, 2004.

101. Pierson, L.S., III, Keppenne, V.D., and Wood, D.W., Phenazine antibiotic biosynthesis in *Pseudomonas aureofaciens* 30-84 is regulated by PhzR in response to cell density, *J. Bacteriol.*, 176, 3966, 1994.

102. Chin-A-Woeng, T.F.C. et al., Description of the colonization of a gnotobiotic tomato rhizosphere by *Pseudomonas fluorescens* biocontrol strain WCS365, using scanning electron microscopy, *Mol. Plant-Microbe Interact.*, 10, 79, 1997.

103. Dong, Y.H. et al., Quenching quorum-sensing-dependent bacterial infection by an N-acyl homoserine lactonase, *Nature*, 411,813, 2001

104. Uroz, S. et al., Novel bacteria degrading N-acylhomoserine lactones and their use as quenchers of quorum-sensing-regulated functions of plant-pathogenic bacteria, *Microbiology*, 149, 1981, 2003.

105. Carlier, A. et al., The Ti plasmid of *Agrobacterium tumefaciens* harbors an *attM*-paralogous gene, *aiiB*, also encoding N-acyl homoserine lactonase activity, *Appl. Environ. Microbiol.*, 69, 4989, 2003.

106. Harman, G.E. et al., *Trichoderma* species — opportunistic, avirulent plant symbionts, *Nat. Rev. Microbiol.*, 2, 43, 2004.

107. Bakker, P.A.H.M. et al., The role of siderophores in potato tuber yield increase by *Pseudomonas putida* in a short rotation of potato, *Neth. J. Plant Pathol.*, 92, 249, 256, 1986.

108. Baron, S.S. and Rowe, J.J., Antibiotic action of pyocyanin, *Antimicrob. Agents. Chemother.*, 20, 814, 1981.

109. Buyer, J.S. and Leong, J., Iron transport-mediated antagonism between plant growth-promoting and plant-deleterious *Pseudomonas* strains, *J. Biol. Chem.*, 261, 791, 1986.

110. Buysens, S.J. et al., Role of siderophores in plant growth stimulation and antagonism by *Pseudomonas aeruginosa* 7NSK2, in *Improving Plant Productivity with Rhizobacteria*, Ryder, M.H. et al., Eds., CSIRO Division of Soils, Adelaide, Australia, 1994, p. 139.

111. Kloepper, J.W., Enhanced plant growth by siderophores produced by plant growth-promoting rhizobacteria, *Nature*, 286, 885, 1980.

112. Lemanceau, P. et al., Effect of pseudobaction 358 production by *Pseudomonas putida* WCS358 on suppression of fusarium wilt of carnations by nonpathogenic *Fusarium oxysporum* Fo47, *Appl. Environ. Microbiol.*, 58, 2978, 1992.

113. Leong, J., Siderophores: their biochemistry and possible role in the biocontrol of plant pathogens, *Annu. Rev. Phytopathol.*, 24, 187, 1986.

114. Loper, J.E., Role of fluorescent siderophore production in biological control of *Pythium ultimum* by a *Pseudomonas fluorescens* strain, *Phytopathology*, 78, 166, 1988.

115. Koster, M. et al., Multiple outer membrane receptors for uptake of ferric pseudobactins in *Pseudomonas putida* WCS385, *Mol. Gen. Genet.*, 248, 735, 1995.

116. Raaijmakers, J.M. et al., Utilization of heterologous siderophores and rhizosphere competence of fluorescent Pseudomonas spp., *Can. J. Microbiol.*, 41, 126, 1995.

117. Raupach, G.S. et al., Induced systemic resistance in cucumber and tomato against cucumber mosaic cucumovirus using plant growth-promoting rhizobacteria (PGPR), *Plant Dis.*, 80, 891, 1996.

118. Wei, G., Kloepper, J.W., and Tuzun, S., Induction of systemic resistance of cucumber to *Colletotrichum orbiculare* by select strains of plant growth-promoting rhizobacteria, *Phytopathology*, 81, 1508, 1991.

119. Duijff, B.J. et al., Siderophore-mediated competition for iron and induced systemic resistance of *Fusarium* wilt of carnation by fluorescent *Pseudomonas* spp., *Neth. J. Plant Pathol.*, 99, 277, 1983.

120. Van Loon, L.C., Bakker, P.A.H.M., and Pieterse, C.M.J., Systemic resistance induced by rhizosphere bacteria, *Annu. Rev. Phytopathol.*, 36, 453, 1998.

121. Knoester, M. et al., Systemic resistance in arabidopsis induced by rhizobacteria requires ethylene-dependent signaling at the site of application, *Mol. Plant-Microbe. Interact*, 8, 720, 1999.

122. Van Wees, S.C.M. et al., Enhancement of induced disease resistance by simultaneous activation of salicylate- and jasmonate-dependent defense pathways in *Arabidopsis thaliana*, *Proc. Natl. Acad. Sci. USA*, 97, 8711, 2000.

123. Van Wees, S.C.M. et al., Differential induction of systemic resistance in *Arabidopsis* by biocontrol bacteria, *Mol.Plant-Microbe Interact.*, 10, 716, 1997.

124. Audenaert, K. et al., Induction of systemic resistance to *Botrytis cinerea* in tomato by *Pseudomonas aeruginosa* 7NSK2: role of salicylic acid, pyochelin, and pyocyanin, *Mol. Plant-Microbe. Interact.*, 15,1147, 2002.

125. Kloepper, J.W., Ryu, C.-M., and Zhang, S. Induced systemic resistance and promotion of plant growth by *Bacillus* spp., *Phytopathology*, 94, 1259, 2004.

126. Leeman, M. et al., Induction of systemic resistance against *Fusarium* wilt of radish by lipopolysaccharides of *Pseudomonas fluorescens*, *Phytopathology*, 85, 1021, 1027, 1995.

127. Dow, M., Newman, M.A., and von Roepenack, E., The Induction and modulation of plant defense responses by bacterial lipopolysaccharides, *Annu. Rev. Phytopathol.*, 38,241, 2000.

128. De Meyer, G. et al., Nanogram amounts of salicylic acid produced by the rhizobacterium *Pseudomonas aeruginosa* 7NSK2 activate the systemic acquired resistance pathway in bean, *Mol. Plant Microbe. Interact.*, 12, 450, 1999.

129. Wäspi, U. et al., Syringolin, a novel peptide elicitor from *Pseudomonas syringae* pv. *syringae* that induces resistance to *Pyricularia oryzae* in rice, *Mol. Plant-Microbe Interact.*, 11, 727, 1998.

130. Ongena, M. et al., Isolation of an N-alkylated benzylamine derivative from *Pseudomonas putida* BTP1 as elicitor of induced systemic resistance in bean, *Mol. Plant-Microbe Interact.*, 18, 562, 2005.

131. Ha, T.T.T. et al., Plant pathogen responses to cyclic lipopeptide surfactants produced by *Pseudomonas fluorescens*, in program and abstract book IOBC-meeting *Multitrophic Interactions in soil*, Wageningen, NL, 2005, p. 14.

131a. Hartmann, A., personal communication, 2005.

131b. Ryu, C.-M. et al., Bacterial volatiles induce systemic resistance in *Arabidopsis, Plant Physiol.*, 134, 1017, 2004.

132. Dekkers, L.C. et al., The *sss* colonization gene of the tomato-*Fusarium oxysporum* f.sp. *radicis-lycopersici* biocontrol strain *Pseudomonas fluorescens* WCS365 can improve root colonization of other wild-type *Pseudomonas* spp. Bacteria, *Mol. Plant Microbe Interact.* 13, 1177, 2000.

133. Felix, G., Regenass, M., and Boller, T., Specific perception of subnanomolar concentrations of chitin fragments by tomato cells: induction of extracellular alkalinization, changes in protein phosphorylation, and establishment of a refractory state, *Plant J.*, 4, 307, 1993.

134. Zipfel, C. et al., Bacterial disease resistance in *Arabidopsis* through flagellin perception, *Nature*, 428, 764, 2004.

135. Gomez-Gomez, L. and Boller, T., FLS2, An LRR receptor-like kinase involved in the perception of the bacterial elicitor flagellin in *Arabidopsis*, *Mol. Cell*, 5,1003, 1011, 2000.

136. Hayashi, F. et al., The innate immune response to bacterial flagellin is mediated by Toll-like receptor 5, *Nature*, 410, 1099, 2001.

137. Lugtenberg, B.J.J., Molecular aspects of biocontrol traits, in *Biology of Plant-Microbe Interactions,* Vol. 4, Tikhonovich, I., Lugtenberg, B.J.J., and Provorov, N., Eds., International Society for Molecular Plant-Microbe Interactions, St. Paul, MN, 2004, p. 310.

138. Nürnberger, T. and Brunner, F., Innate immunity in plants and animals: emerging parallels between the recognition of general elicitors and pathogen-associated molecular patterns, *Curr. Opin. Plant Biol.*, 5, 318, 2002.

138a. Check, W., Innate immunity depends on Toll-like receptors, *ASM News*, 7, 317, 2004.

139. Verhagen, B.W.M. et al., The transcriptome of rhizobacteria-induces systemic resistance in *Arabidopsis, Mol. Plant-Microbe. Interact.* 8, 895, 2004.

140. Van der Ent, S. et al., Transcription factors in roots and shoots of *Arabidopsis* involved in rhizobacteria-induced systemic resistance, in *Multitrophic Interactions in Soil*, Raaijmakers, J.M. and Sikora, R.A., Eds., IOBC/wprs Bulletin, 29, 157, 2006.

141. Curl, E.A. and Truelove, B., *The Rhizosphere,* Springer-Verlag, Berlin Heidelberg, 1986.

142. Duponnois, R. and Garbaye, J., Some mechanisms involved in growth stimulation of ectomycorrhizal fungi by bacteria, *Can. J. Bot.*, 68, 2148, 1990.

143. Frey-Klett, P., Pierrat, J.C., and Garbaye, J., Location and survival of mycorrhiza helper *Pseudomonas fluorescens* during establishment of ectomycorrhizal symbiosis between *Laccaria bicolor* and Douglas fir, *Appl. Environ. Microbiol.*, 63, 139, 1997.

144. Garbaye, J., Helper bacteria: a new dimension to the mycorrhizal symbiosis, *New Phytologist*, 128, 97, 1994.

145. Bianciotto, V. et al., Cellular interactions between arbuscular mycorrhizal fungi and rhizosphere bacteria, *Protoplasma*, 193, 123, 1996.

146. Bianciotto, V. et al., Mucoid Mutants of the Biocontrol Strain *Pseudomonas fluorescens* CHA0 Show Increased Ability in Biofilm Formation on Mycorrhizal and Nonmycorrhizal Carrot Roots, *Mol. Plant-Microbe Interact.*, 14, 255, 2001.

147. Barea, J.M. et al., Impact of arbuscular mycorrhiza formation of *Pseudomonas* strains used as inoculants for biocontrol of soil-borne fungal plant pathogens, *Appl. Environ. Microbiol.*, 64, 2304, 1998.

148. Bolwerk, A. et al., Interactions in the tomato rhizosphere of two *Pseudomonas* biocontrol strains with the phytopathogenic fungus Fusarium oxysporum f. sp. radicis-lycopersici, *Mol.Plant-Microbe Interact.*, 16, 983, 2003.

149. Nelson, E.B. et al., Attachment of *Enterobacter cloacae* to Hyphae of *Pythium ultimum*: possible role in the biological control of pythium preemergence damping-off, *Phytopathology*, 76, 327, 1986.

149a. Kamilova, F. et al., unpublished data, 2005.

150. Lugtenberg, B.J.J. and Kamilova, F.D., Rhizosphere management: microbial manipulation for biocontrol, in *Encyclopedia of Plant and Crop Science*, Marcel Dekker, New York, 2004, p. 1098.

151. Bolwerk, A. et al., Biocontrol of tomato foot and root rot by *Trichoderma* spp. and the role of chitinases, *Mol. Plant-Microbe Interact.*, in press, 2007.

152. Slininger, P.J. and Shea-Andersh, M.A., Proline-based modulation of 2,4-diacetylphloroglucinol and viable cell yields in cultures of *Pseudomonas fluorescens* wild-type and over-producing strains, *Appl. Microbiol. Biotechnol.*, in press, 2006.

153. Duffy, B.K. and Defago, G., Environmental factors modulating antibiotic and siderophore biosynthesis by *Pseudomonas fluorescens* biocontrol strains, *Appl. Environ. Microbiol.*, 65, 2429, 1999.

154. Duffy, B.K. and Defago, G., Zinc improves biocontrol of *Fusarium* crown and root rot of tomato by *Pseudomonas fluorescens* and represses the production of pathogen metabolites inhibitory to bacterial antibiotic biosynthesis, *Phytopathology*, 87, 1250, 1997.

155. Shtark, O.Y. et al., Effect of growth media and wheat exudates components on abtibiotic production by root-colonizing pseudomonads, in *Modern fungicides and antifungal compounds III.*, Dehne, H.-W., Gisi, U., Kuck, K.H., Russell, P.E., and Lyr, H., Eds., AgroConcept GmbH, Bonn, Germany and Verlag Th. Mann GmbH and Co. KG, Gelsenkirchen, Germany, 2002.

156. Van Rij, E.T. et al., Influence of environmental conditions on the production of phenazine-1-carbox-amide by *Pseudomonas chlororaphis* PCL1391, *Mol. Plant-Microbe Interact.*, 17, 557, 2004.

157. Haas, D., Blumer, C., and Keel, C., Biocontrol ability of fluorescent pseudomonads genetically dissected: importance of positive feedback regulation, *Curr. Opin. Biotechnol.*, 11, 290, 2000.

158. Reimmann, C. et al., Posttranscriptional repression of GacS/GacA-controlled genes by the RNA-binding protein RsmE acting together with RsmA in the biocontrol strain *Pseudomonas fluorescens* CHA0, *J. Bacteriol.*, 187, 276, 2005.

159. Aarons, S. et al., A regulatory RNA (PrrB RNA) modulates expression of secondary metabolite genes in *Pseudomonas fluorescens* F113, *J. Bacteriol.*, 182, 3913, 2000.

160. Whistler, C.A. and Pierson, L.S., III, Repression of phenazine antibiotic production in *Pseudomonas aureofaciens* strain 30-84 by RpeA, *J. Bacteriol.*, 185, 3718, 2003.

161. Haas, D., Keel, C., and Reimmann, C., Signal transduction in plant-beneficial rhizobacteria with biocontrol properties, *Antonie van Leeuwenhoek*, 81, 385, 2002.

162. Notz, R. et al., Fusaric acid producing strains of *Fusarium oxysporum* alter 2,4-diacetylphloroglucinol biosynthetic gene expression in *Pseudomonas fluorescens* CHA0 *in vitro* and in the rhizosphere of wheat, *Appl. Environ. Microbiol.*, 68, 2229, 2002.

163. Van Rij, E.T. et al., Influence of fusaric acid on phenazine-1-carboxamide synthesis and gene expression of *Pseudomonas chlororaphis* strain PCL1391, *Microbiology*, 151, 2805, 2005.

164. Koch, B. et al., Lipopeptide production in *Pseudomonas* sp. Strain DSS73 is regulated by components of sugar beet exudate via the Gac two-component regulatory system, *Appl. Environ. Microbiol.*, 68, 4509, 2002.

165. Smith, K.P. and Goodman, R.M., Host variation for interactions with beneficial plant-associated microbes, *Annu. Rev. Phytopathol.*, 37, 473, 1999.

166. Smith, K.P., Handelsman, J., and Goodman, R.M., Genetic basis in plants for interactions with disease-suppressive bacteria, *Proc. Natl. Acad. Sci. USA*, 96, 4786, 1999.

167. Simon, H. et al., Influence of tomato genotype on growth of inoculated and indigenous bacteria in the spermosphere, *Appl. Environ. Microbiol.*, 67, 514, 2000.

168. Lindow, S.E., The use of reporter genes in the study of microbial ecology, *Mol. Ecol.*, 4, 555, 1995.

169. Hansen, L.H. and Sørensen, S.J., The use of whole-cell biosensors to detect and quantify compounds or conditions affecting biological systems, *Microbial Ecol.*, 42, 483, 2001.

170. Casavant, N.C. et al., Site-specific recombination-based genetic system for reporting transient or low-level gene expression, *Appl. Environ. Microbiol.*, 68, 3588, 2002.

171. Beauchamp, C.J., Kloepper, J.W., and Lemke, P.A., Luminometric analyses of plant-root colonization by bioluminescent pseudomonads, *Can. J. Microbiol.*, 39, 434, 1993.

172. Brennerova, M.V. and Crowley, D.E., Direct detection of rhizosphere-colonizing *Pseudomonas* sp using an *Escherichia coli* ribosomal RNA promoter in a Tn7-lux system, *FEMS Microbiol. Ecol.*, 14, 319, 1994.

173. De Weger, L.A. et al., Use of phosphate reporter bacteria to study phosphate limitation in the rhizosphere and in bulk soil, *Mol. Plant-Microbe Interact.*, 7, 32, 1994.

174. Kozdroj, J., Survival of lux-marked bacteria introduced into soil and the rhizosphere of bean (*Phaseolus vulgaris* L), *World J. Microbiol. Biotechnol.*, 12, 261, 1996.

175. Koch, B. et al., Carbon limitation induces sigma-dependent gene expression in *Pseudomonas fluorescens* in soil, *Appl. Environ. Microbiol.*, 67, 3363, 2001.

176. Meikle, A. et al., Matric potential and the survival and activity of a Pseudomonas fluorescens inoculum in soil, *Soil Biol. Biochem.*, 27, 881, 1995.

177. Unge, A. et al., Simultaneous monitoring of cell number and metabolic activity of specific bacterial populations with a dual *gfp-luxAB* marker system, *Appl. Environ. Microbiol.*, 65, 813, 1999.

178. Kragelund, L., Hosbond, C., and Nybroe, O., Distribution of metabolic activity and phosphate starvation response of *lux*-tagged *Pseudomonas fluorescens* reporter bacteria in the barley rhizosphere, *Appl. Environ. Microbiol.*, 63, 4920, 1997.

179. Porteous, F., Killham, K., and Meharg, A., Use of a *lux*-marked rhizobacterium as a biosensor to assess changes in rhizosphere C flow due to pollutant stress, *Chemosphere*, 41, 1549, 2000.

180. Marschner, P. and Crowley, D.E., Physiological activity of a bioluminescent *Pseudomonas fluorescens* (strain 2-79) in the rhizosphere of mycorrhizal and non-mycorrhizal pepper (*Capsicum annuum* L), *Soil Biol. Biochem.*, 28, 869, 1996.

181. Boldt, T.S. et al., Combined use of different Gfp reporters for monitoring single-cell activities of a genetically modified PCB degrader in the rhizosphere of alfalfa, *FEMS Microbiol. Ecol.*, 48, 139, 2004.

182. Ramos, C., Molbak, L., and Molin, S., Bacterial activity in the rhizosphere analyzed at the single-cell level by monitoring ribosome contents and synthesis rates, *Appl. Environ. Microbiol.*, 66, 801, 2000.

183. Loper, J.E. and Lindow S.E., Reporter gene systems useful in evaluating *in situ* gene expression by soil and plant-associated bacteria, in *Manual of Environmental Microbiology*, Hurst, C.J., Knudsen, G.R., McInerney, M.J., Stetzenbach, L.D., and Walter, M.V., Eds., ASM Press, Washington, D.C., 1997, pp. 482–492.

183a. Miller, W.G. et al., Biological sensor for sucrose availability: relative sensitivities of various reporter genes, *Appl. Environ. Microbiol.*, 67, 1308, 2001.

183b. Zhang, Y.-Z. et al., Detecting *lacZ* gene expression in living cells with new lipophilic, fluorogenic - galactosidase substrates, *FASEB J.*, 5, 3108, 1991.

183c. Buell, C.R. and Anderson, A.J., Expression of the *aggA* locus of *Pseudomonas putida in vitro* and *in planta* as detected by the reporter gene, xylE, *Mol. Plant Microbe Interact.*, 6, 331,1993.

184. Camilli, A., Beattie, D.T., and Mekalanos, J.J., Use of genetic recombination as a reporter of gene expression, *Proc. Natl. Acad. Sci. USA*, 91, 2634, 1994.

185. Casavant, N.C. et al., Use of a site-specific recombination-based biosensor for detecting bioavailable toluene and related compounds on roots, *Environ. Microbiol.*, 5, 238, 2003.

186. Vande Broek, A.V. et al., Spatial-temporal colonization patterns of *Azospirillum brasilense* on the wheat root surface and expression of the bacterial nifH gene during association, *Mol. Plant-Microbe Interact.*, 6, 592, 1993.

187. Cleyet-Marel, J.C. et al., Host-specific regulation of nodulation mediated by flavonoid compounds present in plant, *Acta Bot. Gallica*, 143, 521, 1996.

188. Buell, C.R. and Anderson, A.J., Expression of the *aggA* locus of *Pseudomonas putida in vitro* and *in planta* as detected by the reporter gene *xylE*, *Mol. Plant-Microbe Interact.*, 6, 331, 1993.

189. Howie, W.J. and Suslow, T.V., Role of antibiotic biosynthesis in the inhibition of Pythium ultimum in the cotton spermosphere and rhizosphere by *Pseudomonas fluorescens*, *Mol. Plant-Microbe Interact.*, 4, 393, 1991.

190. Georgakopoulos, D.G. et al., Analysis of expression of a phenazine biosynthesis locus of *Pseudomonas aureofaciens* PGS12 on seeds with a mutant carrying a phenazine biosynthesis locus ice nucleation reporter gene fusion, *Appl. Environ. Microbiol.*, 60, 4573, 1994.

191. Lutz, M.P. et al., Signaling between bacterial and fungal biocontrol agents in a strain mixture, *FEMS Microbiol. Ecol.*, 48, 447, 2004.

192. Koch, B. et al., Lipopeptide production in *Pseudomonas* sp strain DSS73 is regulated by components of sugar beet seed exudate via the *gac* two-component regulatory system, *Appl. Environ. Microbiol.*, 68, 4509, 2002.

193. Notz, R. et al., Biotic factors affecting expression of the 2,4-diacetylphloroglucinol biosynthesis gene *phlA* in *Pseudomonas fluorescens* biocontrol strain CHA0 in the rhizosphere, *Phytopathology*, 91, 873, 2001.

194. Seveno, N.A., Morgan, J.A.W., and Wellington, E.M.H., Growth of Pseudomonas aureofaciens PGS12 and the dynamics of HHL and phenazine production in liquid culture, on nutrient agar, and on plant roots, *Microb. Ecol.*, 41, 314, 2001.

195. Siddiqui, I.A. and Shaukat, S.S., *Trichoderma harzianum* enhances the production of nematicidal compounds *in vitro* and improves biocontrol of *Meloidogyne javanica* by *Pseudomonas fluorescens* in tomato, *Lett. Appl. Microbiol.*, 38, 169, 2004.

196. Teplitski, M., Robinson, J.B., and Bauer, W.D., Plants secrete substances that mimic bacterial N-acyl homoserine lactone signal activities and affect population density-dependent behaviors in associated bacteria, *Mol. Plant-Microbe Interact.*, 13, 637, 2000.

197. Jaeger, C.H. et al., Mapping of sugar and amino acid availability in soil around roots with bacterial sensors of sucrose and tryptophan, *Appl. Environ. Microbiol.*, 65, 2685, 1999.

198. Bringhurst, R.M., Cardon, Z.G., and Gage, D.J., Galactosides in the rhizosphere: utilization by *Sinorhizobium meliloti* and development of a biosensor, *Proc. Natl. Acad. Sci. USA*, 98, 4540, 2001.

199. Standing, D., Meharg, A.A., and Killham, K., A tripartite microbial reporter gene system for real-time assays of soil nutrient status, *FEMS Microbiol. Lett.*, 220, 35, 2003.

200. Jensen, L.E. and Nybroe, O., Nitrogen availability to *Pseudomonas fluorescens* DF57 is limited during decomposition of barley straw in bulk soil and in the barley rhizosphere, *Appl. Environ. Microbiol.*, 65, 4320, 1999.

201. Ravnskov, S., Nybroe, O., and Jakobsen, I., Influence of an arbuscular mycorrhizal fungus on *Pseudomonas fluorescens* DF57 in rhizosphere and hyphosphere soil, *New Phytologist*, 142, 113, 1999.

202. Dollard, M.A. and Billard, P., Whole-cell bacterial sensors for the monitoring of phosphate bioavailability, *J. Microbiol. Methods*, 55, 221, 2003.

203. Hojberg, O. et al., Oxygen-sensing reporter strain of *Pseudomonas fluorescens* for monitoring the distribution of low-oxygen habitats in soil, *Appl. Environ. Microbiol.*, 65, 4085, 1999.

204. Joyner, D.C. and Lindow, S.E., Heterogeneity of iron bioavailability on plants assessed with a whole-cell GFP-based bacterial biosensor, *Microbiology*, 146, 2435, 2000.

205. Loper, J.E. and Lindow, S.E., A biological sensor for iron available to bacteria in their habitats on plant surfaces, *Appl. Environ. Microbiol.*, 60, 1934, 1994.

206. Loper, J.E. and Henkels, M.D., Availability of iron to *Pseudomonas fluorescens* in rhizosphere and bulk soil evaluated with an ice nucleation reporter gene, *Appl. Environ. Microbiol.*, 63, 99, 1997.

207. Marschner, P., Crowley, D.E., and Sattelmacher, B., Root colonization and iron nutritional status of a *Pseudomonas fluorescens* in different plant species, *Plant and Soil*, 196, 311, 1997.

208. Brazil, G.M. et al., Construction of a rhizosphere Pseudomonad with potential to degrade polychlorinated-biphenyls and detection of *bph* gene expression in the rhizosphere, *Appl. Environ. Microbiol.*, 61, 1946, 1995.

209. Tom-Petersen, A., Hosbond, C., and Nybroe, O., Identification of copper-induced genes in *Pseudomonas fluorescens* and use of a reporter strain to monitor bioavailable copper in soil, *FEMS Microbiol. Ecol.*, 38, 59, 2001.

210. Petanen, T. et al., Construction and use of broad host range mercury and arsenite sensor plasmids in the soil bacterium *Pseudomonas fluorescens* OS8, *Microb. Ecol.*, 41, 360, 2001.

211. Paterson, E. et al., Effects of elevated CO_2 on rhizosphere carbon flow and soil microbial processes, *Glob. Change Biol.*, 3, 363, 1997.

212. Bassler, B.L., How bacteria talk to each other: regulation of gene expression by quorum sensing, *Curr. Opin. Microbiol.*, 2, 582, 1999.

213. Elasri, M. et al., Acyl-homoserine lactone production is more common among plant-associated *Pseudomonas* spp. than among soilborne Pseudomonas spp., *Appl. Environ. Microbiol.*, 67, 1198, 2001.

214. Ulrich, R.K., Quorum quenching: enzymatic disruption of N-acylhomoserine lactone-mediated bacterial communication in *Burkholderia thailandensis*, *Appl. Environ. Microbiol.*, 70, 6173, 2004.

215. Dong, Y.H. et al., AiiA, an enzyme that inactivates the acylhomoserine lactone quorum sensing signal and attenuates the virulence of *Erwinia carotorova*, *Proc. Natl. Acad. Sci. USA*, 97, 3526, 2000.

216. Dong, Y.H. et al., Identification of quorum quenching N-homoserine lactonases from *Bacillus* species, *Appl. Environ. Microbiol.*, 68, 1754, 2002.

217. Park, S.-Y. et al. Identification of extracellular N-acylhomoserine lactone acylase from a *Streptomyces* sp. and its application to quorum quenching, *Appl. Environ. Microbiol.*, 71, 2632, 2005.

218. Piper, K.R., Beck von Bodman, S., and Farrand, S. K., Conjugation factor of *Agrobacterium tumefaciens* regulates Ti plasmid transfer by autoinduction, *Nature*, 362, 448, 1993.

219. Toth, I.K. et al., Potato plants genetically modified to produce N-acylhomoserine lactones increase susceptibility to soft rot Erwiniae, *Mol. Plant-Microbe Interact.*, 17, 880, 2004.

220. Ochsner, U.A. and Reiser, J., Autoinducer-mediated regulation of rhamnolipid biosurfactant synthesis in *Pseudomonas aeruginosa*, *Proc. Natl. Acad. Sci. USA.*, 92, 6424, 1995.

221. Pierson, L.S., III, Wood, D.W., and Pierson, E.A., Homoserine lactone-mediated gene regulation in plant-associated bacteria, *Annu. Rev. Phytopathol.*, 36, 207, 1998.

222. Chin-A-Woeng, T.F.C., Bloemberg, G.V., and Lugtenberg, B.J.J., Phenazines and their role in biocontrol by *Pseudomonas* bacteria, *New Phytologist*, 157, 503, 2003.

223. Molina, L. et al., Autoinduction in *Erwinia amylovora*: evidence of an acyl-homoserine lactone signal in the fire blight pathogen, *J. Bacteriol.*, 187, 3206, 2005.

224. Smadja, B. et al., Involvement of N-acylhomoserine lactones throughout plant infection by *Erwinia carotovora subsp. atroseptica* (*Pectobacterium atrosepticum*), *Mol. Plant-Microbe Interact.*, 17, 1269, 2004.
225. Calfee, M.W. et al., Solubility and bioactivity of the pseudomonas quinolone signal are increased by a *Pseudomonas aeruginosa*-produced surfactant, *Infect. Immun.*, 73, 878, 2005.
226. Pierson, E.A. et al., Interpopulation signaling via N-acyl-homoserine lactone among bacteria in the wheat rhizosphere, *Mol. Plant-Microbe Interact.*, 11, 1078, 1998.
227. Givskov, M. et al., Eukaryotic interference with homoserine lactone-mediated prokaryotic signalling, *J. Bacteriol.*, 178, 6618, 1996.
228. Rasmussen, T.B. et al., How *Delisea pulchra* furanones affect quorum sensing and swarming motility in *Serratia liquefaciens* MG1, *Microbiology*, 146, 3237, 2000.
229. Rasmussen, T.B. et al., Screening for quorum-sensing inhibitors (QSI) by use of a novel genetic system, the QSI selector, *J. Bacteriol.*, 187, 1799, 2005.
230. Idris, E.E. et al., Use of *Bacillus subtilis* as biocontrol agent. VI. Phytohormone-like action of culture filtrates prepared from plant growth-promoting *Bacillus amyloliquefaciens* FZB24, FZB42, FZB45 and *Bacillus subtilis* FZB37, *J. Plant Dis. Prot.*, 111, 583, 2004
231. Dey, R. et al., Growth promotion and yield enhancement of peanut (*Arachis hypogaea* L.) by application of plant growth-promoting rhizobacteria, *Microbiol. Res.*, 159, 371, 2004.
232. Slininger, P.J., Burkhead, K.D., and Schisler, D.A., Antifungal and sprout regulatory bioactivities of phenylacetic acid, indole-3-acetic acid, and tyrosol isolated from the potato dry rot suppressive bacterium *Enterobacter cloacae* S11, *J. Ind. Microbiol. Biotechnol.*, 31, 517, 2004.
233. Kumar, R.S. et al., Characterization of antifungal metabolite produced by a new strain *Pseudomonas aeruginosa* PUPa3 that exhibits broad-spectrum antifungal activity and biofertilizing traits, *J. Appl. Microbiol.*, 98, 145, 2005.
234. Suzuki, S., He, Y.X., and Oyaizu, H., Indole-3-acetic acid production in *Pseudomonas fluorescens* HP72 and its association with suppression of creeping bentgrass brown patch, *Curr. Microbiol.*, 47, 138, 2003.
235. Deshwal, V.K., Dubey, R.C., and Maheshwari, D.K., Isolation of plant growth-promoting strains of *Bradyrhizobium* (*Arachis*) sp with biocontrol potential against *Macrophomina phaseolina* causing charcoal rot of peanut, *Curr. Sci.*, 84, 443, 2003.
236. Bano, N. and Musarrat, J., Characterization of a new *Pseudomonas aeruginosa* strain NJ-15 as a potential biocontrol agent, *Curr. Microbiol.*, 46, 324, 2003.
237. Pal, K.K. et al., Suppression of maize root diseases caused by *Macrophomina phaseolina*, *Fusarium moniliforme* and *Fusarium graminearum* by plant growth promoting rhizobacteria, *Microbiol. Res.*, 156, 209, 2001.
238. Bedini, S. et al., Pseudomonads isolated from within fruit bodies of *Tuber borchii* are capable of producing biological control or phytostimulatory compounds in pure culture, *Symbiosis*, 26, 223, 1999.
239. Oberhansli, T., Defago, G., and Haas, D., Indole-3-acetic acid (IAA) synthesis in the biocontrol strain CHA0 of *Pseudomonas fluorescens* — role of tryptophan side-chain oxidase, *J. Gen. Microbiol.*, 137, 2273, 1991.
240. Le Floch, G. et al., Impact of auxin-compounds produced by the antagonistic fungus *Pythium oligandrum* or the minor pathogen *Pythium* group F on plant growth, *Plant Soil*, 257, 459, 2003
241. Beyeler, M. et al., Enhanced production of indole-3-acetic acid by a genetically modified strain of *Pseudomonas fluorescens* CHA0 affects root growth of cucumber, but does not improve protection of the plant against *Pythium* root rot, *FEMS Microbiol. Ecol.*, 28, 225, 1999.
242. Kamensky, M. et al., Soil-borne strain IC14 of *Serratia plymuthica* with multiple mechanisms of antifungal activity provides biocontrol of *Botrytis cinerea* and *Sclerotinia sclerotiorum* diseases, *Soil Biol. Biochem.*, 35, 323, 2003.
243. Wang, C.X. et al., Effect of transferring 1-aminocyclopropane-1-carboxylic acid (ACC) deaminase genes into *Pseudomonas fluorescens* strain CHA0 and its gacA derivative CHA96 on their growth-promoting and disease-suppressive capacities, *Can. J. Microbiol.*, 46, 898, 2000.
244. Castang, S. et al., N-Sulfonyl homoserine lactones as antagonist of bacterial quorum sensing, *Bioorg. Med. Chem. Lett.*, 14, 5145, 2004.
245. Fray, R.G., Plants genetically modified to produce N-acylhomoserine lactones communicate with bacteria, *Nat. Biotechnol.*, 17, 1017, 1999.

246. Mae, A. et al., Transgenic plants producing the bacterial pheromone N-acyl-homoserine lactone exhibit enhanced resistance to the bacterial phytopathogen *Erwinia carotovora*, *Mol. Plant-Microbe Interact.*, 14, 1035, 2001.

247. Huang, Z. et al., Transformation of *Pseudomonas fluorescens* with genes for biosynthesis of phenazine-1-carboxylic acid improves biocontrol of rhizoctonia root rot and *in situ* antibiotic production, *FEMS Microbiol. Ecol.*, 49, 243, 2004.

248. Chin-A-Woeng, T.F.C. et al., Introduction of the *phzH* gene of *Pseudomonas chlororaphis* PCL1391 extends the range of biocontrol ability of phenazine-1-carboxylic acid-producing *Pseudomonas* spp. Strains, *Mol. Plant-Microbe Interact.*, 14, 1006, 2001.

249. Hill, D.S. et al., Cloning of genes involved in the synthesis of pyrrolnitrin from *Pseudomonas fluorescens* and role of pyrrolnitrin synthesis in biological control of plant disease, *Appl. Environ. Microbiol.*, 60, 78, 1994.

250. Siddiqui, I.A., Shaukat, S.S., and Khan, A., Differential impact of some *Aspergillus* species on *Meliodogyne javanica* biocontrol by *Pseudomonas fluorescens* strain CHA0, *Lett. Appl. Microbiol.*, 39, 74, 2004.

251. Rabea, E.I. et al., Chitosan as antimicrobial agent: applications and mode of action, *Biomacromolecules*, 4, 1457, 2003.

252. Duke, S.O. et al., ARS research on natural products for pest management, *Pest Manage. Sci.*, 59, 708, 2003.

253. Colbert, S.F. et al., Enhanced growth and activity of a biocontrol bacterium genetically engineered to utilize salicylate, *Appl. Environ. Microbiol.*, 59, 2071, 1993.

254. Savka, M.A. et al., Engineering bacterial competitiveness and persistence in the phytosphere, *Mol. Plant-Microbe Interact.*, 15, 866, 2002.

255. Oger, P., Petit, A., and Dessaux, Y., Genetically engineered plants producing opines alter their biological environment, *Nat. Biotechnol.*, 15, 369, 1997.

256. Savka, M.A. and Ferrand, S., Modification of rhizobacterial populations by engineering bacterial utilization of a novel plant-produced resource, *Nat. Biotechnol.*, 15, 363, 1997.

257. O'Connell, K.P., Goodman, R.M., and Handelsman, J., Engineering the rhizosphere: expressing a bias, *Trends Biotechnol.*, 14, 83, 1996.

258. Fravel, D.R. et al., Formulation of microorganisms to control plant diseases, in *Formulation of Microbial Pesticides*, Burges, H. D., Ed., Kluwer Academic Publishers, Dordrecht, The Netherlands, 1998.

259. Duffy, B., Schouten, A, and Raaijmakers, J.M., Pathogen self-defense: mechanisms to counteract microbial antagonism, *Annu. Rev. Phytopathol.*, 41, 501, 2003.

260. Schouten, A. et al., Defense responses of *Fusarium oxysporum* to 2,4-diacetylphloroglucinol, a broad-spectrum antibiotic produced by *Pseudomonas fluorescens*, *Mol. Plant-Microbe Interact.*, 17, 1201, 2004.

261. Lutz, M.P. et al., Mycotoxigenic *Fusarium* and deoxynivalenol production repress chitinase gene expression in the biocontrol agent *Trichoderma atroviride* P1, *Appl. Environ.Microbiol.*, 69, 3077, 2003.

262. Smith, L.M. et al., Signalling by the fungus *Pythium ultimum* repressess expression of two ribosomal operons with key roles in the rhizosphere ecology of *Pseudomonas fluorescens* F113, *Environ. Microbiol.*, 1, 495, 1999.

263. Fedi, S. et al., Evidence for signaling between the phytopathogenic fungus *Pythium ultimum* and *Pseudomonas fluorescens* F113: *P. ultimum* represses the expression of genes in *P. fluorescens* F113, resulting in altered ecological fitness, *Appl. Environ.Microbiol.*, 63, 4261, 1997.

264. Bloemberg, G.V. and Lugtenberg, B.J.J., Bacterial biofilm on plants: relevance and phenotypic aspects, in *Microbial Biofilms*, Ghannoum, M. and O'Toole, G.A.O., Eds., ASM Press, Washington, D.C., 2004, p. 141.

265. Campbell, R. and Rovira, A.D., The study of the rhizosphere by scanning electron microscopy, *Soil Biol. Biochem.*, 5, 747, 1973.

266. Chin-A-Woeng, T.F.C. et al., Visualisation of interactions of Pseudomonas and Bacillus biocontrol strains, in *Plant Surface Microbiology*, Varma, A. et al., Eds., Springer, Berlin, Germany, 2004, p. 431.

267. Foster, R.C. and Rovira, A.D., Ultrastructure of wheat rhizosphere, *New Phytologist*, 76, 343, 1976.

268. Foster, R.C., The ultrastructure and histochemistry of the rhizosphere, *New Phytologist*, 89, 263, 1981.

269. Bahme, J.B. and Schroth, M.N., Spatial-temporal colonization patterns of a rhizobacterium on underground organs of potato, *Phytopathology*, 77, 1093, 1987.

270. Bowen, G.D. and Rovira, A.D., Microbial colonization of plant roots, *Annu. Rev. Phytopathol.*, 14, 121, 1976.

271. Foster, R.C., The fine structure of epidermal cell mucilages of roots, *New Phytologist*, 91, 727, 1982.

272. Van Elsas, J.D. et al., Bacterial conjugation between pseudomonads in the rhizosphere of wheat, *FEMS Microbiol. Lett.*, 53, 299, 1998.

273. Lagopodi, A.L. et al., Novel aspects of tomato root colonization and infection by *Fusarium oxysporum* f. sp. *radicis-lycopersici* revealed by confocal laser scanning microscopic analysis using the green fluorescent protein as a marker, *Mol. Plant-Microbe Interact.*, 15, 172, 2002.

274. Bolwerk, A. et al., Visualization of interactions between a pathogenic and a beneficial *Fusarium* strain during biocontrol of tomato foot and root rot, *Mol. Plant-Microbe Interact.*, 18, 710, 2005.

275. Van Overbeek, L.S. et al., A polyphasic approach for studying the interaction between Ralstonia solanacearum and potential control agents in the tomato phytosphere, *J. Microbiol. Methods*, 48, 69, 2002.

276. Sigler, W.V. et al., Fate of the biological control agent *Pseudomonas aureofaciens* TX-1 after application to turfgrass, *Appl. Environ. Microbiol.*, 67, 3542, 2001.

277. Rubio, M.B. et al., Specific PCR assays for the detection and quantification of DNA from the biocontrol strain *Trichoderma harzianum* 2413 in soil, *Microb. Ecol.*, 49, 25, 2005.

278. Van Dillewijn, P. et al., Plant-dependent active biological containment system for recombinant rhizobacteria, *Environ. Microbiol.*, 6, 88, 2004.

279. Montesinos, E., Development, registration and commercialization of microbial pesticides for plant protection, *Int. Microbiol.*, 6, 245, 2003.

280. Burges, H.D., *Formulation of Microbial Pesticides*, Kluwer Academic Publishers, Dordrecht, The Netherlands, 1998.

281. Chebotar, V.K. and Kang, U.G., Production and application efficiency of the biopreparations based on rhizobacteria (PGPR), in *Biology of Plant-Microbe Interactions*, Vol. 4., Tikhonovich, I. et al., Eds., International Society for Molecular Plant-Microbe Interactions, St. Paul, MN, 2004, p. 597.

282. Smith, R.S., Rhizobial inoculant technology in North America, in *Biology of Plant-Microbe Interactions*, Vol. 4, Tikhonovich, I., Lugtenberg, B.J.J., and Provorov, N., Eds., International Society for Molecular Plant-Microbe Interactions, St. Paul, MN, 2004, p. 594.

11 Chemical Signals in the Rhizosphere: Root–Root and Root–Microbe Communication

Laura G. Perry, Élan R. Alford, Junichiro Horiuchi, Mark W. Paschke, and Jorge M. Vivanco

CONTENTS

I. INTRODUCTION

The rhizosphere, or soil immediately surrounding plant roots, houses complex communities that include plant roots, soil bacteria and fungi, nematodes, annelids, and arthropods. Within these communities, plant roots interact with competitors, mutualists, parasites, and pathogens.

These interactions are sometimes purely mechanical in nature, such as tissue damage by herbivores or pathogens. Other interactions are mediated entirely by the effects on limiting resources, such as resource uptake, microbial decomposition, microbial nutrient immobilization and transformation, and carbon and nutrient transfer among organisms. However, numerous interactions between plant roots and other organisms are also mediated by chemical signals produced by plants and the organisms with which they interact. These signals operate to induce changes in the behavior, morphology, physiology, or biochemistry of the organisms exposed to them, sometimes with positive consequences for those organisms, and sometimes with negative consequences. As such, these signals constitute a complex, underground chemical system through which plants and soil microbes communicate and influence their biotic and abiotic environment.

A. Chemical Signaling as Belowground Communication

The American Heritage Dictionary [1] defines *communication* as the "exchange of thoughts, messages, or information, as by speech, signals, writing, or behavior." Secondary metabolites released by one organism that alter the behavior, morphology, physiology, or biochemistry of other organisms may be viewed as signals mediating the exchange of messages that evoke particular responses; in other words belowground communication. Such communication may develop through evolutionary pressure on the signaler, involving the release of signals directed at specific organisms to benefit the signaler. Alternatively, such communication may involve organisms that *eavesdrop* on chemical conversations between others to gain information for their own benefit or are *innocent bystanders* inadvertently affected by a chemical message that was sent to affect a different organism. Whittaker and Feeny [2] proposed several terminologies to describe six types of chemical signals. Among chemical signals with interspecific effects (i.e., allelochemicals), allamones benefit the signaler but not the recipient, kairomones benefit the recipient but not the signaler, synomones benefit both the signaler and the recipient, and depressants benefit neither the signaler nor the recipient. Among chemical signals with intraspecific effects, autotoxins negatively affect the producing population; autoinhibitors positively affect the producing population by negatively affecting some of its members, serving as an adaptive mechanism for population control, and pheromones positively affect both the signaler and the receiver, mediating reproduction and social behavior, and communicating the presence of food or danger. All of these chemical signals are distinct from hormones, which serve as chemical signals between cells within organisms. However, hormones may have evolved in multicellular organisms from cell-to-cell signals (i.e., pheromones) in primitive colonies of single-celled organisms.

In this chapter, we describe recent research on chemical-mediated communication in the rhizosphere, including both communication between plants and communication between plants and soil microbes. We begin with a brief review of the presence and behavior of chemical signals in the rhizosphere. Then, we discuss recent discoveries on the biology, chemistry, and mode of action of chemical signals that mediate negative interactions in the rhizosphere, first between plants and then between plants and soil microbes (Figure 11.1). Negative communication between plants includes phytotoxins with interspecific effects (allamones), phytotoxins with intraspecific effects (autotoxins and autoinhibitors), and chemical signals involved in parasitic plant–host interactions (allamones and kairomones). Negative communication between plants and soil microbes includes plant defense compounds (allamones), pathogenic virulence factors (allamones), and pathogenic compounds that elicit plant defenses (kairomones). In addition, we discuss new research on chemical signals that mediate positive interactions between plants and between plants and soil microbes (Figure 11.1). Positive communication between plants includes warning signals that induce plant defenses (kairomones and synomones) and signals involved in root growth into the soil matrix (kairomones, synomones, and autoinhibitors). Positive communication between plants and soil microbes includes chemical signals that mediate mutualistic associations between plant hosts and extracellular plant-growth-promoting rhizobacteria (ePGPR), intracellular nitrogen-fixing bacteria, and mycorrhizal fungi (synomones).

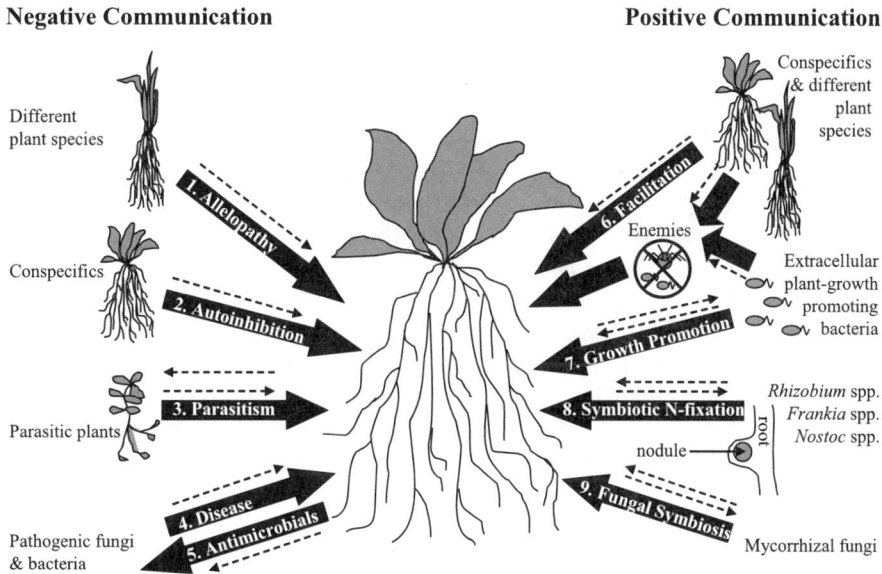

FIGURE 11.1 Modes of chemical-mediated communication between plant roots and between plant roots and soil microbes. Block arrows indicate different types of interactions. Negative interactions mediated by below-ground chemical signals include (1) interspecific interference among plants (i.e., allelopathy), (2) intraspecific interference among plants (i.e., autoinhibition), (3) host infection by parasitic plants, (4) host infection by pathogenic soil microbes, and (5) plant defense against soil pathogens. Positive interactions mediated by belowground chemical signals include (6) facilitation by other plants, which is often mediated by induced herbivore resistance or reduced herbivore populations, (7) growth promotion by extracellular plant-growth-promoting bacteria, which is also often mediated by biological control of soil pathogens, (8) symbiotic relationships between plants and nitrogen-fixing bacteria, and (9) symbiotic relationships between plants and mycorrhizal fungi. Dashed arrows indicate the source and recipient of the chemical signals involved in each interaction. In many cases, both organisms involved in these interactions are known to both send and respond to chemical signals.

B. CHEMICAL SIGNALS IN THE RHIZOSPHERE

Plant roots contribute an estimated 5 to 21% of photosynthetically fixed carbon to the rhizosphere [3]. This, surprisingly, large quantity of carbon is released into the soil in the form of root debris, border cells, and root exudates. The relative quantities of these materials in the rhizosphere are described in detail in Chapter 1 and vary considerably with environmental conditions, plant species, and plant physiological status as discussed in Chapter 2. Root exudates include carbohydrates, mucilage, and a variety of secondary metabolites. Secondary metabolites are low-molecular-weight compounds that are produced by metabolic pathways, other than the primary pathways, involved in carbohydrate, protein, lignin, fat, and energy production. Whereas primary metabolites occur in most or all living organisms, specific secondary metabolites may occur in many or only a few organisms. Thus, secondary metabolites are responsible for much of the chemical diversity of plant root exudates. More than 100,000 secondary metabolites for plants have been identified [4]. These include an array of organic acids, flavonoids, tannins, terpenoids, alkaloids, polyacteylenes, and simple phenolics [5,6] in root exudates.

Secondary metabolites in root exudates have the potential to perform numerous important functions in the rhizosphere. When plant secondary metabolites were first discovered in root exudates, they were presumed to be merely waste products of metabolism disposed of by excretion. However, as research on secondary metabolites has progressed, and the techniques for examining their biochemical effects have become more sophisticated, it has become clear that many plant secondary metabolites may play important roles as chemical signals in the rhizosphere. Specifically, plant secondary metabolites may be involved in interspecific and intraspecific interference between plants, antimicrobial activity, plant induction of disease resistance, attraction and repulsion of pathogens and mutualists, development of symbiotic associations between plants and microbes, and regulation of microbial populations that influence nutrient cycling [6,7,8], and can affect microbial diversity (see Chapter 3).

Organic compounds are also released into the rhizosphere by populations of soil bacteria and fungi with a similarly wide array of potential roles as signals in the rhizosphere (Figure 11.1). Soil microbes produce and secrete secondary metabolites that mediate microbial interactions, including both antibiotics and chemicals involved in microbial associations, such as quorum sensing and biofilm formation [9,10]. Other secondary metabolites produced by soil microbes directly affect plant growth, including phytotoxins, plant hormones, and plant growth promoters [11,12]. Still others reduce plant disease responses, thus facilitating either pathogen infection or infection by symbiotic bacteria or fungi [13].

C. Signal Mobility, Persistence, and Activity in the Rhizosphere

For chemical signals to mediate belowground root–root or root–microbe communication, the signals must be able to travel the distance between organisms, be of sufficient concentration to elicit a response, and persist in an available and active state for sufficient time in the soil to affect the signal recipient. Consequently, plant and microbial compounds that, under artificial conditions, appear to have the potential to act as chemical signals are not necessarily active under rhizosphere conditions.

Several factors that are discussed in detail in Chapter 1 can operate to limit the stability, mobility, concentration, and activity of secondary metabolites in the rhizosphere. First, microbial degradation can rapidly reduce concentrations of at least some secondary metabolites. Soil microbes often rely on plant root exudates as a major source of carbon, including low-molecular-weight organic compounds such as secondary metabolites [14]. Second, some secondary metabolites are chemically unstable and may transform rapidly following exudation. Such chemically unstable secondary metabolites are less likely to play important roles in belowground communication, but might be effective if exuded in conjunction with a stabilizing agent, if the degradation products are stable and active, or if the distance between sender and recipient is small. Third, secondary metabolites can be adsorbed by soil particles resulting in reduced signal mobility and activity. Charged and hydrophobic (e.g., phenolics) secondary metabolites, which include many putative chemical signals, are readily adsorbed by soil particles, particularly in soils with high humic content or cation exchange capacity [15]. Chemical signals that must be taken up by the recipient will not be effective if bound to soil particles. However, signals that induce a biochemical response by activating a receptor on the exterior of the recipient may still be active even when bound to soil particles, so long as the structural components of the compound that are responsible for activating the receptor, remain unbound.

Thus, the mode of action of a chemical signal, in addition to its behavior in soil, must be understood to predict its activity under different soil conditions. Most measurements of secondary metabolite concentrations do not distinguish between organic compounds in soil solution and organic compounds adsorbed to soil particles. Strong extraction procedures may increase the yield of the target signal but may also increase the occurrence of artifacts if the compound is not active when bound to soil particles.

The further a chemical signal must travel through the soil to reach a recipient, the more likely it is to encounter a biotic or abiotic factor that alters its structure, concentration, or activity. Chemical communication between roots, and between roots and soil microbes, is thus most likely to occur in the rhizosphere, where the individuals involved are in close proximity and local microsites of concentrated chemicals signals are most easily generated. Soil microbial populations are often concentrated in the rhizosphere as a result of abundant carbon in plant root exudates [16], leading to a high potential for chemical signals to travel between roots and soil microbes and for microbes to degrade chemical signals. Chemical communication between plants may be most frequent between individuals with roots that are close together. For most of the rhizosphere signals discussed in the literature, little is known about concentrations or activity in soil. In Chapter 1, Uren proposed that the right set of circumstances at the root interface, such as low oxygen conditions generated by root exudation, may allow some secondary metabolites to avoid oxidation and microbial degradation. Such conditions may be necessary for some secondary metabolites to act as chemical signals in the rhizosphere, particularly the many secondary metabolites that are chemically unstable or readily degraded by microbes.

II. NEGATIVE ROOT–ROOT COMMUNICATION

A. ALLELOPATHY

Phytotoxic secondary metabolites have the potential to mediate interspecific plant–plant interference (i.e., allelopathy) by reducing competitor establishment, growth, and survival. To date, hundreds of plant-produced phytotoxins have been identified [17]. These compounds vary substantially in chemical structure, modes of action, and effects on plants [18]. Different phytotoxic allamones inhibit seed germination, root growth, shoot growth, and plant survival with effects on metabolite production, photosynthesis, respiration, membrane transport, and within-plant chemical signaling [6,17,18]. Plant-produced phytotoxins are released into the soil as root exudates, as leachates from live and dead plant tissue, as green leafy volatiles, and as decomposition products from dead plant material [6,18].

Whereas all phytotoxic allamones have the potential to act as chemical signals in the rhizosphere, only root-exuded phytotoxins are involved specifically in root–root communication. Many plants are known to rhizosecrete potent phytotoxins (Figure 11.2). Examples include, but are not limited to, trees such as *Juglans nigra* L. [19], crops such as *Cucumis sativa* L. [20], *Oryza sativa* L. [21], *Sorghum* spp. [22], and *Triticum aestivum* L. [23], and invasive weeds such as *Centaurea maculosa* Lam. [24,25], *C. diffusa* Lam. [26], and *Acroptilon repens* (L.) DC [27]. The modes of action of the phytotoxins associated with many of these examples are well understood [6]. In particular, recent research on two of the species, *C. maculosa* and *Sorghum bicolor*, has elucidated much of the biochemical and genetic basis of allelopathy in these species.

Centaurea maculosa, a perennial forb native to Eurasia, is a highly invasive weed in North American grasslands [28]. The competitive success of *C. maculosa* in North America may be mediated in part by root exudation of a phytotoxin that reduces the growth and survival of North American grassland species [24,25,29–31]. *C. maculosa* roots exude a racemic mixture of (+)-catechin and (−)-catechin (Figure 11.2). Both enantiomers are phytotoxic, although (−)-catechin is substantially more potent than (+)-catechin [24,32]. Treatment of susceptible plants with (−)-catechin or the racemic mixture, (±)-catechin, *in vitro* inhibits root elongation, can result in plant mortality, particularly among dicots, and can reduce shoot length and germination [24,25,30,31]. Perry et al. [31] found that (±)-catechin treatment inhibited root elongation of 13 out of 20 native North American grassland species examined, suggesting that *C. maculosa* (±)-catechin production may contribute to *C. maculosa* invasions of North American grasslands.

Efforts to quantify (±)-catechin production have yielded mixed results. Early studies indicated that young *Centaurea maculosa* plants grown at high densities *in vitro* could produce as much as 80 μg ml^{-1} of (±)-catechin [24]. In contrast, in a more recent study, *C. maculosa* plants grown individually *in vitro* produced a maximum of 2.5 μg ml^{-1} of (±)-catechin [33]. Further, several studies have reported very high soil (±)-catechin concentrations (> 1 mg g^{-1}) in well-established field populations of *C. maculosa* [24,25,34]. However, one recent study failed to detect any

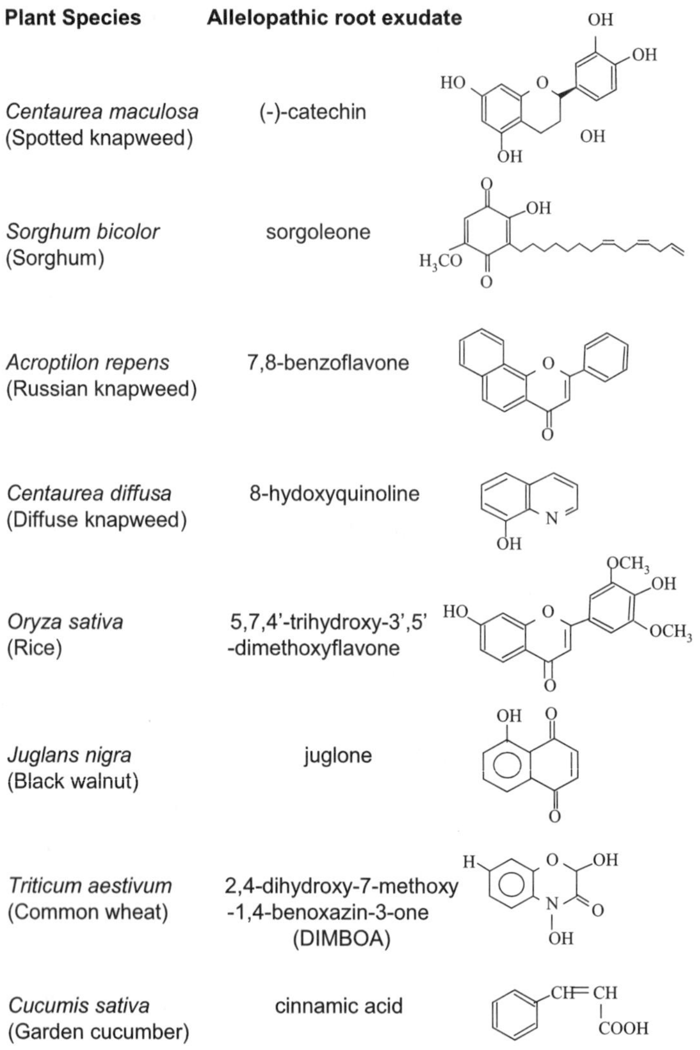

Plant Species	Allelopathic root exudate	
Centaurea maculosa (Spotted knapweed)	(-)-catechin	
Sorghum bicolor (Sorghum)	sorgoleone	
Acroptilon repens (Russian knapweed)	7,8-benzoflavone	
Centaurea diffusa (Diffuse knapweed)	8-hydroxyquinoline	
Oryza sativa (Rice)	5,7,4'-trihydroxy-3',5' -dimethoxyflavone	
Juglans nigra (Black walnut)	juglone	
Triticum aestivum (Common wheat)	2,4-dihydroxy-7-methoxy -1,4-benoxazin-3-one (DIMBOA)	
Cucumis sativa (Garden cucumber)	cinnamic acid	

FIGURE 11.2 Phytotoxic compounds in the root exudates of some allelopathic plants. For some species (e.g., *Centaurea* maculosa, *Sorghum bicolor*, and *Juglans nigra*), the phytotoxin shown makes up a large portion of the root exudates and is the only phytotoxin identified in the root exudates. For other species (e.g., *Oryza sativa*, *Triticum aestivum,* and *Cucumis sativa*), the phytotoxin shown is one of several identified in the root exudates.

(±)-catechin in soil in two *C. maculosa* populations [33], and another study that included a wide range of sites sampled on a number of dates, detected soil (±)-catechin in only one site and on only one sampling date (L.G. Perry, unpublished data). The reasons for these differences in results among studies are uncertain. All studies to date of soil (±)-catechin concentrations have examined bulk soil rather than rhizosphere soil, and therefore they may not have evaluated (±)-catechin concentrations under the most relevant circumstances (see Chapter 1). A better understanding of (±)-catechin dynamics in the rhizosphere is needed to evaluate whether and when (±)-catechin is present at sufficient concentrations in root–root interactions to influence *C. maculosa*'s neighbors.

The exact cellular target of (±)-catechin has not yet been identified, but many of the initial physiological responses to (±)-catechin in susceptible plants are well understood, including several biochemical signals and changes in gene expression that appear to initiate root cell death [25]. In *Arabidopsis thaliana* and *Centaurea diffusa* Lam. roots, cytoplasmic condensation begins in the root tip and root elongation zone within 10 min of (±)-catechin treatment, and then travels up the main axis of the root to mature root tissues within 1 h. Fluorescent viability staining with fluorescein diacetate indicates that cell death begins in the root tip soon after cytoplasmic condensation, within 15 min of (±)-catechin treatment and reaches mature root tissue within 1 h. Bais et al. [25] examined the production of biochemical signals in *A. thaliana* and *C. diffusa* roots immediately after (±)-catechin treatment, including production of reactive oxygen species (ROS) and fluctuation in cytoplasmic calcium and pH. ROS are strong electron donors and are involved in signal transduction processes leading to numerous plant physiological responses to stress, including hypersensitive responses to pathogen infection leading to cell death [35]. Cytoplasmic calcium signals are also often induced by plant stress and are involved in plant responses to high salinity, heavy metals, and oxidative damage [36,37]. Fluctuations in cytoplasmic pH occur in conjunction with signal transduction processes in plants and are involved in regulation of root growth [36,38]. ROS production, cytoplasmic calcium concentrations, and cytoplasmic pH were all strongly influenced by (±)-catechin treatment in susceptible plant roots [25]. ROS signals, visualized using the fluorescent dye 6-carboxy-2′,7′-dichlorodihydrofluorescein diacetate-di(acetoxymethly ester), increased 12-fold in the root tip within 10 sec of (±)-catechin treatment. Following the same pattern as cell death in response to (±)-catechin, ROS production first increased in the root tip, then in the elongation zone, and then traveled in a wave up the main axis of the root. Cytoplasmic calcium concentrations, visualized using the fluorescent dye indol-1, increased substantially in the root tip within 30 sec of (±)-catechin treatment, and then dissipated within 10 min of (±)-catechin treatment. Cytoplasmic pH declined from 7.2 to 5.6 within 15 to 20 min of (±)-catechin treatment, perhaps indicating loss of membrane function in conjunction with cell death. Treating susceptible plant roots with an antioxidant, ascorbic acid, prevented the induction of ROS production, increased cytoplasmic calcium, and cell death [25], suggesting that cell death in response to (±)-catechin may result from oxidative stress similar to hypersensitive responses to pathogen infection [35].

Bais et al. [25] used a 12,000-gene oligoarray to examine gene expression in (±)-catechin-treated *Arabidopsis thaliana* plants relative to untreated *Arabidopsis* plants, 10 min, 1 h, and 12 h after treatment. A number of genes associated with oxidative stress were upregulated within 1 h of (±)-catechin treatment, including glutathione transferase (GST), monooxygenase, lipid transfer protein, heat shock protein, DNA-J protein, and blue copper-binding protein [25], suggesting further that oxidative stress may play an important role in (±)-catechin phytotoxicity. Genes involved in phenylpropanoid and terpenoid phytoalexin pathways, some of which produce enzymes that serve as antioxidants [39], were also upregulated 1 h after treatment. Importantly, 10 other genes were strongly upregulated within the first 10 min of (±)-catechin treatment, and then not upregulated 1 h and 12 h after (±)-catechin treatment, suggesting that these genes may be involved in the initial plant responses to (±)-catechin that lead to cell death [25]. Several of these genes are associated with calcium signaling and oxidative stress, and four are of unknown function. Further examination of the role of these genes in plant responses to (±)-catechin may yield insights into the genetic basis of (±)-catechin phytotoxicity and (±)-catechin resistance.

Another example of plants that rhizosecrete potent phytotoxins is the *Sorghum* genus. The *Sorghum* genus includes important crop and cover crop species such as *S. bicolor* L. and *S. sudanese* (Piper) Stapf, and important weeds such as *S. halepense* (L.) Pers. Most *Sorghum* species rhizosecrete large quantities of the phytotoxin sorgoleone (2-hydroxy-5-methoxy-3-[(8′Z, 11′Z)-8′, 11′, 14′-pentadecatriene]-*p*-benzoquinone) (Figure 11.2). HPLC analyses suggest that sorgoleone is the most abundant compound in *Sorghum* spp. root exudates [40]. Sorgoleone production by *Sorghum* spp. can be very high. Most *Sorghum* species produce between 1.3 to 1.9 mg g^{-1} of fresh roots, whereas *S. halepense* produces as much as 14.8 mg g^{-1} [40]. *Sorghum* spp. also produce a number of other compounds similar to sorgoleone, with similar activities, but in smaller quantities that also may play a role in *Sorghum* spp. allelopathy [40,41]. Sorgoleone appears to be synthesized in the endoplasmic reticulum of *Sorghum* spp. root hairs, and then transported to between the cell wall and plasmalemma of the root hairs prior to exudation [42]. An exogenous application of 80 μg g^{-1} sorgoleone to soil substantially reduces the growth of susceptible plants, indicating that sorgoleone is phytotoxic under relatively realistic conditions [42]. However, *Sorghum* spp. rhizosphere and bulk soil concentrations of sorgoleone have not been determined. Exogenously applied sorgoleone degrades substantially in soil within the first week after application, but can persist at low concentrations in soil for several weeks [42], and might accumulate in soils if continuously exuded at high concentrations. A gene involved in sorgoleone biosynthesis was recently identified [43], which may create opportunities to use transgenic plants to examine directly the importance of sorgoleone production to outcomes of plant interference.

Sorgoleone appears to influence the physiology of susceptible plants in a number of ways. Numerous studies suggest that sorgoleone inhibits electron transport in both photosynthesis and respiration in susceptible plants [6], although one recent study found no effect of sorgoleone on photosynthesis [44]. In addition, sorgoleone inhibits the enzyme hydroxyphenylpyruvate dioxygenase in *Arabidopsis thaliana*, which is involved in carotenoid biosynthesis [45]. Sorgoleone also disrupts photosystem II (PSII), which is responsible for organizing the ligands that bind the pigments and other cofactors involved in photosynthesis, but this effect will be relevant only if sorgoleone is transported *in planta* from soil to leaf tissue [46]. As an electron acceptor, sorgoleone may also disrupt redox reactions in the root plasma membrane [44], although this effect of sorgoleone has not been demonstrated. Finally, sorgoleone substantially reduces H$^+$-ATPase activity in *Zea mays* L. root microsomal membranes [44], as is discussed in Chapter 6. H$^+$-ATPase activity in roots is necessary for maintenance of the electrochemical gradients in the rhizosphere that drive nutrient uptake. Sorgoleone inhibition of H$^+$-ATPase activity likely reduces nutrient and water uptake in susceptible plants, perhaps also accounting for effects of sorgoleone on photosynthesis and respiration [44].

Plant species often vary substantially in susceptibility to particular phytotoxins [31,47]. However, the mechanisms through which many species resist the effects of phytotoxic allomones have not yet been determined. Some plants may resist phytotoxic allomones via sequestration or secretion of the phytotoxins once they are taken up [48]. Other plants detoxify phytotoxins by altering the chemical structure of the compounds enzymatically. For example, a number of plants detoxify benzoxazinoid phytotoxins via glucosylation, or the attachment of one or more sugar moieties to the compound [49]. Two phytotoxic benzoxazinoids, 2,4-dihydroxy-1,4-benzoxazin-3-one (DIBOA) and 2,4-dihydroxy-7-methoxy-1,4-benoxazin-3-one (DIMBOA) are rhizosecreted by both *Secale secale* L. and *Elytrigia repens* (L.) Gould [50,51]. The detoxification mechanisms used by *Zea mays* L. for DIBOA, DIMBOA, and benzoxazolin-2(3*H*)-one (BOA), a degradation product of DIBOA, are well understood. DIBOA, when released into the soil, is quickly converted to BOA. BOA, when incubated with *Z. mays*, undergoes N-glucosylation, ultimately resulting in production of isomeric 1-(2-hydroxyphenylamino)-1-deoxy-β-glucoside 1,2,-carbamate, a less toxic compound [52]. Two glucosyltransferases in *Z. mays*, *Bx8* and *Bx9*, have been identified as highly specific to DIBOA and DIMBOA [53]. Transgenic *Arabidopsis thaliana* plants containing the *Z. mays* genes for *Bx8* and *Bx9* are resistant to DIBOA and DIMBOA, indicating an important role of these enzymes in phytotoxin detoxification [53]. The glucosylation products are apparently released in *Z. mays* root exudates and then may be degraded microbially [52]. Mechanisms of plant resistance to other well-known phytotoxic allomones, such as (±)-catechin and sorgoleone, are not yet understood but may involve similar processes.

Because plants are able to develop resistance to phytotoxins, the importance of allelopathy in structuring plant communities may often be limited to only some interspecific interactions [54]. In particular, the effectiveness of a phytotoxin against a particular species is likely to depend on the historical selection pressures for resistance to the phytotoxin. Plant species with relatively little previous exposure to a phytotoxin are less likely to be resistant and, therefore, more likely to be displaced by the phytotoxin than more experienced species. Thus, allelopathy may be particularly effective for exotic allelopathic plants in their invaded ranges, where the native species have not been exposed previously to the phytotoxic allamone (i.e., the novel weapons hypothesis, *sensu* Callaway and Aschehoug [55]). Several studies comparing the effects of exotic plants and their phytotoxins on congeners, from their native and exotic ranges, have supported the novel weapons hypothesis [56]. The invasive plant *Centaurea diffusa* Lam. reduced the growth of three naive species from its invaded range significantly more than three congeneric experienced species from its native range [55]. Adding activated carbon to the soil to adsorb organic compounds removed this difference in part by improving the growth of the naive species, suggesting that *C. diffusa* allelochemicals had a greater negative effect on the naive species than on the experienced congeners. North American populations of the invasive plant *Alliaria petiolata* (Bieb.) Cavara and Grande also inhibited a naive North American species more than an experienced European congener, but European populations of *A. petiolata* did not affect the species differently, only partially supporting the novel weapons hypothesis [57].

Recent studies have also suggested that plants may rapidly evolve partial resistance to novel phytotoxic allamones following invasions by allelopathic exotic species. Callaway et al. [58] found that offspring from populations of North American grasses that had experienced *Centaurea maculosa* invasion were less inhibited by interactions with *C. maculosa* than offspring from populations of North American grasses that had not encountered *C. maculosa*. Although Callaway et al. [58] did not attempt to rule out maternal effects as a cause of this trend, their results suggest that North American grasses may be evolving resistance to *C. maculosa* allelopathic interference. Similarly, Mealor et al. [59] found that offspring from populations of one North American species that had experienced invasions by *Acroptilon repens* (L.) DC., another exotic species that may be allelopathic [27], were less inhibited by interactions with *A. repens* than offspring from populations that had not previously encountered *A. repens*. Mealor et al. [59] compared offspring from plants grown in a common garden, thus demonstrating evolution of increased resistance to *A. repens* in one species. The apparent importance of historic coexistence in these potentially allelopathic interactions suggests an important role of coevolution in determining effects of chemical communication.

B. AUTOINHIBITION

In addition to the interspecific effects of phytotoxic root exudates, some root exudates have intraspecific effects (i.e., autotoxicity and autoinhibition). Distinguishing between autotoxicity, which negatively affects population dynamics, and autoinhibition, which positively affects population dynamics, is difficult and has not yet been achieved for any plant species. Consequently, we use autoinhibition to refer to both autotoxicity and autoinhibition. Autoinhibition has long been suggested as a mechanism to explain declining yields in agricultural fields planted with the same crop species over many years [60]. Whereas autoinhibition in agricultural crops has often been attributed to phytotoxins in decomposing leaf and stem tissue [60], some phytotoxins responsible for autotoxicity are also released in crop root exudates. For example, *Asparagus officinalis* L. root exudates are toxic to *A. officinalis* seedlings [61] and may increase *A. officinalis* susceptibility to the fungal pathogen *Fusarium oxysporum* spp. *Asparagi* [62]. Similarly, cinnamic acid and perhaps other compounds in *Cucumis sativus* L. root exudates reduce photosynthesis and uptake of several important nutrients by *C. sativus* plants [20,63].

Autoinhibition may also play a role in natural plant communities, particularly among weedy species. Indeed, several invasive plants reported to be allelopathic are also autoinhibitory, including *Amaranthus palmeri* S. Wats. [64,65], *Centaurea maculosa* Lam. [24,34], *Cirsium arvense* (L.) Scop. [60,66], *Elytrigia repens* (L.) Gould [51,67], *Lantana camara* L. [68,69], and *Parthenium hysterophorus* L. [70]. For most of these species, it is not known whether the phytotoxins responsible for allelopathy are the same as those responsible for autoinhibition. Further, as in agricultural examples, most known instances of autoinhibition in natural communities involve phytotoxins in leaf volatiles or decomposing plant residues [60]. However, phytotoxic root exudates involved in allelopathy can also be autoinhibitory. Very high concentrations of (±)-catechin, the phytotoxic root exudate produced by *C. maculosa*, inhibit *C. maculosa* seedling root elongation by as much as 75% [34]. Provided that (±)-catechin is held in solution at high concentrations, it also induces *C. maculosa* seed dormancy, reducing seed germination by as much as 50%.

Whereas autoinhibitory chemicals have been identified in many plants, proving that these chemicals explain the population dynamics of the plants that produce them under natural conditions can be difficult. Experiments to examine the ecological role of autotoxins by manipulating autotoxin behavior or abundance in the field have rarely been done. However, there is evidence to suggest that autoinhibition influences *Centaurea maculosa* population dynamics [34]. In well-established *C. maculosa* populations in North America, *C. maculosa* adults are often widely spaced, separated by unoccupied space or by a few small seedlings. Perry et al. [34] found that adding activated carbon, which adsorbs organic compounds, to soil around adult *C. maculosa* plants in the field increases *C. maculosa* seedling density, suggesting that autoinhibition is one of the factors that limits *C. maculosa* seedling establishment in established populations. Whether (±)-catechin concentrations in *C. maculosa* populations are sufficiently high to explain these field results is uncertain. Early measurements of soil (±)-catechin suggested that (±)-catechin concentrations could be very large [25,71], but more recent studies have been unable to repeat these results (L. G. Perry, unpublished data) [33]. (±)-Catechin in the rhizosphere may still be responsible for *C. maculosa* autoinhibition but may not be detectable in bulk soil, or a different compound may be responsible for *C. maculosa* autoinhibition.

Plants may benefit from autoinhibitory chemical signals in several ways. Adults that produce autoinhibitory root exudates may reduce the establishment of intraspecific competitors [72]. Similarly, seedlings that rhizosecrete autoinhibitors may benefit from reducing interference from establishing siblings or other intraspecific neighbors [73]. In instances where autoinhibitors delay seedling establishment by inducing seed dormancy, autoinhibitors produced by adults may serve as a positive signal to protect their offspring from establishing in areas with intense intraspecific competition [70]. In some cases, autoinhibition also may occur simply because plants produce phytotoxins designed to inhibit interspecific competitors (i.e., allomones) but invest only in partial resistance to the phytotoxins to avoid the metabolic cost of full resistance.

Whereas many cases of autoinhibition have been reported, many allelopathic plants also appear to at least partially avoid the negative effects of their phytotoxins. The mechanisms through which plants resist their own phytotoxins are not fully understood and probably vary considerably among species and phytotoxins. Some plants appear to avoid autoinhibition by sequestering phytotoxins in vacuoles or specialized structures such as trichomes [74], although this mechanism of resistance has not been demonstrated for phytotoxic root exudates. Other allelopathic plants appear to avoid autoinhibition by producing and releasing nontoxic compounds that are then degraded enzymatically, microbially, or oxidatively to produce phytotoxic compounds. For example, *Polygonella myriophylla* (Small) Horton leaves accumulate arbutin, a glycoside of the phytotoxin hydroquinone [75]. Arbutin is not phytotoxic and therefore does not inhibit *P. myriophylla*. However, microbial degradation of arbutin from *P. myriophylla* leaves in the soil releases the phytotoxins hydroquinone and benzoquinone, leading to inhibition of neighboring plants [75]. Glycosides may play a similar role in plant resistance to autoinhibition from phytotoxic root exudates.

C. Host–Parasitic Plant Interactions

Secondary metabolites in root exudates also may be used by parasitic plants to detect potential hosts and induce infection. Communication between parasitic plants and their hosts involves a complex combination of chemical signals from both the host and the parasite that induces the development of structures for transferring water, nutrients, and carbon (Figure 11.3). Approximately 4000 plant species have been identified as facultative or obligate parasites [76]. The signaling pathways involved in parasite establishment and host infection are particularly well understood for

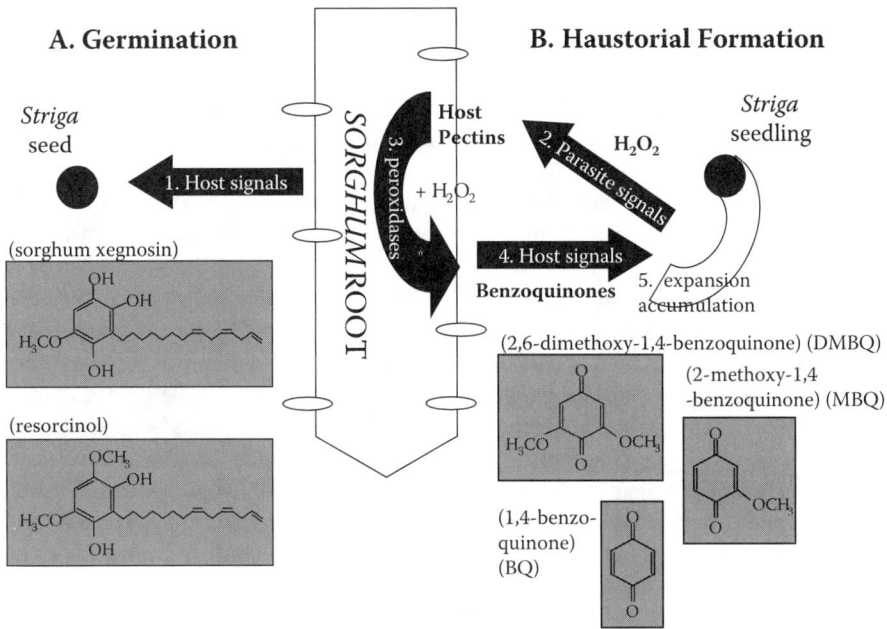

FIGURE 11.3 Chemical signals involved in interactions between the parasitic plant *Striga asiatica* and a common *Striga* host, *Sorghum bicolor*. Chemical signals mediate both *Striga* germination and *Striga* formation of haustoria, a specialized root structure involved in host infection. *Striga* seed germination is induced near to host roots by (1) host root signals such as sorghum xegnosin. Other host root exudates such as resorcinol stabilize sorghum xegnosin. *Striga* haustorial formation is induced by a complex series of signals. (2) *Striga* seedlings release hydrogen peroxide, which (3) activates host peroxidases, transforming host pectins into benzoquinones such as 2,6-dimethyl-1,4-benzoquinone (DMBQ), 2-methoxy-1,4-benzoquinone (MBQ), and 1,4-benzoquinone (BQ). (4) The host benzoquinones act as chemical signals, (5) altering *Striga* gene expression and inducing accumulation of expansin proteins in the root tip, ultimately leading to haustorial formation.

the plants *Striga asiatica* (L.) Kuntze and *S. hermonthica* (Del.) Benth [77]. However, other parasitic members of the Scrophulariaceae family (e.g., *Triphysaria versicolor* Fisch. & C.A. May and *Orobanche* spp.) are thought to participate in similar chemical interactions with their hosts [78].

The *Striga* species are obligate plant parasites unable to survive for more than 5 d after germination without attachment to a host [77]. The *Striga* seeds are extremely small and therefore have limited carbohydrate reserves for initial growth, although they can survive in the soil for decades before germinating [77]. Consequently, it is critical for the survival of the *Striga* seedling that their seeds germinate very near to a potential host. To ensure that germination is limited to areas near potential hosts, *Striga* seeds germinate only when in the sustained presence of relatively high concentrations of particular host root exudates [79]. The many potential hosts of *Striga* do not all produce the same signal for *Striga* germination. Rather, the compounds responsible for inducing *Striga* germination differ among host species and do not even appear to share any particular inducing structure. To date, only one *Striga* germination inducer in host root exudates, sorghum xenognosin (SXSg), has been isolated and characterized (Figure 11.3). Similarities in the chemical behavior of SXSg and other reagents that induce *Striga* germination suggest that *Striga* germination inducers participate in oxidation/reduction reactions that generate aryloxy radical intermediates, which, might in turn, induce germination [77]. However, this proposed mechanism for *Striga* germination induction has not yet been proven.

Interestingly, both SXSg and a *Striga* germination inducer produced by *Zea mays* L. are relatively unstable compounds. *Striga* may benefit from the chemical instability of these germination inducers, because unstable inducers are less likely to accumulate in the soil indicating incorrectly that a host is nearby. SXSg decomposes so rapidly in aqueous solution that it would not be expected to travel the distance from sorghum roots to nearby *Striga* seeds or to be present for sufficient time (10 to 12 h) to induce *Striga* germination [80]. However, SXSg is not as unstable when in *Sorghum* root exudates as it is when in pure solution. Another less abundant compound in *Sorghum* root exudates, resorcinol, stabilizes SXSg sufficiently to allow it to induce *Striga* germination (Figure 11.3) [80]. Resorcinol has no direct effect on *Striga* germination but enhances the effect of SXSg [80]. The role of resorcinol in *Striga* germination emphasizes the potential importance of combinations of chemicals in root exudates.

Once *Striga* germinates, it must quickly produce specialized root structures (i.e., haustoria) necessary to penetrate the epidermis of the host root and connect to the host vascular system. Haustorial development in *Striga* occurs in approximately 1 d and involves cessation of root elongation, radial meristematic swelling, and haustorial hair formation at the bulbous root tip [77]. A complex series of chemical signals (Figure 11.3) and genetic responses are involved in *Striga* haustorial formation to ensure that haustoria develop in the proper location relative to the host [77,81]. Whereas *Striga* germination involves host-derived signals, *Striga* haustorial development appears to begin with a parasite-derived signal. According to the current conceptual model of *Striga* haustorial development, *Striga* begins the sequence of signals with constitutive release of hydrogen peroxide at the parasite root tip. The importance of hydrogen peroxide in haustorial development was demonstrated by treating *Striga* roots with catalase, which degrades hydrogen peroxide and prevents haustorial development [82]. Hosts also produce hydrogen peroxide upon wounding or other forms of stress but do not produce hydrogen peroxide constitutively.

Hydrogen peroxide is a necessary cofactor for many peroxidases [77]. The parasite peroxides induce root peroxidases to act on pectins in the host cell walls resulting in the oxidative release of benzoquinones [81]. Both the host and the parasite produce peroxidases that could generate benzoquinones, but the hosts produce a greater abundance of peroxidases, suggesting that the peroxidases involved may most often originate in the host [82]. The pectins that are reduced to benzoquinones are known to originate in the host, because haustorial development does not occur in the absence of host material, even when hydrogen peroxide and parasite peroxidases are present.

The benzoquinones produced by the peroxidases are thought to induce haustorial development in parasite roots via redox reactions with a receptor in the parasite root. Two lines of evidence suggest that redox reactions are involved. First, the benzoquinones that initiate haustorial development possess similar oxidative reactivity (electromotive potential, or Em, between −280 and +20 mV), which is consistent with the hypothesis that the benzoquinones act as single electron carriers [83]. Second, the semiquinone intermediate involved in the redox reactions would be expected to bind to a redox binding site on the parasite root. To demonstrate that binding was possible between a semi-quinone intermediate and the binding site, Smith et al. [83] generated a synthetic semiquinone intermediate with Em > 20 mV. The synthetic intermediate inhibited haustorial formation, perhaps by binding to the parasite receptor.

The biochemical and physiological processes through which *Striga* haustoria develop following initiation by host benzoquinones are not fully understood. However, the benzoquinones involved are known to directly or indirectly induce expression of two genes coding for protein expansins in *Striga* roots [84]. Benzoquinones without the necessary oxidative reactivity or chemical structure to bind to the parasite receptor block upregulation of the expansins genes, indicating that the specific host benzoquinones that initiate haustorial development are necessary to induce the observed changes in gene expression [84]. Expansins facilitate plant cell growth by disrupting the hydrogen bonds between cellulose microfibrils and the polysaccharide matrix in the cell wall, thus reducing cell wall resistance and allowing cell expansion [85]. The gene saExp3, which codes for the expansin present in *Striga* seedlings during initial root development, is downregulated in response to host benzoquinones, whereas two other genes, saExp1 and saExp2, that code for expansins are upregulated [84]. saExp1 and saExp2 are relatively unusual expansin genes and may play a role in the unique structure of haustoria. Transcripts of saExp1 and saExp2 accumulate in *Striga* root cells in a linear manner to a threshold concentration over the course of several hours. The exact time necessary is dependent on the benzoquinone signal identity and concentration [84]. Once the threshold concentration of expansin transcripts is reached, the root tip begins to swell and haustorial development begins. Interestingly, if the benzoquinone signal is removed temporarily, before accumulation of the expansin transcripts is complete, the transcripts appear to remain present for a few hours, allowing transcript accumulation to continue as though undisturbed when the signal is returned. If, however, the signal is removed for longer periods, signaling definitively that a host root is not present, the expansin transcripts decay and haustorial development must be reinitiated [84].

Additional chemical signaling may occur between plant parasites and their hosts as the process of infection continues. For example, treatment of *Triphysaria versicolor* with host root exudates induces expression of a gene for enzyme asparagine synthetase, which may play a role in facilitating nitrogen transfer from the host to the parasite after infection is complete [86]. Numerous other genes that may be involved in benzoquinone detoxification, signal transduction, and haustorial initiation, as well as genes of unknown function, are also induced in *T. versicolor* upon exposure to the benzoquinones that initiate haustorial development [78,86]. Further examination of these genetic responses to chemical signals in host–parasite interactions will likely yield further insights into the biochemical mechanisms of host infection.

Host resistance to *Striga* infection also appears to be mediated by chemical signals in root exudates. Many nonhosts of parasitic plants appear to resist infection via hypersensitive responses that lead to cell death in the area of infection and block resource supply to the parasite [87,88]. However, at least one nonhost of *S. hermonthica* appears to resist *Striga* infection by producing chemical signals that inhibit haustorial development [89]. When *Striga* attaches to *Tripsacum dactyloides* L., it is able to develop a connection with *T. dactyloides* xylem but haustorial tissue differentiation is impaired. Further, once *Striga* has attempted an attachment to *T. dactyloides* it loses the ability to develop normal secondary haustoria with *Zea mays*, a common *Striga* host. These results suggest that *T. dactyloides* roots produce a chemical signal that causes systemic inhibition of haustorial development in *Striga*. Among *Striga* hosts, there is also considerable variation in susceptibility to *Striga* infection. Some relatively resistant *S. bicolor* genotypes appear to limit *Striga* infection by producing low concentrations of *Striga* germination inducers [90,91]. Others may resist *Striga* infection via low production of the chemical signals involved in haustoria development or production of *Striga* germination and growth inhibitors [59,92,93].

III. NEGATIVE ROOT–MICROBE COMMUNICATION

A. PLANT ANTIMICROBIAL SIGNALS AND RESISTANCE TO SOIL PATHOGENS

The rich diversity of secondary metabolites in plant root exudates is believed to have arisen in part from selection for improved defense against pathogenic microbes and herbivores [94,95]. Numerous antimicrobial plant root exudates are secreted into the soil constitutively with the potential for strong negative effects on microbial populations in the rhizosphere (Figure 11.4). For example, *Ocimum basilicum* L. roots exude rosmarinic acid (α-*o*-caffeoyl-3-4-dihydroxyphenyllactic acid) [96], an antimicrobial agent, that likely allows *O. basilicum* to resist infection by a variety of soilborne microorganisms. Rosmarinic acid (RA) has been observed to damage the cytoskeleton of the pathogenic fungus *Aspergillus niger,* resulting in broken interseptas in the mycelia, flows of nuclei in the hyphal tips, and convoluted cell surfaces [96]. Further, RA increases spatial division and condensation of DNA in the bacterial pathogen *Pseudomonas aeruginosa*, suggesting that *P. aeruginosa* may respond to RA-induced cell damage with rapid cell division [96]. Similarly, *Centaurea maculosa* Lam. roots exude (+)-catechin, which inhibits an array of soilborne bacteria and fungi [24,97], perhaps mediating *C. maculosa* disease resistance. Constitutively-produced root exudates also appear to play a role in *Gladiolus* L. resistance to the fungal pathogen *Fusarium*

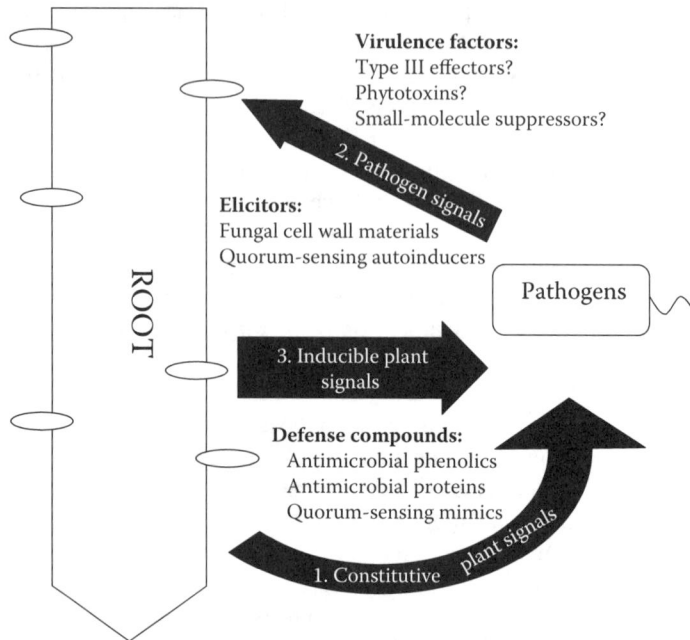

FIGURE 11.4 Chemical-mediated interactions between plant roots and pathogenic soil microbes. (1) Plants rhizosecrete numerous defense compounds constitutively, including antimicrobial phenolics and proteins and quorum-sensing mimics. (2) Pathogenic soil microbes also secrete compounds into the rhizosphere. Some pathogenic signals probably mediate pathogen infection. Others, including fungal cell wall materials and autoinducer compounds involved in quorum sensing, elicit plant defense responses. (3) Plant defense responses to pathogen signals include physiological changes, increased production of constitutively produced antimicrobial compounds, and production of additional antimicrobial signals.

oxysporum f. sp. gladioli. Root exudates from a resistant *Gladiolus* cultivar inhibit microconidial germination of *F. oxysporum*, whereas root exudates from a susceptible cultivar do not affect *F. oxysporum* germination [98]. Root exudates from the resistant cultivar contain greater relative amounts of aromatic phenolic compounds and lower relative amounts of carbonylic and aliphatic compounds than root exudates from the susceptible cultivar, perhaps accounting for their different effects on *F. oxysporum* germination [98].

In addition to antimicrobial phenolic compounds, some plant roots exude antimicrobial proteins into the rhizosphere [99,100]. For example, hairy roots of *Phytolacca americana* L. secrete numerous defense proteins, including a ribosome-inactivating protein, PAP-H, and several pathogenesis-related (PR) proteins (β-1,3-glucanase, chitinase, and protease) [101,102]. Ribosome-inactivating proteins (RIPs) are produced by many higher plants and inhibit protein synthesis in plants and microbes through *N*-glycosidase activity [103,104]. PAP-H, the RIP in *P. americana* root exudates, is active against fungal pathogens, including *Rhizoctonia solani* and *Trichoderma reesei*, which cause root rot [101,102]. The PR proteins in *P. americana* root exudates may facilitate movement of PAP-H into *R. solani* and *T. reesei* cells by damaging the fungal cell walls [101,102], illustrating the potential for rhizosecreted compounds to act in concert in defense against pathogens.

Some root exudates also mimic chemical signals involved in microbial communication, disrupting the organization of microbial populations. Gram-negative and Gram-positive bacteria, including plant pathogens such as *Erwinia* spp., *Pseudomonas* spp., and *Agrobacterium* spp., participate in quorum-sensing behavior in which small diffusible signaling molecules mediate cell–cell communication [10]. The signaling molecules for Gram-negative bacteria are termed *autoinducers*, which are typically acylated homoserine lactones (AHLs), whereas Gram-positive bacteria employ peptide-signaling molecules for quorum sensing [10]. When quorum sensing signals reach a threshold concentration in dense bacteria populations, the signals activate transcription factors that induce specific genes in the bacteria [10]. Thus, bacterial gene expression is controlled by cell density, allowing populations to behave as single units. This mode of density-dependent behavior mediates diverse processes in prokaryotes, including production of virulence factors, bioluminescence, sporulation, swarming, antibiotic biosynthesis, and plasmid conjugal transfer [105]. Interestingly, some higher plants, including *Pisum sativum* L., *Glycine max* (L.) Merr., *Oryza sativa* L., *Solanum lycopersicum* L., *Coronilla* L. spp., and *Medicago truncatula* Gaertner, secrete quorum sensing mimics that stimulate AHL reporters and thus interfere with bacterial communication [106,107]. For example, when the bacteria *Chromobacterium violaceum* is exposed to *P. sativum* root exudates, two processes that are typically regulated by AHLs, antibiotic violacein synthesis and protease activity, are strongly inhibited. However, *C. violaceum* growth is not affected, indicating that the effect of the root exudates is specific to quorum-sensing processes [106,107]. Mathesius et al. [108] used bacterial reporter genes to examine effects of *M. truncatula* root exudates on AHL-regulated processes. Many components of *M. trunculata* root exudates stimulate AHL-regulated bioluminescence, indicating the presence of numerous quorum sensing mimics. The plant compounds that serve as quorum sensing mimics have not been identified but are probably different from known bacterial AHLs, because they possess distinct solvent partitioning properties [108]. Whereas the relationship between bacterial quorum-sensing and plant metabolites is still largely unexplored, plants may use quorum sensing mimics to inhibit pathogenic bacteria that rely on quorum sensing for host infection.

Experiments using activated carbon, which adsorbs root-exuded organic compounds, have indicated that constitutively rhizosecreted compounds can play an important role in plant defense against soilborne microbial pathogens. The model plant *Arabidopsis thaliana* rhizosecretes a variety of potentially antimicrobial compounds, including butanoic acid, *trans*-cinnamic acid, *o*-coumaric acid, *p*-coumaric acid, ferulic acid, *p*-hydroxybenzamide, methyl *p*-hydroxybenzoate, 3-indolepropanoic acid, syringic acid, and vanillic acid [109,110]. These compounds exhibit a wide range of antimicrobial activity against both soilborne bacteria and fungi at the concentrations detected in *A. thaliana* root exudates [109,110] and may protect *A. thaliana* from pathogenic infection. Several strains of the bacterial pathogen *Pseudomonas syringae* have been identified that do not normally infect *A. thaliana*. However, when *A. thaliana* plants inoculated with the nonpathogenic *P. syringae* strains are grown in soil with activated carbon added, they exhibit pathogen colonization, disease symptoms, and plant mortality [111], suggesting that organic compounds in the *A. thaliana* rhizosphere normally prevent these microbial infections.

It has been suggested that root exudation of antimicrobial proteins and secondary metabolites may be particularly concentrated at the root tip [112]. For example, PAP-H, the RIP exuded by *Phytolacca americana* roots, is present within the plant mainly in the cell walls of root border cells [101,102]. Root border cells occur around the root tip and are programmed to detach from roots and continue living in the rhizosphere. Border cells are thought to protect root meristematic tissue from microbial pathogens both by exuding an array of biochemicals that influence microbial behavior and survival [113,114] and by separating from the root and serving as host-specific "decoys" [114,115]. Most plant root infections are initiated in the root elongation zone rather than the root tip, suggesting that plant defenses may be particularly effective at the root tip where border cells are present [116]. Further, *Medicago sativa* L. mutants that express antisense mRNA, leading to slowed cell cycling and substantially reduced border cell production, are significantly more susceptible than the wild type to root tip infection by the fungal pathogen *Nectria haematococca* [117], suggesting an important role of border cells in plant root defense. However, the locations of release for most plant root exudates are unknown and may often occur elsewhere along the root.

B. Microbial Signals and Induced Plant Disease Responses

Chemical signals produced by soilborne microbes often have negative effects on other microbes [9,118], and probably also have negative effects on plant roots including increasing root susceptibility to infection (Figure 11.4). However, most of the information available on the chemical signals involved in plant pathogen virulence has been gained from studies of leaf infection. These aboveground chemical signals include type III effectors, which are proteins injected into plants by pathogens using a needlelike, protein-based structure (i.e., type III secretion), toxins from phytopathogenic bacteria [119,120], host-selective toxins and small molecule suppressors from phytopathogenic fungi [121,122], and suppressors of posttranscriptional gene silencing from plant viruses [123]. Unfortunately, research has not yet been done to determine whether similar chemical signals are produced by soilborne pathogens to increase root susceptibility to infection.

Instead, much of the research on the effects of chemical signals from pathogenic soil microbes has focused on plant detection of potential pathogens and induced defense responses. Plants exposed to chemical signals from soilborne microbes respond by increasing production of constitutively rhizosecreted compounds or by producing novel root exudates, many of which are antimicrobial (Figure 11.4) [24,96,110,124,125]. These induced defense responses allow plants to invest in defense equipment when necessary and to invest in growth, reproduction, or other traits when defenses are unnecessary [126]. Many microbial signals that elicit plant root exudation do so for numerous plant species but typically trigger production of different compounds in different species [110]. The signals that induce root exudation of defense compounds are often highly general, apparently indicating the presence of abundant fungi or bacteria rather than the presence of a particular pathogen. However, specific elicitors that inform particular hosts of particular dangers are also likely to exist in specialized host–pathogen interactions.

Although little is known about belowground induced defense responses, several recent papers have documented antimicrobial and antifungal properties of elicited, rhizosecreted secondary metabolites [24,124]. The most frequently reported plant root elicitors are fungal cell wall materials. When exposed to fungal cell walls or other elicitors, both *Arabidopsis thaliana* and *Ocimum basilicum* L. roots exude an array of antimicrobial compounds not detected in constitutively produced exudates [96,110]. Whereas nonelicited *Arabidopsis* roots secrete 68 compounds, *A. thaliana* roots exposed to chitosan (a component of fungal cell walls), or to cell walls of the fungi *Phytophthora cinnamoni* and *Rhizoctonia solani*, secrete up to 289 compounds [110]. Further, *O. basilicum* hairy roots treated with cell wall material from the fungus *Phytophthora cinnamoni* produce 2.67 times more of the constitutively expressed antimicrobial compound RA than untreated plants [96]. *O. basilicum* RA exudation also increases with *in situ* attack by the pathogen *Pythium ultimum*. Similarly, treatment with hyphae from the pathogenic fungi *R. solani*, *Pythium aphanidermatum*, and *Nectria hematococca* induces production of pigmented naphthoquinones in the epidermal cells of *Lithospermum erythrorhizon* hairy roots, whereas pigment development is normally limited to root hairs and border cells in *L. erythrorhizon* [127]. Several of the elicited pigmented naphthoquinones exhibit strong negative effects on soilborne bacteria and fungi [127], suggesting a role for the compounds in preventing microbial infection.

Autoinducers (i.e., AHLs) involved in Gram-negative bacterial quorum sensing can also alter plant gene expression and root exudation [108]. Treatment of *Medicago truncatula* Gaertner with AHLs from symbiotic (*Sinorhizobium meliloti*) and pathogenic (*Pseudomonas aeruginosa*) bacteria significantly altered the relative abundance of over 150 proteins in *M. truncatula* roots. Among these proteins, several auxin-responsive and flavonoid synthesis proteins increased in response to the AHLs, and treatment with the AHLs activated an auxin-responsive promoter and three chalcone synthase promoters in *Trifolium repens* L. Further, treatment with the AHLs altered the chemical composition of *M. truncatula* root exudates, including inducing novel secretion of a number of quorum-sensing mimics, which may, in turn, disrupt bacterial communication.

C. Microbial Resistance to Plant Antimicrobial Signals

Plant defenses effectively prevent infections by most microbial pathogens. However, despite the many defense strategies employed by plants in the rhizosphere, most plants are susceptible to at least some soil pathogens. Microbial pathogens employ a number of strategies to reduce the efficacy of plant defenses, including enzymatic degradation of antimicrobial secondary metabolites, resistance to particular antimicrobial compounds, and efflux mechanisms that reduce intracellular accumulation of antimicrobial compounds [128]. Some pathogens also degrade host antimicrobial compounds to form products that suppress host defense responses [129], although this strategy has not yet been observed in soilborne pathogens.

In addition, some forms of microbial organization may allow soilborne pathogens to avoid the effects of plant signals. Biofilms, which are communities of bacterial cells that are adhered to one another or to other surfaces and are enclosed within an extracellular polymeric matrix [130], may sometimes act as shields to protect bacteria from antimicrobial root exudates. For example, RA, which is rhizosecreted by *Ocimum basilicum* L., inhibits planktonic cells of the bacterial pathogen *Pseudomonas aeruginosa* [131]. However, *P. aeruginosa* biofilms are resistant to RA [131]. During plant infection, *P. aeruginosa* forms a biofilm on *O. basilicum* roots, allowing it to infect and eventually kill the plant.

Some specialized pathogenic bacteria may also counter plant defense strategies by reducing plant root exudation of antimicrobial compounds. Recently, Bais et al. [111] found that some pathogenic bacteria are capable of blocking plant synthesis or exudation of antimicrobial secondary metabolites. *Arabidopsis thaliana* rhizosecretes a number of antimicrobial secondary metabolites that together inhibit an array of pathogenic bacteria. Interestingly, whereas treatment with nonpathogenic strains of the bacteria *Pseudomonas syringae* increases *Arabidopsis* root exudation of these antimicrobial metabolites, treatment with a pathogenic strain of *P. syringae*, which is relatively resistant to *Arabidopsis* defense compounds, reduces *Arabidopsis* root exudation of these compounds. These results suggest that some plant pathogens may succeed in infecting their hosts through a combination of partial resistance to plant signals and inhibition of plant signal production [111].

IV. POSITIVE ROOT–ROOT COMMUNICATION

A. Herbivore Resistance

A recent study demonstrated that a root-exuded phytotoxin known to inhibit seedling growth of numerous plant species also reduces leaf tissue attractiveness to herbivores [132]. Aphids, which feed on plant sugars from phloem, were significantly less likely to choose to settle on *Hordeum vulgare* L. plants whose roots were exposed to root exudates collected from *Elytrigia repens* (L.) Desv. ex B.D. Jackson than on control *H. vulgare* plants. Aphids also rejected *H. vulgare* plants whose roots were treated with carboline, a known phytotoxin in *E. repens* root exudates. Carboline alone did not influence aphid behavior, indicating that aphid rejection was due to *H. vulgare* responses to carboline rather than to direct effects of the carboline. Aphid behavior upon exposure to odors from *H. vulgare* leaf tissue suggested that treatment with carboline altered the odor of *H. vulgare* leaves making them repellent to aphids. This positive effect of *E. repens* phytotoxins on potential competitors is likely not the purpose for which *E. repens* phytotoxins evolved. Carboline may induce secondary metabolite production in *H. vulgare*, including production of metabolites involved in herbivore resistance. Alternatively, carboline may be involved directly in *E. repens* and *H. vulgare* herbivore resistance. Such positive effects of plant-produced phytotoxins may have the potential to outweigh the negative effects.

In addition, some plants appear to use chemical signals in root exudates to coordinate population-wide herbivore defense responses [133]. Insect herbivory of aboveground plant tissue often induces the release of leaf volatiles that have a variety of functions, including induction of herbivore resistance and mediation of plant–plant and plant–insect communication [134]. However, at the same time, herbivory of aboveground plant tissue can induce root exudation of secondary metabolites. These root exudates may serve as chemical signals, communicating the presence of herbivores among plant neighbors. In particular, root exudates produced by herbivore-damaged plants have been shown to induce leaf volatile production in undamaged neighbors, attracting predators and parasitoids of the herbivores [135,136]. Thus, numerous plants may participate in reducing herbivore populations in response to herbivory on a single plant. For example, when *Vicia faba* L. plants are under attack by pea aphids (*Acyrthosiphon pisum*), the plants release volatiles from their leaves that attract an aphid parasitoid, *Aphidius ervi* [137]. Moreover, when undamaged *V. faba* plants were grown in soil with aphid-infested *V. faba* plants, or in water that previously held the roots of an aphid-infested *V. faba* plant, the undamaged plants became just as attractive to aphid parasitoids [136]. Similar results have also been reported for interactions between *Phaseolus lunatus* L., spider mites (*Tetranychus urticae*), and predatory mites (*Phytoseiulus persimilis*) [135]. Whereas such plant–plant communication is most likely a case of plants "eavesdropping" for their own benefit on the chemical signals of their less-fortunate neighbors, the resulting coordination of herbivore defense responses could have considerable effects on predator behavior and herbivore populations. To date, much of the research on these interactions has focused on the chemical signals involved in the plant–predator communication and in volatile-mediated aboveground plant–plant communication, rather than in belowground plant–plant communication [134]. The specific elicitors in the root exudates of herbivore-damaged plants that lead to responses in undamaged plants have not been identified nor are their biochemical modes of action yet understood. Additionally, most research to date on positive plant–plant interactions has examined plants in agricultural systems. Chemical-mediated communication may be more common and more important in coevolved, natural plant communities.

B. ROOT DETECTION AND NAVIGATION

Ecological theory predicts that in competition for soil resources, plant fitness will be greatest for individuals that concentrate root growth first in unoccupied space, second in space occupied by competitors' roots, and only last in space occupied by their own roots [138].

Accordingly, experimental studies indicate that chemical signals in the rhizosphere may mediate self-detection and detection of obstacles and competitor roots, perhaps facilitating root navigation. For example, chemical signals produced by *Pisum sativum* L. roots appear to limit *P. sativum* root growth in the vicinity of belowground obstacles. *P. sativum* roots growing in the direction of a root-shaped obstacle (a nylon string) either withered or grew shorter than roots growing away from the obstacle [139]. Adding activated carbon, which adsorbs organic compounds, to the soil allowed the roots to grow normally toward the obstacle, suggesting that accumulation of *P. sativum* auto-inhibitors reduced root growth by the obstacle, perhaps as a mechanism for limiting root growth in areas less likely to be favorable for resource acquisition [139].

Further, communication between roots appears to mediate self-detection, perhaps reducing within-plant competition. Both *Buchloe dactyloides* (Nutt.) Engelm. and *Pisum sativum* grew fewer roots when grown with a genetically identical clone than when grown with a clone from a different, conspecific individual [140,141]. Further, in split root experiments, *P. sativum* grew fewer roots in the direction of other roots connected to the same plant than in the direction of roots of other *P. sativum* individuals [140]. These results suggest that roots of these species can distinguish between "self" and "nonself" roots, and limit growth near self roots. In another series of experiments, root elongation rates of the shrub *Ambrosia dumosa* declined considerably on contact with roots of conspecific individuals, but not of other species, perhaps indicating a similar mechanism for avoiding intraspecific competition [142–144].

The signaling mechanisms that drive self-detection and species identification in root–root interactions are not known. Genetically identical clones of *Buchloe dactyloides* that were grown in separate pots for 8 weeks and then reunited did not limit their root growth in response to one another like recently separated clones, suggesting that the mechanism of self-detection was based on physiological rather than genetic identity [141]. The mechanism of *Ambrosia dumosa* conspecific detection also appeared to be based on physiological identity; inhibition of *A. dumosa* root elongation occurred between roots of separated clones from a single individual but did not occur upon contact between roots of a single individual [144]. Addition of activated carbon, which adsorbs organic compounds, to *A. dumosa* soils did not prevent root inhibition, suggesting that organic chemical signals in the rhizosphere were probably not involved [143]. One potential mechanism is the resonant amplification of hormonal or electrical oscillatory signals [140,141]. Such signals might be unique to each individual based on environmental conditions and might be amplified between roots with similar signatures, leading to a signal-based measure of self. However, there is currently little evidence in support of this hypothesis. More research is required to determine the mechanisms of self or nonself detection in plant roots.

V. POSITIVE ROOT–MICROBE COMMUNICATION

Plant roots and soil microorganisms participate in interactions that form a continuum between pathogenic infection and symbiotic mutualism. Colonization of the rhizosphere or plant roots by beneficial soil microorganisms can increase nutrient availability, stimulate plant growth, and increase plant resistance to pathogen infection [145], in exchange for plant sugars, proteins, and secondary metabolites. A plant's ability to distinguish between beneficial and pathogenic root-associated microorganisms is critical to its survival and success. Consequently, interactions between plants and beneficial microbes in the rhizosphere are often highly regulated, involving complex chemical communication between plants and microbes.

The general mechanisms that drive the enormous variety of pathogenic and mutualistic root–microbe interactions, however, are often surprisingly similar. Roots produce signals that are recognized by microbes, and microbes in turn produce signals that initiate infection. Initial recognition of plant root exudates by both beneficial and pathogenic microorganisms usually involves movement of the microbe in response to chemical signals such as organic acids or carbohydrates, an attraction known as *chemotaxis* [146,147]. Electrotaxis, or movement in response to an electrical current, can also play a role in microbial attraction, sometimes overriding chemotaxis, and can be initiated by root exudates [148]. In addition, plants exude lectins (proteins that specifically bind carbohydrates) to attract beneficial microbes. Further, for beneficial bacteria that exist at low concentrations in soils (e.g., many *Rhizobium* spp.), quorum-sensing mechanisms are important for aggregation. As described earlier, quorum sensing, or bacteria cell–cell communication mediated by small diffusible signaling molecules (autoinducers), regulates gene expression in dense bacterial populations [149]. Plants can produce quorum-sensing mimics that may stimulate or repress quorum-regulated behavior [106], potentially affecting gene expression and chemical signal production in both beneficial and pathogenic soil bacteria. Chemical signals involved in early stages of pathogen and mutualist infections are also sometimes similar, supporting the notion that mutualists are simply highly-evolved pathogens. For example, the same flavonoids from *Pisum sativum* L. can initiate root–microbe interactions in both the beneficial nitrogen-fixing *Rhizobium leguminosarium* bv. viciae symbiont by inducing *nod* gene transcription, and in the pathogenic *Nectria haematococca* MP 6 (*Fusarium solani*) relationship by inducing spore germination [150].

Pathogenic and beneficial soil microbes both also induce plant defense responses to infection. Microbial products shared by beneficial and pathogenic microbes, such as chitin, flagellins, and glyoproteins, can trigger plant defenses. Consequently, specialist beneficial soil microbes, like specialist plant pathogens, must resist, evade, or deactivate plant immune responses, including reactive oxygen species (ROS), chitinases, and phytoalexins, to successfully infect the host. Beneficial microbes employ many of the same strategies used by pathogenic microbes to resist plant defenses, including surface and extracellular polysaccharides, antioxidants, degrading or deactivating enzymes [151], and effector proteins delivered by type III secretion systems [147,151,152]. Effector protein production can be induced by root exudates, suggesting a role of plant chemical signals in facilitating infection by beneficial soil microbes [147,151,152]. However, these root exudates may also facilitate some pathogen infections. Despite the substantial overlap in the signals involved in pathogen and mutualist infections of plant roots, in many cases, plants are able to enter into beneficial interactions with soil microbes while avoiding most pathogen infections. The signaling pathways that confer this specificity are well understood for some root–microbe interactions but remain largely unknown for many symbiotic associations.

Studies to examine the biochemical interactions between roots and beneficial soil microbes are most often performed in the laboratory, as microsite variation in soils makes these interactions difficult to observe *in situ*. Therefore, much of the information known about the signals that drive these complex interactions is limited to culturable microorganisms. Culturable bacteria make up less than 1% of the diversity of soil bacteria [153]. Nevertheless, recent research on these few species has provided important insights into the nature of biochemical interactions between plant roots and beneficial soil microbes.

A. Extracellular Plant Growth-Promoting Rhizobacteria

Extracellular plant growth-promoting rhizobacteria (ePGPR) are growth-promoting bacteria that colonize the rhizosphere, the root surface (i.e., rhizoplane), or areas between cells of the root cortex, in contrast to intracellular plant growth-promoting rhizobacteria (iPGPR) that colonize root cells [11]. ePGPR in the rhizosphere are also discussed in Chapter 3. Chemical signals in the rhizosphere may mediate both initiation of plant–ePGPR associations and ePGPR promotion of plant growth. ePGPR movement to the root appears to be mediated in part by root chemical signals, which may attract bacteria, affect bacterial motility, and regulate bacterial attachment to the root (Figure 11.5). The particular signals involved in ePGPR chemotaxis have not yet been identified, although carbohydrates and amino acids in plant root exudates weakly stimulate ePGPR chemotaxis [147]. ePGPR exhibit weaker chemotaxis than iPGPR in response to the same root exudates, suggesting that either different root exudates are involved or that ePGPR are less affected than iPGPR by plant signals [154]. Plant root exudates also influence bacteria flagellar movement by altering bacterial chemosensory pathways [155] and regulate bacterial attachment to the root surface. For example, the major outer membrane protein from *Azospirillum brasilense* has greater affinity for root exudates from cereals than from legumes or tomatoes [156], indicating a role of cereal root exudates in *A. brasilense* attachment.

Colonization by ePGPR can promote plant growth via several mechanisms. Whereas ePGPR can alter nutrient cycling, increase nutrient availability, or moderate environmental stress, many ePGPR also produce chemical signals that increase plant growth (Figure 11.5). ePGPR chemical signals include phytohormones that stimulate plant growth, antimicrobial agents that inhibit plant pathogens, and defense response elicitors that reduce susceptibility to pathogen infection. ePGPR, such as *Rhizobium leguminosarum* and *Azospirillum* spp., produce phytohormones, including auxins, cytokinins, and gibberellins that stimulate plant growth [147,157,158]. Both *R. leguminosarum* and *Azospirillum* spp. can also fix atmospheric dinitrogen, but their effects on plant growth in extracellular associations with plant roots are mainly attributed to production of indole-3-acetic acid (IAA) [147], an auxin that stimulates root cell elongation. ePGPR that produce IAA via the indole-3-acetaldehyde or indole-3-pyruvic pathways can stimulate plant growth, whereas plant pathogens that produce IAA via the indoleacetamide pathway, which occurs exclusively in microorganisms, produce toxic IAA concentrations in plants [12]. Plant root exudates may serve as phytohormone precursors. Addition of tryptophan, a precursor of IAA biosynthesis, is necessary for some bacteria to produce IAA [159]. Mapping tryptophan in soil around *Avena barbata* Pott ex Link roots with biosensor bacteria suggests that tryptophan is released by plants into the rhizosphere [160].

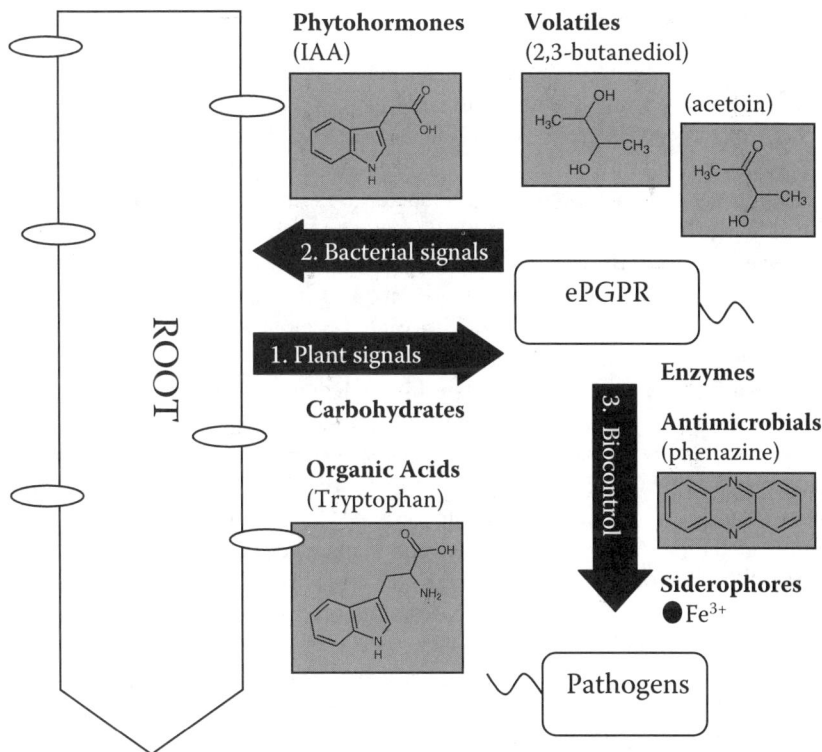

FIGURE 11.5 Mechanisms of chemical communication between plant roots and extracellular plant-growth-promoting bacteria (ePGPR). (1) Roots produce signals that attract bacteria and induce chemotaxis. ePGPR may increase plant growth through direct mechanisms, such as increased nutrient availability and (2) chemical signals to roots; or through indirect mechanisms, such as (3) biological control of plant pathogens, biofilm formation, and activation of plant defenses.

Some ePGPR also produce enzymes and antibiotics that target plant pathogens. For example, *Stenotrophomonas maltophilia* produces an extracellular serine protease that reduces virulence of the fungal pathogen *Pythium ultimum* [161]. Similarly, *Pseudomonas fluorescens* produces anti-fungal compounds, including phenazine [162] and viscosinamide [163]. Other ePGPR, such as *Bacillus* sp. A24 and transformed *P. fluorescens* P3, degrade bacterial quorum-sensing molecules (i.e., AHLs), thereby preventing quorum formation and subsequent infection by bacterial pathogens [164]. Some ePGPR may limit pathogen infection by forming biofilms that physically prevent root access to pathogens [146] or by producing siderophores that may limit iron availability to pathogens, although iron-limited conditions in the rhizosphere are probably rare [165]. The use of ePGPR as biocontrols against pathogens is discussed in Chapter 10. The role of siderophores in the rhizosphere is discussed in Chapter 7.

Finally, some ePGPR promote plant growth by producing signals that activate plant defenses including induced systemic resistance (ISR), thus making plants less susceptible to pathogen infection [128]. For example, bacterial lipopolysaccharide (LPS) production by *Rhizobium etli* induces systemic resistance in *Solanum tuberosum* L. roots, which in turn enhances plant resistance to nematode infection [13]. Similarly, 2,3-butanediol, a volatile produced by the ePGPR *Bacillus subtilis* GB03 and *B. amyloliquefaciens* IN937a, significantly decreases *Arabidopsis thaliana* susceptibility to the bacterial pathogen *Erwinia carotovora* [129]. Experiments with mutant *Arabidopsis* suggest that several redundant pathways may be involved in ePGPR induction of ISR, including the jasmonic acid and ethylene-dependent pathways [129].

B. Rhizobia–Legume Interactions

Symbiotic associations between members of the plant family Fabaceae and the bacteria family Rhizobiaceae are the most studied beneficial plant–microbe interactions and are described in detail in Chapter 9. Members of this group of iPGPR, termed *rhizobia*, enter legume root cells, induce nodule formation, and fix atmospheric dinitrogen in exchange for plant photosynthates. Most plants in the Fabaceae family and bacteria in the Rhizobiaceae family are able to form symbiotic associations (Table 11.1) [166]. However, rhizobia–legume interactions are generally highly specific; most rhizobia species will form nodules with only particular legume species [167].

Flavonoid signals in plant root exudates are involved in a number of processes that mediate rhizobia–legume interactions and can be effective at picomolar concentrations [168]. First, host flavonoids often serve as chemoattractants to the bacteria in the initial host recognition process [169]. Significant differences in the flavonoid profiles of legume root exudates are thought to explain in part rhizobia–host specificity [170]. Second, exposure to host flavonoids can induce changes in rhizobia surface molecules, which may protect rhizobia from host defense signals. For example, host root exudates alter the O-antigen structure of lipopolysaccharides (LPS) in *Rhizobium etli* and *Sinorhizobium fredii* [171], perhaps conferring resistance to host phytoalexins. Third, exposure to host flavonoids, including flavones, flavanones, and isoflavones, activates the transcriptional regulator, *nodD* (LysR family of transcriptional activators) in rhizobia cells [172]. The NodD protein complexes with the flavonoid activator and transcriptionally activates nodulation genes (*nod* genes) involved in symbiosis by binding to genetic elements known as *nod boxes*. Nonflavonoids in some legume root exudates, including aldonic acids, anthocynidins, betaines, chalcones, and phenolics, can also induce *nod* gene expression [172,173]. *nod* Gene-inducing flavonoids are continuously exuded into the rhizosphere, but signals from compatible rhizobia species can increase legume

TABLE 11.1
Root-Associated Symbiotic Dinitrogen-Fixing Bacteria and Plant Hosts

Microsymbiont		Host Plant	
Family	Genera	Family	Genera
Rhizobiaceae	*Allorhizobium*	Fabaceae[a]	*Acacia, Astragalus, Desmodium, Galega, Glycine,*
	Azorhizobium		*Hedysarum, Lablab, Lathyrus, Lens, Leucaena Lotus,*
	Bradyrhizobium		*Lupinus, Medicago, Melilotus, Mimosa, Neptunia,*
	Rhizobium		*Oxytropis, Phaseolus, Pisum, Prosopis, Sesbania,*
	Mesorhizobium		*Trifolium, Trigonella, Vicia, Vigna*
	Sinorhizobium	Ulmaceae	*Parasponia*
Frankiaceae	*Frankia*	Betulaceae	*Alnus*
		Casuarinaceae	*Allocasuarina, Casuarina, Ceuthostoma, Gymnostoma*
		Coriariaceae	*Coriaria*
		Datiscaceae	*Datisca*
		Elaeagnaceae	*Elaeagnus, Hippophae, Shepherdia*
		Myricaceae	*Comptonia, Morella, Myrica*
		Rhamnaceae	*Ceanothus, Colletia, Discaria, Kentrothamnus,*
			Retanilla, Talguenea, Trevoa
		Rosaceae	*Cercocarpus, Chamaebatia, Dryas, Purshia*
Nostocaceae	*Nostoc*	Cycadaceae	*Cycas*
		Stangeriaceae	*Bowenia, Stangeria*
		Zamiaceae	*Ceratozamia, Dioon, Encephalartos, Lepidozamia,*
			Macrozamia, Microcycas, Zamia

[a] Host genera within Fabaceae are representative, not inclusive.

flavonoid exudation [174]. Flavonoid concentrations in soil may also depend on microbial degradation; some flavonoids are easily degraded while others are recalcitrant [175,176]. Both low
constitutive production and microbial degradation may explain why the flavonoids that induce *nod*
gene expression are sometimes not found in legume rhizospheres.

Among the products of rhizobia *nod* genes are a number of lipochitooligosaccharides termed
Nod factors. Nod factors are critical bacterial signals that mediate the process of rhizobia host
infection. These factors differ among rhizobia species in substituents on the glucosamine residues
but share the same function, mediating communication with the host root. Several Nod factor
receptors have been identified recently in model legumes. In *Medicago sativa* L., a nodulation
receptor kinase (NORK) gene has been proposed to function in Nod factor perception and signal
transduction [177]. In *M. truncatula* Gaertner and *Lotus japonicus*, transmembrane serine or
threonine receptorlike kinases (RLKs) containing extracellular LysM domains have been identified
as putative Nod factor receptors [178–180]. In addition to Nod factors, rhizobia also produce
extracellular polysaccharides (e.g., succinoglycan and EPS II), lipopolysaccharides, and type III
system-secreted proteins that are involved in infection thread development [174,181–183].

Rhizobia must evade or modify plant defense responses in order to successfully colonize roots
and form functional nodules. The LPS structure of rhizobia may be important in host recognition
of specific microsymbionts. For example, when purified *Sinorhizobium meliloti* LPS were added
to host (*Medicago sativa* L.) and nonhost (*Nicotiana tabacum* L.) cell cultures, *N. tabacum* produced
characteristic defense responses (alkalinization and oxidative bursts), whereas *M. sativa* did not
[184]. Extracellular polysaccharides (EPS) from rhizobia may also influence host-specificity [185].
Bradyrhizobium japonicum expressing wild type EPS do not induce *Glycine max* (L.) Merr. defense
responses, whereas *B. japonicum* mutants with altered EPS induce host phytoalexin production
similar to pathogenic bacteria [186]. Rhizobial Nod factors (lipochitooligosaccharides) can also
downregulate plant defense responses [187]. In addition to saccharides (i.e., LPS, EPS, and Nod
factors), some rhizobia produce enzymes and type III secreted proteins that moderate host defenses.
B. elkanii produces rhizobitoxine, which inhibits a key enzyme, 1-aminocyclopropane-1-carboxylate
(ACC) synthase, in the ethylene biosynthetic pathway [188,189]. Ethylene is known to inhibit
nodulation upstream of Nod factor reception [190] and is involved in plant defense responses to
pathogens. Further *Rhizobium* sp. NGR234 produces a putative type III effector protein, NopL,
which likely serves as a substrate for plant protein kinases and may alter the expression of plant
defense genes [191].

C. FRANKIA–ACTINORHIZAL INTERACTIONS

Filamentous actinomycetes from the bacterial genus *Frankia* develop intracellular root symbioses
with woody plant hosts from several plant families. As in rhizobia–legume interactions, *Frankia*
organisms induce formation of specialized root structures (nodules) in which they reside and fix
atmospheric dinitrogen in exchange for photosynthates (Table 11.1). The genus *Frankia* comprises Gram-
positive bacteria with high guanine and cytosine content (i.e., high GC) in the Firmicutes phylum.
Like rhizobia, these organisms can survive outside the host as heterotrophic saprophytes [192,193].
Unlike rhizobia, *Frankia* organisms can fix atmospheric dinitrogen outside host nodules [194].
They use two general pathways for host plant infection: intracellular root hair infection and
intercellular penetration [195]. The method of host infection appears to be under host control, as
a single *Frankia* strain uses different infection pathways when nodulating different host species
[195]. The intracellular infection pathway is very similar to the rhizobia–legume infection process;
except that in the actinorhizal symbiosis, bacteria are not released from the infection threads [196].
The bacteria initiate root hair curling, infection threads form, and the microsymbiont travels down
the thread to a newly forming nodule primordia. In contrast, the intercellular infection pathway
occurs when *Frankia* hyphae penetrate the middle lamella between two root epidermal cells. Once
inside the root, the hyphae branch and migrate through intercellular spaces to the root cortex, where
they infect nodule primordia.

Host–microbe specificity among *Frankia* organisms and actinorhizal plants suggests that specific signals likely permit or inhibit symbiotic *Frankia*–host interactions [197]. The initial chemical signals involved in host recognition and infection in these interactions are not known but may include flavonoids or other chemoattractants that induce chemotaxis. Exposure to different flavonones and isoflavonones from *Alnus rubra* Bong. seeds can stimulate or inhibit *Frankia* nodulation in *A. rubra* roots [198]. Similarly, incubation with the flavonol kaempferol inhibits *Frankia* nodulation in *A. glutinosa* (L.) Gaertn. roots [199]. Further, flavonoids from *Casuarina* seeds induce *Frankia* organisms to produce factors that cause root hair deformation [195]. These factors are functionally similar to rhizobia Nod factors (i.e., involved in root hair curling) but structurally divergent [200]. Identification and characterization of the host genes involved in nodulation (i.e., actinorhizins) is underway. Host genes that are significantly upregulated in actinorhizal nodules may have important roles in nodule metabolism, nodule development, or internalization of the microsymbiont [201].

D. Nostoc–Cycad Interactions

Like the genus *Frankia*, members of the Cyanobacteria phylum can fix atmospheric dinitrogen as free-living cells. Cyanobacteria can also form symbiotic associations with diatoms, fungi, lower plants, and higher plants, fixing atmospheric dinitrogen in exchange for host photosynthates [202]. Cyanobacteria infections of plant roots are restricted to interactions between members of the *Nostoc* genus and the gymnosperm families Cycadaceae, Stangeriaceae, and Zamiaceae (Table 11.1). Unlike rhizobia and *Frankia*, *Nostoc* infects specialized host structures that are developed before infection, known as *precoralloid roots*. *Nostoc* persist extracellularly in the host root cortex, embedded in mucilage under microaerobic, nonphotosynthetic conditions. *Nostoc* infection of cycads and other gymnosperms occurs on apogeotropic precoralloid roots. Infection may occur throughout precoralloid root development, but precoralloid roots eventually senesce, after which infection is no longer possible [202]. *Nostoc* invasion transforms the precoralloid roots into coralloids that exhibit geotropic growth and induces development of a cortical layer in the roots. *Nostoc* organisms occupy this cortical layer and spread among the host coralloids from within. The development of coralloid roots and symbiotic *Nostoc* populations is highly synchronized, suggesting that complex chemical signaling may be involved.

Little is known about the chemical signals that regulate *Nostoc*–cycad symbioses. However, some information is available on chemical signals involved in interactions between *Nostoc* and other hosts. In interactions between *Nostoc* and bryophytes, *Gunnera* L., or cycads, the chemical signals involved include chemoattractants and factors that induce formation of *Nostoc* infection units termed *hormogonia* [202,203]. However, root exudates from some nonhosts, including *Oryza sativa* L., also induce chemotaxis and hormogonia development in *Nostoc* [204], suggesting that additional factors are involved in host infection. Some *Nostoc* gene sequences with homologies to genes upregulated in rhizobia during legume infection have also been identified [202]. For example, *Nostoc* collected from *Nostoc*–*Gunnera* L. associations contain gene sequences similar to the NodD binding component of a *nod* promotor, *nod* box, and host-specific *nod* genes [205]. However, whether these genes play a role in *Nostoc* infection is not clear [202].

E. Mycorrhizal Fungi

Symbiotic interactions with mycorrhizal fungi are widespread in the plant kingdom, occurring in more than 80% of all terrestrial plant species. They are characterized by fungal acquisition and translocation of nutrients (particularly phosphorus) from soil to plant roots in exchange for host photosynthates. In addition, mycorrhizae may protect hosts from water stress and pathogen infection [206]. There are several classifications of mycorrhizal symbioses [206]. We will limit our discussion to signaling in plant associations with arbuscular mycorrhizae (AM) and ectomycorrhizae (EM). Further discussion of mycorrhizal associations is provided in Chapter 8.

AM fungi, which form symbioses with a wide range of plant hosts, form appresoria (hyphal swellings) in host roots before entering root epidermal cells [207]. Appresoria formation is induced by purified fragments of plant epidermal cell walls, suggesting a role of host chemical signals in this process [208]. Recently, a strigolactone, 5-deoxy-strigol, rhizosecreted by *Lotus japonicus* (Regel) K. Larsen was identified as one of the chemical signals that induces AM hyphal branching prior to appresoria formation [209]. Exposure to plant root exudates, perhaps particularly to flavonoids, increases spore germination, hyphal growth, and branching in AM fungi. Flavonoids from host root exudates and commercial sources increase spore germination in the AM fungi *Gigaspora margarita* and some *Glomus* species [172]. Further, flavonoid exudation by *Medicago sativa* L. is altered during AM symbioses, increasing and decreasing during different stages of symbiotic development [210]. Signals produced by AM fungi also appear to play a role in the development of AM symbioses. Germinating AM fungal spores of *Gigaspora* spp. and *Glomus intraradices* release an unknown diffusible signal that activates a symbiotic gene in *M. truncatula* Gaertner, which is also induced by *Sinorhizobium* Nod factors [211]. The diffusible signal may be specific to AM fungi; it does not occur in pathogenic fungi [211].

Associations with AM fungi may also influence plant interactions with other soil microbes. For example, plants that form associations with AM fungi also form more associations with bacteria than nonmycorrhizal plants [212]. Increased bacterial colonization on AM plants has been attributed to changes in soil aggregation and increased nutrient availability with mycorrhizal infection but may also be due to effects of hyphal exudates or plant root exudates induced by mycorrhizal infection. For example, some bacterial strains are more attracted to root exudates from mycorrhizal tomato plants than to those of nonmycorrhizal plants [213]. Further, signals from some AM fungi may inhibit soil pathogens in the rhizosphere [212]. For example, crude extracts from the AM species *Glomus intraradices* inhibit conidial germination of the plant pathogen *Fusarium*, although *G. intraradices* extracts also stimulate growth of the mycoparasitic fungi *Trichoderma harzianum* and *Pseudomonas chlororaphis* [214].

Unlike AM symbioses, EM symbioses occur within a limited set of tree and shrub taxa. Host flavonoid signals and hormone balances both appear to be involved in initiating EM symbioses [215]. For example, *Eucalyptus globulus* root exudates contain the flavonol rutin, which increases fungal growth, and the cytokinin zeatin, which stimulates *Pisolithus* hyphal branching, inducing changes in fungal morphology that resemble changes that occur during ectomycorrhizal development [215–217]. Flavonoids and phytohormones, including auxins, cytokinins, abscisic acid, and ethylene, from EM fungi also may be involved in EM infection [218]. Conifers treated with EM fungal exudates, extracts, or synthetic auxins exhibit similar changes in root development to conifers treated with fungal mycelium, suggesting a role of fungal exudates, including phytohormones, in EM infection [215]. In addition, fungal production of the alkaloid hypaphorine increases during EM development and may regulate root development. *Pisolithus*-secreted hypaphorine decreases root hair elongation and induces transient apical swelling in *E. globulus* seedlings, perhaps by counteracting endogenous IAA in the host [219]. Normal root hair elongation can be restored by application of IAA.

As described earlier, fungal products, including those from AM and EM fungi, can elicit plant defense responses [220]. For example, *Picea abies* (L.) Karst. cells elicited by the EM fungi *Amanita muscaria* and *Hebeloma crustuliniforme* produce chitinases, as do *Eucalyptus globulus* Labill. seedlings treated with *Pisolithus* cell free extracts [215]. Similarly, initial contact with the AM fungus *Glomus versiforme* activates defense and stress-related genes in *Medicago truncatula* Gaertner roots [221]. The mechanisms through which mycorrhizal fungi evade plant defense responses and develop effective symbioses have not been determined. A number of plant genes involved in the rhizobia-legume symbiosis are also found in AM-competent plants, suggesting that AM symbioses may have similar modes of signal transduction and genetic regulation to the rhizobia symbioses [222–224]. Further, plants with mutations in genes required for rhizobia symbiosis (e.g., *dmi3*) are also deficient in AM symbioses [224]. However, similarities in mechanisms of AM and rhizobia resistance to host defenses have not yet been demonstrated.

VI. CONCLUDING REMARKS

This chapter provides only a partial review of the full literature on chemical signals and communication in the rhizosphere, highlighting the best-understood examples and most recent discoveries. Plant and soil microbe chemical signals mediate an array of root–root and root–microbe interactions. Most plants and soil microbes appear to participate in at least some modes of chemical-mediated communication in the rhizosphere, suggesting that chemical signals may be a common mechanism for regulating belowground interactions. However, the chemical structures, modes of action, and biological roles of these chemical signals vary considerably among interactions. In particular, the signaling pathways that regulate associations between *Striga* spp. and *Sorghum* spp., and between rhizobia and legumes, highlight the potential biochemical complexity and specificity of belowground root–root and root–microbe communication. However, even for those root and microbe chemical signals with known functions in rhizosphere communication, their modes of action, mechanisms by which they are produced and detected, movement, persistence, and behavior in soil are most often not yet understood [225]. The continued development of increasingly sophisticated techniques to examine chemical signals and the mechanisms through which they operate may speed up the process of identifying and understanding chemical communication among organisms. Future studies must be coupled with experiments conducted under realistic field conditions and with noncultivated species, to evaluate with more certainty, the importance of chemical signals in determining plant and soil community structure. In addition, further research on the evolution of chemical signal production and detection and of cross talk between chemical signals may deepen our understanding of root–root and root–microbe relationships.

ACKNOWLEDGMENTS

Our research on allelochemistry is supported by U.S. Department of Defense SERDP (SI-1388). Junichiro Horiuchi acknowledges financial support from the Japan Society for the Promotion of Science.

REFERENCES

1. *The American Heritage Dictionary of the English Language*, 4th ed., Houghton Mifflin Company, Boston, MA, 2000.
2. Whittaker, R.H. and Feeny, P.P., Allelochemics: chemical interactions between species, *Science,* 171, 757, 1971.
3. Marschner, H., *Mineral Nutrition of Higher Plants*, 2nd ed., Academic Press, London, 1995.
4. Dixon, R.A., Natural products and plant disease resistance, *Nature,* 411, 843, 2001.
5. Flores, H.E., Vivanco, J.M., and Loyola-Vargas, V.M., 'Radicle' biochemistry: the biology of root-specific metabolism, *Trends Plant Sci.,* 4, 220, 1999.
6. Bertin, C., Yang, X.H., and Weston, L.A., The role of root exudates and allelochemicals in the rhizosphere, *Plant Soil,* 256, 67, 2003.
7. Dakora, F.D. and Phillips, D.A., Root exudates as mediators of mineral acquisition in low-nutrient environments, *Plant Soil,* 245, 35, 2002.
8. Sudha, G. and Ravishankar, G.A., Involvement and interaction of various signaling compounds on the plant metabolic events during defense response, resistance to stress factors, formation of secondary metabolites and their molecular aspects, *Plant Cell Tiss. Org. Cult.,* 71, 181, 2002.
9. Haas, D. and Keel, C., Regulation of antibiotic production in root-colonizing *Pseudomonas* spp. and relevance for biological control of plant disease, *Ann. Rev. Phytopathol.,* 41, 117, 2003.
10. Fray, R.G., Altering plant-microbe interaction through artificially manipulating bacterial quorum sensing, *Ann. Bot. — London,* 89, 245, 2002.
11. Gray, E.J. and Smith, D.L., Intracellular and extracellular PGPR: commonalities and distinctions in the plant-bacterium signaling processes, *Soil Biol. Biochem.,* 37, 395, 2005.

12. Patten, C.L. and Glick, B.R., Role of *Pseudomonas putida* indoleacetic acid in development of the host plant root system, *Appl. Environ. Microbiol.,* 68, 3795, 2002.

13. Reitz, M. et al., Lipopolysaccharides of *Rhizobium etli* strain G12 act in potato roots as an inducing agent of systemic resistance to infection by the cyst nematode *Globodera pallida, Appl. Environ. Microbiol.,* 66, 3515, 2000.

14. Darrah, P.R., Models of the rhizosphere.1. Microbial-population dynamics around a root releasing soluble and insoluble carbon, *Plant Soil,* 133, 187, 1991.

15. Varanini, Z. and Pinton, R., Direct versus indirect effects of soil humic substances on plant growth and nutrition, in *The Rhizosphere: Biochemistry and Organic Substances at the Soil-Plant Interface,* Pinton, R., Varanini, Z., and Nannipieri, P., Eds., Marcel Dekker, New York, 2001, p. 141.

16. Foster, R.C., Microenvironments of soil-microorganisms, *Biol. Fertil. Soils,* 6, 189, 1988.

17. Einhellig, F.A., Mechanisms of action of allelochemicals in allelopathy, in *Allelopathy: Organisms, Processes, and Applications,* Inderjit, Dakshini, K.M.M., and Einhellig, F.A., Eds., American Chemical Society, Washington, D.C., 1995, p. 96.

18. Weir, T.L., Park, S.W., and Vivanco, J.M., Biochemical and physiological mechanisms mediated by allelochemicals, *Curr. Opin. Plant Biol.,* 7, 472, 2004.

19. Jose, S. and Gillespie, A.R., Allelopathy in black walnut (*Juglans nigra* L.) alley cropping. I. Spatio-temporal variation in soil juglone in a black walnut-corn (*Zea mays* L.) alley cropping system in the midwestern USA, *Plant Soil,* 203, 191, 1998.

20. Yu, J.Q. and Matsui, Y., Effects of root exudates of cucumber (*Cucumis sativus*) and allelochemicals on ion uptake by cucumber seedlings, *J. Chem. Ecol.,* 23, 817, 1997.

21. Kong, C.H. et al., Release and activity of allelochemicals from allelopathic rice seedlings, *J. Agric. Food Chem.,* 52, 2861, 2004.

22. Nimbal, C.I. et al., Phytotoxicity and distribution of sorgoleone in grain sorghum germplasm, *J. Agric. Food Chem.,* 44, 1343, 1996.

23. Wu, H.W. et al., Allelochemicals in wheat (*Triticum aestivum* L.): variation of phenolic acids in root tissues, *J. Agric. Food Chem.,* 48, 5321, 2000.

24. Bais, H.P. et al., Enantiomeric-dependent phytotoxic and antimicrobial activity of (+/−)-catechin. A rhizosecreted racemic mixture from spotted knapweed, *Plant Physiol.,* 128, 1173, 2002.

25. Bais, H.P. et al., Allelopathy and exotic plant invasion: from molecules and genes to species interactions, *Science,* 301, 1377, 2003.

26. Vivanco, J.M. et al., Biogeographical variation in community response to root allelochemistry: novel weapons and exotic invasion, *Ecol. Lett.,* 7, 285, 2004.

27. Stermitz, F.R. et al., 7,8-Benzoflavone: a phytotoxin from root exudates of invasive Russian knapweed, *Phytochemistry,* 64, 493, 2003.

28. Sheley, R.L., Jacobs, J.S., and Carpinelli, M.F., Distribution, biology, and management of diffuse knapweed (*Centaurea diffusa*) and spotted knapweed (*Centaurea maculosa*), *Weed Technol.,* 12, 353, 1998.

29. Ridenour, W.M. and Callaway, R.M., The relative importance of allelopathy in interference: the effects of an invasive weed on a native bunchgrass, *Oecologia,* 126, 444, 2001.

30. Weir, T.L., Bais, H.P., and Vivanco, J.M., Intraspecific and interspecific interactions mediated by a phytotoxin, (−)-catechin, secreted by the roots of *Centaurea maculosa* (spotted knapweed), *J. Chem. Ecol.,* 29, 2397, 2003.

31. Perry, L.G. et al., Screening of grassland plants for restoration after spotted knapweed invasion, *Rest. Ecol.,* 13, 725, 2005.

32. Veluri, R. et al., Phytotoxic and antimicrobial activities of catechin derivatives, *J. Agric. Food Chem.,* 52, 1077, 7746, 2004.

33. Blair, A.C. et al., New techniques and findings in the study of a candidate allelochemical implicated in invasion success, *Ecol. Lett.,* 8, 1039, 2005.

34. Perry, L.G. et al., Dual role for an allelochemical: (+/−)-catechin from *Centaurea maculosa* root exudates regulates conspecific seedling establishment, *J. Ecol.,* 93, 1126, 2005.

35. Huckelhoven, R. and Kogel, K.H., Reactive oxygen intermediates in plant-microbe interactions: who is who in powdery mildew resistance?, *Planta,* 216, 891, 2003.

36. Fasano, J.M. et al., Changes in root cap pH are required for the gravity response of the *Arabidopsis* root, *Plant Cell,* 13, 907, 2001.

37. Jones, D.L. et al., Effect of aluminum on cytoplasmic Ca^{2+} homeostasis in root hairs of *Arabidopsis thaliana* (L.), *Planta,* 206, 378, 1998.

38. Scott, A.C. and Allen, N.S., Changes in cytosolic pH within *Arabidopsis* root columella cells play a key role in the early signaling pathway for root gravitropism, *Plant Physiol.*, 121, 1291, 1999.

39. Sticher, L., MauchMani, B., and Metraux, J.P., Systemic acquired resistance, *Ann. Rev. Phytopathol.*, 35, 235, 1997.

40. Czarnota, M.A., Rimando, A.M., and Weston, L.A., Evaluation of root exudates of seven sorghum accessions, *J. Chem. Ecol.*, 29, 2073, 2003.

41. Kagan, I.A., Rimando, A.M., and Dayan, F.E., Chromatographic separation and *in vitro* activity of sorgoleone congeners from the roots of *Sorghum bicolor*, *J. Agric. Food Chem.*, 51, 7589, 2003.

42. Czarnota, M.A. et al., Mode of action, localization of production, chemical nature, and activity of sorgoleone: a potent PSII inhibitor in *Sorghum* spp. root exudates, *Weed Technol.*, 15, 813, 2001.

43. Yang, X.H., Scheffler, B.E., and Weston, L.A., SOR1, a gene associated with bioherbicide production in sorghum root hairs, *J. Exp. Bot.*, 55, 2251, 2004.

44. Hejl, A.M. and Koster, K.L., The allelochemical sorgoleone inhibits root H$^+$–ATPase and water uptake, *J. Chem. Ecol.*, 30, 2181, 2004.

45. Meazza, G. et al., The inhibitory activity of natural products on plant p-hydroxyphenylpyruvate dioxygenase, *Phytochemistry*, 60, 281, 2002.

46. Gonzalez, V.M. et al., Inhibition of a photosystem II electron transfer reaction by the natural product sorgoleone, *J. Agric. Food Chem.*, 45, 1415, 1997.

47. Lawrence, T. and Kilcher, M.R., The effect of fourteen root extracts upon germination and seedling length of fifteen plant species, *Can. J. Plant Sci.*, 42, 308, 1962.

48. Inderjit and Duke, S.O., Ecophysiological aspects of allelopathy, *Planta*, 217, 529, 2003.

49. Schulz, M. and Wieland, I., Variation in metabolism of BOA among species in various field communities — biochemical evidence for co-evolutionary processes in plant communities?, *Chemoecology*, 9, 133, 1999.

50. Perez, F.J. and Ormeno-Nunez, J., Weed growth interference from temperate cereals: the effect of a hydroxamic-acids-exuding rye (*Secale cereale* L.) cultivar, *Weed Res.*, 33, 115, 1993.

51. Friebe, A. et al., Phytotoxins from shoot extracts and root exudates of *Agropyron repens* seedlings, *Phytochemistry*, 38, 1157, 1995.

52. Sicker, D. et al., Glycoside carbamates from benzoxazolin-2(3H)-one detoxification in extracts and exudates of corn roots, *Phytochemistry*, 58, 819, 2001.

53. von Rad, U. et al., Two glucosyltransferases are involved in detoxification of benzoxazinoids in maize, *Plant J.*, 28, 633, 2001.

54. Fitter, A., Making allelopathy respectable, *Science*, 301, 1337, 2003.

55. Callaway, R.M. and Aschehoug, E.T., Invasive plants versus their new and old neighbors: a mechanism for exotic invasion, *Science*, 290, 521, 2000.

56. Callaway, R.M. and Ridenour, W.M., Novel weapons: invasive success and the evolution of increased competitive ability, *Front. Ecol. Environ.*, 2, 436, 2004.

57. Prati, D. and Bossdorf, O., Allelopathic inhibition of germination by *Alliaria petiolata* (Brassicaceae), *Am. J. Bot.*, 91, 285, 2004.

58. Callaway, R.M. et al., Natural selection for resistance to the allelopathic effects of invasive plants, *J. Ecol.*, 93, 576, 2005.

59. Mealor, B.A., Hild, A.L., and Shaw, N.L., Native plant community composition and genetic diversity associated with long-term weed invasions, *West. N. Am. Nat.*, 64, 503, 2004.

60. Singh, H.P., Batish, D.R., and Kohli, R.K., Autotoxicity: concept, organisms, and ecological significance, *Crit. Rev. Plant Sci.*, 18, 757, 1999.

61. Young, C.C., Autointoxication in root exudates of *Asparagus officinalis* L., *Plant Soil*, 82, 247, 1984.

62. Nigh, E.L., Jr., Stress factors influencing *Fusarium* infection in asparagus, *Acta Horticult.*, 271, 315, 1990.

63. Yu, J.Q. et al., Effects of root exudates and aqueous root extracts of cucumber (*Cucumis sativus*) and allelochemicals, on photosynthesis and antioxidant enzymes in cucumber, *Biochem. Syst. Ecol.*, 31, 129, 2003.

64. Menges, R.M., Allelopathic effects of Palmer amaranth (*Amaranthus palmeri*) on seedling growth, *Weed Sci.*, 36, 325, 1988.

65. Bradow, J.M. and Connick, W.J., Volatile methyl ketone seed-germination inhibitors from *Amaranthus palmeri* S Wats residues, *J. Chem. Ecol.*, 14, 1617, 1988.

66. Kazinczi, G., Beres, I., and Narwal, S.S., Allelopathic plants. 1. Canada thistle [*Cirsium arvense* (L.) Scop], *Allelopathy J.,* 8, 29, 2001.

67. Bokhari, U.G., Allelopathy among prairie grasses and its possible ecological significance, *Ann. Bot. — London,* 42, 127, 1978.

68. Sahid, I.B. and Sugau, J.B., Allelopathic effect of lantana (*Lantana camara*) and siam weed (*Chromolaena odorata*) on selected crops, *Weed Sci.,* 41, 303, 1993.

69. Arora, R.K. and Kohli, R.K., Autotoxic impact of essential oil extracted from *Lantana camara* L, *Biol. Plant.,* 35, 293, 1993.

70. Picman, J. and Picman, A.K., Autotoxicity in *Parthenium hysterophorus* and its possible role in control of germination, *Biochem. Syst. Ecol.,* 12, 287, 1984.

71. Perry, L.G. et al., Callaway, R.M., Paschke, M.W. and Vivanco, J.M., Dual role for an allelochemical: (+/−)-catechin from *Centaurea maculosa* root exudates regulates conspecific seedling establishment, *J. Ecol.,* 93, 1126–1135, 2005.

72. Schenck, J., Mahall, B.E., and Callaway, R.M., Spatial segregation of roots, *Adv. Ecol.,* 28, 145, 1999.

73. Dyer, A.R., Maternal and sibling factors induce dormancy in dimorphic seed pairs of *Aegilops triuncialis,* *Plant Ecol.,* 172, 211, 2004.

74. Williamson, G.B., Allelopathy, Koch's postulates and neck riddles, in *Perspectives in Plant Competition,* Grace, J.B. and Tilman, D., Eds., Academic Press, London, 1990, p. 143.

75. Weidenhamer, J.D. and Romeo, J.T., Allelochemicals of *Polygonella myriophylla*: chemistry and soil degradation, *J. Chem. Ecol.,* 30, 1067, 2004.

76. Yoder, J.I., Parasitic plant responses to host plant signals: a model for subterranean plant-plant interactions, *Curr. Opin. Plant Biol.,* 2, 65, 1999.

77. Palmer, A.G. et al., Chemical biology of multi-host/pathogen interactions: chemical perception and metabolic complementation, *Ann. Rev. Phytopathol.,* 42, 439, 2004.

78. Matvienko, M., Torres, M.J., and Yoder, J.I., Transcriptional responses in the hemiparasitic plant *Triphysaria versicolor* to host plant signals, *Plant Physiol.,* 127, 272–282, 2001.

79. Chang, M. et al., Chemical-regulation of distance — characterization of the 1st natural host germination stimulant for *Striga asiatica, J. Am. Chem. Soc.,* 108, 7858, 1986.

80. Fate, G.D. and Lynn, D.G., Xenognosin methylation is critical in defining the chemical potential gradient that regulates the spatial distribution in *Striga pathogenesis, J. Am. Chem. Soc.,* 118, 11369, 1996.

81. Keyes, W.J. et al., Signaling organogenesis in parasitic angiosperms: xenognosin generation, perception, and response, *J. Plant Growth Regul.,* 19, 217, 2000.

82. Kim, D.J. et al., On becoming a parasite: evaluating the role of wall oxidases in parasitic plant development, *Chem. Biol.,* 5, 103, 1998.

83. Smith, C.E. et al., A mechanism for inducing plant development: the genesis of a specific inhibitor, *Proc. Natl. Acad. Sci. USA,* 93, 6986, 1996.

84. O'Malley, R.C. and Lynn, D.G., Expansin message regulation in parasitic angiosperms: marking time in development, *Plant Cell,* 12, 1455-1465, 2000.

85. McQueen-Mason, S. and Cosgrove, D.J., Disruption of hydrogen-bonding between plant-cell wall polymers by proteins that induce wall extension, *Proc. Natl. Acad. Sci. USA,* 91, 6574, 1994.

86. Delavault, P. et al., Host-root exudates increase gene expression of asparagine synthetase in the roots of a hemiparasitic plant *Triphysaria versicolor* (Scrophulariaceae), *Gene,* 222, 155, 1998.

87. Gowda, B.S., Riopel, J.L., and Timko, M.P., NRSA-1: a resistance gene homolog expressed in roots of non-host plants following parasitism by *Striga asiatica* (witchweed), *Plant J.,* 20, 217, 1999.

88. Goldwasser, Y. et al., The differential susceptibility of vetch (*Vicia* spp.) to *Orobanche aegyptiaca*: anatomical studies, *Ann. Bot. — London,* 85, 257, 2000.

89. Gurney, A.L. et al., Novel sources of resistance to *Striga hermonthica* in *Tripsacum dactyloides,* a wild relative of maize, *New Phytologist,* 160, 557, 2003.

90. Haussmann, B.I.G. et al., Analysis of resistance to *Striga hermonthica* in diallel crosses of sorghum, *Euphytica,* 116, 33, 2000.

91. Rich, P.J., Grenier, U., and Ejeta, G., *Striga* resistance in the wild relatives of sorghum, *Crop Sci.,* 44, 2221, 2004.

92. Mohamed, A. et al., Hypersensitive response to *Striga* infection in *Sorghum, Crop Sci.,* 43, 1320, 2003.

93. Serghini, K. et al., Sunflower (*Helianthus annuus* L.) response to broomrape (*Orobanche cernua* Loefl.) parasitism: induced synthesis and excretion of 7-hydroxylated simple coumarins, *J. Exp. Bot.,* 52, 2227, 2001.

94. Bais, H.P. et al., How plants communicate using the underground information superhighway, *Trends Plant Sci.,* 9, 26, 2004.

95. Blossey, B. and Hunt-Joshi, T.R., Belowground herbivory by insects: influence on plants and aboveground herbivores, *Ann. Rev. Entomol.,* 48, 521, 2003.

96. Bais, H.P. et al., Root specific elicitation and antimicrobial activity of rosmarinic acid in hairy root cultures of *Ocimum basilicum, Plant Physiol. Biochem.,* 40, 983, 2002.

97. Veluri, R. et al., Phytotoxic and antimicrobial activities of catechin derivatives, *J. Agric. Food Chem.,* 52, 1077, 2004.

98. Taddei, P. et al., Vibrational, H-1-NMR spectroscopic, and thermal characterization of gladiolus root exudates in relation to *Fusarium oxysporum* f. sp *gladioli* resistance, *Biopolymers,* 67, 428, 2002.

99. Shepherd, T. and Davies, H.V., Carbon loss from the roots of forage rape (*Brassica napus* L) seedlings following pulse-labeling with (CO_2)-C^{14}, *Ann. Bot. — London,* 72, 155, 1993.

100. Borisjuk, N.V. et al., Production of recombinant proteins in plant root exudates, *Nat. Biotechnol.,* 17, 466, 1999.

101. Park, S.W. et al., Isolation and characterization of a novel ribosome-inactivating protein from root cultures of pokeweed and its mechanism of secretion from roots, *Plant Physiol.,* 130, 164, 2002.

102. Park, S.W., Stevens, N.M., and Vivanco, J.M., Enzymatic specificity of three ribosome-inactivating proteins against fungal ribosomes, and correlation with antifungal activity, *Planta,* 216, 227, 2002.

103. Nielsen, K. and Boston, R.S., Ribosome-inactivating proteins: a plant perspective, *Ann. Rev. Plant Physiol. Plant Mol. Biol.,* 52, 785, 2001.

104. Vepachedu, R., Bais, H.P., and Vivanco, J.M., Molecular characterization and post-transcriptional regulation of ME1, a type-I ribosome-inactivating protein from *Mirabilis expansa, Planta,* 217, 498, 2003.

105. Swift, S. et al., Quorum sensing: a population-density component in the determination of bacterial phenotype, *Trends Biochem. Sci.,* 21, 214, 1996.

106. Teplitski, M., Robinson, J.B., and Bauer, W.D., Plants secrete substances that mimic bacterial N-acyl homoserine lactone signal activities and affect population density-dependent behaviors in associated bacteria, *Mol. Plant Microbe Interact.,* 13, 637, 2000.

107. Bauer, W.D. and Robinson, J.B., Disruption of bacterial quorum sensing by other organisms, *Curr. Opin. Biotech.,* 13, 234, 2002.

108. Mathesius, U. et al., Extensive and specific responses of a eukaryote to bacterial quorum-sensing signals, *Proc. Natl. Acad. Sci. USA,* 100, 1444, 2003.

109. Walker, T.S. et al., Root exudation and rhizosphere biology, *Plant Physiol.,* 132, 44, 2003.

110. Walker, T.S. et al., Metabolic profiling of root exudates of *Arabidopsis thaliana, J. Agric. Food Chem.,* 51, 2548, 2003.

111. Bais, H.P. et al., Mediation of pathogen resistance by exudation of antimicrobials from roots, *Nature,* 434, 217, 2005.

112. Hawes, M.C. et al., Root caps and rhizosphere, *J. Plant Growth Regul.,* 21, 352, 2002.

113. Hawes, M.C. et al., Function of root border cells in plant health: pioneers in the rhizosphere, *Ann. Rev. Phytopathol.,* 36, 311, 1998.

114. Gunawardena, U. et al., Tissue-specific localization of pea root infection by *Nectria haematococca.* Mechanisms and consequences, *Plant Physiol.,* 137, 1363, 2005.

115. Gunawardena, U. and Hawes, M.C., Tissue specific localization of root infection by fungal pathogens: role of root border cells, *Mol. Plant Microbe Interact.,* 15, 1128, 2002.

116. Foster, R.C., The ultrastructure of the rhizoplane and rhizosphere, *Ann. Rev. Phytopathol.,* 24, 211, 1986.

117. Woo, H.H., Hirsch, A.M., and Hawes, M.C., Altered susceptibility to infection by *Sinorhizobium meliloti* and *Nectria haematococca* in alfalfa roots with altered cell cycle, *Plant Cell Rep.,* 22, 967, 2004.

118. Laville, J. et al., Global control in *Pseudomonas fluorescens* mediating antibiotic-synthesis and suppression of black root-rot of tobacco, *Proc. Natl. Acad. Sci. USA,* 89, 1562, 1992.

119. Collmer, A. et al., Genomic mining type III secretion system effectors in *Pseudomonas syringae* yields new picks for all TTSS prospectors, *Trends Microbiol.,* 10, 462, 2002.

120. Bender, C.L., Alarcon-Chaidez, F., and Gross, D.C., *Pseudomonas syringae* phytotoxins: mode of action, regulation, and biosynthesis by peptide and polyketide synthetases, *Microbiol. Mol. Biol. Rev.*, 63, 266, 1999.

121. Wolpert, T.J., Dunkle, L.D., and Ciuffetti, L.M., Host-selective toxins and avirulence determinants: what's in a name?, *Ann. Rev. Phytopathol.*, 40, 251, 2002.

122. Shiraishi, T. et al., The role of suppressors in determining host-parasite specificities in plant cells, *Int. Rev. Cytol.*, 172, 55, 1997.

123. Moissiard, G. and Voinnet, O., Viral suppression of RNA silencing in plants, *Mol. Plant Pathol.*, 5, 71, 2004.

124. Narasimhan, K. et al., Enhancement of plant-microbe interactions using a rhizosphere metabolomics-driven approach and its application in the removal of polychlorinated biphenyls, *Plant Physiol.*, 132, 146, 2003.

125. Bais, H.P. et al., Exudation of fluorescent beta-carbolines from *Oxalis tuberosa* L. roots, *Phytochemistry*, 61, 539, 2002.

126. Baldwin, I.T., Jasmonate-induced responses are costly but benefit plants under attack in native populations, *Proc. Natl. Acad. Sci. USA*, 95, 8113, 1998.

127. Brigham, L.A., Michaels, P.J., and Flores, H.E., Cell-specific production and antimicrobial activity of naphthoquinones in roots of *Lithospermum erythrorhizon*, *Plant Physiol.*, 119, 417, 1999.

128. van Loon, L.C., Bakker, P., and Pieterse, C.M.J., Systemic resistance induced by rhizosphere bacteria, *Ann. Rev. Phytopathol.*, 36, 453, 1998.

129. Ryu, C.M., Farag, M.A., Hu, C.H., Reddy, M.S., Kloepper, J.W., and Pare, P.W., Bacterial volatiles induce systemic resistance in *Arabidopsis*, *Plant Physiol.*, 134, 1017, 2004.

130. Morrissey, J.P. and Osbourn, A.E., Fungal resistance to plant antibiotics as a mechanism of pathogenesis, *Microbiol. Mol. Biol. Rev.*, 63, 708, 1999.

131. Bouarab, K. et al., A saponin-detoxifying enzyme mediates suppression of plant defences, *Nature*, 418, 889, 2002.

132. Glinwood, R. et al., Change in acceptability of barley plants to aphids after exposure to allelochemicals from couch-grass (*Elytrigia repens*), *J. Chem. Ecol.*, 29, 261, 2003.

133. Chamberlain, K. et al., Can aphid-induced plant signals be transmitted aerially and through the rhizosphere?, *Biochem. Syst. Ecol.*, 29, 1063, 2001.

134. Bruin, J. and Sabelis, M.W., Meta-analysis of laboratory experiments on plant-plant information transfer, *Biochem. Syst. Ecol.*, 29, 1089, 2001.

135. Dicke, M. and Dijkman, H., Within-plant circulation of systemic elicitor of induced defence and release from roots of elicitor that affects neighbouring plants, *Biochem. Syst. Ecol.*, 29, 1075, 2001.

136. Guerrieri, E. et al., Plant-to-plant communication mediating in-flight orientation of *Aphidius ervi*, *J. Chem. Ecol.*, 28, 1703, 2002.

137. Du, Y.J. et al., Identification of semiochemicals released during aphid feeding that attract parasitoid *Aphidius ervi*, *J. Chem. Ecol.*, 24, 1355, 1998.

138. Gersani, M. et al., Tragedy of the commons as a result of root competition, *J. Ecol.*, 89, 660, 2001.

139. Falik, O. et al., Root navigation by self inhibition, *Plant Cell Environ.*, 28, 562, 2005.

140. Falik, O. et al., Self/non-self discrimination in roots, *J. Ecol.*, 91, 525, 2003.

141. Gruntman, M. and Novoplansky, A., Physiologically mediated self/non-self discrimination in roots, *Proc. Natl. Acad. Sci. USA*, 101, 3863, 2004.

142. Mahall, B.E. and Callaway, R.M., Root communication among desert shrubs, *Proc. Natl. Acad. Sci. USA*, 88, 874, 1991.

143. Mahall, B.E. and Callaway, R.M., Root communication mechanisms and intracommunity distributions of 2 mojave desert shrubs, *Ecology*, 73, 2145, 1992.

144. Mahall, B.E. and Callaway, R.M., Effects of regional origin and genotype on intraspecific root communication in the desert shrub *Ambrosia dumosa* (Asteraceae), *Am. J. Bot.*, 83, 93, 1996.

145. Lugtenberg, B.J.J., Deweger, L.A., and Bennett, J.W., Microbial stimulation of plant-growth and protection from disease, *Curr. Opin. Biotech.*, 2, 457, 1991.

146. Bais, H.P., Fall, R., and Vivanco, J.M., Biocontrol of *Bacillus subtilis* against infection of *Arabidopsis* roots by *Pseudomonas syringae* is facilitated by biofilm formation and surfactin production, *Plant Physiol.*, 134, 307, 2004.

147. Somers, E., Vanderleyden, J., and Srinivasan, M., Rhizosphere bacterial signalling: a love parade beneath our feet, *Crit. Rev. Microbiol.*, 30, 205, 2004.

148. van West, P. et al., Oomycete plant pathogens use electric fields to target roots, *Mol. Plant Microbe Interact.,* 15, 790, 2002.

149. Fuqua, W.C., Winans, S.C., and Greenberg, E.P., Quorum sensing in bacteria: the luxr-luxi family of cell density-responsive transcriptional regulators, *J. Bacteriol.,* 176, 269, 1994.

150. Bagga, S. and Straney, D., Modulation of cAMP and phosphodiesterase activity by flavonoids which induce spore germination of *Nectria haematococca* MP VI (*Fusarium solani*), *Physiol. Mol. Plant P.,* 56, 51, 2000.

151. D'Haeze, W. and Holsters, M., Surface polysaccharides enable bacteria to evade plant immunity, *Trends Microbiol.,* 12, 555, 2004.

152. Preston, G.M., Bertrand, N., and Rainey, P.B., Type III secretion in plant growth-promoting *Pseudomonas fluorescens* SBW25, *Mol. Microbiol.,* 41, 999, 2001.

153. Torsvik, V., Goksoyr, J., and Daae, F.L., High diversity in DNA of soil bacteria, *Appl. Environ. Microbiol.,* 56, 782, 1990.

154. Bacilio-Jimenez, M. et al., Chemical characterization of root exudates from rice (*Oryza sativa*) and their effects on the chemotactic response of endophytic bacteria, *Plant Soil,* 249, 271, 2003.

155. Blair, D.F., How bacteria sense and swim, *Ann. Rev. Microbiol.,* 49, 489, 1995.

156. Burdman, S. et al., Purification of the major outer membrane protein of *Azospirillum brasilense*, its affinity to plant roots, and its involvement in cell aggregation, *Mol. Plant Microbe Interact.,* 14, 555, 2001.

157. Sessitsch, A. et al., Advances in *Rhizobium* research, *Crit. Rev. Plant Sci.,* 21, 323, 2002.

158. Bashan, Y., Holguin, G., and de-Bashan, L.E., *Azospirillum*-plant relationships: physiological, molecular, agricultural, and environmental advances (1997–2003), *Can. J. Microbiol.,* 50, 521, 2004.

159. Patten, C.L. and Glick, B.R., Regulation of indoleacetic acid production in *Pseudomonas putida* GR12-2 by tryptophan and the stationary-phase sigma factor RpoS, *Can. J. Microbiol.,* 48, 635, 2002.

160. Jaeger, C.H. et al., Mapping of sugar and amino acid availability in soil around roots with bacterial sensors of sucrose and tryptophan, *Appl. Environ. Microbiol.,* 65, 2685, 1999.

161. Dunne, C. et al., Overproduction of an inducible extracellular serine protease improves biological control of *Pythium ultimum* by *Stenotrophomonas maltophilia* strain W81, *Microbiol.-SGM,* 146, 2069, 2000.

162. Thomashow, L.S. and Weller, D.M., Role of a phenazine antibiotic from *Pseudomonas fluorescens* in biological-control of *Gaeumannomyces graminis* var *tritici*, *J. Bacteriol.,* 170, 3499, 1988.

163. Nielsen, T.H. et al., Viscosinamide, a new cyclic depsipeptide with surfactant and antifungal properties produced by *Pseudomonas fluorescens* DR54, *J. Appl. Microbiol.,* 87, 80, 1999.

164. Molina, L. et al., Degradation of pathogen quorum-sensing molecules by soil bacteria: a preventive and curative biological control mechanism, *FEMS Microbiol. Ecol.,* 45, 71, 2003.

165. Joyner, D.C. and Lindow, S.E., Heterogeneity of iron bioavailability on plants assessed with a whole-cell GFP-based bacterial biosensor, *Microbiol.-UK,* 146, 2435, 2000.

166. Doyle, J.J., Phylogeny of the legume family: an approach to understanding the origins of nodulation, *Ann. Rev. Ecol. Syst.,* 25, 325, 1994.

167. Pueppke, S.G., The genetic and biochemical basis for nodulation of legumes by rhizobia, *Crit. Rev. Biotechnol.,* 16, 1, 1996.

168. Pueppke, S.G. et al., Release of flavonoids by the soybean cultivars McCall and Peking and their perception as signals by the nitrogen-fixing symbiont *Sinorhizobium fredii*, *Plant Physiol.,* 117, 599, 1998.

169. Peters, N.K., Frost, J.W., and Long, S.R., A plant flavone, luteolin, induces expression of *Rhizobium meliloti* nodulation genes, *Science,* 233, 977, 1986.

170. Perret, X., Staehelin, C., and Broughton, W.J., Molecular basis of symbiotic promiscuity, *Microbiol. Mol. Biol. Rev.,* 64, 180, 2000.

171. Lerouge, I. and Vanderleyden, J., O-antigen structural variation: mechanisms and possible roles in animal/plant-microbe interactions, *FEMS Microbiol. Rev.,* 26, 17, 2002.

172. Jain, V. and Nainawatee, H.S., Plant flavonoids: signals to legume nodulation and soil microorganisms, *J. Plant Biochem. Biotechnol.,* 11, 1, 2002.

173. Brencic, A. and Winans, S.C., Detection of and response to signals involved in host-microbe interactions by plant-associated bacteria, *Microbiol. Mol. Biol. Rev.,* 69, 155, 2005.

174. Broughton, W.J., Jabbouri, S., and Perret, X., Keys to symbiotic harmony, *J. Bacteriol.,* 182, 5641, 2000.

175. Rao, J.R. and Cooper, J.E., Rhizobia catabolize Nod gene-inducing flavonoids via C-ring fission mechanisms, *J. Bacteriol.*, 176, 5409, 1994.
176. Leon-Barrios, M. et al., Isolation of *Rhizobium meliloti* Nod gene inducers from alfalfa rhizosphere soil, *Appl. Environ. Microbiol.*, 59, 636, 1993.
177. Endre, G. et al., A receptor kinase gene regulating symbiotic nodule development, *Nature*, 417, 962, 2002.
178. Limpens, E. et al., LysM domain receptor kinases regulating rhizobial Nod factor-induced infection, *Science*, 302, 630, 2003.
179. Madsen, E.B. et al., A receptor kinase gene of the LysM type is involved in legume perception of rhizobial signals, *Nature*, 425, 637, 2003.
180. Radutoiu, S. et al., Plant recognition of symbiotic bacteria requires two LysM receptor-like kinases, *Nature*, 425, 585, 2003.
181. Cheng, H.P. and Yao, S.Y., The key *Sinorhizobium meliloti* succinoglycan biosynthesis gene exoY is expressed from two promoters, *FEMS Microbiol. Lett.*, 231, 131, 2004.
182. Gage, D.J., Infection and invasion of roots by symbiotic, nitrogen-fixing rhizobia during nodulation of temperate legumes, *Microbiol. Mol. Biol. Rev.*, 68, 280, 2004.
183. Hirsch, A.M., Plant-microbe symbioses: a continuum from commensalism to parasitism, *Symbiosis*, 37, 345, 2004.
184. Albus, U. et al., Suppression of an elicitor-induced oxidative burst reaction in *Medicago sativa* cell cultures by *Sinorhizobium meliloti* lipopolysaccharides, *New Phytologist*, 151, 597, 2001.
185. Fraysse, N., Couderc, F., and Poinsot, V., Surface polysaccharide involvement in establishing the rhizobium-legume symbiosis, *Eur. J. Biochem.*, 270, 1365, 2003.
186. Parniske, M., Schmidt, P.E., Kosch, K., and Muller, P., Plant defense responses of host plants with determinate nodules induced by eps-defective exob mutants of *Bradyrhizobium japonicum*, *Mol. Plant Microbe Interact.*, 7, 631, 1994.
187. Shaw, S.L. and Long, S.R., Nod factor inhibition of reactive oxygen efflux in a host legume, *Plant Physiol.*, 132, 2196, 2003.
188. Yasuta, T., Satoh, S., and Minamisawa, K., New assay for rhizobitoxine based on inhibition of 1-aminocyclopropane-1-carboxylate synthase, *Appl. Environ. Microbiol.*, 65, 849, 1999.
189. Yuhashi, K.I., Ichikawa, N., Ezura, H., Akao, S., Minakawa, Y., Nukui, N., Yasuta, T., and Minamisawa, K., Rhizobitoxine production by *Bradyrhizobium elkanii* enhances nodulation and competitiveness on *Macroptilium atropurpureum*, *Appl. Environ. Microbiol.*, 66, 2658, 2000.
190. Oldroyd, G.E.D., Engstrom, E.M., and Long, S.R., Ethylene inhibits the nod factor signal transduction pathway of *Medicago truncatula*, *Plant Cell*, 13, 1835, 2001.
191. Bartsev, A.V. et al., Purification and phosphorylation of the effector protein NopL from *Rhizobium* sp NGR234, *FEBS Lett.*, 554, 271, 2003.
192. Paschke, M.W., Dawson, J.O., and Condon, B.M., *Frankia* in prairie, forest, and cultivated soils of central Illinois, USA, *Pedobiologia*, 38, 546, 1994.
193. Paschke, M.W. and Dawson, J.O., The occurrence of *Frankia* in tropical forest soils of Costa Rica, *Plant Soil*, 142, 63, 1992.
194. Ganthier, D., Diem, H.G., and Dommergues, Y., *In vitro* nitrogen fixation by two actinomycete strains isolated from *Casuarina* nodules, *Appl. Environ. Microbiol.*, 41, 306, 1981.
195. Schwencke, J. and Caru, M., Advances in actinorhizal symbiosis: host plant-*Frankia* interactions, biology, and applications in arid land reclamation. A review, *Arid Land Res. Manage.*, 15, 285–327, 2001.
196. Gualtieri, G. and Bisseling, T., The evolution of nodulation, *Plant Mol. Biol.*, 42, 181, 2000.
197. Wall, L.G., The actinorhizal symbiosis, *J. Plant Growth Regul.*, 19, 167, 2000.
198. Benoit, L.F. and Berry, A.M., Flavonoid-like compounds from seeds of red alder (*Alnus rubra*) influence host nodulation by *Frankia* (Actinomycetales), *Physiol. Plant.*, 99, 588, 1997.
199. Hughes, M. et al., Effects of the exposure of roots of *Alnus glutinosa* to light on flavonoids and nodulation, *Can. J. Bot.*, 77, 1311, 1999.
200. Ceremonie, H., Debelle, F., and Fernandez, M.P., Structural and functional comparison of *Frankia* root hair deforming factor and rhizobia Nod factor, *Can. J. Bot.*, 77, 1293, 1999.
201. Vessey, J.K., Pawlowski, K., and Bergman, B., Root-based N-2-fixing symbioses: legumes, actinorhizal plants, *Parasponia* sp and cycads, *Plant Soil*, 266, 205, 2004.

202. Rai, A.N., Soderback, E., and Bergman, B., Cyanobacterium-plant symbioses, *New Phytologist,* 147, 449, 2000.
203. Bergman, B., Matveyev, A., and Rasmussen, U., Chemical signalling in cyanobacterial-plant symbioses, *Trends Plant Sci.,* 1, 191, 1996.
204. Nilsson, M., Rasmussen, U., and Bergman, B., Competition among symbiotic cyanobacterial *Nostoc* strains forming artificial associations with rice (*Oryza sativa*), *FEMS Microbiol. Lett.,* 245, 139, 2005.
205. Rasmussen, U. et al., A molecular characterization of the *Gunnera-Nostoc* symbiosis: comparison with *Rhizobium-* and *Agrobacterium*-plant interactions, *New Phytologist,* 133, 391, 1996.
206. Sylvia, D.M., Mycorrhizal symbioses, in *Principles and Applications of Soil Microbiology,* 2nd ed., Sylvia, D.M., Fuhrmann, J.J., Hartel, P.G., and Zuberer, D.A., Eds., Pearson Prentice Hall, Upper Saddle River, NJ, 2005, p. 263.
207. Parniske, M., Molecular genetics of the arbuscular mycorrhizal symbiosis, *Curr. Opin. Plant Biol.,* 7, 414, 2004.
208. Nagahashi, G. and Douds, D.D., Appressorium formation by AM fungi on isolated cell walls of carrot roots, *New Phytologist,* 136, 299, 1997.
209. Akiyama, K., Matsuoka, H., and Hayashi, H., Plant sesquiterpenes induce hyphal branching in arbuscular mycorrhizal fungi, *Nature,* 435, 824, 2005.
210. Larose, G. et al., Flavonoid levels in roots of *Medicago sativa* are modulated by the developmental stage of the symbiosis and the root colonizing arbuscular mycorrhizal fungus, *J. Plant Physiol.,* 159, 1329, 2002.
211. Kosuta, S. et al., A diffusible factor from arbuscular mycorrhizal fungi induces symbiosis-specific MtENOD11 expression in roots of *Medicago truncatula, Plant Physiol.,* 131, 952, 2003.
212. Johansson, J.F., Paul, L.R., and Finlay, R.D., Microbial interactions in the mycorrhizosphere and their significance for sustainable agriculture, *FEMS Microbiol. Ecol.,* 48, 1, 2004.
213. Sood, S.G., Chemotactic response of plant-growth-promoting bacteria towards roots of vesicular-arbuscular mycorrhizal tomato plants, *FEMS Microbiol. Ecol.,* 45, 219, 2003.
214. Filion, M., St-Arnaud, M., and Fortin, J.A., Direct interaction between the arbuscular mycorrhizal fungus *Glomus intraradices* and different rhizosphere microorganisms, *New Phytologist,* 141, 525, 1999.
215. Martin, F. et al., Developmental cross talking in the ectomycorrhizal symbiosis: signals and communication genes, *New Phytologist,* 151, 145, 2001.
216. Lagrange, H., Jay-Allgmand, C., and Lapeyrie, F., Rutin, the phenolglycoside from eucalyptus root exudates, stimulates *Pisolithus* hyphal growth at picomolar concentration, *New Phytologist,* 149, 349, 2001.
217. Tagu, D., Lapeyrie, F., and Martin, F., The ectomycorrhizal symbiosis: genetics and development, *Plant Soil,* 244, 97, 2002.
218. Gogala, N., Regulation of mycorrhizal infection by hormonal factors produced by hosts and fungi, *Experimentia,* 47, 331, 1991.
219. Ditengou, F.A., Beguiristain, T., and Lapeyrie, F., Root hair elongation is inhibited by hypaphorine, the indole alkaloid from the ectomycorrhizal fungus *Pisolithus tinctorius*, and restored by indole-3-acetic acid, *Planta,* 211, 722, 2000.
220. Garcia-Garrido, J.M. and Ocampo, J.A., Regulation of the plant defence response in arbuscular mycorrhizal symbiosis, *J. Exp. Bot.,* 53, 1377, 2002.
221. Liu, J.Y. et al., Transcript profiling coupled with spatial expression analyses reveals genes involved in distinct developmental stages of an arbuscular mycorrhizal symbiosis, *Plant Cell,* 15, 2106-2123, 2003.
222. Ane, J.M., Kiss et al., *Medicago truncatula* DMI1 required for bacterial and fungal symbioses in legumes, *Science,* 303, 1364, 2004.
223. Levy, J. et al., A putative Ca^{2+} and calmodulin-dependent protein kinase required for bacterial and fungal symbioses, *Science,* 303, 1361, 2004.
224. Mitra, R.M. et al., A Ca^{2+}/calmodulin-dependent protein kinase required for symbiotic nodule development: gene identification by transcript-based cloning, *Proc. Natl. Acad. Sci. USA,* 101, 4701, 2004.
225. Pinton, R., Varanini, Z., and Nannipieri, P., The rhizosphere as a site of biochemical interactions among soil components, plants and microorganisms, in *The Rhizosphere: Biochemistry and Organic Substances at the Soil-Plant Interface*, Pinton, R., Varanini, Z., and Nannipieri, P., Eds., Marcel Dekker, New York, 2001, p. 1.

12 Modeling the Rhizosphere

Peter R. Darrah and Tiina Roose

CONTENTS

I. INTRODUCTION

The rhizosphere is a highly complex environment where almost every process involves multiple interactions between the soil–root–microbe triumvirate as dealt with in other chapters in this volume document. Understanding rhizospheric processes is therefore extremely challenging, especially, as many of the questions to be studied require quantitative answers. For example, the rate of nitrogen influx into the root system is an important determinant of plant productivity, and yet explaining why this flux differs between different experimental treatments requires a very detailed knowledge of plant physiology, soil physics, chemistry, and biology as it relates to the N cycle [1]. The value of mathematical models is that they do generate quantitative answers and allow the integration of many individual plant and soil processes to predict a single outcome. However, such models require parameters to run the simulations and the experiments to measure them may be extremely difficult to do.

The most striking, indeed the defining, feature of the rhizosphere are gradients of solute concentration extending from the root surface into the surrounding soil. The zone encompassed by this gradient is, in modeling terms, the rhizosphere and it defines the volume of soil that is influenced by the root. This is in stark contrast to the practical definition of the rhizosphere used elsewhere in the volume, that is, the soil adhering to gently shaken roots. To talk about *the* rhizosphere is rather meaningless, as each solute under consideration will have a different gradient encompassing a different soil volume, and strictly, the rhizosphere is the totality of all the solute gradients surrounding a root. Thus there will be gradients of ammonium, nitrate, phosphorus, and each of the other 13 essential mineral nutrients (MN) required by higher plants — most of these gradients will be depletion profiles, that is, the solute concentrations will be lowest at the root surface. There will also be gradients of the 200 or so soluble organic solutes, which roots release into the soil [2] — these will typically be accumulation gradients, that is, concentrations of each solute are highest at the root surface. There may also be gradients in volatile compounds in the gaseous phase, for example, O_2 depletion profiles and CO_2 accumulation profiles due to root respiration, as well as gradients of compounds such as ethylene and other volatile organic compounds. The spatial scale over which these gradients are formed varies tremendously, from a few micrometers to perhaps meters but the typical scale for most soluble inorganic and organic compounds would be the millimeter scale. A typical idealized root rhizosphere is shown in Figure 12.1, which illustrates how much the environment of a rhizosphere bacterium depends on its spatial location in the rhizosphere.

The existence of cylindrical multiple gradients in very close proximity to root surfaces, in part explains the extraordinary difficulty of studying the rhizosphere. Rather few experimental techniques are appropriate for sampling soil with sufficient precision to explore these gradients, and many sampling strategies simply average the gradients and compare these averaged concentrations to those of bulk soil (see Chapter 5 and Chapter 11). Yet in doing so, most of the detail that drives rhizospheric processes is inevitably lost. Where appropriate techniques are used, gradients around roots have been demonstrated (Figure 12.2).

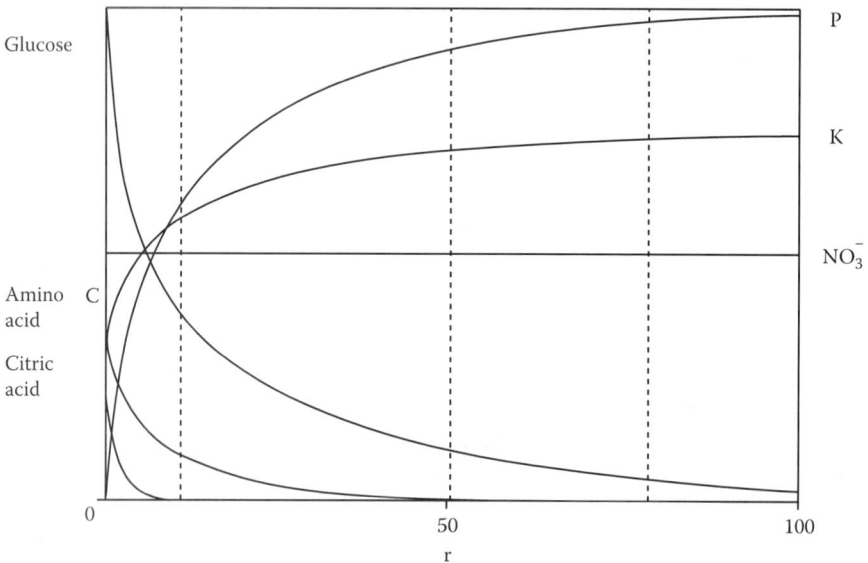

FIGURE 12.1 Diagrammatic representation of the nondimensional profiles of mineral nutrients and organic compounds in the soil surrounding a root. Note that the width of the rhizosphere (arbitrary units) differs for different solutes.

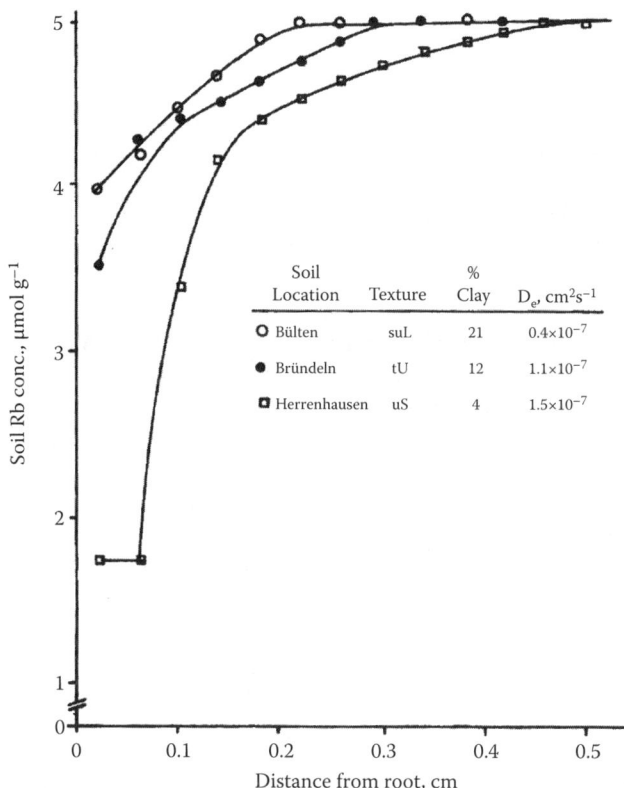

FIGURE 12.2 Gradients in Rb around a maize root as visualized by autoradiography. The depletion of Rb becomes more pronounced with time whereas the overall width of the Rb rhizosphere is largely independent of time. (From Jungk, A. and Claassen, N., *Adv. Agron.*, 61, 53, 1997. With permission from Academic Press.)

The rhizosphere defines the soil volume from which plants extract their MN requirements. It also defines the zone where plant–microbial interactions occur. Each unit length of root has its own unique, dynamic rhizosphere gradients, its own unique flux of nutrients into and out of the root, and its own unique microbial interactions. The whole plant integrates all these unique fluxes into a global nutrient acquisition rate and adjusts its growth rate, physiology, and allocation patterns accordingly [3]. Adjustments may involve major changes, such as a new root architectural form or increased root turnover or may be subtler, involving an increase in uptake capacity at the local rhizosphere scale (see the following text and Chapter 5 and Chapter 6 in this volume). However, such adjustments are likely to have implications at the rhizosphere level as whole-plant adjustments feedback to the activities of individual roots. Any mathematical consideration of the rhizosphere must therefore recognize these two spatial scales: the root scale and the plant scale; and consider how they are interconnected. Ultimately, mathematical models of the rhizosphere will have to consider even larger scales, for example, field, community, and landscape scales, and the final section of this contribution deals with these aspects.

II. MODELING SHORT-RANGE TRANSPORT AND UPTAKE IN SOILS

What processes drive the formation of solute gradients around roots? Why does the rhizosphere exist? It exists for the following two main reasons:

1. Soil is a heterogeneous medium consisting of a solid phase interspersed with a pore space that is filled with both air and water. Because of the high capillary and surface tension forces, the convective mixing of both the air and water phases hardly occurs, and both these phases are relatively immobile in an undisturbed soil. Any concentration imbalances in the mobile phases are therefore dependent largely on diffusion for their resolution.
2. Roots are active biological organs and remove some solutes and deposit others from/into the soil at their surfaces. Such activities generate concentration imbalances.

Rhizosphere modeling, therefore, simply consists of identifying and mathematically describing two types of processes: those responsible for perturbing soil solution equilibria — predominantly, the biological activities of plant roots and soil microorganisms; and processes responsible for restoring soil solution equilibria — predominantly, physical and chemical processes involving transport and chemical phase reactions.

Mathematical modeling of these two processes led to the development of the model [4], which forms the basis of most mechanistic models of nutrient uptake from soil.

A. Diffusion of Solutes

The development of the theory of solute diffusion in soils was largely because of the work of Nye et al. in the late sixties and early seventies, culminating in the essential reference work [5]. They adapted the Fickian diffusion equations for homogeneous media to describe diffusion in a heterogeneous porous medium by taking into account the impedance and chemical reaction processes in the soil.

Fick's law describes the relationship between the flux of a solute (mass per unit surface area per unit time, J_L) and the concentration gradient driving the flux, that is,

$$J_L = -D_L \nabla C_L \qquad (12.1)$$

where the proportionality constant, D_L, is the diffusion coefficient in pure water at the ambient temperature and pressure, and C_L is the concentration of the solute in solution. For a soluble compound diffusing in soil, three additional factors must be taken into account [5]:

1. The cross-sectional area available for diffusion is reduced because the solute can only move in the solution phase.
2. The distance that the compound must travel in moving from a point x_1 to x_2 is not simply the difference between x_1 and x_2. In an unsaturated soil, an individual solute molecule must follow a tortuous path through the moisture films around the soil particles, which increases the effective path length between x_1 and x_2. This introduces a tortuosity or impedance factor.
3. Diffusive movement is only significant in the solution phase and only concentration gradients in the soil solution drive diffusive movement.

To account for these factors, all of which act to reduce the rate of diffusion in soil compared to water, the solution diffusion coefficient must be modified [5] to represent the effective diffusion coefficient in soil:

$$J = -D_L \theta f \nabla C_L \qquad (12.2)$$

where θ is the volumetric moisture content and f defines an empirically determined impedance/tortuosity factor that accounts for the increased path length. C_L defines the soil solution concentration of the solute; some solutes occur entirely or predominantly in the soil solution. This would include most anions in most soils and uncharged organic molecules such as glucose. However, many solutes display positive charge and will interact electrostatically with the negatively charged surfaces found in most soils. Others, such as phosphate, may interact with surfaces via ligand exchange mechanisms. Such inner and outer sphere surface complexes [6], jointly termed *exchangeable solutes* do not contribute directly to diffusional flux and are not considered to be part of soil solution.

In the simplest case, a linear relationship may exist between solutes in soil solution and those on the solid phase such that $C_x = bC_L$, where C_x is the solute concentration on the solid phase and b defines the buffer power. The *buffer power* is hence the ratio between sorbed and solution solute. In perhaps more typical situations, the relationship might be defined by a more complicated, concentration-dependent, relationship such as a Freundlich or Langmuir expression; here, the instantaneous buffer power may be defined as the tangent to the exchange isotherm, that is, $b = dC_x/dC_L$. Some care is needed when defining this buffer power [7].

The conservation equations for a mobile nutrient are [8]

$$\frac{\partial C_x}{\partial t} = d_s \tag{12.3}$$

$$\frac{\partial(\theta C_L)}{\partial t} = \nabla \cdot (D_L f \theta \nabla C_L) - d_s \tag{12.4}$$

If we assume that absorption and desorption of the solute on the soil solid particles follows first-order kinetics, then,

$$d_S = k_a C_L - k_d C_x \quad \text{or} \quad \frac{d_S}{k_d} = \frac{k_a}{k_d} C_L - C_x \tag{12.5}$$

where k_a and k_d are the respective first-order rate constants. If we assume that $1/k_d$ is very small, that is, desorption is very fast relative to the diffusional timescale, then we obtain

$$C_x = \frac{k_a}{k_d} C_L = b C_L \tag{12.6}$$

where b is the buffer power in the preceding equation. By combining Equation 12.3 and Equation 12.4 and using Equation 12.6 we get

$$(b + \theta)\frac{\partial C_L}{\partial t} = \nabla \cdot (D_L f \theta \nabla C_L) \tag{12.7}$$

Thus, the effective diffusion coefficient for solute movement in the soil pore water is given by

$$D_e = \frac{D_L \theta f}{(\theta + b)} \tag{12.8}$$

To describe the diffusion of solutes in the rhizosphere, where concentration gradients change with time, t, as well as space, mass conservation is invoked with the spatial geometry appropriate for the cylindrical root [9]:

$$\frac{\partial C}{\partial t} = \frac{1}{r}\frac{\partial}{\partial r}\left[rD_e \frac{\partial C}{\partial r}\right] \qquad r_a < r < r_b \tag{12.9}$$

Here the spatial coordinate, r, defines the radial position in the cylinder of soil surrounding the root, the root surface is at r_a. This parabolic, second-order partial differential equation (PDE) describes how the concentration of available solute at an arbitrary point in the rhizosphere changes with time, in response to the concentrations on either side of it. Note that this equation is still effectively one-dimensional because concentration gradients along the root length can be neglected on the basis that roots are long and thin, and concentrations around the root are assumed to be radially symmetric. If concentration gradients along the root are thought to be important, then two-dimensional equations and solutions are required [10].

B. CONVECTIVE TRANSPORT

As water moves through the soil pores in response to water potential gradients, it moves with it, the solutes dissolved in soil solution. In a rhizosphere context, water moves radially toward the root to replace water taken up by the roots for transpiration. The flux of solute due to water movement (J_w) is simply the product of the rate of water flow at that point and the concentration in soil solution:

$$J_w = vC_L \tag{12.10}$$

In a cylindrical system, because equal amounts of water are being transported across successive concentric radial increments, the rate of flow of water must vary across the rhizosphere, with the fastest flows at the root surface. It is conventional to use a single parameter for rhizosphere water flow, v_0, which is defined as the water flux (volume per unit area per unit time) at the root surface, r_a.

C. DIFFUSION AND CONVECTION

If convection is included as a transport process in the rhizosphere, Equation 12.9 becomes

$$\frac{\partial C}{\partial t} = \frac{1}{r}\frac{\partial}{\partial r}\left[rD_e \frac{\partial C}{\partial r} + \frac{r_a v_0}{(\theta + b)}C\right] \tag{12.11}$$

Strictly speaking, in this formulation, the effective diffusion coefficient D_e is replaced by an empirical dispersion coefficient D^* to account for the effect of water flow on diffusion, that is, the so-called Taylor diffusion. However, in practice, the rate of transpirational water flow is sufficiently slow that dispersion effects are minimal and Equation 12.11 can be used without error. This is because the Péclet number (see Section F.2) is small. For the same reason, in almost all cases, diffusion is the most important process in moving nutrients to the root and the convection term can be omitted entirely.

D. TRANSFERS BETWEEN AVAILABLE AND UNAVAILABLE FORMS

During the lifetime of a root, considerable depletion of the available MN in the rhizosphere is to be expected. This, in turn, will affect the equilibrium between available and unavailable forms of MN. For example, dissolution of insoluble calcium or iron phosphates may occur, clay-fixed ammonium or potassium may be released and nonlabile forms of P associated with clay and sesquioxide surfaces may enter soil solution [11]. Any or all of these conversions to available forms will act to buffer the soil solution concentrations and reduce the intensity of the depletion curves around the root. However, because they occur relatively slowly, for example, over hours, days, or weeks, they cannot be accounted for in the buffer capacity term and have to be included as separate source ($\partial C/\partial t$) terms in Equation 12.11. Such source terms are likely to be highly soil specific and difficult to measure [12]. Many rhizosphere modelers have chosen to ignore them altogether, either by dealing with soils in which they are of limited importance or by growing plants for relatively short periods of time where their contribution is small. Where such terms have been included, it is common to find first-order kinetic equations being used to describe the rate of interconversion [13].

Another transfer, which is rather more difficult to deal with, is the conversion from organic to inorganic forms of MN [14]. Because these reactions are biologically catalyzed, their simulation depends on a thorough description of the activity of the soil microbial biomass and an understanding of the abiotic control of their activity that is still relatively poorly developed. The microbial biomass in the rhizosphere also changes with root age with their growth fed by root C flow (see Chapter 3 and Chapter 13 in this volume). It might be expected that the rhizosphere biomass would grow exponentially and this introduces a highly nonlinear element in the sink term of the diffusion–convection Equation 12.11. Transfers of N, for example, from organic to inorganic, root-available form are important in many agricultural situations and need to be included in realistic models. Yet when the effects of elevated CO_2 on the rhizosphere were reviewed, it was found that increased C input could inhibit, have no effect on, or stimulate N mineralization [15].

Rates of transfers between different forms will also tend to be different in the rhizosphere compared to bulk soil because of the activity of the roots and microorganisms [16]. Perhaps the most obvious difference is that the pH of the rhizosphere is usually different to that of bulk soil (Figure 12.3). The principal cause of this difference is root uptake of mineral ions in the form of cations and anions: inequalities in the amounts and valence are compensated for by root excretion of H^+ (cation excess) or HCO_3^- (anion excess). These excretions may change the pH of the rhizosphere very substantially [3]. The solubilities of many MN-containing mineral phases are pH dependent, and root-induced pH change is predicted to substantially alter the availability of nutrient ions in the rhizosphere. The ionic form in which nitrogen is taken up (NH_4^+ vs. NO_3^-) has, arguably, the largest effect on the pH of the rhizosphere but the other ions may also be important [3]. Hence, predicting pH change in the rhizosphere is complex and may involve the simultaneous solution of equations describing the uptake of all the macronutrients to determine the imbalances; such a solution has been attempted but without the feedback effect of pH on subsequent macronutrient availability [17].

Similarly, the redox status of the rhizosphere, the abundance of metal-complexing organics, and the activity of the soil microorganisms are all likely to be different in the rhizosphere. The differences are root-induced and many are thought to be plant adaptations to alleviate nutrient stress [16,18]. However, because the processes occur simultaneously, experiments have been unable to determine the adaptive significance of each for the mineral nutrition of plants. Some progress may be possible using mutants deficient in one or more adaptive traits [16] (see Section V). Models have proved very useful in quantifying fluxes between different MN pools and in distinguishing between the importances of different mechanisms for resupplying the rhizosphere; several examples are given in the following paragraphs.

In experiments with lowland rice (*Oryza sativa* L) it was found that roots quickly exhausted available sources of P and subsequently exploited the acid-soluble pool with small amounts deriving from the alkaline-soluble pool [19]. More recalcitrant forms of P were not utilized. The zone of net P depletion was 4 to 6 mm wide and showed accumulation in some P pools giving rather complex concentration profiles in the rhizosphere. Several mechanisms for P solubilization could be invoked in a conceptual model to describe this behavior. However, using a mathematical model with independently measured parameters [20], it was shown that it could be accounted for solely by root-induced acidification. The acidification resulted from H^+ produced during the oxidation of Fe^{2+} by O_2 released from roots into the anaerobic rhizosphere as well as from cation/anion imbalances in ion uptake [19]. Rice was shown to depend on root-induced acidification for more than 80% of its P uptake.

A mathematical model was used to evaluate the significance of organic-acid excretion by roots on P nutrition [21]. The model predicted that observed levels of citrate release could increase P

FIGURE 12.3 Root induced changes in pH in agar surrounding the roots of Fe-deficient tobacco (*Nicotiana tabaccum*) seedling as visualised by the indicator bromocresol purple. (From Hinsinger[131] with kind permission from Kluwer Academic Publishers.)

availability from goethite by 2- to 30-fold, depending on the P-loading of the mineral. The mechanism involved was competitive sorption between citrate and phosphate on the mineral surface. They also found that rhizosphere acidification by citric acid gave a smaller degree of P mobilization (compared to the release at a fixed pH of 5.0) due to pH-dependent interactions of P and citrate. In contrast, a model developed to describe the release of P from Mali rock phosphate predicted that the acidifying effect of root exudation of malic and citric acid was capable of solubilizing more P than rape plants required for growth [22]. These models showed how a common process (exudation of organic acids) could change the availability of an important mineral nutrient by two completely different mechanisms (competitive sorption and dissolution) depending on the insoluble P source. In many soils, roots would come into contact with both P forms.

E. ROOT BOUNDARY CONDITIONS

The preceding equations describe how solutes in the soil will move in response to concentration or water potential gradients. Such gradients form when the rhizosphere is perturbed by the activities of the root including water and MN abstraction and carbon deposition. These activities need to be mathematically described and form one of the two boundary conditions required to solve the initial-value problem.

1. Inner-Boundary Condition

Traditionally, nutrient uptake from solution culture was taken to depend on the concentration of the external mineral nutrient, C_L, the amount of nutrient-absorbing surface, and the kinetics of uptake per unit surface area or unit length of root [23]. The flux of nutrients into the roots, J, is described by one of two functionally equivalent equations

$$J = I_{max} \frac{C_L - C_{min}}{C_L - C_{min} + Km} \quad \text{or} \quad J = I_{max} \frac{C_L}{C_L + Km} - E \qquad (12.12)$$

where I_{max} is the maximum rate of uptake per unit surface area of root, Km is the affinity constant for uptake, C_{min} is the concentration at which efflux and influx exactly balance, and E is the rate of loss of nutrient to the soil per unit surface area. The three parameter values describing uptake are usually obtained from plants grown in solution culture [24]. The amount of root absorbing surface is normally equated with the root radius, that is, the root is viewed as a solid object.

Mass conservation at the rhizoplane means that the diffusive flux toward the root must equal the rate of extraction by the root (Equation 12.12), leading to the boundary condition

$$I_{max} \frac{C_L}{C_L + Km} - E = D_L \theta f \frac{\partial C}{\partial r} \qquad r = r_a \qquad (12.13)$$

The root surface at r_a is usually taken to be a fixed point. The effect of root thickening on uptake, where the root surface becomes a moving boundary has also been modeled [25].

2. Root Hairs and Root Structure

The classic Barber–Cushman model treats the root surface as a smooth solid cylinder. Yet many experimental studies have shown that root hairs are important for the uptake of some nutrients, for example, P [26,27]. Various mathematical models for root hairs have been used [5,28,29], all of which differ slightly in the way in which root hairs are modeled. Most authors conclude that root hairs make a substantial contribution to uptake, particularly for relatively immobile nutrients.

Textbook descriptions of roots stress that the root cortex is porous with the internal volume termed the apoplast, yet the Barber–Cushman model treats the root simply as an absorbing surface. There are some indications that inclusion of a diffusive-convective pathway within the root volume may affect the predicted uptake of MN [30]. Because higher root-surface concentrations can penetrate further into the root before becoming depleted, any additional uptake sites within the cortex become exposed to nutrients. This means that uptake is more sensitive to concentration than Equation 12.12 suggests, because I_{max} becomes an essentially concentration-dependent parameter.

For single-nutrient simulations at physiologically realistic soil solution concentrations, the apoplasmic pathway can be neglected because most of the nutrient is taken up by the epidermal cell layer with little further apoplasmic penetration. However, because some nutrients compete for uptake sites [31], the internal uptake pathway may become important in multinutrient models. It may also become important when high concentrations of potentially toxic elements such as Al^{3+} accumulate at the root surface.

3. Outer-Boundary Condition

Several boundary conditions have been used to prescribe the outer limit of an individual rhizosphere, ($r = r_b$). For low root densities, it has been assumed that each rhizosphere extends over an infinite volume of soil; in the model, r_b is set sufficiently large that the soil concentration at r_b is never altered by the activity in the rhizosphere. The majority of models assume that the outer limit is approximated by a fixed value that is calculated as a function of the maximum root density found in the simulation, under the assumption that the roots are uniformly distributed in the soil volume. Each root can then extract nutrients only from this finite soil cylinder. Hoffland [32] recognized that the outer limit would vary as more roots were formed within the simulated soil volume and periodically recalculated r_b from the current root density. This recalculation thus resulted in existing roots having a reduced r_b. New roots were assumed to be formed in soil with an initial solute concentration equal to the average concentration present in the cylindrical shells stripped away from the existing roots. The effective boundary equation for all such assumptions is the same:

$$\frac{\partial C}{\partial r} = 0 \quad 0 < t < \infty, \quad r = r_b \tag{12.14}$$

4. Interroot Competition

The use of a fixed boundary at r_b implicitly recognizes the fact that roots compete with each other for nutrients, but is based on the assumption that roots are parallel and uniformly distributed. However, roots growing in soil may have a more clustered distribution as they may grow preferentially down cracks or fissures in the soil fabric [33]. Clustering would mean that some roots would have a smaller r_b, whereas for others it would be larger than for the uniform case. This problem has been approached by assigning a "Voronoi polygon" to each root and considering each polygon to equate to a cylinder of radius r_b of the same area [34]. Uptake by the root system could then be simulated for solving the Barber–Cushman model for the population of cylinders so obtained. This study reported that the effect of clustering was small [34], whereas Reference 35 found a rather larger effect. A similar approach with "Thiessen areas" and worst-case situations for clustering with slash pine found that the regular distribution assumption was appropriate for K but led to an overestimate of P uptake [33].

Tinker and Nye [5] provided some useful guidelines for deciding when interroot competition would become an issue. They concluded that parallel roots do not really compete provided that the distance between roots is greater than the diffusion length, $(D_e t)^{1/2}$. For the standard parameter set for maize, some critical distances are given in Table 12.1 for selected nutrients and for different times. A typical root density for maize is 1 cm cm^{-2}, giving $r_b = 0.5$ cm for evenly spaced roots. Inspection of Table 12.1 reveals that interroot competition is generally unimportant for K and P even where experiments are simulated over 3 months. However, interroot competition for nitrate cannot be ignored, and models assuming uniform root distributions would be likely to overestimate nitrate fluxes. Long simulations might also overestimate K fluxes if significant clustering of roots

---

TABLE 12.1
Standard Parameter Values for the Uptake of Mineral Nutrients by Maize

Mineral Nutrient	C_0 μmol cm^{-3}	b	I_{max} μmol cm^{-2} sec^{-1}	Km μmol cm^{-3}	$\frac{a^2(\theta+b)}{D\theta}$ sec	λ
NO$_3^-$-N	5.0	1	1.0×10^{-5}	0.025	578	0.04
K	0.046	39	3.0×10^{-5}	1.4×10^{-2}	17467	14
S	0.1	2	3.0×10^{-7}	1.0×10^{-2}	1020	0.06
P	2.9×10^{-3}	239	3.26×10^{-6}	5.8×10^{-3}	106355	24
Mg	1.0×10^{-3}	1.2	4.0×10^{-6}	0.15	666	88
Ca	0.8×10^{-3}	156	1.0×10^{-6}	4.0	69466	27

Note: The soil is representative of a silt loam. Parameters are defined in the text. Values common to all MN are θ = 0.3, f = 0.3, D$_L$ = 1.0 × 10^{-5}.

occurred experimentally. However, they [5] also concluded that even when interroot competition did occur, its effects on uptake were not very great compared to a regular distribution (see Section V).

F. Solution of the Continuity Equations

Mathematical modeling of nutrient uptake requires the solution of Equation 12.11, subject to the boundary conditions imposed by Equation 12.13 and Equation 12.14. The initial state of the system (concentration distribution) must also be defined (initial-value problem). There are various methods of solution.

1. Numerical Approximation

There are a variety of methods suitable for the solution of second-order PDE but the method most widely used seems to be the finite-difference technique [9]. For simplicity, the solution is illustrated for the case of linear geometry. Imagine a volume of soil in which a one-dimensional concentration gradient of a solute has developed. At an arbitrary point, x, in this volume, we can identify 3 points located at $x - \Delta x$, x, and $x + \Delta x$ with different concentrations designated C_{i-1}, C_i, and C_{i+1} respectively (see Figure 12.4). We need to predict how the solute concentration changes over time in the shaded volume bounded by the planes at L and R. To do this, we assume that the Δx distance is sufficiently short that the concentration gradient between any adjacent points can be approximated by a straight line. In this case, the amount of solute crossing the plane at L in a short period of time, Δt, is given by Fick's law, that is,

$$F_{i-1} = -D_e \frac{C_i - C_{i-1}}{\Delta x} \tag{12.15}$$

and the amount crossing the plane at $x = $ R is given by

$$F_i = -D_e \frac{C_{i+1} - C_i}{\Delta x} \tag{12.16}$$

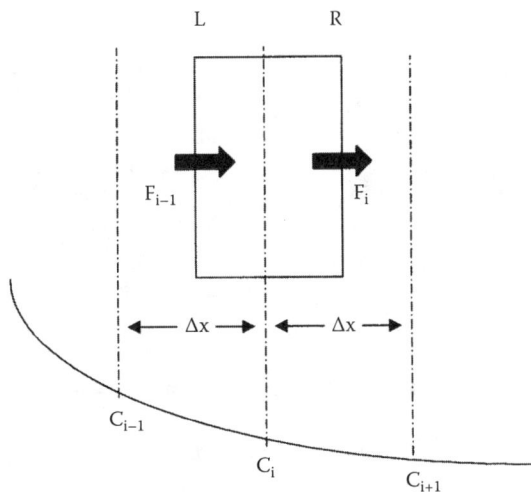

FIGURE 12.4 Schematic representation of a small section of a diffusion profile illustrating the application of Fick's law to determine the concentration change in the central volume element as a result of the fluxes (F) across the two planes at L and R.

The net change in solute in the compartment is then

$$\frac{F_i - F_{i-1}}{\Delta x} = \frac{D_e}{\Delta x^2}(C_{i-1} - 2C_i + C_{i+1}) \tag{12.17}$$

Mass must be conserved, so the two fluxes must cause the concentration in the compartment to change. If we define the new concentration in the compartment at time $t + \Delta t$ as $C_i^{t+\Delta t}$, then we can calculate this new concentration directly as

$$C_i^{t+\Delta t} = C_i + \frac{D_e \Delta t}{\Delta x^2}(C_{i-1} - 2C_i + C_{i+1}) \tag{12.18}$$

where the last term on the right-hand side can be evaluated either at the current time point t giving an explicit finite difference scheme, or at an intermediate time point, $t > = t + \nabla t$, to give an implicit scheme requiring the solution of a tridiagonal system of linear equations [36].

This then provides a physical derivation of the finite-difference technique and shows how the solution to the differential equations can be propagated forward in time, from knowledge of the concentration profile, at a series of mesh points. Algebraic derivations of the finite-difference equations can be found in most textbooks on numerical analysis [36]. There are a variety of finite-difference approximations ranging from the fully explicit method (illustrated previously) via Crank–Nicolson and other weighted implicit forward schemes to the fully implicit backward method, which can be used to solve the equations. The methods tend to increase in stability and accuracy in the order given. The difference scheme for the cylindrical geometry appropriate for a root is

$$C_i^{t+\Delta t} = C_i^t + \frac{2D_e \Delta t}{\left(r_{i+\frac{1}{2}} + r_{i-\frac{1}{2}}\right)\Delta r^2}\left(r_{i+\frac{1}{2}}C_{i+1}^t - \left(r_{i+\frac{1}{2}} + r_{i-\frac{1}{2}}\right)C_i^t + r_{i-\frac{1}{2}}C_{i-1}^t\right) \tag{12.19}$$

where the notation, $r_{i \pm 1/2}$, indicates the radial position midway between successive mesh points [36].

It must be remembered that all of the techniques involve numerical approximation. If the grid of mesh points is too coarse (violating the assumption of approximate linearity between adjacent mesh points) or the time step is too large (violating the assumption that $\partial C/\partial t \cong (C^{t+\Delta t} - C^t)/\Delta t$), then the solution will be inaccurate, but a solution will still be calculated. The general condition on stability for the explicit iteration scheme given earlier is that $D_e \Delta t/\Delta r^2 < 0.5$, which is quite restrictive [36]. It is therefore important when using these numerical approximations that appropriate steps are taken to check the accuracy of the solution. At its simplest, this would involve running the simulations for different combinations of Δx and Δt to ensure that the solutions are not significantly affected by the magnitude of these variables. The simulations should also be checked to ensure that mass is conserved. For rhizosphere simulations, where concentrations change most rapidly at the root surface, it may be helpful to use a nonuniform mesh spacing with points more clustered toward the root–soil interface. Mass conservation schemes and stability criteria are discussed in Morton and Mayers [36].

Numerical techniques are iterative and require considerable computer processing power. With modern desktop computers, this is usually not an issue, and solutions of root uptake over days or weeks typically take a few seconds to generate. However, for some strongly nonlinear problems, such as the development of rhizosphere microbial populations (Section III), where the increase in microbial biomass may be exponential over time, processing time may become an important issue.

Although most authors have used the finite-difference method, the finite-element method has also been used, for example, a two-dimensional finite element model incorporating shrinkable subdomains was used to describe interroot competition to simulate the uptake of N from the rhizosphere [37]. It included a nitrification submodel and found good agreement between observed and predicted uptake by onion on a range of soil types. However, whereas a different method of solution was used, the assumptions and the equations solved were still based on the Barber–Cushman model.

2. Analytical Solution

A problem with the solution of initial-value differential equations is that they always have to be solved iteratively from the defined initial conditions. Each time a parameter value is changed, the solution has to be recalculated from scratch. When simulations involve uptake by root systems with different root orders and, hence, many different root radii, the calculations become prohibitive. An alternative approach is to try to solve the equations analytically, allowing the calculation of uptake at any time directly. This has proved difficult because of the nonlinearity in the boundary condition, where the uptake depends on the solute concentration at the root–soil interface. Another approach is to seek relevant model simplifications that allow approximate analytical solutions to be obtained.

For analytical solutions, it is more convenient to work with nondimensional forms of the diffusion equations. We choose the following nondimensional substitutions: the time coordinate, t is replaced by the nondimensional parameter, t^*, and a is the root radius

$$t^* = \frac{tD_L f\theta}{a^2(\theta + b)}$$

the radial coordinate, r is also replaced by the nondimensional parameter, r^*

$$r^* = \frac{r}{a} \tag{12.20}$$

and the concentration variable, C is replaced by the nondimensional variable C^*

$$C^* = \frac{C}{Km} \tag{12.21}$$

Substituting for t, r, and C by t^*, r^*, and C^* into Equation 12.17 to Equation 12.19 gives the following equations:

$$\frac{\partial C^*}{\partial t^*} - Pe\frac{1}{r^*}\frac{\partial C^*}{\partial r^*} = \frac{1}{r^*}\frac{\partial}{\partial r}\left(r^*\frac{\partial C^*}{\partial r^*}\right) \tag{12.22}$$

$$\frac{\partial C^*}{\partial r^*} + PeC^* = \lambda\frac{C^*}{1+C^*} - \varepsilon \quad \text{at } r = 1 \tag{12.23}$$

$$C^* = C^*_\infty \quad t > 0, \, r \rightarrow \infty \tag{12.24}$$

$$C^* = C^*_\infty \quad t = 0, \, 1 < r < \infty \tag{12.25}$$

where the introduced dimensionless parameters are defined by

$$P_e = \frac{aV}{D_L f\theta} \tag{12.26}$$

$$\lambda = \frac{I_{max}a}{D_L fKm\theta} \tag{12.27}$$

$$\varepsilon = \frac{Ea}{D_L fKm\theta} \tag{12.28}$$

$$C_\infty^* = \frac{C_0}{Km} \tag{12.29}$$

These substitutions replace the eight dimensional parameters in the original equations by the four compound nondimensional parameters as shown earlier. The parameter P_e is the Péclet number [38] and indicates the relative importance of convection compared to diffusion. The advantage of this formulation becomes obvious when typical parameter values are substituted into the equations.

For all the essential nutrient ions, the diffusion coefficient, D_L, is essentially the same with a value of around 10^{-5} cm^2 sec^{-1}, whereas the water flux at the root surface is typically of the order 10^{-7}cm sec^{-1} for soils at around field capacity. The tortuosity factor typically scales with the volumetric moisture content over quite a wide range of moisture content, that is, $f \approx \theta$. As the soil becomes drier, the water flux will decline much faster than the tortuosity factor due to the typically log–linear relationship between the hydraulic conductivity and the moisture content [5]. Therefore, the maximum value of the Péclet number (Pe) will be

$$Pe = a\frac{10^{-7}}{0.3 \cdot 10^{-5}} \tag{12.30}$$

Typical root radii range from 0.0005 to 0.06 cm that yield Péclet values of the order 10^{-7} to 10^{-3}. Hence, in Equation 12.20, the Péclet number is always very small in comparison to the diffusion term and it can be neglected without error in almost all cases. This, in turn, implies that diffusion is the overwhelmingly dominant process responsible for moving nutrient ions around the rhizosphere. This mathematical conclusion contrasts with the tabulated estimates of the relative importance of mass-flow and diffusion for supplying ions to roots (for example, see Table 4.4, on p. 96, of Reference [23]), which implies that diffusion is an unimportant process for the supply of nitrogen, calcium, and magnesium. Here, the contribution of mass flow is calculated simply as the product of the rate of water flow and the average solution concentration of the ion. If this amount exceeds the quantity taken up by the plant, then mass flow is deemed to fulfill the nutrient demand giving the rather misleading impression that diffusion is unimportant. In reality, diffusive transport is still the most important transport mechanism but here it acts to move ions away from the root where they are accumulating because the plant demand for water exceeds that for N, Ca, or Mg. Without this diffusion, these ions would accumulate to potentially toxic levels, and the osmotic functioning of the root would be compromised.

In practice, the Péclet number can always be ignored in the diffusion–convection equation. It can also be ignored in the root boundary condition unless $C_\infty > \lambda/Pe$ or $\lambda < Pe$. Inspection of the table of standard parameter values (Table 12.2) shows that this is never the case for realistic soil and root conditions. Inspection of Table 12.2 also reveals that the term relating to nutrient efflux,

---

TABLE 12.2
Critical Root Distances for Three Mineral
Nutrients for Different Periods of Uptake

Days	NO_3^--N	K	P
1	0.24	0.04	0.02
30	1.34	0.24	0.1
90	2.32	0.42	0.17

Note: If neighboring roots are separated by more than the critical distance, then interroot competition can be neglected.

ε, can also be ignored because $\varepsilon < Pe \ll 1$. Roose et al. (2001) [8] investigated the solution of this simplified system of equations in various asymptotic limits ($\lambda \gg 1$, $C_\infty \gg 1$, etc.) and found a solution that was valid for all values of λ and C_∞, and at all times larger than the natural diffusive timescale, which typically ranges from a few seconds to a few hours (Table 12.1). Subsequently, a more direct solution for the nutrient flux at the root surface was derived [8], again valid for all values of λ and C_∞ which is given as follows:

$$F^*(t^*) = \cfrac{2\lambda C_\infty^*}{1 + C_\infty^* + \dfrac{\lambda}{2}\ln(4e^{-\gamma}t^* + 1) + \left[4C_\infty^* + \left\{1 - C_\infty^* + \dfrac{\lambda}{2}\ln(4e^{-\gamma}t^* + 1)\right\}^2\right]^{\frac{1}{2}}} \tag{12.31}$$

where γ is Euler's constant ($\cong 0.5772$). This equation gives the instantaneous flux into unit surface area of plant root at any time point in the simulation period. The cumulative uptake can easily be calculated by numerical integration of Equation 12.31*. Comparison of the numerical and analytical solutions for a range of nutrient parameters showed very close agreement indicating that the asymptotic approximations were valid.

The nutrient concentration around the root is given by

$$C^*(t^*) = C_\infty^* - \frac{F^*(t^*)}{2}E_1\left(\frac{r^{*2}}{4t^*}\right) \tag{12.32}$$

$$E_1(x) = \int_x^\infty \frac{e^{-y}}{y}dy \tag{12.33}$$

---

* The equations are required in dimensional form to deal with experimental data. The dimensional equivalents of Equation 12.31, Equation 12.32, and Equation 12.33 are:

$$F(t) = \frac{2F_m C_\infty^*}{1 + C_\infty^* + L(t) + \sqrt{4C_\infty^* + \left[1 - C_\infty^* + L(t)\right]^2}}$$

$$C_L(r,t) = C_{L,0} - \frac{2C_{L,0}\lambda}{1 + C_\infty^* + L(t) + \sqrt{4C_\infty^* + \left[1 - C_\infty^* + L(t)\right]^2}}E_1\left(\frac{(\phi + b)}{\phi D}\frac{r^2}{4t}\right)$$

$$L(t) = \frac{\lambda}{2}\ln\left(4e^{-\gamma}\frac{\phi D}{(\phi + b)a^2}t + 1\right)$$

An important advantage of the analytical approach is that it allows more sophisticated sensitivity analyses to be carried out. A traditional sensitivity analysis [23] shows how step changes in a single-model parameter changes the cumulative uptake into the plant after a certain time. A typical plot is shown for P in Figure 12.5 and reveals that uptake is most sensitive to changes in root radius with changes in E and v_0 having a negligible effect (as predicted earlier). However, these types of analyses can only present a single snapshot of the importance of the parameters at a particular time. They may also be slightly misleading in that they imply that the response is linear over the interval where the parameter values are halved or doubled; there is no reason to believe that the responses are in fact linear. Figure 12.6 shows sensitivity analyses of flux (F) of P into the root over time for similar parameter values as Figure 12.5, but calculated for infinitely small parameter changes (e.g., dF/da, where a is any given parameter). If the value at a particular time is negative, then this indicates that the flux decreases as the parameter value increases, whereas conversely, a positive value indicates that the flux increases as the parameter increases. The figure illustrates how different parameters dynamically affect uptake during the uptake period. Initially, the root radius has a very strong influence on uptake but this effect declines as the solution concentration at the root surface falls close to zero. Conversely, factors affecting the diffusion coefficient (θ, f) become more important over time as uptake becomes increasingly controlled by transport processes. Km is seen to have a transitory effect on flux during the period that soil solution concentrations of P are relatively high at the root surface.

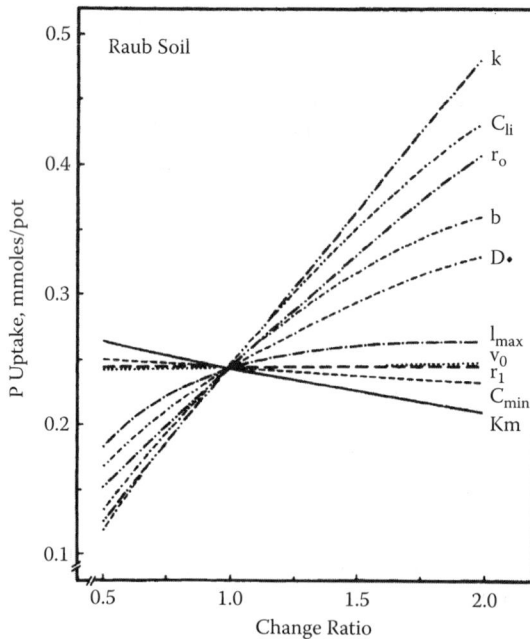

FIGURE 12.5 A sensitivity analysis for the Barber–Cushman model for the uptake of P by maize in Raub soil. The sensitivity was analyzed by halving and doubling each parameter value in turn while keeping all other parameters at their standard values. (From Silberbush, M. and Barber, S.A., *Plant Soil*, 74, 93, 1983. With permission from Kluwer Academic Publishers.)

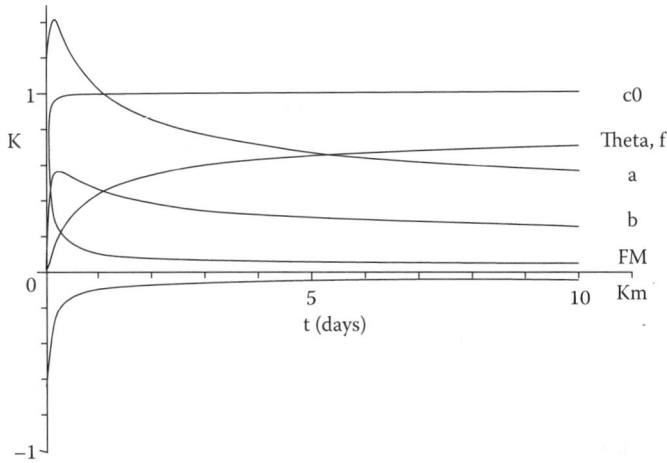

FIGURE 12.6 Sensitivity analysis of maize seedlings to some model parameter values during the first 10 d of uptake. The curves show how phosphorus flux (F) into the roots responds to differential perturbation to the parameters, a, (i.e., $\delta F/\delta a$).(Model parameters are given in Table 12.2.)

3. Other Analytical Approximations

One of the earliest approximations studied is to assume that the solute concentration is small or rapidly becomes small in comparison to the affinity constant for uptake. This then allows the nonlinear Michaelis–Menten equation to be approximated by:

$$I_{max}\frac{C_L - C_{min}}{C_L - C_{min} + Km} \cong \frac{I_{max}}{Km}C_L \tag{12.34}$$

which is linear in the concentration term and can be solved analytically. The ratio, I_{max}/Km was termed the *root absorbing power* [5]; the reference also discussed the implications of this simplification for the accuracy of prediction.

Another commonly used approximation for which there is an analytical solution is to assume that the root acts as a zero sink for uptake. Here, the solute concentration at the root surface is taken to be zero and uptake is therefore completely controlled by the diffusive flux to the root [22,39,40]. The implicit assumption is that root uptake is very rapid in comparison to resupply by transport, and hence the root very rapidly depletes the solute concentration at the root surface to zero and maintains it there. The validity of this assumption depends on the value of λ, and it is inapplicable, unless λ is greater than or about 10^8. For such large λ, there is a nondimensional critical time (t_c) after which it is reasonable to assume a zero sink [8,40]. Approximate values of t_c are

$$t_c \approx \frac{\pi C_\infty^2}{4\lambda^2} \ll 1 \quad \text{if} \quad \lambda \gg C_\infty \gg 1$$

or   (12.35)

$$t_c \approx \frac{1}{\lambda^2} \ll 1 \quad \text{if} \quad \lambda \gg 1;\ C_\infty \ll 1$$

Using the standard parameter values for maize presented in Table 12.2, the dimensional critical time after which a zero-sink approximation is valid are 180 sec for P and 89 sec for K. The zero-sink assumption is not valid at all for NO_3^-–N ($\lambda = 0.04$).

Such an analysis indicates that the zero-sink assumption must be used with extreme caution if accurate flux calculations are required at the local root level. Potassium for example is close to the limiting value of λ for the zero-sink assumption to be fulfilled and simulations with larger roots or larger buffer powers could well lead to inaccurate simulation results. Any zero-sink model involving nitrate should be treated with some suspicion. The zero-sink assumption is also widely used in root architecture models (see following sections).

G. Validation of the Local-Scale Model of Nutrient Uptake

The measurement of spatial gradients around roots at a resolution sufficient to provide an acceptable test of model predictions is very challenging and few studies have attempted it. Figure 12.7 shows the measured and predicted depletion of K for 4-d-old rape seedlings grown as a planar mat in contact with a column of soil. This method, although not representative of rhizosphere geometry, at least allows adequate spatial resolution to be obtained by sectioning the soil column. An alternative method for nutrients with radioisotopes is to scan the film density of autoradiographs. Figure 12.2 shows the results obtained from a cross-section of a maize root growing in a sandy soil with ^{86}Rb. These and other published studies (reviewed in Tinker and Nye [5] and Jungk and Claassen [13]) generally show good agreement between observed and predicted gradients; however, only rarely have such approaches been used to validate local-scale and plant-scale models in the same experiment.

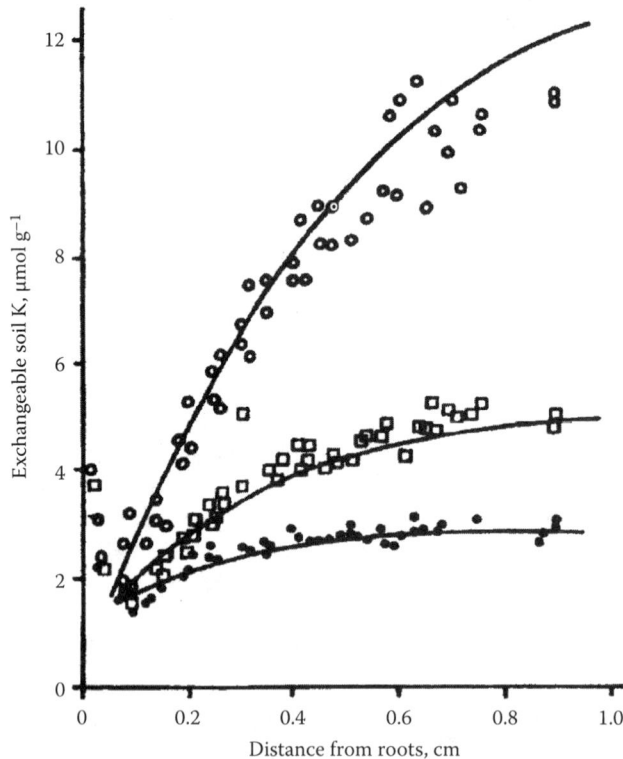

FIGURE 12.7 Comparison of experimental and modeled potassium depletion in the soil close to a planar mat of rape roots for three soil K^+ levels. The modeled lines were calculated using the Barber–Cushman model. (From Claassen, N. et al., *Plant Soil*, 95, 209, 1986. With permission from Kluwer Academic Publishers.)

III. MODELING THE DYNAMICS OF MICROBIAL POPULATIONS
AROUND THE ROOT

The rhizosphere was first defined [41] as a zone of increased proliferation of nitrogen-fixing bacteria associated with the roots of legumes. This definition was later widened to encompass increased proliferation of any microorganisms in the soil around any living plant root [42]. Many scientists still reserve the term rhizosphere for this microbiological phenomenon. Carbon from root exudates, sloughed root cells, and other root debris is thought to fuel this enhanced microbial growth, although it is possible that some species may respond to the altered mineral nutrient, O_2, or CO_2 concentrations or to other perturbations in the physicochemical environment surrounding the root. So far, modelers have concentrated on describing the microbial response in terms of a microbial biomass response rather than a species-specific response. This mostly reflects the uncertainty about the niche requirements and the identities and competitive abilities of the vast majority of microbes that may inhabit the rhizosphere. The exception to this rule is where the population dynamics of root pathogens have been modeled [43,44].

Newman and Watson [45] first applied mathematical modeling to the population dynamics of rhizosphere microorganisms growing on root exudates. Later, their equations were modified [10,46] to account for microbial death, C recycling, and the deposition of insoluble C. The basis of these models is the description of microbial growth using classical bacterial growth equations coupled with transport equations describing how microbial growth substrates are transported from the root surface through the soil. Microbial growth is generally described in terms of biomass carbon, and it is assumed that a single specific growth rate can describe the response of all the microbial species in the rhizosphere [47]. The specific growth rate of total soil biomass, μ (h^{-1}) is given by [46]:

$$\mu = \mu_{max} \frac{C}{C + K_s} - ZY\left(1 - \frac{M_0}{M}\right) \tag{12.36}$$

where C is the concentration of carbon growth substrate in soil, M_0 and M are the initial and current microbial biomass, u_{max} is the maximum specific growth rate, and K_s is the Monod affinity constant. Some carbon substrate is required for maintenance of the existing biomass, where Z is the maintenance requirement and Y is the growth yield. Any biomass dying because of insufficient substrate (i.e., $\mu < 0$) is assumed to enter a necromass pool, which can be decomposed to recycle carbon for further growth. The growth of microbial biomass is then described by:

$$\frac{\partial M}{\partial t} = \mu M \tag{12.37}$$

Therefore, if the carbon substrate is present at sufficiently high concentration anywhere in the rhizosphere (i.e., $\mu \approx \mu_{max}$), the microbial biomass will increase exponentially. Most models have considered the microbes to be immobile and so Equation 12.37 can be solved independently for each position in the rhizosphere provided the substrate concentration is known. This, in turn, is simulated by treating substrate carbon as the diffusing solute in Equation 12.36. The substrate consumption by microorganisms is considered as a sink term in the diffusion equation (Equation 12.11).

The outer-boundary condition is that given by Equation 12.14. The inner-boundary condition must describe the flux of carbon from the root into soil as a function of time. In the earlier models, this flux was either treated as time invariant or assumed to be maximal at the root tip and decline linearly over time, corresponding to a reduction in carbon efflux with root age [45,46]. Solution of these models revealed that the population dynamics of microbial biomass around roots was surprisingly complex (Figure 12.8). In the early stages of rhizosphere formation, the influence of the root extended several millimeters from the surface. As the root matured, the extent of the rhizosphere influence diminished

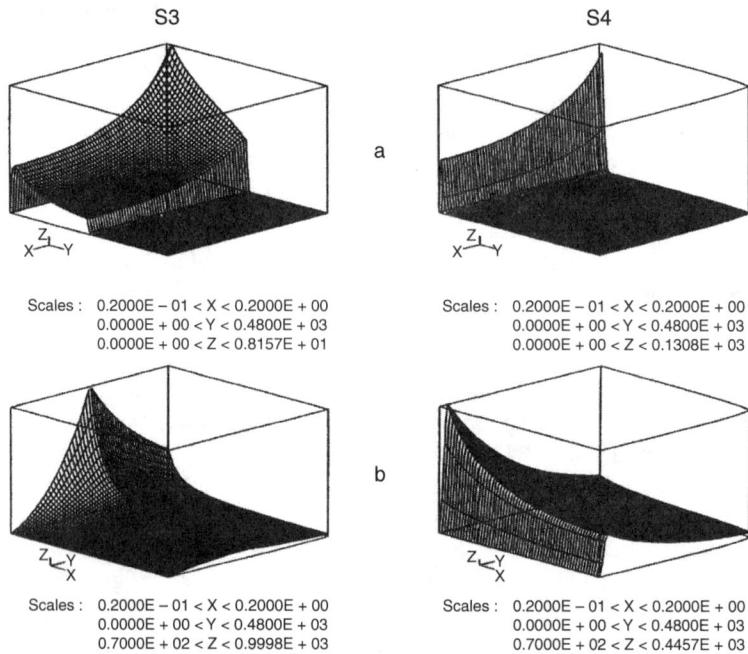

S3   S4

a

Scales : 0.2000E − 01 < X < 0.2000E + 00   Scales : 0.2000E − 01 < X < 0.2000E + 00
0.0000E + 00 < Y < 0.4800E + 03   0.0000E + 00 < Y < 0.4800E + 03
0.0000E + 00 < Z < 0.8157E + 01   0.0000E + 00 < Z < 0.1308E + 03

b

Scales : 0.2000E − 01 < X < 0.2000E + 00   Scales : 0.2000E − 01 < X < 0.2000E + 00
0.0000E + 00 < Y < 0.4800E + 03   0.0000E + 00 < Y < 0.4800E + 03
0.7000E + 02 < Z < 0.9998E + 03   0.7000E + 02 < Z < 0.4457E + 03

FIGURE 12.8 The simulated distribution of (a) soluble carbon and (b) microbial biomass in the rhizosphere of maize over a period of 10 d. The X-axis represents distance (cm) from the root surface and the Y-axis represents time (h). A uniform exudation rate (S3) is compared to the situation where the same amount of exudate is released in the first 24 h (S4). Note that the Z-axes have different scales although both represent μg C cm^{-3}. (From Darrah, P.R., *Plant Soil*, 133, 187, 1991. With permission from Kluwer Academic Publishers.)

as the population closer to the root consumed all the root exudation, hence depriving outer shells of a source of substrate. However, the intensity of the rhizosphere effect close to the root increased enormously with the ratio of $M:M_0$ reaching a peak of 14 [46]. The model revealed that the time course of root exudation had profound consequences for rhizosphere population dynamics. In Figure 12.8a, the root was assumed to exude C at a constant rate over a 10-d period while in Figure 12.8b, the same amount of C was exuded but all within a 1-d period. In turn, these different patterns of biomass proliferation could have important impacts on root functioning: for example [3], roots in Fe-deficient soils release phytosiderophores from near the root-tip to chelate and transport Fe. These organic compounds can also serve as carbon-substrates for rhizosphere microorganisms; the impact of microbial consumption on Fe nutrition will depend on the size and speed of response of the microbial community around the root tip [48]. These simulations suggest that the spatial and temporal patterns of exudation could have a major effect on root acquisition efficiency.

More recently, it was realized that rhizosphere carbon dynamics are more complex than these early models envisaged. Jones and Darrah [49–51] showed that roots actively scavenge their root exudates and that the reuptake of exudates was selective. In most situations, sugars and amino acids were scavenged by roots whereas organic acid exudates were not [52]. The authors also found that exudation losses were largely passive. This active involvement of the root essentially involves substrate-specific competition between the root and the rhizosphere biomass. This more complex situation was modeled using the root boundary condition [53]:

$$P(C_{cyt} - C_L) - V_{max} \frac{C_L}{C_L + K_m} = D_L f \frac{\partial C}{\partial r} \quad r = r_a \qquad (12.38)$$

where C_{cyt} and C_L are the exudate-carbon concentrations in the root cytoplasm and soil solution, respectively, P describes the permeability of the root to exudates. The second term describes the reuptake of C in terms of Michaelis–Menten kinetics. It was found that, in sterile soil, exudation losses declined over time and eventually reached a pseudo steady state where the reuptake of carbon matched the efflux. When microorganisms were included in the simulation model, total carbon losses increased because of microbial consumption, but roots were nevertheless surprisingly good competitors for their exuded carbon, being capable of reacquiring about 50% of the their exudation losses. A sensitivity analysis of the model (Figure 12.9) revealed that the exudate loss was most dependent on the factors controlling passive loss and active reuptake, that is, C_{cyt} and V_{max}, respectively. Exudation was surprisingly insensitive to the growth rate of microorganisms, implying that the rate of microbial accumulation had little effect on efflux. These and similar findings have revealed that roots are capable of exerting subtle effects on the carbon spectrum that are made available in the soil. This may then mean that the microbial composition of the rhizosphere is to some extent controlled by the root.

The importance of including soil-based parameters in rhizosphere simulations has been emphasized [54]. They used a time-dependent exudation boundary condition and a layer model to predict how introduced bacteria would colonize the root environment from a seed-based inoculum. They explicitly included pore-size distribution and matric potential as determinants of microbial growth rate and diffusion potential. Their simulations showed that the total number of bacteria in the rhizosphere and their vertical colonization were sensitive to the matric potential of the soil. Soil structure and pore-size distribution was also predicted to be a key determinant of the competitive success of a genetically modified microorganism introduced into the soil [55]. The Scott [54] model also demonstrated that the diffusive movement of root exudates was an important factor in determining microbial abundance. Results from models that ignore the spatial nature of the rhizosphere

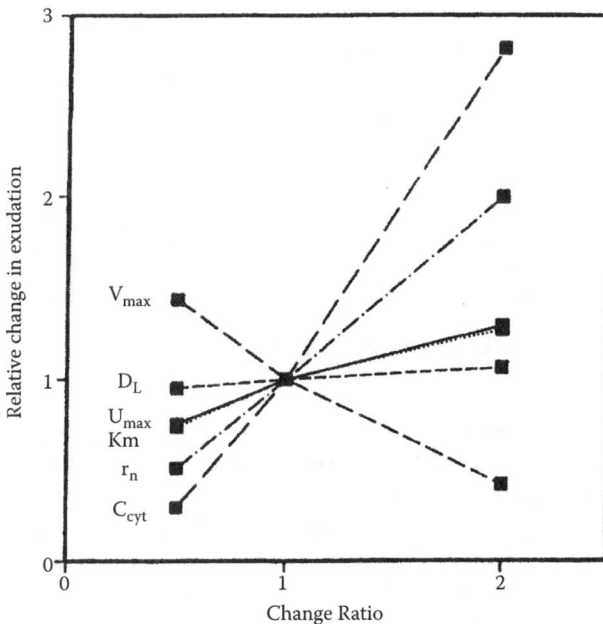

FIGURE 12.9 A sensitivity analysis for exudation losses by maize roots. The sensitivity was analyzed by halving and doubling each parameter value, in turn, while keeping all other parameters at their standard values. Parameters shown on the graph are defined in the text. (From Darrah, P.R., *Plant Soil*, 187, 265, 1996. With permission from Kluwer Academic Publishers.)

and treat exudate concentration as a spatially averaged parameter [15] should therefore be treated with some caution.

A major challenge is to move from such biomass-based models to consider the dynamics of individual species. Bacterial novelty is everywhere that researchers look, especially in soil, and it is no longer considered novel to report the discovery of novel organisms [56]. Much of the novelty is associated with uncultivatable bacteria and hence their physiology and ecology is unknown, for example, the diverse but abundant acidobacteria found in many soils [57]. Frequently, analysis of natural environments uncovers new taxa unrelated to previously known bacteria [58,59]. Continuing advances in molecular microbial ecology [60] have allowed the diversity of soil communities to be explored and has revealed a startling level of diversity. A catalogue of all the microbial species present in a soil sample has not yet been achieved and statistical approaches are currently used to estimate biodiversity from sample data. Reported diversities are around 10^3 to 10^5 species per g-Kg sample, whereas a ton of soil could theoretically contain three million different species [61]. The only direct experimental method for estimating diversity is based on DNA reassociation kinetics and reported 6000 species (forest soil), 3500 to 8800 species (pasture soil), and 140 to 350 species (arable soil) for samples of around 100 g [62].

The distribution of bacterial cells from a topsoil often shows relatively strong spatial autocorrelation with most of the variance occurring at the micrometer to millimeter scale [63,64] with evidence of bacterial patches and gradients. The evidence of spatial structure as opposed to random noise in soil ecology has recently been reviewed [65], and biological properties have been shown to have structured spatial distributions ranging from micrometer to kilometer scales with, for example, microbial biomass distribution being strongly correlated at the meter scale with the growth of individual plants.

A common assumption in models is to assume homogeneity in distribution but this recent work on diversity both in species and in space runs counter to this assumption. This has profound implications for rhizosphere models. The apex of the root is largely sterile [66], suggesting that rhizosphere colonizers do not comigrate with the root, but colonize from the surrounding soil so each new root can be thought of as an independent, random sampling of the bacteria in the soil. But if the pool of possible colonizers is very large and spatially structured at several scales, then the chances are that each new rhizosphere would be a unique subsample of the community pool. Selection pressures on each unique community may subsequently reduce the heterogeneity between them, but there seems to be few theoretical grounds to assume that the microbiology of individual rhizospheres of the same plant are reproducible. Modelers are therefore faced by multiple instances of very large numbers of very diverse organisms of generally unknown potential — not surprisingly, progress in modeling detailed rhizosphere microbiology has been slow!

IV. MODELING UPTAKE AT THE PLANT SCALE

The objective of the local-scale simulation of uptake into roots is normally to allow prediction of nutrient acquisition by the whole plant. This is true even where the objective is to validate the local-scale model because of the difficulties of measuring experimentally the cylindrical gradients around plant roots to the necessary precision and resolution to validate adequately the local-scale model. Typically, the models are applied to growing plants with continually expanding root systems. Nutrient acquisition at the plant scale is then achieved by integrating root uptake at the local scale and root growth. Barber [23] reviewed the general approach. He made the following assumptions about the growth of maize root systems:

1. Roots grew exponentially over the experimental period. The growth rate of the roots (g) was calculated as $\ln(L_t - L_0)/(t_1 - t_0)$, where L_t denotes the total root length at the end of the experiment at t_1 and L_0 the initial root length at the start of the experiment at t_0. In a typical experiment, t_0 would occur 6 d after germination, and t_1, some 2 to 3 weeks later.

2. A single average root radius could approximate the radius of the roots for the entire experimental period. To calculate this radius, the total root length and the total root weight at harvest was measured. The latter was converted into a root volume, V_t, and the following equation applied:

$$r = \sqrt{\frac{V_t}{\pi L_t}} \qquad (12.39)$$

The cumulative uptake into whole plants (N_t) could then be calculated as a function of time using the equation:

$$N_t = L_0 \int_0^T J_L dt + L_0 \int_0^T \frac{de^{gt}}{dt} \int_0^{T-t} J_L \, dt \, dt \qquad (12.40)$$

which integrates the flux into the root, J_L, (expressed on a unit length basis) by an exponentially growing root system over time, T. The main feature of this model is the exponential growth rate of the root system with a constant coefficient, g, and it is therefore hereafter referred to as the uptake with convection and diffusive model with constant growth or UDC-GC model.

This equation has been used, mainly by the original authors, to compare predicted and observed uptake by a range of plant species. A typical outcome for maize seedlings up to 22 d old is shown in Figure 12.10a. Generally, these models have shown good correlation between observed and predicted uptakes in both pot and field experiments. It is true that the imposition of a constant growth rate forces the model to accumulate nutrient and therefore some degree of correlation between observed and predicted uptake must be expected if intermediate harvest data is used. Moreover, examination of the regression coefficients often indicates that there is substantial under- or overprediction of plant uptake, indicating a systematic error in the formulation of the models. Without an independent test of the local-scale model, it is impossible to be sure whether this systematic bias derives from the local scale or from the plant scale, as both scales make assumptions that may not be appropriate.

A. THE ASSUMPTION OF CONSTANT ROOT RADIUS

One obvious deficiency of the treatment of root growth in the UDC-GC model is the assumption of a constant root radius. Root systems consist of several orders of root, with each order branching off the parent root and having a smaller diameter. Therefore, one would predict that the average root radius would decrease over time as the root system consisted of proportionately more higher-order roots. Because local-scale uptake of nutrients is highly sensitive to root radius (see Figure 12.5), the average root radius simplification can be expected to be rather inaccurate.

Recent developments have led to the formulation of mathematical models capable of describing the three-dimensional architecture of plant roots growing in soil [67–70]. Most of the models operate by applying sets of rules, for example, those that define the distance between an existing lateral branch and a new branch, or the branching angle and orientation for each root order, or by imposing the rules of fractal geometry. Most are also capable of visually displaying the spatial distribution of the evolving root system (see Figure 12.11). The application of different branching parameters leads to the development of different root architectures that are particularly efficient for the exploitation of large soil volumes (herringbone architecture) or for the intensive exploitation of small soil volumes (dichotomous architecture). Model simulations [71] suggest that the herringbone topology is efficient for acquiring rather mobile nutrients such as nitrate, whereas the much more immobile nutrients like P are most efficiently exploited by a

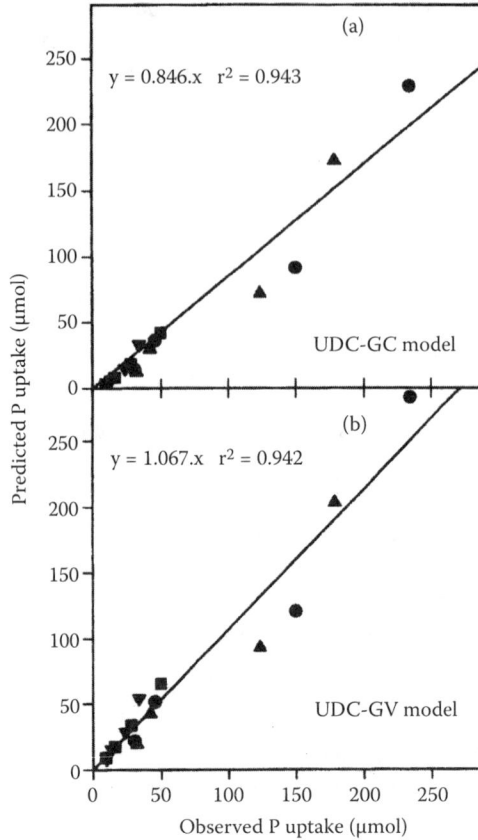

FIGURE 12.10 Relationship between observed and predicted P uptake by two tomato cultivars grown in soil of differing P supply. Uptake was predicted by the original (a) Barber–Cushman model or (b) by the UDC-GV model. (From Darrah, P.R., *Inherent Variation in Plant Growth: Physiological Mechanisms and Ecological Consequences*, Lambers, H., Poorter, H., and van Vuuren, M.M.I., Eds., Backhuys Publishers, 1998, p. 159. With permission from Backhuys Publishers.)

dichotomous form. Here, efficiency is measured by comparing the carbon cost of construction to the amount of nutrient acquired. Virtual root systems can now be inspected online [72] and the general field of morphological modeling of plants was recently reviewed [73]. Root system models have recently been combined with nutrient uptake models to model depletion zones around individual roots [74].

Architectural models explicitly specify the distribution of roots in space. An alternative approach, which is also useful for rhizosphere studies, is the continuum approach where only the amount of roots per unit soil volume is specified. Rules are defined that specify how roots propagate in the vertical and horizontal dimensions, and root propagation is usually viewed as a diffusive phenomenon, that is, root proliferation favors unexploited soil. This defines the exploitation intensity per unit volume of soil and, under the assumption of even distribution, provides the necessary information for the integration step mentioned earlier. Acock and Pachepsky [75] provide an excellent review of the different assumptions made in the various continuum models formulated and show how such models can explain root distribution data relating to chrysanthemum.

In one of the more recent continuum models [76], an approach first suggested by Leonardo da Vinci in the 15th century, from his observations that the cross-sectional area (CSA) of the main

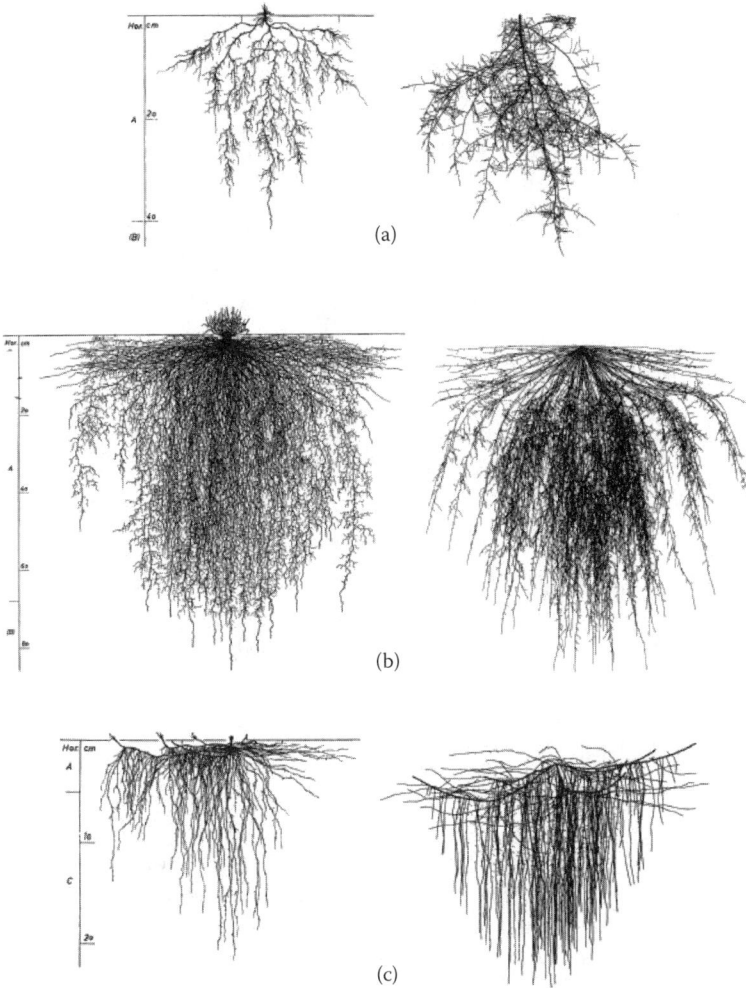

FIGURE 12.11 Observed and simulated root systems of a) *Arabidopsis thaliana*, b) *Lolium multiflorum*, and c) *Achillea millefolium*. Simulations (on the right-hand side) were generated by *RootTyp*. (From Pages, L. et al., *Plant Soil*, 258, 103, 2004. With permission from Springer Science and Business Media.)

stem of a tree equaled the cross-sectional area of the tree branches, was used. The more generalized rule was considered:

$$d^\Delta = \alpha\left(d_1^\Delta + d_2^\Delta\right)$$

where d_1 and d_2 represent diameter values before and after branching, respectively, Δ is the diameter exponent, and α is a proportionality factor that allows for unequal branch diameters to be formed. By assuming that each segment of root had to attain a minimum length before branching could occur, and assuming that there was a minimum root diameter below which no further branching could occur, a computer program was developed that could produce very realistic computer representations of real roots [76].

Maize roots typically have 3 to 4 orders with zero-order roots having the largest diameters and being formed from the seed and shoot [68]. Each order can be divided into three regions: a basal nonbranching region (l_b), a branched zone (l), and an apical nonbranching zone (l_a). New branches

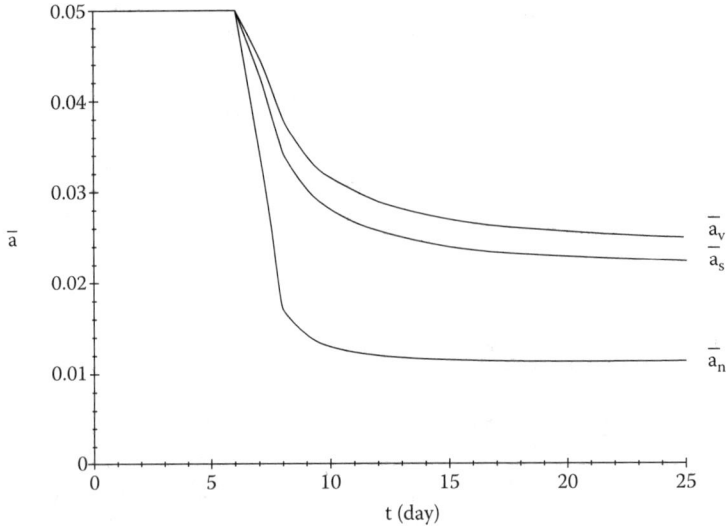

FIGURE 12.12 The predicted changes in average root radius of maize root systems during 20 d of growth. The subscripts on the curves denote differing ways of calculating the average (by volume, by surface area, or by root number). (From Roose, T. et al., *J. Math. Biol.*, 42, 347, 2001.)

can form on this root if $l > l_a + l_b$ and these branches arise at regular intervals defined by the internodal distance, l_n. Hence, a root of length l has a maximum number of branches given by $[(l - l_a - l_b)/l_n]$. The lengths l_a, l_b, and l_n depend primarily on the branch order, but could also depend on age, nutritional status, or other physiological or environmental factors. Roose et al. [8] extended an existing theory of variable root elongation [68] and constructed a model describing the length and time development of the populations of roots of different orders in the form of a hyperbolic PDE. Each root order was also considered to have a maximum attainable length and the elongation rates of roots were assumed to decline as this length was approached. The average root radius could then be calculated as a function of time during the development of a maize seedling and these values could be compared with the fixed, average value used in the classical approach mentioned earlier. The results are shown in Figure 12.12, which demonstrates the highly dynamic nature of root radius during the early development of seedlings where the volume-averaged root radius halves in magnitude during the first few weeks of growth. The changes suggest that the approach of estimating root radius as a single fixed point calculated from the root size distribution at the end of the experiment is likely to yield inaccurate cumulative flux estimates.

Once the root size distribution had been successfully modeled, the total uptake by the plant could then be predicted [8] by integrating the flux equations for each root order separately and then summing the contributions of each root order. Figure 12.13 shows the relative contributions of each root order to total uptake over a 20-d period. Initially, all uptake is by growing zero-order roots that do not reach their maximum lengths until about 50 d. From around day 6, first-order roots start to be produced and these are responsible for the increase in uptake observed at this time. The simulations shown are for phosphate uptake where it is possible to assume that interroot competition between individual roots is unimportant. Second-order roots are relatively unimportant in the timescale simulated but would start to have a major impact on uptake at a later stage. These simulations can also be compared with the classical UDC-GC approach and these simulations are shown in Figure 12.14; it is obvious that the Barber–Cushman method of radius averaging significantly underestimates the uptake.

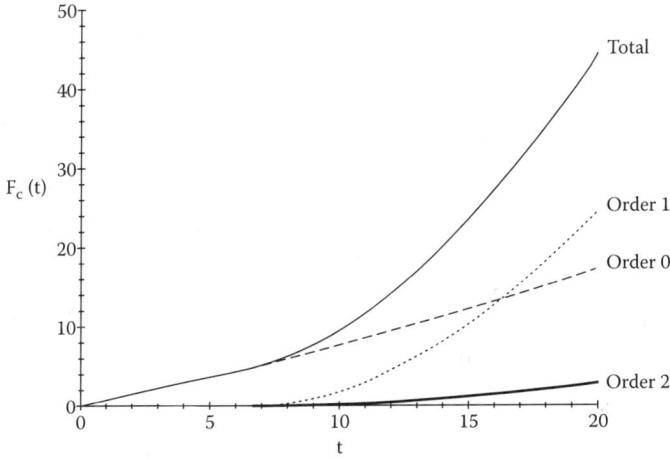

FIGURE 12.13 The predicted contribution of roots of different order to the uptake of phosphorus by maize over a 20-d period. Parameter values are given in Table 12.2.

B. THE ASSUMPTION OF EXPONENTIAL ROOT GROWTH WITH A CONSTANT COEFFICIENT

The local-scale model predicts nutrient uptake by plants as a function of the nutrient concentration in the rhizosphere; it should be equally accurate for both high and low nutrient concentrations provided the buffer capacity term has been described accurately. It has indeed been used to validate the uptake model for a range of soil types with wide ranges of nutrient availability [23]. However, this introduces a complication at the plant scale. If the rates of uptake at the plant scale are low, then it might be reasonable to expect that the rate of plant growth might also decline as a consequence of nutrient deficiency. In fact, in any situation where the yield potential of the plant is constrained by a limiting nutrient, which by itself is the nutrient being simulated, the assumption of constant exponential root growth seems *a priori* to be unacceptable. In these nutrient-limited plants grown from seed, one might, instead, predict a gradual decline in plant (and root) growth

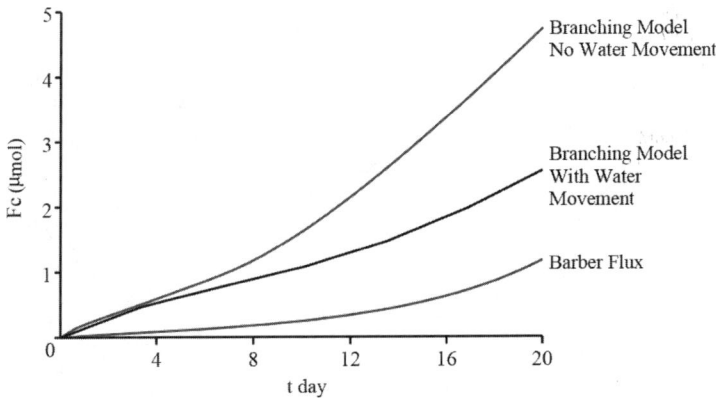

FIGURE 12.14 Predicted cumulative P uptake (F_c) by maize according to three models. The Barber flux curve uses the minimal model with a single average root radius and constant water content. The topmost curve introduces a root branching model into the minimal model above that partitions the root system into three orders of root with different radii. The middle curve includes different radii and allows water content to fluctuate in the soil profile in response to rainfall and drainage. (From Roose, T. and Fowler, A.C., *J. Theor. Biol.*, 228, 173, 2004. With permission of Elsevier.)

with time, as the seed reserves become diluted by growth, and the uptake mechanisms cannot keep pace with the MN demands of growth. The lack of feedback mechanisms between cumulative nutrient acquisition and subsequent growth seem to be one of the main deficiencies of the classical approach.

Inclusion of feedback between uptake and growth obviously requires knowledge of the relationship between MN content and growth, and such relationships form the basis of many whole-plant growth models [77]. Whole-plant growth models (recently reviewed [78,79]) typically relate aboveground parameters such as a photosynthetically active radiation, humidity, leaf area, and net assimilation rate to predict rates of carbon fixation. Their belowground submodels are usually empirically based and predict water and MN uptake. When combined with developmentally controlled descriptions of resource allocation and ontogeny, these models aim to predict biomass or grain yield. Such models are typically management tools, used for guidance on when and where to apply fertilizer or irrigation and most are "models-without-roots." Such empirically based models are of limited use in the current context. Other models [80] specify the nutrient flux as a parameter and then model the allocation of this resource and photosynthate within the plant. There is obviously considerable scope for amalgamating this approach with the mechanistic modeling of nutrient uptake.

One approach that has been used [29] is to try to assess the potential importance of feedback, while keeping the description of the whole plant model as simple as possible. The relative addition rate technique (RAR) is an experimental procedure [81] that allows plants to be maintained at a constant exponential growth rate in nutrient-limited conditions. This is achieved by supplying nutrients at an exponentially increasing rate such that the supply of nutrients is always just sufficient to maintain a constant tissue concentration of this growth-limiting nutrient. If the tissue concentration departs from this constant value, then the plant growth rate changes. In a series of papers (see Reference 82 for review), the authors derived a linear relationship between internal nutrient concentration and relative growth rate or RAR, extending from zero-growth at some threshold internal concentration to the maximum specific growth rate at a concentration termed optimal. Above this optimal internal concentration, plants grew at the maximum specific growth rate. This then gives the simplest possible relationship between nutrient uptake and subsequent plant growth (Figure 12.15). The only modification required to solve the equations is to allow the growth rate, g, to depend on internal nutrient concentration in Equation 12.35. This approach was used to

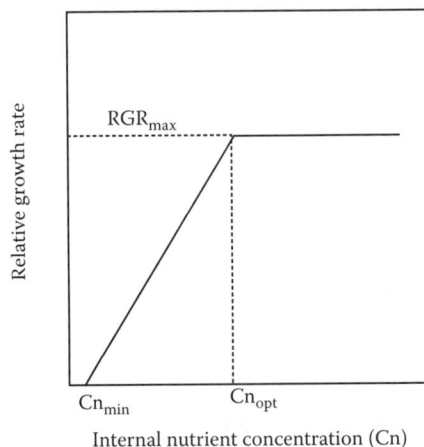

FIGURE 12.15 The relationship between relative growth rate (RGR) and internal nutrient concentration (Cn). Growth ceases at nonzero Cn (at Cn_{min}) and the maximum rate of growth is reached at Cn_{opt}. (From Ingestad, T. and Ågren, G.I., *Plant Soil*, 168–169, 15, 1995. With permission.)

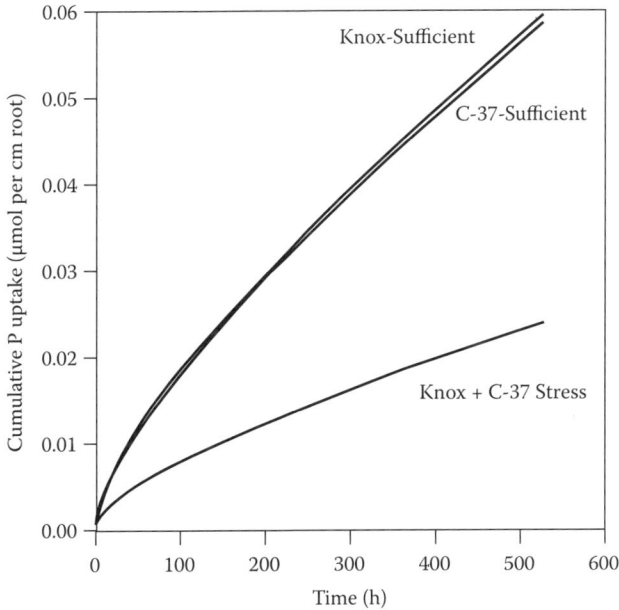

FIGURE 12.16 Cumulative phosphorus uptake curves (per unit root length) predicted for two tomato cultivars (Knox and C37) grown in a soil fertilized to two P levels. (From Darrah, P.R., *Inherent Variation in Plant Growth: Physiological Mechanisms and Ecological Consequences*, Lambers, H., Poorter, H., and van Vuuren, M.M.I., Eds., Backhuys Publishers, 1998, p. 159. With permission from Backhuys Publishers.)

reanalyze some data [83] on the growth of young tomatoes in soils of differing P status. These data had previously been analyzed using the UDC-GC approach, with a fixed growth rate calculated from the root system at the end of the experimental period. Figure 12.16 shows the local-scale-predicted P uptake for the two tomato cultivars in the P-sufficient and P-deficient soils and shows that the deficient soil is capable of delivering P to the plant at a much lower rate than the sufficient soil. Therefore, unless the plants are indulging in luxury uptake in both cases, one would expect the growth rates of the plants in the different conditions to be different. This was indeed found and the UDC-GC model was validated [83] with postexperimentally determined average growth rates of 0.0096 and 0.0085 h^{-1} for the sufficient and deficient soils, respectively; note that these are the constants in an equation describing exponential growth and so the differences reveal substantially different growth potentials. However, the seedlings were pregerminated and maintained under identical conditions prior to starting the experiments. Therefore, it is reasonable to assume that immediately after transplanting, all the seedlings were growing at the same rate irrespective of soil treatment. In the new analysis [29], a constant growth rate was used for all seedlings initially, and these growth rates were allowed to vary as a consequence of their P uptake history. Figure 12.17 shows the predicted growth rates for both cultivars in both soil treatments over time and reveals an approximate halving in growth rate over the experimental period for the deficient soil. However, when the predicted uptake was compared with the experimentally observed uptake (Figure 12.10b), the correlation coefficients were identical for the UDC-GC and UDC-GV (growth variable) models, and the accuracy of the predictions was actually better for the UDC-GV model.

What the UDC-GV model has shown is that plants grown in nutrient-limited conditions could downregulate their growth rate in response to nutrient deficiency and still produce predictions that validate the model at least as well as the original analysis. It is extremely unlikely that the very simple and direct feedback between growth and uptake used in the UDC-GV model is correct. For example, there is likely to be some time delay in response to deficiency and probably some internal

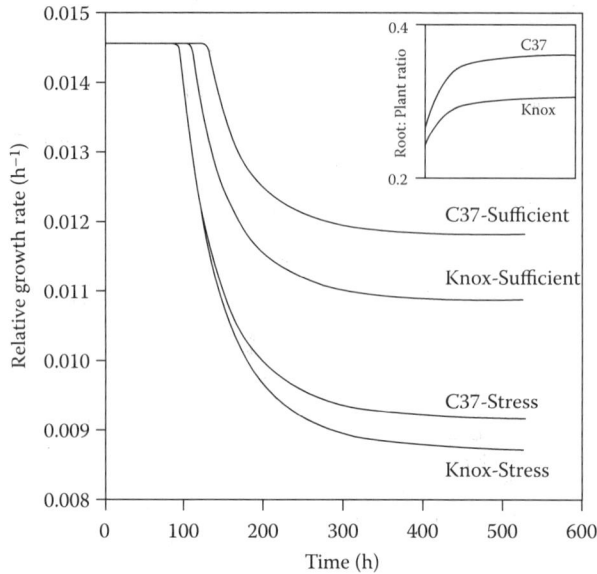

FIGURE 12.17 Predicted changes in the relative growth rates of two tomato cultivars grown under sufficient and insufficient levels of phosphorus. The inset shows the changes in root mass: plant mass ratio of the cultivars in the low-P soil. (From Darrah, P.R., *Inherent Variation in Plant Growth: Physiological Mechanisms and Ecological Consequences*, Lambers, H., Poorter, H., and van Vuuren, M.M.I., Eds., Backhuys Publishers, 1998, p. 159. With permission from Backhuys Publishers.)

reallocation of P to enable growth to be maintained at a higher rate for longer. This then raises the question of whether agreement between observed and predicted values at the plant scale can be considered to provide adequate validation at the root scale. If two models with very different root dynamics produce similar results, how does one judge which is correct?

The most complete mechanistic model to include both local- and plant-scale processes is *ecosys* [84,85]. This includes a Barber–Cushman mineral nutrient submodel, a mineralization by microbial biomass submodel, and explicit models mycorrhizae and root growth. It also includes a whole-plant growth model based on the functional equilibrium concept [86]. The model is driven by around 50 parameters and requires supercomputing facilities to execute. Validation is primarily at the plant scale. The model has been used to investigate P uptake in barley in two soils, over the whole growing season [87]. Reasonable agreement was found between observed and predicted variables such as water soluble P in the soil, root length density with depth, shoot P, and shoot dry matter. The correspondence was much better for plant variables than soil variables. *Ecosys* represents the first serious attempt to employ fully mechanistic principles to construct an agronomic management model [88].

V. PLANT UPTAKE IN HETEROGENEOUS ENVIRONMENTS

Soils are spatially variable at all scales from the micro- to the landscape. Variability on the field scale has led to the development of precision agriculture [89–91] where fertilizers or other agro-chemicals are applied at different rates at different locations to compensate for local patchiness. Within the rooting zone of an individual plant or a small group of plants, spatial heterogeneity is also present [92,93]. The most obvious is a general decline of nutrient content with depth that is often matched with a similar decline in rooting density. Drew [94] conducted a much-cited experiment where the roots of barley were exposed to an artificially provided patch of nitrogen.

The plants responded by proliferating extensively in the patch and reducing root investment in the unamended soil. A second response to nutrient patches is a stimulation of uptake per unit of root (increased I_{max}). The proliferation response seems to be largely unspecific whereas the I_{max} response seems to be much more nutrient-specific and the magnitude of the response seems related to the concentration of the nutrient [95,96].

A list of the responses of 27 uncultivated forbs and grasses, differing in inherent maximum growth rates, to localized nutrient patches has been compiled [97]. It was found that fast-growing species displayed more plasticity within root systems and responded more quickly to patches than slow-growing species. However, the authors pointed out that root proliferation within a patch cannot always be expected to result in greater nutrient capture. In fact, for a mobile ion like nitrate, such proliferation may have little effect on nitrate inflows and with isolated plants, there is only a weak link between proliferation and nitrate capture. Recently, it has been shown [98] that proliferation confers an advantage only when roots are competing for the N resource in N-limited plant communities. It is also suggested that attempting to increase the proliferation response in crops grown in well-fertilized monocultures would have little benefit [98].

Components of the underlying sensory mechanism have now been identified [99]. However, the signal transduction pathway remains uncertain with some evidence for the involvement of auxin in both N and P responses [100–102] whereas others have found auxin-deficient *Arabidopsis* mutants behave like wild-type plants [103,104]. Encouragingly, *Arabidopsis* does show a proliferation response [105], and the sequencing of the rice and maize genomes should allow rapid progress in this area [106]. It is already clear that the signaling pathways controlling root development in *Arabidopsis* are very complex [107]. However, it seems clear that both short-range (e.g., local nitrate availability) and long-range (e.g., plant nitrogen status) stimuli are important in modifying root architecture [101] (see also Chapter 5 in this volume). The development and regulation of root system architecture is further complicated by the involvement of microorganisms [108] and soil animals [109], which can markedly change root morphology indirectly or directly via phytohormones. Some progress is being made in incorporating this molecular detail into plant architectural models [110].

Jackson and Caldwell [111] used the Barber–Cushman model to examine the significance of localized nutrient patches and root proliferation for whole plant uptake by *Agropyron desertorum*. They divided the simulated soil volume into 25 soil cells, either all at the same nutrient concentrations (homogeneous) or with one cell with an elevated nutrient concentration (heterogeneous). In both cases, the total amount of nutrient in the soil volume was the same. Roots were either assumed to exploit the soil volume with evenly spaced roots (nonplastic) or were allowed to proliferate preferentially in the enriched cell (plastic). Their simulations showed that plastic roots in the heterogeneous environment acquired more N and P than in the homogeneous state while nonplastic roots acquired less in the heterogeneous state compared to the homogeneous. They also simulated the response to the 3-fold (P) and 12-fold (NO_3^--N) variations observed experimentally in a 0.25-m² area in the field. Preferential proliferation in nutrient-rich patches was estimated to increase acquisition by 28% for P and 61% for N where roots were allowed to colonize unoccupied soil (Figure 12.18). They attributed these gains to increased root surface area in the patches together with higher patch concentrations allowing faster uptake per unit surface area of root. However, when roots were allowed to proliferate in patches already containing some roots, the gains for nitrate largely disappeared as the new roots essentially competed for N with the existing roots.

In a further paper [112], the effect of a highly heterogeneous environment on nutrient uptake was simulated with *Agropyron desertorum*. Nutrients patches were allocated using Monte Carlo techniques, and the volume was divided into 10 or 1000 such patches. Root density per cell was constant, and adaptive root proliferation was therefore not examined. The results showed that root acquisition of P and NO_3^--N was always lower when the nutrients were unevenly distributed. However, the reduction in uptake could be largely offset by allowing the uptake capacity per unit surface area of root (I_{max}) to increase with increased nutrient concentration (i.e., within nutrient-rich

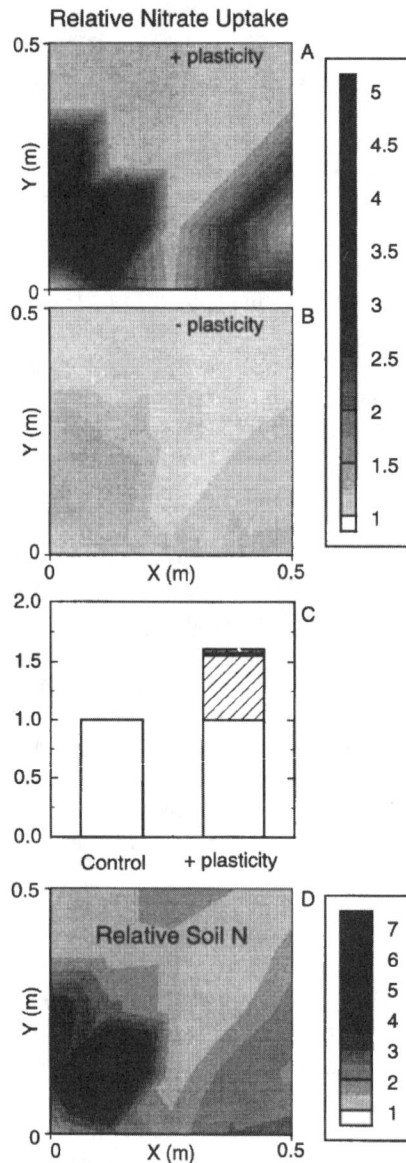

FIGURE 12.18 Simulated uptake of nitrate from a patchy soil. The relative nitrate distribution from field data is shown in D. In A, the plant roots were allowed to respond in a plastic fashion to local resource availability while B shows the nonplastic response: both plots show the relative nitrate uptake over the 0.5-m^2 area. The total nitrate acquired in each plot is shown in C. Plastic roots acquired 61% more nitrate during the 2-d simulation. (From Jackson, R.B. and Caldwell, M.M., *J. Ecol.*, 84, 891, 1996.) (With permission from Blackwell Science.)

patches). Such modification of uptake capacity has been shown to occur for many plant species [113] as well as for *A. desertorum* in direct response to nutrient concentration [111]. Localization of P in nutrient patches increased the predicted uptake by maize even without invoking root proliferation or altered uptake kinetics [114]. In this case, the benefit was because of a highly nonlinear buffer power that increased the soil solution concentration of P allowing a much more efficient uptake of P per unit root surface area within patches. In the Jackson and Caldwell [111] study, the buffer term was essentially linear, which minimized the uptake efficiency gains. Such studies illustrate the power of mathematical models to resolve apparently contradictory results.

Gleeson and Fry [115] developed a model that simulated optimal foraging strategies for roots with respect to root proliferation. Uptake was not simulated mechanistically. They assumed that the root system would operate by investing roots in patches such that the marginal return (differential uptake rate divided by differential root allocation rate) would be equal across all patches and that marginal return was correlated with patch quality. Container-grown *Sorghum vulgare* fine-root biomass showed behavior consistent with the foraging model; however, much of the overall root behavior remained unexplained by the model. Recently, statistically distinct foraging patterns [116] were observed in a study of 55 plant species.

VI. WATER AND NUTRIENT UPTAKE

The nutrient uptake models described earlier assume that the soil water content is constant. However, clearly, because plants do take up water, water uptake and movement should be taken into account when calculating nutrient movements in the soil. There has been very little research focused on simultaneous nutrient and water uptake. Roose and Fowler developed a model [117] for water uptake by root branching structures, which they subsequently linked with nutrient uptake to assess fertilizer movement in the soil and the impact it had on nutrient uptake [118]. Here we briefly outline their approach.

Water inside the roots flows along the xylem tubes and this movement is well characterized by the Poiseuille law, that is, the axial flux of water is given by

$$q_z = -k_z \left[\frac{\partial p_r}{\partial z} - \rho g \right] \tag{12.41}$$

$$k_z = \sum_i \frac{\pi n_i R_i}{8\mu} \tag{12.42}$$

where k_z is the axial conductivity of the root and depends on the number of the xylem vessels n_i of radius R_i and μ, the viscosity of water, where the index i refers to the size class of the xylem vessels; p_r is the water pressure inside the xylem vessels. Water enters the xylem vessels by flowing through the cortex in a radial direction. This flow is modeled by assuming that the flux is proportional to the water pressure difference between the soil and the xylem vessels, that is,

$$q_r = k_r (p - p_r) \tag{12.43}$$

where p is the water pressure in the soil pores and k_r is root tissue radial conductivity.

The conservation of water in the single root means

$$2\pi a k_r (p - p_r) = -k_z \frac{\partial^2 p_r}{\partial z^2} \tag{12.44}$$

This is a single equation for the root internal pressure p_r. This equation needs two boundary conditions. We assume that there is no fluid flux through the tip of the root

$$\frac{\partial p_r}{\partial z} - \rho g = 0 \qquad Z = L \tag{12.45}$$

and at the base of the root we will prescribe a fluid pressure

$$p_r = P \qquad Z = 0 \tag{12.46}$$

By nondimensionalizing these equations with scales $p_r \sim |P|$, $p \sim |P|$, and $z \sim L$, we find that there are only two dimensionless groupings of the parameters that determine the solution. The dimensionless equations are

$$\kappa^2 (p - p_r) = -\frac{\partial^2 p_r}{\partial z^2}, \qquad \begin{matrix} Z = 0 \\ Z = 1 \end{matrix} \tag{12.47}$$

$$p_r = 1 \tag{12.48}$$

$$\frac{\partial p_r}{\partial z} = \omega \tag{12.49}$$

where

$$\omega = \frac{\rho g L}{|P|} \tag{12.50}$$

$$\kappa^2 = \frac{2 \pi a k_r L^2}{k_z} \tag{12.51}$$

For most roots, $\omega \ll 1$, that is, the gravity effects on root internal pressure are small. However, depending on the root radius and length, the parameter κ can be either small or large. For maize first-order lateral branches, $\kappa \gg 1$ meaning that it is very easy for water to flow into small thin roots, and the dominant impedance to water flow in these roots comes from axial conductivity. Therefore, for the most part these roots do not take up water apart for a very small thin region near the base, that is, the leading order solution when $\kappa \gg 1$ is given by

$$p_r(z) \approx p(z) + [-1 - p(0)] e^{-\kappa z} \tag{12.52}$$

When $\kappa \ll 1$, then most of the water uptake is dominated by the soil water pressure profile, but the internal root pressure is approximately equal to the pressure at the base of the root. Thus, even though the water saturation and uptake modeling by the root system is very complicated, using the scaling presented earlier, the authors [117] were able to identify that for most agricultural crops, the side branches do not participate significantly in water uptake. Instead, it is the thickest roots that take up most of the water. This fact enabled the development of a more tractable model for water uptake by root branching structures than it had been previously thought possible [117].

The relative soil moisture content in the soil, S, from the Darcy–Richard's equation is given by

$$\phi \frac{\partial S}{\partial t} = \nabla \cdot [D_0 D(S) \nabla S - K_s k(S) \hat{\mathbf{k}}] - F_w(S, z, t) \tag{12.53}$$

where ϕ is the porosity of the soil, D_0 and K_s are the reference water diffusivity and conductivity, respectively, and $D(S)$ and $k(S)$ mathematically represent the reduction in diffusivity and conductivity of water with changes in relative water saturation; F_w is a spatially heterogeneous sink term (functional forms of $D(S)$, $k(S)$ and F_w are available [117]).

The equation for nutrient conservation in the soil with variable moisture content is given by

$$\frac{\partial}{\partial t}[(b+\phi S)c] + \nabla \cdot [c\mathbf{u}] = \nabla \cdot [D_f \phi^{d+1} S^{d+1} \nabla c] - F(c,S,t) \qquad (12.54)$$

The essential task was then to identify what the functional form for the distributed nutrient uptake sink term should be [118]. By scaling $z \sim L, t \sim bL^2/D_0, c \sim K_m$, where L is the length of the root system, we get the following dimensionless equations

$$\delta \frac{\partial S}{\partial t} = \frac{\partial}{\partial z}\left[D(S)\frac{\partial S}{\partial z} - \varepsilon k(S) \right] - F_w \qquad (12.55)$$

$$(1+\delta S)\frac{\partial c}{\partial t} - \left[D(S)\frac{\partial S}{\partial z} - \varepsilon k(S) \right]\frac{\partial c}{\partial z} = D_\varepsilon \frac{\partial}{\partial z}\left[S^{1+d}\frac{\partial c}{\partial z} \right] - F + F_w c \qquad (12.56)$$

Here, the most important parameter is $\delta = \phi/b$, which mathematically represents the timescale of water movement in comparison to nutrient movement. For highly buffered nutrients like phosphate and potassium, $b \gg 1$ and therefore $\delta \ll 1$. Thus, on the timescale of nutrient movement, the water saturation is only varying slowly in time and, therefore, the analytic approximations for nutrient uptake given by Equation 12.31 can still be used. This separation of water and nutrient movement timescales is very useful because even properties like the speed of fertilizer movement in the soil can be calculated analytically [118].

Comparison between the model calculations with and without water movement and the Barber flux calculation is shown in Figure 12.14. Both root architecture and water movement can lead to large differences in the model predictions for phosphate uptake. Thus, more detailed experimental data on the spatiotemporal evolution of root branching structures and water movement are needed to improve on the current estimates of nutrient uptake from the soil.

VII. FUTURE PROSPECTS

Most mineral nutrient uptake rhizosphere models have considered resource capture by isolated plants or by monocultures. Elsewhere in the ecological literature are models dealing with interspecies interactions and particularly competition. For example, one of the main theories for nutrient competition by roots assumes that the mechanism operates by depletion of an average soil solution concentration [118] (concentration reduction theory). When this theory was tested with a cellular automaton model [119], aimed at approximating the diffusion of nitrate in soil, the model indicated that competition relied on superior interception, that is, competition depended on localized supply, not globally averaged concentration. Similarly, an interspecies competition model [120] including spatially-explicit nutrient uptake and transport, found that space occupancy by roots was more important than their sink strength at low nutrient diffusivities but the two became increasingly comparable as diffusivity increased. As yet, the treatment of nutrient transport in such models is rather simplified, but the marriage of such models with mechanistic models of the rhizosphere could provide powerful tools to aid in the understanding of belowground competition in natural communities. This is important because belowground competition is predicted to be much more symmetrical than aboveground competition for light and may therefore act to preserve biodiversity in natural communities [121,122].

Many studies have shown that the rhizosphere microbial community is different from the bulk soil community and differs between plant species (see also Chapter 3 and Chapter 13 of this volume).

Microbial rhizosphere models have thus far largely dealt with the community as a single entity using biomass as the state variable. But RAPD and DGGE analysis of whole-community DNA extracts revealed bacterial successions that correlated with the succession of different grass species [123]. Future models may therefore have to attempt to understand the rhizosphere at the species level. This will involve understanding why particular species are selected as rhizosphere colonists, how microbial species interact between themselves, their population dynamics and how their activities influence, and are influenced by, the root. However, the interpretation of taxonomic diversity as measured by molecular methods is uncertain and the analysis gives little idea of the functional capabilities of the community. Chloroform fumigation [124,125] and reconstruction of communities in sterile soil [126] were used to manipulate the diversity of the soil community. In fumigated soils, functions such as thymidine incorporation, response to added nutrients, and decomposition of plant residues increased with decreased biodiversity whereas nitrification, denitrification, and methane oxidation decreased. In the reconstructed communities, there was consistent trend in similar functions with biodiversity change. There was some evidence that low biodiversity correlated with a reduction in functional stability. Most available evidence suggests that there is no predictable relationship between diversity and function in soils [127,128]; yet it is the functional aspects of the microbial community, which are of importance to plant functioning.

Global climate change is predicted to have major consequences for plant growth and this will provide a fresh impetus for the mathematical modeling of the rhizosphere [129,130].

REFERENCES

1. de Willigen, P., Nitrogen turnover in the soil crop system — comparison of 14 simulation models, *Fert. Res.,* 27, 141, 1991.
2. Curl, E.A. and Truelove, B., *The Rhizosphere,* Adv. Ser. Agric. Sci. 15, Springer-Verlag, Berlin, 1986.
3. Marschner, H., *Mineral Nutrition of Higher Plants,* Academic Press, 1995.
4. Barber, S.A. and Cushman, J.H., Nitrogen uptake model for agricultural crops, in *Modeling Waste Water Renovation-Land Treatment,* Iskander, I., Ed., Wiley-Interscience, New York, pp. 382–404.
5. Nye, P.H. and Tinker, P.B., *Solute Movement in the Soil-Root System (Studies in Ecology),* Vol. 4, Blackwell Scientific Publications, Oxford, 1977.
6. Sposito, G., *The Chemistry of Soils,* Oxford University Press, 1989.
7. van Rees, K.C.J., Comerford, N.B., and Rao, P.S.C., Defining soil buffer power: implications for ion diffusion and nutrient uptake modeling, *Soil Sci. Soc. Am. J.,* 54, 1505, 1990.
8. Roose, T., Fowler, A.C., and Darrah, P.R., A mathematical model of plant nutrient uptake, *J. Math. Biol.,* 42, 347, 2001.
9. Crank, J., *The Mathematics of Diffusion,* Clarendon Press, Oxford, 1975.
10. Darrah, P.R., Models of the rhizosphere. 2. A quasi 3-dimensional simulation of the microbial-population dynamics around a growing root releasing soluble exudates, *Plant Soil,* 138, 147, 1991.
11. Lindsay, W.L., *Chemical Equilibria in Soils,* Wiley-Interscience, Wiley, New York, 1979.
12. Barrow, N.J., The reaction of plant nutrients and pollutants with soil, *Aust. J. Soil Res.,* 27, 475, 1989.
13. Jungk, A. and Claassen, N., Ion diffusion in the soil-root system, *Adv. Agron.,* 61, 53, 1997.
14. Bosatta, E. and Agren, G.I., Theoretical analyses of carbon and nutrient dynamics in soil profiles, *Soil Biol. Biochem.,* 28, 1523, 1996.
15. Cheng, W., Rhizosphere feedbacks in elevated CO_2, *Tree Physiol.,* 19, 313, 1999.
16. Hinsinger, P., How do plant roots acquire mineral nutrients? Chemical processes involved in the rhizosphere, *Adv. Agron.,* 64, 225, 1998.
17. Bouldin, D.R., A multiple ion uptake model, *J. Soil. Sci.,* 40, 309, 1989.
18. Schmidt, W., Mechanisms and regulation of reduction-based iron uptake in plants, *New Phytologist,* 141, 1, 1999.
19. Kirk, G.J.D. and Saleque, M.A., Root-induced solubilization of phosphate in the rhizosphere of lowland rice, *New Phytologist,* 129, 325, 1995.
20. Kirk, G.J.D. and Saleque, M.A., Solubilization of phosphate by rice plants growing in reduced soil — prediction of the amount solubilized and the resultant increase in uptake, *Eur. J. Soil Sci.,* 46, 247, 1995.

21. Geelhoed, J., Findenegg, G.R., and van Riemsdijk, W.H., Availability to plants of phosphate adsorbed onto goethite: experiment and simulation, *Eur. J. Soil Sci.,* 48, 473, 1997.

22. Hoffland, E., Quantitative-evaluation of the role of organic-acid exudation in the mobilization of rock phosphate by rape, *Plant Soil,* 140, 279, 1992.

23. Barber, S.A., *Soil Nutrient Bioavailability: A Mechanistic Approach,* Wiley, New York, 1995.

24. Claassen, N. and Barber, S.A., A method for characterizing the relationship between nutrient concentration and flux into roots of intact plants, *Plant Physiol.,* 54, 564, 1974.

25. Reginato, J.C., Tarzia, D.A., and Cantero, A., On the free-boundary problem for the Michealis-Menten absorption model for root growth. 2. High concentrations, *Soil Sci.,* 152, 63, 1991.

26. Gahoonia, T.S., Care D., and Nielsen, N.E., Root hairs and phosphorus acquisition of wheat and barley cultivars, *Plant Soil,* 191,181, 1997.

27. Gahoonia, T.S. and Nielsen, N.E., Variation in root hairs of barley cultivars doubled soil phosphorus uptake, *Euphytica,* 98, 177, 1997.

28. Itoh, S. and Barber, S.A., A numerical solution of whole plant uptake for soil-root uptake including root hairs, *Plant Soil,* 70, 403, 1983.

29. Darrah, P.R., Interactions between root exudates, mineral nutrition and plant growth, in *Inherent Variation in Plant Growth: Physiological Mechanisms and Ecological Consequences,* Lambers, H., Poorter, H., and van Vuuren, M.M.I., Eds., Backhuys Publishers, 1998, p. 159.

30. Darrah, P.R., The rhizosphere and plant nutrition: a quantitative approach, *Plant Soil,* 155, 3, 1993.

31. Epstein, E., *Mineral Nutrition of Plants: Principles and Perspectives,* Wiley, New York, 1972.

32. Hoffland, E. et al., Simulation of nutrient-uptake by a growing root-system considering increasing root density and interroot competition, *Plant Soil,* 124, 149, 1990.

33. Comerford, N.B., Porter, P.S., and Escamilla, J.A., Use of Theissen areas in models of nutrient uptake. I. Ecosystems, *Soil Sci. Soc. Am. J.* 58, 210, 1994.

34. Barley, K.P., The configuration of the root system in relation to nutrient uptake, *Adv. Agron.,* 22, 159, 1970.

35. Baldwin, J.P., Nutrient Uptake by Competing Roots in Soil, D.Phil. thesis, University of Oxford, 1972.

36. Morton, K.W. and Mayers, D.F., *Numerical Solution of Partial Differential Equations,* Cambridge University Press, 1994.

37. Abbes, C., Robert, J.L., and Parent, L.E., Mechanistic modeling of coupled ammonium and nitrate uptake by onions using the finite element method, *Soil Sci. Soc. Am. J.,* 60, 1160, 1996.

38. Fowler, A.C., *Mathematical Models in the Applied Sciences,* Cambridge University Press, 1997.

39. Geelhoed, J.S., Sipko, L.J.M., and Findenegg, G.R., Modelling zero sink nutrient uptake by roots with root hairs from soil: comparison of two models, *Soil Sci.,* 162, 544, 1997.

40. Greenwood, D.J. and Karpinets, T.V., Dynamic model for the effects of K-fertilizer on crop growth, K-uptake and soil-K in arable cropping. 1. Description of the model, *Soil Use Manage.,* 134, 178, 1997.

41. Hiltner, L., Uber neuere Erfahrungen und Problem auf dem Gebeit der Bodenbakteriologie und unter besonderer Berucksichtigung der Grundungung und Brache, *Arb. Dtsch. Landwirt. Ges.,* 98, 59, 1904.

42. Lynch, J.M., Introduction: some consequences of microbial rhizosphere competence for plant and soil, in *The Rhizosphere,* Lynch, J.M., Ed., Wiley series in Ecol. and Appl. Micro., Wiley-Interscience, New York, 1990, p. 1.

43. Gilligan, C.A., Mathematical models of infection, in *The Rhizosphere,* Lynch, J.M., Ed., Wiley series in Ecol. and Appl. Micro., Wiley-Interscience, New York, 1990, p. 207.

44. Gibson, G.J., Gilligan, C.A., and Kleczkowski, A., Predicting variability in biological control of a plant-pathogen system using stochastic models, *Proc. R. Soc. B.,* 266, 1743, 1999.

45. Newman, E.I. and Watson, A., Microbial abundance in the rhizosphere: a computer model, *Plant Soil,* 48, 17, 1977.

46. Darrah, P.R., Models of the rhizosphere 1. Microbial population dynamics around a root releasing soluble and insoluble carbon, *Plant Soil,* 133, 187, 1991.

47. Van der Werf, H. and Verstraete, W., Estimation of active soil microbial biomass by mathematical analysis of respiration curves: development and verification of the model, *Soil Biol. Biochem,* 19, 253, 1987.

48. Romheld, V., The role of phytosiderophores in acquisition of iron and other micronutrients in graminaceous species: an ecological approach, *Plant Soil,* 130, 127, 1991.

49. Jones, D.L. and Darrah, P.R., Resorption of organic-components by roots of *Zea mays* and its consequences in the rhizosphere. 1. Resorption of ^{14}C labeled glucose, mannose and citric-acid, *Plant Soil,* 143, 259, 1992.

50. Jones, D.L. and Darrah, P.R., Re-sorption of organic-compounds by roots of *Zea mays* L. and its consequences in the rhizosphere. 2. Experimental and model evidence for simultaneous exudation and re-sorption of soluble C compounds, *Plant Soil,* 153, 47, 1993.
51. Jones, D.L. and Darrah, P.R., Influx and efflux of amino-acids from *Zea-mays* L roots and their implications for N-nutrition and the rhizosphere, *Plant Soil,* 156, 87, 1993.
52. Jones, D.L. and Darrah, P.R., Influx and efflux of organic-acids across the soil-root interface of *Zea-mays* L and its implications in rhizosphere C flow, *Plant Soil,* 173, 103, 1995.
53. Darrah, P.R., Rhizodeposition under ambient and elevated CO_2 levels, *Plant Soil,* 187, 265, 1996.
54. Scott, E.M. et al., A mathematical model for dispersal of bacterial inoculants colonizing the wheat rhizosphere, *Soil Biol. Biochem.,* 27, 1307, 1995.
55. van der Hoeven, N., van Elsas, J.D., and Heijnen, C.E., A model based on soil structural aspects describing the fate of genetically modified bacteria in soil, *Ecol. Modelling,* 89, 161, 1996.
56. Forney, L.J., Zhou, X., and Brown, C.J., Molecular microbial ecology: land of the one-eyed king, *Curr. Opin. Microbiol.,* 7, 210, 2004.
57. Kuske, C.R., Barns, S.M., and Busch, J.D., Diverse uncultivated bacterial groups from soils of the arid southwestern United States that are present in many geographic regions, *Appl. Environ. Microbiol.,* 63, 3614, 1997.
58. Ueda, T., Suga, Y., and Matsuguchi, T., Molecular phylogenetic analysis of a soil microbial community in a soybean field, *Eur. J. Soil Sci.,* 46, 415, 1995.
59. Venter, J.C. et al., Environmental genome shotgun sequencing of the Sargasso Sea, *Science,* 304, 66, 2004.
60. Handelsman, J. and Smalla, K., Conversations with the silent majority, *Curr. Opin. Microbiol.,* 6, 271, 2003.
61. Curtis, T.P. and Sloan, W.T., Prokaryotic diversity and its limits: microbial community structure in nature and implications for microbial ecology, *Curr. Opin. Microbiol.,* 7, 221, 2004.
62. Torsvik, V., Ovreas, L., and Thingstad, T.F., Prokaryotic diversity — magnitude, dynamics, and controlling factors, *Science,* 296, 1064, 2002.
63. Nunan, N. et al., *In situ* spatial patterns of soil bacterial populations, mapped at multiple scales, in an arable soil, *Microb. Ecol.,* 44, 296, 2002.
64. Nunan, N. et al., Spatial distribution of bacterial communities and their relationships with the micro-architecture of soil, *FEMS Microbiol. Ecol.,* 44, 203, 2003.
65. Ettema, C.H. and Wardle, D.A., Spatial soil ecology, *Trends Ecol. Evol.,* 17, 177, 2002.
66. Farrar, J. et al., How roots control the flux of carbon to the rhizosphere, *Ecology,* 84, 827, 2003.
67. Diggle, A.J., ROOTMAP — a model in three-dimensional coordinates of the growth and structure of fibrous root systems, *Plant Soil,* 105, 169, 1988.
68. Pagès, L., Jourdan, M.D., and Picard, D., A simulation model of the three-dimensional architecture of the maize root system, *Plant Soil,* 119, 147, 1989.
69. Pages, L. et al., Root type: a generic model to depict and analyse the root system architecture, *Plant Soil,* 258, 103, 2004.
70. Nielsen, K.L. and Lynch, J., Fractal analysis of bean root systems, *Agron. Abstr.,* 150, 1, 1994.
71. Fitter, A.H., An architectural approach to the comparative ecology of plant root systems, *New Phytologist,* 106, 61, 1987.
72. Room, P.M., Hanan, J.S., and Prusinkiewicz, P., Virtual plants: new perspectives for ecologists, pathologists and agricultural scientists, *Trends Plant Sci.,* 1, 33, 1996.
73. Michalewicz, M.T., *Plants to Ecosystems (Advances in Computational Life Sciences),* Vol. 1, CSIRO Publishing, Melbourne, 1997.
74. Dunbabin, V., Rengel, Z., and Diggle, A.J., Simulating form and function of root systems: efficiency of nitrate uptake is dependent on root system architecture and the spatial and temporal variability of nitrate supply, *Funct. Ecol.,* 18, 204, 2004.
75. Acock, B. and Pachepsky, Y.A., Convective-diffusive model of two-dimensional root growth and proliferation, *Plant Soil,* 180, 231, 1996.
76. Spek, L.Y., Generation and visualization of root-like structures in a three-dimensional space, *Plant Soil,* 197, 9, 1997.
77. van Noordwijk, M. and van de Geijn, S.C., Root, shoot and soil parameters required for process-oriented models of crop growth limited by water or nutrients, *Plant Soil,* 183, 1, 1996.

78. Hopmans, J.W. and Bristow, K.L., Current capabilities and future needs of root water and nutrient uptake modelling, *Adv. Agron.*, 77, 103, 2002.

79. Wang, E.L. and Smith, C.J., Modelling the growth and water uptake function of plant root systems: a review, *Aust. J. Agric. Res.*, 55, 501, 2004.

80. Lemaire, G. and Millard, P., An ecophysiological approach to modelling resource fluxes in competing plants, *J. Exp. Bot.*, 50, 15, 1999.

81. Ingestad, T., Relative addition rate and external concentration; driving variables used in plant nutrition research, *Plant Cell Environ.*, 5, 443, 1982.

82. Ingestad, T. and Ågren, G.I., Plant nutrition and growth: basic principles, *Plant Soil*, 168–169, 15, 1995.

83. Fontes, P.C.R., Barber, S.A., and Wilcox, G.E., Prediction of phosphorus uptake by two tomato cultivars growing under sufficient and insufficient phosphorus soil conditions using a mechanistic mathematical model, *Plant Soil*, 94, 87, 1986.

84. Grant, R.F., The distribution of water and nitrogen in the soil-crop system: a simulation study with validation from a winter wheat field trial, *Fert. Res.*, 27, 199, 1991.

85. Grant, R.F., Simulation model of soil compaction and root growth. I Model development, *Plant Soil*, 150, 15, 1993.

86. Thornley, J.H.M., A balanced quantitative model for root:shoot ratios in vegetative plants, *Ann. Bot.*, 36, 431, 1972.

87. Grant, R.F. and Robertson, J.A., Phosphorus uptake by root systems: mathematical modelling in ecosys, *Plant Soil*, 188, 279, 1997.

88. Grant, R.F. et al., Mathematical modelling of phosphorus losses from land application of hog and cattle manure, *J. Environ. Qual.*, 33, 210, 2004.

89. Cook, S.E. and Bramley, R.G.V., Precision agriculture — opportunities, benefits and pitfalls of site-specific crop management in Australia, *Aus. J. Exp. Agric.*, 38, 753, 1998.

90. Sylvester-Bradley, R. et al., An analysis of the potential for precision farming in Northern Europe, *Soil Use Manage.*, 15, 1, 1999.

91. van Niel, T.G. and McVicar, T.R., Current and potential uses of optical remote sensing in rice-based irrigation systems: a review, *Aust. J. Agric. Res.*, 55, 155, 2004.

92. Robertson, G.P. et al., Spatial variability in a successional plant community: patterns of nitrogen availability, *Ecology*, 69, 1517, 1988.

93. Robertson, G.P., Crum, J.R., and Ellis, B.G., The spatial variability of soil resources following long term disturbance, *Oecologia*, 96, 451, 1993.

94. Drew, M.C., Comparison of the effects of a localized supply of phosphate, nitrate, ammonium and potassium on the growth of the seminal root system, and the shoot, in barley, *New Phytologist*, 75, 479, 1975.

95. Robinson, D., Variation, co-ordination and compensation in root systems in relation to soil variability, *Plant Soil*, 187, 57, 1996.

96. Hodge, A., The plastic plant: root responses to heterogeneous supplies of nutrients, *New Phytologist*, 162, 9, 2004.

97. Robinson, D. and van Vuuren, M.M.I., Responses of wild plants to nutrient patches in relation to growth rate, in *Inherent Variation in Plant Growth: Physiological Mechanisms and Ecological Consequences*, Lambers, H., Poorter, H., and van Vuuren, M.M.I., Eds., Backhuys Publishers, 1998, p. 237.

98. Robinson, D. et al., Plant root proliferation in nitrogen-rich patches confers competitive advantage, *Proc. R. Soc. B*, 266, 431, 1999.

99. Zhang, H.M. and Forde, B.G., An *Arabidopsis* MADS box gene that controls nutrient-induced changes in root architecture, *Science*, 279, 407, 199.

100. Al-Ghazi, Y. et al., Temporal responses of Arabidopsis root architecture to phosphate starvation: evidence for the involvement of auxin signalling, *Plant Cell Environ.*, 26, 1053 2003.

101. Forde, B.G., Local and long-range signaling pathways regulating plant responses to nitrate, *Ann. Rev. Plant Biol.*, 53, 203, 2002.

102. Zhang, H.M. and Forde, B.G., Regulation of Arabidopsis root development by nitrate availability, *J. Exp. Bot.*, 51, 51, 2000.

103. Linkohr, B.I. et al., Nitrate and phosphate availability and distribution have different effects on root system architecture of Arabidopsis, *Plant J.*, 29, 751, 2002.

104. Williamson, L.C. et al., Phosphate availability regulates root system architecture in Arabidopsis, *Plant Phys.*, 126, 875, 2001.

105. Casimiro, I. et al., Dissecting Arabidopsis lateral root development, *Trends Plant Sci.*, 8, 165, 2003.

106. Hochholdinger, F., From weeds to crops: genetic analysis of root development in cereals, *Trends Plant Sci.*, 9, 42, 2004.

107. Casson, S.A. and Lindsey, K., Genes and signalling in root development, *New Phytologist*, 158, 11, 2003.

108. Persello-Cartieaux, F., Nussaume, L., and Robaglia, C., Tales from the underground: molecular plant-rhizobacteria interactions, *Plant Cell Environ.*, 26, 189, 2003.

109. Bonkowski, M., Protozoa and plant growth: the microbial loop in soil revisited, *New Phytologist*, 162, 617, 2004.

110. Prusinkiewicz, P., Modeling plant growth development, *Curr. Opin. Plant Biol.*, 7, 79, 2004.

111. Jackson, R.B. and Caldwell, M.M., Integrating resource heterogeneity and plant plasticity: modelling nitrate and phosphate uptake in a patchy soil environment, *J. Ecol.*, 84, 891, 1996.

112. Ryel, R.J. and Caldwell, M.M., Nutrient acquisition from soils with patchy nutrient distributions as assessed with simulation models, *Ecology*, 79, 2735, 1998.

113. Clarkson, D.T., Factors affecting mineral nutrient uptake by plants, *Ann. Rev. Plant Physiol.*, 36, 77, 1984.

114. Kovar, J.L. and Barber, S.A., Reasons for differences among soils in placement of phosphorus for maximising predicted uptake, *Soil. Sci. Soc. Am. J.*, 53, 1733, 1989.

115. Gleeson, S.K. and Fry, J.E., Root proliferation and marginal patch value, *Oikos*, 79, 387, 1997.

116. Levang-Brilz, N. and Bondini, M.E., Growth rate, root development and nutrient uptake of 55 plant species from the Great Plains Grasslands, USA, *Plant Ecol.*, 165, 117, 2003.

117. Roose, T. and Fowler, A.C., A model for water uptake by plant roots, *J. Theor. Biol.*, 228, 155, 2004.

118. Roose, T. and Fowler, A.C., A mathematical model for water and nutrient uptake by plant root systems, *J. Theor. Biol.*, 228, 173, 2004.

119. Tilman, D., On the meaning of competition and the mechanisms of competitive superiority, *Funct. Ecol.*, 1, 304, 1987.

120. Craine, J.M., Fargione, J., and Sugita, S., Supply pre-emption, not concentration reduction, is the mechanism of competition for nutrients, *New Phytologist*, 166, 933, 2005.

121. Raynaud, X. and Leadley, P.W., Soil characteristics play a key role in modelling nutrient competition in plant communities, *Ecology*, 85, 2200, 2004.

122. Casper, B.B., Scvhenk, H.J., and Jackson, R.B., Defining a plant's below ground zone of influence, *Ecology*, 84, 2313, 2003.

123. Kang, S.H. and Mills, A.L., Soil bacterial community structure changes following disturbance of the overlying plant community, *Soil Sci.*, 169, 55, 2004.

124. Griffiths, B.S. et al., The relationship between microbial community structure and functional stability, tested experimentally in an upland pasture soil, *Microb. Ecol.*, 47, 104, 2004.

125. Griffiths, B.S. et al., Ecosystem response of pasture soil communities to fumigation-induced microbial diversity reductions: an examination of the biodiversity-ecosystem function relationship, *Oikos*, 90, 279, 2000.

126. Griffiths, B.S. et al., An examination of the biodiversity-ecosystem function relationship in arable soil microbial communities, *Soil Biol. Biochem*, 33, 1713, 2001.

127. Bardgett, R.D., Causes and consequences of biological diversity in soil, *Zoology*, 105, 367, 2002.

128. Nannipieri, P. et al., Microbial diversity and soil functions, *Eur. J. Soil Sci.*, 54, 655, 2003.

129. van Noordwijk, M. et al., Global change and root function, *Glob. Change Biol.*, 4, 759, 1998.

130. Pendall, E. et al., Below-ground process responses to elevated CO_2 and temperature: a discussion of observations, measurement methods, and models, *New Phytologist*, 162, 311, 2004.

131. Hinsinger, P., Plassard, C., Tang, C.X., and Jaillard, B., Origins of root-mediated pH changes in the rhizosphere and their responses to environmental constraints: a review. *Plant Soil*, 248: 43, 2003.

132. Silberbush, M. and Barber, S.A., Sensitivity of simulated phosphorus uptake to parameters used by a mathematical model, *Plant Soil*, 74, 93, 1983.

133. Claassen, N., Syring, K.M., and Jungk, A., Verification of a mathematical model by simulating potassium uptake from soil, *Plant Soil*, 95, 209, 1986.

134. Lynch, J. and Nielsen, K.L., Simulation of root architecture, in *Plant Roots: The Hidden Half*, 2nd ed., Waisel, Y., Eshel, A., and Kafkafi, U., Eds., Marcel Dekker, New York, 1996, p. 251.

13 Methodological Approaches to the Study of Carbon Flow and the Associated Microbial Population Dynamics in the Rhizosphere

Johannes A. van Veen, J. Alun W. Morgan, and John M. Whipps

CONTENTS

I. INTRODUCTION

Rhizodeposition, i.e., the release of carbon compounds from roots, is a general phenomenon of plant roots. The compounds lost from different species, or even cultivars, can vary markedly in quality and quantity over time and space.

Details of the types and quantities of compounds lost from the root are discussed in Chapter 1.

Rhizodeposition describes the total carbon (C) transfer from plant roots to soil, and it comprises exudates (small molecules such as organic acids, amino acids, and sugars), secretions (such as enzymes), lysates from dead cells, and mucilage [1]. The release of plant-derived carbon compounds commences during seed imbibition, often as water-soluble exudates and gases, and continues as the plant grows [2]. Cells are lost from the root cap continuously as the root grows through soil accompanied by losses of exudates and polysaccharide gel. The zone just behind the root tip appears to be the site of maximum exudation with both the zone of elongation and the cell junctions also indicated as localized regions of greater exudation in some species [3–5]. Eventually, whole roots or parts of root systems die, and this root loss with CO_2 derived from roots and the accompanying microbiota respiration are the major sources of rhizodeposition.

Such differences in the amount and type of rhizodeposition that occur on the root with time result in concomitant variations in microbial populations in the rhizosphere, within the root, on the surface of the root, and in the soil adjacent to the root. The general microbial population changes and specific interaction of individual compounds from specific plants or groups of plants with individual microbial species are covered in detail in Chapter 3 and Chapter 11. Consequently, this chapter is restricted to consideration of methodologies used to study carbon flow and associated microbial population dynamics in the rhizosphere, drawing on specific plant–microbe examples only when required.

II. METHODS FOR THE STUDY OF RHIZOSPHERE CARBON FLOW IN ARTIFICIAL MEDIA

Experiments to examine rhizodeposition can vary markedly in scale and complexity, depending on the information required, the equipment available, and the plants concerned (see also Chapter 2). To avoid the complexity of the soil, experiments to study exudates and other material released from young roots of plants grown in nutrient solution culture, sometimes with sand or other solid support systems, are the simplest and are carried out in the laboratory under controlled conditions. Under these conditions, rhizodeposition can be measured chemically by growing plants in solution culture. The type of sampling employed reflects the target of the experiment and can range from detailed analyses of a single class of chemicals to much more general observations. For example, in sterile solution culture, dicarboxylic acids were quantified in one study of exudates of wheat and flax [6], and isoflavonoids in exudates of white lupin were measured in another [7]. In contrast, changes in gross classes of compounds, such as low- and high-molecular-weight materials and particulate matter, have been examined in exudates derived from maize grown axenically in solution culture [8]. The ability to change and control the composition of the nutrient solution and the relatively small size of the microcosms used enables manipulation of environmental variables and time-course studies of rhizodeposition to be made relatively easily. The influences of nutrient availability, pH, temperature, anoxia, light intensity, CO_2 concentration, and specific microorganisms have all been examined within a range of plant species [1,9–12]. Besides enabling the identification of the effects of a range of variables on rhizodeposition, plant culture solution studies have also helped to highlight the reabsorption of organic compounds by roots (see Chapter 12), causing a reappraisal of the quantification of carbon loss [13–17]. However, it is well known that the flow of carbon from the roots into the rhizosphere depends on a large number of factors including the presence of microbiota, soil particles, and the age of the plant [18–20]. Therefore, studies with young plants grown in solution cultures can only provide specific information on rhizodeposition under these artificial conditions.

Plants grown for longer periods in solid supports, such as sand or soil, represent the next level of complexity, and, although other techniques are available, carbon flow through the plant to the roots and into the surrounding environment is mostly measured with the use of carbon isotope tracers. The simplest system devised comprises a microcosm where seedlings with roots sandwiched between MF-Millipore™ membranes are in contact with agar-containing plant nutrients [21]. The shoots can be exposed to the tracer, and the loss of the tracer into the agar via the roots is then monitored. The design enables the effect of microorganisms on rhizodeposition to be examined easily [22] but really lacks the complexity of substratum to allow data obtained to be related to rhizodeposition in soil.

The next level of sophistication is to use microcosms containing sand with nutrient solution to provide the physical conditions of soil but without its inherent complexity. Such systems have been used to measure rhizodeposition, ranging from exudates to total carbon budgets as well as specific microbe–plant interactions. For example, soluble fractions of root exudates from 5- and 7-week-old mycorrhizal and nonmycorrhizal maize seedlings were collected by percolating the sand sub-stratum in a tube microcosm with distilled water [23]. Similarly, forage rape (*Brassica napus*) was grown in a syringe-based microcosm containing sand and nutrient solution, and via a series of root zone flushing systems, amino acids released over specified time intervals were collected and quantified chemically with plants up to 60 d old [24]. Such microcosms were adapted to enable ^{14}C-CO_2 pulse-chase experiments to be carried out with forage rape and full carbon budgets were obtained [25].

Other sand-based systems using ^{14}C- labeling procedures have been used to produce carbon budgets for *Festuca ovina* and *Plantago lanceolata* seedlings [26], and white lupin (*Lupinus albus*) [27]. Significantly, ^{14}C-CO_2 labeling of proteoid roots of white lupin under phosphate-deficient conditions showed that high levels of dark fixation of ^{14}C-CO_2 by the roots took place, and that 66% of this root-fixed carbon was exuded from the roots [27]. Clearly, dark fixation of CO_2 by roots and subsequent rhizodeposition is an area that deserves further study in the future.

Sand-based systems can gradually be made to mimic soils by addition of other solid materials or nutrient sources used in agriculture. For example, rock phosphate and mica were added to sand microcosms containing mycorrhizal or nonmycorrhizal pine (*Pinus sylvestris*) and beech (*Fagus sylvaticus*) to investigate the effect of added microorganisms on rhizodeposition [28]. The effluents from the tube microcosms were collected continuously, and after 1 and 2 years the tubes were harvested. Estimates of total organic carbon, as well as of sugars, amino acids, organic acids, and phenolics, were made. This work clearly demonstrates that, depending on the plant species involved, sand-based microcosms can support plant growth suitable for rhizodeposition studies for long periods of time, if required.

Although it has not been applied for the study of carbon flows in the rhizosphere, the microcosm system developed by Lugtenberg and colleagues offers a new opportunity for detailed studies of interactions between plants and microbes [29]. In this system, sterile germinated seedlings are aseptically placed in a 10 cm long quartz sand column moisturized with a plant nutrient solution without added carbon source. Prior to the placement in the columns, the seedlings are bacterized by incubating them in a suspension of specific bacteria to be tested for their ability to colonize roots and to suppress plant pathogens. This gnotobiotic system works well for a number of plants, including tomato, radish, potato, wheat, and grass. The advantages of the system over the use of soil are better reproducibility due to the use of the plant growth medium and higher numbers of the test bacteria on the root due to the absence of competition from indigenous soil bacteria. More detailed information on these interesting results obtained with this system is described in Chapter 10.

III. MEASUREMENT OF RHIZODEPOSITION IN SOIL

Despite the advantages that sand-based systems provide for estimating rhizodeposition, studies in soil are required to obtain realistic information of the process under natural conditions. These are far more technically demanding and are heavily biased toward the use of tracer methods. Yet, other approaches have also been applied. In particular, in studies on the carbon flow in plant–soil systems at conditions of elevated atmospheric CO_2- concentrations, unlabeled CO_2 in different atmospheric concentrations have also been used. In these studies, detailed measurements of changes in root biomass and constitution of CO_2 fluxes from the root–soil compartment provided valuable information about the impact of elevated atmospheric CO_2- concentrations on carbon dynamics in terrestrial ecosystems [30].

Yet, in most studies C-tracers have been used to follow the fate of photosynthetically fixed C through the plant into the surrounding soil and to discriminate against the vast soil organic-C pool. In the past, the radioactive isotope ^{14}C was used most frequently as a C-tracer, in more recent studies the nonradioactive isotope ^{13}C has often been applied as well to avoid the intrinsic risk associated with radioactive tracers [31]. The main reason for the use of ^{14}C was the much greater sensitivity of measuring radioactive tracers over nonradioactive tracers. However, recent developments in isotope ratio mass spectrometry (IRMS) [31] have increased the sensitivity of nonradioactive tracer measurements, considerably. For instance, laser ablation IRMS could potentially be applied to study carbon flow in the rhizosphere at the microscale level [32]. Furthermore, a number of integrated technologies have been developed recently, through which measurements of carbon flow and carbon-utilizing microbial community can be combined, enabling a complete picture of the impact of plant-derived carbon on the functioning of the rhizosphere ecosystem (Section V of this chapter). This has led to an increase in the use of nonradioactive ^{13}C tracers in rhizodeposition studies. Although organic compounds have been used, in most rhizodeposition studies the carbon tracer has been applied in the form of CO_2 to include the entire carbon allocation process in the plant and in the surrounding environment. In the laboratory, the C tracer can be supplied to shoots either in short pulses or continuously, and the carbon flow can be monitored. In the field, due to technical limitations, only C tracer pulse labeling procedures are possible. In this section we will first deal with some examples of the use of ^{14}C with emphasis on more recent work related to carbon sequestering issues, and then we will provide an overview of the work using ^{13}C as a tracer. In the latter, we will provide examples of the use of ^{13}C tracers after addition of ^{13}C-labeled compounds (mainly CO_2) and natural abundance studies. A more extensive review including technical details on the use of ^{13}C and ^{14}C (as well as ^{11}C) in carbon flow studies in plant–soil systems is given by Kuzyakov and Domanski [33].

A. $^{14}CO_2$-LABELING TECHNIQUES FOR CARBON FLOW STUDIES

Two different approaches to the use of $^{14}CO_2$ labeling have been applied to measure carbon flow within plant–soil systems, i.e., pulse-chase and continuous labeling. The $^{14}CO_2$-labeling apparatus can range from a polyethylene bag [34] to more complex Plexiglas canopies [35,36] to highly sophisticated dedicated growth cabinet facilities such as the experimental soil–plant–atmosphere system in The Netherlands [37,38]. Nevertheless, the key feature in all these facilities is the separation of the shoot compartment where the $^{14}CO_2$ is released from the root compartment, thus allowing the spread of the ^{14}C through the plant and into the rhizosphere to be monitored.

Each of these approaches has advantages and disadvantages, and these have been reviewed [39]. Essentially, pulse labeling results in the labeling of nonstructural labile carbon pools, whereas continuous labeling homogeneously labels the whole plant. Pulse labeling is relatively quick and easy to carry out and allows plant responses to rapid changes in environmental conditions to be determined, which is impossible with continuous labeling. The pulse-labeling approach has been applied in a large variety of studies under laboratory conditions with wheat and maize [35,40,41], grasses [40,42–44], and with trees [45,46]. ^{14}C-CO_2 pulse labeling has also been applied in field studies [36,47–52], mainly, to determine the C flow in arable cropping systems. To be able to quantify the carbon flow from the pulse-labeling data, not only the flux of C to the roots needs to be assessed but also the fluxes of organic and inorganic release of root-derived material. C fluxes have to be calculated from data on shoot and root biomass, and from data on ^{14}C distribution at different development stages [41,50]. Swinnen et al. [51] extrapolated the ^{14}C distribution curves obtained after labeling wheat plants in the field from the first labeling date (elongation stage) down to crop emergence and from the last labeling date (dough ripening stage) up to crop harvest using different extrapolation procedures. Their results showed that whereas the maximum shoot growth rate occurred around ear emergence, the flux of C to the roots had a maximum around tillering. Over the entire growing season, shoot growth amounted to 5730 kg C ha y^{-1} and 2310 ± 90 kg C ha y^{-1} was translocated belowground. Of this 920 ± 150 kg C ha y^{-1} was lost in root respiration and 500 ± 120 kg C ha y^{-1} was released as young rhizodeposits, which are defined as organic materials released from the roots within 19 d after assimilation. An interesting procedure was suggested by Michunas and Lauenroth [53] to estimate the belowground net primary production after

[14]C pulse-labeling from ^{14}C isotope decay assessments due to *in situ* decomposition. They considered the loss of ^{14}C to be the inverse of production, and they calculated turnover times by regression to time of complete loss of the isotope from the system. They noticed a two-phase loss of ^{14}C over a 13-year period in a short-grass prairie system. Pulse labeling has also been applied to study the impact of air polluting ozone on the assimilation and translocation of ^{14}C-CO_2 in adult pine trees [54]. Pulses of both ozone and ^{14}C-CO_2 were applied in a box covering a branch of the trees at different times during the growing season, and the ^{14}C in different fractions in needles, in the rest of the branch, and in the tree were measured. These studies indicated that short term exposure to ozone retarded the translocation of photosynthates from the needles to the branches and the roots in a similar fashion in both juvenile and mature trees. Whereas, pulse labeling allows determination of carbon flux to the root, continuous labeling gives only cumulative data regarding carbon transport. Nevertheless, continuous labeling is the only way to determine total flux of carbon through the root–soil system. Consequently, different qualities of information are provided by both labeling procedures.

One of the most intriguing questions related to the dynamics of carbon in plant–soil system is related to the relative contribution of plant- vs. microbe-produced CO_2 in the root environment. Obviously, when ^{14}C-CO_2 is released by microorganisms the plant-derived carbon had been involved in the stimulation of microbial activity, whereas plant-derived ^{14}C-CO_2 results from carbon that has not entered the rhizosphere system. Earlier studies had attempted to compare sterile with nonsterile systems, but these worked for only short periods before sterility was lost [55]. In other studies, microbial inhibitors had been added to nutrient solutions, but side effects on the plant and on nontarget microorganisms could not be ruled out [56], and the lack of soil as substratum compounded the problem [57]. Therefore, ^{14}C-based systems to address this problem in soil have been devised. Johansson [58] calculated gross rhizodeposition in continuously ^{14}C-labeled plants from the stable ^{14}C residue remaining in the soil after long-term incubation. The former is calculated from the latter through use of a stabilization factor, which is estimated by comparison with decomposition of known organic materials (glucose, grass shoots, and roots). Thus, C from microbial respiration is the difference between the calculated gross rhizodeposition and that measured in soil immediately after plant growth. Cheng et al. [59,60] involved saturating the soil with glucose immediately before [14]CO_2 labeling to eliminate ^{14}C microbial respiration. Finally, Swinnen [61] injected model ^{14}C-labeled rhizodeposits (glucose, root extract, or root cell wall material) into the rooted soil of an unlabeled plant, simultaneously with the [14]CO_2 pulse labeling of a similar but separate plant. Hence, the assumption is made that microbial transformation of the introduced compounds is representative of the transformation of ^{14}C-labeled rhizodeposits in the ^{14}C-CO_2 pulse-labeled system. Kuzyakov [62] concluded from a comparative study on the feasibility of these methods that the methods that have been suggested to separate root respiration and microbial respiration of rhizodeposits in nonsterile soil show different results. Each is based on many assumptions and has certain shortcomings. It remains unclear whether the different results are method-inherent or reflect environmental and experimental conditions, i.e., different plants, soils, equipment, and environmental conditions, etc.

One practical aspect of the procedure for monitoring carbon flow following ^{14}C labeling is the need to separate roots from the soil for analysis. Incomplete removal of roots can lead to an overestimation of rhizodeposition, but overzealous washing of soil may lead to leaching of ^{14}C or loss of fine roots. This problem has been examined in detail for wheat and barley and correction procedures for these errors have been developed [63].

[14]C also provides excellent possibilities for following the fate of root released carbon into the microbial biomass. Generally, chloroform fumigation is carried out to measure the microbial biomass and the microbial content, including the tracer ^{14}C, which is measured using direct extraction, incubation, or centrifugation methods [38,40,43,47,64,65].

Most recently ^{14}C-CO_2 has mainly been applied to assess the impact of elevated atmospheric CO_2 concentrations on the carbon flow in plant–soil systems in order to estimate the carbon sequestering capacity of terrestrial ecosystems. Paterson et al. [66] used ^{14}C-CO$_2$ pulse labeling to investigate the impacts of CO$_2$ concentration on plant growth, dry matter partitioning, and rhizodeposition as affected by photon flux density (PFD) and growth matrix. Plants were grown at two CO$_2$ concentrations: 450 (low CO$_2$) and 720 μmol mol^{-1} (high CO$_2$). Their results showed that plant responses to CO$_2$ are potentially affected by PFD and by feedbacks from the growth matrix. Van Ginkel et al. [67,68] applied a continuous ^{14}C-CO$_2$ labeling treatment for up to 115 d to investigate the effect of elevated atmospheric CO$_2$ concentrations on the carbon flow in perennial grass plants and then into the rhizosphere. As observed by others, they found that the root biomass was greater at elevated CO$_2$, and in a follow-up experiment the roots of plants grown at elevated CO$_2$ appeared to be decomposed more slowly than the roots grown at ambient CO$_2$.

B. ^{13}CO$_2$-LABELING TECHNIQUES FOR CARBON FLOW STUDIES

As mentioned before, ^{13}C is gradually replacing ^{14}C as the main tracer in carbon flow studies in plant–soil systems, because it is naturally occurring, stable, nonradioactive, and is less discriminated against than ^{14}CO$_2$ during photosynthesis. Because of its nonradioactive nature, it can easily be combined with other methodologies, and some rather exciting new approaches have been developed in recent years, integrating measurements of the flow of carbon in the rhizosphere with assessments of the associated responses of the microbial community. In Section V.C a short description of these combined approaches will be given.

Basically, ^{13}C is used in two ways in carbon flow studies: (1) it has been applied as either enriched or depleted ^{13}C-CO$_2$ in pulse-labeling procedures and (2) use is made of the differences in natural abundance of ^{13}C in plant systems dominated by either C4 or C3 plants. Concerning the latter, the natural abundance of ^{13}C is higher in C4 compared with C3 species [69], and by growing C4 plants on soils previously exposed only to C3 plants, or *vice versa*, this difference may be used to study carbon partitioning in plants and the rhizosphere.

Butler et al. [70] demonstrated the effectiveness of using ^{13}C enrichment labeling to carbon flows from plants at different stages of growth. They used a ^{13}C pulse-chase labeling procedure to examine the flow of photosynthetically fixed ^{13}C into the microbial biomass of the bulk and rhizosphere soils of greenhouse-grown annual ryegrass (*Lolium multiflorum* Lam.). The temporal dynamics of rhizosphere C flow through the microbial biomass was assessed by labeling plants, either during the transition between active root growth and rapid shoot growth (labeling period 1), or 9 d later during the rapid shoot growth stage (labeling period 2). Within 24 h of labeling, more than 10% of the ^{13}C retained in the plant–soil system resided in the soil. Turnover of ^{13}C through the microbial biomass was faster in rhizosphere soil than in bulk soil, and faster in labeling period 1 than labeling period 2.

Whereas this study was conducted in the greenhouse, successful ^{13}C-enrichment pulse labeling studies have also been carried out in the field. For example, using this approach, it appeared that pasture plants allocated more carbon to roots under low-soil fertility and high-grazing intensity [71]; that liming increased soil carbon throughput in an upland grassland [31] up to the level that more C was lost through the soil than has been gained via photosynthetic assimilation over a period of 8 years [72]; and that mycorrhizal fungi provide a rapid and important pathway of carbon flow from plants back into the atmosphere [73].

The use of ^{13}C-depleted CO$_2$ in a free air CO$_2$ enrichment experiment in a mature deciduous forest allowed for tracing the carbon transfer from tree crowns to the rhizosphere of 100- to 120-year-old trees [74]. During the first season of CO$_2$ enrichment, the CO$_2$ released from soil originated substantially from concurrent assimilation. The small contribution of recent carbon in fine roots suggests a much slower fine root turnover than is often assumed. The spatial variability of δ^{13}C in soil air showed relationships to aboveground tree types such as conifers vs. broad-leaved trees. Depleted ^{13}C-CO$_2$ labeling was also applied by Andrews et al. [75] to determine the relative contributions of root and soil heterotrophic respiration to total soil respiration *in situ*. In the free-air CO$_2$

enrichment (FACE) facility, in the Duke University forest plots, U.S., of an undisturbed loblolly pine (*Pinus taeda* L.) forest was fumigated with CO_2 that was strongly depleted in ^{13}C. By measuring the depletion of ^{13}C-CO_2 in the soil system, they found that the rhizosphere contribution to soil CO_2 reflected the distribution of fine roots in the soil and that late in the growing season roots contributed 55% of total soil respiration at the surface.

Carbon flow, including rhizosphere respiration and original soil carbon decomposition, can also be assessed on the basis of measurements of the differences in the natural abundance of ^{13}C [76]. The principle of the ^{13}C natural tracer method is based on the differences in ^{13}C:^{12}C ratios (often reported in $\delta^{13}C$ values) between plants with the C3 photosynthetic pathway, whose mean $\delta^{13}C$ is −27‰, and plants with the C4 pathway, whose mean $\delta^{13}C$ is −12‰ [77]. Similarly, there are subsequent differences in the $\delta^{13}C$ values between soil organic matters derived from the two types of plants. Soil organic matter derived from C4 plants (C4-derived soil) such as continuous corn fields has $\delta^{13}C$ values ranging from −12 to −14‰, whereas $\delta^{13}C$ values of soil organic matter derived from cold and temperate forests (C3-derived soil) range from −24 to −29‰. If one grows C3 plant in a C4-derived soil, the carbon entering the soil via roots will have a different $\delta^{13}C$ value than the $\delta^{13}C$ value of the soil. Natural abundance methodologies are often used in combination with planned previous manipulation of the system to obtain pools of different $\delta^{13}C$. In terms of carbon flows, $\delta^{13}C$ natural abundances have been successfully used, for instance, to determine the contribution of roots to soil carbon storage [78,79] and the exudation of carbon from growing roots [80]. The only limitation to the use of natural abundance methods is that the $\delta^{13}C$ signature between the carbon pools of interest must be sufficiently different to be able to track carbon movement. For example, Cheng et al. [81] developed a new method for measuring nondestructively, tree root respiration *in situ* under natural forest conditions. To obtain sufficient differences in $\delta^{13}C$ values between the rhizodeposits and the native soil organic matter, three kinds of tree growth media were investigated. The first medium was surface soil from under long-term wire grass patches (C4 soils) near to the longleaf pine plantation. This wire grass soil was significantly enriched with ^{13}C ($\delta^{13}C$ was −17.83‰ from wiregrass soil and −27.56‰ for pine roots, i.e., approx. 10‰ higher). The second one was soil that was obtained from a site where C4 grasses dominated. It had a $\delta^{13}C$ of −14.6‰. The third medium was a mixture of acid-washed sand and burnt vermiculite that was used as reference.

In most studies only one tracer was used. However, a combination of ^{13}C and ^{14}C could provide extra information on carbon flow processes. For instance, Johansson [82] combined ^{13}C and ^{14}C continuous labeling to assess the impact of cutting on the carbon flow in *Festuca* spp. Kuzyakov and Cheng [83,84] determined the effects of shading wheat and maize on rhizosphere respiration and rhizosphere priming of soil organic matter decomposition by using a natural abundance ^{13}C tracer method and ^{14}C pulse labeling simultaneously. Both ^{13}C natural abundance and ^{14}C pulse labeling techniques used gave similar estimates of root-derived CO_2 during the whole observation period, and the application of combined tracer methods showed that the cultivation of the wheat led to increasing decomposition intensity of soil organic matter (priming effect), whereas maize showed a negative priming effect.

A final, highly specialized procedure for monitoring carbon flow in plants involves the use of ^{11}C-CO_2. The use of this positron gamma-emitting isotope of C, with a half-life of 20.3 min, allowed several physiological parameters of mycorrhizal and nonmycorrhizal plants of *Panicum coloratum* to be measured simultaneously in real time [85]. However, the technical problems associated with using ^{11}C-CO_2, particularly the ability to produce this isotope of carbon and handling it with such a short half-life are likely to limit this approach to specialized facilities.

IV. ANALYSIS OF CARBON FLOW COMPOSITION

The significance of carbon flow for the dynamics of microbial populations cannot be assessed without considering the composition of the carbon compounds released by the plant. Starting with the information provided by Vančura in the 1960s (for example, see Reference 86 and Reference 87) numerous studies have been performed to analyze the composition of the organic compounds released from the roots. The main emphasis has always been on the simple, water-soluble root exudates, although the bulk of the root releases are probably organic complexes derived from plant cell walls and other structural cell materials. However, the precise analysis of the root exudates has proven to be hard, if not impossible; this has led to the so-called *great rhizosphere frustration* [88]. The main reason is the fact that it is very difficult to differentiate between root exudates and compounds released by microorganisms present in the rhizosphere. It has been proven that microorganisms play a significant role in the exudation process such that the quality and quantity of root exudation under axenic conditions can be significantly different from the exudation under natural conditions [89].

Improved and more accessible technologies for chemical analysis have allowed for a better understanding of the variety of compounds released by roots, but they have not enabled us to unravel the composition of the mix of compounds continuously provided by the roots for microbial utilization. Among the techniques commonly used to analyze the exudate composition are HPLC, TLC, and LC [90,91]. Czarnota et al. [92] used both TLC and HPLC to analyze the composition of root exudates of different sorghum accessions and found that within each accession variation existed in the amount of exudates produced and the chemical constituents in each of the accessions. The average exudate production ranged from 0.5 to 14.8 mg/g root. The exudates were obtained from plants grown on a so-called *capillary mat system* on which seeds were allowed to germinate and after 5 to 14 d roots were harvested from the screens and exudates were collected. That system was similar to the Kuchenbuch model rhizosphere system developed in the 1980s to study the dynamics of nutrients in the immediate vicinity of roots and later used for the study of microbial processes, including gene transfer in the rhizosphere [93–95].

More sophisticated technologies such as gas chromatography–mass spectrometry (GC–MS) and pyrolysis-field ionization-MS have also been applied to determine the composition of exudates. Fan et al. [96] applied ^1H and ^{13}C multidimensional nuclear magnetic resonance (NMR) and silylation GC-MS to analyze the exudates composition of barley and in particular the presence of phytometallophores. The advantages of this approach are: (1) minimal sample preparation and reduced net analysis time, (2) structure-based analysis for universal detection and identification, and (3) simultaneous analysis of a large number of constituents in a complex mixture. Quantification of all major root exudate components of barley using these methods showed a sevenfold increase in total exudation under moderate iron deficiency. Total quantities of exudates per gram of root remained unchanged as iron deficiency increased, but the relative quantity of carbon allocated to phytosiderophores increased to approximately 50% of the total exudates in response to severe iron deficiency. Melnitchouk et al. [97] investigated the composition and diurnal dynamics of water-soluble, root-derived substances and products, and their interaction with sandy soil were investigated in maize plants (*Zea mays* L.) by pyrolysis–field ionization mass spectrometry (Py–FIMS). Rhizodeposition was larger during day than during the night, and the composition of these deposits was different. The largest differences in the Py–FI mass spectra resulted from signals of amino acids (aspartic acid, asparagine, glutamic acid, leucine, isoleucine, hydroxyproline, and phenylalanine) and carbohydrates, in particular pentoses, which were exuded in the photosynthetic period. Other compounds detected in the Py–FI mass spectra were interpreted as constituents of rhizodeposits (lipids, suberin, fatty acids) or products of the interaction of rhizodeposits and microbial metabolites with stable soil organic matter (lignin dimers and alkylaromatics).

Another approach to study the flow of carbon and the response of the microbial community in the rhizosphere is the use of molecular tools and modified organisms. Bacterial cells harbor environmentally responsive genomic elements which, when fused to an appropriate reporter gene, will react toward the presence of an appropriate effector molecule [98]. Van Overbeek and Van Elsas [99] used Tn5-B20 (*lacZ* as reporter gene) transcriptional fusion mutants of *Pseudomonas fluorescens* R2f to study the response of the bacterium to wheat root exudates. Several mutants showed β-galactosidase activity under the influence of wheat root exudates. In one such mutant,

RIWE8, gene expression was specifically induced by proline but not by 125 other substrates. This mutant also showed reporter gene induction, albeit to a lesser extent, by exudates of maize and grass roots but not by that of clover roots, which are known not to contain proline. De Weger et al. [100] used *lacZ* to evaluate phosphate limitation of rhizosphere bacteria. The same group of Lugtenberg used a similar approach to determine the main exudate components involved in root colonization by rhizobacteria [29]. They developed auxotrophic mutants of *Pseudomonas fluorescens*, which were impaired in growth on specific compounds, such as monosaccharides and organic acids, which were supposed to be major components of tomato root exudates. The ability of the mutants to compete with the wild type for root colonization was an indication of the importance of the utilization of the different exudates components for rhizosphere competence. Their results indicate that organic acids are the nutritional basis for tomato root colonization by these bacteria, whereas sugars obviously play a much smaller role.

A strategy for quantification of specific root exudates on the basis of this approach is the use of green fluorescent protein (GFP) as reporter gene fused to a responsive genomic element [98]. This has successfully been tested for the detection of sugars using a GFP-producing *Erwinia herbicola*. More biosensors have been developed [101,102]. As these biosensors are inherently specific, only small fractions of the microbially relevant root exudates can be detected.

Besides modified microbes, also modified plants might conceptually be suitable for the evaluation of the impact of root exudates on the rhizosphere community. As early as 1978, Petit et al. suggested to take advantage of the close relationship existing between a plant and its associated microflora to engineer plant root exudation (see Reference 103). This should provide the microorganism of interest with a selective advantage that may help establish it in the rhizosphere, a strategy later termed *biased rhizosphere* or *artificial symbiosis*. Several recently published studies on plants modified in the release of root exudates have originally been developed for the assessment of the risks of genetically modified plants on soil ecosystems (see reviews by Bruinsma et al. [104] and Kowalchuk et al. [105]). The most well-known example of this approach is the use of modified tobacco plants capable of releasing substantial amounts of opine-derived compounds in the rhizosphere [103,106]. In these studies, transgenic *Lotus* plants producing two opine, namely mannopine and nopaline, were used to characterize the microbial communities directly influenced by the modification of root exudation. The results showed that opine utilizers represent a large community in the rhizosphere of opine-producing transgenic *Lotus*. This community is composed of at least 12 different bacterial species, one third of which are able to utilize mannopine and two thirds nopaline. Opine utilizers are diverse belonging to the Gram-positive and Gram-negative bacteria. The authors concluded that transgenic plants with engineered exudation constitute an excellent tool to isolate and characterize specific microbial populations.

Finally, although the basic concept of the rhizosphere effect, i.e., the stimulation of microbial growth and biological activity by the release of organic root-derived compounds, has not been altered, we get more and more aware of the fact that the influence of the root on the growth and activity of microorganisms in the soil might also result from effects of specific compounds released by plants on the activity and the behavior of specific microorganisms [107]. There has been a tremendous effort in all fields of biology, during the last two or three decades, to analyze and comprehend the exchange of (chemical) signals that take place among organisms. Regarding the rhizosphere, the signaling between host plants and symbiotic nitrogen-fixing bacteria, and plants and pathogens are among the best-studied processes (e.g., 108–110, Chapter 11). Specific signaling compounds also play a role in the contact between root pathogens and their hosts; specific compounds within the root exudates can stimulate germination of fungal propagules and subsequent directed growth of the mycelium of a range of fungal pathogens [111]. There are many more processes in the rhizosphere that are regulated or initiated by the exchange of more or less specific signals between plants and the rhizosphere microflora. We also know that some of the key microbial processes involved in these interactions are under the control of quorum sensing [112], that is, the density-dependent regulation of gene expression in microbial cells. In both Gram-negative and

Gram-positive bacteria many different quorum-sensing systems have been developed. In particular, the N-acylated homoserine lactone system found in many Proteobacteria has extensively been studied. Acylated homoserine lactones (AHLs) quorum-sensing assessments are commonly based on the use of reporter gene constructs in *Escherichia coli* or *Agrobacterium tumefaciens* [112]. It is also possible to monitor the expression of an AHL-controlled phenotype, such as the production of protease, antibiotics, or pigments. The most commonly used strains in this respect are mutants of *Chromobacterium violaceum* for which the formation of the purple pigment violacein is dependent on the production of AHL by other bacteria [113].

Other assays are based on chemical analysis of the signal. It is beneficial to understand what the objective of the study is and to know the abilities of particular methods to achieve those aims. Rice et al. [114] provide detailed information on these assays. With respect to the dynamics of microbial populations in the rhizosphere, it is important to realize that, recently, evidence was provided that certain plants are releasing compounds that interfere with bacterial quorum sensing [115], presumably determining the outcome of the interactions between plants and rhizosphere bacteria.

V. METHODS FOR THE STUDY OF MICROBIAL POPULATION DYNAMICS ASSOCIATED WITH RHIZOSPHERE CARBON FLOW

The flow of carbon compounds from roots will influence microbial populations in various ways. The interactions among the microbial populations in the rhizosphere and between these populations and plants are key processes for plant and soil health.

The main obstacles to increasing our knowledge on the size, the dynamics, and the activity of microbial communities in the rhizosphere are methodological. Microbial cell enumeration techniques and identification procedures are often difficult or tedious, and the collection of relevant samples or the simulation of natural conditions in the laboratory can be problematical. However, in the last two decades, the development of molecular approaches for the study of microbial populations has significantly contributed to solving the problems related to the identification and the dynamics of microbial communities, in particular of the so-called unculturable majority of the microbial rhizosphere community. Genomics approaches, which are gradually applicable in soil ecology studies, are promising in shedding more light on the activities of the wide variety of microorganisms in soil and rhizosphere. Considering the vast array of techniques presented in the literature, only a selection of these methods is discussed in this review. After discussing recent progress in both the traditional culturing techniques and the molecular tools for detection and identification of both culturable and unculturable microorganisms, we will focus on these techniques that integrate the assessment of carbon flow dynamics and related responses of the microbial community in the rhizosphere. An excellent reference of the wide variety of methodologies available for microbiological studies, including descriptions of the details of the practical application of the many different techniques, is the recently published *Molecular Microbial Ecology Manual* [116].

A. CULTURE-DEPENDENT TECHNOLOGIES

Traditionally, since the first formulation of the concept of the rhizosphere by pioneer German agronomist and plant physiologist Lorenz Hiltner in 1904 [117], the agar plate enrichment methodology has been used to determine microbial populations in the rhizosphere. Based on this methodology, a wealth of information has become available since then and some of the main concepts describing rhizosphere processes have been developed. Of great importance for the understanding of the rhizosphere was the formulation of the R/S-ratio concept by Katznelson et al. [118]. The *R/S ratio* is defined as the number of microorganisms in the rhizosphere (R) to the number in the bulk soil (S). So, it expresses the response of specific microbial species to the root and the root released compounds. The R/S ratio measured by isolation and culture for unicellular bacteria is generally between 10 and 20, but the figure can be greater than 100. For other groups of organisms the R/S ratio decreases in the following order: unicellular bacteria > actinomycetes and fungi > microfauna [119]. Increased numbers of microorganisms in the rhizosphere have also been observed microscopically [120]. Another culture-dependent approach is the measuring of the rate of colony development on culture media in order to discriminate between so-called r and K strategists (fast-growing opportunists and slow-growing specialists, respectively), which was developed by De Leij et al. [121]. This method is still used frequently as the information obtained by this rather simple approach is unique and highly relevant for understanding of the soil and rhizosphere ecosystem. On the basis of this methodology, the ecophysiological (EP) index can be calculated. The EP index is calculated on the basis of the proportion of the total number of bacteria that appear per unit time, and the EP index thus expresses the evenness in the distribution of fast and slow-growing bacteria. Recent applications of this approach are by Garbeva et al. [122], where the EP index was used to obtain information on the dynamics of microbial populations in soil and rhizosphere of fields under different arable regimes over a period of four growing seasons. Another recent example concerned the study by Ruiz Palomino et al. [123] on the seasonal diversity changes in the culturable rhizobacterial communities of alder. For technical details the reader is to refer to Insam and Goberna [124].

To overcome the need for isolating individual colonies and then profiling their metabolic capacity, Garland and Mills [125] developed a method to profile the whole population. Cell suspensions taken from samples are placed directly on media with single carbon sources and a metabolic indicator (tetrazolium). If any members of the population have the ability to utilize the carbon source, a positive reaction will be recorded. In this way the carbon source utilization profile of the total population would be determined. Initially, a simple bacterial identification system (of Biolog, Inc.) was used to profile samples [126], but since then the types of carbon sources used have been broadened to include those likely to be found in the environment under study. Yet, Biolog plates are still frequently used to generate community-level physiological profiles (CLPP). Some recent examples of the width of the application of the CLPP methodology, including the use of Biolog plates, are given by Bucher and Lanyon [127] in their evaluation of soil management practices; by Selmants et al. [128] who assessed the changes in the microbial functions induced by Red alder (*Alnus rubra*) in conifer forests, and by Ritz et al. [129] who measured the spatial structure in microbial properties in upland grassland.

Biolog-based approaches to assess the metabolic profile of a community have several drawbacks. It is culture dependent, changes in the composition of microbial communities may occur during incubation, the contribution of fungi cannot be assessed, and, for reproducible results, the replicates need to be inoculated with equally dense inocula [130]. Degens and Harris [131] developed an alternative *in situ* approach to determine the metabolic profile of soil microbial communities. The method utilizes differences between the substrate induced respiration (SIR) responses of microbial communities and simple organic compounds to quantify catabolic diversity. They argued that the approach provided a reasonably rapid and simple method to assess the catabolic diversity of microbial communities without extracting or culturing organisms from soils and, so, to overcome some of the main drawbacks of the Biolog approach.

It was obvious from the early information that was obtained when comparing the results of agar plate counts with direct microscopic counts (for example, see Reference 132), that plate counts greatly underestimate the size of populations in natural systems (reviewed by Amann et al. [133]). From the classic paper of Torsvik et al. [134], the huge diversity of the microbial community in soil began to be recognized. In that paper, an estimate of 4600 distinct genomes per gram of soil was determined by the reassociation time of total community deoxyribonucleic acid (DNA) compared with a standard curve of reassociation kinetics of a known number of cultured genomes. Clearly, culture-based methodology is inadequate to serve the needs of microbial ecologists seeking to describe the diversity of bacterial communities in environmental samples.

Although genomics approaches might in future enable the study of physiological properties of specific microorganisms *in vivo*, at present we still need isolates for these detailed studies.

Janssen et al. [135] used different diluted growth media to maximize the isolation of bacteria from soils. They were able to isolate a number of bacteria that represent the first known isolates of groups of soil bacteria belonging to a novel, largely uncultured, lineages within the divisions Actinobacteria, Acidobacteria, Proteobacteria, and Verrucomicrobia. Species belonging to those lineages have been observed to be dominant members of the rhizosphere community by molecular methods [136]. Davis et al. [137] noticed that some media, which are traditionally used for soil microbiological studies, returned low viable counts and did not result in the isolation of members of rarely isolated groups. In contrast, newly developed media and use of increased incubation times of up to 3 months resulted in high viable counts and in the isolation of many members of rarely isolated groups. Uncommon culture media have also been used to isolate nonthermophilic archaea from the rhizosphere [138]. It turned out that mesophilic soil crenarchaeotes are found associated with plant roots, and it is the first evidence for growth of nonthermophilic crenarchaeotes in culture. Interestingly, De Ridder-Duine et al. [139] observed that the number of bacteria that could be isolated from rhizosphere soil relative to the total count by direct microscopy was much larger than the relative number of bacteria that could be isolated from bulk soil.

B. CULTURE-INDEPENDENT TECHNOLOGIES

All cell culture-based methodologies are essentially limited when studying whole populations, as the dominant proportion of the microbial community of soil, rhizosphere, and rhizoplane, and other environments cannot be cultured on standard laboratory media as mentioned before. To obtain information on the composition and activity of the nonculturable fraction and to aid the study of the culturable fraction, direct detection methods are needed.

Direct microscopic analysis of microbial cells can provide information on cell size, numbers, and biomass [140]. Modifications of this basic method can be used to obtain additional information on cell viability and activity, or to identify types of microorganisms present. Table 13.1 outlines a range of these methods. For example, the identification of individual cells may be possible using one of a number of immunologically based methods. Fluorescently labeled antibodies are commonly used in conjunction with microscopy to allow the study of cells in soil (Table 13.1). In most cases, the results are affected by background fluorescence levels and the nonspecific binding of all antibodies to nontarget cells or particles. Specialized confocal microscopes that use a laser light

---

TABLE 13.1
Examples of Microscope-Based Methods for the Detection of Cells in Environmental Samples

Target	Stain	Comments	References
DNA	Ethidium bromide, DAPI	Stains all cells	142
RNA/DNA	Acridine orange	May be affected by membrane changes	143
RNA	Fluorescent oligonucleotides	16S rRNA probes, specific for a species, genera, or kingdom	133
Proteins	FITC	General stain	144
Membrane potential	Rhodamine 123	Vital stain	145
Outer cell wall structures	Fluorescent antibodies	Specific antibodies must be available	140,146
Others	Molecular probes — stain all, vital stain	Vital stains specially useful for total viable cell counts	147
Viable cells	Direct cell count	Detection of viable cells by elongation	148
Esterases	Fluorescein diacetate	Vital stain	149
Respiration systems	INT, CTC (tetrazolium compounds)	Vital stain	142,150
Polysaccharide	Calcafluor white M23	Direct staining	143

source can reduce these problems and may also penetrate a small distance into the sample to view cells below the surface [141]. This is a particularly useful feature for observing microbial films, including those that develop around roots and soil particles. The use of another laser-based system, the flow cytometer, may also overcome some of the methodological limitations that prevent the analysis of a large number of samples by direct microscopy.

1. Nucleotide-Based Biomarkers

The most frequently used techniques for the assessment of the composition of microbial communities are based on the analysis of ribosomal RNA, rRNA, extracted from the communities in soil and rhizosphere. The rRNA molecules comprise highly conserved sequence domains interspersed with variable regions. The earliest attempts to analyze the diversity of naturally occurring microbial populations relied upon direct extraction and sequencing of 5S rRNA-gene from environmental samples [151]. The limited length of the 5S rRNA molecule of approximately 120 nucleotides means that there is limited scope for high-resolution phylogenetic analyses based on 5S rRNA gene sequences. The development of robust and simple DNA cloning techniques and the polymerase chain reaction (PCR) have, however, allowed higher resolution analyses of more complex communities using small subunit (SSU; 16S for prokaryotes and 18S for eukaryotes) rRNA sequence analysis. The SSU rRNA molecule is approximately 13 times longer than the 5S rRNA and thus contains considerably more information.

The starting point for this and other related procedures is the extraction of nucleic acids of sufficient quality to permit activity of the enzymes used in subsequent procedures. This is not a trivial matter and requires extensive experience. Frostegård et al. [152] examined several approaches to improve the extraction, purification, and quantification of DNA derived from as large a portion of the soil microbial community as possible. Lysis was carried out by grinding, sonication, thermal shocks, and chemical treatments of the soil. Grinding increased the DNA yield compared with the yield obtained without any lysis treatment, but none of the subsequent treatments clearly increased the DNA yield. Griffiths et al. [153] described a method for the extraction of total nucleic acids from soil, i.e., DNA and RNA. Of great importance is that prior to nucleic acid extraction all solutions and glassware are rendered RNase free and only certified RNase- and DNase-free plasticware are used. Griffiths et al. [153] extracted nucleic acids from a sample of 0.5 g (fresh weight) soil involving a bead beating procedure. The extracted DNA is subjected to PCR amplification using "universal" primers or primers designed to amplify rRNA genes from a particular group of organisms. The PCR product can then either be cloned or analyzed or the DNA fragment is resolved electrophoretically. For the latter approach, a large number of techniques are available. Relevant to rhizosphere studies are techniques, such as ribosomal intergenic spacer analysis (RISA) [154,155], denaturing gradient gel electrophoresis (DGGE) [136,139,156–159], temperature gradient gel electrophoresis (TGGE) [160–162], single strand conformation polymorphism (SSCP) [163,164], internal transcribed spacer–restriction fragment length polymorphism (ITS-RFLP) [165–167], random amplified polymorphic DNA (RAPD) [168,169], amplified ribosomal DNA restriction analysis (ARDRA) [170–172], and Terminal Restriction Fragment Length Polymorphism (T-RFLP) [173–175]. Differences in electrophoretic profiles between samples reflect differences in community composition and abundance of individual microbial populations in a community [176]. Although the fingerprint obtained from an environmental sample does not reveal the taxonomic composition of a microbial community, phylogenetic information about particular community members may be obtained by isolation and sequence analysis of bands of interest.

Community analyses based on PCR have a number of steps that may introduce biases [176]. Cell structure varies among taxonomic groups with some microbes being more easily disrupted than others in the extraction process. In addition, the copy number of rRNA genes present within the genomes of different organisms can vary significantly. When analyzing known specific species this can be taken into account, but when analyzing a highly diverse community from soil, this is inherently a major drawback for quantitative analyses. Environmental variables may affect DNA extraction and PCR. Inhibition of PCR by environmental compounds has been reviewed by Wilson et al. [177]. Methods for sample collection and DNA extraction must take into account such factors as coextraction of humic substances from soil and low bacterial cell density in some environments, and at the same time optimize lysis of structurally different cells. Niemi et al. [178] demonstrated that soil bacterial community profiles differed depending on the DNA extraction and purification method utilized. Methods that include mechanical lysis using a bead beater were found to yield the most consistent results.

In fact, the implication of these biases and limitations is that quantitative analyses should be considered with great care. Yet, in recent years, quantitative PCR (qPCR, also referred to as *real-time PCR*) has emerged as a promising tool for studying soil microbial communities [179]. qPCR is based on the real-time detection of a reporter molecule whose fluorescence increases as PCR product accumulates during each amplification cycle. Fierer et al. [179] described a qPCR-based approach to assessing soil microbial community structure and to quantify the abundances of the dominant groups of bacteria and fungi found in the soil environment.

As mentioned before, these PCR-based community analyses are at most times using the ribosomal RNA operon, typically the 16S rRNA gene (for bacteria) and 18S (for fungi). Another RNA-based marker used to design specific primer sets for *Paenibacillus* is *rpoB*, the RNA polymerase beta subunit [180]. Other marker genes have been suggested as well. When specific traits or functional characteristics are under investigation, phylogenetic markers, other than the rRNA genes, can be used to characterize microbial communities. For soil microorganisms, functions associated with nitrogen metabolism have been widely used for community analysis. The phylogenetic markers used in these studies include a structural nitrogenase gene (*nifH*) [181–183]. When a particular function is restricted to specific bacterial taxa, 16S rRNA sequence may be used to differentiate these community members. For instance, this approach has been used to study autotrophic ammonia oxidizers in a variety of environments [184].

It will not be possible to describe the wide variety of PCR-based techniques in detail. Excellent reviews have been published for instance by Head et al. [151], Kent and Triplett [176], Singh et al. [185], Kowalchuk et al. [116], Van Elsas et al. [186], and Nannipieri and Smalla [187]. Here, we will only briefly deal with some recent developments in the DGGE methodology that probably has been used most frequently in recent studies on the dynamics of microbial populations in the rhizosphere [136,158,159,188–191]. DGGE was originally developed as a technique for the detection of single point mutations in a variety of genes in particular for medical purposes [192–194]. Muyzer et al. [195] adapted the method for the detection of bacteria in environmental samples. The latter adapted techniques including the separation of PCR-amplified DNA fragments on the basis of their base composition. For reasons mentioned before, SSU (16S and 18S) rDNA fragments were the main target for the DGGE analysis, although other DNA-based markers have been used, such as the ITS for the detection of fungal species, structural genes for key enzymes involved in N_2 fixation [183], 2,4-diacetylphloroglucinol (DAPG) production [196], and chitinases [197]. As the PCR product migrates through a gel containing a chemical gradient of a denaturing compound, such as formamide or urea, the two DNA strands begin to melt at the denaturant concentration specific for the DNA composition. Thus, PCR products with slightly different base compositions melt at different locations within the gel. The position and number of each band on a gel reflects the sequence diversity within the PCR product and ideally represents specific microbial species. The PCR-DGGE analysis of soil samples with bacterial PCR primers thus yields bacterial community profiles that may represent numerically the dominant members of the bacterial populations [160]. However, there are several problems that conceptually restrict the use or interpretation of PCR-DGGE of soil DNA [198]. These problems are partially related to the biases inherent to the use of DNA or RNA extractions from soil and to the use of PCR as mentioned before, and partially they are specific for the DGGE method. For instance, single bacterial types might produce more than one band in the DGGE profiles due to the presence of several copies of the 16S rRNA gene with slightly differing sequences [199], or, conversely, bands at similar positions may originate from different sequences with similar melting behavior.

Yet, over the years, strong evidence has been provided that microbial DGGE analysis of PCR amplified soil and rhizosphere DNA constitutes a reproducible and robust strategy for the assessment of the diversity of microbial communities with an emphasis on numerically dominant types. We have to realize that the detection limits for specific microbial groups are fairly high, and that minority groups are easily missed by the strategy based on total bacterial and fungal primers that targets whole microbial communities. To assess the prevalence and diversity of such minority groups, the use of group-specific primers is a prerequisite. Therefore, a variety of primer sets for specific microbial groups, including rhizosphere relevant groups, have been developed and tested for many different environments. These include primer sets for *Pseudomonas* [e.g., 200,201] antibiotic-producing *Pseudomonas* spp. [196], *Bacillus* spp. [202], actinomycetes [203], *Burkholderia* spp. [204], N_2-fixing bacteria [183], total fungi [158,161], Gigasporaceae [205], wood inhabiting fungi [206], *Glomus* spp. [207], and *Paenibacillus* spp. [180,208].

DGGE has also been applied in combination with other approaches. One of the most interesting, recently developed combinations is that of DGGE and stable isotope probing (SIP), which will be dealt with later in this chapter. The technique has been used to analyze the culturable fraction of rhizosphere samples, either after isolation on agar plates or in Biolog or related plates. Lin et al. [209] applied CLPP based on the use of Biolog GN plates to detect the effects of a genetically modified bacterial inoculant on microbial communities associated with rice seedlings. The potential for utilization of substrates of the Biolog system by these communities remained largely unchanged after the inoculation of the genetically modified bacterium [210]. However, significant differences in the utilization of selected substrates were observed between the control and inoculated soils, as evidenced by the application of PCR-DGGE to the bacterial communities inhabiting selected wells of the Biolog plates. Duineveld et al. [188] compared DGGE patterns obtained from DNA extracted directly from rhizosphere soil with that obtained after extraction of DNA from microbial colonies from agar plates. The DGGE patterns of the DNA of both sources differed significantly.

Duineveld et al. [136] used PCR and reverse transcriptase (RT) PCR to amplify 16S/rDNA and 16S rRNA, respectively, and the products were subjected to DGGE. Prominent DGGE bands were excised and sequenced to gain insight into the identities of the total PCR and active RT-PCR bacterial populations present. DGGE analysis of RT-PCR products detected a subset of bands visible in the rDNA-based analysis, indicating that some dominantly detected bacterial populations did not have high levels of metabolic activity. The sequences detected by the RT-PCR approach were however derived from a wide taxonomic range, suggesting that activity in the rhizosphere was not determined at broad taxonomic levels but rather was a strain- or species-specific phenomenon. This approach by analyzing RNA extracts from soil and rhizosphere was also applied in other studies to detect the active fraction of the microbial community [136,190]. The use of rRNA as a marker for the active microbial fraction, as presumed in these studies, should be considered with great care as the correlation between the rRNA content of a cell and its physiological state is rather weak, and the ribosomal copies per species in a community differs considerably.

2. Other Biomarkers

Ergosterol is a predominant sterol limited to the true fungi [211], and it is possible to use measurements of ergosterol content as an indicator of fungal biomass [212–214] in soil. The content of ergosterol in fungal membranes is species-dependent and can also vary with physiological state. Factors such as age, developmental stage, and general growth conditions can all cause variations in ergosterol levels. Joergensen [215] determined fungal and microbial biomass by ergosterol and fumigation–extraction, respectively, in bulk grassland soil, rhizosphere, and root material, to quantify the contribution of these three microbial fractions into the total soil microbial biomass and to soil organic matter. About 75% of the total ergosterol was found in the bulk soil fraction, 11% in the rhizosphere, and 14% in the root material. Klamer et al. [168] assessed changes in relative fungal biomass in the soil and root fractions of a scrub oak forest soil due to elevated atmospheric CO_2 by the ergosterol technique. In the bulk soil and root fractions, a significantly increased level of ergosterol was detected in the elevated CO_2 treatments relative to ambient controls. They combined the fungal biomass measurement by ergosterol with community composition analysis using T-RFLP analysis of the ITS region.

Volatile fungal metabolites can also be used as indicators of fungal growth in samples, such as stored cereals and wheat. Metabolites such as 3-octanone, *l*-octen-3-*ol*, and 3-methyl-*l*-butanol, 3-methylfuran, or total concentrations of a group of compounds, such as carbonyl compounds, have been used for this purpose [216–219]. In general, the production of a volatile metabolite that is to be used as a biomarker must not change with substrate type or level — an essential feature that needs to be addressed in such studies. The metabolite *l*-octen-3-01 is produced during the breakdown of lipids; its quantity can vary and is dependent on the lipid content of the substrate [219]. This limits its use as a biomarker of biomass.

A group of chemicals especially useful as biomarkers are fatty acids. These have been of great value in determining bacterial phylogeny and also provide a useful set of features for characterizing strains [220]. Specific fatty acids, especially phospholipids, which are the major constituents of the membranes of all living cells (with the exception of archaea), have the following useful properties: (1) they are degraded rapidly following cell death, (2) they are not found in storage lipids or in anthropogenic contaminants, and (3) they usually have a comparatively rapid turnover [221]. Bacteria also contain phospholipids as a relatively constant proportion of their biomass. This makes the analysis of the phospholipid fraction of mixed communities a useful measure of the viable cellular biomass and complementary to other traditional methods such as enzyme profile, muramic acid levels, and total adenosine triphosphate (ATP) [222]. Lipid markers can be recovered from isolates and environmental samples by a single-phase chloroform–methanol extraction, fractionation of the lipids on columns containing silicic acid, and derivatization prior to analysis by capillary GC-MS.

Quantitative estimates of microbial and community structure by means of analysis of the phospholipid fraction have been performed on soil and rhizosphere [70,223–228]. The method is applicable to the study of mixed populations of varying degrees of complexity and is relatively straightforward to perform. Changes in the environment that result in an alteration in the physiology of a microorganism can sometimes be detected in differences in the lipid profile. This factor is often mentioned when using fatty acid profiles for identification purposes, where growth conditions must be kept constant. However, these changes may be exploited to determine the metabolic status of an organism and the environmental conditions it encounters [229]. Changes in fatty acid composition of phospholipids have been observed as bacteria enter a nonculturable but viable condition [230], where cells are unable to grow on laboratory media but are able to take up vital stains. This may be linked with a survival strategy where alterations are occurring to a cell to ensure the efficient uptake of nutrients or for strengthening the membrane.

C. Analysis of Microbial Responses to Carbon Flow

In 1998, Boschker et al. [231] published a paper in which they proposed a method to link directly microbial populations to specific biochemical processes by ^{13}C labeling of PLFA biomarkers. They applied this approach in aquatic sediments and were able to identify the bacteria involved in sulfate reduction coupled to acetate oxidation and methane oxidation. Their results demonstrated the power of this new approach, of which the authors stated that it "could potentially provide a wealth of information on many of the natural microbial populations involved in the main biogeochemical cycles. In principle, all processes that involve the biological consumption and incorporation of carbon by organisms could be studied using label additions, including autotrophic as well as heterotrophic processes. As shown in this study, ^{13}C-labeling greatly extends the use of PLFA analysis in natural environments. As the full spectrum of the labeled PLFA can be used, researchers are no longer restricted to studying specific biomarkers that are only found in a limited number of genera. By directly linking activity with identity, this approach can greatly extend our knowledge about how the huge diversity of the microbial world relates to microbial processes in nature." Since then a large number of studies have been published in which a combined measurement of biomarkers and isotopes, in particular the stable isotope ^{13}C, was used to trace the fate of C substrates in microbial communities from an array of environments, including the rhizosphere; these approaches are commonly referred to as *stable isotope probing* (SIP). Butler et al. [70] used a ^{13}C pulse-chase labeling procedure to examine the incorporation of rhizodeposition into individual PLFAs in the bulk and rhizosphere soils of greenhouse-grown annual ryegrass. Treonis et al. [232] combined microbial community PLFA analyses with an *in situ* stable isotope $^{13}CO_2$ labeling approach to identify microbial groups actively involved in assimilation of root-derived C in limed grassland soils. They concluded that: "^{13}C stable isotope pulse-labeling technique paired with analyses of PLFA microbial biomarkers shows promise for *in situ* investigations of microbial function in soils." Pelz et al. [228] determined the flow of new and native plant-derived C in the rhizosphere of an agricultural field during one growing season. They used the natural differences in $\delta^{13}C$ abundances of different carbon sources as tracer and measured total amounts and $\delta^{13}C$ values of a number of relevant C fractions, including PLFA. The $\delta^{13}C$ values of the dominant soil PLFA showed wide ranges, suggesting that the microbial community utilized different pools as C sources during the season, and they concluded that the $\delta^{13}C$ values of PLFA, therefore, enabled the analysis of the metabolically active populations.

Analogous to the aforementioned approaches of combining stable isotopes labeling with biomarker measurement, Radajewski et al. [233] showed that ^{13}C-DNA, produced during the growth of specific microbial groups on a ^{13}C-enriched carbon source, can be resolved from ^{12}C-DNA by density gradient centrifugation. DNA isolated from the target group of microorganisms can be characterized taxonomically and functionally by gene probing and sequence analysis. They applied this methodology to investigate methanol-utilizing microorganisms in soil and demonstrated the involvement of members of two phylogenetically distinct groups of eubacteria. The ability to isolate an entire copy of the genome of microorganisms involved in the metabolism of a substrate makes SIP unique and provides advantages over other methods when attempting to link metabolic activity with taxonomic identity. Another valuable feature of SIP is that it enables genes involved in a particular function to be identified. The SIP approach has potential limitations, especially in relation to its sensitivity. Perhaps the single most important factor that might affect the identification of the target microorganisms is the dilution of a labeled substrate before its assimilation and incorporation. In the case of carbon, simultaneous growth on an unlabeled (^{12}C) substrate will dilute the proportion of ^{13}C that is incorporated into DNA. The same holds, to even a much larger extent, for the use of ^{13}C-labeled CO_2 when assessing the incorporation of root-derived compounds into the microbial community of the rhizosphere. Such dilution will reduce the proportion of DNA that becomes isotopically labeled, making it less easy to target those microorganisms that are primary users of a substrate. Another important factor is the fate of the isotope, as ^{13}C-labeled intermediates may be assimilated by other, nonprimary organisms. Production of the corresponding ^{12}C-labeled compounds and trophic interactions would, however, dilute these modified substrates in a complex environment such as soil. With further development of SIP, it is estimated that as little as 20% incorporation of a ^{13}C-labeled substrate into DNA will be sufficient to resolve ^{13}C-DNA from ^{12}C-DNA [217]. In future other isotopes might be used in SIP as well, including ^{15}N and ^2H.

SIP has been applied to trace ^{13}C-labeled plant photosynthate into phylogenetic groups of microbial taxa in the rhizosphere, permitting an examination of the link between soil microbial diversity and carbon flow *in situ*. Ostle et al. [234] used ^{13}C-CO_2 tracer to measure the incorporation of recently assimilated plant C into soil microbial RNA and DNA pools as a means to determine the turnover of the "active" rhizosphere community. This also required the development of a method for the extraction, purification, and preparation of small samples of soil DNA and RNA (<5 μg C) for isotope analysis. They used the method of Griffiths et al. [153] for extraction of DNA and RNA and included an extra purification step in order to quantify the ^{13}C signal in pure DNA and RNA. Results showed that both soil DNA and RNA rapidly incorporated recent photosynthate, with most ^{13}C found in the "active" microbial RNA fraction reflecting higher rates of microbial RNA turnover. Following this first approach to apply the SIP methodology to assess carbon flow from plants into the microbial community, Griffiths et al. [235] tested the feasibility of SIP to detect functional differences in microbial communities, utilizing recently fixed plant photosynthate in moisture perturbed grassland turfs. One of the main goals of this study was to find out, whether or not ^{13}C deposited in soil from pulse-labeled plants can be used to identify microbes utilizing plant exudates. Measurements of pulse-derived ^{13}C incorporated into soil RNA over 2 months showed that, at maximum, isotopic values represented only a 0.1 to 0.2 ^{13}C atom percentage increase over natural abundance levels and were found to be insufficient for the application of RNA-SIP. These findings reveal that in this experimental system the microbial uptake of labeled carbon from plant exudates is low, and that the methodology requires further optimization for application of SIP to natural plant–soil systems. This was partly achieved by a study to determine the effect of liming on the structure of the microbial community in the rhizosphere involved in the utilization of root exudates [72]. Here ^{13}C-CO$_2$ pulse labeling was applied followed by RNA-SIP. Prior to the ^{13}C-CO$_2$ pulse labeling, ^{13}C-labeled *Pseudomonas fluorescens* were inoculated in soil to assess the detection limit of detection by SIP. It appeared that the limit is between 10^5 and 10^6 cells per gram of soil. After extraction, the ^{12}C and ^{13}C RNA were separated by density gradient, isopycnic ultracentrifugation [236]. After centrifugation, the contents of the centrifugation tubes were fractionated into 12 fractions and the δ-^{13}C values were determined. Two distinct groups of fractions could be isolated from the tubes, containing ^{13}C and ^{12}C labeled RNA. The RNA of each of the fractions were analyzed by DGGE after reversed transcription applying primer sets for bacterial sequences, archaea, and fungi, and using a nested PCR approach targeting all three microbial groups. The more successful analysis in comparison, the work of Griffiths et al. [235] was considered to be due to minor modifications improving the separation efficiency. For example, ultracentrifugation was carried out with a vertical rotor, considered most efficient for isopycnic separation and RT-PCR amplification targeting large amplicons, followed by nested amplification using primers yielding shorter sequences that may have improved DGGE analysis. Further improvements in separation efficiency would, however, be valuable. Sensitivity could also be improved for investigation of utilization of root exudates by addition of ^{13}C compounds directly to soil, but this will eliminate effects of spatial interactions between plant roots and the rhizosphere microbial community that are critical in competition for C flux from plants to soil.

A significant strength of the SIP technique is that the ^{13}C-enriched DNA will contain the entire genome of each functionally active microbe of the community. Cloning large fragments of the labeled DNA using bacterial artificial chromosome (BAC) vectors may facilitate genome-level analysis of uncultivated microbes that can be associated with specific metabolic and ecological functions (see following text).

Another combined technology that holds great promise for detailed studies of the response of microbial populations to the flow of carbon from roots is the combination of fluorescent *in situ* hybridization (FISH) technology coupled with ^{14}C microautoradiography. FISH allows the phylogenetic identification of bacteria in natural environments using fluorescent-group-specific phylogenetic probes (targeting rRNA) and fluorescence microscopy [185]. Combining FISH with microautoradiography, after application of radioactive tracers and their uptake by individual cells, can be used to detect and quantify the active population utilizing a specific substrate. The combination of FISH-microautoradiography has been used successfully in several studies, such as in substrate utilization by Proteobacteria present in an activated sludge system [237,238] and in various natural environments [239,240]. Conceptually, this method could be used to study plant–microbial interactions, with a few modifications. Pulse labeling of plants with $^{14}CO_2$, followed by FISH-microautoradiography analyses of the rhizospheric soil could reveal the identity of bacteria utilizing the root released ^{14}C labeled compounds. The functional role of root-exudate-utilizing bacteria could be assigned by using probes designed to target functional genes. The major pitfall with this technique is the specificity of the applied probe. Only the particular organism or the group of organisms, for which the probe was designed, would be revealed when they utilized the tracer. Thus, it would not identify all organisms that have utilized root exudates. Furthermore, it is a time-consuming and labor-intensive technique. Technological aspects of FISH have been discussed recently [241].

VI. FINAL REMARKS AND THE FUTURE GENOMICS OUTLOOK

At present, we cannot explain plant–microbe interactions in the rhizosphere with sufficient confidence as to the specificity of these interactions and the detailed mechanisms involved. Therefore, we are not yet able to make sufficiently accurate predictions of the impact of changes in the environment, such as those caused by global change on these interactions, and so, on the key ecosystem functions resulting from the interactions. The main constraints for the improvement of our knowledge remain the insufficient methodology, in spite of the great progress made in the last two decades. No single method will elucidate all interactions, but a combination of new physicochemical and molecular techniques and genomics can answer some specific questions. For example, following the combination of FISH and microautoradiography, or SIP and DGGE, the combined use of isotope microarrays that are not only able to detect more microbes and a better expression of genes in an ecosystem than ever before and SIP will be one of the most exciting developments in rhizosphere research for the near future. The use of SIP in combination with microarrays will allow identification of microbes utilizing plant C exudates and estimations of the levels of expression of bacterial genes in the rhizosphere in an unprecedented way. Another combination of SIP with metagenomic approaches using BAC libraries will help to determine structure–function relationships of rhizosphere microorganisms. Metagenomics using libraries of BAC clones provides culture-independent genomic analysis of microbial communities [242]. In essence, DNA extracted from samples can be archived in a BAC library. BAC vectors can be used to propagate very large DNA fragments (around 100 kb), which would otherwise be unstable [243]. Because of these large DNA fragments in BAC clone libraries, the chances of finding phylogenetic and functional markers in the same clone are considerably greater than with standard cloning vectors, especially when the 16S rRNA gene might be located close to the functional gene [244]. However, even when phylogenetic and functional markers are detectable on the same clone, only a minor part (i.e., 1 to 1.5% = average BAC fragment of 40 kb of an average genome size of a bacterium of 4 to 6 Mb) of the functions of the identified organism can be determined. Therefore, at present, metagenomic approaches are suitable to mining ecosystems for unknown genes [245], but they do not allow for analyses of overall microbial functioning.

REFERENCES

1. Grayston, S.J., Vaughan, D., and Jones, D., Rhizosphere carbon flow in trees, in comparison with annual plants: the importance of root exudation and its impact on microbial activity and nutrient availability, *Appl. Soil Ecol.,* 5, 29, 1997.
2. Nelson, E.B., Microbial dynamics and interactions in the spermosphere, *Annu. Rev. Phytopathol.,* 42, 271, 2004.
3. McDougall, B.M. and Rovira, A.D., Sites of exudation of ^{14}C-labelled compounds from wheat roots, *New Phytologist,* 69, 999, 1970.
4. Bowen, G.D. and Rovira, A.D., Are modelling approaches useful in rhizosphere ecology?, in *Modern Methods in the Study of Microbial Ecology,* Rosswall, T., Ed., *Bull. Ecol. Res. Com. (Stockholm),* 17, 443, 1973.

5. Trofymow, J.A., Coleman, D.C., and Cambardella, C., Rates of rhizodeposition and ammonium depletion in the rhizosphere of axenic oat roots, *Plant Soil*, 97, 333, 1987.
6. Szmigielska, A.M. et al., Determination of low molecular weight dicarboxylic acids in root exudates by gas chromatography, *J. Agric. Food Chem.*, 43, 956, 1995.
7. Gagnon, H., Tahara, S., and Ibrahim, R.K., Biosynthesis, accumulation and secretion of isoflavonoids during germination and development of white lupin (*Lupinus albus* L.), *J. Exp. Bot.*, 46, 609, 1995.
8. Chaboud, A. and Rougier, M., Effect of root density in incubation medium on root exudate composition of axenic maize seedlings, *J. Plant Physiol.*, 131, 602, 1991.
9. Shepherd, T. and Davies, H.V., Patterns of short-term amino acid accumulation and loss in the root-zone of liquid-cultured forage rape (*Brassica napus* L.), *Plant Soil*, 158, 99, 1994.
10. Ohwaki, Y. and Sugahara, K., Active extrusion of protons and exudation of carboxylic acids in response to iron deficiency by roots of chickpea (*Cicer arietinum* L.), *Plant Soil*, 189, 49, 1997.
11. Hoffland, E., Quantitative evaluation of the role of organic acid exudation in the mobilization of rock phosphate by rape, *Plant Soil*, 140, 279, 1992.
12. Mozafar, A., Duss, F., and Oertli, J.J., Effect of *Pseudomonas fluorescens* on the root exudates of two tomato mutants differently sensitive to Fe chlorosis, *Plant Soil*, 144, 167, 1992.
13. Jones, D.L. and Darrah, P.R., Re-sorption of organic compounds by roots of *Zea mays* L. and its consequences in the rhizosphere: I. Re-sorption of ^{14}C labelled glucose, mannose and citric acid, *Plant Soil*, 143, 259, 1992.
14. Jones, D.L. and Darrah, P.R., Re-sorption of organic compounds by roots of *Zea mays* L. and its consequences in the rhizosphere: II. Experimental and model evidence for simultaneous exudation and resorption of soluble C compounds, *Plant Soil*, 153, 47, 1993.
15. Jones, D.L. and Darrah, P.R., Amino-acid influx at the soil-root interface of *Zea mays* L. and its implications in the rhizosphere, *Plant Soil*, 143, 1, 1994.
16. Jones, D.L. and Darrah, P.R., Influx and efflux of organic acids across the soil-root interface of *Zea mays* L. and its implications in rhizosphere C flow, *Plant Soil*, 173, 103, 1995.
17. Miihling, K.H., Schubert, S., and Mengel, K., Mechanism of sugar retention by roots of intact maize and field bean plants, *Plant Soil*, 155/156, 99, 1993.
18. Martin, J.K., Factors influencing the loss of organic carbon from wheat roots, *Soil Biol. Biochem.*, 12, 551, 1977.
19. Hütsch, B.W., Augustin, J., and Merbach, W., Plant rhizodeposition — an important source for carbon turnover in soils, *J. Plant Nutr. Soil Sci.*, 165, 397, 2002.
20. Nguyen, C., Rhizodeposition of organic C by plants: mechanisms and controls, *Agronomie*, 23, 375, 2003.
21. Meharg, A.A. and Killham, K., A novel method of quantifying root exudation in the presence of soil microflora, *Plant Soil*, 33, 111, 1991.
22. Meharg, A.A. and Killham, K., Loss of exudates from the roots of perennial ryegrass inoculated with a range of micro-organisms, *Plant Soil*, 170, 345, 1995.
23. Posta, K., Marschner, H., and Romheld, V., Manganese reduction in the rhizosphere, of mycorrhizal and non-mycorrhizal maize, *Mycorrhiza*, 5,119, 1994.
24. Shepherd, T. and Davies, H.V., Effect of exogenous amino acids, glucose and citric acid on the patterns of short-term accumulation and loss of amino acids in the root-zone of sand-cultured forage rape (*Brassica napus* L.), *Plant Soil*, 158, 111, 1994.
25. Shepherd, T. and Davies, H.V., Carbon loss from the roots of forage rape (*Brassica napus* L.) seedlings following pulse-labelling with $^{14}CO_2$, *Annu. Bot.*, 72, 155, 1993.
26. Hodge, A., Grayston, S.J., and Ord, B.G., A novel method for characterisation and quantification of plant root exudates, *Plant Soil*, 184, 97, 1996.
27. Johnson, J.F. et al., Root carbon dioxide fixation by phosphorus-deficient *Lupinus albus;* contribution to organic acid exudation by proteoid roots, *Plant Physiol.*, 112, 19, 1996.
28. Leyval, C. and Berthelin, J., Rhizodeposition and net release of soluble organic compounds by pine and beech seedlings inoculated with rhizobacteria and ectomycorrhizal fungi, *Biol. Fertil. Soils*, 15, 259. 1993.
29. Lugtenberg, B.J.J., Dekkers, L., and Bloemberg, G.V., Molecular determinants of rhizosphere colonization by *Pseudomonas, Annu. Rev. Phytopathol.*, 39, 461, 2001.
30. Janssens, I.A. et al., Elevated atmospheric CO_2 increases fine root production, respiration, rhizosphere respiration and soil CO_2 efflux in Scots pine seedlings, *Glob. Change Biol.*, 4, 871, 1998.

31. Staddon, P.L., Carbon isotopes in functional soil ecology, *Trends Ecol. Evol.*, 19, 148, 2004.
32. Bruneau, P.M.C. et al., Determination of rhizosphere ^{13}C pulse signals in soil thin sections by laser ablation isotope ratio mass spectrometry, *Rapid Commun. Mass Spectrom.*, 16, 2190, 2002.
33. Kuzyakov, Y. and Domanski, G., Carbon input by plants into the soil. Review, *J.Plant Nutr. Soil Sci.*, 163, 421, 2000.
34. Huang, B., North, G.B., and Nobel, P.S., Soil sheaths, photosynthate distribution to roots, and rhizosphere water relations for *Opuntiaficus-indica*, *Int. J. Plant Sci.*, 154, 425, 1993.
35. Holland, J.N., Cheng, W., and Crossley, D.A., Herbivore-induced changes in plant carbon allocation: assessment of below-ground C fluxes using carbon-14, *Oecologia*, 107, 87, 1996.
36. Jensen, B., Rhizodeposition by ^{14}CO$_2$-pulse-labelled spring barley grown in small field plots on sandy loam, *Soil Biol. Biochem.*, 25, 1553, 1993.
37. Gorissen, A. et al., ESPAS-an advanced phytotron for measuring carbon dynamics in a whole plant-soil system, *Plant Soil*, 179, 81, 1996.
38. Van Ginkel, J.H., Gorissen, A., and Van Veen, J.A., Carbon and nitrogen allocation in *Lolium perenne* in response to elevated atmospheric CO$_2$ with emphasis on soil carbon dynamics, *Plant Soil*, 188, 299, 1997.
39. Merharg, A.A., A critical review of labeling techniques used to quantify rhizosphere carbon-flow, *Plant Soil*, 166, 55, 1994.
40. Paterson, E., Rattray, E.A.S., and Killham, K., Effect of elevated atmospheric CO$_2$ concentration on C-partitioning and rhizosphere C-flow for three plant species, *Soil Biol. Biochem.*, 28,195, 1996.
41. Kiselle, K.W. et al., Method for ^{14}C-labelling maize field plots and assessment of label uniformity within plots, *Commun. Soi Sci. Plant Anal.*, 30, 1759, 1999.
42. Meharg, A.A. and Killham, K., Carbon distribution within the plant and rhizosphere for *Lolium perenne* subjected to anaerobic soil conditions, *Soil Biol. Biochem.*, 22, 643, 1990.
43. Rattray, E.A.S., Paterson, E., and Killham, K., Characterisation of the dynamics of C-partitioning within *Lolium perenne* and to the rhizosphere microbial biomass using ^{14}C pulse chase, *Biol. Fertil. Soils*, 19, 280, 1995.
44. Newton, P.C.D. et al., Plant growth and soil processes in temperate grassland communities at elevated CO$_2$, *J. Biogeog.*, 22, 235, 1995.
45. Rouhier, H. et al., Effect of elevated CO$_2$ on carbon and nitrogen distribution within a tree *(Castanea sativa* Mill.)-soil system, *Plant Soil*, 162, 281, 1994.
46. Rouhier, H. et al., Carbon fluxes in the rhizosphere of sweet chestnut seedlings (*Castanea sativa*) grown under two atmospheric CO$_2$ concentrations: ^{14}C partitioning after pulse labelling, *Plant Soil*, 180, 101, 1996.
47. Xu, J.G. and Juma, N.G., Above- and below-ground transformation of photosynthetically fixed carbon by two barley (*Hordeum vulgare* L.) cultivars in a typic cryoboroll, *Soil Biol. Biochem.*, 25, 1263, 1993.
48. Gregory, P.J. and Atwell, B.J., The fate of carbon in pulse-labelled crops of barley and wheat, *Plant Soil*, 136, 205, 1991.
49. Swinnen, J., Van Veen, J.A., and Merckx, R., ^{14}C pulse-labelling of field-grown spring wheat: an evaluation of its use in rhizosphere carbon budget estimations, *Soil Biol. Biochem.*, 25,161, 1994.
50. Swinnen, J., Van Veen, J.A., and Merckx, R., Rhizosphere carbon fluxes in field- grown spring wheat: model calculations based on ^{14}C partitioning after pulse-labelling, *Soil Biol. Biochem.*, 26, 171, 1994.
51. Swinnen, J., Van Veen, J.A., and Merckx, R., Root decay and turnover of rhizodeposits in field-grown winter wheat and spring barley estimated by ^{14}C pulse-labelling, *Soil Biol. Biochem.*, 27, 211, 1995.
52. Swinnen, J., Van Veen, J.A., and Merckx, R., Carbon fluxes in the rhizosphere of winter wheat and spring barley with conventional vs. integrated farming, *Soil Biol. Biochem.*, 27, 811, 1995.
53. Michunas, D.G. and Lauenroth, W.K., Belowground primary production by carbon isotope decay and long-term root biomass dynamics, *Ecosystems*, 4, 139, 2001.
54. Smeulders, S.M. et al., Effects of short-term ozone exposure of mature and juvenile Douglas firs [*Pseudotsuga menziesii* (Mirb.) Franco], *New Phytologist*, 129, 45, 1995.
55. Whipps, J.M. and Lynch, J.M., Substrate flow and utilization in the rhizosphere of cereals, *New Phytologist*, 95, 605, 1983.
56. Landi, L. et al., Effectiveness of antibiotics to distinguish the contributions of fungi and bacteria to net nitrogen mineralization, nitrification and respiration, *Soil Biol. Biochem.*, 25, 1771, 1993.

57. Helal, H.M. and Sauerbeck, D., Short term determination of the actual respiration carbon flow and rate of intact plant roots, in *Plant Roots and Their Environment,* McMichael, B.L. and Persson, H., Eds., Elsevier, Amsterdam, The Netherlands, 1981, p. 88.

58. Johansson, G., Below-ground carbon distribution in barley (*Hordeum vulgare* L.) with and without nitrogen fertilization, *Plant Soil,* 144, 93, 1992.

59. Cheng, W. et al., *In situ* measurement of root respiration and soluble C concentrations in the rhizosphere, *Soil Biol. Biochem.,* 25, 1189, 1993.

60. Cheng, W. et al., Investigating short-term carbon flows in the rhizospheres of different plant species, using isotopic trapping, *Agron. J.,* 86, 782, 1994.

61. Swinnen, J., Evaluation of the use of a model rhizodeposition technique to separate root and microbial respiration in soil, *Plant Soil,* 165, 89, 1994.

62. Kuzyakov, Y., Separating microbial respiration of exudates from root respiration in non-sterile soils: a comparison of four methods, *Soil Biol. Biochem.,* 34, 1621, 2002.

63. Swinnen, J., Van Veen, J.A., and Merckx, R., Losses of ^{14}C from roots of pulse-labelled wheat and barley during washing from soil, *Plant Soil,* 166, 93, 1994.

64. Norton, J.M., Smith, J.L., and Firestone, M.K., Carbon flow in the rhizosphere of ponderosa pine seedlings, *Soil Biol. Biochem.,* 22, 449, 1990.

65. Martin, M.K. and Merckx, R., The partitioning of photosynthetically fixed carbon within the rhizosphere of mature wheat, *Soil Biol. Biochem.,* 24, 1147, 1992.

66. Paterson, E. et al., Carbon partitioning and rhizosphere C-flow in *Lolium perenne* as affected by CO_2 concentration, irradiance and below-ground conditions, *Glob. Change Biol.,* 5, 669, 1999.

67. Van Ginkel, J.H., Gorissen, A., and Van Veen, J.A., Carbon and nitrogen allocation in *Lolium perenne* in response to elevated atmospheric CO_2 with emphasis on soil carbon dynamics, *Plant Soil,* 188, 299, 1997.

68. Van Ginkel, J.H., Whitmore, A.P., and Gorissen, A., *Lolium perenne* grasslands may function as a sink for atmospheric carbon dioxide, *J. Environ. Qual.,* 28, 1580, 1999.

69. Mary, B., Mariotti, A., and Morel, J.L., Use of ^{13}C variations at natural abundance for studying the biodegradation of root mucilage, roots and glucose in soil, *Soil Biol. Biochem.,* 24, 1065, 1992.

70. Butler, J.L. et al., Distribution and turnover of recently fixed photosynthate in ryegrass rhizospheres, *Soil Biol. Biochem.,* 36, 371, 2004.

71. Stewart, D.P.C. and Metherell, A.K., Carbon (^{13}C) uptake and allocation in pasture plants following field pulse-labelling, *Plant Soil,* 210, 61, 1999.

72. Rangel-Castro, J.I. et al., Stable isotope probing analysis of the influence of liming on root exudate utilization by soil microorganisms, *Environ. Microbiol.,* 7, 2100, 2005.

73. Johnson, S.D. et al., *In situ* $^{13}CO_2$ pulse labeling of upland grassland demonstrates a rapid pathway of carbon flux from arbuscular mycorrhizal mycelia to the soil, *New Phytologist,* 153, 327, 2002.

74. Steinmann, K. et al. Carbon fluxes to the soil in a mature temperate forest assessed by ^{13}C isotope tracing, *Oecologia,* 141, 489, 2004.

75. Andrews, J.A. et al., Separation of root respiration from total soil respiration using carbon 13 labeling during free-air carbon dioxide enrichment (FACE), *Soil Sci. Soc. Am. J.,* 63, 1429, 1999.

76. Cheng, W., Measurement of rhizosphere respiration and organic matter decomposition using natural ^{13}C, *Plant Soil,* 183, 263, 1996.

77. Smith, B.N. and Epstein, S., Two categories of $^{13}C/^{12}C$ ratios for higher plants, *Plant Physiol.,* 47, 380, 1971.

78. Balesdent, J. and Balabane, M., Major contribution of roots to soil carbon storage inferred from maize cultivated soils, *Soil Biol. Biochem.,* 28, 1261, 1996.

79. Collins, H.P. et al., Soil carbon dynamics in corn-based agroecosystems: results from carbon-13 natural abundance, *Soil Sci. Soc. Am. J.,* 63 584, 1999.

80. Qian, J.H. and Doran, J.W., Available carbon released from crop roots during growth as determined by carbon-13 natural abundance, *Soil Sci. Soc. Am. J.,* 60, 828, 1996.

81. Cheng, W. et al., Measuring tree root respiration using ^{13}C natural abundance: rooting medium matters, *New Phytologist,* 167, 297, 2005.

82. Johansson, G., Carbon distribution in grass (*Festuca pratensis* L.) during regrowth after cutting- utilization of stored and newly assimilated carbon, *Plant Soil,* 151, 11, 1993.

83. Kuzyakov, Y. and Cheng, W., Photosynthesis controls of rhizosphere respiration and organic matter decomposition, *Soil Biol. Biochem.,* 33, 1915, 2001.

84. Kuzyakov, Y. and Cheng, W., Photosynthesis controls of CO$_2$ efflux from maize rhizosphere, *Plant Soil*, 263, 85, 2004.
85. Wang, G.M. et al., Carbon partitioning patterns of mycorrhizal versus non-mycorrhizal plants: real-time dynamic measurements using [11]CO$_2$, *New Phytologist*, 112, 489, 1989.
86. Vančura, V., Root exudates of plants: I. Analysis of root exudates of barley and wheat in their initial phases of growth, *Plant Soil*, 21, 231, 1964.
87. Vančura, V. and Hovadk, A., Root exudates of plants: II. Composition of root exudates of some vegetables, *Plant Soil*, 22, 21, 1965.
88. Van Veen, J.A., The rhizosphere-historical and future perspectives from a microbiologist's viewpoint, in *Rhizosphere 2004-Perspectives and Challenges — A Tribute to Lorenz Hiltner*, Hartmann, A. et al., Eds., GSF, Neuherberg, Germany, 2005, p. 29.
89. Curl, E.A. and Truelove, B., *The Rhizosphere*, Springer Verlag, Berlin, 1985.
90. Kravchenko, L.V. et al., Root exudates of tomato plants and their effect on the growth and antifungal activity of *Pseudomonas* strains, *Microbiology*, 72, 37, 2003.
91. Ponce, M.A. et al., Flavonoids from shoots, roots and root exudates of *Brassica alba*, *Phytochemistry*, 65, 3131, 2004.
92. Czarnota, M.A., Rimando, A.M., and Weston, L.A., Evaluation of root exudates of seven sorghum accessions, *J. Chem. Ecol.*, 29, 2073, 2003.
93. Kuchenbuch, R. and Jungk, A., A method for determining concentration profiles at the soil-root interface by thin slicing rhizospheric soil, *Plant Soil*, 68, 391, 1982.
94. Dijkstra, A.F. et al., A soil chamber for studying the bacterial distribution in the vicinity of roots, *Soil Biol. Biochem.*, 19, 351, 1987.
95. Van Elsas, J.D., Trevors, J.T., and Starodrub, M.E., Bacterial conjugation between pseudomonads in the rhizosphere of wheat, *FEMS Microbiol. Ecol.*, 53, 299, 1988.
96. Fan, T.W.-M. et al., Comprehensive analysis of organic ligands in whole root exudates using nuclear magnetic resonance and gas chromatography-mass spectrometry, *Anal. Biochem.*, 251, 57, 1997.
97. Melnitchouk, A. et al., Qualitative differences between day-and night-time rhizodeposition in maize (*Zea mays* L.) as investigated by pyrolysis-field ionization mass spectrometry, *Soil Biol. Biochem.*, 37, 155, 2005.
98. Farrar, J. et al., How roots control the flux of carbon to the rhizosphere, *Ecology*, 84, 827, 2003.
99. Van Overbeek, L.S. and Van Elsas, J.D., Root exudates-induced promoter activity in *Pseudomonas fluorescens* mutants in the wheat rhizosphere, *Appl. Environ. Microbiol.*, 61, 890, 1995.
100. De Weger, L.A. et al., Use of phosphate-reporter bacteria to study phosphate limitation in the rhizosphere and in bulk soil, *MPMI*, 7, 32, 1994.
101. Joyner, D. and Lindow, S.E., Heterogeneity of iron bioavailability on plants assessed with a whole-cell GFP-based bacterial biosensor, *Microbiology*, 146, 2435, 2000.
102. Bringhurst, R.M., Cardon, Z.G., and Gage, D.J., Galactosides in the rhizosphere: utilization by *Sinorhizobium meliloti* and development of a biosensor, *PNAS*, 98, 4540, 2001.
103. Oger, P., Petit, A., and Dessaux, Y., Genetically engineered plants producing opines alter their biological environment, *Nat. Biotechnol.*, 15, 369, 1997.
104. Bruinsma, M., Kowalchuk, G.A., and Van Veen, J.A., Effects of genetically modified plants on microbial communities and proceses in soil, *Biol. Fertil. Soils*, 37, 329, 2003.
105. Kowalchuk, G.A., Bruinsma, M., and van Veen, J.A., Assessing responses of soil microorganisms to GM plants, *Trends Ecol. Evol.*, 18, 403, 2003.
106. Oger, P.M. et al., Engineering root exudation of *Lotus* toward the production of two novel carbon compounds leads to the selection of distinct microbial populations in the rhizosphere, *Microb. Ecol.*, 47, 96, 2004.
107. Clarke, H.R.G., Leigh, J.A., and Douglas, C.J., Molecular signals in the interactions between plants and microbes, *Cell*, 71, 191, 1992.
108. Bladergroen, M.R. and Spaink, H.P., Genes and signal molecules involved in the rhizobia- Leguminoseae symbiosis, *Curr. Opin. Plant Biol.*, 1, 353, 1998.
109. Spaink, H.P., Root nodulation and infection factors produced by rhizobial bacteria, *Annu. Rev. Microbiol.*, 54, 257, 2000.
110. Perret, X., Staehelin, C., and Broughton, W.J., Molecular basis of symbiotic promiscuity, *Microbiol. Mol. Biol. Rev.*, 64, 180, 2000.

111. Nelson, E.B., Exudate molecules initiating fungal responses to seeds and roots, in *The Rhizosphere and Plant Growth* Keister, D.L. and Cregan, P.B., Eds., Kluwer Academic Publishers, The Netherlands, 1991, p. 197.

112. Pierson, L.S., Wood, D.W., and Von Bodman, S., Quorum sensing in plant–associated bacteria, in *Cell-Cell Signaling in Bacteria*, Dunny, G.M. and Winans, S.C., Eds., ASM Press, Washington, D.C., 1999, p. 101.

113. McClean, R.J.C. et al., Evidence of autoinducer activity in naturally occurring biofilms, *FEMS Microbiol. Lett.*, 154, 259, 1997.

114. Rice, S.A. et al., Detection of homoserine lactone quorum sensing signals, in *Molecular Microbial Ecology Manual*, 2nd ed., Kowalchuk, G.A., Eds., Kluwer Acad. Publ., The Netherlands, 2004, p. 1629.

115. Teplitski, M., Robinson, J.B., and Bauer, W.D., Plants secrete substances that mimic bacterial N-acyl homoserine lactone signal activities and affect population density dependent behavior in associated bacteria, *MPMI*, 13, 637, 2000.

116. Kowalchuk, G.A. et al., Eds., *Molecular Microbial Ecology Manual*, 2nd ed., Kluwer Acad. Publ., Dordrecht, The Netherlands, 2004.

117. Hiltner, L., Uber Neuere Erfahrungen und Probleme auf dem Gebiet der Bodenbakteriologie und unter besonderer Berucksichtigung der Grundungung und Brache, *Arb. Dtsch. Landwirt. Ges.*, 98, 59, 1904.

118. Katznelson, H., Lochhead, A.G., and Timonin, M.I., Soil microorganisms and the rhizosphere, *Bot. Rev.*, 14, 543, 1948.

119. Katznelson, H., Nature and importance of the rhizosphere, in *Ecology of Soil-Borne Plant Pathogens-Prelude to Biological Control*, Baker, K.F. and Snyder, W.C., Eds., University of California Press, Berkeley, CA, 1965, p. 187.

120. Foster, R.C., Microenvironments of soil microorganisms, *Biol. Fertil. Soils*, 6, 189, 1988.

121. De Leij, F.A.A.M., Whipps, J.M., and Lynch, J.M., The use of colony development for the characterization of bacterial communities in soil and on roots, *Microb. Ecol.*, 27, 81, 1993.

122. Garbeva, P. et al., Effect of above-ground plant species on soil microbial community structure and its impact on suppression of *Rhizoctonia solani* AG3, *Environ. Microbiol.*, 8, 233, 2006.

123. Ruiz Palomino, M. et al., Seasonal diversity changes in alder (*Alnus glutinosa*) culturable rhizobacterial communities throughout a phenological cycle, *Appl. Soil Ecol.*, 29, 215, 2005.

124. Insam, H. and Goberna, M., Community level physiological profiles (Biolog substrate use tests) of environmental samples, in *Molecular Microbial Ecology Manual*, 2nd ed., Kowalchuk, G.A. et al., Eds., Kluwer Acad. Publ., Dordrecht, The Netherlands, 2004, p. 853.

125. Garland, J.L. and Mills A.L., Classification and characterisation of heterotrophic microbial communities on the basis of patterns of community-level sole carbon-source utilisation, *Appl. Environ. Microbiol.*, 57, 2351, 1991.

126. Garland, J.L., Patterns of potential C source utilization by rhizosphere communities, *Soil Biol. Biochem.*, 28, 223, 1996.

127. Bucher, A.E. and Lanyon, L.E., Evaluating soil management with microbial community-level physiological profiles, *Appl. Soil Ecol.*, 29, 59, 2005.

128. Selmants, P.C. et al., Red alder (*Alnus rubra*) alters community-level soil microbial function in conifer forests of the Pacific Northwest, USA, *Soil Biol. Biochem.*, 37, 1860, 2005.

129. Ritz, K. et al., Spatial structure in soil chemical and microbiological properties in an upland grassland, *FEMS Microbiol. Ecol.*, 49, 191, 2004.

130. Nannipieri, P. et al., Microbial diversity and soil functions, *Eur. J. Soil Sci.*, 54, 655, 2003.

131. Degens, B.P. and Harris, J.A., Development of a physiological approach to measuring the catabolic diversity of soil microbial communities, *Soil Biol. Biochem.*, 29, 1309, 1997.

132. Skinner, F.A., Jones, P.C.T., and Mollison, J.E., A comparison of a direct- and a plate-counting technique for the quantitative estimation of soil micro-organisms, *J. Gen. Microbiol.*, 6, 261, 1952.

133. Amann, R.I., Ludwig, W., and Schleifer, K.H., Phylogenetic identification and *in situ* detection of individual microbial cells without cultivation, *Microbiol. Rev.*, 59, 143, 1995.

134. Torsvik, V., Goksoyr, J., and Daae, F.L., High diversity in DNA of soil bacteria, *Appl. Environ. Microbiol.*, 56, 782, 1990.

135. Janssen, P.H. et al., Improved culturability of soil bacteria and isolation in pure culture of novel members of the divisions *Acidobacteria*, *Actinobacteria*, *Proteobacteria*, and *Verrucomicrobia*, *Appl. Environ. Microbiol.*, 68, 2391, 2002.

136. Duineveld, B.M. et al., Analysis of bacterial communities in the rhizosphere of chrysanthemum via Denaturing Gradient Gel Electrophoresis of PCR-amplified 16S rRNA as well as DNA fragments coding for 16S rRNA, *Appl. Environ. Microbiol.*, 67, 172, 2001.
137. Davis, K.E.R., Joseph, S.J., and Janssen, P.H., Effects of growth medium, inoculum size, and incubation time on culturability and isolation of soil bacteria, *Appl. Environ. Microbiol.*, 71, 826, 2005.
138. Simon, H.M. et al., Cultivation of mesophilic soil crenarchaeotes in enrichment cultures from plant roots, *Appl. Environ. Microbiol.*, 71, 4751, 2005.
139. De Ridder-Duine, A.S. et al., Rhizosphere bacterial community composition in natural stands of *Carex arenaria* (sand sedge) is determined by bulk soil community composition, *Soil Biol. Biochem.*, 37, 349, 2005.
140. Bloem, J. et al., Microscopic methods for counting bacteria and fungi in soil, in *Methods in Applied Soil Microbiology and Biochemistry*, Alef, K. and Nannipieri, P., Eds., Academic Press, New York, 1995, p. 162.
141. Bloem, J., Veninga, M., and Shepherd, J., Fully automatic determination of soil bacterium numbers, cell volumes, and frequencies of dividing cells by confocal laser-scanning microscopy and image-analysis, *Appl. Environ. Microbiol.*, 61, 926, 1995.
142. Yu, W. et al., Optimal staining and sample storage time for direct microscopic enumeration of total and active bacteria in soil with two fluorescent dyes, *Appl. Environ. Microbiol.*, 61, 3367, 1995.
143. Scheu, S. and Parkinson, D., Changes in bacterial and fungal biomass-C, bacterial and fungal biovolume and ergosterol content after drying, remoistening and incubation of different layers of cool temperate forest soils, *Soil Biol. Biochem.*, 26, 1515, 1994.
144. Diaper, J.P. and Edwards, C., Survival of *Staphylococcus aureus* in lake water monitored by flow cytometry, *Microbiology*, 140, 35, 1994.
145. Morgan, J.A.W., Rhodes, G., and Pickup, R.W., Survival of non-culturable *Aeromonas salmonicida* in lake water, *Appl. Environ. Microbiol.*, 59, 874, 1993.
146. Schloter, M., Assmus, B., and Hartmann, A., The use of immunological methods to detect and identify bacteria in the environment, *Biotechnol. Adv.* 13, 75, 1995.
147. Haughland, R.P., *Handbook of Fluorescent Probes and Research Chemicals,* 5th ed., Molecular Probes, Eugene, Oregon, 1994.
148. Kogure, K., Simidu, U., and Taga, N., A tentative direct microscopic method for counting living marine bacteria, *Can. J. Microbiol.*, 25, 415, 1979.
149. Lundgren, B., Fluorescein diacetate as a stain of metabolically active bacteria in soil, *Oikos*, 36, 17, 1981.
150. Zimmermann, A., Iturriaga, R., and Becker-Kirk, J., Simultaneous determination of the total number of aquatic bacteria and number thereof involved in respiration, *Appl. Enivron. Microbiol.*, 36, 926, 1978.
151. Head, I.M., Saunders, J.R., and Pickup, R.W., Microbial evolution, diversity, and ecology: a decade of ribosomal RNA analysis of uncultivated microorganisms, *Microbiol. Ecol*, 35, 1, 1998.
152. Frostegård, Å. et al., Quantification of bias related to the extraction of DNA directly from soils, *Appl. Environ. Microbiol.*, 65, 5409, 1999.
153. Griffiths, R.I. et al., Rapid method for coextraction of DNA and RNA from natural environments for analysis of ribosomal DNA- and rRNA-based microbial community composition, *Appl. Environ. Microbiol,*. 66, 5488, 2000.
154. Baudoin, E., Benizri, E., and Guckert, A., Impact of artificial root exudates on the bacterial community structure in bulk soil and maize rhizosphere, *Soil Biol. Biochem.*, 35, 1183, 2003.
155. Benizri, E. et al., Replant diseases: Bacterial community structure and diversity in peach rhizosphere as determined by metabolic and genetic fingerprinting, *Soil Biol. Biochem.*, 37, 1738, 2005.
156. Viebahn, M. et al., Assessment of differences in ascomycete communities in the rhizosphere of field-grown wheat and potato, *FEMS Microbiology Ecology*, 53, 245, 2005.
157. Salles, J.F., Van Veen, J.A., and Van Elsas, J.D., Multivariate analyses of *Burkholderia* species in soil: Effect of crop and land use history, *Appl. Environ. Microbiol.*, 70, 4012, 2004.
158. Marcial Gomes, N.C. et al., Dynamics of fungal communities in bulk and maize rhizosphere soil in the tropics, *Appl. Environ. Microbiol.*, 69, 3758, 2003.
159. Kowalchuk, G.A. et al., Effects of above-ground plant species composition and diversity on the diversity of soil-borne microorganisms, *Antonie van Leeuwenhoek*, 81, 509, 2002.

160. Heuer, H. and Smalla, K., Application of denaturing gradient gel electrophoresis (DGGE) and tem-
      perature gradient gel electrophoresis (TGGE) for studying soil microbial communities, in *Modern
      soil microbiology*, Van Elsas, J.D., Trevors, J.T., and Wellington, E.M.H., Eds., Marcel Dekker,
      New York, 1997, 353.
161. Smit, E. et al., Analysis of fungal diversity in the wheat rhizosphere by sequencing of cloned PCR-
      amplified genes encoding 18S rRNA and temperature gradient gel electrophoresis, *Appl. Environ.
      Microbiol.*, 65, 2614, 1999.
162. Gomes, N.C.M. et al., Bacterial diversity of the rhizosphere of maize (*Zea mays*) grown in tropical
      soil studied by temperature gradient gel electrophoresis, *Plant Soil*, 232, 167, 2001.
163. Schwieger, F. and Tebbe, C.C., A new approach to utilize PCR-single-strand-conformation polymor-
      phism for 16S rRNA gene-based microbial community analysis, *Appl. Environ. Microbiol*, 64, 4870,
      1998.
164. Sliwinski, M.K. and Goodman, R.M., Comparison of Crenarchaeal consortia inhabiting the rhizo-
      sphere of diverse terrestrial plants with those in bulk soil in native environments, *Appl. Environ.
      Microbiol.*, 70, 1821, 2004.
165. Uroz, S. et al., Novel bacteria degrading N-acylhomoserine lactones and their use as quenchers of
      quorum-sensing-regulated functions of plant-pathogenic bacteria, *Microbiology*, 149, 1981, 2003.
166. Mounler, E. et al., Influence of maize mucilage on the diversity and activity of the denitrifying
      community, *Environ. Microbiol.* 6, 301, 2004.
167. Berg, G. et al., Endophytic and ectophytic potato-associated bacterial communities differ in
      structure and antagonistic function against plant pathogenic fungi, *FEMS Microbiol. Ecol.*, 51,
      215, 2005.
168. Klamer, M. et al., Influence of elevated CO_2 on the fungal community in a coastal scrub oak forest
      soil investigated with terminal-restriction fragment length polymorphism analysis, *Appl. Environ.
      Microbiol.*, 68, 4370, 2002.
169. Graff, A. and Conrad, R., Impact of flooding on soil bacterial communities associated with poplar
      (*Populus* sp.) trees, *FEMS Microbiol. Ecol.*, 53, 401, 2005.
170. Frey-Klett, P. et al., Ectomycorrhizal symbiosis affects functional diversity of rhizosphere fluorescent
      pseudomonads, *New Phytologist*, 165, 317, 2005.
171. Di Battista-Leboeuf, C. et al., Distribution of *Pseudomonas* sp. populations in relation to maize root
      location and growth stage, *Agronomie*, 23, 441, 2003.
172. Leeflang, P. et al., Effects of *Pseudomonas putida* WCS358r and its genetically modified phenazine
      producing derivative on the *Fusarium* population in a field experiment, as determined by 18S rDNA
      analysis, *Soil Biol. Biochem.*, 34, 1021, 2002.
173. Tiquia, S.M. et al., Effects of mulching and fertilization on soil nutrients, microbial activity and
      rhizosphere bacterial community structure determined by analysis of TRFLPs of PVR-amplified 16S
      rRNA genes, *Appl. Soil Ecol.*, 21, 3, 2002.
174. Brodie, E., Edwards, S., and Clipson, N., Soil fungal community structure in a temperate upland
      grassland soil, *FEMS Microbiol. Ecol.*, 44, 105, 2003.
175. Edel-Hermann, V. et al., Terminal restriction fragment length polymorphism analysis of ribosomal
      RNA genes to assess changes in fungal community structure in soils, *FEMS Microbiol. Ecol.*, 47,
      397, 2004.
176. Kent, A.D. and Triplett, E.W., Microbial communities and their interactions in soil and rhizosphere
      ecosystems, *Annu. Rev. Microbiol.*, 56, 211, 2002.
177. Wilson, M.J., Weightman, A.J., and Wade, W.G., Applications of molecular ecology in the character-
      ization of uncultured microorganisms associated with human disease, *Rev. Med. Microbiol.*, 8, 91,
      1997.
178. Niemi, R.M., Extraction and purification of DNA in rhizosphere soil samples for PCR-DGGE analysis
      of bacterial consortia, *J. Microbiol. Methods*, 45, 155, 2001.
179. Fierer, N. et al., Assessment of soil microbial community structure by use of taxon-specific quantitative
      PCR assays, *Appl. Environ. Microbiol.*, 71, 4117, 2005.
180. Da Mota, F.F. et al., Assessment of the diversity of *Paenibacillus* species in environmental samples
      by a novel rpoB-based PCR-DGGE method, *FEMS Microbiol. Ecol.*, 53, 317, 2005.
181. Mergel, A. et al., Relative abundance of denitrifying and dinitrogen-fixing bacteria in layers of a forest
      soil, *FEMS Microbiol. Ecol.*, 36, 33, 2001.

182. Poly, F. et al., Comparison of *nifH* gene pools in soils and soil microenvironments with contrasting properties, *Appl. Environ. Microbiol.*, 67, 2255, 2001.

183. Lovell, C.R. et al., Molecular analysis of diazotroph diversity in the rhizosphere of the smooth cordgrass, *Spartina alterniflora, Appl. Environ. Microbiol.*, 66, 3814, 2000.

184. Kowalchuk, G.A. and Stephen, J.R., Ammonia-oxidizing bacteria: a model for molecular microbial ecology, *Annu. Rev. Microbiol.*, 55, 485, 2001.

185. Singh, B.K. et al., Unravelling rhizosphere-microbial interactions: opportunities and limitations, *Trends Microbiol.*, 12, 386, 2004.

186. Van Elsas, J.D., Trevors, J.T., and Wellington, E.M.H, Eds., *Modern Soil Microbiology,* Marcel Dekker, New York, 1997, p. 683.

187. Nannipieri, P. and Smalla, K., Eds., *Nucleic Acids and Proteins in Soil,* Springer Verlag, Berlin, in press.

188. Duineveld, B.M. et al., Analysis of the dynamics of bacterial communities in the rhizosphere of the chrysanthemum via denaturing gradient gel electrophoresis and substrate utilization patterns, *Appl. Environ. Microbiol.*, 64, 4950, 1998.

189. Marschner, P. and Timonen, S., Interactions between plant species and mycorrhizal colonization on the bacterial community composition in the rhizosphere, *Appl. Soil Ecol.*, 28, 23, 2005.

190. Nicol, G.W., Glover, L.A., and Prosser, J.I., Molecular analysis of methanogenic archaeal communities in managed and natural upland pasture soils, *Glob. Change Biol.*, 9, 1451, 2003.

191. Marschner, P. et al., Soil and plant specific effects on bacterial community composition in the rhizosphere, *Soil Biol. Biochem.*, 33, 1437, 2001.

192. Lerman, L.S. et al., Sequence-determined DNA separations, *Annu. Rev. Biophys. Bioeng.*, 13, 399, 1984.

193. Lerman, L.S., Detecting sequence changes in a gene, *Somatic Cell Mol. Genet.*, 13, 419, 1987.

194. Myers, R.M. and Lerman, L.S., Detection and localization of single base changes by denaturing gradient gel electrophoresis, *Methods Enzymol.*, 155, 501, 1987.

195. Muyzer, G., De Waal, E.C., and Uitterlinden, A.G., Profiling of complex microbial populations by denaturing gradient gel electrophoresis analysis of polymerase chain reaction-amplified genes coding for 16S rRNA, *Appl. Environ. Microbiol.*, 59, 695, 1993.

196. Bergsma-Vlami, M. et al., Assessment of genotypic diversity of antibiotic-producing *Pseudomonas* species in the rhizosphere by denaturing gradient gel electrophoresis, *Appl. Environ. Microbiol.*, 71, 993, 2005.

197. Williamson, N., Brian, P., and Wellington, E.M.H., Molecular detection of bacterial and streptomycete chitinases in the environment, *Antonie van Leeuwenhoek*, 78, 315, 2000.

198. Gelsomino, A. et al., Assessment of bacterial community structure in soil by polymerase chain reaction and denaturing gradient gel electrophoresis, *J. Microbiol. Methods*, 38, 1, 1999.

199. Nübel, U. et al., Sequence heterogeneities of genes encoding 16 rRNAs in *Paenibacillus polymyxa* detected by temperature gradient gel electrophoresis, *J. Bacteriol.*, 178, 5636, 1996.

200. Gyamfi, S. et al., Effects of transgenic glufosinate-tolerant oilseed rape and the associated herbicide application on eubacterial and *Pseudomonas* communities in the rhizosphere, *FEMS Microbiol.Ecol.*, 41, 181, 2002.

201. Reiter, B. et al., Endophytic *Pseudomonas* spp. populations of pathogen-infected potato plants analysed by 16S rDNA- and 16S rRNA-based denaturating gradient gel electrophoresis, *Plant Soil,* 257, 397, 2003.

202. Garbeva, P., Van Veen, J.A., and Van Elsas, J.D., Predominant *Bacillus* spp. in agricultural soil under different management regimes detected via PCR-DGGE, *Microb. Ecol.*, 45, 302, 2003.

203. Heuer, H. et al., Analysis of actinomycete communities by specific amplification of genes encoding 16S rRNA and gel-electrophoretic separation in denaturing gradients, *Appl. Environ. Microbiol.*, 63, 3233, 1997.

204. Salles, J.F., De Souza, F.A., and Van Elsas, J.D., Molecular method to assess the diversity of *Burkholderia* species in environmental samples, *Appl. Environ. Microbiol.*, 68, 1595, 2002.

205. De Souza, F.A. et al., PCR-denaturing gradient gel electrophoresis profiling of inter- and intraspecies 18S rRNA gene sequence heterogeneity is an accurate and sensitive method to assess species diversity of arbuscular mycorrhizal fungi of the genus *Gigaspora, Appl. Environ. Microbiol.*, 70, 1413, 2004.

206. Vainio, E.J. and Hantula, J., Direct analysis of wood-inhabiting fungi using denaturing gradient gel electrophoresis of amplified ribosomal DNA, *Mycol. Res.*, 104, 927, 2000.

207. Ma, W.K., Siciliano, S.D., and Germida, J.J., A PCR-DGGE method for detecting arbuscular mycorrhizal fungi in cultivated soils, *Soil Biol. Biochem.*, 37, 1589, 2005.
208. Araújo Da Silva, K.R. et al., Application of a novel *Paenibacillus*-specific PCR-DGGE method and sequence analysis to assess the diversity of *Paenibacillus* spp. in the maize rhizosphere, *J. Microbiol. Methods*, 54, 213, 2003.
209. Lin, M. et al., Effect of an *Alcaligenes faecalis* inoculant strain on bacterial communities in flooded soil microcosms planted with rice seedlings, *Appl. Soil Ecol.*, 15, 211, 2000.
210. Smalla, K. et al., Analysis of Biolog GN substrate utilization patterns by microbial communities, *Appl. Environ. Microbiol.*, 64, 1220–1225, 1998.
211. Nylund, J.E. and Wallander, H., Ergosterol analysis as a means of quantifying mycorrhizal biomass, in *Methods in Microbiology*, Norris, J.R., Read, D.J., and Varma, A.K., Eds., Academic Press, London, 1992, p. 77.
212. Grant, W.D. and West, A.W., Measurement of ergosterol, diaminopimelic acid and glucosamine in soil: evaluation as indicators of microbial biomass, *J. Microbiol. Methods*, 6, 47, 1986.
213. Gessner, M.O. and Chauvet, E., Ergosterol-to-biomass conversion factors for aquatic hyphomycetes, *Appl. Environ. Microbiol.*, 59, 502, 1993.
214. Bååth, E., Estimation of fungal growth rates in soil using ^{14}C-acetate incorporation into ergosterol, *Soil Biol. Biochem.*, 33, 2011, 2001.
215. Joergensen, R.G., Ergosterol and microbial biomass in the rhizosphere of grassland soils, *Soil Biol. Biochem.*, 32, 647, 2000.
216. Borjesson, T., Stollman, U., and Schnurer, J., Volatile metabolites produced by six fungal species compared with other indicators of fungal growth on cereal grains, *Appl. Environ. Microbiol.*, 58, 2599, 1992.
217. Abramson, D., Sinha, R.N., and Mills, J.T., Mycotoxin and odour formation in moist cereal grain during granary storage, *Cereal Chem.*, 57, 346, 1980.
218. Tuma, D. et al., Odour volatiles associated with microflora in damp ventilated and non-ventilated bin-stored bulk wheat, *Int. J. Food Microbiol.*, 8, 11, 1989.
219. Wurzenberger, M. and Grosch, W., The enzymatic oxidative breakdown of linoleic acid in mushrooms (*Psalliota bispora*), *Z. Lebensm. Unters. Forsch.*, 175, 186, 1982.
220. Lechevalier, M.P., Lipids in bacterial taxonomy: a taxonomist's view, *Crit. Rev. Microbiol.*, 5, 109, 1976.
221. Zelles, L. et al., Signature fatty acids in phospholipids and lipopolysaccharides as indicators of microbial biomass and community structure in agricultural soils, *Soil Biol. Biochem.*, 24, 317, 1992.
222. White, D.C., Analysis of microorganisms in terms of quantity and activity in natural environments. Microbes in their natural environments, *Soc. Gen. Microbiol. Symp.*, 34, 37, 1983.
223. Bååth, E., Frostegård, A., and Fritze, H., Soil bacterial biomass, activity, phospholipid fatty acid pattern, and pH tolerance in area polluted with alkaline dust deposition, *Appl. Environ. Microbiol.*, 58, 4026, 1992.
224. Frostegård, A. and Bååth, E., The use of phospholipid fatty-acid analysis to estimate bacterial and fungal biomass in soil, *Biol. Fertil. Soils*, 22, 59, 1996.
225. Lindahl, A.V. et al., Phospholipid fatty acid composition of size fractionated indigenous soil bacteria, *Soil Biol. Biochem.*, 29, 1565, 1997.
226. Diab el Arab, H.G., Vilich, V., and Sikora, R.A., The use of phospholipids fatty acid (Pl-FA) in the determination of rhizosphere specific microbial communities (RSMC) of two wheat cultivars, *Plant Soil*, 228, 291, 2001.
227. Innes, L., Hobbs, P.J., and Bardgett, R.D., The impacts of individual plant species on rhizosphere microbial communities in soils of different fertility, *Biol. Fertil. Soils*, 40, 7, 2004.
228. Pelz, O. et al., Microbial assimilation of plant-derived carbon in soil traced by isotope analysis, *Biol. Fertil. Soils*, 41, 153, 2005.
229. Frostegård, A., Tunlid, A., and Bååth, E., Phospholipid fatty acid composition, biomass, and activity of microbial communities from two soil types experimentally exposed to different heavy-metals, *Appl. Environ. Microbiol.*, 59, 3605, 1993.
230. Morgan, J.A.W. and Pickup, R.W., Activity of microbial peptidases, oxidases and esterases in lake waters of varying trophic status, *Can. J. Microbiol.*, 39, 795, 1993.

231. Boschker, H.T.S. et al., Direct linking of microbial processes by ^{13}C-labelling of biomarkers, *Nature*, 392, 801, 1998.
232. Treonis, A.M. et al., Identification of groups of metabolically-active rhizosphere microorganisms by stable isotope probing of PLFAs, *Soil Biol. Biochem.*, 36, 533, 2004.
233. Radajewski, S. et al., Stable-isotope probing as a tool in microbial ecology, *Nature*, 403, 646, 2000.
234. Ostle, N. et al., Active microbial RNA turnover in a grassland soil estimated using a ^{13}CO$_2$ spike, *Soil Biol. Biochem.*, 35, 877, 2003.
235. Griffiths, R.I. et al., ^{13}CO$_2$ pulse labeling of plants in tandem with stable isotope probing: methodological considerations for examining microbial function in the rhizosphere, *J. Microbiol. Methods*, 58, 119, 2004.
236. Manefield, M. et al., RNA stable isotope probing, a novel means of linking microbial community function to phylogeny, *Appl. Environ. Microbiol.*, 68, 5367, 2002.
237. Lee, N. et al., Combination of fluorescent *in situ* hybridization and microautoradiography — a new tool for structure and function analyses in microbial ecology, *Appl. Environ. Microbiol.*, 65, 1289, 1999.
238. Gray, N.D. and Head, I.M., Linking genetic identity and function in communities of uncultured bacteria, *Environ. Microbiol.*, 3, 481, 2001.
239. Gray, N.D. et al., Use of combined microautoradiography and fluorescence *in situ* hybridization to determine carbon metabolism in mixed natural communities of uncultured bacteria from the genus *Achromatium*, *Appl. Environ. Microbiol.*, 66, 4518, 2000.
240. Cottrell, M.T. and Kirchman, D.L., Natural assemblage of marine proteobacteria and members of the *Cytophaga-Flavobacter* cluster consuming low- and high-molecular-weight dissolved organic matter, *Appl. Environ. Microbiol.*, 66, 1692, 2000.
241. Bouvier, T. and del Giorgio, P.A., Factors influencing the detection of bacterial cells using fluorescence *in situ* hybridization (FISH): a quantitative review of published reports, *FEMS Microbiol. Ecol.*, 44, 3, 2003.
242. Riesenfeld, C.S., Schloss, P.D., and Handelsman, J., Metagenomics: genomic analysis of microbial communities, *Annu. Rev. Microbiol.*, 38, 525, 2004.
243. Rondon, M.R. et al., Cloning the soil metagenome: a strategy for accessing the genetic and functional diversity of uncultured microorganisms, *Appl. Environ. Microbiol.*, 66, 2541, 2000.
244. Liles, M.R. et al., A census of rRNA genes and linked genomic sequences within a soil metagenomic library, *Appl. Environ. Microbiol.*, 69, 2684, 2003.
245. Streit, W.R. and Schmitz, R.A., Metagenomics-the key to the uncultured microbes, *Curr. Opin. Microbiol.*, 7, 492, 2004.

14 Gene Flow in the Rhizosphere

Anne Mercier, Elisabeth Kay, Timothy M. Vogel, and Pascal Simonet

CONTENTS

I. INTRODUCTION

The fundamental role of horizontal gene flow in bacterial adaptation and evolution has been supported by the complete genome sequence analysis of more than 200 bacteria. Bioinformatics tools are able to identify recently acquired genes and detect the mosaic patterns of some bacterial genes such as those encoding heavy metal or antibiotic resistance, pathogenic traits, etc. These *in silico* data strongly suggest that bacterial genomes result, in part, from an incorporation of DNA originating from close bacterial species and also from organisms belonging to phylogenetically remote taxa [1–10].

Various mechanisms are involved in DNA transfer among bacteria including transduction, natural transformation, and conjugation [1,11–13]. In addition, DNA exchange was also reported to be mediated by phage-like structures [14], protoplast fusion [15], and transposition [16–18]. Up to thousands of kilobases can be transferred between two bacterial cells in a single transfer event, but the amount of transferred DNA varies over a wide range depending on the transfer mechanism involved. Conjugation is considered to be the most efficient mechanism by transferring the largest amount of DNA [19,20].

Numerous studies have examined the translocation of DNA across bacterial membranes and determined the fate of the incoming DNA in the bacterial cytoplasm [13,21,22]. Circular plasmids can be maintained if they are able to replicate autonomously in the new host. The fate of chromosomal or plasmidic linear DNA depends on the bacterial host genetic recombination systems, which select between integration of new DNA in the bacterial genome and its degradation [13,22]. Studies have been carried out to determine the recombination potential expressed by bacteria and the possible involvement of gene transfer and recombination as an evolutionary force [4,10,22–24]. However, the actual role of horizontal gene transfer cannot be evaluated without considering the different biotic and abiotic parameters of the various environments in which bacteria develop and how they regulate gene flow between bacteria. In complex and heterogeneous environments, such as soil, favorable conditions for gene transfer include those for bacterial growth and cell contact and also for DNA, phage persistence, and integrity. Based on these criteria, the rhizosphere (soil surrounding plant roots), spermosphere (soil surrounding germinated seeds), and the "residue-sphere" (interface between decaying plant material and soil matrix) were identified as possible hot-spots for gene transfer [25–30]. Microcosms and open environment-based experiments have been developed to determine which environmental factors control the occurrence of transduction, natural transformation and conjugation-mediated gene transfer in these soil-related environments and the potential impact on soil microbial community and ecosystems [23,31–36]. Most of these studies concentrate on only one mechanism, usually, conjugation and natural transformation but rarely consider the three mechanisms of gene transfer at the same time in the same soil environment.

In this chapter, the three main mechanisms involved in gene acquisition by bacteria will be presented. The specific purpose is to demonstrate how environmental parameters and bacteria themselves are involved in gene flow regulation between bacteria and other organisms. This review helps to evaluate whether the rhizosphere, the nutrient-rich ecosystem between plant roots and the soil, can be considered as a specific hot-spot for intra and interspecies and interkingdom gene transfer.

II. TRANSDUCTION: GENE TRANSFER FROM PHAGES TO BACTERIA IN THE RHIZOSPHERE

A. TRANSDUCTION IN BACTERIA AND REGULATION

Over the past two decades, the potential for bacterial gene exchange in natural environments through transduction (bacteriophage-mediated gene transfer) has been established [37–40]. Studies have demonstrated that both chromosomal and plasmid DNA can be successfully transduced in natural environments that support bacterial multiplication in soil [39], fresh water [37,41], sewage, activated-sludge [42,43] and plant leaf surfaces [44]. Moreover, the current phage-sequencing initiatives are improving our understanding of the complex evolution and ecology of phages [45–48]. The bacteriophages are able to infect almost all (if not all) bacteria tested to date, and have the ability to mediate bacterial gene transfer in various environments. Transduction has been observed with a wide range of bacteria, such as *Escherichia coli* [49,50], *Erwinia* [51], *Pseudomonas* [44,52], *Rhizobium* [53], *Bacillus* [54,55], and *Streptomyces* [56]. Although the host spectrum of certain bacteriophages is restricted to only one bacterial species (sensitivity group of a bacteriophage) due to a high specificity to the bacterial surface receptors, the transfer of genetic material via transduction could also be detected between bacteria of the same species and the same genus [12,57,58]. Bacteriophages can carry large bacterial DNA and survive under adverse conditions that eliminate bacterial populations. Important bacterial DNA, such as virulence factors, can be preserved in bacteriophages until a host for lysogenic conversion is reintroduced into the environmental niche. Bacteriophages can also spread DNA directly to an entire bacterial population, eliminating the need for clonal expansion of a specific population. Thus, bacteriophages are involved in the evolution of bacterial pathogens and consequently in pathogen–host interactions by providing mechanisms to counteract the host barriers [38,59,60]. The dynamic interactions of viruses with their hosts may contribute significantly to the genetic diversity and composition of microbial populations [1]. However, in most transduction studies, bacteriophages are considered only as molecular cloning tools for a better understanding bacterial genetics [61].

1. Multiplication Cycle of Bacteriophages

The survival of bacteriophages depends on the presence of compatible bacterial hosts. A highly specific interaction is established between the fixation sites carried by the phage and the sites on the recipient bacterial envelope. The bacteriophage DNA is then injected inside the bacterium where it will be replicated by the cellular machinery of the bacterium. The phage genes express viral proteins and then the assembly of new viral particles takes place. The mature viruses are released into the environment after bacterial lysis, and can infect other bacteria. Two types of phages are distinguished according to their multiplication cycle. The multiplication cycle of the virulent phages leads to the production of new phages immediately after the infection of the bacterial cell and to their release in the medium after the lysis of the bacterium. The DNA of the temperate phages (or lysogens) such as the lambda phage can determine the immediate production of new phages (similar to virulent phages) or to enter in a repressed state in which a single copy of the DNA (or prophage) will be replicated as part of the bacterial genome. Later, the prophage will be able to produce phages by the phenomenon of spontaneous induction.

2. Types of Transduction

During the development of a phage, errors can occur and lead to the distinction of two types of transduction. The nature of the transferred DNA will depend on these modes of transduction:

- The generalized transduction involves moderate phages and certain virulent phages. During the lytic process, the phages synthesize an endonuclease that partially degrades the bacterial genome. Bacterial chromosomal or plasmidic DNA fragments could be randomly encapsulated with the DNA of the bacteriophage. Bacterial lysis will lead to a mixture of virulent phages and transducers phages that could be adsorbed into a new recipient bacterium. The survival of the transduced DNA is dependent on its integration by recombination into the genome of the recipient bacteria.
- The specialized or restricted transduction is rare and characteristic of temperate phages. The errors occur when the prophage is excised from the bacterial chromosome. Specialized transduction results from an illegitimate recombination between sequences nonhomologous to the phage and bacterial DNA located in the immediate vicinity of the integration site of the prophage [32]. The quantity of bacterial DNA that can be encapsulated would be very limited.

B. Biotic and Abiotic Rhizospheric Factors Affecting Transduction

Currently, the importance of the gene transfer by transduction in the soil environments cannot be evaluated due to the lack of data. However, some *in vitro* and *in situ* studies identified several environmental conditions that can affect transduction efficiency by acting on the persistence of bacteriophages in soil, bacterial hosts densities, and phage–bacterial cells interactions.

1. Persistence of Phages and Bacterial Hosts in Soil

Bacterial gene transfer by transduction under natural conditions is modulated by many biotic parameters, such as the density and the physiological state of the bacterial hosts, the adsorption of the bacteriophage on the bacterial cell surface, its multiplication, and its propagation. To survive, the bacteriophage must infect bacteria and use the metabolic and cellular machinery of its host for its multiplication and propagation in the environment [62]. However, natural habitats are often oligotrophic and the energy resources undergo important variations. Bacteria are generally found in a starvation state that limits metabolism and bacterial multiplication. When nutrients are scarce, the replication of phages inside the host cell is modified, the latent periods are longer, and the virulence of the lytic phages is reduced [63]. However, results from Schrader et al. [64] indicate that starvation does not protect bacterial hosts from bacteriophage infection and suggest that bacteriophages would be responsible for significant bacterial mortality under most natural ecosystems. Consequently, the addition of nutrients to soil has been studied with the objectives to analyze their influence on transduction efficiency [50]. Addition of phages, recipient bacteria, and a rich medium into a sterile soil leads to an increase in the number of recipient cells and of transductants by three to four orders of magnitude. Results have also shown that, when the phages and the bacterial hosts are inoculated in the presence of salt water, the increase is more progressive, but the percentage of transductants in the bacterial recipient population is the same in both sterile and nonsterile soil microcosms. Zeph et al. [50] conclude that the addition of nutrients in nonsterile soil increases the densities of phages and bacteria, but does not affect the transduction efficiency. According to results from Reanney and Marsch [65], propagation of the phages in the environment is directly related to bacterial cellular density and the frequencies of transduction will be higher with a massive bacterial colonization [44]. Wiggins and Alexander [66] showed that the bacteriophages did not affect number and activity of the bacteria in the environment, when the density of bacterial population is lower than a threshold of approximately of 10^4 CFU/g in natural ecosystems. However, according to results from Ashelford et al. [39], predation of bacteria by viruses will be an important factor in controlling and stimulating the growth of bacterial populations in soil. Their results also support the importance of bacteriophages for mediating gene transfer in soil [39]. Transduction may affect the course of bacterial evolution by maintaining genetic material in bacterial gene pools that would otherwise be lost due to negative fitness [67].

2. Bacteria and Phage Physiological State and Interactions

Abiotic parameters such as temperature, pH, moisture, and the presence of surfaces and their chemistry can affect the physiological state of recipient bacterial cells and bacteriophages. For example, variations in temperature affect the metabolism of bacteria and can also modify the phage attachment to their bacterial hosts [68]. Variations in water content, cations, and pH can also influence the adsorption of the phages on particles or surfaces in the environment, and thus, modify the phage–bacterial host interaction [35]. Soil properties, especially the presence of clay (such as montmorillonite and kaolinite) can also favor the persistence of bacteriophages, the interaction between bacteriophages and bacteria, and thus, can enhance transduction [54,69]. Phage adsorption on clay is due to electrostatic interactions and is dependent on presence of cations such as calcium or magnesium [54,69,70]. Moreover, results from Hurst et al. [71] and Gerba et al. [70] suggest that bacteriophages persistence in the soil is strongly related to the presence of cations (e.g., Al^{3+}) that bridge between clay negative charges and viruses. However, phage adsorption is not correlated to the cation exchange capacity of the clays [54].

The adsorption of phages on clay would protect the viral genome against the attack of soil proteases and nucleases, other soil degrading factors, or even against UV [54,55,69,72]. However, under certain conditions, the protective role of clay is denied — soil desiccation produces a quicker decrease of virus survival in clay soil than in sandy soils. The remaining water will be tightly bound to clay particles decreasing biological activities [73]. Results from Vettori et al. [55] have shown that survival of the phage was increased by adsorption on clays and that the adsorbed phage maintained its ability to transduce bacterial cells for at least 30 days after the preparation of the clay–phage complex. Moreover, transduction by the clay–phage complex was primarily the result of the phages detaching from the clays in the presence of host cells [54]. Both physical separation and reduced mobility potentially inhibit genetic interactions between phage and bacteria, and thus, would be responsible for the low transduction frequencies observed in soil [37], as compared to aquatic environments.

C. METHODS AND PREDICTIONS OF TRANSDUCTION GENE TRANSFER IN SOIL

For years, transduction was not considered as a relevant mechanism for gene transfer in the environment. The bacteriophage concentration in the natural habitats was estimated to be too low for frequent contact between viruses and bacteria. However, the use of powerful tools such as electron and epifluorescence microscopy led to the direct observation of bacteriophages in oceans [74–76] and freshwater [77], marine sediments [37,76], sewage, and activated sludge [42]. Bacteriophage density up to 10^8/ml of water was reported in marine environments. This represents a phage: bacteria ratio of about 10 [39]. In soil, detection by epifluorescence microscopy is difficult due to the nonspecific interactions between soil fine particles and stains used to detect bacteriophages [39]. Until recently, the only method to study soil bacteriophages was the plating technique — the spreading out of a soil suspension against a suspension of indicator bacteria. According to this technique, the number of plaque forming units (pfu) was in a range from 0 to 4×10^4/g of soil [52,65]. However, the recent application of transmission electron microscopy (TEM) to soil samples has provided new data on the average number of phage particles in soil that was up to 1.5×10^7/g, which is at least 350-fold more than the highest number estimated on Petri dishes [39]. Moreover, this technique still underestimated the actual bacteriophage concentration that could reach 1.5×10^8/g of soil, which is equivalent to 4% of the total bacterial population in rhizosphere [39].

These new data on the unexpectedly high bacteriophage concentration led to speculation on their potential role in soil such as bacterial density control and gene transfer enhancement.

According to the process of viral propagation previously described, bacteriophages could be considered as efficient vectors for genetic information exchange. However, the level to which transduction occurs in soil remains very difficult to assess in spite of preliminary experiments, which demonstrate that genes were transferred from one bacterium to another via bacteriophages [62].

Factors modulating transduction efficiency in soil remain unexplored [55,78]. Most studies on transduction were carried out in soil microcosms seeded with recipient bacteria and bacteriophage particles as donor DNA or lysogenic cells with antibiotic-resistant genes or other markers to positively select bacterial transductants [49,50]. Colony hybridization, plasmid or gene DNA restriction analysis, or Southern hybridization could be used to confirm that the gene was actually transferred from the donor to the recipient bacterium [39]. However, in most studies, transduction frequency remains below the detection limit leading to new methods to increase the transfer rate, such as a UV treatment on transducing lysates [37] or engineering the recipient bacteria to favor homologous recombination with the donor DNA [35].

Transduction was monitored in sterile and nonsterile soil microcosms using an *Escherichia coli*–phage P1 transduction system [50]. These experiments concluded that transduction occurred in the days following inoculation, without providing the frequency. The detected transductants resulted either from individual transduction events or from a multiplication of initial recombinant clones. Results from Germida and Khachatourians [49] confirmed that phage P1 could transduce *E. coli in situ* although the reported frequency of 10^{-6} should be considered cautiously. Transduction of indigenous soil bacteria was not unequivocally demonstrated [33]. Transduction-mediated gene transfer was also monitored on plant leaves by using *Pseudomonas aeruginosa* and its generalized transducing bacteriophage F116 as a model system [44]. Transfers were detected when donor and recipient bacteria were inoculated on the same leaf and onto adjacent plants, when contact was possible through high-density planting.

III. NATURAL TRANSFORMATION: POSSIBLE OCCURRENCE IN THE RHIZOSPHERE

A. Natural Transformation in Bacteria and Regulation

Natural genetic transformation is defined as the active uptake of extracellular DNA by the specialized bacteria that develop a physiological state called competence state [20,23,79]. The natural transformation process is regulated by coordinated functions encoded by a set of genes present throughout the genome of competent bacteria [23,80]. Natural transformation-mediated gene transfer was demonstrated to occur in various environments including soil [23,36,33], plant tissues [81,82], seawater, and marine sediments [83,84]. A wide range of DNA can be taken up by competent bacteria, including autonomously replicating plasmid molecules, or chromosomal DNA fragments from bacteria belonging to the same species or to phylogenetically less related taxa that can be integrated into the host genome by homologous or illegitimate recombination, respectively [23,84–87].

1. Natural Transformation Determinants

Natural transformation has been studied extensively in some bacterial models. The natural transformation pathway has been defined after most genes were identified and function of the corresponding proteins characterized [20,88–96]. Competence proteins are involved in four interconnected steps including competence induction, DNA binding, DNA uptake, and the inheritable integration of the incoming DNA [20,23,92].

The physiological state of competence is restricted to the naturally transformable bacterial species and, depending on the bacteria, can be reached during the late exponential growth phase, the stationary phase, or is constitutive [23,97,98]. Metabolic changes and modifications of cell surface properties induced during the competence state are often transitory [23,79,98]. Among the huge soil bacterial diversity only few species have been identified (Table 14.1) to have a genetically encoded natural transformation machinery to actively take up and incorporate free DNA [23,32,124]. However, the natural transformable species present in soil encompass a wide range of taxonomic and trophic bacterial groups including saprophyte bacteria, plant and animal symbionts, and pathogens [23]. Complete genome sequence analysis [93,125] and experimental studies [122,126–128] provide insight into the natural transformation potential of most soil bacteria.

Genes encoding homologs of some components of the DNA-uptake machinery have been detected in completely sequenced genomes of several bacteria. This sequence homology as well as conservation of the genetic organization over evolutionary time suggests that numerous bacteria could develop competence and that these genes might have spread among bacteria by horizontal gene transfer [93,112,129]. However, another hypothesis suggests a change of function, such as in *E. coli*, where homologs of these proteins are involved in the uptake of homospecific and heterospecific extracellular DNA as a source of carbon and energy [130]. Moreover, natural transformation-mediated gene transfer could also be used for genome evolution of functions such as DNA repair [79,131–135].

TABLE 14.1
List of Some Naturally Transformable Bacteria Species Isolated from Soil Environment

Species Present in Soil Environment	Taxonomic Group	References
Photolithotrophic		
Synechocystis sp. PCC 6803[a]	Cyanobacteria	99
Chemolithotrophic		
Thiobacillus thiopurans[b]	*β*-Proteobacteria	100
Heterotrophic		
Acinetobacter calcoaceticus[a]	*γ*- Proteobacteria	101
Acinetobacter sp. BD413 (or strain ADP1)	*γ*- Proteobacteria	102
Azotobacter vinelandii[b]	*γ*- Proteobacteria	103
Pseudomonas stutzeri[a]	*γ*- Proteobacteria	104, 105
Ralstonia solanacearum	*β*- Proteobacteria	106
Bacillus subtilis[a]	Low GC Gram+	107
Mycobacterium smegmatis	High GC Gram+	108, 109
Deinococcus radiodurans	*Deinococcus/Thermus*	110
Thermus thermophilus[a] (and *Thermus* spp.)	*Deinococcus/Thermus*	80, 111, 112
Methylotropic		
Methylobacterium organophilum[b]	α- Proteobacteria	113
Archaebacteria		
Methanobacterium thermoautotrophicum[b]	Archaea	114
Clinical pathogenic species		
Campylobacter jejuni[a]	ε- Proteobacteria	115
Campylobacter coli	ε- Proteobacteria	116
Helicobacter pylori[a]	ε- Proteobacteria	117–118
Legionella pneumophila[a] (serogroup1)	*γ*- Proteobacteria	119
Moraxella spp.[a]	*γ*- Proteobacteria	120, 121
Suspected natural transformable bacteria [c]		
Agrobacterium tumefaciens	α- Proteobacteria	122
Pseudomonas fluorescens	*γ*- Proteobacteria	122
Escherichia coli	*γ*- Proteobacteria	93
Listeria monocytogenes	Low GC Gram+	93
Lactococcus lactis	Low GC Gram+	123

[a] Competence regime identified and completed by molecular analyses of the natural transformation machinery.

[b] No recent publication about natural transformation of the species considered.

[c] Competence regime nonidentified, but *in silico* detection of a natural transformation machinery or *in situ* evidence of natural transformation capability.

Source: From Mercier, A. et al., in *Nucleic Acids and Proteins in Soil (Soil Biology)*, Vol. 8, Nannipieri, P. and Smalla, K., Eds., Springer-Verlag, Berlin, 2006, chap. 15. With permission.

2. Natural Transformation Gene Transfer Regulation: The Obstacle Race of the Penetrating Extracellular DNA

Uptake and incorporation of extracellular DNA by competent bacteria are strongly regulated. In some bacteria, the first regulation step involves a sequence-based specific binding of the DNA to the cell wall prior to uptake [136]. The requirement for uptake signal sequences (USS), which are small motifs overrepresented throughout the genome that bind to DNA receptors contributes to the specific uptake of DNA from related species [136]. However, natural transformation with heterologous chromosomal DNA was also detected in some pathogenic bacteria that exhibited USS [137].

Moreover, other bacteria such as *Bacillus subtilis* and *Acinetobacter* sp. take up DNA from homologous and heterologous sources as efficiently indicating that, apparently, these bacteria do not limit the incoming DNA flow [23]. DNA is taken up as single strand, so restriction systems that recognize and cleave double strand DNA would be inefficient with the transforming DNA and would not limit natural transformation in most competent bacteria. However, recent results mainly with *Pseudomonas stutzeri* indicate that the restriction mechanisms could act as a barrier to transformation in some bacteria [138–140].

Inheritable perpetuation of incoming DNA in the host genome requires its integration by homologous, homeologous, or illegitimate recombination [23]. Production of nonmatching and mismatches due to the increase in the DNA sequence divergence between donor and recipient DNA leads successively to the nonactivation of the recombinational RecA protein and elimination of the DNA by enzymes of the methyl-mismatch repair system (MMR). This MMR system acts as a strong inhibitor of interspecies recombination, and thus contributes to the maintenance of equilibrium between generation of genetic diversity and an acceptable level of stability and integrity of the genetic information [22,141–147].

On the other hand, the DNA damage-dependent inducible SOS system, through overproduction of recombinational proteins, could act as a regulator for increasing recombination during natural gene transfer [22]. Moreover, illegitimate and homology-facilitated illegitimate recombination (HFIR), based on presence on one side of extracellular DNA by a short homologous sequence (anchor), can lead to the integration of heterologous DNA [85,87,148,149]. The homologous recombination also facilitates plasmid transfer by chromosomal homology [150].

B. Biotic and Abiotic Rhizospheric Factors Affecting Natural Transformation

1. Persistence of Extracellular DNA in Soil

Occurrence of natural transformation in the environment depends on presence of nondegraded and biologically active extracellular DNA. This DNA is released by living and dead microorganisms and by senescent or decaying plant and animal material. In soil, natural transformation-mediated gene transfer was not considered initially as ecologically relevant until it was demonstrated that most of the DNA released in soil is rapidly degraded, but a significant part of the extracellular DNA escaped physical, chemical, and enzymatic degradation depending on environmental conditions [23,151–155]. Soil is a large reservoir of nondegraded extracellular DNA containing up to 40 μg of extracellular DNA per gram [23,156]. Mechanisms of DNA persistence in soil were investigated in different soil types including silt loam, loamy sand or clay and silty clay soils, under sterile and nonsterile conditions [151,156,157] and by using pure chromosomal and plasmidic DNA [151,153,158] and bacterial lysate [159]. Recombinant DNA from genetically modified plants was also used to monitor the fate of plant DNA under field conditions [160] but also in soil microcosms inoculated with ground plant material [153], plant leaves [154], decaying plant material [155], and pollen [161].

Persistence and integrity of most extracellular DNA in soil are due to a reversible adsorption onto soil components including sand particles, clay minerals such as montmorillonite, illite, kaolinite, and humic compounds [162–164]. Interactions between extracellular DNA and soil components depend mainly on the size and conformation of chromosomal and plasmidic DNA rather than on the base composition and presence of blunt or cohesive ends of DNA fragments [165–168] and on soil parameters such as temperature, pH, concentration, and cation valence [23,151,169,170]. However, DNA persistence in soil could also be due to an efficient adsorption of nucleases onto clay particles leading to a physical separation between enzymes and their DNA substrates [164] as well as to a protection by cell debris to which the released DNA remains bound [159].

These different factors explain the slow turnover of extracellular DNA in soil, which is confirmed by the PCR detection of DNA sequences after months and even years of incubation in soil

[153,154,158,162,171,172]. However, a downward movement of the persisting extracellular DNA have been suggested, probably transported by water-satured soil and groundwater [172,173]. Moreover, the ability of this persisting DNA to transform bacteria remains to be confirmed.

2. Bacterial Competence Development in Soil

Genetic transformation frequency of bacteria in soil is directly related to competence gene expression. The physiological state of bacteria in soil and, consequently, their ability to develop competence is directly related to abiotic parameters such as pH, moisture level, concentration of mono- and bivalent cations, nutrient input and biotic parameters, such as bacterial cell density, growth rate, competence factor concentration, transformation period, competition with indigenous bacterial communities, and stress [23]. However, among the naturally transformable bacteria identified so far in soil, few of them have been shown to express competence under natural soil conditions [174]. Moreover, competence was rapidly lost after inoculation in an oligotrophic soil of an *in vitro* prepared competent *Acinetobacter* sp. BD413 [175]. Competence in this bacterium was found to be induced only when nutrients were artificially added to soil microcosms. Competence development required a high phosphate level, and natural transformation occurred at a higher frequency in a silt loam soil than in a loamy sand soil. Competence of *Acinetobacter* sp. BD413 in soil was also stimulated by various organic compounds (glucose and the high-P salts) naturally found in the rhizosphere of crop plants [176].

C. Methods and Predictions of Natural Transformation-Mediated Gene Transfer in the Rhizosphere

1. Methods to Investigate Natural Transformation-Mediated Gene Transfer in Soil

Detection of significant amounts of extracellular DNA in soil was initially considered as the definite proof that natural transformation is a mechanism that soil bacteria use to acquire new DNA [23]. However, the firm and experimental demonstration, that the DNA released by indigenous soil organisms can be used *in situ* by indigenous bacteria as a source of new genetic information, is still missing. Various techniques were developed to extract DNA from soil [23]. These methods are very useful to quantify the amount of extracellular DNA present in soil, to determine its size and its origin, and even to find new genes. The soil-extracted DNA solution was also used to transform competent bacteria *in vitro* to determine the biological potential of this DNA [155,158,177]. If these genes should originate from bacteria phylogenetically, remote from the recipient strain, the homologous recombination-mediated integration might fail. Moreover, these transformation tests with extracted and purified DNA inoculated to *in vitro* growing bacteria would not indicate the DNA availability and frequency for bacteria *in situ*. Even inoculating the competent bacteria into the soil to capture *in situ* the indigenous DNA suffers from the bias related to the different localizations of inoculated and indigenous bacteria and consequently their DNA [178]. Experiments in which DNA is inoculated in soil are also far from the natural conditions in which DNA is gradually released by decaying biological material. Indigenous and inoculated DNA colonizes different niches [178]. Finally, one of the most promising approaches to study gene transfer in soil is the use of genetically modified plants because their DNA that can be easily tracked even after transformation into bacteria. The specific recombinant DNA of transgenic plants is useful for monitoring its fate in soil by PCR or molecular hybridization. For example, its persistence (as extracellular) for months and even years have been demonstrated several times [153,154,158,162, 171,179]. The plant transgene can also be specifically engineered with bacterial DNA sequences flanking one or several marker genes to improve recombination in a bacterial host genome and detection of potential recombinant bacteria [180]. Based on this strategy, a marker rescue system was developed recently by several groups. It combines the presence of the same marker gene in donor and recipient organisms with, however a deletion in the recipient DNA. This deletion is complemented after recombination with the donor DNA [161,181–184]. Finally, the interest of transgenic plants to address fundamental questions about gene transfer is increased with those in which the transgene is cloned in the chloroplast genome (transplastomic plants) because the higher copy number of the transgene increases the transfer potential to bacteria and facilitates transgene detection [185].

A better understanding of genetic transformation in soil requires considering the bacterial potential. Analysis of complete genome sequences in which competence genes were detected demonstrates that the transformability property might be shared by a wider range of bacteria than initially thought [93,125]. The fundamental role that gene transfer would play in bacterial evolution and adaptation justifies checking the numerous isolates in worldwide strain collections for natural transformability [86]. In addition to the *in vitro* studies in which different transformation protocols can be tested, transformation tests can also be carried out in soil microcosms, where competence might develop more easily [122]. New molecular tools have been developed to address these questions based on the use of broad host range plasmids, containing several marker genes to select transformants among the bacterial community background and even to detect them *in situ* (even those belonging to nonculturable bacteria), when green fluorescent protein (GFP) or other fluorescence genes are used. Conditions for competence development in soil are also to be investigated. Only a few models including *Bacillus subtilis*, *Acinetobacter* sp. and *Pseudomonas stutzeri*, selected based on their transformation potential *in vitro*, have been studied for competence development in soil [23,83,180]. However, such studies cannot be without considering the spatial heterogeneity of soil [186,187]. Soil bacteria develop microcolonies in soil and on aggregates and thus might have the required conditions for competence development locally. However, specific experimental conditions must be developed to consider bacteria in these microniches and their ability to release and take up extracellular DNA.

Finally, several reports indicate that bacterial transformation including for bacteria that are not naturally transformable can be chemically or electrically induced in the environment [188,189]. A better understanding of the involvement of gene transfer in bacterial evolution and adaptation will not be complete without determining to what extent these passive mechanisms contribute to the uptake of DNA by bacteria.

2. Natural Transformation Gene Transfer in the Rhizosphere

The occurrence of natural transformation by bacteria in the rhizosphere remains only speculative. The relatively low turnover of extracellular DNA in soil, explained by its adsorption on soil components [158,162–164,171], supports the existence of a DNA pool for transformation of bacteria. On the other hand, bacteria spend most of their time in soil in a dormancy state that is not compatible with competence development [23]. Conjugation between inoculated and indigenous bacteria was demonstrated *in situ* [190,191], but not natural transformation of indigenous bacteria. Competence development is probably the main factor limiting DNA acquisition by soil bacteria. However, conclusions drawn from bulk soil experiments cannot be extrapolated to what might occur in the rhizosphere. For example, specific plant exudates detected in the rhizosphere such as sugars, amino acids, and some organic acids strongly stimulate bacterial growth and were demonstrated to increase natural transformation-mediated gene transfer frequency [176]. With a total direct count up to 10^9 bacteria per gram of soil [39], the rhizosphere, a rich nutrient, and organic matter environment would be less refractory for competence development than the bulk soil. In addition to other plant-related environments such as the plant tissues colonized by bacterial pathogens [192,193] and the residuesphere (decaying plant material) [155], the rhizosphere has the theoretical potential to be a hot-spot for gene transfer including by natural transformation. Moreover, competent rhizospheric bacteria develop in contact not only with the DNA released by the various bacteria and fungi but also those released by decaying root cells, thus providing favorable conditions for intra and interspecies gene transfer and interkingdom exchange when transgenic plants are considered (see Section V). However, these tightly bound soil aggregates remain difficult to access when significant volumes are necessary to detect rare events that might explain the lack of data on natural transformation in these ecosystems.

IV. CONJUGATION: A CONFIRMED GENE TRANSFER MECHANISM IN THE RHIZOSPHERE

A. CONJUGATION IN BACTERIA AND REGULATION

Conjugation is an active and genetically regulated gene transfer process requiring close contact between bacteria. Complex physiological and genetic interactions lead to the unidirectional transfer of DNA from donor to recipient bacteria [194]. Some conjugative retrotransfer events are also described, leading to the capture by the original bacterial host of a conjugative element bringing DNA (either chromosomal markers or plasmids) from the mating recipient bacterium [195–197].

The genes necessary for conjugation are only present in conjugative plasmids and transposons [198–200]. These two conjugative elements can involve the mobilization, as well as the cointegrate formation of chromosomal or plasmid sequences [11]. The widespread occurrence of conjugative plasmids and transposons has been detected in bacterial populations from various environments including soil [26], seed surfaces [201,202], crown gall tumors [203], the phytosphere [204], sediments, biofilms, activated sludge, wastewater [205], marine environment [206], animals ecosystems like insect larvae or porcine feces [207], and human gut [208]. Bacterial donor and recipient cells may belong to the same species or to different genera or even kingdoms [1,19]. However, the efficiency of the conjugative transfer is controlled at various levels, mainly by plasmid incompatibility group [209], plasmid-encoded traits, restriction–modification systems [210], DNA divergence and recombinational systems [22]. The persistence of conjugative mobile elements over evolutionary time has not been completely characterized, but is mainly determined during bacterial division and by bacterial growth rates [211].

1. Conjugative Plasmids

Conjugation is mainly driven by conjugative plasmids containing gene sequences necessary for carrying out replication and for their transfer from one bacterial cell to another. Conjugal plasmids are assigned to incompatibility (inc) groups, which prevent closely related plasmids to coexist stably in a same bacterium [209]. These conjugative plasmids carry a set of genes (named *tra* genes), which encode the functions necessary to establish contact between cells. In proteobacteria, a pilus allows attachment of the plasmid-containing donor bacterium to another cell (the recipient), which exhibits a receptor for the pilus. The retraction of the pilus draws together the donor and the recipient bacteria. Several donor bacteria can extend pili at the same time and thus, converge together on the recipient bacteria, favoring aggregation of bacterial cells in clusters. Conjugation in Gram-positive bacteria is slightly different without the initial formation of bridges or pores between donor and recipient cells. Potential recipient bacteria excrete substances that prompt donor bacteria to produce proteins able to bring bacteria together. After aggregation, donor and recipient bacteria form the pores required for the conjugative plasmid transfer.

The conjugative plasmids can facilitate the transfer of nonconjugative transfer origin (*ori*T) containing plasmids coexisting in the bacterial donor cell. The mobilization of the IncQ plasmid RSF1010 by conjugative IncP1 plasmids such as RP4 is one of the best described examples [212]. Finally, a nonconjugative and nonmobilizable plasmid can also be transferred, consecutively by the fusion and the cointegration in a conjugative plasmid, in presence or not of insertion elements or transposons that facilitate the fusion. Plasmids are also vectors for chromosomal DNA fragments that integrate into the conjugative plasmids and are transferred to the recipient cell. After transfer, plasmids can recircularize and be maintained as autoreplicative elements in the recipient cell, and also be involved in a recombination process for integration in the recipient host genome.

These promiscuous plasmids transfer genetic information between unrelated species such as proteobacteria and Gram-positive bacteria, yeast cells, and plants [213–215]. The nonconjugative mobilizable IncQ plasmids are often used as mobilizable cloning vectors due to their broad host range spectrum including most of proteobacteria and several Gram-positive bacteria such as *Streptomyces, Mycobacterium,* and *Synechococccus.*

The conjugative transfer of plasmids is the main mechanism responsible for the rapid dissemination of antibiotic resistance genes in unrelated pathogenic bacterial species [216]. However, this mechanism is also involved in the transfer of xenobiotic catabolic genes such as those involved in degradation of polychlorinated biphenyl compounds or in the transformation of mercury or other heavy metals [217–219]. Moreover, many virulence-associated determinants are located on plasmids and are continually evolving through genetic exchange with other replicons [60].

2. Conjugative Transposons

The second type of conjugation requires the presence of a conjugative transposon [198,199] and occurs mainly between Gram-positive bacteria. However, the presence of conjugative transposons has been detected in proteobacteria, for example, in the pathogen *Neisseria meningitidis.* In this case, the DNA could penetrate by natural transformation or be transported by a conjugative plasmid [220].

Conjugative transposons differ from traditional transposons by their capacity to be transferred by conjugation without causing the doubling of the insertion site within the DNA target [199]. Moreover, conjugative transposons are excised in a DNA circular intermediate form. The conjugative transposon is transferred and integrated from a donor cell to the chromosome of a recipient bacterium. Nevertheless, the retrotransfer of a conjugative transposon (or plasmid) in the genome of the original bacterial host is possible [197]. Conjugative transposons may also facilitate plasmid mobilization and cointegrate formation. All the conjugative transposons described so far exhibit at least one antibiotic resistance gene, mainly genes conferring resistance to tetracycline, chloramphenicol, and kanamycin [198]. They are involved in the transmission of antibiotic resistance genes to pathogenic Gram-positive bacteria deprived of plasmids. They are transferred easily among a broad range of bacterial species [198,221,222]. The contact between bacterial donor and recipient cells would start the transposon excision in the donor bacterium. Then, the circular form of the excised transposon is transferred into the bacterial recipient cell and integrated [199].

B. Biotic and Abiotic Factors Affecting Conjugation in the Rhizosphere

1. Cell Density and Physiological Status of Bacterial Cells

Biotic and abiotic factors affecting regulation of gene transfer by conjugation have been identified under *in vitro* conditions [223]. These results were confirmed in soil microcosms [224], in the rhizosphere [25,225], on the surface of the roots, and in the phytosphere [202], although in some cases, conjugation was modulated differently in nature than in the laboratory [201]. Soil is often described as being more refractory to conjugation-mediated gene transfer than marine environments mainly because of a higher heterogeneity. However, high cell densities can be reached locally, such as on the surface of roots or in soil microaggregates, thus providing close contact between bacterial cells.

An active metabolism of donor and recipient bacteria is also required for conjugation, and the influence of nutrient amendment on conjugation frequency was studied in sterile and nonsterile soil microcosms [25,26,226,227]. The addition of organic matter to the soil sample significantly increased conjugative transfer frequency. These studies suggest that conjugation genetic transfers between bacteria are strongly influenced in the rhizosphere by the availability of growth substrates and the bacterial growth rate. However, other studies showed that a direct addition of root exudates to sterile sand microcosms did not modify the frequency of gene transfer in spite of a significant increase in bacterial metabolic activity [228]. The direct inoculation of these compounds did not simulate their gradual release by roots and the conjugation-mediated genetic exchange might be stimulated only when a bacterial metabolic activity threshold was reached. Similarly, the basic metabolic activity that *Pseudomonas putida* cells develop on sterile leaves would not limit conjugation [229].

2. Requirements for Conjugation in Soil

Independent of the conjugative element itself and of the bacterial host spectrum, horizontal gene transfer efficiency in the environment depends on other biotic and abiotic parameters.

Abiotic factors, such as temperature, moisture, pH [226,230], nutritive conditions [231,232], concentration of mono and bivalent cations, or soil structure [226,231] influence conjugative gene transfer by affecting cell-to-cell contact and host-cell physiology. Environmental conditions can directly induce conjugation by activating plasmid-specific properties. For example, the presence of host plants and a high density of *Agrobacterium tumefaciens* cells (quorum sensing control) induced expression of the Ti plasmid-borne *tra* genes [194]. Abiotic parameters, such as soil structure, also influence conjugation. The soil matrix can be considered on one hand as a barrier separating donor and recipient bacteria, yet on the other hand, bacterial growth and cell-to-cell contact are increased in presence of clay minerals (montmorillonite, bentonite) on which bacteria adsorb and develop [226,231].

C. Methods and Predictions of Conjugative Gene Transfer in the Rhizosphere

1. Methods to Investigate Conjugation-Mediated Gene Transfer in Soil

Conjugation is considered as being the most efficient gene transfer mechanism in the environment and also the easiest one to investigate. Thus, conjugation was the first mechanism of bacterial gene transfer to be studied extensively as a way for bacteria to disseminate genetic material in the environment. Most of the experimental studies conducted in soil were carried out to estimate transfer frequency between donor and recipient bacteria [27–29,233].

Initial experiments involved inoculation of soil by donor and recipient bacteria; the transconjugants being specifically selected based on the appropriate combination of marker genes between plasmids, donor, and recipient cells. However, additional experiments requiring an efficient counterselection system demonstrated that introduced donor bacteria could also transfer their plasmids to indigenous bacteria [234].

Selection of transconjugants can be based on the expression of antibiotic resistance genes, but alternate systems involving the use of auxotrophic donor and recipient bacteria could lead to prototrophic recombinants after transfer. These transconjugants were found to be more competitive than their parental auxotrophic isolates in the rhizosphere [26,191].

Studies have also been developed to determine the prevalence and putative role of plasmids present in indigenous bacterial populations in *in situ* genetic mobilization [190,235,236]. A triparental exogenous method based on coinoculation of a mixed soil bacterial community by a strain containing the mobilizable plasmid and a selectable recipient strain, led to the isolation of mobilizing plasmids from polluted soils [236] and from the rhizosphere [190]. Moreover, among the self-transmissible plasmids isolated in the wheat rhizosphere, van Elsas et al. [190] have characterized some cryptic gene-mobilizing plasmids (with no antibiotic and heavy-metal resistance), which might capture and disseminate beneficial genes in natural bacterial populations [190].

2. Conjugative Transfer in the Rhizosphere

According to several microcosm-based studies the nutrient-rich plant root system would stimulate bacterial growth efficiently. These conditions could stimulate mobilization and conjugation-mediated gene transfer compared to the bulk soil [25,26,29,176,237]. For example, the wheat rhizosphere obtained in artificial soil under gnotobiotic conditions or in a second step with natural, nonsterile soil microcosms allowed the mobilizing IncP plasmid by root-colonizing *Pseudomonas* spp. at a higher frequency than in bulk soil [26]. In spite of differences between microcosm and field studies, these data implied that conjugation could occur at a significant frequency between indigenous bacteria in the environment. These results confirm the data obtained by van Elsas et al. [25] who detected the conjugative transfer of the RP4 plasmid between *Pseudomonas* spp. inoculated in the wheat rhizosphere even though these transfers remained undetected in the bulk soil. Transfer frequency decreased when the distance between bacteria and the roots increased [25]. Finally, the same RP4 plasmid was transferred at a higher frequency between a *Pseudomonas fluorescens* strain to a *Serratia* sp. isolate in the rhizosphere of a watery plant than in the nonplanted soil [228] although the artificial addition of root exudates in the bulk soil increased the transfer frequency up to that of the rhizosphere [228]. In spite of this last result, it appeared that conjugation would be stimulated in the rhizosphere, more by the physical presence of roots that contributes to the grouping of bacteria and favors cell-to-cell contact than by the release of nutrients [228]. The potential for gene transfer in the rhizosphere is also confirmed by the demonstration that retrotransfer can occur [197].

However, transfer to indigenous bacteria remains difficult to detect, based the last results by Hirkala and Germida [191] who failed to demonstrate that plasmid-borne atrazine degradation and tellurite resistance genes were transferred from an inoculated *Pseudomonas putida* to the indigenous soil microflora in a bulk soil or in the canola (*Brassica napus*) rhizosphere. Moreover, transconjugants remained undetected even in the presence of atrazine, which was expected to increase the fitness of transconjugants over nonatrazine degrading bacteria, although the donor *P. putida* population remained high based on its capacity to degrade environmental pollutants [191].

Using a direct approach (triparental exogenous isolation), van Elsas et al. [190] have isolated self-transmissible plasmids with IncQ plasmid-mobilizing capacity from a wheat rhizosphere. The mobilization observation confirmed the natural conjugal gene flow in the Proteobacteria communities in soil and the rhizosphere. In this same study, three cryptic gene-mobilizing plasmids (containing no selectable traits) have been characterized by phenotypic and molecular methods. The prevalence of one of these plasmids in soil and the rhizosphere has been demonstrated without identifying their reservoir and their role in mobilizing the IncQ plasmids *in situ* [190].

V. GENE TRANSFER FROM GENETICALLY MODIFIED PLANTS TO RHIZOSPHERIC BACTERIA

The possible transfer of transgenes from genetically modified organisms, such as plants, to bacteria is a modern consequence of the genomic adaptability potential that is conferred to bacteria by horizontal gene transfer mechanisms. Plant transgene contain pieces of DNA, which may behave physically like any other naturally occurring plant DNA fragments in the soil environment, but the recombinant genes could differ in several ways from native genes with respect to their likelihood for transfer to bacteria, genome integration, and expression in the bacterial host. During the DNA recombination process in bacteria, the procaryotic origin of these new plant genes could help them overcome the bacterial genetic barriers that prevent all but a few rare plant genes to be transferred from plants to bacteria [238,239]. The main barrier to gene transfer would not be a physical one (degradation and or unavailability of the DNA) or a physiological one (inability of bacteria to take up DNA) but a genetic barrier that prevents dissimilar exogenous DNA, including plant DNA to be integrated in the recipient genome. Presence of one or more (among the 25) marker genes, which mainly confers resistance to antibiotics or herbicides, and originates from bacterial and viral sources that have been successfully used for plant transformation [240,241], increases dramatically the probability that this DNA will be transferred to bacteria. Moreover, the use of diverse natural plant-associated symbiotic bacteria that can be modified to mediate gene transfer to achieve plant transformation [241] could represent a potential factor of interkingdom gene dissemination.

The antibiotic resistance genes, initially used as selective markers for plant construction [240,242], persist in plant genomes and provide a possible source of transforming DNA for environmental bacteria [30,31,158,192,243,244] and to human pathogenic bacteria [245,246]. The results by Kay et al. [183], demonstrating that a marker gene from a transplastomic plant could be transferred *in planta* to bacteria possessing the appropriate DNA homologous sequences, confirms the availability of plant DNA for endophyte bacteria and the fundamental role of recombination in regulating gene transfer. Moreover, the risk of transgene transfer would increase when transplastomic plants are considered (presence of the transgene in the chloroplast genome) due to the transgene copy number that is increased from 10 in traditional and nuclear modified transgenic plants to more than 10,000 in transplastomic plants [183,185].

However, detection of transgene transfer from plants to bacteria outside the greenhouse conditions is hampered by the problem of isolating transformants. Even presence of antibiotic resistance genes in transgenic plants is not enough to differentiate between transformants that would have acquired whole or part of the transgene from plants and those that naturally contain the same antibiotic resistance gene. The only way to demonstrate that plant DNA could have been transferred to bacteria would be to detect isolated bacteria containing the transgenic plant specific sequences flanked by the bacteria's own DNA. Such an objective requires treating soil samples of an appropriate size to deal with the low transfer frequency of bacteria in soil unless the transgene increases the fitness of the transformant in soil. However, there is no report in the literature of the successful isolation of such recombinant bacteria.

The numerous and successive events that must occur before a plant gene is expressed by a recombinant bacterial cell explains the low transfer frequency and justifies the separate study of each step of the gene transfer to determine the limiting factors. During the first step, DNA must be released by the plant. The recent study by Ceccherini et al. [155] demonstrated that most (98%) of the DNA was degraded during plant decay. Only a minor fraction of the plant DNA would be released into the soil. The second step would be the movement of either the plant DNA or bacteria to establish contact. Data indicate that endophyte bacteria, including pathogens or opportunistic bacteria are those for which exposure to plant DNA will be the strongest [193]. On the other hand, plant DNA was also detectable in soil for years, but these studies still failed to demonstrate that this persisting DNA could be still involved in a transformation process. There are also clear reports indicating that bacteria can reach competence when developing in plant tissues [193], but the location, density, and metabolism of bacteria in soil microenvironments are still almost unexplored and their ability to develop competence totally unknown. Data indicate that gene transfer frequency could vary strongly between the most favorable conditions such as those provided by a plant colonized by a pathogen and the least favorable such as those under heterogeneous and oligotrophic conditions encountered in a bulk soil where gene transfer frequency would remain too low for transformant detection [23].

Finally, the involvement of conjugation or transduction mechanisms in gene transfer from genetically engineered plants to bacteria is theoretically possible but might occur at very low frequencies and has not been studied at the experimental level.

Impact of transgenic plants on soil bacteria extends, however, far beyond the possible acquisition of the transgene by bacteria and includes the direct interactions between transgenic plants and the soil microbial community. Recent interest in these interactions is related to the property of some transgenic plants to release compounds in the soil that are expected to modify the soil microbial community structure by selectively stimulating the growth of adapted microorganisms [247–249]. Such an effect was found with opine-producing transgenic *Lotus* plants that led to an increase of the population level of opine-degrading rhizospheric bacteria [249]. However, most of the recent experiments concluded that transgenic plants have a limited impact on their plant-associated microorganisms, including nitrogen-fixing bacteria, mycorrhizal fungi, and endophytic microbiota [247,248]. The alteration of microbial diversity in association with transgenic plants remains difficult to estimate due to a lack of method sensitivity and to the presence of variable and transient effects including transgenic plant species, transgene insertion, field site, sampling data, and other environmental factors that influence soil- and plant-associated microbial communities.

VI. CONCLUSIONS

The recent analysis of complete genomic sequences confirmed preliminary data about the strong influence that horizontal gene transfer has had in shaping bacterial genomes. In addition to active mechanisms such as transduction, transformation, and conjugation, gene transfer between bacteria could occur passively after cell envelopes were permeabilized by chemical or physical environmental factors. These different mechanisms have been studied for their occurrence in various environments, including water, sediments, animal and human guts, and soils. The biotic and abiotic parameters specifically encountered in rhizospheric soil, such as high nutrient content and strong cell concentration, provide much more favorable gene transfer conditions than bulk soil. The challenge for the future is to develop the experimental tools needed to explore the rhizospheric ecosystems with a sufficient precision to determine if they could be hot-spots for gene transfer and thus explain the high level of laterally transferred genes detected in bacterial genome sequences.

REFERENCES

1. Ochman, H., Lawrence, J.G., and Groisman, E.A., Lateral gene transfer and the nature of bacterial innovation, *Nature*, 405, 299, 2000.
2. Ochman, H., Lerat, E., and Daubin, V., Examining bacterial species under the specter of gene transfer and exchange, *Proc. Natl. Acad. Sci. USA*, 102, 6595, 2005.
3. Daubin, V., Gouy, M., and Perriére, G., A phylogenomic approach to bacterial phylogeny: evidence of a core of genes sharing a common history, *Genome Res.*, 12, 1080, 2002.
4. Daubin, V., Lerat, E., and Perriére, G., The source of laterally transferred genes in bacterial genomes, *Genome Biol.*, 4, R57, 2003.
5. Jain, R. et al., Horizontal gene transfer in microbial genome evolution, *Theor. Popul. Biol.*, 61, 489, 2002.
6. Garcia-Vallvé, S. et al., HGT-DB: a database of putative horizontally transferred genes in prokaryotic complete genomes, *Nucl. Acids Res.*, 31, 187, 2003.
7. Kurland, C.G., Canback, B., and Berg, O.G., Horizontal gene transfer: a critical view, *Proc. Natl. Acad. Sci.* USA, 100, 9658, 2003.
8. Nakamura, Y. et al., Biased function of horizontally transferred genes in prokaryotic genomes, *Nat. Genet.*, 36, 760, 2004.
9. Dufraigne, C. et al., Detection and characterization of horizontal transfers in prokaryotes using genomic signature, *Nucl. Acids Res.*, 33, e6, 2005.
10. Gogarten, J.P. and Townsend, J.P., Horizontal gene transfer, genome innovation and evolution, *Nat. Rev. Microbiol.*, 3, 679, 2005.
11. Yin, X. and Stotzky, G., Gene transfer among bacteria in natural environments, *Adv. Appl. Microbiol.*, 45, 153, 1997.
12. Frost, L.S. et al., Mobile genetic elements: the agents of open source evolution, *Nat. Rev. Microbiol.*, 3, 722, 2005.
13. Thomas, C.M. and Nielsen, K.M., Mechanisms of, and barriers to, horizontal gene transfer between bacteria, *Nat. Rev. Microbiol.*, 3, 711, 2005.
14. Lang, A.S. and Beatty, J.T., Genetic analysis of a bacterial genetic exchange element: the gene transfer agent of *Rhodobacter capsulatus*, *Proc. Natl. Acad. Sci. USA*, 97, 859, 2000.
15. Gokhale, D.V., Puntambekar, U.S., and Deobagkar, D.N., Protoplast fusion: a tool for intergeneric gene transfer in bacteria, *Biotechnol. Adv.*, 11, 199, 1993.
16. Shapiro, J.A., Transposable elements as the key to a 21st century view of evolution, *Genetica*, 107, 171, 1999.
17. Bennett, P.M., Genome plasticity: insertion sequence elements, transposons and integrons and DNA rearrangement, *Methods Mol. Biol.*, 266, 71, 2004.

18. Rowe-Magnus, D.A. and Mazel, D., Integrons: natural tools for bacterial genome evolution, *Curr. Opin. Microbiol.,* 4, 565, 2001.
19. Amabile-Cuevas, C.F. and Chicurel, M.E., Bacterial plasmids and gene flux, *Cell,* 70, 189, 1992.
20. Dubnau, D., DNA uptake in bacteria, *Annu. Rev. Microbiol.,* 53, 217, 1999.
21. Palmen, R., Driessen, A.J., and Hellingwerf, K.J., Bioenergetic aspects of the translocation of macromolecules across bacterial membranes, *Biochim. Biophys. Acta,* 1183, 417, 1994.
22. Matic, I., Taddei, F., and Radman, M., Genetic barriers among bacteria, *Trends Microbiol.,* 4, 69, 1996.
23. Lorenz, M.G. and Wackernagel, W., Bacterial gene transfer by natural genetic transformation in the environment, *Microbiol. Rev.,* 58, 563, 1994.
24. Brown, J.R., Ancient horizontal gene transfer, *Nat. Rev.,* 4, 121, 2003.
25. van Elsas, J.D., Trevors, J.T., and Starodub, M.E., Bacterial conjugation between pseudomonads in the rhizosphere of wheat, *FEMS Microbiol. Ecol.,* 54, 299, 1988.
26. Troxler, J. et al., Conjugative transfer of chromosomal genes between fluorescent pseudomonads in the rhizosphere of the wheat, *FEMS Microbiol. Ecol.,* 63, 213, 1997.
27. de Lipthay, J.R., Barkay, T., and Sorensen, S.J., Enhanced degradation of phenoxyacetic acid in soil by horizontal transfer of the *tfd*A gene encoding a 2,4-dichlorophenoxyacetic acid dioxygenase, *FEMS Microbiol. Ecol.,* 35, 75, 2001.
28. Sengelov, G., Kowalchuk, G.A., and Sorensen, S.J., Influence of fungal-bacterial interactions on bacterial conjugation in the residuesphere, *FEMS Microbiol. Ecol.,* 31, 39, 2000.
29. Sengelov, G. et al., Effect of genomic location on horizontal transfer of a recombinant gene cassette between *Pseudomonas* strains in the rhizosphere and spermosphere of barley seedlings, *Curr. Microbiol.,* 42, 160, 2001.
30. Nielsen, K.M., van Elsas, J.D., and Smalla, K., Dynamics, horizontal transfer and selection of novel DNA in bacterial populations in the phytosphere of transgenic plants, *Ann. Microbiol.,* 51, 79, 2001.
31. Nielsen, K.M. et al., Horizontal gene transfer from transgenic plants to terrestrial bacteria — a rare event?, *FEMS Microbiol. Rev.,* 22, 79, 1998.
32. Davison, J., Genetic exchange between bacteria in the environment, *Plasmid,* 42, 73, 1999.
33. Dröge, M., Pühler, A., and Selbitschka, W., Horizontal gene transfer among bacteria in terrestrial and aquatic habitats as assessed by microcosm and field studies, *Biol. Fertil. Soils,* 29, 221, 1999.
34. Paul, J.H., Microbial gene transfer: an ecological perspective, *J. Mol. Microbiol. Biotechnol.,* 1, 45, 1999.
35. Timms-Wilson, T.M. et al., Quantification of gene transfer in soil and the phytosphere, in *Manual of Environmental Microbiology,* 2nd ed., Hurst, C.J. et al., Eds., ASM Press, Washington, D.C., 2002, p. 648.
36. Mercier, A., Kay, E., and Simonet, P. Horizontal gene transfer by natural transformation in soil environment, in *Nucleic Acids and Proteins in Soil (Soil Biology),* Vol. 8, Nannipieri, P. and Smalla, K., Eds., Springer-Verlag, Berlin, 2006, chap. 15.
37. Jiang, S.C. and Paul, J.H., Gene transfer by transduction in the marine environment, *Appl. Environ. Microbiol.,* 64, 2780, 1998.
38. Miao, E.A. and Miller, S.I., Bacteriophages in the evolution of pathogen–host interactions, *Proc. Natl. Acad. Sci. USA,* 96, 9452, 1999.
39. Ashelford, K.E., Day, M.J., and Fry, J.C., Elevated abundance of bacteriophage infecting bacteria in soil, *Appl. Environ. Microbiol.,* 69, 285, 2003.
40. Canchaya, C., Fournous, G., and Brussow, H., The impact of prophages on bacterial chromosomes, *Mol. Microbiol.,* 53, 9, 2004.
41. Wichels, A. et al., Bacteriophage diversity in the North Sea, *Appl. Environ. Microbiol.,* 64, 4128, 1998.
42. Ewert, D.L. and Paynter, M.J., Enumeration of bacteriophages and host bacteria in sewage and the activated-sludge treatment process, *Appl. Environ. Microbiol.,* 39, 576, 1980.
43. Jensen, E.C. et al., Prevalence of broad-host-range lytic bacteriophages of *Sphaerotilus natans, Escherichia coli,* and *Pseudomonas aeruginosa, Appl. Environ. Microbiol.,* 64, 575, 1998.
44. Kidambi, S.P., Ripp, S., and Miller, R.V., Evidence for phage-mediated gene transfer among *Pseudomonas aeruginosa* strains on the phylloplane, *Appl. Environ. Microbiol.,* 60, 496, 1994.
45. Hendrix, R.W., Bacteriophages genomics, *Curr. Opin. Microbiol.,* 6, 506, 2003.
46. Miller, E.S. et al., Bacteriophage T4 genome, *Microbiol. Mol. Biol. Rev.,* 67, 86, 2003.
47. Levesque, C. et al., Genomic organization and molecular analysis of virulent bacteriophage 2972 infecting an exopolysaccharide-producing *Streptococcus thermophilus* strain, *Appl. Environ. Microbiol.,* 71, 4057, 2005.

48. Paul, J.H. and Sullivan, M.B., Marine phage genomics: what have we learned?, *Curr. Opin. Biotechnol.,* 16, 299, 2005.
49. Germida, J.J. and Khachatourians, G.G., Transduction of *Escherichia coli* in soil, *Can. J. Microbiol.,* 34, 190, 1988.
50. Zeph, L.R., Onaga, M.A., and Stotzky, G., Transduction of *Escherichia coli* by bacteriophage P1 in soil, *Appl. Environ. Microbiol.,* 54, 1731, 1988.
51. Chatterjee, A.K. and Brown, M.A., Generalized transduction in the enterobacterial phytopathogen *Erwinia chrysanthemi, J. Bacteriol.,* 143, 1444, 1980.
52. Campbell, J.I.A., Albrechtsen, M., and Sorensen, J., Large *Pseudomonas* phages isolated from barley rhizosphere, *FEMS Microbiol. Ecol.,* 18, 63, 1995.
53. Mendum, T.A., Clark, I.M., and Hirsch, P.R., Characterization of two novel *Rhizobium leguminosarum* bacteriophages from a field release site of genetically-modified rhizobia, *Antonie Van Leeuwenhoek,* 79, 189, 2001.
54. Vettori, C. et al., Interaction between bacteriophage PBS1 and clay minerals and transduction of *Bacillus subtilis* by clay-phage complexes, *Environ. Microbiol.,* 1, 347, 1999.
55. Vettori, C., Gallori, E., and Stotzky, G., Clay minerals protect bacteriophage PBS1 of *Bacillus subtilis* against inactivation and loss of transducing ability by UV radiation, *Can. J. Microbiol.,* 46, 770, 2000.
56. Burke, J., Schneider, D., and Westpheling, J., Generalized transduction in *Streptomyces coelicolor, Proc. Natl. Acad. Sci. USA,* 98, 6289, 2001.
57. Marrero, R., Young, F.E., and Yasbin, R.E., Characterization of interspecific plasmid transfer mediated by *Bacillus subtilis* temperate bacteriophage SP02, *J. Bacteriol.,* 160, 458, 1984.
58. Sunshine, M.G. et al., Mutation of the *htr*B gene in a virulent *Salmonella typhimurium* strain by intergeneric transduction: strain construction and phenotypic characterization, *J. Bacteriol.,* 179, 5521, 1997.
59. Dobrindt, U. and Hacker, J., Plasmids, phages and pathogenicity islands in relation to bacterial protein toxins: impact on the evolution of microbes, in *The Comprehensive Sourcebook of Bacterial Protein Toxins,* Aloof, J.E. and Freer, J.H., Eds., Academic Press, 1999, p. 3.
60. Dobrindt, U. and Hacker, J., Whole genome plasticity in pathogenic bacteria, *Curr. Opin. Microbiol.,* 4, 550, 2001.
61. Sambrook, J., Fritsch, E.F., and Maniatis, T., Molecular cloning: a laboratory manual, 2nd ed., Cold Spring Harbor Laboratory Press, Cold Spring Harbor, New York, 1989.
62. Ashelford, K.E. et al., Seasonal population dynamics and interactions of competing bacteriophages and their host in the rhizosphere, *Appl. Environ. Microbiol.,* 66, 4193, 2000.
63. Kokjohn, T.A., Sayler, G.S., and Miller, R.V., Attachment and replication of *Pseudomonas aeruginosa* bacteriophages under conditions simulating aquatic environments, *J. Gen. Microbiol.,* 137, 661, 1991.
64. Schrader, H.S. et al., Bacteriophage infection and multiplication occur in *Pseudomonas aeruginosa* starved for 5 years, *Can. J. Microbiol.,* 43, 1157, 1997.
65. Reanney, D.C. and Marsh, S.C.N., The ecology of viruses attacking *Bacillus stearothermophilus* in soil, *Soil Biol. Biochem.,* 5, 399, 1973.
66. Wiggins, B.A. and Alexander, M., Minimum bacterial density for bacteriophage replication: implications for significance of bacteriophages in natural ecosystems, *Appl. Environ. Microbiol.,* 49, 19, 1985.
67. Miller, R.V., Environmental bacteriophage-host interactions: factors contribution to natural transduction, *Antonie Van Leeuwenhoek,* 79, 141, 2001.
68. Seeley, N.D. and Primrose, S.B., The effect of temperature on the ecology of aquatic bacteriophages, *J. Gen. Virol.,* 46, 87, 1980.
69. Ripp, S. and Miller, R.V., Effects of suspended particulates on the frequency of transduction among *Pseudomonas aeruginosa* in a freshwater environment, *Appl. Environ. Microbiol.,* 61, 1214, 1995.
70. Gerba, C.P. et al., Quantitative assessment of the adsorptive behavior of viruses to soils, *Environ. Sci. Technol.,* 15, 940, 1981.
71. Hurst, C.J., Gerba, C.P., and Cech, I., Effects of environmental variables and soil characteristics on virus survival in soil, *Appl. Environ. Microbiol.,* 40, 1067, 1980.
72. Yin, X., Zeph, L.R., and Stotzky, G., A simple method for enumerating bacteriophages in soil, *Can. J. Microbiol.,* 43, 461, 1997.
73. Straub, T.M., Pepper, I.L., and Gerba, C.P., Persistence of viruses in desert soils amended with anaerobically digested sewage sludge, *Appl. Environ. Microbiol.,* 58, 636, 1992.
74. Bergh, O. et al., High abundance of viruses found in aquatic environments, *Nature,* 340, 467, 1989.

75. Hara, S., Terauchi, K., and Koike, I., Abundance of viruses in marine waters: assessment by epifluorescence and transmission electron microscopy, *Appl. Environ. Microbiol.*, 57, 2731, 1991.
76. Paul, J.H. et al., Distribution of viral abundance in the reef environment of Key Largo, Florida, *Appl. Environ. Microbiol.*, 59, 718, 1993.
77. Hennes, K.P. and Simon, M., Significance of bacteriophages for controlling bacterioplankton growth in a mesotrophic lake, *Appl. Environ. Microbiol.*, 61, 333, 1995.
78. Saye, D.J. et al., Transduction of linked chromosomal genes between *Pseudomonas aeruginosa* strains during incubation *in situ* in a freshwater habitat, *Appl. Environ. Microbiol.*, 56, 140, 1990.
79. Solomon, J.M. and Grossman, A.D., Who's competent and when: regulation of natural genetic competence in bacteria, *Trends Genet.*, 12, 150, 1996.
80. Friedrich, A., Hartsch, T., and Averhoff, B., Natural transformation in mesophilic and thermophilic bacteria: identification and characterization of novel, closely related competence genes in *Acinetobacter* sp. strain BD413 and *Thermus thermophilus* HB27, *Appl. Environ. Microbiol.*, 67, 3140, 2001.
81. Bertolla, F. et al., During infection of its host, the plant pathogen *Ralstonia solanacearum* naturally develops a state of competence and exchanges genetic material, *Mol. Plant-Microbe Interact.*, 12, 467, 1999.
82. Kay, E. et al., Intergeneric transfer of chromosomal and conjugative plasmid genes between *Ralstonia solanacearum* and *Acinetobacter* sp. BD413, *Mol. Plant Microbe Interact.*, 16, 74, 2003.
83. Paul, J., Thurmond, J.M., and Frischer, M.E., Gene transfer in marine water column and sediment microcosms by natural plasmid transformation, *Appl. Environ. Microbiol.*, 57, 1509, 1991.
84. Kriz, P. et al., Microevolution through DNA exchange among strains of *Neisseria meningitidis* isolated during an outbreak in the Czech Republic, *Res. Microbiol.*, 150, 273, 1999.
85. de Vries, J. and Wackernagel, W., Integration of foreign DNA during natural transformation of *Acinetobacter* sp. by homology-facilitated illegitimate recombination, *Proc. Natl. Acad. Sci.*, 99, 2094, 2002.
86. Sikorski, J., Teschner, N., and Wackernagel, W., Highly different levels of natural transformation are associated with genomic subgroups within a local population of *Pseudomonas stutzeri* from soil, *Appl. Environ. Microbiol.*, 68, 865, 2002.
87. Meier, P. and Wackernagel, W., Mechanisms of homology-facilitated illegitimate recombination for foreign DNA acquisition in transformable *Pseudomonas stutzeri, Mol. Microbiol.*, 48, 1107, 2003.
88. Dreiseikelmann, B., Translocation of DNA across bacterial membranes, *Microbiol. Rev.*, 58, 293, 1994.
89. Facius, D., Fussenegger, M., and Meyer, T.F., Sequential action of factors involved in natural competence for transformation of *Neisseria gonorrhoeae*, *FEMS Microbiol. Lett.*, 137, 159, 1996.
90. Tortosa, P. et al., Specificity and genetic polymorphism of the *Bacillus* competence quorum-sensing system, *J. Bacteriol.*, 183, 451, 2001.
91. Jarmer, H. et al., Transcriptome analysis documents induced competence of *Bacillus subtilis* during nitrogen limiting conditions, *FEMS Microbiol. Lett.*, 206, 197, 2002.
92. Chen, I. and Dubnau, D., DNA transport during transformation, *Front. Biosci.*, 8, s544, 2003.
93. Claverys, J.P. and Martin, B., Bacterial "competence" genes: signatures of active transformation, or only remnants?, *Trends Microbiol.*, 11, 161, 2003.
94. Hahn, J. et al., Transformation proteins and DNA uptake localize to the cell poles in *Bacillus subtilis, Cell*, 122, 59, 2005.
95. Kidane, D. and Graumann, P.L., Intracellular protein and DNA dynamics in competent *Bacillus subtilis* cells, *Cell*, 122, 73, 2005.
96. Sun, Y.H. et al., Identification and characterization of genes required for competence in *Neisseria meningitidis, J. Bacteriol.*, 187, 3273, 2005.
97. Majewski, J. et al., Barriers to genetic exchange between bacterial species: *Streptococcus pneumoniae* transformation, *J. Bacteriol.*, 182, 1016, 2000.
98. MacFadyen, L.P. et al., Competence development by *Haemophilus influenzae* is regulated by the availability of nucleic acid precursors, *Mol. Microbiol.*, 40, 700, 2001.
99. Yoshihara, S. et al., Mutational analysis of genes involved in pilus structure, motility and transformation competency in the unicellular motile cyanobacterium *Synechocystis* sp. PCC 6803, *Plant Cell Physiol.*, 42, 63, 2001.
100. Yankofsky, S.A. et al., Genetic transformation of obligately chemolithotrophic *thiobacilli, J. Bacteriol.*, 153, 652, 1983.
101. Palmen, R. et al., Physiological characterization of natural transformation in *Acinetobacter calcoaceticus, J. Gen. Microbiol.*, 139, 295, 1993.

102. Young, D.M., Parke, D., and Ornston, L.N., Opportunities for genetic investigation afforded by *Acinetobacter baylyi*, a nutritionally versatile bacterial species that is highly competent for natural transformation, *Annu. Rev. Microbiol.*, 59, 519, 2005.

103. Page, W.J. and Grant, G.A., Effect of mineral iron on the development of transformation competence in *Azotobacter vinelandii*, *FEMS Microbiol. Lett.*, 41, 257, 1987.

104. Carlson, C.A. et al., *Pseudomonas stutzeri* and related species undergo natural transformation, *J. Bacteriol.*, 153, 93, 1983.

105. Meier, P. et al., Natural transformation of *Pseudomonas stutzeri* by single-stranded DNA requires type IV pili, competence state and *comA*, *FEMS Microbiol. Lett.*, 207, 75, 2002.

106. Bertolla, F. et al., Conditions for natural transformation of *Ralstonia solanacearum*, *Appl. Environ. Microbiol.*, 63, 4965, 1997.

107. Dubnau, D., Genetic competence in *Bacillus subtilis*, *Microbiol. Rev.*, 55, 395, 1991.

108. Norgard, M.V. and Imaeda, T., Physiological factors involved in the transformation of *Mycobacterium smegmatis*, *J. Bacteriol.*, 133, 1254, 1978.

109. Bhatt, A. et al., Plasmid transfer from *Streptomyces* to *Mycobacterium smegmatis* by spontaneous transformation, *Mol. Microbiol.*, 43, 135, 2002.

110. Fuchs, P., Agostini, H., and Minton, K.W., Defective transformation of chromosomal markers in DNA polymerase I mutants of the radioresistant bacterium *Deinococcus radiodurans*, *Mutat. Res.*, 309, 175, 1994.

111. Koyama, Y. et al., Genetic transformation of the extreme thermophile *Thermus thermophilus* and of other *Thermus* spp., *J. Bacteriol.*, 166, 338, 1986.

112. Friedrich, A. et al., Molecular analyses of the natural transformation machinery and identification of pilus structures in the extremely thermophilic bacterium *Thermus thermophilus* strain HB27, *Appl. Environ. Microbiol.*, 68, 745, 2002.

113. O'Connor, M., Wopat, A., and Hanson, R.S., Genetic transformation in *Methylobacterium organophilum*, *J. Gen. Microbiol.*, 98, 265, 1977.

114. Worrell, V.E. et al., Genetic transformation system in the archaebacterium *Methanobacterium thermoautotrophicum* Marburg, *J. Bacteriol.*, 170, 653, 1988.

115. Wiesner, R.S., Hendrixson, D.R., and DiRita, V.J., Natural transformation of *Campylobacter jejuni* requires components of a type II secretion system, *J. Bacteriol.*, 185, 5408, 2003 (Erratum in *J. Bacteriol.*, 185, 6493).

116. Wang, Y. and Taylor, D.E., Natural transformation in *Campylobacter* species, *J. Bacteriol.*, 172, 949, 1990.

117. Hofreuter, D. et al., Genetic competence in *Helicobacter pylori*: mechanisms and biological implications, *Res. Microbiol.*, 151, 487, 2000.

118. Smeets, L.C. and Kusters, J.G., Natural transformation in *Helicobacter pylori*: DNA transport in an unexpected way, *Trends Microbiol.*, 10, 159, 2002.

119. Stone, B.J. and Kwaik, Y.A., Natural competence for DNA transformation by *Legionella pneumophila* and its association with expression of type IV pili, *J. Bacteriol.*, 181, 1395, 1999.

120. Juni, E., Heym, G.A., and Newcomb, R.D., Identification of *Moraxella bovis* by qualitative genetic transformation and nutritional assays, *Appl. Environ. Microbiol.*, 54, 1304, 1988.

121. Luke, N.R. et al., Expression of type IV pili by *Moraxella catarrhalis* is essential for natural competence and is affected by iron limitation, *Infect. Immunol.*, 72, 6262, 2004.

122. Demanèche, S. et al., Natural transformation of *Pseudomonas fluorescens* and *Agrobacterium tumefaciens* in soil, *Appl. Environ. Microbiol.*, 67, 2617, 2001.

123. Bolotin, A. et al., The complete genome sequence of the lactic acid bacterium *Lactococcus lactis* ssp. *lactis* IL1403, *Genome Res.*, 11, 731, 2001.

124. de Vries, J. and Wackernagel, W., Microbial horizontal gene transfer and the DNA release from transgenic crop plants, *Plant Soil*, 266, 91, 2005.

125. Davidsen, T. et al., Biased distribution of DNA uptake sequences towards genome maintenance genes, *Nucl. Acids Res.*, 32, 1050, 2004.

126. Baur, B. et al., Genetic transformation in freshwater: *Escherichia coli* is able to develop natural competence, *Appl. Environ. Microbiol.*, 62, 3673, 1996.

127. Bauer, F., Hertel, C., and Hammes, W.P., Transformation of *Escherichia coli* in foodstuffs, *Syst. Appl. Microbiol.*, 22, 161, 1999.

128. Tsen, S.D. et al., Natural plasmid transformation in *Escherichia coli, J. Biomed. Sci.,* 9, 246, 2002.

129. Chen, I. and Gotschlich, E.C., ComE, a competence protein from *Neisseria gonorrhoeae* with DNA-binding activity, *J. Bacteriol.,* 183, 3160, 2001.

130. Finkel, S.E. and Kolter, R., DNA as a nutrient: novel role for bacterial competence gene homologs, *J. Bacteriol.,* 183, 6288, 2001.

131. Michod, R.E., Wojciechowski, M.F., and Hoelzer, M.A., DNA repair and the evolution of transformation in the bacterium *Bacillus subtilis, Genetics,* 118, 31, 1988.

132. Mongold, J.A., DNA repair and the evolution of transformation in *Haemophilus influenzae, Genetics,* 132, 893, 1992.

133. Redfield, R.J., Evolution of natural transformation: testing the DNA repair hypothesis in *Bacillus subtilis* and *Haemophilus influenzae, Genetics,* 133, 755, 1993.

134. Redfield, R.J., Schrag, M.R., and Dean, A.M., The evolution of bacterial transformation: sex with poor relations, *Genetics,* 146, 27, 1997.

135. Tonjum, T. et al., Transformation and DNA repair: linkage by DNA recombination, *Trends Microbiol.,* 12, 1, 2004.

136. Smith, H.O., Gwinn, M.L., and Salzberg, S.L., DNA uptake signal sequences in naturally transformable bacteria, *Res. Microbiol.,* 150, 603, 1999.

137. Kroll, J.S. et al., Natural genetic exchange between *Haemophilus* and *Neisseria*: intergeneric transfer of chromosomal genes between major human pathogens, *Proc. Natl. Acad. Sci. USA,* 95, 12381, 1998.

138. Ando, T. et al., Restriction-modification system differences in *Helicobacter pylori* are a barrier to interstrain plasmid transfer, *Mol. Microbiol.,* 37, 1052, 2000.

139. Aras, R.A. et al., *Helicobacter pylori* interstrain restriction modification diversity prevents genome subversion by chromosomal DNA from competing strains, *Nucl. Acids Res.,* 30, 5391, 2002.

140. Berndt, C., Meier, P., and Wackernagel, W., DNA restriction is a barrier to natural transformation in *Pseudomonas stutzeri* JM300, *Microbiology,* 149, 895, 2003.

141. Strätz, M., Mau, M., and Timmis, K.N., System to study horizontal gene exchange among microorganisms without cultivation of recipients, *Mol. Microbiol.,* 22, 207, 1996.

142. Taddei, F. et al., Genetic variability and adaptation to stress, *EXS,* 83, 271, 1997.

143. Nielsen, K.M., Barriers to horizontal gene transfer by natural transformation in soil bacteria, *APMIS Suppl.,* 84, 77, 1998.

144. Young, D.M. and Ornston, L.N., Functions of the mismatch repair gene *mut*S from *Acinetobacter* sp. strain ADP1, *J. Bacteriol.,* 183, 6822, 2001.

145. Townsend, J.P. et al., Horizontal acquisition of divergent chromosomal DNA in bacteria: effects of mutator phenotypes, *Genetics,* 164, 13, 2003.

146. Meier, P. and Wackernagel, W., Impact of *mut*S inactivation on foreign DNA acquisition by natural transformation in *Pseudomonas stutzeri, J. Bacteriol.,* 187, 143, 2005.

147. Prunier, A.L. and Leclercq, R., Role of *mut*S and *mut*L genes in hypermutability and recombination in *Staphylococcus aureus, J. Bacteriol.,* 187, 3455, 2005.

148. Prudhomme, M., Libante, V., and Claverys, J.P., Homologous recombination at the border: insertion-deletions and the trapping of foreign DNA in *Streptococcus pneumoniae, Proc. Natl. Acad. Sci. USA,* 99, 2100, 2002.

149. de Vries, J., Herzfeld, T., and Wackernagel, W., Transfer of plastid DNA from tobacco to the soil bacterium *Acinetobacter* sp. by natural transformation, *Mol. Microbiol.,* 53, 323, 2004.

150. Lopez, P. et al., Facilitation of plasmid transfer in *Streptococcus pneumoniae* by chromosomal homology, *J. Bacteriol.,* 150, 692, 1982.

151. Romanowski, G. et al., Persistence of free plasmid DNA in soil monitored by various methods, including a transformation assay, *Appl. Environ. Microbiol.,* 58, 3012, 1992.

152. Trevors, J.T., DNA in soil: adsorption, genetic transformation, molecular evolution and genetic microchip, *Antonie Van Leeuwenhoek.,* 70, 1, 1996.

153. Widmer, F., Seidler, R.J., and Watrud, L.S., Sensitive detection of transgenic plant marker gene persistence in soils microcosms, *Mol. Ecol.,* 5, 603, 1996.

154. Widmer, F. et al., Quantification of transgenic plant marker gene persistence in soil microcosms, *Mol. Ecol.,* 6, 1, 1997.

155. Ceccherini, M.T. et al., Degradation and transformability of DNA from transgenic leaves, *Appl. Environ. Microbiol.,* 69, 673, 2003.

156. Frostegard, A. et al., Quantification of bias related to the extraction of DNA directly from soils, *Appl. Environ. Microbiol.*, 65, 5409, 1999.

157. Zhou, J., Bruns, M.A., and Tiedje, J.M., DNA recovery from soil of diverse composition, *Appl. Environ. Microbiol.*, 62, 316, 1996.

158. Gebhard, F. and Smalla, K., Monitoring field releases of genetically modified sugar beets for persistence of transgenic plant DNA and horizontal gene transfer, *FEMS Microbiol. Ecol.*, 28, 261, 1999.

159. Nielsen, K.M., Smalla, K., and van Elsas, J.D., Natural transformation of *Acinetobacter* sp strain BD413 with cell lysates of *Acinetobacter* sp., *Pseudomonas fluorescens,* and *Burkholderia cepacia* in soil microcosms, *Appl. Environ. Microbiol.*, 66, 206, 2000.

160. Paget, E. et al., The fate of recombinant plant DNA in soil, *Eur. J. Soil. Biol.*, 34, 81, 1998.

161. Meier, P. and Wackernagel, W., Monitoring the spread of recombinant DNA from field plots with transgenic sugar beet plants by PCR and natural transformation of *Pseudomonas stutzeri, Transgenic Res.*, 12, 293, 2003.

162. Paget, E., Jocteur-Monrozier, L., and Simonet, P., Adsorption of DNA on clay minerals: protection against Dnase I and influence on gene transfer, *FEMS Microbiol. Lett.*, 97, 31, 1992.

163. Crecchio, C. and Stotzky, G., Binding of DNA on humic acids: effect on transformation of *Bacillus subtilis* and resistance to DNAse, *Soil Biol. Biochem.*, 30, 1061, 1998.

164. Demanèche, S. et al., Evaluation of biological and physical protection against nuclease degradation of clay-bound plasmid DNA, *Appl. Environ. Microbiol.*, 67, 293, 2001.

165. Gallori, E. et al., Transformation of *Bacillus subtilis* by DNA bound on clay in non-sterile soil, *FEMS Microbiol. Ecol.*, 15, 119, 1994.

166. Poly, F. et al., Differences between linear chromosomal and supercoiled plasmid DNA in their mechanisms and extent of adsorption on clay minerals, *Langmuir,* 16, 1233, 2000.

167. Pietramellara, G. et al., Effect of molecular characteristics of DNA on its adsorption and binding on homoionic montmorillonite and caolinite, *Biol. Fertil. Soil,* 33, 410, 2001.

168. Demanèche, S. et al., Influence of plasmid conformations and replication or homologous replication-based integration mechanisms on natural transformation of *Acinetobacter* sp., *Ann. Microbiol.*, 52, 61, 2002.

169. Romanowski, G., Lorenz, M.G., and Wackernagel, W., Adsorption of plasmid DNA to mineral surfaces and protection against DNase I, *Appl. Environ. Microbiol.*, 57, 1057, 1991.

170. Khanna, M. and Stotzky, G., Transformation of *Bacillus subtilis* by DNA bound on montmorillonite and effect of DNase on the transforming ability of bound DNA, *Appl. Environ. Microbiol.*, 58, 1930, 1992.

171. Romanowski, G., Lorenz, M.G., and Wackernagel, W., Use of polymerase chain reaction and electroporation of *Escherichia coli* to monitor the persistence of extracellular plasmid DNA introduced into natural soils, *Appl. Environ. Microbiol.*, 59, 3438, 1993.

172. Agnelli, A. et al., Distribution of microbial communities in a forest soil profile investigated by microbial biomass, soil respiration and DGGE of total and extracellular DNA, *Soil Biol. Biochem.*, 36, 859, 2004.

173. Potè, J. et al., Fate and transport of antibiotic resistance genes in satured soil columns, *Eur. J. Soil Biol.*, 39, 65, 2004.

174. Sikorski, J. et al., Natural genetic transformation of *Pseudomonas stutzeri* in a non-sterile soil, *Microbiology,* 144, 569, 1998.

175. Nielsen, K.M., Bones, A.M., and van Elsas, J.D., Induced natural transformation of *Acinetobacter calcoaceticus* in soil microcosms, *Appl. Environ. Microbiol.*, 63, 3972, 1997.

176. Nielsen, K.M. and van Elsas, J.D., Stimulatory effects of compounds present in the rhizosphere on natural transformation of *Acinetobacter* sp. BD413 in soil, *Soil Biol. Biochem.*, 33, 345, 2001.

177. Nielsen, K.M. et al., Natural transformation and availability of transforming DNA to *Acinetobacter calcoaceticus* in soil microcosms, *Appl. Environ. Microbiol.*, 63, 1945, 1997.

178. van Veen, J.A., van Overbeek, L.S., and van Elsas, J.D., Fate and activity of microorganisms introduced into soil, *Microbiol. Mol. Biol. Rev.*, 61, 121, 1997.

179. Kloos, D.U. et al., Inducible cell lysis system for the study of natural transformation and environmental fate of DNA released by cell death, *J. Bacteriol.*, 176, 7352, 1994.

180. de Vries, J., Meier, P., and Wackernagel, W., The natural transformation of the soil bacteria *Pseudomonas stutzeri* and *Acinetobacter* sp. by transgenic plant DNA strictly depends on homologous sequences in the recipient cells, *FEMS Microbiol. Lett.*, 195, 211, 2001.

181. de Vries, J. and Wackernagel, W., Detection of *nptII* (kanamycin resistance) genes in genome of transgenic plants by marker-rescue transformation, *Mol. Gen. Genet.*, 257, 606, 1998.
182. Nielsen, K.M., van Elsas, J.D., and Smalla, K., Transformation of *Acinetobacter* sp. strain BD413 (pFG4ΔnptII) with transgenic plant DNA in soil microcosms and effects of kanamycin on selection of transformants, *Appl. Environ. Microbiol.*, 66, 1237, 2000.
183. Kay, E. et al., *In situ* transfer of antibiotic resistance genes from transgenic (transplastomic) tobacco plants to bacteria, *Appl. Environ. Microbiol.*, 68, 3345, 2002.
184. de Vries, J. et al., Spread of recombinant DNA by roots and pollen of transgenic potato plants, identified by highly specific biomonitoring using natural transformation of an *Acinetobacter* sp., *Appl. Environ. Microbiol.*, 69, 4455, 2003.
185. Daniell, H. et al., Containment of herbicide resistance through genetic engineering of the chloroplast genome, *Nat. Biotechnol.*, 16, 345, 1998.
186. Dechesne, A. et al., A novel method for characterizing the microscale 3D spatial distribution of bacteria in soil, *Soil Biol. Biochem.*, 35, 1537, 2003.
187. Pallud, C. et al., Modification of spatial distribution of 2,4-dichlorophenoxyacetic acid degrader microhabitats during growth in soil columns, *Appl. Environ. Microbiol.*, 70, 2709, 2004.
188. Demanèche, S. et al., Laboratory-scale evidence for lightning-mediated gene transfer in soil, *Appl. Environ. Microbiol.* 67, 3440, 2001.
189. Cérémonie, H. et al., Isolation of lightning-competent soil bacteria, *Appl. Environ. Microbiol.*, 70, 6342, 2004.
190. van Elsas, J.D. et al., Isolation, characterization, and transfer of cryptic gene-mobilizing plasmids in the wheat rhizosphere, *Appl. Environ. Microbiol.*, 64, 880, 1998.
191. Hirkala, D.L. and Germida, J.J., Field and soil microcosm studies on the survival and conjugation of a *Pseudomonas putida* strain bearing a recombinant plasmid, pADPTel., *Can. J. Microbiol.*, 50, 595, 2004.
192. Bertolla, F. and Simonet, P., Horizontal gene transfers in the environment: natural transformation as a putative process for gene transfers between transgenic plants and microorganisms, *Res. Microbiol.*, 150, 375, 1999.
193. Kay, E. et al., Opportunistic colonization of *Ralstonia solanacearum*-infected plants by *Acinetobacter* sp. and its natural competence development, *Microb. Ecol.* 43, 291, 2002.
194. Zatyka, M. and Thomas, C.M., Control of genes for conjugative transfer of plasmids and other mobile elements, *FEMS Microbiol. Rev.*, 21, 291, 1998.
195. Sia, E.A., Kuehner, D.M., and Figurski, D.H., Mechanism of retrotransfer in conjugation: prior transfer of the conjugative plasmid is required, *J. Bacteriol.*, 178, 1457, 1996.
196. Szpirer, C. et al., Retrotransfer or gene capture: a feature of conjugative plasmids, with ecological and evolutionary significance, *Microbiology*, 145, 3321, 1999.
197. Ronchel, M.C., Ramos-Diaz, M.A., and Ramos, J.L., Retrotransfer of DNA in the rhizosphere, *Environ. Microbiol.*, 2, 319, 2000.
198. Clewell, D.B., Flannagan, S.E., and Jaworski, D.D., Unconstrained bacterial promiscuity: the Tn916-Tn1545 family of conjugative transposons, *Trends Microbiol.*, 3, 229, 1995.
199. Scott, J.R. and Churchward, G.G., Conjugative transposition, *Annu. Rev. Microbiol.*, 49, 367, 1995.
200. Burrus, V. and Waldor, M.K., Shaping bacterial genomes with integrative and conjugative elements, *Res. Microbiol.*, 155, 376, 2004.
201. Lilley, A.K. et al., *In situ* transfer of an exogenously isolated plasmid between *Pseudomonas* spp. in sugar beet rhizosphere, *Microbiology*, 140, 27, 1994.
202. Lilley, A.K. and Bailey, M., The acquisition of indigenous plasmids by a genetically marked Pseudomonad population colonizing the sugar beet phytosphere is related to local environmental conditions, *Appl. Environ. Microbiol.*, 63, 1577, 1997.
203. Zhang, L. et al., Agrobacterium conjugation and gene regulation by N-acyl-L-homoserine lactones, *Nature*, 362, 446, 1993.
204. Bjorklof, K. et al., High frequency of conjugation versus plasmid segregation of RP1 in epiphytic *Pseudomonas syringae* populations, *Microbiology*, 141, 2719, 1995.
205. Bale, M.J., Fry, J.C., and Day, M.J., Plasmid transfer between strains of *Pseudomonas aeruginosa* on membrane filters attached to river stones, *J. Gen. Microbiol.*, 133, 3099, 1987.
206. Dahlberg, C., Bergstrom, M., and Hermansson, M., *In situ* detection of high levels of horizontal plasmid transfer in marine bacterial communities, *Appl. Environ. Microbiol*, 64, 2670, 1998.

207. Kruse, H. and Sorum, H., Transfer of multiple drug resistance plasmids between bacteria of diverse origins in natural microenvironments, *Appl. Environ. Microbiol.,* 60, 4015, 1994.

208. Balis, E. et al., Indications of *in vivo* transfer of an epidemic R plasmid from *Salmonella enteritidis* to *Escherichia coli* of the normal human gut flora, *J. Clin. Microbiol.,* 34, 977, 1996.

209. Kues, U. and Stahl, U., Replication of plasmids in gram-negative bacteria, *Microbiol. Rev.,* 53, 491, 1989.

210. Schafer, A. et al., Cloning and characterization of a DNA region encoding a stress-sensitive restriction system from *Corynebacterium glutamicum* ATCC 13032 and analysis of its role in intergeneric conjugation with *Escherichia coli, J. Bacteriol.,* 176, 7309, 1994.

211. Simonsen, L., Dynamics of plasmid transfer on surfaces, *J. Gen. Microbiol.,* 136, 1001, 1990.

212. van Elsas, J.D., Nikkel, M., and van Overbeek, L.S., Detection of plasmid RP4 transfer in soil and rhizosphere, and the occurrence of homology to RP4 in soil bacteria, *Curr. Microbiol.,* 19, 375, 1989.

213. Trieu-Cuot, P. et al., *In vivo* transfer of genetic information between gram-positive and gram-negative bacteria, *EMBO J.,* 4, 3583, 1985.

214. Heinemann, J.A. and Sprague, G.F., Transmission of plasmid DNA to yeast by conjugation with bacteria, *Methods Enzymol.,* 194, 187, 1991.

215. Nishikawa, M., Suzuki, K., and Yoshida, K., DNA integration into recipient yeast chromosomes by trans-kingdom conjugation between *Escherichia coli* and *Saccharomyces cerevisiae, Curr. Genet.,* 21, 101, 1992.

216. Davies, J., Inactivation of antibiotics and the dissemination of resistance genes, *Science,* 264, 375, 1994.

217. de Rore, H. et al., Evolution of heavy metal resistant transconjugants in a soil environnement with a concomitant selective pressure, *FEMS Microbiol. Ecol.,* 15, 71, 1994.

218. Bogdanova, E.S. et al., Horizontal spread of *mer* operons among gram-positive bacteria in natural environments, *Microbiology,* 144, 609, 1998.

219. Top, E.M. et al., Enhancement of 2,4-dichlorophenoxyacetic acid (2,4-D) degradation in soil by dissemination of catabolic plasmids, *Antonie Van Leeuwenhoek,* 73, 87, 1998.

220. Stephens, D.S. et al., Insertion of Tn916 in *Neisseria meningitidis* resulting in loss of group B capsular polysaccharide, *Infect. Immun.,* 59, 4097, 1991.

221. Salyers, A.A. and Shoemaker, N.B., Chromosomal gene transfer elements of the *Bacteroides* group, *Eur. J. Clin. Microbiol. Infect. Dis.,* 11, 1032, 1992.

222. Salyers, A.A. et al., Conjugative transposons: an unusual and diverse set of integrated gene transfer elements, *Microbiol. Rev.,* 59, 579, 1995.

223. Smit, E., Wolters, A., and van Elsas, J.D., Determination of plasmid transfer frequency in soil: consequences of bacterial mating on selective agar media, *Curr. Microbiol.,* 21, 151, 1990.

224. Kinkle, B.K. et al., Plasmids pJP4 and r68.45 can be transferred between populations of Bradyrhizobia in non sterile soil, *Appl. Environ. Microbiol.,* 59, 1762, 1993.

225. Pearce, D.A., Bazin, M.J., and Lynch, J.M., Substrate concentration and plasmid transfer frequency between bacteria in a model rhizosphere, *Microb. Ecol.,* 40, 57, 2000.

226. Richaume, A., Angle, J.S., and Sadowsky, M.J., Influence of soil variables on *in situ* plasmid transfer from *Escherichia coli* to *Rhizobium fredii, Appl. Environ. Microbiol.,* 55, 1730, 1989.

227. Top, E. et al., Gene escape model: transfer of heavy metal resistance genes from *Escherichia coli* to *Alcaligenes eutrophus* on agar plates and in soil samples, *Appl. Environ. Microbiol.,* 56, 2471, 1990.

228. Kroër, N. et al., Effect of root exudates and bacterial metabolic activity on conjugal gene transfer in the rhizosphere of a marsh plant, *FEMS Microbiol. Ecol.,* 25, 375, 1998.

229. Normander, B. et al., Effect of bacterial distribution and activity on conjugal gene transfer on the phylloplane of the bush bean (*Phaseolus vulgaris*), *Appl. Environ. Microbiol.,* 64, 1902, 1998.

230. Lafuente, R. et al., Influence of environmental factors on plasmid transfer in soil microcosms, *Curr. Microbiol.,* 32, 213, 1996.

231. van Elsas, J.D., Govaert, J.M., and van Veen, J.A., Transfer of plasmid pFT30 between bacilli in soil as influenced by bacterial dynamics and soil conditions, *Soil Biol. Biochem.,* 19, 639, 1987.

232. Clerc, S. and Simonet, P., Efficiency of the transfer of a pSAM-derivative plasmid between two strains of *Streptomyces lividans* in conditions ranging from agar slants to non-sterile soil microcosms, *FEMS Microbiol. Ecol.,* 21, 157, 1996.

233. Desaint, S. et al., Genetic transfer of the *mcd* gene in soil, *J. Appl. Microbiol.,* 95, 102, 2003.

234. Richaume, A. et al., Influence of soil type on the transfer of plasmid RP4p from *Pseudomonas fluorescens* to introduced recipient and indigenous bacteria, *FEMS Microbiol. Ecol.*, 101, 281, 1992.

235. Powell, B.J. et al., Demonstration of *tra*1 plasmid activity in bacteria indigenous to the phyllosphere of sugarbeet: gene transfer to a recombinant pseudomonad, *FEMS Microbiol. Ecol.*, 12, 195, 1993.

236. Top, E. et al., Exogenous isolation of mobilizing plasmids from polluted soils and sludges, *Appl. Environ. Microbiol.*, 60, 831, 1994.

237. Vilas-Boas, L.A. et al., Survival and conjugation of *Bacillus thuringiensis* in a soil microcosm, *FEMS Microbiol. Ecol.*, 31, 255, 2000.

238. Brown, J.R. and Doolittle, W.F., Gene descent, duplication, and horizontal transfer in the evolution of glutamyl- and glutaminyl-tRNA synthetases, *J. Mol. Evol.*, 49, 485, 1999.

239. Brinkman, F.S. et al., Evidence that plant-like genes in *Chlamydia* species reflect an ancestral relationship between Chlamydiaceae, cyanobacteria, and the chloroplast, *Genome Res.*, 12, 1159, 2002.

240. Scutt, C.P., Zubko, E., and Meyer, P., Techniques for the removal of marker genes from transgenic plants, *Biochimie*, 84, 1119, 2002.

241. Broothaerts, W. et al., Gene transfer to plants by diverse species of bacteria, *Nature*, 433, 629, 2005.

242. Flavell, R.B. et al., Selectable marker genes: safe for plants?, *Biotechnology*, 10, 141, 1992.

243. Gebhard, F. and Smalla, K., Transformation of *Acinetobacter* sp. strain BD413 by transgenic sugar beet DNA, *Appl. Environ. Microbiol.*, 64, 1550, 1998.

244. de Vries, J. and Wackernagel, W., Microbial horizontal gene transfer and crop plants, *Plant Soil*, 266, 91, 2004.

245. Dröge, M., Pühler, A., and Selbitschka, W., Horizontal gene transfer as a biosafety issue: a natural phenomenon of public concern, *J. Biotechnol.*, 64, 75, 1998.

246. Normark, B.H. and Normark, S., Evolution and spread of antibiotic resistance, *J. Intern. Med.*, 252, 91, 2002.

247. Azevedo, J.L. and Araujo, W.L., Genetically modified crops: environmental and human health concerns, *Mutat. Res.*, 544, 223, 2003.

248. Dunfield, K.E. and Germida, J.J., Impact of genetically modified crops on soil- and plant-associated microbial communities, *J. Environ. Qual.*, 33, 806, 2004.

249. Oger, P.M. et al., Engineering root exudation of *Lotus* toward the production of two novel carbon compounds leads to the selection of distinct microbial populations in the rhizosphere, *Microb. Ecol.*, 47, 96, 2004.

Index

A

Abcisic acid (ABA), 143
ABC transporters, 32
ACC deaminase gene, 280
Acidification, *See* Rhizosphere acidification
Acid phosphatase, 11, 41–43, 122, 123, *See also* Phosphatases
Acinetobacter spp., 408, 409
Aconitase, 39, 53, 141
Acroptilon repens, 305
Actinomycetes, 89
Actinorhizal plants, *Frankia* symbiosis, 319–320
Acyl-homoserone lactones (AHLs), 272, 273, 275, 280, 311, 317, 380
Acyrthosiphon pisum, 314
Aerobactin, 190
Aeromonas caviae, 85
Agar sheets, 27
Agricultural practices, 94
    cropping systems, 83, 94
    fertilization, 92, 94, 117, 118
    pesticides, 96
Agrobacterium spp.
    quorum sensing, 310, 380
    siderophore production, 183
Agrobacterium tumefaciens, 246, 413
Agropyron desertorum, 361
Alcaligenes siderophore production, 183
Aldonic acid, 241
ALF4, 137
Alkaline phosphatase, 123, 124, *See also* Phosphatases
Allamones, 298
Allelochemicals, 9, 11–12, 301–305, *See also* Phytotoxins; Root exudates; *specific chemicals*
    categories, 298
    detoxification mechanisms, 304
    invasive plants and, 305
    plant resistance, 304–306
    reactive oxygen species, 303
Alnus rubra, 320
Alternaria alternate, 273
Aluminum (Al) toxicity, 166
    Al form and, 9
    carboxylate-permeable channels and, 57, 166
    exudate persistence and, 8
    genotypic differences in tolerance, 53
    H^+-ATPase and, 156
    organic acids and, 13, 51–54
    root environmental "sensing," 115
Amanita muscaria, 321
Amaranthus palmeri, 305
Ambrosia dumosa, 314–315

Aminoacid permease, 220
Amino acids
    active retrieval mechanisms, 30
    bioreporters and, 277
    diffusion, 31
    exudation response to Cd toxicity, 54
    exudation under hypoxic conditions, 55
    light intensity and exudation, 93
    microbial effects on recovery, 28
    microbial uptake, 118–119
    mycorrhizal nitrogen nutrition, 220
    natural roles in rhizosphere, 278
    N rhizodeposition estimates, 76
    plant uptake and dynamics in soil, 118–120
    pyrolysis-field ionization MS, 378
    Rhizobia acid tolerance and, 257–258
    rhizosphere competence and biocontrol, 271
    soil diffusion rate, 120
    temperature effects on exudation, 54
    transport systems, 30, 118–119, 158, 160
Ammonia, 85, 271
Ammoniacal syndrome, 120
Ammonia monooxygenase, 126
Ammonium (NH_4^+) fertilization, 92, 117
Ammonium (NH_4^+) uptake, 43–44, 120
    cation transport system, 162–163
    mycorrhizal fungi and, 163, 220–221
    rhizosphere acidification and, 120
    transport systems, 221
Amplified ribosomal DNA restriction analysis (ARDRA), 269, 383
Anaerobic conditions, 91–92
Anion carriers, 158–159
Anion channels, *See* Ion channels
Anion-exchange resins, 26, 27, 29
Antibiotic resistance genes, 412, 414–415
Antibiotics
    biocontrol agents, 83–84, 191, 271–272, 275–276, 278
    border cell secretions, 33
    microbial synthesis, 83–84
    phenolic exudates, 41
    preventing biodegradation during exudate extraction, 29
Antimicrobial exudates, 309–314, *See also* Biocontrol; Phytotoxins; Root-microbe communication
Aphidius ervi, 314
Aphids, 227, 313, 314
Apoplast, 339
Appresoria formation, 321
Apyrases, 122
Aquaporins, 165–166
Arabidopsis (including *A. thaliana*)
    allelochemical effects, 7, 303–304
    antimicrobial secretions, 311, 312, 313